IARC MONOGRAPHS

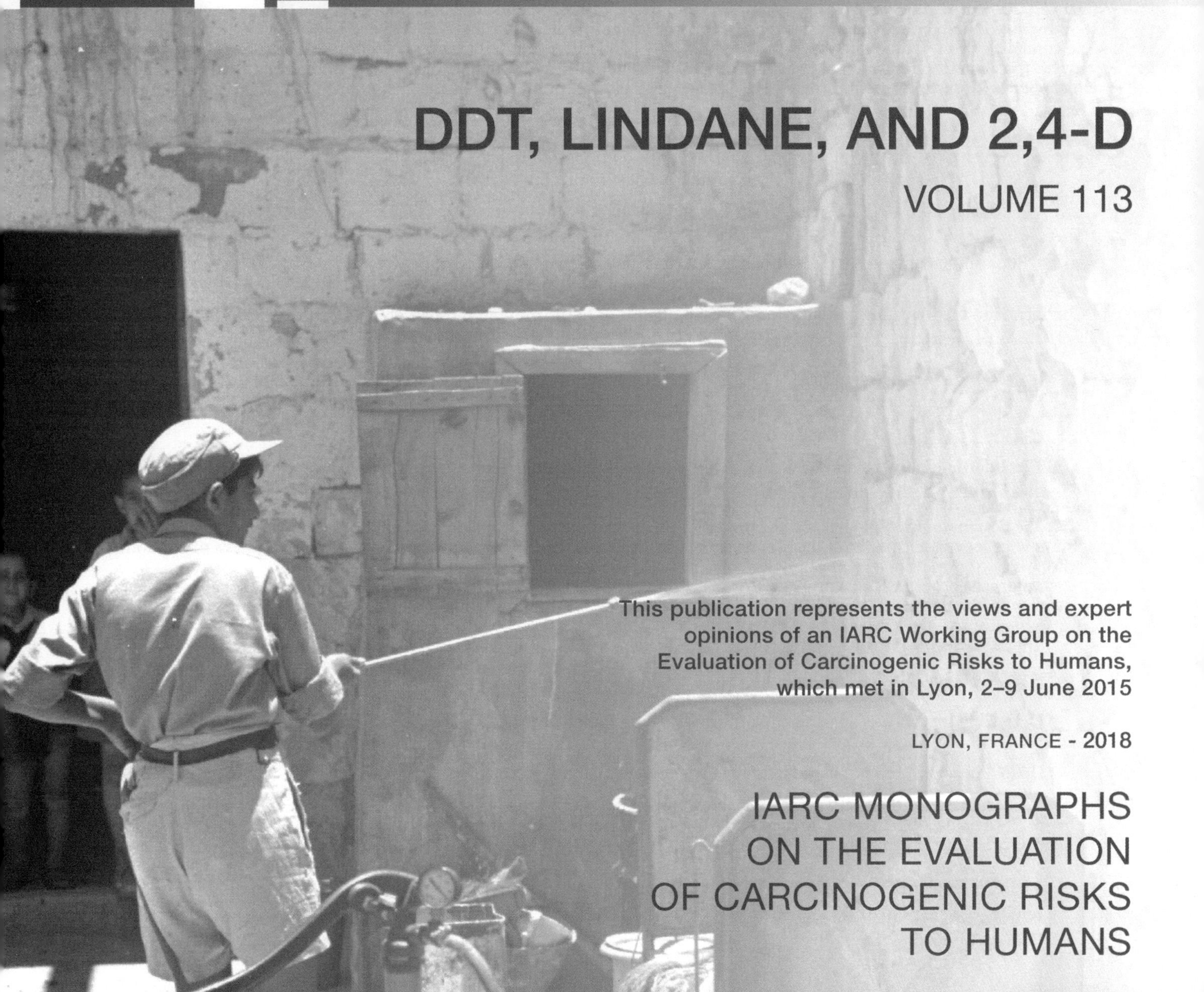

DDT, LINDANE, AND 2,4-D

VOLUME 113

This publication represents the views and expert opinions of an IARC Working Group on the Evaluation of Carcinogenic Risks to Humans, which met in Lyon, 2–9 June 2015

LYON, FRANCE - 2018

IARC MONOGRAPHS ON THE EVALUATION OF CARCINOGENIC RISKS TO HUMANS

International Agency for Research on Cancer

World Health Organization

IARC MONOGRAPHS

In 1969, the International Agency for Research on Cancer (IARC) initiated a programme on the evaluation of the carcinogenic risk of chemicals to humans involving the production of critically evaluated monographs on individual chemicals. The programme was subsequently expanded to include evaluations of carcinogenic risks associated with exposures to complex mixtures, lifestyle factors and biological and physical agents, as well as those in specific occupations. The objective of the programme is to elaborate and publish in the form of monographs critical reviews of data on carcinogenicity for agents to which humans are known to be exposed and on specific exposure situations; to evaluate these data in terms of human risk with the help of international working groups of experts in carcinogenesis and related fields; and to indicate where additional research efforts are needed. The lists of IARC evaluations are regularly updated and are available on the Internet at http://monographs.iarc.fr/.

This programme has been supported since 1982 by Cooperative Agreement U01 CA33193 with the United States National Cancer Institute, Department of Health and Human Services. Additional support has been provided since 1986 by the European Commission Directorate-General for Employment, Social Affairs, and Inclusion, initially by the Unit of Health, Safety and Hygiene at Work, and since 2014 by the European Union Programme for Employment and Social Innovation "EaSI" (2014–2020) (for further information please consult: http://ec.europa.eu/social/easi). Support has also been provided since 1992 by the United States National Institute of Environmental Health Sciences, Department of Health and Human Services. The contents of this volume are solely the responsibility of the Working Group and do not necessarily represent the official views of the United States National Cancer Institute, the United States National Institute of Environmental Health Sciences, the United States Department of Health and Human Services, or the European Commission.

Published by the International Agency for Research on Cancer,
150 cours Albert Thomas, 69372 Lyon Cedex 08, France

On-line publication, December 2017

Distributed by WHO Press, World Health Organization, 20 Avenue Appia, 1211 Geneva 27, Switzerland
(tel.: +41 22 791 3264; fax: +41 22 791 4857; email: bookorders@who.int).

Corrigenda to the IARC Monographs are published online at http://monographs.iarc.fr/ENG/Publications/corrigenda.php
To report an error, please contact: editimo@iarc.fr

Co-funded by the European Union

IARC Library Cataloguing in Publication Data

DDT, Lindane, and 2,4-D / IARC Working Group on the Evaluation of Carcinogenic Risks to Humans (2015: Lyon, France)

(IARC monographs on the evaluation of carcinogenic risks to humans ; volume 113)

1. Carcinogens 2. Neoplasms – chemically induced 3. DDT – toxicity 4. Lindane – toxicity
5. 2,4-Dichlorophenoxyacetic Acid – toxicity 6. Herbicides – toxicity 7. Insecticides – toxicity
8. Occupational exposure

I. International Agency for Research on Cancer II. Series

ISBN 978-92-832-0151-9 (NLM Classification: W1)
ISSN 1017-1606

PRINTED IN FRANCE

CONTENTS

NOTE TO THE READER

The term 'carcinogenic risk' in the *IARC Monographs* series is taken to mean that an agent is capable of causing cancer. The *Monographs* evaluate cancer hazards, despite the historical presence of the word 'risks' in the title.

Inclusion of an agent in the *Monographs* does not imply that it is a carcinogen, only that the published data have been examined. Equally, the fact that an agent has not yet been evaluated in a *Monograph* does not mean that it is not carcinogenic. Similarly, identification of cancer sites with *sufficient evidence* or *limited evidence* in humans should not be viewed as precluding the possibility that an agent may cause cancer at other sites.

The evaluations of carcinogenic risk are made by international working groups of independent scientists and are qualitative in nature. No recommendation is given for regulation or legislation.

Anyone who is aware of published data that may alter the evaluation of the carcinogenic risk of an agent to humans is encouraged to make this information available to the Section of IARC Monographs, International Agency for Research on Cancer, 150 cours Albert Thomas, 69372 Lyon Cedex 08, France, in order that the agent may be considered for re-evaluation by a future Working Group.

Although every effort is made to prepare the *Monographs* as accurately as possible, mistakes may occur. Readers are requested to communicate any errors to the Section of IARC Monographs, so that corrections can be reported in future volumes.

LIST OF PARTICIPANTS

Members [1]

Michael Alavanja

Division of Cancer Epidemiology & Genetics
National Cancer Institute
Rockville, MD
USA

Maarten C. Bosland

College of Medicine
University of Illinois at Chicago
Chicago, IL
USA

Mariano E. Cebrian Garcia [unable to attend]

Department of Toxicology
National Polytechnic Research Institute
Mexico City
Mexico

Rajendra S. Chhabra

Toxicology Consulting Services
Raleigh, NC
USA

Weihsueh A. Chiu (Subgroup Chair, Mechanisms)

College of Veterinary Medicine
Texas A&M University
College Station, TX
USA

Pierluigi Cocco (Subgroup Chair, Cancer in Humans)

Department of Public Health
University of Cagliari
Monserrato
Italy

[1] Working Group Members and Invited Specialists serve in their individual capacities as scientists and not as representatives of their government or any organization with which they are affiliated. Affiliations are provided for identification purposes only. Invited Specialists do not serve as Meeting Chair or Subgroup Chair, draft text that pertains to the description or interpretation of cancer data, or participate in the evaluations.
Each participant was asked to disclose pertinent research, employment, and financial interests. Current financial interests and research and employment interests during the past 4 years or anticipated in the future are identified here. Minor pertinent interests are not listed and include stock valued at no more than US$ 1000 overall, grants that provide no more than 5% of the research budget of the expert's organization and that do not support the expert's research or position, and consulting or speaking on matters not before a court or government agency that does not exceed 2% of total professional time or compensation. All grants that support the expert's research or position and all consulting or speaking on behalf of an interested party on matters before a court or government agency are listed as significant pertinent interests.

Anneclaire De Roos

Department of Environmental and Occupational Health
Drexel University School of Public Health
Philadelphia, PA
USA

Shoji Fukushima

Japan Bioassay Research Center
Japan Industrial Safety & Health Association
Hadano
Japan

Pascal Guénel

Center for Research in Epidemiology and Population Health
National Institute of Health & Medical Research (INSERM)
Villejuif
France

Ronald A. Herbert (Subgroup Chair, Cancer in Experimental Animals)

Division of the National Toxicology Program
National Institute of Environmental Health Sciences
Research Triangle Park, NC
USA

Manolis Kogevinas (Overall Chair)

Centre for Research in Environmental Epidemiology (CREAL)
Municipal Institute of Medical Research (IMIM)
Barcelona
Spain

Michele La Merrill

Department of Environmental Toxicology
University of California at Davis
Davis, CA
USA

Marcelo L. Larramendy

National Scientific and Technical Research Council (CONICET)
National University of La Plata
La Plata
Argentina

Lizbeth Teresita Lopez Carrillo

Center of Public Health Research
National Institute of Public Health
Cuernavaca
Mexico

Francis L. Martin [unable to attend]

Lancaster University
Bailrigg
England

Saloshni Naidoo

School of Nursing and Public Health
University of KwaZulu-Natal
Durban
South Africa

Tippawan Prapamontol

Research Institute for Health Sciences (RIHES)
Chiang Mai University
Chiang Mai
Thailand

David M. Reif

Department of Biological Sciences
North Carolina State University
Raleigh NC
USA

Deodutta Roy

Robert Stempel College of Public Health & Social Work
Florida International University
Miami, FL
USA

Lesley Rushton [2]

Faculty of Medicine
Imperial College London
London
England

Thomas Sanderson

INRS-Institut Armand-Frappier Research Centre
Laval (Québec)
Canada

Martyn T. Smith

Berkeley Institute of the Environment and Superfund Research Program
School of Public Health
University of California
Berkeley, CA
USA

Bernard W. Stewart

Cancer Control Program
South Eastern Sydney Public Health Unit and Faculty of Medicine
University of New South Wales
Sydney NSW
Australia

Kent Thomas

National Exposure Research Laboratory
United States Environmental Protection Agency
Research Triangle Park, NC
USA

Roel Vermeulen (Subgroup Chair, Exposure)

Institute for Risk Assessment Sciences (IRAS)
University of Utrecht
Utrecht
The Netherlands

Mary S. Wolff

Icahn School of Medicine at Mount Sinai
New York, NY
USA

Invited Specialists

None

[2] Lesley Rushton has been member of an ad hoc Scientific Advisory Group on Epidemiology of the European Crop Protection Association to provide advice on future epidemiological research. She received a small honorarium and reimbursement of travel expenses.

Representatives

Amira Ben Amara [unable to attend] [4]

National Agency of Sanitary and Environmental Control of Products (ANCSEP)
Montplaisir, Tunis
Tunisia

Silvia de Oliveira Santos Cazenave [5]

Brazilian Health Surveillance Agency (ANVISA)
Brasilia
Brazil

Marie Odile Rambourg Schepens [6]

Regulated Products Department
French Agency for Food, Environment and Occupational Health & Safety (ANSES)
Maisons-Alfort
France

Camila Queiroz Moreira [7]

Brazilian Health Surveillance Agency (ANVISA)
Brasilia
Brazil

Deborah M. Winn [8]

National Cancer Institute
Bethesda, MD
USA

Observers [9]

James S. Bus [10]

Center for Toxicology and Mechanistic Biology
Exponent
Midland, MI
USA

[4] Amira Ben Amara attended as a Representative of the National Agency of Sanitary and Environmental Control of Products, Tunisia.
[5] Silvia de Oliveira Santos Cazenave attended as an Representative for the Brazilian Health Surveillance Agency (ANVISA).
[6] Marie Odile Rambourg Schepens attended as a Representative of the French Agency for Food, Environment and Occupational Health & Safety (ANSES).
[7] Camila Queiroz Moreira attended as a Representative for the Brazilian Health Surveillance Agency (ANVISA).
[8] Deborah Winn attended as a Representative of the National Cancer Institute, USA.
[9] Each Observer agreed to respect the Guidelines for Observers at *IARC Monographs* meetings. Observers did not serve as Meeting Chair or Subgroup Chair, draft any part of a *Monograph*, or participate in the evaluations. They also agreed not to contact participants before the meeting, not to lobby them at any time, not to send them written materials, and not to offer them meals or other favours. IARC asked and reminded Working Group Members to report any contact or attempt to influence that they may have encountered, either before or during the meeting.
[10] James Bus attended as an Observer for the Industry Task Force II on 2,4-D Research Data, which is an industry consortium whose members hold registrations for the active ingredient 2,4-D. The Task Force members include Dow AgroSciences LLC, Nufarm, Ltd. and Agro-Gor Corporation, a United States corporation jointly owned by Albaught, Inc. (USA) and PBI-Gordon Corp. (USA). He retired in 2013 as a toxicologist from The Dow Chemical Co., a manufacturer of 2,4-D, and now works for Exponent, Inc. to provide toxicological expertise on 2,4-D. He has interacted with national and international regulatory agencies and in USA-based litigation, all concerning 2,4-D. He holds significant stock in The Dow Chemical Co.

Béatrice Fervers [unable to attend] [11]

Unité Cancer et Environnement
Centre Léon Bérard
Lyon
France

Julie E. Goodman [12]

Gradient
Cambridge, MA
USA

Steve A. McMaster [13]

Industry Task Force II on 2,4-D Research Data
Raleigh, NC
USA

IARC/WHO Secretariat

Lamia Benbrahim-Tallaa
Véronique Bouvard
Richard Brown, WHO Geneva
Fatiha El Ghissassi
Yann Grosse
Neela Guha
Kathryn Guyton
Michael Korenjak
Maria E. Leon Roux
Dana Loomis (*Responsible Officer*)
Heidi Mattock (*Editor*)
Kurt Straif (*Head of Programme*)
Jiri Zavadil

Administrative Assistance

Marieke Dusenberg
Sandrine Egraz
Michel Javin
Brigitte Kajo
Helene Lorenzen-Augros

Production Team

Elisabeth Elbers
Fiona Gould
Solène Quennehen

[11] Beatrice Fervers attended as an Observer for the Centre Léon Bérard, France.

[12] Julie Goodman attended as an Observer for the Industry Task Force II on 2,4-D Research Data, which is an industry consortium whose members hold registrations for the active ingredient 2,4-D. The Task Force members include Dow AgroSciences LLC, Nufarm, Ltd. and Agro-Gor Corporation, a United States corporation jointly owned by Albaught, Inc. (USA) and PBI-Gordon Corp (USA). She is employed by Gradient, which has received funding from the Industry Task Force II on 2,4-D Research Data.

[13] Steve McMaster attended as an Observer for the Industry Task Force II on 2,4-D Research Data (Task Force), which is an industry consortium whose members hold registrations for the active ingredient 2,4-D. The Task Force members include Dow AgroSciences LLC, Nufarm, Ltd, and Agro-Gor Corporation, a United States corporation jointly owned by Albaught, Inc. (USA) and PBI-Gordon Corp. (USA) Steve McMaster is employed by Dow AgroSciences LLC and is the Technical Chairman of the Industry Task Force II on 2,4-D Research Data. He holds significant stock in The Dow Chemical Co.

PREAMBLE

The Preamble to the IARC Monographs describes the objective and scope of the programme, the scientific principles and procedures used in developing a Monograph, the types of evidence considered and the scientific criteria that guide the evaluations. The Preamble should be consulted when reading a Monograph or list of evaluations.

A. GENERAL PRINCIPLES AND PROCEDURES

1. Background

Soon after IARC was established in 1965, it received frequent requests for advice on the carcinogenic risk of chemicals, including requests for lists of known and suspected human carcinogens. It was clear that it would not be a simple task to summarize adequately the complexity of the information that was available, and IARC began to consider means of obtaining international expert opinion on this topic. In 1970, the IARC Advisory Committee on Environmental Carcinogenesis recommended '... that a compendium on carcinogenic chemicals be prepared by experts. The biological activity and evaluation of practical importance to public health should be referenced and documented.' The IARC Governing Council adopted a resolution concerning the role of IARC in providing government authorities with expert, independent, scientific opinion on environmental carcinogenesis. As one means to that end, the Governing Council recommended that IARC should prepare monographs on the evaluation of carcinogenic risk of chemicals to man, which became the initial title of the series.

In the succeeding years, the scope of the programme broadened as *Monographs* were developed for groups of related chemicals, complex mixtures, occupational exposures, physical and biological agents and lifestyle factors. In 1988, the phrase 'of chemicals' was dropped from the title, which assumed its present form, *IARC Monographs on the Evaluation of Carcinogenic Risks to Humans*.

Through the *Monographs* programme, IARC seeks to identify the causes of human cancer. This is the first step in cancer prevention, which is needed as much today as when IARC was established. The global burden of cancer is high and continues to increase: the annual number of new cases was estimated at 10.1 million in 2000 and is expected to reach 15 million by 2020 (Stewart & Kleihues, 2003). With current trends in demographics and exposure, the cancer burden has been shifting from high-resource countries to low- and medium-resource countries. As a result of *Monographs* evaluations, national health agencies have been able, on scientific grounds, to take measures to reduce human exposure to carcinogens in the workplace and in the environment.

The criteria established in 1971 to evaluate carcinogenic risks to humans were adopted by the Working Groups whose deliberations resulted in the first 16 volumes of the *Monographs* series. Those criteria were subsequently updated by further ad hoc Advisory Groups (IARC, 1977, 1978, 1979, 1982, 1983, 1987, 1988, 1991; Vainio et al., 1992; IARC, 2005, 2006).

The Preamble is primarily a statement of scientific principles, rather than a specification of working procedures. The procedures through which a Working Group implements these principles are not specified in detail. They usually involve operations that have been established as being effective during previous *Monograph* meetings but remain, predominantly, the prerogative of each individual Working Group.

2. Objective and scope

The objective of the programme is to prepare, with the help of international Working Groups of experts, and to publish in the form of *Monographs*, critical reviews and evaluations of evidence on the carcinogenicity of a wide range of human exposures. The *Monographs* represent the first step in carcinogen risk assessment, which involves examination of all relevant information to assess the strength of the available evidence that an agent could alter the age-specific incidence of cancer in humans. The *Monographs* may also indicate where additional research efforts are needed, specifically when data immediately relevant to an evaluation are not available.

In this Preamble, the term 'agent' refers to any entity or circumstance that is subject to evaluation in a *Monograph*. As the scope of the programme has broadened, categories of agents now include specific chemicals, groups of related chemicals, complex mixtures, occupational or environmental exposures, cultural or behavioural practices, biological organisms and physical agents. This list of categories may expand as causation of, and susceptibility to, malignant disease become more fully understood.

A cancer 'hazard' is an agent that is capable of causing cancer under some circumstances, while a cancer 'risk' is an estimate of the carcinogenic effects expected from exposure to a cancer hazard. The *Monographs* are an exercise in evaluating cancer hazards, despite the historical presence of the word 'risks' in the title. The distinction between hazard and risk is important, and the *Monographs* identify cancer hazards even when risks are very low at current exposure levels, because new uses or unforeseen exposures could engender risks that are significantly higher.

In the *Monographs*, an agent is termed 'carcinogenic' if it is capable of increasing the incidence of malignant neoplasms, reducing their latency, or increasing their severity or multiplicity. The induction of benign neoplasms may in some circumstances (see Part B, Section 3a) contribute to the judgement that the agent is carcinogenic. The terms 'neoplasm' and 'tumour' are used interchangeably.

The Preamble continues the previous usage of the phrase 'strength of evidence' as a matter of historical continuity, although it should be understood that *Monographs* evaluations consider studies that support a finding of a cancer hazard as well as studies that do not.

Some epidemiological and experimental studies indicate that different agents may act at different stages in the carcinogenic process, and several different mechanisms may be involved. The aim of the *Monographs* has been, from their inception, to evaluate evidence of carcinogenicity at any stage in the carcinogenesis process, independently of the underlying mechanisms. Information on mechanisms may, however, be used in making the overall evaluation (IARC, 1991; Vainio et al., 1992; IARC, 2005, 2006; see also Part B, Sections 4 and 6). As mechanisms of carcinogenesis are elucidated, IARC convenes international scientific conferences to determine whether a broad-based consensus has emerged on how specific mechanistic data can be used in an evaluation of human carcinogenicity. The results of such conferences are reported in IARC Scientific Publications, which, as long as they still reflect the current state of scientific knowledge, may guide subsequent Working Groups.

Although the *Monographs* have emphasized hazard identification, important issues may also involve dose–response assessment. In many cases, the same epidemiological and experimental studies used to evaluate a cancer hazard can also be used to estimate a dose–response relationship. A *Monograph* may undertake to estimate dose–response relationships within the range of the available epidemiological data, or it may compare the dose–response information from experimental and epidemiological studies. In some cases, a subsequent publication may be prepared by a separate Working Group with expertise in quantitative dose–response assessment.

The *Monographs* are used by national and international authorities to make risk assessments, formulate decisions concerning preventive measures, provide effective cancer control programmes and decide among alternative options for public health decisions. The evaluations of IARC Working Groups are scientific, qualitative judgements on the evidence for or against carcinogenicity provided by the available data. These evaluations represent only one part of the body of information on which public health decisions may be based. Public health options vary from one situation to another and from country to country and relate to many factors, including different socioeconomic and national priorities. Therefore, no recommendation is given with regard to regulation or legislation, which are the responsibility of individual governments or other international organizations.

3. Selection of agents for review

Agents are selected for review on the basis of two main criteria: (a) there is evidence of human exposure and (b) there is some evidence or suspicion of carcinogenicity. Mixed exposures may occur in occupational and environmental settings and as a result of individual and cultural habits (such as tobacco smoking and dietary practices). Chemical analogues and compounds with biological or physical characteristics similar to those of suspected carcinogens may also be considered, even in the absence of data on a possible carcinogenic effect in humans or experimental animals.

The scientific literature is surveyed for published data relevant to an assessment of carcinogenicity. Ad hoc Advisory Groups convened by IARC in 1984, 1989, 1991, 1993, 1998 and 2003 made recommendations as to which agents should be evaluated in the *Monographs* series. Recent recommendations are available on the *Monographs* programme web site (http://monographs.iarc.fr). IARC may schedule other agents for review as it becomes aware of new scientific information or as national health agencies identify an urgent public health need related to cancer.

As significant new data become available on an agent for which a *Monograph* exists, a re-evaluation may be made at a subsequent meeting, and a new *Monograph* published. In some cases it may be appropriate to review only the data published since a prior evaluation. This can be useful for updating a database, reviewing new data to resolve a previously open question or identifying new tumour sites associated with a carcinogenic agent. Major changes in an evaluation (e.g. a new classification in Group 1 or a determination that a mechanism does not operate in humans, see Part B, Section 6) are more appropriately addressed by a full review.

4. Data for the *Monographs*

Each *Monograph* reviews all pertinent epidemiological studies and cancer bioassays in experimental animals. Those judged inadequate or irrelevant to the evaluation may be cited but not summarized. If a group of similar studies is not reviewed, the reasons are indicated.

Mechanistic and other relevant data are also reviewed. A *Monograph* does not necessarily cite all the mechanistic literature concerning the agent being evaluated (see Part B, Section 4). Only those data considered by the Working Group to be relevant to making the evaluation are included.

With regard to epidemiological studies, cancer bioassays, and mechanistic and other relevant data, only reports that have been published or accepted for publication in the openly available scientific literature are reviewed. The same publication requirement applies to studies originating from IARC, including meta-analyses or pooled analyses commissioned by IARC in advance of a meeting (see Part B, Section 2c). Data from government agency reports that are publicly available are also considered. Exceptionally, doctoral theses and other material that are in their final form and publicly available may be reviewed.

Exposure data and other information on an agent under consideration are also reviewed. In the sections on chemical and physical properties, on analysis, on production and use and on occurrence, published and unpublished sources of information may be considered.

Inclusion of a study does not imply acceptance of the adequacy of the study design or of the analysis and interpretation of the results, and limitations are clearly outlined in square brackets at the end of each study description (see Part B). The reasons for not giving further consideration to an individual study also are indicated in the square brackets.

5. Meeting participants

Five categories of participant can be present at *Monograph* meetings.

(a) The Working Group

The Working Group is responsible for the critical reviews and evaluations that are developed during the meeting. The tasks of Working Group Members are: (i) to ascertain that all appropriate data have been collected; (ii) to select the data relevant for the evaluation on the basis of scientific merit; (iii) to prepare accurate summaries of the data to enable the reader to follow the reasoning of the Working Group; (iv) to evaluate the results of epidemiological and experimental studies on cancer; (v) to evaluate data relevant to the understanding of mechanisms of carcinogenesis; and (vi) to make an overall evaluation of the carcinogenicity of the exposure to humans. Working Group Members generally have published significant research related to the carcinogenicity of the agents being reviewed, and IARC uses literature searches to identify most experts. Working Group Members are selected on the basis of (a) knowledge and experience and (b) absence of real or apparent conflicts of interests. Consideration is also given to demographic diversity and balance of scientific findings and views.

(b) Invited Specialists

Invited Specialists are experts who also have critical knowledge and experience but have a real or apparent conflict of interests. These experts are invited when necessary to assist in the Working Group by contributing their unique knowledge and experience during subgroup and plenary discussions. They may also contribute text on non-influential issues in the section on exposure, such as a general description of data on production and use (see Part B, Section 1). Invited Specialists do not serve as meeting chair or subgroup chair, draft text that pertains to the description or interpretation of cancer data, or participate in the evaluations.

(c) Representatives of national and international health agencies

Representatives of national and international health agencies often attend meetings because their agencies sponsor the programme or are interested in the subject of a meeting. Representatives do not serve as meeting chair or subgroup chair, draft any part of a *Monograph*, or participate in the evaluations.

(d) Observers with relevant scientific credentials

Observers with relevant scientific credentials may be admitted to a meeting by IARC in limited numbers. Attention will be given to achieving a balance of Observers from constituencies with differing perspectives. They are invited to observe the meeting and should not attempt to influence it. Observers do not serve as meeting chair or subgroup chair, draft any part of a *Monograph*, or participate in the evaluations. At the meeting, the meeting chair and subgroup chairs may grant Observers an opportunity to speak, generally after they have observed a discussion. Observers agree to respect the Guidelines for Observers at *IARC Monographs* meetings (available at http://monographs.iarc.fr).

(e) The IARC Secretariat

The IARC Secretariat consists of scientists who are designated by IARC and who have relevant expertise. They serve as rapporteurs and participate in all discussions. When requested by the meeting chair or subgroup chair, they may also draft text or prepare tables and analyses.

Before an invitation is extended, each potential participant, including the IARC Secretariat, completes the WHO Declaration of Interests to report financial interests, employment and consulting, and individual and institutional research support related to the subject of the meeting. IARC assesses these interests to determine whether there is a conflict that warrants some limitation on participation. The declarations are updated and reviewed again at the opening of the meeting. Interests related to the subject of the meeting are disclosed to the meeting participants and in the published volume (Cogliano et al., 2004).

The names and principal affiliations of participants are available on the *Monographs* programme web site (http://monographs.iarc.fr) approximately two months before each meeting. It is not acceptable for Observers or third parties to contact other participants before a meeting or to lobby them at any time. Meeting participants are asked to report all such contacts to IARC (Cogliano et al., 2005).

All participants are listed, with their principal affiliations, at the beginning of each volume. Each participant who is a Member of a Working Group serves as an individual scientist and not as a representative of any organization, government or industry.

6. Working procedures

A separate Working Group is responsible for developing each volume of *Monographs*. A volume contains one or more *Monographs*, which can cover either a single agent or several related agents. Approximately one year in advance of the meeting of a Working Group, the agents to be reviewed are announced on the *Monographs* programme web site (http://monographs.iarc.fr) and participants are selected by IARC staff in consultation with other experts. Subsequently, relevant biological and epidemiological data are collected by IARC from recognized sources of information on carcinogenesis, including data storage and retrieval systems such as PubMed. Meeting participants who are asked to prepare preliminary working papers for specific sections are expected to supplement the IARC literature searches with their own searches.

Industrial associations, labour unions and other knowledgeable organizations may be asked to provide input to the sections on production and use, although this involvement is not required as a general rule. Information on production and trade is obtained from governmental, trade and market research publications and, in some cases, by direct contact with industries. Separate production data on some agents may not be available for a variety of reasons (e.g. not collected or made public in all producing countries, production is small). Information on uses may be obtained from published sources but is often complemented by direct contact with manufacturers. Efforts are made to supplement this information with data from other national and international sources.

Six months before the meeting, the material obtained is sent to meeting participants to prepare preliminary working papers. The working papers are compiled by IARC staff and sent, before the meeting, to Working Group Members and Invited Specialists for review.

The Working Group meets at IARC for seven to eight days to discuss and finalize the texts and to formulate the evaluations. The objectives of the meeting are peer review and consensus. During the first few days, four subgroups (covering exposure data, cancer in humans, cancer in experimental animals, and mechanistic and other relevant data) review the working papers, develop a joint subgroup draft and write summaries. Care is taken to ensure that each study summary is written or reviewed by someone not associated with the study being considered. During the last few days, the Working Group meets in plenary session to review the subgroup drafts and develop the evaluations. As a result, the entire volume is the joint product of the Working Group, and there are no individually authored sections.

IARC Working Groups strive to achieve a consensus evaluation. Consensus reflects broad agreement among Working Group Members, but not necessarily unanimity. The chair may elect to poll Working Group Members to determine the diversity of scientific opinion on issues where consensus is not readily apparent.

After the meeting, the master copy is verified by consulting the original literature, edited and prepared for publication. The aim is to publish the volume within six months of the Working Group meeting. A summary of the outcome is available on the *Monographs* programme web site soon after the meeting.

B. SCIENTIFIC REVIEW AND EVALUATION

The available studies are summarized by the Working Group, with particular regard to the qualitative aspects discussed below. In general, numerical findings are indicated as they appear in the original report; units are converted when necessary for easier comparison. The Working Group may conduct additional analyses of the published data and use them in their assessment of the evidence; the results of such supplementary analyses are given in square brackets. When an important aspect of a study that directly impinges on its interpretation should be brought to the attention of the reader, a Working Group comment is given in square brackets.

The scope of the *IARC Monographs* programme has expanded beyond chemicals to include complex mixtures, occupational exposures, physical and biological agents, lifestyle factors and other potentially carcinogenic exposures. Over time, the structure of a *Monograph* has evolved to include the following sections:

Exposure data

Studies of cancer in humans

Studies of cancer in experimental animals

Mechanistic and other relevant data

Summary

Evaluation and rationale

In addition, a section of General Remarks at the front of the volume discusses the reasons the agents were scheduled for evaluation and some key issues the Working Group encountered during the meeting.

This part of the Preamble discusses the types of evidence considered and summarized in each section of a *Monograph*, followed by the scientific criteria that guide the evaluations.

1. Exposure data

Each *Monograph* includes general information on the agent: this information may vary substantially between agents and must be adapted accordingly. Also included is information on production and use (when appropriate), methods of analysis and detection, occurrence, and sources and routes of human occupational and environmental exposures. Depending on the agent, regulations and guidelines for use may be presented.

(a) General information on the agent

For chemical agents, sections on chemical and physical data are included: the Chemical Abstracts Service Registry Number, the latest primary name and the IUPAC systematic name are recorded; other synonyms are given, but the list is not necessarily comprehensive. Information on chemical and physical properties that are relevant to identification, occurrence and biological activity is included. A description of technical products of chemicals includes trade names, relevant specifications and available information on composition and impurities. Some of the trade names given may be those of mixtures in which the agent being evaluated is only one of the ingredients.

For biological agents, taxonomy, structure and biology are described, and the degree of variability is indicated. Mode of replication, life cycle, target cells, persistence, latency, host response and clinical disease other than cancer are also presented.

For physical agents that are forms of radiation, energy and range of the radiation are included. For foreign bodies, fibres and respirable particles, size range and relative dimensions are indicated.

For agents such as mixtures, drugs or lifestyle factors, a description of the agent, including its composition, is given.

Whenever appropriate, other information, such as historical perspectives or the description of an industry or habit, may be included.

(b) Analysis and detection

An overview of methods of analysis and detection of the agent is presented, including their sensitivity, specificity and reproducibility. Methods widely used for regulatory purposes are emphasized. Methods for monitoring human exposure are also given. No critical evaluation or recommendation of any method is meant or implied.

(c) Production and use

The dates of first synthesis and of first commercial production of a chemical, mixture or other agent are provided when available; for agents that do not occur naturally, this information may allow a reasonable estimate to be made of the date before which no human exposure to the agent could have occurred. The dates of first reported occurrence of an exposure are also provided when available. In addition, methods of synthesis used in past and present commercial production and different methods of production, which may give rise to different impurities, are described.

The countries where companies report production of the agent, and the number of companies in each country, are identified. Available data on production, international trade and uses are obtained for representative regions. It should not, however, be inferred that those areas or nations are necessarily the sole or major sources or users of the agent. Some identified uses may not be current or major applications, and the coverage is not necessarily comprehensive. In the case of drugs, mention of their therapeutic uses does not necessarily represent current practice nor does it imply judgement as to their therapeutic efficacy.

(d) Occurrence and exposure

Information on the occurrence of an agent in the environment is obtained from data derived from the monitoring and surveillance of levels in occupational environments, air, water, soil, plants, foods and animal and human tissues. When available, data on the generation, persistence and bioaccumulation of the agent are also included. Such data may be available from national databases.

Data that indicate the extent of past and present human exposure, the sources of exposure, the people most likely to be exposed and the factors that contribute to the exposure are reported. Information is presented on the range of human exposure, including occupational and environmental exposures. This includes relevant findings from both developed and developing countries. Some of these data are not distributed widely and may be available from government reports and other sources. In the case of mixtures, industries, occupations or processes, information is given about all agents known to be present. For processes, industries and occupations, a historical description is also given, noting variations in chemical composition, physical properties and levels of occupational exposure with date and place. For biological agents, the epidemiology of infection is described.

(e) Regulations and guidelines

Statements concerning regulations and guidelines (e.g. occupational exposure limits, maximal levels permitted in foods and water, pesticide registrations) are included, but they may not reflect the most recent situation, since such limits are continuously reviewed and modified. The absence of information on regulatory status for a country should not be taken to imply that that country does not have regulations with regard to the exposure. For biological agents, legislation and control, including vaccination and therapy, are described.

2. Studies of cancer in humans

This section includes all pertinent epidemiological studies (see Part A, Section 4). Studies of biomarkers are included when they are relevant to an evaluation of carcinogenicity to humans.

(a) Types of study considered

Several types of epidemiological study contribute to the assessment of carcinogenicity in humans — cohort studies, case–control studies, correlation (or ecological) studies and intervention studies. Rarely, results from randomized trials may be available. Case reports and case series of cancer in humans may also be reviewed.

Cohort and case–control studies relate individual exposures under study to the occurrence of cancer in individuals and provide an estimate of effect (such as relative risk) as the main measure of association. Intervention studies may provide strong evidence for making causal inferences, as exemplified by cessation of smoking and the subsequent decrease in risk for lung cancer.

In correlation studies, the units of investigation are usually whole populations (e.g. in particular geographical areas or at particular times), and cancer frequency is related to a summary measure of the exposure of the population to the agent under study. In correlation studies, individual exposure is not documented, which renders this kind of study more prone to confounding. In some circumstances, however, correlation studies may be more informative than analytical study designs (see, for example, the *Monograph* on arsenic in drinking-water; IARC, 2004).

In some instances, case reports and case series have provided important information about the carcinogenicity of an agent. These types of study generally arise from a suspicion, based on clinical experience, that the concurrence of two events — that is, a particular exposure and occurrence of a cancer — has happened rather more frequently than would be expected by chance. Case reports and case series usually lack complete ascertainment of cases in any population, definition or enumeration of the population at risk and estimation of the expected number of cases in the absence of exposure.

The uncertainties that surround the interpretation of case reports, case series and correlation studies make them inadequate, except in rare instances, to form the sole basis for inferring a causal relationship. When taken together with case–control and cohort studies, however, these types of study may add materially to the judgement that a causal relationship exists.

Epidemiological studies of benign neoplasms, presumed preneoplastic lesions and other end-points thought to be relevant to cancer are also reviewed. They may, in some instances, strengthen inferences drawn from studies of cancer itself.

(b) Quality of studies considered

It is necessary to take into account the possible roles of bias, confounding and chance in the interpretation of epidemiological studies. Bias is the effect of factors in study design or execution that lead erroneously to a stronger or weaker association than in fact exists between an agent and disease. Confounding is a form of bias that occurs when the relationship with disease is made to appear stronger or weaker than it truly is as a result of an association between the apparent causal factor and another factor that is associated with either an increase or decrease in the incidence of the disease. The role of chance is related to biological variability and the influence of sample size on the precision of estimates of effect.

In evaluating the extent to which these factors have been minimized in an individual study, consideration is given to several aspects of design and analysis as described in the report of the study. For example, when suspicion of carcinogenicity arises largely from a single small study, careful consideration is given when interpreting subsequent studies that included these data in an enlarged population. Most of these considerations apply equally to case–control, cohort and correlation studies. Lack of clarity of any of these aspects in the reporting of a study can decrease its credibility and the weight given to it in the final evaluation of the exposure.

First, the study population, disease (or diseases) and exposure should have been well defined by the authors. Cases of disease in the study population should have been identified in a way that was independent of the exposure of interest, and exposure should have been assessed in a way that was not related to disease status.

Second, the authors should have taken into account — in the study design and analysis — other variables that can influence the risk of disease and may have been related to the exposure of interest. Potential confounding by such variables should have been dealt with either in the design of the study, such as by matching, or in the analysis, by statistical adjustment. In cohort studies, comparisons with local rates of disease may or may not be more appropriate than those with national rates. Internal comparisons of frequency of disease among individuals at different levels of exposure are also desirable in cohort studies, since they minimize the potential for confounding related to the difference in risk factors between an external reference group and the study population.

Third, the authors should have reported the basic data on which the conclusions are founded, even if sophisticated statistical analyses were employed. At the very least, they should have given the numbers of exposed and unexposed cases and controls in a case–control study and the numbers of cases observed and expected in a cohort study. Further tabulations by time since exposure began and other temporal factors are also important. In a cohort study, data on all cancer sites and all causes of death should have been given, to reveal the possibility of reporting bias. In a case–control study, the effects of investigated factors other than the exposure of interest should have been reported.

Finally, the statistical methods used to obtain estimates of relative risk, absolute rates of cancer, confidence intervals and significance tests, and to adjust for confounding should have been clearly stated by the authors. These methods have been reviewed for case–control studies (Breslow & Day, 1980) and for cohort studies (Breslow & Day, 1987).

(c) Meta-analyses and pooled analyses

Independent epidemiological studies of the same agent may lead to results that are difficult to interpret. Combined analyses of data from multiple studies are a means of resolving this ambiguity, and well conducted analyses can be considered. There are two types of combined analysis. The first involves combining summary statistics such as relative risks from individual studies (meta-analysis) and the second involves a pooled analysis of the raw data from the individual studies (pooled analysis) (Greenland, 1998).

The advantages of combined analyses are increased precision due to increased sample size and the opportunity to explore potential confounders, interactions and modifying effects that may explain heterogeneity among studies in more detail. A disadvantage of combined analyses is the possible lack of compatibility of data from various studies due to differences in subject recruitment, procedures of data collection, methods of measurement and effects of unmeasured co-variates that may differ among studies. Despite these limitations, well conducted combined analyses may provide a firmer basis than individual studies for drawing conclusions about the potential carcinogenicity of agents.

IARC may commission a meta-analysis or pooled analysis that is pertinent to a particular *Monograph* (see Part A, Section 4). Additionally, as a means of gaining insight from the results of multiple individual studies, ad hoc calculations that combine data from different studies may be conducted by the Working Group during the course of a *Monograph* meeting. The results of such original calculations, which would be specified in the text by presentation in square brackets, might involve updates of previously conducted analyses that incorporate the results of more recent studies or de-novo analyses. Irrespective of the source of data for the meta-analyses and pooled analyses, it is important that the same criteria for data quality be applied as those that would be applied to individual studies and to ensure also that sources of heterogeneity between studies be taken into account.

(d) Temporal effects

Detailed analyses of both relative and absolute risks in relation to temporal variables, such as age at first exposure, time since first exposure, duration of exposure, cumulative exposure, peak exposure (when appropriate) and time since cessation of exposure, are reviewed and summarized when available. Analyses of temporal relationships may be useful in making causal inferences. In addition, such analyses may suggest whether a carcinogen acts early or late in the process of carcinogenesis, although, at best, they allow only indirect inferences about mechanisms of carcinogenesis.

(e) Use of biomarkers in epidemiological studies

Biomarkers indicate molecular, cellular or other biological changes and are increasingly used in epidemiological studies for various purposes (IARC, 1991; Vainio et al., 1992; Toniolo et al., 1997; Vineis et al., 1999; Buffler et al., 2004). These may include evidence of exposure, of early effects, of cellular, tissue or organism responses, of individual susceptibility or host responses, and inference of a mechanism (see Part B, Section 4b). This is a rapidly evolving field that encompasses developments in genomics, epigenomics and other emerging technologies.

Molecular epidemiological data that identify associations between genetic polymorphisms and interindividual differences in susceptibility to the agent(s) being evaluated may contribute to the identification of carcinogenic hazards to humans. If the polymorphism has been demonstrated experimentally to modify the functional activity of the gene product in a manner that is consistent with increased susceptibility, these data may be useful in making causal inferences. Similarly, molecular epidemiological studies that measure cell functions, enzymes or metabolites that are thought to be the basis of susceptibility may provide evidence that reinforces biological plausibility. It should be noted, however, that when data on genetic susceptibility originate from multiple comparisons that arise from subgroup analyses, this can generate false-positive results and inconsistencies across studies, and such data therefore require careful evaluation. If the known phenotype of a genetic polymorphism can explain the carcinogenic mechanism of the agent being evaluated, data on this phenotype may be useful in making causal inferences.

(f) Criteria for causality

After the quality of individual epidemiological studies of cancer has been summarized and assessed, a judgement is made concerning the strength of evidence that the agent in question is carcinogenic to humans. In making its judgement, the Working Group considers several criteria for causality (Hill, 1965). A strong association (e.g. a large relative risk) is more likely to indicate causality than a weak association, although it is recognized that estimates of effect of small magnitude do not imply lack of causality and may be important if the disease or exposure is common. Associations that are replicated in several studies of the same design or that use different epidemiological approaches or under different circumstances of exposure are more likely to represent a causal relationship than isolated observations from single studies. If there are inconsistent results among investigations, possible reasons are sought (such as differences in exposure), and results of studies that are judged to be of high quality are given more weight than those of studies that are judged to be methodologically less sound.

If the risk increases with the exposure, this is considered to be a strong indication of causality, although the absence of a graded response is not necessarily evidence against a causal relationship. The demonstration of a decline in risk after cessation of or reduction in exposure in individuals or in whole populations also supports a causal interpretation of the findings.

Several scenarios may increase confidence in a causal relationship. On the one hand, an agent may be specific in causing tumours at one site or of one morphological type. On the other, carcinogenicity may be evident through the causation of multiple tumour types. Temporality, precision of estimates of effect, biological plausibility and coherence of the overall database are considered. Data on biomarkers may be employed in an assessment of the biological plausibility of epidemiological observations.

Although rarely available, results from randomized trials that show different rates of cancer among exposed and unexposed individuals provide particularly strong evidence for causality.

When several epidemiological studies show little or no indication of an association between an exposure and cancer, a judgement may be made that, in the aggregate, they show evidence of lack of carcinogenicity. Such a judgement requires first that the studies meet, to a sufficient degree, the standards of design and analysis described above. Specifically, the possibility that bias, confounding or misclassification of exposure or outcome could explain the observed results should be considered and excluded with reasonable certainty. In addition, all studies that are judged to be methodologically sound should (a) be consistent with an estimate of effect of unity for any observed level of exposure, (b) when considered together, provide a pooled estimate of relative risk that is at or near to unity, and (c) have a narrow confidence interval, due to sufficient population size. Moreover, no individual study nor the pooled results of all the studies should show any consistent tendency that the relative risk of cancer increases with increasing level of exposure. It is important to note that evidence of lack of carcinogenicity obtained from several epidemiological studies can apply only to the type(s) of cancer studied, to the dose levels reported, and to the intervals between first exposure and disease onset observed in these studies. Experience with human cancer indicates that the period from first exposure to the development of clinical cancer is sometimes longer than 20 years; latent periods substantially shorter than 30 years cannot provide evidence for lack of carcinogenicity.

3. Studies of cancer in experimental animals

All known human carcinogens that have been studied adequately for carcinogenicity in experimental animals have produced positive results in one or more animal species (Wilbourn et al., 1986; Tomatis et al., 1989). For several agents (e.g. aflatoxins, diethylstilbestrol, solar radiation, vinyl chloride), carcinogenicity in experimental animals was established or highly suspected before epidemiological studies confirmed their carcinogenicity in humans (Vainio et al., 1995). Although this association cannot establish that all agents that cause cancer in experimental animals also cause cancer in humans, it is biologically plausible that agents for which there is *sufficient evidence of carcinogenicity* in experimental animals (see Part B, Section 6b) also present a carcinogenic hazard to humans. Accordingly, in the absence of additional scientific information, these agents are considered to pose a carcinogenic hazard to humans. Examples of additional scientific information are data that demonstrate that a given agent causes cancer in animals through a species-specific mechanism that does not operate in humans or data that demonstrate that the mechanism in experimental animals also operates in humans (see Part B, Section 6).

Consideration is given to all available long-term studies of cancer in experimental animals with the agent under review (see Part A, Section 4). In all experimental settings, the nature and extent of impurities or contaminants present in the agent being evaluated are given when available. Animal species, strain (including genetic background where applicable), sex, numbers per group, age at start of treatment, route of exposure, dose levels, duration of exposure, survival and information on tumours (incidence, latency, severity or multiplicity of neoplasms or preneoplastic lesions) are reported. Those studies in experimental animals that are judged to be irrelevant to the evaluation or judged to be inadequate (e.g. too short a duration, too few animals, poor survival; see below) may be omitted. Guidelines for conducting long-term carcinogenicity experiments have been published (e.g. OECD, 2002).

Other studies considered may include: experiments in which the agent was administered in the presence of factors that modify carcinogenic effects (e.g. initiation–promotion studies, co-carcinogenicity studies and studies in genetically modified animals); studies in which the end-point was not cancer but a defined precancerous lesion; experiments on the carcinogenicity of known metabolites and derivatives; and studies of cancer in non-laboratory animals (e.g. livestock and companion animals) exposed to the agent.

For studies of mixtures, consideration is given to the possibility that changes in the physicochemical properties of the individual substances may occur during collection, storage, extraction, concentration and delivery. Another consideration is that chemical and toxicological interactions of components in a mixture may alter dose–response relationships. The relevance to human exposure of the test mixture administered in the animal experiment is also assessed. This may involve consideration of the following aspects of the mixture tested: (i) physical and chemical characteristics, (ii) identified constituents that may indicate the presence of a class of substances and (iii) the results of genetic toxicity and related tests.

The relevance of results obtained with an agent that is analogous (e.g. similar in structure or of a similar virus genus) to that being evaluated is also considered. Such results may provide biological and mechanistic information that is relevant to the understanding of the process of carcinogenesis in humans and may strengthen the biological plausibility that the agent being evaluated is carcinogenic to humans (see Part B, Section 2f).

(a) Qualitative aspects

An assessment of carcinogenicity involves several considerations of qualitative importance, including (i) the experimental conditions under which the test was performed, including route, schedule and duration of exposure, species, strain (including genetic background where applicable), sex, age and duration of follow-up; (ii) the consistency of the results, for example, across species and target organ(s); (iii) the spectrum of neoplastic response, from preneoplastic lesions and benign tumours to malignant neoplasms; and (iv) the possible role of modifying factors.

Considerations of importance in the interpretation and evaluation of a particular study include: (i) how clearly the agent was defined and, in the case of mixtures, how adequately the sample characterization was reported; (ii) whether the dose was monitored adequately, particularly in inhalation experiments; (iii) whether the doses, duration of treatment and route of exposure were appropriate; (iv) whether the survival of treated animals was similar to that of controls; (v) whether there were adequate numbers of animals per group; (vi) whether both male and female animals were used; (vii) whether animals were allocated randomly to groups; (viii) whether the duration of observation was adequate; and (ix) whether the data were reported and analysed adequately.

When benign tumours (a) occur together with and originate from the same cell type as malignant tumours in an organ or tissue in a particular study and (b) appear to represent a stage in the progression to malignancy, they are usually combined in the assessment of tumour incidence (Huff et al., 1989). The occurrence of lesions presumed to be preneoplastic may in certain instances aid in assessing the biological plausibility of any neoplastic response observed. If an agent induces only benign neoplasms that appear to be end-points that do not readily undergo transition to malignancy, the agent should nevertheless be suspected of being carcinogenic and requires further investigation.

(b) Quantitative aspects

The probability that tumours will occur may depend on the species, sex, strain, genetic background and age of the animal, and on the dose, route, timing and duration of the exposure. Evidence of an increased incidence of neoplasms with increasing levels of exposure strengthens the inference of a causal association between the exposure and the development of neoplasms.

The form of the dose–response relationship can vary widely, depending on the particular agent under study and the target organ. Mechanisms such as induction of DNA damage or inhibition of repair, altered cell division and cell death rates and changes in intercellular communication are important determinants of dose–response relationships for some carcinogens. Since many chemicals require metabolic activation before being converted to their reactive intermediates, both metabolic and toxicokinetic aspects are important in determining the dose–response pattern. Saturation of steps such as absorption, activation, inactivation and elimination may produce nonlinearity in the dose–response relationship (Hoel et al., 1983; Gart et al., 1986), as could saturation of processes such as DNA repair. The dose–response relationship can also be affected by differences in survival among the treatment groups.

(c) Statistical analyses

Factors considered include the adequacy of the information given for each treatment group: (i) number of animals studied and number examined histologically, (ii) number of animals with a given tumour type and (iii) length of survival. The statistical methods used should be clearly stated and should be the generally accepted techniques refined for this purpose (Peto et al., 1980; Gart et al., 1986; Portier & Bailer, 1989; Bieler & Williams, 1993). The choice of the most appropriate statistical method requires consideration of whether or not there are differences in survival among the treatment groups; for example, reduced survival because of non-tumour-related mortality can preclude the occurrence of tumours later in life. When detailed information on survival is not available, comparisons of the proportions of tumour-bearing animals among the effective number of animals (alive at the time the first tumour was discovered) can be useful when significant differences in survival occur before tumours appear. The lethality of the tumour also requires consideration: for rapidly fatal tumours, the time of death provides an indication of the time of tumour onset and can be assessed using life-table methods; non-fatal or incidental tumours that do not affect survival can be assessed using methods such as the Mantel-Haenzel test for changes in tumour prevalence. Because tumour lethality is often difficult to determine, methods such as the Poly-K test that do not require such information can also be used. When results are available on the number and size of tumours seen in experimental animals (e.g. papillomas on mouse skin, liver tumours observed through nuclear magnetic resonance tomography), other more complicated statistical procedures may be needed (Sherman et al., 1994; Dunson et al., 2003).

Formal statistical methods have been developed to incorporate historical control data into the analysis of data from a given experiment. These methods assign an appropriate weight to historical and concurrent controls on the basis of the extent of between-study and within-study variability: less weight is given to historical controls when they show a high degree of variability, and greater weight when they show little variability. It is generally not appropriate to discount a tumour response that is significantly increased compared with concurrent controls by arguing that it falls within the range of historical controls, particularly when historical controls show high between-study variability and are, thus, of little relevance to the current experiment. In analysing results for uncommon tumours, however, the analysis may be improved by considering historical control data, particularly when between-study variability is low. Historical controls should be selected to resemble the concurrent controls as closely as possible with respect to species, gender and strain, as well as other factors such as basal diet and general laboratory environment, which may affect tumour-response rates in control animals (Haseman et al., 1984; Fung et al., 1996; Greim et al., 2003).

Although meta-analyses and combined analyses are conducted less frequently for animal experiments than for epidemiological studies due to differences in animal strains, they can be useful aids in interpreting animal data when the experimental protocols are sufficiently similar.

4. Mechanistic and other relevant data

Mechanistic and other relevant data may provide evidence of carcinogenicity and also help in assessing the relevance and importance of findings of cancer in animals and in humans. The nature of the mechanistic and other relevant data depends on the biological activity of the agent being considered. The Working Group considers representative studies to give a concise description of the relevant data and issues that they consider to be important; thus, not every available study is cited. Relevant topics may include toxicokinetics, mechanisms of carcinogenesis, susceptible individuals, populations and life-stages, other relevant data and other adverse effects. When data on biomarkers are informative about the mechanisms of carcinogenesis, they are included in this section.

These topics are not mutually exclusive; thus, the same studies may be discussed in more than one subsection. For example, a mutation in a gene that codes for an enzyme that metabolizes the agent under study could be discussed in the subsections on toxicokinetics, mechanisms and individual susceptibility if it also exists as an inherited polymorphism.

(a) Toxicokinetic data

Toxicokinetics refers to the absorption, distribution, metabolism and elimination of agents in humans, experimental animals and, where relevant, cellular systems. Examples of kinetic factors that may affect dose–response relationships include uptake, deposition, biopersistence and half-life in tissues, protein binding, metabolic activation and detoxification. Studies that indicate the metabolic fate of the agent in humans and in experimental animals are summarized briefly, and comparisons of data from humans and animals are made when possible. Comparative information on the relationship between exposure and the dose that reaches the target site may be important for the extrapolation of hazards between species and in clarifying the role of in-vitro findings.

(b) Data on mechanisms of carcinogenesis

To provide focus, the Working Group attempts to identify the possible mechanisms by which the agent may increase the risk of cancer. For each possible mechanism, a representative selection of key data from humans and experimental systems is summarized. Attention is given to gaps in the data and to data that suggests that more than one mechanism may be operating. The relevance of the mechanism to humans is discussed, in particular, when mechanistic data are derived from experimental model systems. Changes in the affected organs, tissues or cells can be divided into three non-exclusive levels as described below.

(i) Changes in physiology

Physiological changes refer to exposure-related modifications to the physiology and/or response of cells, tissues and organs. Examples of potentially adverse physiological changes include mitogenesis, compensatory cell division, escape from apoptosis and/or senescence, presence of inflammation, hyperplasia, metaplasia and/or preneoplasia, angiogenesis, alterations in cellular adhesion, changes in steroidal hormones and changes in immune surveillance.

(ii) Functional changes at the cellular level

Functional changes refer to exposure-related alterations in the signalling pathways used by cells to manage critical processes that are related to increased risk for cancer. Examples of functional changes include modified activities of enzymes involved in the metabolism of xenobiotics, alterations in the expression of key genes that regulate DNA repair, alterations in cyclin-dependent kinases that govern cell cycle progression, changes in the patterns of post-translational modifications of proteins, changes in regulatory factors that alter apoptotic rates, changes in the secretion of factors related to the stimulation of DNA replication and transcription and changes in gap–junction-mediated intercellular communication.

(iii) Changes at the molecular level

Molecular changes refer to exposure-related changes in key cellular structures at the molecular level, including, in particular, genotoxicity. Examples of molecular changes include formation of DNA adducts and DNA strand breaks, mutations in genes, chromosomal aberrations, aneuploidy and changes in DNA methylation patterns. Greater emphasis is given to irreversible effects.

The use of mechanistic data in the identification of a carcinogenic hazard is specific to the mechanism being addressed and is not readily described for every possible level and mechanism discussed above.

Genotoxicity data are discussed here to illustrate the key issues involved in the evaluation of mechanistic data.

Tests for genetic and related effects are described in view of the relevance of gene mutation and chromosomal aberration/aneuploidy to carcinogenesis (Vainio et al., 1992; McGregor et al., 1999). The adequacy of the reporting of sample characterization is considered and, when necessary, commented upon; with regard to complex mixtures, such comments are similar to those described for animal carcinogenicity tests. The available data are interpreted critically according to the end-points detected, which may include DNA damage, gene mutation, sister chromatid exchange, micronucleus formation, chromosomal aberrations and aneuploidy. The concentrations employed are given, and mention is made of whether the use of an exogenous metabolic system in vitro affected the test result. These data are listed in tabular form by phylogenetic classification.

Positive results in tests using prokaryotes, lower eukaryotes, insects, plants and cultured mammalian cells suggest that genetic and related effects could occur in mammals. Results from such tests may also give information on the types of genetic effect produced and on the involvement of metabolic activation. Some end-points described are clearly genetic in nature (e.g. gene mutations), while others are associated with genetic effects (e.g. unscheduled DNA synthesis). In-vitro tests for tumour promotion, cell transformation and gap–junction intercellular communication may be sensitive to changes that are not necessarily the result of genetic alterations but that may have specific relevance to the process of carcinogenesis. Critical appraisals of these tests have been published (Montesano et al., 1986; McGregor et al., 1999).

Genetic or other activity manifest in humans and experimental mammals is regarded to be of greater relevance than that in other organisms. The demonstration that an agent can induce gene and chromosomal mutations in mammals in vivo indicates that it may have carcinogenic activity. Negative results in tests for mutagenicity in selected tissues from animals treated in vivo provide less weight, partly because they do not exclude the possibility of an effect in tissues other than those examined. Moreover, negative results in short-term tests with genetic end-points cannot be considered to provide evidence that rules out the carcinogenicity of agents that act through other mechanisms (e.g. receptor-mediated effects, cellular toxicity with regenerative cell division, peroxisome proliferation) (Vainio et al., 1992). Factors that may give misleading results in short-term tests have been discussed in detail elsewhere (Montesano et al., 1986; McGregor et al., 1999).

When there is evidence that an agent acts by a specific mechanism that does not involve genotoxicity (e.g. hormonal dysregulation, immune suppression, and formation of calculi and other deposits that cause chronic irritation), that evidence is presented and reviewed critically in the context of rigorous criteria for the operation of that mechanism in carcinogenesis (e.g. Capen et al., 1999).

For biological agents such as viruses, bacteria and parasites, other data relevant to carcinogenicity may include descriptions of the pathology of infection, integration and expression of viruses, and genetic alterations seen in human tumours. Other observations that might comprise cellular and tissue responses to infection, immune response and the presence of tumour markers are also considered.

For physical agents that are forms of radiation, other data relevant to carcinogenicity may include descriptions of damaging effects at the physiological, cellular and molecular level, as for chemical agents, and descriptions of how these effects occur. 'Physical agents' may also be considered to comprise foreign bodies, such as surgical implants of various kinds, and poorly soluble fibres, dusts and particles of various sizes, the pathogenic effects of which are a result of their physical presence in tissues or body cavities. Other relevant data for such materials may include characterization of cellular, tissue and physiological reactions to these materials and descriptions of pathological conditions other than neoplasia with which they may be associated.

(c) Other data relevant to mechanisms

A description is provided of any structure–activity relationships that may be relevant to an evaluation of the carcinogenicity of an agent, the toxicological implications of the physical and chemical properties, and any other data relevant to the evaluation that are not included elsewhere.

High-output data, such as those derived from gene expression microarrays, and high-throughput data, such as those that result from testing hundreds of agents for a single end-point, pose a unique problem for the use of mechanistic data in the evaluation of a carcinogenic hazard. In the case of high-output data, there is the possibility to overinterpret changes in individual end-points (e.g. changes in expression in one gene) without considering the consistency of that finding in the broader context of the other end-points (e.g. other genes with linked transcriptional control). High-output data can be used in assessing mechanisms, but all end-points measured in a single experiment need to be considered in the proper context. For high-throughput data, where the number of observations far exceeds the number of end-points measured, their utility for identifying common mechanisms across multiple agents is enhanced. These data can be used to identify mechanisms that not only seem plausible, but also have a consistent pattern of carcinogenic response across entire classes of related compounds.

(d) Susceptibility data

Individuals, populations and life-stages may have greater or lesser susceptibility to an agent, based on toxicokinetics, mechanisms of carcinogenesis and other factors. Examples of host and genetic factors that affect individual susceptibility include sex, genetic polymorphisms of genes involved in the metabolism of the agent under evaluation, differences in metabolic capacity due to life-stage or the presence of disease, differences in DNA repair capacity, competition for or alteration of metabolic capacity by medications or other chemical exposures, pre-existing hormonal imbalance that is exacerbated by a chemical exposure, a suppressed immune system, periods of higher-than-usual tissue growth or regeneration and genetic polymorphisms that lead to differences in behaviour (e.g. addiction). Such data can substantially increase the strength of the evidence from epidemiological data and enhance the linkage of in-vivo and in-vitro laboratory studies to humans.

(e) Data on other adverse effects

Data on acute, subchronic and chronic adverse effects relevant to the cancer evaluation are summarized. Adverse effects that confirm distribution and biological effects at the sites of tumour development, or alterations in physiology that could lead to tumour development, are emphasized. Effects on reproduction, embryonic and fetal survival and development are summarized briefly. The adequacy of epidemiological studies of reproductive outcome and genetic and related effects in humans is judged by the same criteria as those applied to epidemiological studies of cancer, but fewer details are given.

5. Summary

This section is a summary of data presented in the preceding sections. Summaries can be found on the *Monographs* programme web site (http://monographs.iarc.fr).

(a) Exposure data

Data are summarized, as appropriate, on the basis of elements such as production, use, occurrence and exposure levels in the workplace and environment and measurements in human tissues and body fluids. Quantitative data and time trends are given to compare exposures in different occupations and environmental settings. Exposure to biological agents is described in terms of transmission, prevalence and persistence of infection.

(b) Cancer in humans

Results of epidemiological studies pertinent to an assessment of human carcinogenicity are summarized. When relevant, case reports and correlation studies are also summarized. The target organ(s) or tissue(s) in which an increase in cancer was observed is identified. Dose–response and other quantitative data may be summarized when available.

(c) Cancer in experimental animals

Data relevant to an evaluation of carcinogenicity in animals are summarized. For each animal species, study design and route of administration, it is stated whether an increased incidence, reduced latency, or increased severity or multiplicity of neoplasms or preneoplastic lesions were observed, and the tumour sites are indicated. If the agent produced tumours after prenatal exposure or in single-dose experiments, this is also mentioned. Negative findings, inverse relationships, dose–response and other quantitative data are also summarized.

(d) Mechanistic and other relevant data

Data relevant to the toxicokinetics (absorption, distribution, metabolism, elimination) and the possible mechanism(s) of carcinogenesis (e.g. genetic toxicity, epigenetic effects) are summarized. In addition, information on susceptible individuals, populations and life-stages is summarized. This section also reports on other toxic effects, including reproductive and developmental effects, as well as additional relevant data that are considered to be important.

6. Evaluation and rationale

Evaluations of the strength of the evidence for carcinogenicity arising from human and experimental animal data are made, using standard terms. The strength of the mechanistic evidence is also characterized.

It is recognized that the criteria for these evaluations, described below, cannot encompass all of the factors that may be relevant to an evaluation of carcinogenicity. In considering all of the relevant scientific data, the Working Group may assign the agent to a higher or lower category than a strict interpretation of these criteria would indicate.

These categories refer only to the strength of the evidence that an exposure is carcinogenic and not to the extent of its carcinogenic activity (potency). A classification may change as new information becomes available.

An evaluation of the degree of evidence is limited to the materials tested, as defined physically, chemically or biologically. When the agents evaluated are considered by the Working Group to be sufficiently closely related, they may be grouped together for the purpose of a single evaluation of the degree of evidence.

(a) Carcinogenicity in humans

The evidence relevant to carcinogenicity from studies in humans is classified into one of the following categories:

***Sufficient evidence of carcinogenicity*:**

The Working Group considers that a causal relationship has been established between exposure to the agent and human cancer. That is, a positive relationship has been observed between the exposure and cancer in studies in which chance, bias and confounding could be ruled out with reasonable confidence. A statement that there is *sufficient evidence* is followed by a separate sentence that identifies the target organ(s) or tissue(s) where an increased risk of cancer was observed in humans. Identification of a specific target organ or tissue does not preclude the possibility that the agent may cause cancer at other sites.

***Limited evidence of carcinogenicity*:**

A positive association has been observed between exposure to the agent and cancer for which a causal interpretation is considered by the Working Group to be credible, but chance, bias or confounding could not be ruled out with reasonable confidence.

***Inadequate evidence of carcinogenicity*:**

The available studies are of insufficient quality, consistency or statistical power to permit a conclusion regarding the presence or absence of a causal association between exposure and cancer, or no data on cancer in humans are available.

***Evidence suggesting lack of carcinogenicity*:**

There are several adequate studies covering the full range of levels of exposure that humans are known to encounter, which are mutually consistent in not showing a positive association between exposure to the agent and any studied cancer at any observed level of exposure. The results from these studies alone or combined should have narrow confidence intervals with an upper limit close to the null value (e.g. a relative risk of 1.0). Bias and confounding should be ruled out with reasonable confidence, and the studies should have an adequate length of follow-up. A conclusion of *evidence suggesting lack of carcinogenicity* is inevitably limited to the cancer sites, conditions and levels of exposure, and length of observation covered by the available studies. In addition, the possibility of a very small risk at the levels of exposure studied can never be excluded.

In some instances, the above categories may be used to classify the degree of evidence related to carcinogenicity in specific organs or tissues.

When the available epidemiological studies pertain to a mixture, process, occupation or industry, the Working Group seeks to identify the specific agent considered most likely to be responsible for any excess risk. The evaluation is focused as narrowly as the available data on exposure and other aspects permit.

(b) Carcinogenicity in experimental animals

Carcinogenicity in experimental animals can be evaluated using conventional bioassays, bioassays that employ genetically modified animals, and other in-vivo bioassays that focus on one or more of the critical stages of carcinogenesis. In the absence of data from conventional long-term bioassays or from assays with neoplasia as the end-point, consistently positive results in several models that address several stages in the multistage process of carcinogenesis should be considered in evaluating the degree of evidence of carcinogenicity in experimental animals.

The evidence relevant to carcinogenicity in experimental animals is classified into one of the following categories:

***Sufficient evidence of carcinogenicity*:**

The Working Group considers that a causal relationship has been established between the agent and an increased incidence of malignant neoplasms or of an appropriate combination of benign and malignant neoplasms in (a) two or more species of animals or (b) two or more independent studies in one species carried out at different times or in different laboratories or under different protocols. An increased incidence of tumours in both sexes of a single species in a well conducted study, ideally conducted under Good Laboratory Practices, can also provide *sufficient evidence*.

A single study in one species and sex might be considered to provide *sufficient evidence of carcinogenicity* when malignant neoplasms occur to an unusual degree with regard to incidence, site, type of tumour or age at onset, or when there are strong findings of tumours at multiple sites.

***Limited evidence of carcinogenicity*:**

The data suggest a carcinogenic effect but are limited for making a definitive evaluation because, e.g. (a) the evidence of carcinogenicity is restricted to a single experiment; (b) there are unresolved questions regarding the adequacy of the design, conduct or interpretation of the studies; (c) the agent increases the incidence only of benign neoplasms or lesions of uncertain neoplastic potential; or (d) the evidence of carcinogenicity is restricted to studies that demonstrate only promoting activity in a narrow range of tissues or organs.

***Inadequate evidence of carcinogenicity*:**

The studies cannot be interpreted as showing either the presence or absence of a carcinogenic effect because of major qualitative or quantitative limitations, or no data on cancer in experimental animals are available.

***Evidence suggesting lack of carcinogenicity*:**

Adequate studies involving at least two species are available which show that, within the limits of the tests used, the agent is not carcinogenic. A conclusion of *evidence suggesting lack of carcinogenicity* is inevitably limited to the species, tumour sites, age at exposure, and conditions and levels of exposure studied.

(c) Mechanistic and other relevant data

Mechanistic and other evidence judged to be relevant to an evaluation of carcinogenicity and of sufficient importance to affect the overall evaluation is highlighted. This may include data on preneoplastic lesions, tumour pathology, genetic and related effects, structure–activity relationships, metabolism and toxicokinetics, physicochemical parameters and analogous biological agents.

The strength of the evidence that any carcinogenic effect observed is due to a particular mechanism is evaluated, using terms such as 'weak', 'moderate' or 'strong'. The Working Group then assesses whether that particular mechanism is likely to be operative in humans. The strongest indications that a particular mechanism operates in humans derive from data on humans or biological specimens obtained from exposed humans. The data may be considered to be especially relevant if they show that the agent in question has caused changes in exposed humans that are on the causal pathway to carcinogenesis. Such data may, however, never become available, because it is at least conceivable that certain compounds may be kept from human use solely on the basis of evidence of their toxicity and/or carcinogenicity in experimental systems.

The conclusion that a mechanism operates in experimental animals is strengthened by findings of consistent results in different experimental systems, by the demonstration of biological plausibility and by coherence of the overall database. Strong support can be obtained from studies that challenge the hypothesized mechanism experimentally, by demonstrating that the suppression of key mechanistic processes leads to the suppression of tumour development. The Working Group considers whether multiple mechanisms might contribute to tumour development, whether different mechanisms might operate in different dose ranges, whether separate mechanisms might operate in humans and experimental animals and whether a unique mechanism might operate in a susceptible group. The possible contribution of alternative mechanisms must be considered before concluding that tumours observed in experimental animals are not relevant to humans. An uneven level of experimental support for different mechanisms may reflect that disproportionate resources have been focused on investigating a favoured mechanism.

For complex exposures, including occupational and industrial exposures, the chemical composition and the potential contribution of carcinogens known to be present are considered by the Working Group in its overall evaluation of human carcinogenicity. The Working Group also determines the extent to which the materials tested in experimental systems are related to those to which humans are exposed.

(d) Overall evaluation

Finally, the body of evidence is considered as a whole, to reach an overall evaluation of the carcinogenicity of the agent to humans.

An evaluation may be made for a group of agents that have been evaluated by the Working Group. In addition, when supporting data indicate that other related agents, for which there is no direct evidence of their capacity to induce cancer in humans or in animals, may also be carcinogenic, a statement describing the rationale for this conclusion is added to the evaluation narrative; an additional evaluation may be made for this broader group of agents if the strength of the evidence warrants it.

The agent is described according to the wording of one of the following categories, and the designated group is given. The categorization of an agent is a matter of scientific judgement that reflects the strength of the evidence derived from studies in humans and in experimental animals and from mechanistic and other relevant data.

Group 1: The agent is carcinogenic to humans.

This category is used when there is *sufficient evidence of carcinogenicity* in humans. Exceptionally, an agent may be placed in this category when evidence of carcinogenicity in humans is less than *sufficient* but there is *sufficient evidence of carcinogenicity* in experimental animals and strong evidence in exposed humans that the agent acts through a relevant mechanism of carcinogenicity.

Group 2.

This category includes agents for which, at one extreme, the degree of evidence of carcinogenicity in humans is almost *sufficient*, as well as those for which, at the other extreme, there are no human data but for which there is evidence of carcinogenicity in experimental animals. Agents are assigned to either Group 2A (*probably carcinogenic to humans*) or Group 2B (*possibly carcinogenic to humans*) on the basis of epidemiological and experimental evidence of carcinogenicity and mechanistic and other relevant data. The terms *probably carcinogenic* and *possibly carcinogenic* have no quantitative significance and are used simply as descriptors of different levels of evidence of human carcinogenicity, with *probably carcinogenic* signifying a higher level of evidence than *possibly carcinogenic.*

Group 2A: The agent is probably carcinogenic to humans.

This category is used when there is *limited evidence of carcinogenicity* in humans and *sufficient evidence of carcinogenicity* in experimental animals. In some cases, an agent may be classified in this category when there is *inadequate evidence of carcinogenicity* in humans and *sufficient evidence of carcinogenicity* in experimental animals and strong evidence that the carcinogenesis is mediated by a mechanism that also operates in humans. Exceptionally, an agent may be classified in this category solely on the basis of *limited evidence of carcinogenicity* in humans. An agent may be assigned to this category if it clearly belongs, based on mechanistic considerations, to a class of agents for which one or more members have been classified in Group 1 or Group 2A.

Group 2B: The agent is possibly carcinogenic to humans.

This category is used for agents for which there is *limited evidence of carcinogenicity* in humans and less than *sufficient evidence of carcinogenicity* in experimental animals. It may also be used when there is *inadequate evidence of carcinogenicity* in humans but there is *sufficient evidence of carcinogenicity* in experimental animals. In some instances, an agent for which there is *inadequate evidence of carcinogenicity* in humans and less than *sufficient evidence of carcinogenicity* in experimental animals together with supporting evidence from mechanistic and other relevant data may be placed in this group. An agent may be classified in this category solely on the basis of strong evidence from mechanistic and other relevant data.

Group 3: The agent is not classifiable as to its carcinogenicity to humans.

This category is used most commonly for agents for which the evidence of carcinogenicity is *inadequate* in humans and *inadequate* or *limited* in experimental animals.

Exceptionally, agents for which the evidence of carcinogenicity is *inadequate* in humans but *sufficient* in experimental animals may be placed in this category when there is strong evidence that the mechanism of carcinogenicity in experimental animals does not operate in humans.

Agents that do not fall into any other group are also placed in this category.

An evaluation in Group 3 is not a determination of non-carcinogenicity or overall safety. It often means that further research is needed, especially when exposures are widespread or the cancer data are consistent with differing interpretations.

Group 4: The agent is probably not carcinogenic to humans.

This category is used for agents for which there is *evidence suggesting lack of carcinogenicity* in humans and in experimental animals. In some instances, agents for which there is *inadequate evidence of carcinogenicity* in humans but *evidence suggesting lack of carcinogenicity* in experimental animals, consistently and strongly supported by a broad range of mechanistic and other relevant data, may be classified in this group.

(e) Rationale

The reasoning that the Working Group used to reach its evaluation is presented and discussed. This section integrates the major findings from studies of cancer in humans, studies of cancer in experimental animals, and mechanistic and other relevant data. It includes concise statements of the principal line(s) of argument that emerged, the conclusions of the Working Group on the strength of the evidence for each group of studies, citations to indicate which studies were pivotal to these conclusions, and an explanation of the reasoning of the Working Group in weighing data and making evaluations. When there are significant differences of scientific interpretation among Working Group Members, a brief summary of the alternative interpretations is provided, together with their scientific rationale and an indication of the relative degree of support for each alternative.

References

Bieler GS, Williams RL (1993). Ratio estimates, the delta method, and quantal response tests for increased carcinogenicity. *Biometrics*, 49:793–801. doi:10.2307/2532200 PMID:8241374

Breslow NE, Day NE (1980). Statistical methods in cancer research. Volume I - The analysis of case-control studies. *IARC Sci Publ*, 32:5–338. PMID:7216345

Breslow NE, Day NE (1987). Statistical methods in cancer research. Volume II–The design and analysis of cohort studies. *IARC Sci Publ*, 82:1–406. PMID:3329634

Buffler P, Rice J, Baan R et al. (2004). Workshop on mechanisms of carcinogenesis: contributions of molecular epidemiology. Lyon, 14–17 November 2001. Workshop report. *IARC Sci Publ*, 157:1–27. PMID:15055286

Capen CC, Dybing E, Rice JM, Wilbourn JD (1999). Species differences in thyroid, kidney and urinary bladder carcinogenesis. Proceedings of a consensus conference. Lyon, France, 3–7 November 1997. *IARC Sci Publ*, 147:1–225. PMID:10627184

Cogliano V, Baan R, Straif K et al. (2005). Transparency in IARC Monographs. *Lancet Oncol*, 6:747. doi:10.1016/S1470-2045(05)70380-6

Cogliano VJ, Baan RA, Straif K et al. (2004). The science and practice of carcinogen identification and evaluation. *Environ Health Perspect*, 112:1269–1274. doi:10.1289/ehp.6950 PMID:15345338

Dunson DB, Chen Z, Harry J (2003). A Bayesian approach for joint modeling of cluster size and subunit-specific outcomes. *Biometrics*, 59:521–530. doi:10.1111/1541-0420.00062 PMID:14601753

Fung KY, Krewski D, Smythe RT (1996). A comparison of tests for trend with historical controls in carcinogen bioassay. *Can J Stat*, 24:431–454. doi:10.2307/3315326

Gart JJ, Krewski D, Lee PN et al. (1986). Statistical methods in cancer research. Volume III–The design and analysis of long-term animal experiments. *IARC Sci Publ*, 79:1–219. PMID:3301661

Greenland S (1998). Meta-analysis. In: Rothman KJ, Greenland S, editors. Modern epidemiology. Philadelphia: Lippincott Williams & Wilkins, pp. 643–673.

Greim H, Gelbke H-P, Reuter U et al. (2003). Evaluation of historical control data in carcinogenicity studies. *Hum Exp Toxicol*, 22:541–549. doi:10.1191/0960327103ht394oa PMID:14655720

Haseman JK, Huff J, Boorman GA (1984). Use of historical control data in carcinogenicity studies in rodents. *Toxicol Pathol*, 12:126–135. doi:10.1177/019262338401200203 PMID:11478313

Hill AB (1965). The environment and disease: Association or causation? *Proc R Soc Med*, 58:295–300. PMID:14283879

Hoel DG, Kaplan NL, Anderson MW (1983). Implication of nonlinear kinetics on risk estimation in carcinogenesis. *Science*, 219:1032–1037. doi:10.1126/science.6823565 PMID:6823565

Huff JE, Eustis SL, Haseman JK (1989). Occurrence and relevance of chemically induced benign neoplasms in long-term carcinogenicity studies. *Cancer Metastasis Rev*, 8:1–22. doi:10.1007/BF00047055 PMID:2667783

IARC (1977). IARC Monographs Programme on the Evaluation of the Carcinogenic Risk of Chemicals to Humans. Preamble (IARC Intern Tech Rep No. 77/002).

IARC (1978). Chemicals with sufficient evidence of carcinogenicity in experimental animals – IARC Monographs Volumes 1–17 (IARC Intern Tech Rep No. 78/003).

IARC (1979). Criteria to select chemicals for IARC Monographs (IARC Intern Tech Rep No. 79/003).

IARC (1982). Chemicals, industrial processes and industries associated with cancer in humans (IARC Monographs, volumes 1 to 29). *IARC Monogr Eval Carcinog Risk Chem Hum Suppl*, 4:1–292.

IARC (1983). Approaches to classifying chemical carcinogens according to mechanism of action (IARC Intern Tech Rep No. 83/001).

IARC (1987). Overall evaluations of carcinogenicity: an updating of IARC Monographs volumes 1 to 42. *IARC Monogr Eval Carcinog Risks Hum Suppl*, 7:1–440. PMID:3482203

IARC (1988). Report of an IARC Working Group to Review the Approaches and Processes Used to Evaluate the Carcinogenicity of Mixtures and Groups of Chemicals (IARC Intern Tech Rep No. 88/002).

IARC (1991). A consensus report of an IARC Monographs Working Group on the Use of Mechanisms of Carcinogenesis in Risk Identification (IARC Intern Tech Rep No. 91/002).

IARC (2004). Some drinking-water disinfectants and contaminants, including arsenic. *IARC Monogr Eval Carcinog Risks Hum*, 84:1–477. PMID:15645577

IARC (2005). Report of the Advisory Group to Recommend Updates to the Preamble to the IARC Monographs (IARC Intern Rep No. 05/001).

IARC (2006). Report of the Advisory Group to Review the Amended Preamble to the IARC Monographs (IARC Intern Rep No. 06/001).

McGregor DB, Rice JM, Venitt S (1999). The use of short-and medium-term tests for carcinogens and data on genetic effects in carcinogenic hazard evaluation. Consensus report. *IARC Sci Publ*, 146:1–18. PMID:10353381

Montesano R, Bartsch H, Vainio H et al., editors(1986). Long-term and short-term assays for carcinogenesis—a critical appraisal. *IARC Sci Publ*, 83:1–564. PMID:3623675

OECD (2002). Guidance notes for analysis and evaluation of chronic toxicity and carcinogenicity studies (Series on Testing and Assessment No. 35), Paris: OECD.

Peto R, Pike MC, Day NE et al. (1980). Guidelines for simple, sensitive significance tests for carcinogenic effects in long-term animal experiments. *IARC Monogr Eval Carcinog Risk Chem Hum Suppl*, 2:Suppl: 311–426. PMID:6935185

Portier CJ, Bailer AJ (1989). Testing for increased carcinogenicity using a survival-adjusted quantal response test. *Fundam Appl Toxicol*, 12:731–737. doi:10.1016/0272-0590(89)90004-3 PMID:2744275

Sherman CD, Portier CJ, Kopp-Schneider A (1994). Multistage models of carcinogenesis: an approximation for the size and number distribution of late-stage clones. *Risk Anal*, 14:1039–1048. doi:10.1111/j.1539-6924.1994.tb00074.x PMID:7846311

Stewart BW, Kleihues P, editors (2003). World cancer report, Lyon: IARC.

Tomatis L, Aitio A, Wilbourn J, Shuker L (1989). Human carcinogens so far identified. *Jpn J Cancer Res*, 80:795–807. doi:10.1111/j.1349-7006.1989.tb01717.x PMID:2513295

Toniolo P, Boffetta P, Shuker DEG et al. (1997). Proceedings of the workshop on application of biomarkers to cancer epidemiology. Lyon, France, 20–23 February 1996. *IARC Sci Publ*, 142:1–318. PMID:9410826

Vainio H, Magee P, McGregor D, McMichael A (1992). Mechanisms of carcinogenesis in risk identification. IARC Working Group Meeting. Lyon, 11–18 June 1991. *IARC Sci Publ*, 116:1–608. PMID:1428077

Vainio H, Wilbourn JD, Sasco AJ et al. (1995). [Identification of human carcinogenic risks in IARC monographs] *Bull Cancer*, 82:339–348. PMID:7626841

Vineis P, Malats N, Lang M et al., editors (1999). Metabolic polymorphisms and susceptibility to cancer. *IARC Sci Publ*, 148:1–510. PMID:10493243

Wilbourn J, Haroun L, Heseltine E et al. (1986). Response of experimental animals to human carcinogens: an analysis based upon the IARC Monographs programme. *Carcinogenesis*, 7:1853–1863. doi:10.1093/carcin/7.11.1853 PMID:3769134

GENERAL REMARKS

This one-hundred-and-thirteenth volume of the *IARC Monographs* contains evaluations of the carcinogenic hazard to humans of three pesticides: DDT, lindane, and 2,4-D. DDT (1,1′-(2,2,2-trichloro-ethylidene)bis(4-chlorobenzene)) and lindane (γ-hexachlorocyclohexane) belong to the family of organochlorine insecticides and 2,4-D (2,4-dichlorophenoxyacetic acid) is a chlorophenoxy herbicide. Organochlorine pesticides were accorded priority for evaluation in the *IARC Monographs* programme by an Advisory Group that met in 2014 (Straif et al., 2014). A systematic and objective approach using chemoinformatics, database integration, and automated text mining (Guha et al., 2016) also informed selection of the agents evaluated in this volume. A summary of the findings of this volume appears in *The Lancet Oncology* (Loomis et al., 2015).

DDT and lindane

DDT is one of the best known and most studied chemicals of environmental concern. It was introduced as an insecticide in the 1940s and came into widespread use for disease vector control and agriculture after the Second World War. DDT was an important tool in national and international efforts to eradicate malaria, including the WHO Global Programme for Malaria Eradication; spraying with DDT has been credited with helping to reduce the worldwide burden of malaria. However, by the 1960s concerns began to emerge because of the environmental persistence of DDT and its adverse effects on wildlife, and by the 1970s experimental data from studies in experimental animals began to suggest that organochlorine pesticides, including DDT, might have carcinogenic activity. Based largely on these data, the carcinogenicity of DDT was reviewed by the Working Group early in the history of the *IARC Monographs* in Volume 5, *Some organochlorine pesticides* (IARC, 1974), and again in Supplement 7 and Volume 53 (IARC, 1987; 1991). In these evaluations, data from experimental animals provided *sufficient evidence* of carcinogenicity, while the data from humans provided *inadequate evidence* (IARC, 1987; 1991), resulting in a classification of *possibly carcinogenic to humans* (Group 2B). Most uses of DDT other than limited indoor spraying for malaria control are now severely restricted because of its persistence and environmental effects. Nevertheless, DDT is still detectable in the environment, in food, and in the blood and adipose tissue of people and animals worldwide.

Lindane is the γ isomer of hexachlorocyclohexane; it is the only isomer of that series with insecticidal properties, although other forms, notably the more stable α and β isomers, can be present in technical-grade lindane and are sometimes measured as surrogate indicators of exposure to lindane. Like DDT, lindane was commercialized as an insecticide in the 1940s and is now largely banned due to its toxicity. It was used mainly in agriculture, with use peaking in the 1950s. Lindane was first evaluated for carcinogenicity in Volume 5 of the *IARC Monographs*, and was re-evaluated in Volume 20 and Supplement 7; in all of these evaluations, lindane was considered within the class of hexachlorocyclohexanes, which were classified as *possibly carcinogenic to humans* (Group 2B).

Although active use of DDT and lindane has greatly diminished, research into their carcinogenicity has continued, and new epidemiological and mechanistic data have become available since these compounds were last evaluated by the Working Group. This new research and continuing exposure to DDT justify re-evaluation of the carcinogenicity of both pesticides.

2,4-D

Like the other pesticides evaluated in this volume, 2,4-D was introduced in the 1940s and saw increasing use in the ensuing decades. By the 1960s, it was one of the most widely used herbicide active ingredients. Besides its use in agriculture, during the war in Viet Nam 2,4-D was mixed with another chlorophenoxy herbicide, 2,4,5-T (2,4,5-trichlorophenoxyacetic acid) to produce agent orange that was used extensively as a defoliant by the United States military. 2,4-D continues to be used in substantial quantities, primarily in agriculture, and mixtures of 2,4-D and other active ingredients, including glyphosate, have recently been approved for use to combat weeds that are resistant to other herbicides.

The carcinogenicity of 2,4-D was first evaluated by the Working Group in Volume 15 of the *IARC Monographs* (IARC, 1977) and in Supplement 4 (IARC, 1982), when the evidence for carcinogenicity in humans and animals was found to be *inadequate*. When next evaluated in Volume 41 and Supplement 7 (IARC, 1986; 1987), 2,4-D was considered with the class of chlorophenoxy herbicides, which included several other compounds in addition to 2,4-D and 2,4,5-T. By that time, it had been recognized that commercial preparations of chlorophenoxy herbicides, particularly 2,4,5-T, were frequently contaminated with dibenzodioxins and dibenzofurans, so the evaluations of the chemical class were largely focused on the dioxins, which were subsequently classified as *carcinogenic to humans* (Group 1). Since those evaluations, the dioxin content of chlorophenoxy herbicide formulations has been reduced and epidemiological studies of 2,4-D independently from dioxin and related herbicides have been conducted. In the present re-evaluation the Working Group sought to use the available data on 2,4,-D to disentangle its effects from those of associated agents. Data on 2,4,5-T, dioxin or the class of chlorophenoxy herbicides alone were therefore not considered.

Evaluation of pesticides

In evaluating the agents in this volume, the Working Group took into account several challenges that are particular to research on the carcinogenicity of pesticides.

As intentionally toxic substances, pesticides are subject to licensing and regulation, and these are generally directed towards active pesticide ingredients. For this reason, most experimental studies of pesticides evaluate these active ingredients as single substances. However, many pesticides, including lindane and 2,4-D, are marketed as commercial formulations that contain other substances in addition to the active ingredients. Consequently, epidemiological studies on pesticides almost always concern people who have been occupationally or environmentally exposed to commercial products, rather than to single substances. The Working Group considered such differences in the epidemiological and experimental data in reaching an overall evaluation.

The assessment of exposure to pesticides presents further challenges. The epidemiological studies reviewed in this volume included both studies of cancer among workers exposed occupationally to pesticides and studies of cancer risks associated with pesticide exposures in the population at large. These exposures may be assessed using one of two general approaches based either on questionnaires or on quantitative measurement of pesticides or their metabolites in biological samples from study subjects. Both types of exposure data were available for DDT and lindane, while only questionnaire-based assessments were available for 2,4-D. Although biologically based measurements are objective and quantitative, they are not necessarily superior to traditional methods, as they are potentially subject to bias related to inter- and intra-individual variability. The Working Group took both types of data into account in its evaluations.

In its evaluation of data on the mechanisms of cancer, the Working Group continued the procedures introduced in Volume 112 for assessing the strength of evidence with respect to 10 key characteristics of carcinogens (Smith et al., 2016) and of reviewing data from large-scale toxicity testing programmes (IARC, 2017).

References

Guha N, Guyton KZ, Loomis D, Barupal DK (2016). Prioritizing chemicals for risk assessment using chemoinformatics: examples from IARC Monographs on pesticides. *Environ Health Perspect*, 124(12):1823–9. doi:10.1289/EHP186 PMID:27164621

IARC (1974). Some organochlorine pesticides. *IARC Monogr Eval Carcinog Risk Chem Man*, 5:1–241. Available from: http://monographs.iarc.fr/ENG/Monographs/vol1-42/mono5.pdf

IARC (1977). Some fumigants, the herbicides 2,4-D and 2,4,5-T, chlorinated dibenzodioxins and miscellaneous industrial chemicals. *IARC Monogr Eval Carcinog Risk Chem Man*, 15:1–354. Available from: http://monographs.iarc.fr/ENG/Monographs/vol1-42/mono15.pdf PMID:330387

IARC (1982). Chemicals, industrial processes and industries associated with cancer in humans (IARC Monographs, volumes 1 to 29). *IARC Monogr Eval Carcinog Risk Chem Hum Suppl*, 4:1–292. Available from: http://monographs.iarc.fr/ENG/Monographs/suppl4/index.php PMID:6963265

IARC (1986). Some halogenated hydrocarbons and pesticide exposures. *IARC Monogr Eval Carcinog Risk Chem Hum*, 41:1–407. Available from: http://monographs.iarc.fr/ENG/Monographs/vol1-42/mono41.pdf PMID:3473020

IARC (1987). Overall evaluations of carcinogenicity: an updating of *IARC Monographs* volumes 1 to 42. *IARC Monogr Eval Carcinog Risks Hum Suppl*, 7:1–440. Available from: http://monographs.iarc.fr/ENG/Monographs/suppl7/index.php PMID:3482203

IARC (1991). Occupational exposures in insecticide application, and some pesticides. *IARC Monogr Eval Carcinog Risks Hum*, 53:5–586. Available from http://monographs.iarc.fr/ENG/Monographs/vol53/index.php PMID:1688189

IARC (2017). Some organophosphate insecticides and herbicides. *IARC Monogr Eval Carcinog Risks Hum*, 112:1–452. Available from: http://monographs.iarc.fr/ENG/Monographs/vol112/index.php

Loomis D, Guyton K, Grosse Y, El Ghissasi F, Bouvard V, Benbrahim-Tallaa L et al.; International Agency for Research on Cancer Monograph Working Group, IARC, Lyon, France (2015). Carcinogenicity of lindane, DDT, and 2,4-dichlorophenoxyacetic acid. *Lancet Oncol*, 16(8):891–2. doi:10.1016/S1470-2045(15)00081-9 PMID:26111929

Smith MT, Guyton KZ, Gibbons CF, Fritz JM, Portier CJ, Rusyn I et al. (2016). Key characteristics of carcinogens as a basis for organizing data on mechanisms of carcinogenesis. *Environ Health Perspect*, 124(6):713–21. PMID:26600562

Straif K, Loomis D, Guyton K, Grosse Y, Lauby-Secretan B, El Ghissassi F et al. (2014). Future priorities for the IARC Monographs. *Lancet Oncol*, 15(7):683–4. doi:10.1016/S1470-2045(14)70168-8

DDT

1. Exposure Data

1.1. Identification of the agent

1.1.1 Nomenclature

The term "DDT" (dichlorodiphenyltrichloroethane) refers to *para,para′*-DDT – 1,1′-(2,2,2-trichloro-ethylidene)bis (4-chloro benzene). The structure of DDT permits several different isomeric forms, such as *ortho,para′*-DDT (1-chloro-2-[2,2,2-trichloro-1-(4-chlorophenyl) ethyl]benzene). The term DDT is also applied to commercial products consisting predominantly of *p,p′*-DDT, but also containing smaller amounts of other compounds (WHO, 1989).

Chem. Abstr. Serv. Name: 1,1′-(2,2,2-trichloroethylidene)-bis(4-chlorobenzene)

Chem. Abstr. Serv. Reg. No.: 50-29-3

Preferred IUPAC Name: 1-chloro-4-[2,2,2-trichloro-1-(4-chlorophenyl)ethyl]benzene

Synonyms: dichlorodiphenyltrichloroethane; DDT; 1,1,1-trichloro-2,2-bis(*para*-chlorophenyl)ethane; α,α-bis(*para*-chlorophenyl)-β,β,β-trichloroethane; 1,1-bis(*para*-chlorophenyl)-2,2,2-trichloroethane; 2,2-bis(*para*-chlorophenyl)-1,1,1-trichloroethane; 4,4′-DDT; 4,4′-dichlorodiphenyltrichloroethane; *p,p′*-DDT; *para,para′*-dichlorodiphenyltrichloroethane; OMS 0016; 1,1,1-trichloro-2,2-bis(4,4′-dichlorodiphenyl)ethane; 2,2,2-trichloro-1,1-bis(4-chlorophenyl) ethane; trichlorobis(4′-chlorophenyl)ethane; TbisC-ethane; benzene, 1,1′-(2,2,2-trichloroethylidene)-bis(4-chloro-).

Trade Names: DDT has been used in many commercial product formulations. It is not possible to prepare an exhaustive list of all trade names that have been or are being used currently. Trade names specified below are presented as examples. Agritan; Arkotine; Benzochloryl; Bovidermol; Chlorophenothane; Chlorphenotoxum; Clofenotane; Detoxan; Dicophane; Didigam; Didimac; Estonate; Genitox; Gesafid; Parachlorocidum; Pentachlorin; Penticidum.

From NCBI (2015) and ChemIDplus (2015) unless otherwise specified.

The World Health Organization (WHO) specification for technical-grade DDT intended for use in public health programmes requires that the product contains a minimum of 70% *p,p′*-DDT (WHO, 2009). Technical-grade DDT typically also contains smaller amounts of other compounds such as *o,p′*-DDT, *p,p′*-DDD (1-chloro-4-[2,2-dichloro-1-(4-chlorophenyl) ethyl]benzene, also known as TDE), *o,p′*-DDE (1,1-dichloro-2,2-bis(*p*-chlorophenyl) ethylene) (WHO, 1989). These other isomeric forms may add to the pesticide action of DDT since it is known that *o,p′*-DDT has insecticidal properties (Worthing & Walker, 1987). The structure and full chemical names of *p,p′*-DDT and its major metabolites are presented in Fig. 1.1. Technical-grade DDT has been formulated in almost every conceivable form, including solutions in xylene and petroleum distillates, emulsifiable concentrates, water-wettable powders, granules, aerosols, smoke candles, and charges for vapourizers and lotions (WHO, 1989).

1.1.2 Structural and molecular formulae, and relative molecular mass

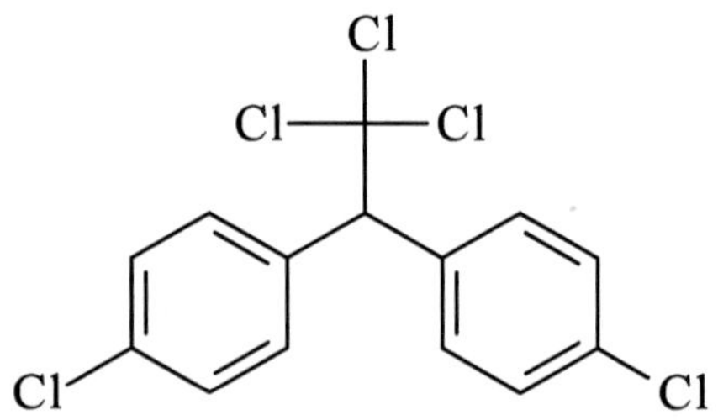

Structure from NIH (2015)

Molecular formula: $C_{14}H_9Cl_5$

Relative molecular mass: 354.49

1.1.3 Chemical and physical properties of the pure substance

See NCBI (2015)

Description: Colourless crystals or white powder, odourless or with weak aromatic odour

Solubility: Practically insoluble in water (0.055 mg/100 mL at 25 °C); soluble in acetone (58 g/100 mL), benzene (78 g/100 mL), benzyl benzoate (42 g/100 mL), carbon tetrachloride (45 g/100 mL), chlorobenzene (74 g/100 mL), cyclohexanone (116 g/100 mL), 95% alcohol (2 g/100 mL), ethyl ether (28 g/100 mL), and other organic solvents.

Octanol/water partition coefficient: log Pow = 7.48 (WHO, 1989)

Conversion factor for airborne concentrations: mg/m^3 = 14.5 × ppm

1.2 Production and use

1.2.1 Production

The history of the development of DDT as a pesticide has been documented (Casida & Quistad, 1998; Jarman & Ballschmiter, 2012). Following the discovery of the insecticidal properties of DDT in 1939, DDT became a commonly used agent for insect control worldwide (both for public health and for agricultural uses), until agricultural uses were curtailed in various countries from the early 1970s (see Section 1.5 regulations).

DDT is made industrially by condensing chloral hydrate with chlorobenzene in the presence of sulfuric acid. To prepare *o,p'*-DDT, an excess of chlorobenzene is condensed with 1-(2-chlorophenyl)-2,2,2-trichloroethanol in the presence of a mixture of 96% sulfuric acid and 25% oleum at 60 °C (Brooks, 1974).

Most DDT production can be assumed to have been technical-grade material that included 15–21% of *o,p'*-DDT, up to 4% of *p,p'*-DDD, and up to 1.5% of 1-(*p*-chlorophenyl)-2,2,2-trichloroethanol (Metcalf, 1955).

In the USA, peak production of DDT was reported in the early 1960s, with 56 000 tonnes produced in 1960 and 82 000 tonnes in 1962; DDT was registered for use on 334 agricultural commodities (Metcalf, 1955). Production then declined and by 1971 had dipped to about 2000 tonnes, shortly before DDT was banned (ATSDR, 2002).

In China, the cumulative production of DDT over 30 years until 1983 was reported to be 0.4 million tonnes, estimated to then account for 20% of the cumulative worldwide production (about 2 million tonnes) (ATSDR, 2002; Xin et al., 2011). During 2000–2003, average annual production in China was approximately 4500 tonnes and there were no DDT imports to China from other counties (Lin et al., 2009).

Fig. 1.1 Structure of *p,p'*-DDT and its major metabolites

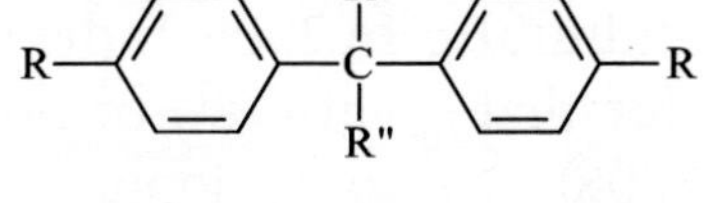

Basic structure of *p,p'*-DDT related compounds

Name	Chemical name	R	R'	R"
DDT	1,1'-(2,2,2-trichloroethylidene)- bis[4-chlorobenzene]	- Cl	- H	- CCl_3
DDE	1,1'-(2,2-dichloroethenylidene)- bis[4-chlorobenzene]	- Cl	None	= CCl_2
DDD (also referred to as TDE)	1,1'-(2,2-dichloroethylidene)- bis[4-chlorobenzene]	- Cl	- H	- $CHCl_2$
DDMU	1,1'-(2-chloroethenylidene)- bis[4-chlorobenzene]-	- Cl	None	= CHCl
DDMS	1,1'-(2-chloroethylidene)- bis[4-chlorobenzene]	- Cl	- H	- CH_2Cl
DDNU	1,1'-bis(4-chlorophenyl)ethylene	- Cl	None	= CH_2
DDOH	2,2-bis(4-chlorophenyl)ethanol	- Cl	None	- CH_2OH
DDA	2,2-bis(4-chlorophenyl)- acetic acid	- Cl	- H	- COOH

Adapted from WHO (1989)

As assessed in 2007, DDT was produced in three countries - India, China and the Democratic People's Republic of Korea - with by far the largest amount then produced in India (van den Berg, 2009). Currently, DDT is produced only in India, from where it is exported as a pure (also called "technical") product, or as a commercially formulated product to other, mostly African, countries (UNEP, 2010; UNEP/WHO 2014).

[The Working Group noted that despite it being recognized that DDT being produced by many countries during the latter part of the twentieth century, no information on DDT production was available for most of these countries.]

1.2.2 Use

DDT has been, and is, primarily used as a pesticide. DDT is also used as an intermediate in the production of the pesticide dicofol (Kelthane), and in antifouling paint.

(a) Worldwide use as a pesticide

DDT was initially used by the military in World War II to control malaria, typhus, body lice, and bubonic plague. DDT was a key tool in malaria eradication efforts in Italy and the USA. In Italy, cases of malaria decreased from 400 000 in 1946 to virtually none in 1950. DDT was also used to manage an epidemic of typhus in Italy and Germany in 1943–1944 (Casida & Quistad, 1998; Jarman & Ballschmiter, 2012).

Historically, in addition to its public health uses, growers used DDT on a variety of food crops worldwide. In USA, some of the crops to which DDT was applied included beans, cotton, soybeans, sweet potatoes, peanuts, cabbage, tomatoes, cauliflower, brussel sprouts, corn, and other crops. DDT was also used to treat livestock and in buildings for pest control (EPA, 1975a).

The usage of DDT in the former Soviet Union was intensive in the 1950s and 1960s, and continued until the early 1990s. The total such use of DDT from 1946 till 1990 is estimated to have been between 250 000 and 520 000 tons [254 000 and 528 000 tonnes] (Li et al., 2006).

Having been introduced as an agricultural chemical in the 1950s, DDT was used for that purpose in China until 1983 (Cai et al., 2008).

DDT is one of the 12 insecticides, and the only organochlorine compound, currently recommended by WHO for use in indoor residual spraying for disease vector control (WHO, 2011a). Currently, DDT represents some 71% of the global annual amount of insecticides used for vector control. In 2005, an estimated 5000 tonnes of DDT was used for disease vector control (Table 1.1). DDT is used mostly for malaria control, but 19% of the global share is sprayed to control transmission of leishmaniasis by sandflies (van den Berg et al., 2012).

As assessed in 1996, use of DDT and malathion was credited with reducing incident malaria in India to "only" 2 to 3 million cases per year (Sharma et al., 1996). In 2005, DDT was shown to be still useful for residual spraying in India, particularly in areas where the vectors are endophilic and not resistant, and its continued use in India was endorsed (Gunasekaran et al., 2005).

DDT has been used in China as an antimalarial agent, specifically in provinces where standing water in rice fields favours the breeding of *Anopheles anthropophagus* as a principal vector. Residual spraying of houses and cattle sheds was shown to reduce malarial prevalence (Xu et al., 1998). China has now stopped the use of DDT (UNEP/WHO, 2014).

In South Africa, spraying of DDT from 1945 to 1995 was highly successful in controlling the vectors of malaria, *Anopheles funestus* and *Anopheles arabiensis*, without emergent resistance. DDT was replaced by pyrethroids in 1995, but after 4 years, the incidence of malaria increased at least fourfold due to vector resistance to pyrethroids. In 2000, DDT was reintroduced and the incidence of malaria subsequently declined (Curtis, 2002). Currently, DDT is still used in South Africa and in a few other African countries for indoor residual spraying (Table 1.1). In 2009, India remained the largest user of DDT, with 78% of global use, the remainder being used in Ethiopia (15%), Mozambique (2%), Namibia (1.8%), South Africa (1.4%), Zimbabwe (1.2%), and Zambia, Madagascar, Eritrea, Swaziland, Uganda, and Mauritius (each < 1% of global use). Trends in the use of DDT have been associated with, and in some instances determined by, resurgence in malaria, as variously documented worldwide from the 1930s to the 2000s (Cohen et al., 2012). While the use of DDT declined in India in the 2000s, several countries in Africa, including Mozambique, Zambia, and Zimbabwe, significantly increased their DDT use until 2008 (Fig. 1.2; WHO, 2011b; van den Berg, 2009).

(b) Use as a chemical intermediate

DDT is used as a feedstock for the production of dicofol (Kelthane), an acaricide used for controlling mites that attack cotton, fruit trees, and vegetables. Dicofol is usually synthesized from technical DDT, which is first chlorinated and then hydrolysed to produce dicofol. DDT and reaction intermediates may remain in the dicofol product as impurities (Gillespie et al., 1994; Qiu et al., 2005).

In an assessment for the European Union in 2004, a single producer of dicofol within the European Union was identified; production was also indentified in China, Brazil, and India, with largest production volume (≈2000 tonnes/year) occurring in China (Ministerie van VROM, 2004).

Dicofol was introduced in China in the 1970s and, after discontinuation of DDT as an agricultural chemical, dicofol production became the major use of DDT. Use of DDT as an intermediate in dicofol production accounted for approximately 80% of the chemical used in China in 2002–2004, antimalarial usage and production of antifouling paint accounting for the balance. Difocol produced in China was reported to have

Table 1.1 Annual global production and use of DDT for vector control, in 2003, 2005, 2007, and 2009

Country	Quantity (1000 kg active ingredient)				Comment	Source
	2003	2005	2007	2009[c]		
DDT production						
China[a]	450	490	NA	NA	For export	Pd
India[b]	4100	4250	4495	NA	For malaria and leishmaniasis	Pd, WS, Dc
DPRK	NA	NA	5	NA	> 155 tonnes for use in agriculture	UNITAR
Global production	> 4550	> 4740	> 4500	NA		
DDT use						
Cameroon	0	0	0	NA	Plan to pilot in 2009	WHO
China	0	0	0	NA	Discontinued use in 2003	SC
Eritrea	13	15	15	12	Epidemic-prone areas	Qu, WHO
Ethiopia	272	398	371	781	Epidemic-prone areas	WHO, WS
Gambia	0	0	NA	NA	Reintroduction in 2008	Dc
India	4444	4253	3413	3399	For malaria and leishmaniasis	WHO, Dc
DPRK	NA	NA	5	NA	> 155 metric tonnes used in agriculture	UNITAR
Madagascar	42	0	0	NA	Plan to resume use in 2009	Qu
Malawi	0	0	0	NA	Plan to pilot in 2009	WHO
Mauritius	1	1	< 1	0.4	To prevent malaria introduction	Qu
Morocco	1	1	0	NA	For occasional outbreaks	Qu
Mozambique	0	308	NA	104	Reintroduction in 2005	WHO
Myanmar	1	1	NA	0.2	Phasing out	WS
Namibia	40	40	40	92	Long-term use	WHO
Papua New Guinea	NA	NA	0	NA	No recent use reported	SC
South Africa	54	62	66	70	Reintroduction in 2000	Qu, WHO
Sudan	75	NA	0	NA	No recent use reported	Qu, WHO
Swaziland	NA	8	8	7	Long-term use	WHO
Uganda	0	0	NA	NA	High Court prohibited use, 2008	SC, Dc
Zambia	7	26	22	NA	Reintroduction in 2000	WS, Qu, WHO
Zimbabwe	0	108	12	61	Reintroduction in 2004	WHO
Global use	> 4953	> 5219	> 3950	5127		

Dc, direct communication with national authorities; DDT, dichlorodiphenyltrichloroethane; DPRK, Democratic People's Republic of Korea; NA, not available; Pd, project proposals submitted to the Global Environment Facility; Qu, questionnaires on DDT by the Secretariat of the Stockholm Convention completed by national authorities; SC, documents published by the Secretariat; WS, workshop presentations by country delegates in the context of the Stockholm Convention. Further information was obtained from the World Health Organization (WHO) and United Nations Institute for Training and Research (UNITAR) reports, as indicated

[a] The figure for 2005 was extrapolated from the total production; in addition to production for vector control; DDT is produced for dicofol manufacture (≈3800 metric tonnes per year) and for antifoulant paints (≈200 metric tonnes per year)

[b] DDT is also produced for dicofol manufacture (≈280 metric tonnes per year)

[c] The data for 2009 were taken from WHO (2011b)

Adapted from van den Berg (2009)

Fig. 1.2 Trends in the global use of DDT, 2000–2009

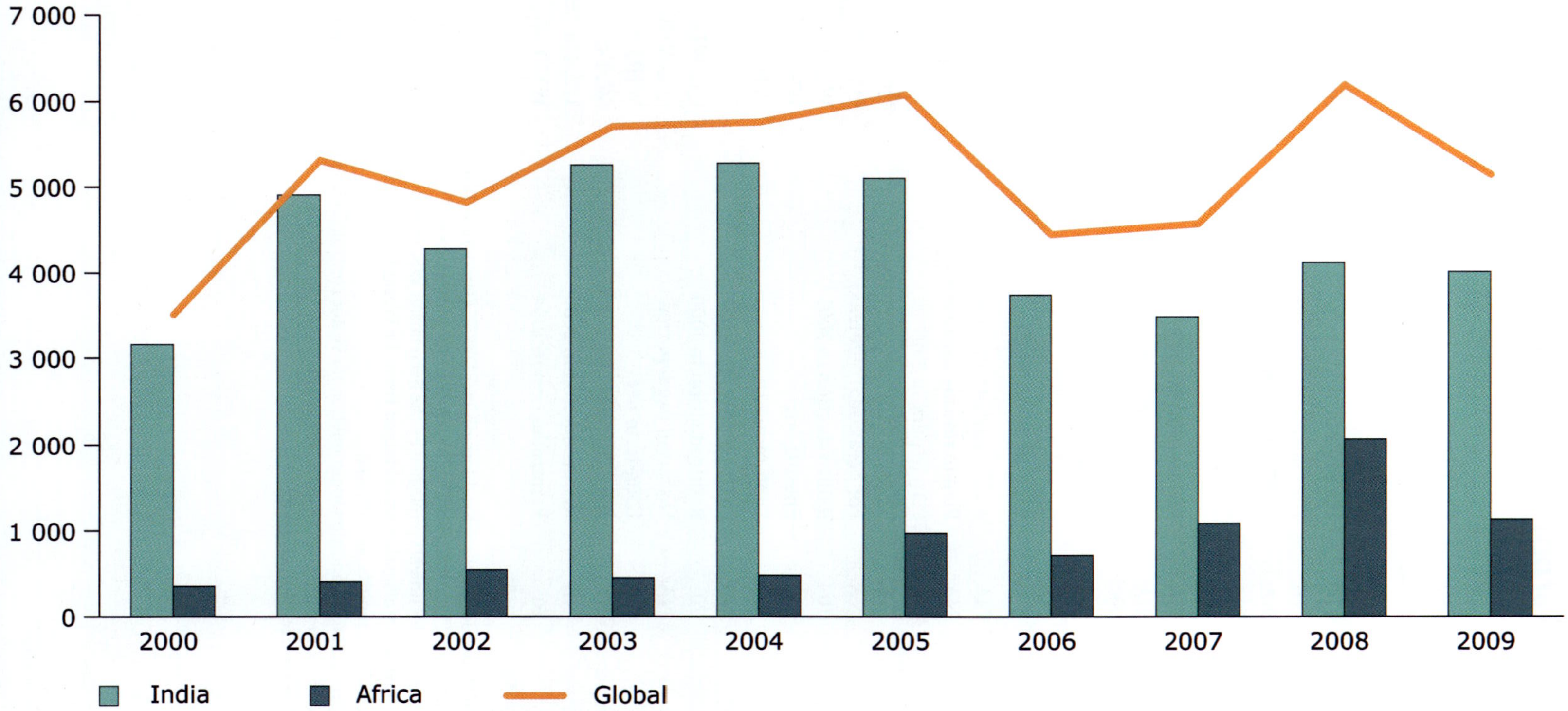

Figure compiled with data from van den Berg et al. (2012) 'Global trends in the use of insecticides to control vector-borne diseases', *Environ. Health Perspect*, (120):577–582 AND reproduced from Bouwman et al. (2013). DDT: fifty years since Silent Spring. Chapter 11. Late lessons from early warnings: science, precaution, innovation. Brussels, Belgium: European Environment Agency; pp. 272–91.

a high DDT content (on average, 20%) (Qiu et al., 2005; Global Environment Facility, 2006).

From the 1950s until 2009, DDT was used as an additive to antifouling paint produced in China (Xin et al., 2011). During 2000–2003, production of antifouling paint accounted for 4% of the DDT produced in China (Lin et al., 2009). By 2014, all uses of DDT in China were withdrawn; the only reported use of DDT as an intermediate in dicofol production was in India (UNEP/WHO, 2015).

1.3 Measurement and analysis

Methods for the analysis of organochlorine pesticides (OCPs) in a variety of environmental, biota, food, and human biological matrices have been well developed during the past several decades. These methods appear in both academic articles and in the standard operational procedures from various environmental and health agencies. The United States Environmental Protection Agency (EPA) (EPA, 2015a), the European Committee for Standardization (CEN/EU), the Japanese Industrial Standards (JIS), and agencies in other industrialized countries have developed comprehensive protocols for pesticide analysis. In particular, the EPA has posted an online index of pesticide analytical methods.

The analysis of DDT, its metabolites, and other OCPs involves extraction, clean-up, and instrumental analysis based on gas chromatography (GC) or gas chromatography-mass spectrometry (GC-MS) (EPA, 2015a). The available sample extraction and clean-up methods described above have generally provided recoveries of target OCP analytes in the range of 50–130%, depending on the sample matrix and analyte concentration. Regarding detection, gas chromatography-electron capture detection (GC-ECD) was widely used to quantify DDT, DDE, and other related isomers and metabolites in many media, especially in the 1970s until the 1990s, and is still a low-cost routine technique for most OCPs. One concern with GC-ECD methods is the potential for interfering non-target chemicals leading to misidentification or incorrect quantitation. GC-MS methods are now widely used have been used to identify and quantify DDT, DDE, and other related isomers and metabolites in many media. These include single quadruple MS detectors running in electron ionization mode with target analyses monitored by selective ion monitoring and GC coupled with high-resolution MS (HRMS) (Barr et al., 2003). These detection methods increase the confidence in confirmative analysis by decreasing matrix interferences, improving selectivity and, in the case of tandem MS and HRMS, showing higher signal-to-noise ratio, therefore improving the detection limits and selectivity.

1.4 Occurrence and exposure

The following summary of the literature is not a systematic review, but provides a descriptive narrative of the occurrence of and exposure to DDT and its metabolites in the environment and among the occupational and general population globally. Because DDT is a synthetic product that consists of variable quantities of several isomers, the term "ΣDDT" used here represents the sum of all isomers, total DDT, as reported by the authos.

DDT is a persistent organochlorine pesticide that together with its metabolites can have a half-life in soil of up to 30 years and in aquatic life of up to 150 years. DDT and its important metabolites for exposure assessment, DDE and DDD, are chemicals not known to occur naturally in the environment (see their structure, Fig. 1.1).

Historically, DDT was released to the environment during its production, formulation, and extensive use as a pesticide in agriculture and malaria-control applications. Although DDT was banned for use in many countries after 1972, it is still being used in some areas of the world for vector-control purposes (ATSDR, 2002).

Occupational exposures to DDT may occur in pesticide-manufacturing workers, commercial pesticide applicators, and historically occurred in farmers. Bystander exposure occurs during DDT application in homes for malaria control, if workers from manufacturing plants take home contaminated clothing, and historically occurred from field application in agriculture. DDT is widely distributed in the environment even decades after its use ended in most countries. General population exposures occur primarily through dietary intake, and as a result of contact with DDT in environmental media, including indoor environments. Where DDT remains in use for malaria control, there is the potential for associated human exposures. DDT and its metabolites are fairly ubiquitous in the atmosphere as a result of spraying operations in areas of the world where it is still used (ATSDR, 2002).

1.4.1 Occupational exposure

Detailed information on occupational exposure is given in Table 1.2

(a) Pesticide manufacturing

No studies reporting exposure measurements for workers manufacturing DDT pesticides were available to the Working Group. However, DDT has been reported to be used in the synthesis of the insecticide dicofol in China since 1978. Analytical studies have shown that DDT compounds, including *p,p′*-DDT and *p,p′*-DDE, may be contaminants in technical grades of difocol (Risebrough et al., 1986) with the potential for DDT to be released from difocol products. In a Chinese factory producing difocol with a closed-loop production system, ΣDDT was measured in workshop air (range, 1.88–17.53 μg/m^3) and worker inhalation doses were estimated (ΣDDT, 0.38–3.51 μg/kg bw per day) (Li et al., 2014a). Occupational hygiene assessment was conducted to assess the risks of workers involved in the demolition of a plant in which dicofol was manufactured between 1978 and 2004. Concentrations of ΣDDT were measured on process equipment surfaces and ranged between 5.4 and 37.5 g/m^2 (Luo et al., 2014).

(b) Agriculture

In North Carolina, USA, between 1995 and 1996, African-American farmers, most of whom were no longer working on a farm, presented with mean plasma DDE concentrations of 11.4 ng/mL (Cooper et al., 2004). In Bolivia, between 2010 and 2011, serum *p,p′*-DDE concentrations of 19.7 ng/mL (4788.7 ng/g lipid) were reported in agricultural workers (Mercado et al., 2013). In India, in a study carried out in 2009 among 30 agriculture, sheep-breeding and wool-shearing workers in rural Bangalore, the mean blood concentration of ΣDDT was 10.6 ng/mL (Dhananjayan et al., 2012).

(c) Malaria control

In Tzaneen, South Africa, median ΣDDT concentrations of 83.3 μg/g of lipid were reported among malaria-control workers (Dalvie et al., 2004a). In 2008 in northern Uganda, malaria-control workers had mean DDE/DDT concentrations in plasma collected post-spraying that ranged empirically from 24 to 128 ppb, with a mean of 77 ppb (Bimenya et al., 2010). In 1994, in the state of Veracruz, Mexico, workers employed to control vectors for malaria and dengue presented mean ΣDDT concentrations of 104.5 μg/g lipids in abdominal adipose tissue (Rivero-Rodriguez et al., 1997). In Mato Grosso, Brazil, between 1999 and 2000, malaria-control sprayers had median serum concentrations of ΣDDT, DDT, and *p,p′*-DDE of 135, 23.3, and 107.3 ng/mL, respectively (Dores et al., 2003). In the Amazon region of Brazil, malaria-control workers in 1997 had serum ΣDDT and *p,p′*-DDE concentrations of 231.5 and 156.9 ng/mL, respectively, levels that were much higher than those found in the general population in 2001, in which concentrations were

Table 1.2 Occupational exposure to DDT and its metabolites

Occupation	Country, collection date	Sampling matrix	Exposure		Comments	Reference
			Level [a]	Range		
Pesticide manufacture, closed-system, dicofol production	China, year NR	Air	NR	ΣDDT, 1.88–17.53 μg/m^3	Dicofol production with DDT as intermediate Estimated inhalation exposure range of 0.38–3.51 μg/kg bw per day	Li et al. (2014a)
Dicofol manufacture, demolition workers	China, around 2009	Dust on production equipment	NR	ΣDDT, 5.4–37.5 g/m^2	The factory produced difocol with DDT intermediate between 1978 and 2004	Luo et al. (2014)
Farmers	USA, North Carolina, 1995–1996	Plasma	*p,p′*-DDE, 7.7 ng/mL (median)	*p,p′*-DDE, 0.6–77.4 ng/mL		Cooper et al. (2004)
Farmers	Bolivia, Santa Cruz, 2010–2011	Serum	*p,p′*-DDE, 4788.7 ng/g lipid (median)	*p,p′*-DDE, 1197–35 131 ng/g lipid (25th–75th percentiles)		Mercado et al. (2013)
Agriculture and sheep wool-associated jobs	India, Bangalore city, Karmataka, (rural) 2009	Blood	*p,p′*-DDD, 3.01 ± 0.19; *p,p′*-DDE, 5.67 ± 1.21; *p,p*-DDT, 3.81 ± 1.51 ΣDDT, 10.6 ± 2.15 ng/mL	*p,p′*-DDD, < LOD–5.65; *p,p′*-DDE, < LOD–32.1; *p,p*-DDT, < LOD–48.3 ΣDDT, 6.72–51.7 ng/mL		Dhananjayan et al. (2012)
Malaria control, 1946–1950	Italy, Sardinia, 2000	Serum	DDT, 47; *p,p′*-DDE, 396 ng/g lipid (median)	DDT, 33–74. *p,p′*-DDE, 157–1045 ng/g lipid		Cocco et al. (2004)
Malaria-control workers, DDT spraying, mixing	South Africa, Tzaneen, Limpopo year NR	Serum	DDE, 52.4; DDD, 0.74; DDT, 28.2 ΣDDT, 83.3 μg/g lipid	DDE, 1.1–273.6; DDD, 0–3.1; DDT, 0.3–67.1 ΣDDT, 1.4–315 μg/g lipid	"DDT," "DDD", and "DDE" are used for the sum of the two respective *p,p′* and *o,p′* isomers	Dalvie et al. (2004a)
Malaria-control DDT weighing and spraying	Mexico, Veracruz, 1994	Abdominal adipose tissue	ΣDDT, 104.48 *p,p′*-DDT, 31; *o,p′*-DDT, 2.1; *p,p′*-DDE, 60.98; *p,p′*-DDD, 0.95 μg/g lipid (geometric mean)	ΣDDT, 10.56–665.56 *p,p′*-DDT, 0.72–344.98; *o,p′*-DDT, 0.07–29.74; *p,p′*-DDE, 9.57–298.42; *p,p′*-DDD, ND-3.51 μg/g lipid		Rivero-Rodriguez et al. (1997)
Malaria control DDT spraying	Brazil, Para state, 1997–2001	Serum	1997: ΣDDT, 231.5; *p,p′*-DDE, 156.9 2001: ΣDDT, 50.4; *p,p′*-DDE, 39.4 ng/mL	1997: ΣDDT, 5.3–3839.8; *p,p′*-DDE, 4.6–513.8 2001: ΣDDT, 3.3–357.9; *p,p′*-DDE, 2.3–284.1 ng/mL		Ferreira et al. (2011)

Table 1.2 (continued)

Occupation	Country, collection date	Sampling matrix	Exposure		Comments	Reference
			Level [a]	Range		
Malaria control DDT spraying	Brazil, Mato Grosso, 1999–2000	Serum	ΣDDT, 135.5; DDT, 23.3; *p,p′*-DDE, 107.3 ng/mL (median)	ΣDDT, 7.5–875.5; DDT, ND-476; *p,p′*-DDE, 7.5–518.5 ng/mL		Dores et al. (2003)
Malaria control DDT spraying	Uganda, Lango, 2008	Plasma	DDE/DDT, 77 ppb	DDE/DDT, 24–128 ppb	The mean value reported is a mean DDE/DDT; plasma samples analysed 6 months after one round of DDT spraying	Bimenya et al. (2010)
Electronic waste and fishery industry, e-waste cycling and fishery	China, Guangdong province (Guiyu town, Haojiang district), 2005	Serum	ΣDDTs: e-waste, 600; fishery, 2300 ng/g lipid (median)	ΣDDTs: e-waste, 210–1800; fishery, 380–5100 ng/g lipid	Exposure to ΣDDT was dominant in both e-waste and fishery industries compared with PCBs and other OCPs	Bi et al. (2007)
Textile	China, Anhui 1996–1998	Serum, preconception	ΣDDT, 27.9, *p,p′*-DDT, 1.42, *o,p′*-DDT, 0.16, *p,p′*-DDE, 26.24, *o,p′*-DDE, 0.09, *p,p′*-DDD, 0.21 ng/g lipid (median)	ΣDDT, 5.52–113.3, *p,p′*-DDT, 0.37–13.12, *o,p′*-DDT, 0.04–1.49, *p,p′*-DDE, 4.76–97.54, *o,p′*-DDE, 0.03–1.07, *p,p′*-DDD, 0.05–0.96 ng/g lipid	ΣDDT was positively associated with the risk of subsequent early pregnancy losses	Venners et al. (2005)

ΣDDT, total DDT corresponding to the sum of *o,p′*-DDE, *p,p′*-DDE, *o,p′*-DDD, *p,p′*-DDD, *o,p′*-DDT, *p,p′*-DDT; DDE, dichlorodiphenyldichloroethylene; DDT, dichlorodiphenyltrichloroethane; LOD, limit of detection; NR, not reported; OCPs, organochlorine pesticides; PCBs, polychlorinated biphenyl; ppb, parts per billion

[a] Arithmetic mean, except where otherwise stated

50.4 and 39.4 ng/mL, respectively (Ferreira et al., 2011). In Sardinia, Italy, workers who had participated in the 1946–1950 antimalarial campaign presented in 2000 with median serum DDT and *p,p'*-DDE concentrations of 47 and 396 ng/g lipid, respectively (Cocco et al., 2004).

(d) Other occupational exposures

In Guangdong province, China, in 2005, serum ΣDDT concentrations among electronic waste and fishery workers ranged from 210 to 5100 ng/g lipid (Bi et al., 2007). In Anhui, China, between 1996 and 1998, serum ΣDDT concentrations among female textile workers in China ranged from 5.52 to 113.3 ng/g lipid (Venners et al., 2005).

1.4.2 Environmental occurrence

Detailed information on environmental occurrence is given in Table 1.3.

(a) Water

The release of DDT to surface water still occurs in countries that rely on DDT for malaria control near open water. DDT also enters surface water as a result of dry and wet deposition from the atmosphere and direct gas transfer, contributing to the loading in rivers, deep wells, lakes, and oceans.

Concentrations of *p,p'*-DDE, *p,p'*-DDD, *p,p'*-DDT in water from the Nile delta ranged from 35 to 67, 19 to 33, and 24 to 31 ng/L respectively between 1995 and 1997 (Abbassy et al.,1999). In 2011 in South Africa, ΣDDT and residues were measured in Johannesburg in surface water (i.e. river and dam) as 14 day-measurement samples taken in spring and summer, and ranged from 0.026 to 0.549 ng/L (Amdany et al., 2014). Surface water sampled twice in 2012 from the Konya Basin in Turkey contained ΣDDT at concentrations ranging from not detected (ND) to 47 ng/L; *p,p'*-DDE represented the major metabolite measured, with concentrations ranging from ND to 37 ng/L (Aydin et al., 2013). Water samples taken monthly between 2009 and 2011 from the Karun river in Khuzestan Province, in the Islamic Republic of Iran, contained ΣDDT at mean concentrations of 32.2ng/L, with *o',p'*-DDT and *p,p'*-DDT concentrations of 26 and 10.2 ng/L, respectively (Behfar et al., 2013). DDT concentrations in surface-water samples taken between August 1989 and October 1991 from southern Asia and Oceania ranged from 1.3 pg/L to up to 5 orders of magnitude higher (120 ng/L) measured in India (Iwata et al., 1994). Much higher ΣDDT concentrations were reported in a 20-year follow-up study in India. ΣDDT concentrations ranged from ND to < 163 × 106 ng/L at 45 surface-water sites, and from ND to 75 × 103 ng/L at 15 ground-water sites. Highest ΣDDT concentrations were measured in 2005 in the Ganges river, India (Sharma et al., 2014).

In the USA, no DDE was detected in 3251 samples obtained from 2001 to 2003 as a statistically selected, nationally representative sampling of small drinking-water systems (EPA, 2008). In the USA National Water Quality Assessment Program from 1992–2001, *p,p'*-DDE was detected in only 5.75% of 2013 samples from 83 agricultural surface water sites with only 0.29% being > 10 ng/L; and 2.02% detection frequency in 812 samples from 30 urban surface-water sites with none being > 10 ng/L (USGS, 2006). In central Poland, between 2002 and 2003, DDT concentrations in drinking-water ranging from 10.6 to 166 ng/L were reported in the vicinity of orchard areas (Badach et al., 2007). Little information was available in the open literature for Latin American countries or other European countries.

In Limpopo, South Africa, where DDT is still used for indoor residual spraying to control malaria vectors, ΣDDT was detected in potable water (tap water) with concentrations ranging from 600 to 7600 ng/L, while 83% of ΣDDT concentrations in exposed areas were < 2000 ng/L. In contrast, in control areas none of the tap-water samples contained residues of

Table 1.3 Environmental occurrence of DDT and its metabolites

Location, collection date	Sampling matrix	Exposure		Comments	Reference
		Level[a]	Range		
South Africa, Johannesburg, 2011	Surface water	NR	ΣDDT, 0.026–0.549 ng/L	Measurements were done for 14 days in spring and in summer; concentrations of isomers also reported by site	Amdany et al. (2014)
Egypt, Nile Delta, 1995–1997	Surface water	NR	*p,p'*-DDE, 35–67, *p,p'*-DDD, 19–33 *p,p'*-DDT, 24–31 ng/L	Levels were reported for different seasons, 1995–1997	Abbassy et al. (1999)
Islamic Republic of Iran, Karun river in Khuzestan Province, August 2009 to March 2011	Surface water	ΣDDT, 36.2; *o,p'*-DDT, 26; *p,p'*-DDT, 10.2 ng/L	ΣDDT, 5.9–90.3; *o,p'*-DDT, ND–87.3; *p,p'*-DDT, 2.4–28 ng/L	There was significant correlation between *o,p'*-DDT and ΣDDT	Behfar et al. (2013)
Turkey, Konya closed basin, March and August 2012	Surface water	NR	ΣDDT, ND–47; *p,p'*-DDE, ND– 37; *p,p'*-DDD, ND–5; *p,p'*-DDT, ND–5 ng/L	Measurements taken in March and August; slightly higher levels of ΣDDT were found in March	Aydin et al. (2013)
Pakistan, River Chenab, Punjab, January–March 2013	Surface water	ΣDDT, 9.07 ng/L	ΣDDT, 1.90–20.6 ng/L		Mahmood et al. (2014)
Eastern and southern Asia and Oceania, August 1989 to October 1991	Surface water	NR	ΣDDT: India: 0.87–120, Thailand: 0.23–2.5, Viet Nam: 0.29–25; Malaysia: 1.7; Indonesia: 0.19–0.27; Solomon Islands: 0.062–21; Japan: 0.0065–0.016; Taiwan, China: 0.0095–0.19; Australia: 0.0013–1.1 ng/L		Iwata et al. (1994)
India, year NR NA	Surface water, ground water	NR	ΣDDT: Surface water: ND–16 367 × 10^3 Ground water: ND–75 × 103 ng/L	> 45 surface water sites, 15 ground water sites, approximately 20-year data; highest ΣDDT levels were measured in 2005 in the Ganges river	Sharma et al. (2014)
Poland, Warka-Grojec region Spring and autumn 2002–2003	Drinking-water	NR	DDT, 10.6–166.9 ng/L	Higher DDT levels were measured in autumn than in spring, and a higher percentage of samples was found contaminated in autumn (47.1%) than in spring (29.4%)	Badach et al. (2007)
South Africa, Limpopo Province February 2008	Drinking-water	NR	ΣDDT: exposed area: 600–7600 ng/L	83% of ΣDDT levels in exposed area: < 2000 ng/L; control area: < 500 ng/L [b]	Van Dyk et al. (2010)

Table 1.3 (continued)

Location, collection date	Sampling matrix	Exposure		Comments	Reference
		Level[a]	Range		
Brazil, Cidade dos Meninos 2002–2003	Soil and sediments, water, pasture and vegetables	ΣDDT: soil and sediments: 188 × 10^3; pasture: 57.2 × 10^3; vegetables: 5100; water: NR ng/g or ng/L	ΣDDT: soil and sediments: < 30 × 10^3 to 465 × 10^3 ; water: ND–120; pasture: 9180–189 × 10^3; ΣDDT: vegetables: 3.22–7.01 ng/g or ng/L		Brilhante & Franco (2006)
China, Taihu lake 2006	Surface sediment	*p,p'*-DDT, 1.26, *p,p'*-DDD, 0.36, *p,p'*-DDE, 0.14, ΣDDT, 53.9 ng/g dw	ΣDDT, 0.25–375 ng/g dw	DDTs were detected in all sediment samples; benthic organisms were also employed for DDT environmental exposure levels	Zhao et al. (2009)
South Africa, Lake Sibaya, KwaZulu-Natal January–March 2012	Surface sediment	NR	ΣDDT, 0.8–123; *p,p'*-DDE, 1.8–42.8; *p,p'* DDD, 1.8–74.1; *p,p'*-DDT, < LOD–6.2 ng/g		Humphries (2013)
Thailand, tributaries of Mae Klong river; Mae Klong river basin 2003–2005	Surface sediment	NA	ΣDDT: Klong river: 80–1830; Klong river basin: < 1–6780 ng/g dw	DDT detection reflected a recent contamination in the study area	Poolpak et al. (2008)
Pakistan, River Chenab, Punjab January–March 2013	Sediment	ΣDDT, 40.3 ng/g	ΣDDT, 5.84–89.8 ng/g		Mahmood et al. (2014)
China Beibu Gulf and its tributary rivers 2010	Sediment	*p,p'*-DDT, 6.43; *o,p'*-DDT, 1.6; *p,p'*-DDE, 4.33; *p,p'*-DDD, 9.39; ΣDDT, 21.8 ng/g dw	*p,p'*-DDT, 0.19–44.1; *o,p'*-DDT, 0.07–10.1; *p,p'*-DDE, 0.08–19.2; *p,p'*-DDD, 0.16–52.2; ΣDDT, 0.59–126 ng/g dw	Concentrations of DDTs were higher than those reported in the sediments from other regions of the world	Xu et al. (2013b)
Eastern and southern Asia and Oceania August 1989 to October 1991	Sediment	NR	ΣDDT: India: 8.0–450; Thailand: 4.8–170; Viet Nam: 0.37–790; Malaysia: 1.8; Indonesia: 3.4–42; Papua New Guinea: 4.7–130; Solomon Islands: 9.3–750; Japan: 2.5–12; Australia: < 0.01–1700 ng/g dw		Iwata et al. (1994)
India	Sediment, soil	NR	ΣDDT: sediment: < 0.01–128 600, soil: 34–903 ng/g	Sediment from 20 sites; soil from 15 mostly agricultural areas; approximately 20-year data	Sharma et al. (2014)

Table 1.3 (continued)

Location, collection date	Sampling matrix	Exposure		Comments	Reference
		Level[a]	Range		
Bahrain, Oman, Qatar and the United Arab Emirates (UAE) 2000 & 2001	Coastal sediment	NR	ΣDDT: Bahrain: 0.088–0.430; Oman: 0.7×10^{-3} to 0.0852; Qatar: 0.63×10^{-3} to 0.0367; UAE: ND–0.0519 ng/g dw		de Mora et al. (2005)
Canada Toronto 2002	Soil and sediment	NR	ΣDDT: soil: 1–18; sediment: 0.2–472 ng/g		Wong et al. (2009b)
Mexico, Culiacan Valley, Sinaloa 2007	Agricultural sediment	NR	ΣDDT, 1.1–25.95 ng/g dw	The Culiacan Valley is an extensive agricultural region characterized by a variety of crops with high-yield production; measurements were taken in the agricultural drainage system of the valley; highest measured concentration among the DDT-related compounds was for *p,p′*-DDE (20.19 ng/g dw)	García de la Parra et al. (2012)
Jordan, Humrat Al-Sahn, Jordan Valley 1998	Soil	NR	*p,p′*-DDE, ND–0.46; *p,p′*-DDD, ND–0.16; *o,p′*-DDT, ND–0.11; *p,p′*-DDT, ND–4.05 ppm	Mean values reported for five different sites	Al-Mughrabi & Qrunfleh (2002)
Brazil, Itirapina region at the Northeastern part of Sao Paulo 2005	Soil	*p,p′*-DDE, 5.16; *o,p′*-DDT, 0.47; *p,p′*-DDT, 0.5; *p,p′*-DDD, 0.48 ng/g dw	*p,p′*-DDE, 2.05–8.8; *o,p′*-DDT, 0.05–1.69; *p,p′*-DDT, 0.03–1.12; *p,p′*-DDD, < 0.01–1.23; ΣDDT, 0.12–11.01 ng/g dw		Rissato et al. (2006)
China, Beijing-Tianjin-Hebei Economic Zone and Bohai Bay Rim city Year NA	Surface soils	ΣDDT, 73.9 ng/g	ΣDDT, ND–2417 ng/g	Industrial area where technical DDT was highly produced	Li et al. (2011)
USA, Texas Year NR	Soil	NR	DDE, Palmview, ND–60; San Benito, 2–60; Harlingen, 1–20; McAllen, 10–50 ng/g	Soils were collected from elementary school yards in cities/towns in the state of Texas	Miersma et al. (2003)
South Africa, Limpopo February 2008	Outside soil	ΣDDT: exposed area: 25; control area: 21 ng/g dw	ΣDDT: exposed area: 5.7–59; control area: 2.1–93 ng/g dw		Van Dyk et al. (2010)

Table 1.3 (continued)

Location, collection date	Sampling matrix	Exposure		Comments	Reference
		Level[a]	Range		
Eastern and southern Asia and Oceania August 1989 to October 1991	Air	NA	ΣDDT: India: 46–12 000; Thailand: 35–2600; Viet Nam: 1700–2400; Solomon Islands: 1300; Japan: 75; Taiwan, China: 230; Australia, 8.8–22 pg/m^3		Iwata et al. (1994)
Global 2005	Air	NA	*p,p′*-DDE: Delhi: 3600–6600; California: 210–460; Canary Islands: 190–250 pg/m^3	Reported from the first year of the Global Atmospheric Passive Sampling (GAPS) Network Highest and seasonably variable concentrations were detected at two agricultural sites near Delhi, India	Poo et al. (2009)
Mexico, Chiapas and Monterrey, and other Mexico sites 2005–2006	Air	ΣDDT, 558 pg/m^3	ΣDDT: southern Mexico sites: 239–2360; central sites: 15–750; Monterrey: 15; Chiapas: 2360 pg/m^3		Wong et al. (2009a)
USA, Canada, Mexico, and Belize 2000–2001	Air	NR	*p,p′*-DDT, 0.12–360; *p,p′*-DDE, 0.04–378; *p,p′*-DDD, 0.11–100 pg/m^3	The highest levels of DDT-related compounds were found in Mexico and Belize, which were about an order of magnitude higher than in the USA and southern Canada, and more than two orders of magnitude higher than in samples from the Arctic region	Shen et al. (2005)
China, Shanghai 2008–2011	Air	ΣDDT: air–gas phase: 4.78; air–particulate: 9.13 pg/m^3 (median)	ΣDDT: air–gas phase: ND–142.2, air–particulate: ND–120 pg/m^3	Estimated total daily uptake ΣDDT from food, dust and air 79.4 ng/day for children and 131.1 ng/day for adults > 95% from food	Yu et al. (2012)
South Africa, Limpopo Province February 2008	Indoor air	ΣDDT: exposed area: 3900; control area: 10 ng/m^3	ΣDDT: exposed area: 1100–8800; control area: 1.5–41 ng/m^3		Van Dyk et al. (2010)
China, Shanghai 2008–2011	Dust	ΣDDT: indoor dust: 29.8; outdoor dust: 5.7 ng/g (median)	ΣDDT: indoor dust: 0.15–179, outdoor dust: 0.16–107 ng/g	Estimated total daily uptake of ΣDDT from food, dust, and air was 79.4 ng/day for children and 131.1 ng/day for adults; > 95% from food	Yu et al. (2012)
South Africa, Limpopo Province February 2008	Floor dust	ΣDDT: in exposed area: 1200; in control area: 1.8 μg/m^2	ΣDDT: in exposed area: 8.3–4800; in control area: 0.1–8.9 μg/m^2		Van Dyk et al. (2010)

Table 1.3 (continued)

Location, collection date	Sampling matrix	Exposure		Comments	Reference
		Level[a]	Range		
Singapore 2005	House dust	ΣDDT, 14 ng/g (median)	ΣDDT, < LOD–770 ng/g	ΣDDT represented *p,p′*-DDE, *p,p′*-DDD and *p,p′*-DDT	Tan et al. (2007)
Mexico, Chihuahua Year NR	House dust	NR	DDT, 1–9587; DDE, 1–797 ng/ g	Chihuahua is a north Mexican state where DDT was sprayed several years previously for malaria-vector control	Díaz-Barriga-Martinez et al. (2012)
USA, California 2001–2006	Carpet dust	*p,p′*-DDE, 9.4; *p,p′*-DDT, 16 ng/g (geometric mean)	NA	Exposure assessment to DDT and other organochlorines using residential carpet dust	Ward et al. (2009)
USA, Iowa, Los Angeles county, Detroit, and Seattle 1999–2001	Carpet dust	*p,p′*-DDE, 43; *p,p′*-DDT, 343 ng/g	NA		Colt et al. (2004)

[a] Arithmetic mean, unless otherwise stated

[b] DDT is still used in specific areas of South Africa for indoor residual spray to control malaria vectors. In this study, indoor air, floor dust, outside soil, and drinking-water were sampled for measurement of DDT and related compounds in summer, 2 months after indoor residual spraying

ΣDDT, total DDT corresponding to the sum of *o,p′*-DDE, *p,p′*-DDE, *o,p′*-DDD, *p,p′*-DDD, *o,p′*-DDT, and *p,p′*-DDT; DDE, dichlorodiphenyldichloroethylene; DDT, dichlorodiphenyltrichloroethane; LOD, detection limit; dw, dry weight; NA, not applicable; ND, not determined; NR, not reported

ΣDDT at above the detection limit of 500 ng/L (Van Dyk et al., 2010).

(b) Soil and sediment

DDT and other persistent OCPs are found in soil and sediment samples from all regions of the globe, and residues are widely distributed in all types of soil (Table 1.3). For example in 2003, soils from elementary school yards in cities/towns within the state of Texas, USA, were reported to contain DDE at concentrations between 1 and 60 ng/g (Miersma et al., 2003). Soil sampled in 1998 from Humrat Al-Sahn in Jordan had levels of *p,p′*-DDE, *p,p′*-DDD, *o,p′*-DDT, and *p,p′*-DDT which ranged from ND to 0.46 ppm, ND to 0.16 ppm, ND to 0.11 ppm, and ND to 4.05 ppm, respectively (Al-Mughrabi & Qrunfleh, 2002). In Canada, in 2002, ΣDDT concentrations ranging from 1.0 to 18 ng/g were detected in soil in suburban Toronto (Wong et al., 2009b). In Sao Paulo, Brazil, ΣDDT concentrations in soil ranged from 0.12 to 11.01 ng/g dry weight (Rissato et al., 2006). In a review by Sharma et al. (2014), mean ΣDDT concentrations ranged from 34–903 ng/g in soil at 15 mainly agricultural sites in India. DDT concentrations exceeding 100 ng/g have been reported in sediment from several countries in Asia; the highest concentrations of ΣDDT, > 100 000 ng/g, were measured in India (Iwata et al., 1994; Sharma et al., 2014).

(c) Residential dust

In Chihuahua, Mexico, DDT and DDE concentrations quantified in household dust ranged from 1 to 9587 ng/g and 1 to 797 ng/g, respectively (Díaz-Barriga Martinez et al., 2012). In Singapore, the median ΣDDT concentration in 31 samples of house dust was 14 ng/g dust, with a range of below the limit of detection to 770 ng/g dust (Tan et al., 2007). In a study in California, USA, in 2001–2006, geometric mean concentrations of *p,p′*-DDE and *p,p′*-DDT detected in samples of carpet dust were 9.4 and 16 ng/g dust, respectively (Ward et al., 2009). In carpet-dust samples collected from the homes of 513 control subjects in Detroit, Iowa, Los Angeles, and Seattle, USA, between 1999 and 2001, mean concentrations of *p,p′*-DDE and *p,p′*-DDT were 43 and 343 ng/g dust (Colt et al., 2004).

(d) Air

A study on the atmospheric distribution of OCPs in North America between 2000 and 2001 reported that concentrations were in the range of 0.12–360 pg/m^3 for DDT, 0.04–378 pg/m^3 for DDE, and 0.11–100 pg/m^3 for DDD; the highest levels of DDT-related substances found in Mexico and Belize which about an order of magnitude higher than in the USA and southern Canada, and more than two orders of magnitude higher than in samples from the Arctic region (Shen et al., 2005). A study from the Global Atmospheric Passive Sampling Network reported that DDE concentrations were below the limit of detection at most sites. DDE concentrations at a background site in the Canary Islands in 2005 were between 190 and 250 pg/m^3, and at a rural site in California, USA, were in the range of 210–460 pg/m^3, but higher levels (3600–6600 pg/m^3) were detected at two agricultural sites near Delhi, India (Pozo et al., 2009). In Mexico, the highest concentration of ΣDDT was found in Chiapas (2360 pg/m^3) and the lowest in Monterrey (15 pg/m^3). In general terms, levels measured between 2005 and 2006 tended to be higher in the southern and central parts of Mexico (Wong et al., 2009a). In Limpopo, South Africa, in 2008, the mean ΣDDT concentration measured in indoor air was 3900 ng/m^3 (Van Dyk et al., 2010).

Between 1989 and 1991, ΣDDT concentrations were measured in air samples from urban and estuarine areas of eastern and southern Asia and Oceania; highest concentrations were measured in India (46–12 000 pg/m^3), followed by Thailand (35–2600 pg/m^3) and Viet Nam (1700–2400 pg/m^3) (Iwata et al., 1994).

(e) Food

DDT and its metabolites have been detected in many foods and in many countries. Generally, ΣDDT concentrations were higher in food containing more fat (see Table 1.4).

In Limpopo, South Africa, in 2008, mean ΣDDT concentrations were measured in vegetables (mean ΣDDT, 43 ng/g), chicken meat (mean ΣDDT, 700 ng/g), chicken fat (mean ΣDDT, 240 × 104 ng/g), and chicken liver (mean ΣDDT, 1600 ng/g) (Van Dyk et al., 2010). In Ethiopia between February and April 2008, ΣDDT mean concentrations in fish ranged from 0.89 to 172 ng/g wet weight (Deribe et al., 2013). In Ethiopia in 2010, mean ΣDDT concentrations in cows' milk samples from three study sites ranged from 269 to 477 ng/g milk fat (Gebremichael et al., 2013).

Dairy products in Jordan were analysed for DDT and its metabolites between 2001 and 2007, and only *p,p′*-DDE was detected. Mean *p,p′*-DDE concentrations ranged from 0.006 to 0.064 mg/kg fat (Salem et al., 2009). Edible fish from the Shadegan marshes in the south-western part of the Islamic Republic of Iran sampled in 2007 had a mean concentration of ΣDDT of 330 ng/g lipid weight, with levels ranging from 43 to 1590 ng/g lipid weight (Davodi et al., 2011).

The main source of human exposure to DDT in the Americas and Europe is dietary consumption of contaminated meat, fish, poultry, and dairy products. In the United States Total Diet Study for the years 1986–1991, mean dietary intakes of ΣDDT ranged from 0.009 to 0.0448 µg/kg bw per day across age and sex groups (Gunderson, 1995).

The Canadian Total Diet Study in 1998 reported that DDE was present in 25.8% of composite samples, with a mean level of 1.2 ng/g wet weight (ww) of prepared food; and reported values in dairy products of 0.71–3.48 ng/g ww, meat and meat products of 0.37–1.14 ng/g ww, and fish and fish products of 0.43–6.69 ng/g ww (Rawn et al., 2004). In Mexico, a ΣDDT concentration of 1.53 ng/g in cows' milk was reported in Chiapas in in 2011 (Gutiérrez et al., 2012), while a DDT concentration of 0.27 ng/g milk fat was reported in Hidalgo in 2010 (Gutiérrez et al., 2013). Also in Mexico, in Veracruz, a ΣDDT concentration of 539 ng/g lipid was reported in bovine meat (Pardío et al., 2012). On the east coast of Brazil, ΣDDT concentrations of 0.93 and 2.47 ng/g w/w were reported in blue shark and swordfish, respectively (de Azevedo e Silva et al., 2007).

The European Union report on pesticide residues in food indicated that only 368 out of 53 493 samples from 27 countries contained DDT or its metabolites, mostly in cows' milk and swine meat (EFSA, 2015).

In Asia, dietary exposure from food consumption, e.g. vegetables, fruits, and food of animal origin and animal products, was conducted to monitor dietary exposures and establish whether DDT concentrations had declined after DDT use was banned. In China, Yu et al. (2013) report dietary intake of DDTs (unspecified) for three time periods before (1970 and 1992) and after the ban (2005–2007) on DDT use on food crops. The range in estimated mean intakes across two cities and five age groups was 125–240 µg/kg per day in the 1970s, 21.6–50.2 µg/kg per day in 1992, and 2.69–4.95 µg/kg per day in 2005–2007. The decrease in estimated intakes between the 1970s and 2005/2007 was almost two orders of magnitude. In the 2007 Chinese Total Diet Study, *p,p′*-DDT was not detected in any foods, while the mean concentration of *p,p′*-DDE ranged from 0 to 13.3 ng/g across six food groups, with a maximum value of 53.4 ng/g in aquatic food products (Zhou et al., 2012). Marine fish were found to contain DDT at higher concentrations than freshwater fish in 2013 in China (Fang et al., 2015). Between 2008 and 2011, Yu et al. (2012) estimated the daily uptake of ΣDDTs from food, dust, and air to be 79.4 ng per day for children and 131.1 ng per day for adults, with > 95% of intake from food consumption. Chung et al. (2008) in 2005 reported the estimated daily

Table 1.4 Exposure to DDT and its metabolites in food

Location, collection date	Sampling matrix	Exposure		Comments	Reference
		Level[a]	Range		
Ethiopia Jimma, Asendabo, Serbo March–May 2010	Cows' milk	ΣDDT, 389 ng/g milk fat	ΣDDT, 269–477 ng/g milk fat		Gebremichael et al. (2013)
Mexico Hidalgo 2008- 2010	Cows' milk	*p,p'*-DDT, 2008: 0.22; 2010: 0.27 ng/g milk fat	NR		Gutiérrez et al. (2013)
Mexico Chiapas January to December 2011	Cows' milk	ΣDDT, 1.53 ng/g milk fat	NR		Gutiérrez et al. (2012)
India Haryana 1992–1993 and 1998–1999	Cows' milk	1992–1993: *p,p'*-DDT, 297.8; *p,p'*-DDE, 27.6; ΣDDT, 514 ng/mL 1998–1999: *p,p'*-DDT, 8; *p,p'*-DDE, 20.4; ΣDDT, 36.7 ng/mL	1992–1993: *p,p'*-DDT, 58.2–674.6; *p,p'*-DDE, 6.2–146.9; ΣDDT, 119.9–989.9 ng/mL 1998–1999: *p,p'*-DDT, < LOD–78; *p,p'*-DDE, < LOD–231.1; ΣDDT, 1.7–286.4 ng/mL	ΣDDT concentrations in cows' milk significantly declined (92.8%) from 1992–1993 to 1998–1999	Kaushik et al. (2011)
China Beijing and Shenyang 2005–2007	Cows' milk	Beijing: *p,p'*-DDT, 1.26 ± 1.83; *p,p'*-DDE, 0.804 ± 0.471 ng/g ww Shenyang: *p,p'*-DDT, 0.403 ± 0.264; *p,p'*-DDE, 0.575 ± 0.426 ng/g ww	NR		Tao et al. (2008)
China Beijing and Shenyang 2005–2007	Pork, chicken and crucian fish	*p,p'*-DDE: Beijing: pork: 1.19 ± 0.20; chicken: 0.54 ± 0.25; crucian fish: 2.98 ± 3.32 Shenyang: pork: 0.21 ± 0.1; chicken: 0.14 ± 0.04; crucian fish: 0.59 ± 0.20 ng/g	NR		Tao et al. (2008)

Table 1.4 (continued)

Location, collection date	Sampling matrix	Exposure		Comments	Reference
		Level[a]	Range		
South Africa Limpopo February 2008	Vegetables, chicken muscle, chicken fat; chicken liver	DDT: vegetables: 43; chicken muscle: 700; chicken fat: 240 000; chicken liver: 1600 ng/g	DDT: vegetables: 10–150; chicken muscle: 69–1400; chicken liver: 52–5600; chicken fat: 99–1 300 000 ng/g		Van Dyk et al. (2010)
Portugal October 2002-July 2003	Fish	*p,p′*-DDT: sardine: 30.1; mackerel: 109.9 *p,p′*-DDD: sardine: 3; mackerel: 51.9 ng/g ww	*p,p′*-DDT: sardine: 663.1; mackerel: 1540.2 *p,p′*-DDD: sardine: 22.3; mackerel: 852.2 ng/g ww (maximum)		Campos et al. (2005)
USA Aleutian Islands of Alaska 2004	Sockeye salmon	DDE, 6.9 ng/g w/w (median)	DDE, 0–56 ng/g w/w	Note that in this study the author stated that *p,p′*-DDE coeluted with PCB-85 under the gas chromatography conditions used for the analyses	Hardell et al. (2010a)
Brazil East Coast 2001	Fish	ΣDDT: blue shark: 0.93; swordfish: 2.47 ng/g w/w	ΣDDT: blue shark: 0.4–2.1; swordfish: 0.15–10.53 ng/g w/w		de Azevedo e Silva et al. (2007)
Mexico Veracruz Year NR	Bovine muscle	ΣDDT: A: 493.1; B: 539.8; C: 1121.7; D: 445.1 ng/g lipid	NR	Livestock originated from four (A–D) extensive breeding stock on commercial bovine producing farms located in the south-western agrarian zone of Veracruz. Between 28.6% and 66.7% of the samples were positive for DDT and/or its metabolites	Pardío et al. (2012)
China Hong Kong Special Administrative Region 2005	Food	ΣDDT, 0.145 μg/kg bw per day	NR	The mean and high EDIs among secondary school student consumers were 0.145 and 0.291 μg/kg bw per day Seafood (39%) and cereal and cereal products (20%) represented the major sources of dietary exposure to DDT and its metabolites Foodstuffs: cereal and cereal products, vegetables, fruits, meat, poultry, egg and their products, seafood, dairy products	Chung et al. (2008)
China Northern metropolis 2013	Vegetable and fish	NA	ΣDDT: vegetable: ND–10.4, fish: 0.77–25.0 ng/g fresh weight	EDI of ΣDDT from vegetable; fresh waterfish and marine fish were 1.13–10.4, 2–40.3 and 2.99–20.2 ng/kg bw per day, respectively	Fang et al. (2015)

Table 1.4 (continued)

Location, collection date	Sampling matrix	Exposure		Comments	Reference
		Level[a]	Range		
Thailand 1989–1996	Food	DDT levels in 1989 and 1996: Meat and milk: 10– < 2; poultry and eggs: 15–5; fats and oils: 35–ND ng/g or ng/L	NR	In Thai total diet, highest DDT levels were found in meat and milk, poultry and eggs and fats and oils; these concentrations decreased in 1989–1996. The dietary daily intake of DDT showed a gradual decline in 1989–1996 from 0.042 to 0.005 µg/kg bw per day	Vongbuddhapitak et al. (2002)
Japan Fukuoka 1992–1993	Food	Fish and fish products: *p,p'*-DDE, 4.74; ΣDDT, 7.58 Meat and meat products: *p,p'*-DDE, 2.87; ΣDDT, 5.25 Eggs: *p,p'*-DDE, 1.28; ΣDDT, 1.28 Cheese: *p,p'*-DDE, 2.24; ΣDDT, 2.28 µg/kg	Fish and fish products: *p,p'*-DDE, < LOD–24.8; ΣDDT, < LOD–40.64 Meat and meat products: *p,p'*-DDE, 1.04–6.54; ΣDDT, 1.04–15.06 µg/kg	*p,p'*-DDE was widely detected in the foods analysed, and levels were high in foods high in fat. Fish, fish products, meat, eggs, milk and milk products were analysed as the major sources of ΣDDT. The EDI in Japan from food was 1.42 µg/day per person	Nakagawa et al. (1995)
Brazil Ponta Grossa lake 2008	Fresh-water fish	ΣDDT: liver: 105; muscle: 29.9 ng/g dw	ΣDDT: liver: 2–641. muscle: ND–210 ng/g dw		Bussolaro et al. (2012)
Jordan 2001–2007	Dairy products	*p,p'*-DDE: milk: 27; butter: 9; cheese: 64; labneh: 6; yoghurt: 32 µg/kg fat	Milk: 5–70, butter: 8–10; cheese: 5–430; labneh: 5–6; yoghurt: 6–60 µg/kg fat	*o,p'*-DDD, *p,p'*-DDD, *o,p'*-DDE, *o,p'*-DDT, *p,p'*-DDT were all ND	Salem et al. (2009)
Islamic Republic of Iran Shadegan Marshes October & November 2007	Fish	ΣDDT, 330 ± 335 ng/g lipid weight	ΣDDT, 43–1590 ng/g lipid weight	Concentrations of DDE > DDD > DDT	Davodi et al. (2011)
Islamic Republic of Iran Fereydoon-kenar, Wildlife Refuge Winter 2008	Liver and muscle tissue of birds	Liver: DDE, 19; DDT, 9.3 Muscle: DDE, 396.8; DDT, 2 ng/g ww	Liver: DDE, < LOQ–89; DDT, < LOQ-28 Muscle: DDE, 0.5–2199; DDT, < LOQ–7.5 ng/g ww	Values are the means of results from three different birds: pintail, common teal and maillard DDE = sum of *o,p'*-DDE and *p-p'*-DDE; DDT = sum of *o,p'*-DDT and *p-p'*-DDT	Rajaei et al. (2010)
Canada Yukon 1998	Food	*p,p'*-DDE, 1.2 ng/g ww	*p,p'*-DDE: dairy: 0.71–3.48. meat: 0.37–1.14; fish: 0.43–6.69 ng/g ww	Canadian total diet study; 25.8% of the food samples were positive for *p,p'*-DDE compared with 28.9% in 1992–1996	Rawn et al. (2004)

Table 1.4 (continued)

Location, collection date	Sampling matrix	Exposure		Comments	Reference
		Level[a]	Range		
Ethiopia Lake Ziway, Rift Valley February–April 2008	Fish	ΣDDT, 18.02 ng g ww	ΣDDT, 0.89–171.96 ng/g ww	Mean reported for four different fish species. Levels reported for *p,p′*-DDT, *p,p′*-DDE and *p,p′*-DDD	Deribe et al. (2013)
USA 1986–1991	Food	NA	ΣDDT, 0.009–0.0448; *p,p′*-DDT, 0.0004–0.0011; *p,p′*-DDE, 0.0082–0.0441 µg/kg bw per day	Total diet study estimated across eight age groups	Gunderson (1995)
China Beijing, Shenyang 1970s, 1992, 2005–2007	Food	NA	1970s: 125–240; 1992: 21.6–50.2 2005–2007: 2.69–5.88 µg/kg per day	Temporal trend dietary intake	Yu et al. (2013)
China Shanghai 2008–2011	Food	ΣDDT (range of median): fish: 1.86–126.6; shellfish: 0.59–8.34; livestock: 0.2–0.5; poultry: 0.35–0.38 ng/g ww (median)	ΣDDT: fish: 0.51–340.1; shellfish: 0.1–13.2; livestock: 0.2–0.98; poultry: 0.06–0.93 ng/g ww	Estimated total daily uptake of ΣDDT from food, dust, and air was 79.4 ng/day for children and 131.1 ng/day for adults; > 95% from food	Yu et al. (2012)
China 2007	Food	*p,p′*-DDE: aquatic foods and aquatic food products, 13.3; meat and meat products, 3.65; eggs and egg products, 0.96; milk and milk products, 0.46; vegetables and vegetable products, 0.25 ng/g	*p,p′*-DDE: aquatic foods and aquatic food products, 1.21–53.4; meat and meat products, 0–13.3; eggs and egg products, 0–6.59; milk and milk products, 0–1.36; vegetables and vegetable products, 0–2.12 ng/g	Chinese total diet study; estimated average dietary exposure to ΣDDT was 0.016 µg/kg bw per day; authors stated that this represented a significant decrease compared with the past *p,p′*-DDE was the only DDT-related compound detected in food samples; aquatic foods and aquatic food products, meat and meat products and eggs and egg products were the major dietary sources of *p,p′*-DDE	Zhou et al. (2012)

ΣDDT, total DDT corresponding to the sum of *o,p′*-DDE, *p,p′*-DDE, *o,p′*-DDD, *p,p′*-DDD, *o,p′*-DDT, and *p,p′*-DDT; DDE, dichlorodiphenyldichloroethylene; DDT, dichlorodiphenyltrichloroethane; dw, dry weight; EDI, estimated daily intake; LOD, limit of detection; LOQ, limit of quantification; NA, not applicable; ND, not detected; NR, not reported; PTDI, the provisional tolerable daily intake; w/w, weight per weight; ww, wet weight

[a] Arithmetic mean, unless otherwise stated

intake of DDT and its metabolites, finding that the mean and maximum estimated daily intakes among secondary school students, 0.145 and 0.291 μg/kg bw per day, fell below the provisional tolerable daily intake of 10 μg/kg bw per day established by the Joint FAO/WHO Expert Committee on Food Additives (JECFA) (Food and Agriculture Organization of the United Nations/World Health Organization). However, the reported average estimated daily intakes were one to three orders of magnitude higher than those reported in Australia (FSANZ, 2003), New Zealand (NZFSA, 2005), Japan (Maitani, 2004), and Thailand (Vongbuddhapitak et al., 2002).

1.4.3 Exposure in the general population

DDT and/or its metabolites, mainly *p,p′*-DDE, have been measured in serum in the general population (in adults and children) in all parts of the world. Infants can also be exposed to these compounds, which have been detected in cord blood, placenta, and breast milk. ΣDDT has been shown to accumulate in adipose tissue and has also been detected in hair. Detailed biological measurements of DDT and its metabolites in the general population are given in Table 1.5.

Globally there are strong downward trends in population levels of plasma or serum *p,p′*-DDT and *p,p′*-DDE due to the ban of most uses of DDT in many countries (Fig. 1.3).

In populations living in homes where there has been indoor residual spraying for malaria control, the corresponding decrease is less and current levels are about 60 times higher than in comparable areas not subject to malaria-vector control (Ritter et al., 2011).

In breast milk, total DDT concentrations of up to 380 000 ng/g lipid have been reported in India in 1982 and in Zimbabwe in 1993–1995 (Ramachandran et al., 1984; Chikuni et al., 1997). ΣDDT concentrations in breast milk have tended to decrease globally, but remain high in some countries (Table 1.5).

The United Nations Environment Programme (UNEP) and WHO (UNEP/WHO, 2013) have presented the results of a global survey on concentrations of persistent organic pollutants in human milk. Large global differences with respect to contamination by DDT and its metabolites during 2000–2012 were apparent; the five countries with higher levels of total DDT in ascending order were Haiti, India, Solomon Islands, Tajikistan, and Ethiopia, which reported levels well above 20 000 ng/g lipid. Lower levels were reported for the Nordic countries.

1.4.4 Exposure assessment in epidemiological studies on DDT

The key epidemiological studies evaluated in this monograph can be categorized as studies of occupational exposure in farmers and commercial applicators, and studies involving the general population using questionnaire- or biological-based exposure assessments to DDT. The present section provides an assessment of the strengths and weaknesses of the exposure assessment and assignment methods used in these studies. The absence of similar detailed discussions for other epidemiological investigations should not be construed to suggest that these studies are inferior. In fact, in many ways these studies have improved on pesticide-exposure assessment when compared with earlier studies.

(a) Occupational exposure

In studies of occupational exposure to DDT among farmers and commercial applicators, exposure assessment has relied on either retrospective reporting of DDT use and similar reporting of factors potentially affecting exposure, or on job-title and task review to determine the circumstances of DDT exposure (Cocco et al., 2005; Alavanja et al., 2014). In both types of study, semiquantitative exposure scores were derived using exposure algorithms.

Table 1.5 Biological measurements of DDT and its metabolites in the general population

Location, collection date	Sampling matrix	Exposure		Comments	Reference
		Level[a]	Range		
USA 1994–1995	Serum	DDE, 0.637 ± 0.125 µg/g lipid	NR	Pretreatment concentrations in NHL patients (lipid-adjusted)	Baris et al. (2000)
USA North Carolina 1978–1982 and folllow up in 2003–2004	Serum	DDE: 1978–1982: 8.5; follow-up 2003–2004: 1.2 µg/L (median)	DDE: 1978–1982: 1.3–41; follow-up: 0.01–11 µg/L	Baseline: 1978–1982 (pregnant women) and follow-up organochlorine levels were compared	Vo et al. (2008)
USA Akwesasne Mohawk Nation 1996–2000	Serum	*p,p′*-DDE, 0.45 ± 0.35 ppb	NR	Male adolescents (age, 10 to < 17 years)	Schell et al. (2014)
Mexico Mexico City 1994–1996	Serum	DDE, 505.46; *p,p′*-DDT, 84.53 ng/g lipid	DDE, 0.004–4361.7; *p,p′*-DDT, 0.012–1262.7 ng/g lipid	Control women participating in a breast cancer case–control study	López-Carrillo et al. (1997)
Mexico Morelos 1999	Serum	*p,p′*-DDE, 21.8 ± 2.58; *p,p′*-DDT, 2.9 ± 2.84 ng/mL (geometric mean)	NR	Women of childbearing age	López-Carrillo et al. (2001)
Mexico Mexico City 1990–1995	Serum	*p,p′*-DDE, 2,51; *p,p′*-DDT, 0.23 µg/g lipid	*p,p′*-DDE, 0.97–6.05. *p,p′*-DDT, 0.04–0.33 µg/g lipid	Control women participating in a breast cancer case–control study	Romieu et al. (2000)
Mexico Chiapas 2002–2003	Serum	*p,p′*-DDE, 2.7; *p,p′*-DDT, 0.3 µg/g lipid (median)	NR	Levels in serum from mothers at delivery; this population was exposed for almost 40 years: DDT was used for agriculture until 1991 and for malaria control until 1998	Cupul-Uicab et al. (2010)
Mexico Sonora 2009	Serum	*p,p′*-DDE, 1.24; *p,p′*-DDT, 0.38 µg/L	*p,p′*-DDE, 0.25–10.3; *p,p′*-DDT, 0.25–1 µg/L	Children aged 6–12 years *p,p′*-DDE in serum was found in 100% of the children whereas *p,p′*-DDT was found in 25%	Meza-Montenegro et al. (2013)
Brazil Rio de Janeiro 2003–2004	Serum	Men: *p,p′*-DDE, 8.32; *o,p′*-DDT, 0.30; *p,p′*-DDT, 3.09; *p,p′*-DDD, 0.61 Women: *p,p′*-DDE, 9.64; *o,p′*-DDT, 0.42; *p,p′*-DDT, 3.2; *p,p′*-DDD, 0.66 ng/mL (median)	P25–P75 men: *p,p′*-DDE, 2.86–21.9; *o,p′*-DDT: < LOD–0.89; *p,p′*-DDT, 0.94–6.96; *p,p′*-DDD, 0.19–1.34 P25–P75 women: *p,p′*-DDE, 3.45–28.9; *o,p′*-DDT, < LOD–1.10; *p,p′*-DDT, 1.03–7.59; *p,p′*-DDD, 0.21–1.41 ng/mL	Levels in men and women in a heavily contaminated rural area in Brazil	Freire et al. (2013)

Table 1.5 (continued)

Location, collection date	Sampling matrix	Exposure		Comments	Reference
		Level[a]	Range		
Saudi Arabia Riyadh NR	Serum	ΣDDT: diabetics: 18.3 ± 1.4; non-diabetics: 11.8 ± 1.3 ng/mL	NR	The study compared serum concentrations of DDT and its metabolites in diabetic and non-diabetic individuals	Al-Othman et al. (2015)
Tunisia Bizerte June 2011 to May 2012	Serum	*p,p*′-DDE, 168.8 ± 158; *p,p*′-DDT, 24.3 ± 18.8; ΣDDT, 213.1 ± 160 ng/g lipid	Maximum: *p,p*′-DDE, 950.4; *p,p*′-DDT, 71.5; ΣDDT, 994.6 ng/g lipid		Ben-Hassine et al. (2014)
Egypt Port Said region July 1999 to July 2000	Serum	DDE, 31 ng/g	DDE, 12–44 ng/g		Ahmed et al. (2002)
Benin Republique Borgou region 2011	Serum	*p,p*′-DDT, 32; *p,p*′-DDE, 607.2 ng/g total lipid (geometric mean)	*p,p*-DDT, 22.8–45; *p,p*′-DDE, 453–813.9 ng/g total lipid		Azandjeme et al. (2014)
South Africa KwaZulu November 1986 to November 1987	Serum	Range of means: *p,p*′-DDE, 103.4–127.1; *p,p*′-DDT, 31.4–47.5; ΣDDTs 140.9–174.6 µg/L	NR		Bouwman et al. (1994)
South Africa KwaZulu-Natal Province 2008	Serum	o,*p*′-DDE, 9; *p,p*′-DDE, 3840; *o,p*′-DDD, 8; *p,p*′-DDD, 26; *o,p*′-DDT; 168; *p,p*′-DDT, 2194 ng/g lipid (geometric mean)	o,*p*′-DDE, 7–10; *p,p*′-DDE, 3008–4902; *o,p*′-DDD, 6–9; *p,p*′-DDD, 20–32; *o,p*′-DDT, 127–221; *p,p*′-DDT, 1706–2823. ng/g lipid	Data reported for women living in malaria-endemic areas who were admitted for delivery at the local hospital	Channa et al. (2012)
South Africa Limpopo 2008	Serum	*p,p*′-DDE, 5900; *p,p*′-DDD, 1500; *o,p*′-DDD, 1500; ΣDDT, 7300 ng/g lipid	*p,p*′-DDE, 1200–23 000; *p,p*′-DDD; 800–3800; *o,p*′-DDD, 300–2700; ΣDDT, 1300–23 000 ng/g lipid	Sampling was done in February 2008 during the summer season 2 months after the IRS process was completed	Van Dyk et al. (2010)
Belgium Flanders 2003–2004	Serum	*p,p*′-DDE: boys, 104; girls, 84 ng/g lipid (median)	*p,p*′-DDE 10th-90th percentile: boys: 47–404; girls: 39–247 ng/g lipid	Serum measurements in adolescents (aged 14–15 years)	Den Hond et al. (2011)
France Ille-et-Vilaine Côte d'Or 2005–2007	Serum	*p,p*′-DDE, 84.8 ng/g lipid (median)	*p,p*′-DDE, 51.9–131.3 ng/g lipid	Women CECILE Study	Bachelet et al. (2011)
France 2006–2007	Serum	*p,p*′-DDE, 88.7; *p,p*′-DDT, 3.6 ng/g lipid (geometric mean)	*p,p*′-DDE, 72.5–108.6; *p,p*′-DDT, 2.6–5.1 ng/g lipid	French National Nutrition and Health Study	Saoudi et al. (2014)

Table 1.5 (continued)

Location, collection date	Sampling matrix	Exposure		Comments	Reference
		Level[a]	Range		
Spain Barcelona 2002–2006	Serum	2002: *p,p′*-DDE, 491.2; *p,p′*-DDT, 37.2; 2006: *p,p′*-DDE, 233.6; *p,p′*-DDT, 20.3 ng/g lipid (geometric mean)	2002: *p,p′*-DDE, 421.3–572.7; *p,p′*-DDT, 31.9–43.5 2006: *p,p′*-DDE, 206.8–263.8; *p,p′*-DDT, 17.9–22.9 ng/g lipid		Porta et al. (2012)
Spain Ribera d'Ebre and Menorca 1997–1999 and 2001–2012	Serum	*p,p′*-DDE: Ribera d'Ebre: cord: 0.86; 4 years: 0.75; 14 years: 0.12 Menorca: cord, 1.03; 4 years, 0.81; 14 years, 0.33 ng/mL (median)	*p,p′*-DDE (25th–75th percentile): Ribera d'Ebre: cord, 0.50–1.68; 4 years, 0.38–1.32; 14 years, 0.09–0.19 Menorca: cord: 0.57–1.94; 4 years, 0.44–1.77; 14 years, 0.23–0.58 ng/mL	Follow-up of organochlorine compounds serum concentration in children from birth until adolescence. Three measurements: at birth (cord), age 4 years, and age 14 years	Gascon et al. (2015)
Spain Asturias, Navarra, Guipúzcoa, Murcia and Granada) 1992–1996	Serum	DDE, 822.1 ng/g lipid (geometric mean)	DDE, 779.2–867.2 ng/g lipid	Participants in the EPIC cohort; Murcia and Granada showed higher levels with geometric mean for *p,p*-DDE of 1287.2 and 957.4 ng/g lipid, respectively	Jakszyn et al. (2009)
Japan Akita prefecture, a rural area in northern Japan 1999	Serum	*p,p′*-DDE + *p,p′*-DDT, 6.3 ng/mL	*p,p′*-DDE + *p,p′*-DDT, 0.9–31 ng/mL		Hanaoka et al., (2002)
the Russian Federation Chapaevsk 2003–2005	Serum	*p,p′*-DDE median, 287 ng/g lipid	*p,p′*-DDE 10th-90th percentile, 122–866 ng/g lipid	Serum measurements in Russian boys aged 8 and 9 years; association with growth was evaluated	Burns et al. (2012)
Thailand a highland village at the north of Chiang Mai 2003–2004	Serum	*p,p′*-DDE, 4013, *p,p′*-DDD, 390.5; *p,p′*-DDT, 628.7 ng/g lipid (geometric mean)	*p,p′*-DDE, 1325–12 683; *p,p′*-DDD, 212–1162; *p,p′*-DDT, 225–3085 ng/g lipid	Measurements in adult men	Asawasinsopon et al. (2006b)

Table 1.5 (continued)

Location, collection date	Sampling matrix	Exposure		Comments	Reference
		Level[a]	Range		
Greenland, Sweden, Poland, Ukraine Warsaw, Kharkiv June 2002 to May 2004	Serum	*p,p′*-DDE: Sweden: 820; Ukraine: 620; Poland: 360. Greenland: 300 ng/g lipid (median)	*p,p′*-DDE: Sweden: 92–14 000; Ukraine: 230–1800; Poland: 100–1100. Greenland: 26–1700 ng/g lipid	Part of the INUENDO Collaborative Project; measurements in women	Axmon et al. (2006)
Republic of Korea Uljin county 2006	Serum	*p,p′*-DDE, 376 ± 290.7; *p,p′*-DDD, 5.7 ± 3.7; *p,p′*-DDT, 23.8 ± 12.1; *o,p′*-DDT, 3.2 ± 2.8 ng/g lipid	NA		Son et al. (2010)
India Ahmedabad urban area Year NR	Serum	*p,p′*-DDE, 20.74; *o,p′*-DDT, 0.99; *p,p′*-DDD, 1.6; *p,p′*-DDT, 7.65; ΣDDT, 29.63 ng/mL (median)	p,*p′*-DDE, 10.43–38.33; *o,p′*-DDT, 0.42–2.41; *p,p′*-DDD, 0.77–4.43; *p,p′*-DDT, 3.66–24.06; ΣDDT, 21.17–54.47 ng/mL	Measurements in adult men	Bhatnagar et al. (2004)
India Delhi 1982	Serum	ΣDDT, 390 ng/mL (geometric mean)	ΣDDT, ND–4610 ng/mL		Ramachandran et al. (1984)
Thailand nationwide 2011	Serum	Males: *p,p′*-DDE, 1539; *p,p′*-DDT, 135 Females: *p,p′*-DDE, 1547; *p,p′*-DDT, 133 ng/g lipid (geometric mean)	Males: *p,p′*-DDE, 1242–1837; p,*p′*-DDT, 116–164 Females: *p,p′*-DDE, 1293–1806; *p,p′*-DDT, 112–147 ng/g lipid		Teeyapant et al. (2014)
Egypt Cairo and Nile Delta NR	Serum	Rural women: DDE, 17.3; ΣDDT, 18.3 Urban women: DDE, 9.7; ΣDDT, 9.9 ppb (geometric mean)	Rural women: DDE, 0–142.1; ΣDDT, 0–144.8 Urban women: DDE, 0.7–59.8; ΣDDT, 0.7–61.8 ppb	Interestingly, women with low DDE serum levels had breast-fed their children for an average of 18 months; women with no lactation history had much higher organochlorine levels than women who breast-fed	Soliman et al. (2003)
Republic of Korea Seoul, Pyuncheon, Anson and Jeju 2011	Serum	*p,p′*-DDE, 57.4; *p,p′*-DDT, 5.2; ΣDDT, 64.4 ng/g lipid (median)	25th–75th percentile: *p,p′*-DDE, 38.8–78.9; *p,p′*-DDT, 2.94–8.99; ΣDDT, 42.2–92.4 ng/g lipid	Pregnant women (n = 138) from five university hospitals; blood samples were collected on the day before delivery	Kim et al. (2013a)
New Zealand Nationwide 1996–1997	Serum	*p,p′*-DDE: 15–24 years: 646; 25–34 years: 771; 35–49 years: 1060; 50–64 years: 1310; 65+: 1780 ng/g lipid	NR	General trend of increasing concentration with age, and no consistent differences between the sexes, or between people of Maori and non-Maori ethnicity	Bates et al. (2004)

Table 1.5 (continued)

Location, collection date	Sampling matrix	Exposure		Comments	Reference
		Level[a]	Range		
Peru Trujillo 2004–2005	Serum, cord blood	*p,p'*-DDE: 1st trimester: 581; 2nd trimester: 486; 3rd trimester: 418; cord: 383 *p,p'*-DDT: 1st trimester: 39; 2nd trimester: 32; 3rd trimester: 29; cord: 20 ng/g lipid (geometric mean)	*p,p'*-DDE: 1st trimester: 373–906; 2nd trimester: 310–762; 3rd trimester: 255–686; cord: 225–652 *p,p'*-DDT: 1st trimester: 26–59; 2nd trimester: 20–51; 3rd trimester: 18–45; cord: 12–33 ng/g lipid	Samples collected once per pregnancy trimester and at delivery in the cord blood; the detection of DDT and DDE in cord serum suggested substantial transfer of these compounds from mother to fetus	Adetona et al. (2013)
Saudi Arabia Al-Kahrj 2005–2006	Serum, cord blood, placental tissue	Cord: *p,p'*-DDE, 0.197; *p,p'*-DDD, 0.005; *p,p'*-DDT, 0.005 µg/L Maternal serum: *p,p'*-DDE, 0.551; *p,p'*-DDD, 0.002; *p,p'*-DDT, 0.008 µg/L Placenta: *p,p'*-DDE, 10.17; *p,p'*-DDD, 7.042; *p,p'*-DDT, 29.62 µg/kg dw	Cord: *p,p'*-DDE, 0–19.95; *p,p'*-DDD, 0–5.08; *p,p'*-DDT, 0–2.97 µg/L Maternal serum: *p,p'*-DDE, 0–29; *p,p'*-DDD, 0–0.90; *p,p'*-DDT, 0–3.4 µg/L Placenta: *p,p'*-DDE, 0–314, *p,p'*-DDD, 0–223; *p,p'*-DDT, 0–2038 µg/kg dw	Detection of DDT, DDD, and DDE in cord serum and placenta suggested substantial transfer of these compounds from mother to fetus	Al-Saleh et al. (2012)
Thailand Chiang Mai Province 2003–2004	Maternal and cord sera	Maternal/cord serum: *p,p'*-DDE, 1191/742; *p,p'*-DDD, 104/89.1; *p,p'*-DDT, 123/77.1 ng/g lipid (geometric mean)	Maternal/cord serum: *p,p'*-DDE, 58.3–7981/81.3–4265, *p,p'*-DDD, 16.8–527/24.1–309, *p,p'*-DDT, 18.0–1067/21.5–660 ng/g lipid	Comparison of concentrations of DDT and metabolites in maternal and cord serum	Asawasinsopon et al. (2006a)
Thailand Chiang Mai, Chiang Dao district 2003–2004	Maternal and cord sera	Maternal/cord sera: *p,p'*-DDE, 1793/1255; *p,p'*-DDT, 145/102; *p,p'*-DDD, 152/145 ng/g lipid (geometric mean)	Maternal/cord sera: *p,p'*-DDE, 208–27 169/146–23 185; *p,p'*-DDT, 11.6–2003/17–2363; *p,p'*-DDD, 11.1–2984/28.6–1800 ng/g lipid	Pregnant women from an agricultural and former malaria-endemic area; cord serum levels of *p,p'*-DDE, *p,p'*-DDT, and *p,p'*-DDD were approximately 70%, 62%, and 79% of maternal serum levels, respectively	Sapbamrer et al. (2008)
China Shanxi province 2005–2007	Placenta	Neural tube defects/ controls: Σ*o,p'*-DDTs, 4.3/2.7; Σ*p,p'*-DDTs, 55/59; ΣDDTs, 60/61 ng/g lipid (median)	Σ*o,p'*-DDTs: 2–7.6; Σ *p,p'*-DDTs, 31–85; ΣDDTs, 35–98 ng/g lipid	Placental concentrations given for newborn infants with neural tube defects compared with healthy (control) newborn infants	Ren et al. (2011)

Table 1.5 (continued)

Location, collection date	Sampling matrix	Exposure		Comments	Reference
		Level[a]	Range		
Italy Brescia, an urban polluted area 2006	Serum placenta adipose tissue	*p,p'*-DDE: serum: 112.3; placenta: 62.5; adipose tissue: 202 ng/g lipid (median)	*p,p'*-DDE, 5th–95th percentile: serum: 42–377; placenta: 24–226; adipose tissue: 76–730 ng/g lipid		Bergonzi et al. (2009)
Nicaragua Rio Atoya basin NR	plasma and cord blood	*p,p'*-DDE: maternal plasma: 7.12; cord blood: 6.39 ng/g	*p,p'*-DDE: maternal plasma: 0–35.23; cord blood: 0–9.35 ng/g	Women at delivery	Dorea et al. (2001)
USA New York 1991–1998	Plasma	DDE, 5.83 ng/mL	DDE, 0.02–23.67 ng/mL	Women with cystic disease (not lipid -adjusted)	Blackwood et al. (1998)
Mexico Quintana Roo State 2007	Plasma	*p,p'*-DDT, 2206; *p,p'*-DDE, 7828 ng/g lipid (geometric mean)	*p,p'*-DDT, 463.5–9046.3; *p,p'*-DDE, 490.8–57 712.4 ng/g lipid	Children aged 6–12 years living in three communities	Trejo-Acevedo et al. (2013)
Mexico Chiapas 2012	Plasma	*p,p'*-DDE, 24.66; *p,p'*-DDT, 14.71 ng/mL (geometric mean)	*p,p'*-DDE, 1.1–222.6; *p,p'*-DDT, 6.37–29.66 ng/mL	Middle-aged men and women; a high proportion of participants were engaged in agricultural work	Ruiz-Suárez et al. (2014)
South Africa Limpopo Province 2010–2011	Plasma	*p,p'*-DDT: unsprayed villages:0.31; non-DDT IRS homes: 1.4; DDT-IRS homes: 2.6 μg/L *p,p'*-DDE: unsprayed villages: 1.7; non-DDT IRS homes: 7.95; DDT-IRS homes: 8.5 μg/L (median)	*p,p'*-DDT: unsprayed villages: 0.11–0.86; non-DDT IRS homes: 0.5–3; DDT-IRS homes: 1.1–6.6 μg/L *p,p'*-DDE: unsprayed villages: 0.7–5.5; non-DDT IRS homes: 3.4–12; DDT-IRS homes: 4.65–18 μg/L	Median and ranges reported for DDT and DDE in three different exposure settings	Whitworth et al. (2014)
South Africa 7 sites NR	Plasma	Range of means for sites: *p,p'*-DDE, 41.1–5178; *p,p'*-DDT, 1.9–1797 ng/g lipid (geometric mean)	*p,p'*-DDE,15–14 482; *p,p'*-DDT, 0.9–5278 ng/g lipid	Measurements in maternal plasma at delivery from seven different communities: rural, urban, industrial, fishing, mining, coastal endemic malaria, inland endemic malaria; highest levels of DDTs were measured in the coastal malaria site (Indian Ocean) with geometric means of 5178 ng/g lipid and 1797 ng/g lipid for *p,p'*-DDE and *p,p'*-DDT, and 1966 ng/g lipid and 726 ng/g lipid for *p,p'*-DDE and *p,p'*-DDT in inland endemic malaria site	Röllin et al. (2009)

Table 1.5 (continued)

Location, collection date	Sampling matrix	Exposure		Comments	Reference
		Level[a]	Range		
South Africa Limpopo Province Year NR	Plasma	*p,p′*-DDT, 109 200 ± 106 600; *p,p′*-DDE, 246 200 ± 218 500 ng/g lipid	NR	Measurements in young men from a malaria area	de Jager et al. (2012)
Norway Vestvagoy in north-western coast of Norway 1997	Plasma	*p,p′*-DDE, 0.936 ng/g lipid (median)	*p,p′*-DDE, 0.15–5.075 ng/g lipid	Women	Furberg et al. (2002)
Central America Year NR	Serum	*p,p′*-DDE + *p,p′*-DDT, 3.25 ng/mL (geometric mean)	*p,p′*-DDE + *p,p′*-DDT, 0.3–64 ng/mL	Measurements in children from several countries in Central America, except Mexico; by far the highest levels were found in children living in Mexico, with levels of *p,p′*-DDE+ *p,p′*-DDT up to 390 ng/mL and a geometric mean of 50 ng/mL	Pérez-Maldonado et al. (2010)
Brazil Sao Paolo State 2007–2008	Blood	*p,p′*-DDE, 280; *p,p′*-DDT, 18 ng/g lipid (median)	*p,p′*-DDE, 126–645 ng/g lipid	Measurements in women at delivery	Rudge et al. (2012)
India Dibrugarb and Nagaon districts, Assam state. 2009–2010	Blood	ΣDDT: Dibrugarb: 417; Nagaon: 743 ng/mL (median)	ΣDDT: Dibrugarb: 4–4718; Nagaon: 7–9906 ng/mL		Mishra et al. (2011)
Sweden 1993–2007	Blood; adipose tissue	DDE: blood: 211; adipose tissue: 555 ng/g lipid (median)	DDE: blood: 16–3725; adipose tissue: 41–3900 ng/g lipid	Measurements in control individuals participating in different cancer studies; during 1993–2007, an annual decrease of 13.5% was found for DDE concentrations in both tissues	Hardell et al. (2010b)
Jordan Jordan University Hospital 1996	Adipose tissue	ΣDDT: 0–14 years: 2.6; 15–29 years: 3.9; 30–44 years: 3.8; 45–59 years: 4.6; 60 years and over: 4.6 ppm	ΣDDT: 0–14 years: 0.36–9.94; 15–29 years: 0.52–16.36; 30–44 years: 0.28–11.04; 45–59 years: 0.11–12.88; 60 years and over: 0.80–7.68 ppm	Means and ranges also reported for individual metabolites (*o,p′*-DDT, *p,p′*-DDT; *o,p′*-DDE, *p,p′*-DDE; *o,p′*-DDD, *p,p′*-DDD) in each age category and by sex Men stored higher amounts of ΣDDT than women; highest levels of ΣDDT were in men (4.3 ppm) and women (5 ppm) aged ≥ 60 years	Alawi et al. (1999)
India Delhi 1982	Adipose tissue	ΣDDT, 15 430 ng/g (geometric mean)	320–380 000 ng/g		Ramachandran et al. (1984)

Table 1.5 (continued)

Location, collection date	Sampling matrix	Exposure		Comments	Reference
		Level[a]	Range		
Denmark NR 1993 –1997	Adipose tissue	DDE, 505 ng/g lipid (median)	DDE, 17–6693 ng/g lipid	Measurements in postmenopausal women. Adipose tissue concentrations of DDE were consistently positively associated with age and the consumption of fish with high fat content. Total lifetime duration of lactation had an inverse relationship	Vaclavik et al. (2006)
Mexico Veracruz 2009	Adipose tissue	Control/pregnant women: *p,p'*-DDE, 695 /421; *p,p'*-DDT, 67/42; ΣDDT, 873/474 ng/g lipid (median)	Control/pregnant women: *p,p'*-DDE, 12–6007/34–5220; *p,p'*-DDT, 5–675/6–1695 ng/g lipid	Measurements in control and pregnant women	Herrero-Mercado et al. (2010)
Poland Poznan NR	Adipose tissue	ΣDDT, 773 ng/g lipid (median)	ΣDDT, 258–3570 ng/g lipid	Values are concentrations in control women from a breast cancer study	Ociepa-Zawal et al. (2010)
USA New York 1991–1998	Breast cyst fluid/plasma	DDE: breast cyst fluid: 1.98; plasma: 4.83 ng/mL	DDE: breast cyst fluid: 0.111–2.35; plasma: 0.02–9.52 ng/mL	Women with cystic disease; highest levels found in one woman: 25.27 and 23.67 ng/mL DDE in breast cyst fluid and plasma, respectively	Blackwood et al. (1998)
Colombia Bogota NR	Breast milk	*p,p'*-DDE, 203 ng/g lipid (milk fat)	*p,p'*-DDE, 17–14 948 ng/g lipid (milk fat)	Measurements in breastfeeding mothers	Rojas-Squella et al. (2013)
Tunisia Bizerte 2010	Breast milk	*p,p'*-DDE, 371.2; *p,p'*-DDD, 92; *p,p'*-DDT, 271.2; ΣDDT, 805.9 ng/g lipid (median)	*p,p'*-DDE, 73.3–3470.8; *p,p'*-DDD, 10.8–1701.6; *p,p'*-DDT, 27.8–2147.3; ΣDDT, 125.8–4574.8 ng/g lipid		Ben-Hassine et al. (2012)
Tunisia 12 different regions 2003–2005	Breast milk	*p,p'*-DDE, 676; *p,p'*-DDD, 92; *p,p'* DDT, 256; ΣDDT, 1931 ng/g lipid	*p,p'*-DDE, 3–6800; *p,p'*-DDD, 2–2461; *p,p'*-DDT, 1–2499; ΣDDT, 8–7060 ng/g lipid		Ennaceur et al. (2008)
Islamic Republic of Iran Southern coast of the Caspian sea 2006	Breast milk	ΣDDT, 2554 ng/g lipid	ΣDDT, 70–18 370 ng/g lipid		Behrooz et al. (2009)
South Africa KwaZulu-Natal April–November 2002	Breast milk	ΣDDT, 83.22 ng/mL	ΣDDT, 3.52–1537.73 ng/mL	Mean value calculated based on mean values reported from six subgroups; *p,p'*-DDE, *p,p'*-DDD and *p,p'*-DDT levels also reported	Bouwman et al. (2006)

Table 1.5 (continued)

Location, collection date	Sampling matrix	Exposure		Comments	Reference
		Level[a]	Range		
South Africa Thohoyando, a rural district in the Limpopo province April–November 2004	Breast milk	*p,p'*-DDE, 4580; *p,p'*-DDT, 1600; ΣDDT, 6320 ng/g lipid	*p,p'*-DDE, 98–13 600; *p,p'*-DDT, 20–6000; ΣDDT, 162–20 200 ng/g lipid		Darnerud et al. (2011)
South Africa KwaZulu-Natal & Limpopo 2008	Breast milk	ΣDDT at three sites, 18 000, 11 000 & 9500 ng/g lipid (milk fat)	ΣDDT, 106–140 000 ng/g lipid	Measurements from women in three DDT-sprayed villages	Bouwman et al. (2012)
South Africa Thohoyando, Limpopo Province 2004	Breast milk	NR	*p,p'*-DDE, 1–14 508; *p,p'*-DDD, ND–5901; ΣDDT, ND–8504 ng/g lipid	Values reported for 10 sites in Thohoyando area	Okonkwo et al. (2008)
Ethiopia Jimma, Asendabo, Serbo March–May 2010	Breast milk	Mean range: *p,p'*-DDE, 2520 – 4760; *p,p'*-DDD, 320–390; *p,p'*-DDT, 3550–12 200; ΣDDT, 6420–17 170 ng/g lipid	NR	Values reported for each of the three areas in which annual spraying with DDT for malaria control was common	Gebremichael et al. (2013)
Zimbabwe Seven sites in Kariba area February 1993 to April 1995	Breast milk	Mean range: *p,p'*-DDE, 1176–13 606; *p,p'*-DDT, 250–9080; ΣDDT, 1607–25 259 ng/g lipid	*p,p'*-DDE: 77–182 523; *p,p'*-DDT, 0–53 000; ΣDDT, 85–380 580 ng/g lipid	Vector-control programmes, agricultural activities, and possibly dietary habits (since DDT is also found in the urban population) were the main contributing factors towards high levels of DDT and DDT metabolites in breast milk	Chikuni et al. (1997)
Czech Republic Nationwide 2005–2009	Breast milk	*p,p'*-DDT, 7600; *p,p'*-DDE, 234 000 ng/g lipid (median)	NR	The Human Biomonitoring Project; preastfeeding primiparas *p,p'*-DDT and *p,p'*-DDE showed a downward trend, with median values decreasing respectively from 41 and 455 mg/kg milk fat in 1996, to 7.6 and 234 mg/kg milk fat in 2009	Cerná et al. (2012)
Cambodia Phnom Penh city (urban area), Meanchey (suburban area) 1999–2000	Breast milk	ΣDDT: urban area: 1100; suburban area: 860 ng/g lipid (median)	ΣDDT: urban area: 310–11 000, suburban area: 360–3800 ng/g lipid		Kunisue et al. (2004b)

Table 1.5 (continued)

Location, collection date	Sampling matrix	Exposure		Comments	Reference
		Level[a]	Range		
China Shanghai 2011–2012	Breast milk	*p,p′*-DDT, 6.4; *p,p′*-DDE, 207; ΣDDT, 221 ng/ g lipid (median)	*p,p′*-DDT, ND–160; *p,p′*-DDE, 42.6–1604; ΣDDT, 52–1643 ng/g lipid	ΣOCPs in this study were much lower than those in human breast milk samples collected in 2002 and 2007	Lu et al. (2015)
China Dalian and Shenyang 2002	Breast milk	Dalian: *p,p′*-DDE, 2000; *p,p′*-DDT, 130; ΣDDT, 2100 Shenyang: *p,p′*-DDE, 830; *p,p′*-DDT, 40; ΣDDT, 870 ng/g lipid	Dalian: *p,p′*-DDE, 710–5300; *p,p′*-DDT, 45–380; ΣDDT, 780–5400 Shenyang: *p,p′*-DDE, 110–3100; *p,p′*-DDT, 12–140; ΣDDT, 140–3200 ng/g lipid		Kunisue et al. (2004a)
Japan Osaka prefecture 1972–1998	Breast milk	1972: DDE, 1686; DDT, 538 1982: DDE, 2446; DDT, 171 1992: DDE, 510; DDT, 18.2 1998: DDE, 270; DDT, 17.8 ng/g lipid	1972: DDE, 640–2630; DDT, 130–1380 1982: DDE, 580–8967; DDT, 82–415 1992: DDE, 102–1318; DDT, 4.5–444 1998: DDE, 77–997; DDT, 4.3–122.7 ng/g lipid	Compared with peak levels found in 1972 (100%), concentrations of DDT and DDE in breast milk fell to about 3% and 16% in 1998, respectively	Konishi et al. (2001)
China Hong Kong SAR and Guangzhou 1999, 2000	Breast milk	Guangzhou: *p,p′*-DDE, 2850; *p,p′*-DDT, 700 Hong Kong SAR: *p,p′*-DDE, 2480; *p,p′*-DDT, 390 ng/g lipid	NA	In Hong Kong SAR, mean *p,p′*-DDT and *p,p′*-DDE levels (390 and 2480 ng/g lipid) from the present study (1999) were considerably lower than those reported 14 years previously (1985) (i.e. 2170 and 11 670 ng/g lipid)	Wong et al. (2002)
Indonesia Jarkata, Purwakarta, Bogor and Lampung 2001–2003	Breast milk	Jakarta: *p,p′*-DDE, 140; *p,p′*-DDT, 10; ΣDDT, 160 Purwakarta: *p,p′*-DDE, 430; *p,p′*-DDT, 17; ΣDDT, 440 Bogor: *p,p′*-DDE, 780; *p,p′*-DDT, 16; ΣDDT, 820 Lampung: *p,p′*-DDE, 860; *p,p′*-DDT, 8; ΣDDT, 910 ng/g lipid (median)	*p,p′*-DDE, 14–12 000; *p,p′*-DDT, < 2–2400; ΣDDT, 18–15 000 ng/g lipid	Study sites were Jakarta (major dumping site of municipal wastes, urban), Purwakarta (agriculture, rural), Bogor (city, suburban) and Lampung (coastal area, rural)	Sudaryanto et al. (2006)

Table 1.5 (continued)

Location, collection date	Sampling matrix	Exposure		Comments	Reference
		Level[a]	Range		
Japan Fukuoka prefecture 2001–2004	Breast milk	Primipara: *p,p′*-DDE, 330; *p,p′*-DDT, 13; ΣDDT, 340 Multipara: *p,p′*-DDE, 220; *p,p′*-DDT, 10; ΣDDT, 230 ng/g lipid	Primipara: *p,p′*-DDE, 32–1100; *p,p′*-DDT, < 0.6–44; ΣDDT, 34–1100 Multipara: *p,p′*-DDE, 29–1100; *p,p′*-DDT, < 0.6–30; ΣDDT, 33–1100 ng/g lipid		Kunisue et al. (2006)
Republic of Korea Seoul and Pyungchon (residential area), Ansan (industrial) and Jeju (rural) 2011	Breast milk	*p,p′*-DDE, 106; *p,p′*-DDT, 7.11; ΣDDT, 114 ng/g lipid	*p,p′*-DDE, < LOQ–375; *p,p′*-DDT, < LOQ–51.7; ΣDDT, < LOQ–392 ng/g lipid		Lee et al. (2013)
Thailand Chiang Mai Province 1998	Breast milk	ΣDDT, 14 960 ng/g lipid (geometric mean)	ΣDDT, 3370–46 880 ng/g lipid	Participants were Hmong hill-tribe mothers living in an area formerly using DDT for malaria-vector control and in agriculture Estimated ΣDDT median daily intake among primipara infants was 63.2 µg/kg bw per day	Stuetz et al. (2001)
Viet Nam Hanoi and Hochiminh 2000–2001	Breast milk	Hanoi: *p,p′*-DDE, 1900; *p,p′*-DDT, 170; ΣDDT, 2100 Hochiminh: *p,p′*-DDE, 2000; *p,p′*-DDT, 265; ΣDDT, 2300 ng/g lipid	Hanoi: *p,p′*-DDE, 420–6300; *p,p′*-DDT, 34–6900; ΣDDT, 480–6900 Hochiminh: *p,p′*-DDE, 340–16 000; *p,p′*-DDT, 100–1000; ΣDDT, 440–17 000 ng/g lipid		Minh et al. (2004)
Thailand Chiang Mai City District 2002	Breast milk	*p,p′*-DDT, 600; *p,p′*-DDE, 3900; ΣDDT, 4800 ng/g lipid (median)	*p,p′*-DDT, 0–5300; *p,p′*-DDE, 700–24 700; ΣDDT, 900–30 100 ng/g lipid		Zimmermann et al. (2005)
Australia Melbourne and 12 other regions Melbourne: 1993; Melbourne plus other regions: 2002–2003	Breast milk	1993: *p,p′*-DDE, 280; *p,p′*-DDT, 12 2002–3: *p,p′*-DDE, 279; *p,p′*-DDT, 7 ng/g lipid (median)	NR		Mueller et al. (2008)

Table 1.5 (continued)

Location, collection date	Sampling matrix	Exposure		Comments	Reference
		Level[a]	Range		
Serbia South Bačka, Voyvodina 1983–2009	Breast milk	1983: ΣDDT, 3630 2009: ΣDDT, 108 ng/g lipid	1983: ΣDDT, 440–9730 2009: ΣDDT, 34–189 ng/g lipid		Vukavić et al. (2013)
China Beijing and Shenyang 2005–2007	Breast milk	Beijing: *p,p'*-DDT, 4.95; *p,p'*-DDE, 169; ΣDDT, 183 Shenyang: *p,p'*-DDT, 4.48; *p,p'*-DDE, 117; ΣDDT, 154 ng/g lipid	Beijing: *p,p'*-DDT, 1.21–17; *p,p'*-DDE, 30.2–1010; ΣDDT, 34.7–1050 Shenyang: *p,p'*-DDT, ND–14.6; *p,p'*-DDE, 15.65–763; ΣDDT, 18.74–833 ng/g lipid	Significant correlation between human milk concentration and daily dietary intake of DDTs; the dietary intake could explain 22% of the variation of DDTs in human milk	Tao et al. (2008)
Turkey Ankara (suburban area) Year NR	Breast milk	DDT, 223 ng/kg lipid	DDT, < LOD–1265.7 ng /g lipid		Yalçin et al. (2014)
Islamic Republic of Iran Ahvaz and Noushahr November 2007 to January 2008	Hair	ΣDDT, 23 ng/g	ΣDDT, 0.8–305 ng/g		Dahmardeh Behrooz et al. (2012)

ΣDDT, total DDT corresponding to the sum of *o,p'*-DDE, *p,p'*-DDE, *o,p'*-DDD, *p,p'*-DDD, *o,p'*-DDT, and *p,p'*-DDT; ADI, acceptable daily intake; DDE, dichlorodiphenyldichloroethylene; DDT, dichlorodiphenyltrichloroethane; dw, dry weight; EPIC, European Prospective Investigation into Cancer and Nutrition; EDI, estimated daily intake; IRS, indoor residual spraying; LOD, limit of detection; NA, not applicable; ND, not detected; NHL, non-Hodgkin lymphoma; NR, not reported; OCP, organochlorine pesticides; SAR, Special Administrative Region

[a] Arithmetic mean, unless otherwise stated

Fig. 1.3 Temporal trends in human biomonitoring data for DDT

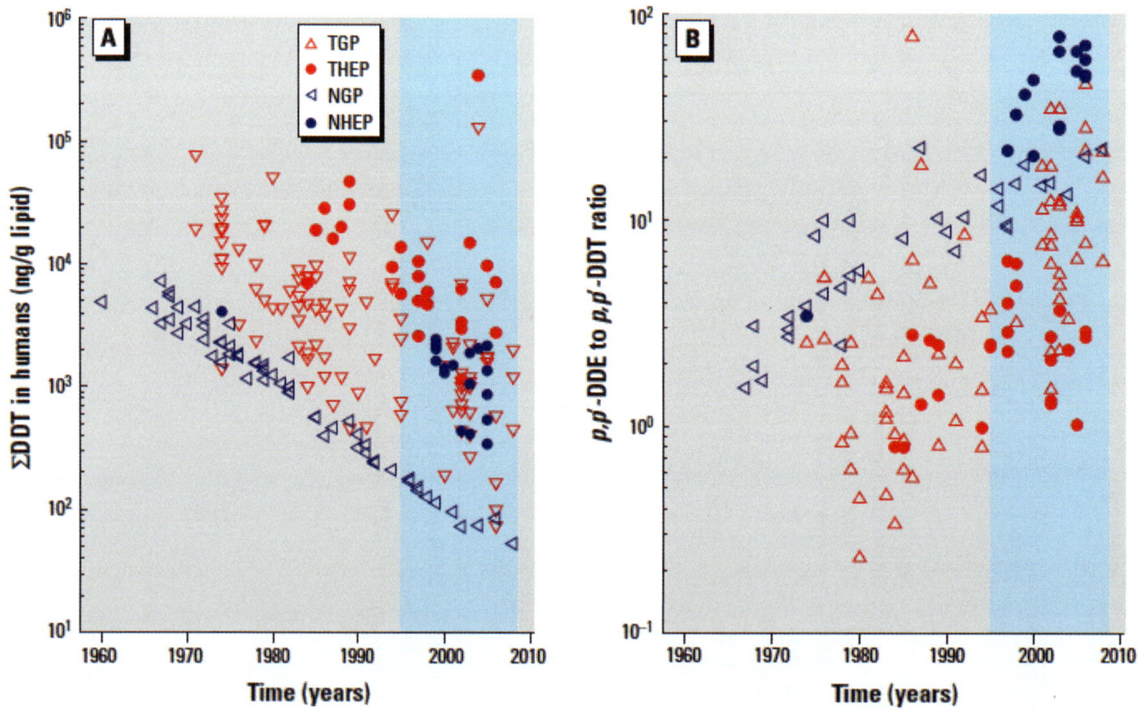

(A) Trends in ΣDDT. (B) Trends in the *p,p′*-DDE to *p,p′*-DDT ratio. Blue shaded area marks the time period investigated in the integrated exposure assessment (1995–2008).
Reproduced from Ritter et al. (2011), *Environmental Health Perspectives.*

The Agricultural Health Study (AHS) of farmers and licensed pesticide applicators in Iowa and North Carolina, USA, estimated exposures to 50 pesticides, including DDT. Pesticide exposure information was ascertained from two phases of questionnaire administration. In the first phase (1993–1997), ever/never use of pesticides was ascertained for all applicators; detailed information on lifetime use of DDT use was obtained for 25 291 applicators from a self-administered questionnaire during phase 1. A follow-up questionnaire ascertained detailed pesticide information for the 5 years since enrolment in the study. For participants who did not complete the follow-up questionnaire, a multiple imputation procedure was used to impute exposure to specific pesticides. Information about factors potentially affecting exposure, such as spraying techniques and use of personal protective equipment, was also obtained by questionnaire. A semiquantitative exposure-assessment based on this information was developed in which estimated intensity was combined with years and annual frequency of use (Dosemeci et al., 2002). Cumulative exposure matrices of ever/never use, lifetime days of use, and intensity-weighted lifetime days of use and were created (Alavanja et al., 2014). The intensity score was based on development of an a priori exposure intensity algorithm. Several validity evaluations of the exposure assessment process were carried out. These included: (i) assessment of the reliability of reporting agricultural factors by requiring completion of the enrolment questionnaires twice, approximately 1 year apart; (ii) confirmatory checks correlating the years in which a pesticide was reportedly used with dates of registered use of that particular pesticide; and (iii) comparison of the exposure algorithm with external exposure data. Agreement of reporting of ever/never use of specific pesticides and application practices was high, and generally ranged from 70% to > 90%. Agreement was lower (typically 50–60%) for duration or frequency of use of specific pesticides (Blair et al., 2000). The confirmatory checks on reported usage of specific pesticides established that the majority of respondents provided plausible responses for decade of first use and total duration of use (Hoppin et al., 2002). The exposure-intensity algorithm was evaluated in three studies (Coble et al., 2005; Acquavella et al., 2006; Thomas et al., 2010). When combined, these studies showed that the AHS algorithm had the capacity to separate the upper tertiles of exposure intensity from the lower.

[The Working Group noted that the AHS has collected detailed information on pesticide use and practices and through validation studies has shown this data to be appropriate for estimating historical exposure to pesticides. However, the validity studies were based on information reported at the time the exposure surveys were completed and would not necessarily reflect the recall of information for all aspects, in particular frequency and duration of use. The assessment of DDT exposure in the AHS is based on the baseline questionnaire (1993–1997) and relies on historical recall. Although DDT may be a pesticide the use of which is more readily recalled than others because of wide recognition and environmental and health concerns, the validity of recalled information is nonetheless unknown. Moreover, due to the persistent nature of DDT, it is likely that a certain proportion of farmer exposure is attributable to non-application circumstances, such as re-entry and contaminated work and home environments. Exposures related to non-application days may be collinear with reported lifetime days of application, but if not the non-application exposures could contribute to marked measurement misclassification. For example, Bakke et al. (2009) detected elevated urinary 2,4-D concentrations in corn farmers relative to non-farmers even on days when the farmers had not applied 2,4-D. These levels, although an order of magnitude lower than those recorded when applying 2,4-D, suggest a contribution of non-application days to cumulative annual exposure. This consideration is relevant when accounting for the fact that active applications by most farmers amount to only a few days during the growing season. Given the more persistent nature of DDT, it is reasonable to assume that similar concerns are in play for this pesticide.]

In the study of Cocco et al. (2005), based on workers exposed to DDT during antimalarial operations in Sardinia, Italy, estimated exposure was based on occupation held at the time of the antimalarial operations. Relevant occupations were those involving either direct exposure to DDT (i.e. applicators) and those involving the likelihood of bystander exposure (i.e. inspectors, warehouse workers, drivers). Dermal and inhalation exposure was estimated using the EUROPOEM model for applicators (van Hemmen, 2001), while for bystander situations only dermal exposure was estimated using the algorithm developed by Krebs et al. (2000).

[The Working Group noted that categorization specifying unexposed, directly exposed and bystander situations is appropriately based on job information. However, estimation of quantitative levels of exposure using the EUROPOEM model, whose scenarios are based on best practice (daily cumulative exposure range, 54–140 400 μg), might have underestimated the exposure to DDT from work practices as they occurred in 1946–1950 in the antimalarial campaigns in Sardinia. The validity of the assumption of no inhalation exposure for bystanders cannot be determined.]

(b) General population

In relevant studies based upon the general population, DDT exposure was assessed either by relying on data from questionnaires and/or biological monitoring.

(i) Questionnaire-based approaches

The studies relying on questionnaires collected varying information on jobs held, agricultural practices, and pesticide use. [The Working Group noted that these studies have employed standard accepted techniques in environmental and occupational epidemiology. As in all studies, self-reported exposure data are prone to inaccurate or biased recall that can lead to exposure misclassification. Due to the absence of confirmatory data on recall and exposure assignment, the validity of these exposure assessments could not be evaluated.]

(ii) Biomonitoring approaches

A number of relevant studies have relied on measurement of *p,p′*-DDE and/or *p,p′*-DDT in serum or plasma or in adipose tissue (see Table 1.4). The implicit rationale for monitoring levels of DDT and its metabolites in serum and adipose is the assumption that those levels would reflect past exposure and be indicative of differences in total exposures between individuals. However, this assumption has been questioned on the basis of long-term *p,p′*-DDT toxicokinetics and metabolism. Studies involving human ingestion of DDT have shown that *o,p′*-DDT is rapidly metabolized and excreted, and that when *p,p′*-DDT is compared with *p,p′*-DDE, the former is more rapidly metabolized and excreted, with the latter being the most persistent compound of this series (Morgan & Roan, 1974). In addition to chronic ingestion of *p,p′*-DDE from the diet and exposures from other environmental media, inter-individual differences in capacity to metabolize DDT may further complicate the interpretation of serum levels of *p,p′*-DDE. Thus, *p,p′*-DDE concentrations in human serum may not accurately reflect past exposure to *p,p′*-DDT, particularly in samples obtained decades after direct exposure to DDT (Perry et al., 2005; Wolff et al., 2005; Cohn et al., 2007). In addition, there are strong trends in population levels of *p,p′*-DDT and *p,p′*-DDE due to the ban of most uses of DDT (Fig. 1.3A; Ritter et al., 2011). The ratio of *p,p′*-DDE to *p,p′*-DDT is changing over time, with ratios around 1 recorded before 1980 and ratios of around 10 recorded in more recent time (2000–2010), the exception being populations living in homes with indoor residual spraying of DDT for disease-vector control (Ritter et al., 2011). This indicates that for many populations, direct exposures to DDT are unlikely and that current levels are predominantly attributable to *p,p′*-DDE in the diet and other environmental media (Fig. 1.3B).

[The Working Group noted that the ratio of *p,p′*-DDE to *p,p′*-DDT employed in the evaluated key epidemiological studies tended to range from 6 to 36, with lower ratios being reported for studies using measurements recorded in the more distant past, and higher ratios corresponding to more recent times. Studies with a higher ratio may indicate an earlier exposure and/or a predominant contribution of *p,p′*-DDE through diet and other environmental media. The Working Group also noted the high levels of *p,p′*-DDT and *p,p′*-DDE in the Persson et al. (2012) study for which no clear explanation was evident.]

1.5 Regulation

Regulation of pesticides typically involves restrictions inherent to registration or licensing for use and measures to limit exposure in an occupational and wider community context; the regulation of DDT is overwhelmingly concerned with its use being prohibited. Prohibition in many countries acknowledged, DDT is approved for use in limited circumstances and is therefore appropriately regulated as a pesticide in that context. Moreover, many countries in which DDT is banned have adopted regulations concerning residues, and may specify limits in respect of occupational exposure.

1.5.1 International agreements

The Stockholm Convention on Persistent Organic Pollutants, signed in 2001, provided initially for the elimination of 12 chemicals, one of which was DDT (UNEP, 2002b). The Convention now addresses 22 chemicals. Within the convention, DDT is addressed in Annex B: Restriction, Part II (acceptable purpose – disease vector control in accordance with WHO recommendations), which has been published separately (Bouwman et al., 2013) and which, among other things, provides for establishment of a DDT register, and agreement between the parties to restrict usage of DDT and identify alternatives, and to report on and evaluate those circumstances where DDT usage continues to be approved.

International instruments variously addressing DDT also include the Basel Convention on the Control of Transboundary Movements of Hazardous Waste and Their Disposal, and the Rotterdam Convention on the Prior Informed Consent Procedure for Certain Hazardous Chemicals and Pesticides in International Trade (Matthews et al., 2011).

Vector control is essential to malaria control, and as such is recognized by WHO, specifically with reference to 12 insecticides recommended for indoor residual spraying, including DDT – the only organochlorine compound in this context – until suitable alternatives are available (WHO, 2011a).

1.5.2 Transnational and national regulations

Based on the increasing amount of data on its toxicity, environmental persistence, bioaccumulation and potential for transboundary movement, restrictions on the use of DDT were adopted in different countries from the early 1970s (Bouwman et al., 2013).

Sweden was the first country to ban the use of DDT, beginning in January 1970 (Turusov et al., 2002).

In 1972, the EPA issued a cancellation order for DDT in the USA (EPA, 2015b). Prior to restrictions being adopted, DDT was variously registered in the USA, including application to livestock. The United States Department of Agriculture (USDA) had previously cancelled Federal registrations as of 1970 for several specified uses of DDT products, including on food crops, livestock and wood products (EPA, 1975b).

In the United Kingdom, use of DDT was severely restricted in 1964, and all use was prohibited from 1984 (Gillespie et al., 1994). In the former Soviet Union, the production and use of DDT in agriculture was officially banned in 1969–1970, but use for public health purposes (against mosquitoes, malarial plasmodia, fleas, lice, and ticks) was permitted until 1989 (Turusov et al., 2002).

From 1983, technical DDT was banned as an agricultural pesticide in China. As reported in 2005, China had requested specific exemptions for DDT under the Stockholm Convention (Wong et al., 2005). A national ban on use of DDT as an additive to antifouling paint was implemented in 2009 in China (Lin et al., 2009). By 2014, all uses of DDT in China were withdrawn (UNEP/WHO, 2014, 2015).

Granted usage of DDT for vector control in India, strict regulations to prevent diversion of DDT to agricultural uses are advocated (Gunasekaran et al., 2005).

As assessed in 2009, many countries that use DDT have inadequate legislation or lack capacity to implement or enforce regulations on pesticide management (Li et al., 2006; van den Berg, 2009; Aliyeva et al., 2013).

Despite bans on production and use, relevant national authorities may specify occupational exposure levels for DDT. In the USA, for example, the National Institute of Occupational Safety and Health (NIOSH) has issued a recommended exposure level (REL, 0.5 mg/m³ time-weighted average, TWA) and the American Conference of Governmental Industrial Hygienists (ACGIH) has put in place a threshold limit value (TLV, 1 mg/m³ TWA for skin) for DDT (CDC, 2015), and such measures remain relevant to jurisdictions in which DDT continues to be used.

For the European Commission, maximum residue limits (MRL) for DDT and related compounds are specified for 378 products, almost all of which are 0.05 mg/kg, extending as high as 1 mg/kg for some products, e.g. coffee beans, spices, meat, and edible offal (European Commission, 2015).

In several countries regulations exist with respect to DDT and related substances in commercial dicofol. In the European Union, USA, and Canada, the limit for such content is 0.1% (Ministerie van VROM, 2004). Dicofol formulations sold in the United Kingdom were found to conform to this requirement (Gillespie et al., 1994). The use of dicofol in China has been banned on tea plant and vegetables. Having been reported at levels of 10% or more in the past, the DDT impurity in dicofol is now required by Chinese regulation to be no more than 0.5% of technical dicofol (Qiu et al., 2005).

2. Cancer in Humans

2.1 Cohort studies

Associations between exposure to DDT and several types of cancer have been extensively investigated in epidemiological studies. The pertinent studies reviewed by the Working Group are summarized here according to study design and cancer site. Several meta-analyses were also available to the Working Group. These are reviewed with the case–control studies for the cancers concerned, as most of the included studies had that design.

2.1.1 Cancer of the breast

See Table 2.1

The relationship between serum concentration of DDE [1,1-dichloro-2,2-bis(*p*-chlorophenyl) ethylene], a main metabolite of DDT [1,1′-(2,2,2-trichloro-ethylidene)bis (4-chlorobenzene).], and cancer of the breast has been evaluated in case–control analyses of 10 prospective cohort studies (Wolff et al., 1993; Krieger et al., 1994; Hunter et al., 1997; Høyer et al., 1998; Dorgan et al., 1999; Helzlsouer et al., 1999; Ward et al., 2000; Raaschou-Nielsen et al., 2005; Cohn et al., 2007; Iwasaki et al., 2008; Cohn et al., 2015). These studies were performed in the USA, Denmark, Norway, and Japan. The duration of follow-up ranged from 1 month (Wolff et al., 1993) to 54 years (Cohn et al., 2015) after blood collection.

In addition, the relationship between DDT and breast cancer was evaluated in some of these studies (Høyer et al., 1998; Dorgan et al., 1999; Høyer et al., 2000; Ward et al., 2000; Høyer et al., 2002; Raaschou-Nielsen et al., 2005; Cohn et al., 2007; Iwasaki et al., 2008; Cohn et al., 2015).

The New York University Women's Health Study enrolled a cohort of 14 290 women from New York City, USA, between 1985 and 1991; these women donated a 30 mL blood sample while attending a mammography screening clinic (Wolff et al., 1993). During this period, women who were diagnosed with cancer of the breast 1–6 months after entry into the study were defined as cases. Controls were selected at random from all cohort members who were alive and free of cancer at the time of the cancer diagnosis in a case patient, matched on age at entry, number of blood donations, menopausal status, and day of menstrual cycle at the time of blood collection. Serum DDE and DDT concentrations were determined in a total of 58 cases and 171 controls.

Table 2.1 Cohort studies of cancer of the breast and exposure to DDT and its metabolites

Reference, location enrolment/follow-up period, study design	Population size, description, exposure assessment method	Organ site	Exposure category or level	Exposed cases/deaths	Risk estimate (95% CI)	Covariates	Comments
Wolff et al. (1993) New York, USA Enrolment, 1985–1991 Nested case–control	Cases: 58 women aged 35–65 yrs with breast cancer diagnosed 1–6 mo after entry Controls: 171 cohort members alive and free of cancer at the time of case diagnosis Exposure assessment method: personal monitoring; GC-ECD	Breast	DDE, 0.5–3.2 ng/mL	6	1.00	First-degree relative with cancer, lifetime lactation, age at first full-term pregnancy	From all prospective studies that have been performed so far, this is the only one that included prevalent cases Strengths: control for confounders, including lactation Limitations: prevalent cases; only one clinic as a source of women; no information on reluctant participants and losses in follow-up; no information on whether DDE measurements were performed before or after cancer treatment began; no information on lipid-adjusted DDE levels
			3.2–5.2	6	1.67		
			5.2–7.5	16	4.37		
			7.5–11.9	8	2.31		
			11.9–44.3	13	3.68		
Wolff et al. (2000a) New York Enrolment 1985–1991; follow-up to 1994 Nested case–control	Cases: 110 Controls: 213 Exposure assessment method: personal monitoring; GC-ECD	Breast	DDE: < 664 ng/g lipid	31	1.00	Age at menarche, number of full-term pregnancies, age at first full-term pregnancy, family history of breast cancer, lifetime history of lactation, height), BMI, BMI-menopausal status interaction	Stratification by ER status showed no significant higher GM of DDE among controls Strengths: only incident cases included (see Wolff et al., 1993) Limitations: only women with 3 blood samples were included; short follow-up
			664–1172	30	0.81 (0.35–1.87)		
			1173–1934	24	0.6 (0.26–1.38)		
			> 1934	25	1.3 (0.51–3.35)		
			Trend-test *P* value: 0.99				

Table 2.1 (continued)

Reference, location enrolment/ follow-up period, study design	Population size, description, exposure assessment method	Organ site	Exposure category or level	Exposed cases/ deaths	Risk estimate (95% CI)	Covariates	Comments
Krieger et al. (1994) California, USA Enrolment 1964–1971; follow-up until 1990 Nested case–control study	Cases: race-stratified random sample of 150 cases from the cohort Controls: 150 women free of cancer up to the time of matched case patient's diagnosis, matched by race, date of joining the programme, year, and age at examination and length of follow up Exposure assessment method: personal monitoring; GC-ECD	Breast	5.33–29.68 DDE ng/mL	NR	1.00	BMI, age at menarche, ever vs never pregnant, menopausal status at time of case patient's diagnosis of breast cancer plus variables matched by design	Strengths: serum samples taken up to 26 yrs before diagnosis before DDT prohibition in USA Limitations: small sample size by race
			29.69–49.60	NR	1.29 (0.67–2.47)		
			49.61–149.50	NR	1.33 (0.68–2.62)		
			Trend-test *P* value: 0.431				
Hunter et al. (1997) 11 USA states Enrolment 1976, follow-up June 1992 Nested case–control study	Cases: 236 incident cases with no previous cancer Controls: 236, matched by year of birth, menopausal status, month, time of day and fasting status at blood sampling and postmenopausal hormone use Exposure assessment method: personal monitoring; GC-ECD; plasma levels of DDE measured after adjustment for plasma cholesterol concentrations	Breast	DDE, ppb			History of breast cancer in a mother or sister, history of benign breast disease, age at menarche, number of children and age at birth of first child, duration of lactation, BMI, plus variables matched by design	Strengths: well-designed study; control of key confounders; low attrition rate (5%) Limitations: DDE levels were measured 2–3 yrs before the end of follow-up
			≤ 2.78	61	1.00		
			> 2.78–4.54	54	0.80 (0.45–1.43)		
			> 4.54–6.26	35	0.47 (0.25–0.9)		
			> 6.26–9.46	43	0.74 (0.40–1.36)		
			> 9.46	43	0.72 (0.37–1.4)		
			Trend-test *P* value: 0.47				

Table 2.1 (continued)

Reference, location enrolment/ follow-up period, study design	Population size, description, exposure assessment method	Organ site	Exposure category or level	Exposed cases/ deaths	Risk estimate (95% CI)	Covariates	Comments
Laden et al. (2001a) 11 USA states Enrolment 1976, follow-up to 1994 Nested case–control study	Cases: 372; as in Hunter et al. (1997) plus 143 cases of invasive postmenopausal cancer diagnosed 1992–1994 Controls: 372 cancer-free women matched on year of birth, menopausal status, date and fasting status at blood collection and post-menopausal hormone use Exposure assessment method: personal monitoring; GC-ECD	Breast	DDE, µg/g lipid			History of breast cancer in mother or a sister, a history of benign breast disease, age at menarche, BMI at blood draw, number of children and age at birth of first child, and duration of lactation	Strengths: 98% completeness of follow-up Limitations: mostly premenopausal women
			0.007–0.427	88	1.00		
			0.428–0.703	85	0.95 (0.59–1.53)		
			0.708–0.955	48	0.51 (0.31–0.86)		
			0.955–1.441	83	0.91 (0.57–1.47)		
			1.466–6.054	68	0.82 (0.49–1.37)		
Dorgan et al. (1999) Columbia, Missouri Blood donation 1977 and 1987; follow-up to 1989 Nested case–control study	Cases: 105 histologically confirmed breast cancer cases up to 1989 Controls: 208; alive and free of cancer at the age of the case's diagnosis, matched on age, date of blood draw, history of benign breast disease and number of blood draws Exposure assessment method: personal monitoring; GC-ECD; serum cholesterol and triglycerides were measured	Breast	DDE, ng/g lipid			Height, weight, BMI, parity, age at menarche, menopausal status, exogenous estrogen use, history of breast cancer among first-degree relatives, education, and number of packs of cigarettes smoked per day	Strengths: up to 9.5 yrs of follow-up Limitations: study population were volunteers; relatively small sample size; no adjustment for lactation
			31–1377	33	1.0		
			1378–2355	32	0.9 (0.5–1.7)		
			2356–3500	14	0.4 (0.2–0.8)		
			3501–20 667	26	0.8 (0.4–1.5)		
			DDT, ng/g lipid				
			0–180	29	1.0		
			181–292	29	1.0 (0.5–2.0)		
			293–467	33	1.1 (0.6–2.1)		
			468–1724	14	0.4 (0.2–1.0)		
			Trend-test *P* value: 0.65				

Table 2.1 (continued)

Reference, location enrolment/ follow-up period, study design	Population size, description, exposure assessment method	Organ site	Exposure category or level	Exposed cases/ deaths	Risk estimate (95% CI)	Covariates	Comments
Helzlsouer et al. (1999) Washington County, Maryland, USA Two cohorts enrolled in 1974 and 1989; follow up until 1994 Nested case–control study	Cases: cohort 1: 235; cohort 2: 105; county cancer registry Controls: cohort 1: 235; cohort 2: 105; women within the cohorts not diagnosed with cancer, matched by sex, race, age, menopausal status, date of blood donation Exposure assessment method: personal monitoring; serum DDE by GC-ECD	Breast	DDE ng/g (1974):			History of breast cancer, BMI at age 20 yrs or current age at menarche, age at first birth, duration of lactation plus matching variables	Strengths: serum collected up to 20 yrs before diagnosis; adjustment by key confounders Limitations: sample size limited for subgroup analyses
			< 1017.19	49	1.00		
			1017.20–1425.39	61	1.24 (0.72–2.13)		
			1425.40–1864.57	47	0.96 (0.55–1.67)		
			1864.58–2446.69	42	0.86 (0.49–1.51)		
			2446.70–10 795.91	36	0.73 (0.40–1.32)		
			DDE ng/g (1989):				
			< 816.3	38	1.00		
			816.4–1595.1	44	1.18 (0.65–2.13)		
			1595.2–10 065.6	23	0.58 (0.29–1.17)		

Table 2.1 (continued)

Reference, location enrolment/ follow-up period, study design	Population size, description, exposure assessment method	Organ site	Exposure category or level	Exposed cases/ deaths	Risk estimate (95% CI)	Covariates	Comments
Cohn et al. (2007) Oakland, California, USA Enrolment between 1959–1967; follow-up to 1998 Nested case–control study	Cases: 129 from state cancer registry and vital records Controls: 129; matched to cases on birth year Exposure assessment method: personal monitoring; no lipid adjustment	Breast	*p,p*′-DDE, µg/L			None	Strengths: exposure measurements from stored serum samples reflecting early life exposures; long follow-up Limitations: small sample size (main results are based on 94 case–control pairs); no discussion of the effect of adjustment by correlated metabolites; no information on risk factors between the time of the pregnancy and development of breast cancer
			≤ 35.23	NR	1.0		
			35.23–58.49	NR	1.5 (0.8–2.6)		
			> 58.49	NR	1.1 (0.6–2.0)		
		Breast	*p,p*′-DDT, µg/L			None	
			≤ 8.09	NR	1.0		
			8.09–13.90	NR	1.4 (0.7–2.7)		
			> 13.90	NR	1.6 (0.8–3.0)		
		Breast	Age < 14 yrs in 1945			*p,p*′DDT, *p,p*′DDE and *o,p*′DDT, and women aged < 14 yrs in 1945	
			p,p′-DDE, µg/L				
			≤ 35.23	NR	1.0		
			> 35.23–58.49	NR	1.5 (0.6–3.4)		
			> 58.49	NR	0.9 (0.3–3.0)		
			p,p′-DDT, µg/L				
			< 8.09	NR	1.0		
			8.09–13.90	NR	2.5 (1.0–6.3)		
			> 13.90	NR	5.2 (1.4–19.1)		
Cohn et al. (2015) Oakland, California, USA 1959–1967 Nested case–control study	Cases: 103 from state cancer registry & vital records Controls: 315; cohort Exposure assessment method: personal monitoring	Breast	Concentration in mother's serum			Maternal cholesterol and triglycerides, maternal overweight in early pregnancy and maternal history of breast cancer	See Cohn et al. (2007) for details Strengths: long follow-up; DDT was measured during mothers'pregnancy Limitations: breast cancer risk factors in daughters were not considered
			Q1 DDE	NR	1.0		
			Q2	NR	1.3		
			Q3	NR	1.1		
			Q4	NR	1.3		
			Q1 DDT	NR	1.0		
			Q2	NR	1.9		
			Q3	NR	1.5		
			Q4	NR	2.2		
			Trend-test *P* value: 0.074				

Table 2.1 (continued)

Reference, location enrolment/follow-up period, study design	Population size, description, exposure assessment method	Organ site	Exposure category or level	Exposed cases/deaths	Risk estimate (95% CI)	Covariates	Comments
Høyer et al. (2000) Copenhagen, Denmark First enrolment 1976–1978; follow-up until December 1992 Nested case–control study	Cases: 143 from Danish cancer registry Controls: 274 age-matched controls from the cohort Exposure assessment method: personal monitoring; GC-ECD; total lipid content was calculated	Breast	I DDE ng/g	NR	1.0	Weight changes	Parity, weight and HRT reported to be confounders but were not included in the final model Strengths: up to 8 yrs of follow-up; average concentration of two DDE measurements with a 5- yr interval was used Limitations: possibly inadequate control for breast-cancer risk factors; limited power, particularly for subgroups
			II	NR	1.0 (0.5–2.0)		
			III	NR	0.8 (0.4–1.6)		
			IV	NR	1.4 (0.7–2.8)		
			I DDT ng/g	NR	1.0		
			II	NR	1.3 (0.4–4.5)		
			III	NR	2.1 (0.6–7.0)		
			IV	NR	3.6 (1.1–12.2)		
Raaschou-Nielsen et al. (2005) Denmark Enrolment 1993–1997; follow-up to 2000 Nested case–control study	Cases: 409 from Danish cancer registry Controls: 409, cancer-free at age of case diagnosis matched by age, HRT & postmenopausal status Exposure assessment method: personal monitoring; gluteal adipose tissue; analysis by GC-MS	Breast	DDE, µg/kg lipids				
			15–282	100	1.0	Education, BMI, alcohol, number of childbirths, age at first delivery, lactation, HRT, history of benign breast disease	Strengths: largest prospective study with in adipose tissue measurements; ER status was determined Limitations: only postmenopausal women
			283–507	100	1.0 (0.7–1.5)		
			508–903	100	0.9 (0.6–1.4)		
			904–6693	100	0.7 (0.5–1.2)		
			DDT, µg/kg lipids				
			6–14	100	1.0		
			14–20	100	0.8 (0.5–1.3)		
			20–31	100	1.4 (0.9–2.3)		
			31–159	100	0.6 (0.3–1.0)		

Table 2.1 (continued)

Reference, location enrolment/ follow-up period, study design	Population size, description, exposure assessment method	Organ site	Exposure category or level	Exposed cases/ deaths	Risk estimate (95% CI)	Covariates	Comments
Ward et al. (2000) Norway Enrolment 1991; follow-up to 1993 Nested case–control study	Cases: 150; random sample of eligible cases in the Norwegian cancer registry Controls: 150; alive and cancer-free at case diagnosis, matched by date of sample collection and date of birth Exposure assessment method: personal monitoring; high-resolution GC/high-resolution isotope dilution MS	Breast	Q1 DDE ng/g lipid	NR	1.0	None specified	Strengths: lipid-adjusted DDE values; ER status assessed Limitations: no BMI nor menopausal status information available; limited power; CIs not reported
			Q2	NR	0.7		
			Q3	NR	1.0		
			Q4	NR	1.2		
			Q1 DDT ng/g lipid	NR	1.0		
			Q2	NR	0.2		
			Q3	NR	0.5		
			Q4	NR	0.3		
Iwasaki et al. (2008) Japan Enrolment 1990–1993, follow-up to 2002 Nested case–control study	Cases: 139 from cancer registries, death certificates, hospitals Controls: 278 cohort members matched by age, public health centre area, area, date and time of blood collection, time of day of blood collection, and menopausal status Exposure assessment method: personal monitoring; GC isotope-dilution MS	Breast	DDE, ng/mL			Age at menarche, menopausal status at baseline, number of births, age at first birth, height (continuous), BMI, alcohol consumption	Strengths: up to 10 yrs of follow-up; control for confounders Limitations: no information on lactation; limited power; no lipid adjustment
			2.50	25	1.0		
			4.83	32	1.01 (0.47–2.19)		
			7.58	36	1.24 (0.60–2.53)		
			14.41	46	1.48 (0.70–3.13)		
			DDT, ng/mL				
			0.50	40	1.00		
			0.89	29	0.65 (0.32–1.32)		
			1.36	25	0.56 (0.26–1.22)		
			2.24	45	0.99 (0.47–2.08)		
			Trend-test *P* value: 0.25				

BMI, body mass index; CI, confidence interval; DDE, dichlorodiphenyldichloroethylene; DDT, dichlorodiphenyltrichloroethane; ER, estrogen receptor; GC-ECD, gas chromatography with electron capture detector; GC-MS, gas chromatography with mass spectrometry; GM, geometric mean; HRT, hormone replacement therapy; yr, year

Mean concentrations of DDE were statistically higher for cases of breast cancer than for control subjects (DDE in cases, 11.0 ± 9.1 ng/mL; DDE in controls, 7.7 ± 6.8 ng/mL; $P = 0.031$). After adjustment for first-degree family history of breast cancer, lifetime lactation, and age at first full-term pregnancy, the relative risk (RR) of cancer of the breast was found to increase by fourfold for an elevation in serum DDE concentrations from 2.0 ng/mL (10th percentile) to 19.1 ng/mL (90th percentile). [A major limitation of this study was the inclusion of prevalent cases. DDE measurements were not adjusted for lipids.]

In the same cohort, with an extended follow-up to 1994, only incident cases were considered. Cases with at least three annual blood samples were included and serum DDE concentrations were adjusted for lipids; 110 cases and 123 controls were included in the analysis (Wolff et al., 2000a). Geometric mean DDE concentration was 6.95 ng/mL among cases and 7.27 ng/mL among controls. The results did not confirm a significant increased risk of breast cancer in relation to DDE (odds ratio, OR, Q4 versus Q1, 1.30; 95% confidence interval, CI, 0.31–3.35). The authors also assessed whether changes of DDE over time were related to risk of cancer of the breast. The median half-life of DDE was estimated from the consecutive measurements to be 13 years among cases and 15 years among controls [reported to be non-significant, data not shown] and the risk of breast cancer was reported not to be associated with DDE half-life (data not shown). [A limitation of this study was that only women with three or more annual blood samples were included, and no key information about non-included women from the original cohort was provided. In addition, the follow-up was relatively short and the analysis thus had limited precision.]

Krieger et al. (1994) performed a nested case–control study among women in northern California, USA, who were members of the Kaiser Permanente Medical Care Program and who underwent a health examination, including giving a sample of blood between 1964 and 1969 and were followed up until 1990. Among the 2097 patients identified with cancer of the breast, 150 cases were randomly selected (50 white, 50 black, and 50 Asian) and matched to 150 controls by race, age, date of entry, and date of follow-up. Mean serum DDE concentration was 43.3 ± 25.9 ng/mL among cases, and 43.1 ± 23.7 ng/mL among controls. After adjustment for reproductive factors, menopausal status, and body mass index (BMI), no significant association was seen between risk of cancer of the breast and serum DDE concentrations for all subjects (OR, 1.33; 95% CI, 0.68–2.62; for the third versus first tertile of concentration), nor in ethnic subgroups: whites (OR, 2.38; 95% CI, 0.54–10.64); blacks (OR, 3.85; 95% CI, 0.93–16.05); or Asians (OR, 0.71; 95% CI, 0.23–2.18). [This was a well-designed study with a long follow-up period. Relevant confounders were adjusted. However, only 150 of more than 2000 available cases were selected for measurement of DDE and thus power was limited, especially for subgroup analyses.]

The Nurses' Health Study was established in 1976 and included more than 120 000 registered nurses in 11 states of the USA; participants were followed by questionnaire every 2 years and 32 826 provided a blood sample between 1989 and 1990. Findings on the association between cancer of the breast and plasma DDE concentrations were reported from follow-ups to 1992 (Hunter et al., 1997) and 1994 (Laden et al., 2001a). In the first follow-up, a total of 236 cases diagnosed within 3 years of blood collection were included. The same number of controls was selected from the cohort and matched by year of birth, menopausal status, month in which blood samples was returned, time of day blood sample was drawn, fasting status at blood sampling and hormone use among postmenopausal women. After adjusting for reproductive variables, BMI, and familial history of breast cancer and/or benign breast disease, the risk of cancer of the breast tended to be lower and not statistically significant, among women with higher serum concentrations of DDE (Hunter et al., 1997). An extended follow-up to 1994 included 372 case–control pairs. Medians DDE concentrations were 768 ng/g lipid among cases, and 817 ng/g lipid among controls. The relative risks of breast cancer were below unity, but not statistically significant, among all groups of women compared with those with lowest levels of DDE (Laden et al., 2001a). [This was a well-designed study with good control for most relevant confounders, including reproductive factors and family history of breast cancer, and larger sample size compared with previous studies. Blood samples were taken after the start of follow-up, but 2–3 years before diagnosis for cases.]

Serum samples were obtained in 1976 from a cohort of 7712 women aged ≥ 20 years who participated in the Copenhagen City Heart Study and provided information and a blood sample (Høyer et al., 1998). Case ascertainment was achieved by linkage to the Danish cancer registry up to 1993. For each case, two women who were free of breast cancer and alive at the time of diagnosis and matched for age and date of examination were selected from the rest of the cohort; 240 cases with a valid serum sample and 447 age-matched controls were included in the analysis. Several potential confounders were tested, including reproductive variables, alcohol consumption, and physical activity, but only full-term pregnancies and weight were retained in final models, which showed no association between quartiles of DDE or DDT and breast cancer.

Participants in the same cohort study were invited for a second examination 5 years after recruitment; 155 cases and 274 controls from the previous study who had two serum sample available were included in a further study (Høyer et al., 2000). No significant association was found between breast cancer and the average concentration of *p,p′*-DDE in the two time periods; however, a 3.6-fold (95% CI, 1.1–12.2) risk of breast cancer was observed among women in the upper quartile of average *p,p′*-DDT serum concentration, with a significant trend. Elevated but non-significant odds ratios were observed for total DDT (OR, 2.4; 95% CI, 0.7–7.8 in the upper quartile). [Parity, weight, and use of hormone replacement therapy use were identified as confounders by the authors, but these were not included in the final model: it is possible that control for confounding was inadequate.]

Within the same cohort, a total of 161 cases with ER status information and 318 matched controls who were free of breast cancer were included in an analysis according to ER status (Høyer et al., 2001). ER status did not modify the association with breast cancer.

Finally, paraffin-embedded tumour-tissue specimens were retrieved for 162 cases and 316 controls from the same cohort and analysed for *p53* (*TP53*) tumour-suppressor gene mutation status (Høyer et al., 2002). No measure of DDT or DDE was associated with breast cancer, regardless of the presence of *p53* mutation. [Several analyses were carried out using data from this Danish study, but power was limited, particularly for subgroups.]

In another study in the USA, 7224 female volunteers donated blood to the Columbia, Missouri Breast Cancer Serum Bank between 1977 and 1987; active follow-up continued until 1989 (Dorgan et al., 1999). Among these women, 105 were diagnosed with histologically confirmed cancer of the breast, and two controls for each were selected, matched to each case on age, date of blood sampling, and history of benign breast disease at the time of enrolment. No association was found between risk of cancer of the breast and lipid-corrected concentrations of DDT or DDE. [This study had limited precision due to relatively small sample size. There was no adjustment for lactation history.]

Residents of Maryland, USA, who had participated in one of two studies conducted in 1974 and 1989 to obtain blood samples for a serum bank (the CLUE I and CLUE II studies) were invited to participate in a case–control study (Helzlsouer et al., 1999). Participants were followed up until 1994 by linkage with the county cancer registry. Of the 346 cases of cancer of the breast diagnosed, valid measurements of DDE were available for 340 cases, which were matched to 340 participating women without cancer of the breast by age, menopausal status, date of blood collection, and study. Taking into account relevant confounders, no association was found between breast cancer and DDE, including after stratifying for menopausal status, estrogen-receptor (ER) status, or polymorphism in *GSTM1*, *GSTT1*, *GSTP1*, *COMT*, or *CYP17*. [Although the sample size was adequate for the main analysis, it was limited for subgroup analyses.]

The JANUS Serum Bank contains serum samples collected between 1973 and 1991 from almost 300 000 individuals undergoing routine health examinations in Norway. Cases of cancer of the breast were identified among 25 431 women who worked outside the home or lived on farms and were followed until 1993 through linkage with the Norwegian cancer registry (Ward et al., 2000). From the 272 cases diagnosed during this period, 150 women with a blood sample taken 2 or more years before diagnosis were randomly selected; an equal number of controls were matched to cases by date of sample collection and date of birth. Mean lipid-adjusted serum DDE concentrations were 1230 ng/g lipid among cases and 1260 ng/g lipid among controls. No association between DDE or DDT concentration and breast cancer was found. [This study was well designed, but had limited precision and did not consider confounding by BMI and menopausal status.]

Between 1993 and 1997, 29 875 Danish women aged 50 to 64 years were enrolled in a prospective study of diet and cancer and followed until December 2000 through linkage with Danish cancer registry (Raaschou-Nielsen et al., 2005). During this period, 409 women were diagnosed with postmenopausal cancer of the breast; each was matched to one randomly selected control-matched by age, postmenopausal status, and use of hormone replacement therapy, and *p,p′*-DDE and *p,p′*-DDT were measured in adipose tissue biopsies. Median DDE concentrations were 476.7 µg/kg lipids among cases and 507.1 µg/kg among controls. No association was found between concentrations of DDE or DDT and risk of cancer of the breast in the whole data set; however, a statistically significant inverse association with DDE was observed when the analysis was restricted to ER-negative (ER–) cases (OR, 0.1; 95% CI, 0.0–0.5 in the highest exposure group). [This was the largest nested case–control study with measurements in adipose tissue rather than serum. The inverse association of DDE among women with ER– tumours does not have a clear interpretation.]

The association of incident breast cancer with young adults' exposure to DDT before it was banned was investigated in a nested case–control study of California residents who provided serum samples for the Child Health and Development Studies between 1959 and 1967 (Cohn et al., 2007). These women would have been mostly aged < 20 years when DDT use peaked. Cases were 129 women who developed cancer of the breast before age 50 years, identified by linkage to the California cancer registry and California vital status records. An equal number of controls from the cohort were matched to cases on birth year. The median time from blood draw to diagnosis was 17 years, and mean age of cases at diagnosis was 44 years. No associations were reported between serum DDE or DDT concentrations and breast cancer in unadjusted analyses. Significant positive associations were found with *p,p′*-DDT after adjustment for *o,p′*-DDT and *p,p′*-DDE. These associations were strongest in the subset of women aged < 14 years in 1945. [This study was notable for providing data on early-life exposure to DDT during a time when exposures were likely to have been higher. However, although breast-cancer risk factors including race, blood lipids, BMI, reproductive history, and breast feeding were evaluated, most models did not adjust for these factors. In addition, the Working Group considered that adjustment for multiple DDT congeners could introduce bias due to correlations among related metabolites, thus the non-adjusted results were taken to be more valid.]

In a further study based on this cohort, the incidence of cancer of the breast in 9300 daughters of women who provided blood samples in the original study was examined in relation to the mothers' prenatal exposure to DDT (Cohn et al., 2015). The methods of case and control selection were similar to those used in the parent study (Cohn et al., 2007). Daughters of the original participants were followed until 2012; 103 cases diagnosed before age 52 years and 315 controls provided serum samples and were included. Median *p,p*′-DDT concentrations were 13.18 ng/mL among cases and 12.98 ng/mL among controls. Mothers' perinatal serum concentration of *o,p*′-DDT was significantly associated with risk of breast cancer in the daughters in models adjusted for maternal lipids, overweight, and history of breast cancer (OR for Q4 versus Q1, 2.8; $P = 0.007$; P for trend, 0.053). Weaker positive and non-significant positive associations were observed for *p,p*′-DDT and DDE. [A major limitation of this study was that risk factors for breast cancer in daughters were not taken into account. As in the parent study, the Working Group gave less weight to models adjusted for multiple DDT metabolites.]

A total of 24 226 women aged 40–69 years in the Japan Public Health Center-based Prospective Study who responded to the baseline questionnaire and provided blood in 1990–1995 were followed until December 2002; 144 incident cases of breast cancer were identified (Iwasaki et al., 2008). Two matched controls for each case were selected from the cohort, and plasma DDT and DDE concentrations were measured for cases and controls. Median DDE concentrations were 7.04 ng/mL among cases versus 6.08 ng/mL among controls). After adjusting for reproductive variables, BMI, and alcohol consumption, the relative risk in the highest compared with the lowest quartile of *p,p*′-DDE was 1.48 (95% CI, 0.7–3.1) and for *p,p*′-DDT it was 0.99 (95% CI, 0.47–2.08). [This study was well-designed and considered most relevant confounders for cancer, but had limited power.]

2.1.2 Non-Hodgkin lymphoma

See Table 2.2

In all the cohort studies described below, a nested case–control analysis was used to investigate the relationship between non-Hodgkin lymphoma (NHL) and exposure to DDT.

Seventy-four cases of NHL (ICD-8 200 or 202) identified during follow-up from 1975 to 1994 of the CLUE I cohort from Washington County, Maryland, USA and 147 controls matched by race, sex, age and study-related factors were included in a nested case–control study (Rothman et al., 1997). Serum samples were collected before diagnosis. Four DDT-related compounds were measured and used to estimate total DDT concentration corrected for total lipids. Median concentrations of total lipid corrected DDT were 3150 ng/g lipid in cases and 2770 ng/g lipid in controls. There was no association between risk of NHL and quartiles of total DDT concentration (OR, 1.2; 95% CI, 0.5–3.0 for the fourth quartile compared with the first quartile) with adjustment, in addition to matching variables, for education, cigarette smoking, occupational exposure to suspected risk factors for NHL, and serum polychlorinated biphenyls (PCBs). [This study had the advantage that serum was collected at baseline before diagnosis.]

A later study using the same data set reported an analysis of the effect of *p,p*′-DDE on the risk of NHL found a slight increase in risk with the

Table 2.2 Cohort studies on cancers of the lympho-haematopoietic system and exposure to DDT and its metabolites

Reference, location, enrolment/ follow-up period, study-design	Population size, description, exposure assessment method	Organ site	Exposure category or level	Exposed cases/ deaths	Risk estimate (95% CI)	Covariates	Comments
Rothman et al. (1997) Washington County, Maryland, USA 1975–94 Nested case–control study	Cases: 74 from county cancer registry Controls: 147 matched on race, sex, birthdate & study variables Exposure assessment method: personal monitoring; lipid-adjusted serum concentrations	NHL (200, 202)	ng/g lipid			PCBs	Strengths: large study; serum collected at baseline before diagnosis Limitations: no data on diet so could not evaluate influence of DDT from foods; limited precision
			180–1740	14	1.0		
			1760–2660	16	1.1 (0.4–2.7)		
			2690–4020	19	1.1 (0.4–2.7)		
			4140–20 500	25	1.2 (0.5–3.0)		
			Trend-test *P* value: 0.87				
Engel et al. (2007) USA 1975–94 Nested case–control study	Cases: 74; as in Rothman et al. (1997) Controls: 147 Exposure assessment method: personal monitoring; lipid-adjusted adjusted serum concentration	NHL (200, 202)	*p,p′*-DDE, median, ng/g lipid			Age, race, state, sex, educational level, smoking	See Rothman et al. (1997) for details
			912.2	17	1.0		
			1616.2	17	0.9 (0.4–2.2)		
			2443.8	14	0.8 (0.3–2.0)		
			4475.0	26	1.5 (0.7–3.2)		
			Early follow-up, 0–12 yrs				
			912.2	7	1.0		
			1616.2	11	1.6 (0.5–5.3)		
			2443.8	7	0.8 (0.2–2.8)		
			4475.0	15	2.1 (0.7–6.3)		
			Late follow-up: 13–19 yrs:				
			912.2	10	1.0		
			1616.2	6	0.4 (0.1–1.6)		
			2443.8	7	0.9 (0.2–3.6)		
			4475.0	11	1.1 (0.3–3.4)		

Table 2.2 (continued)

Reference, location, enrolment/ follow-up period, study-design	Population size, description, exposure assessment method	Organ site	Exposure category or level	Exposed cases/ deaths	Risk estimate (95% CI)	Covariates	Comments
Engel et al. (2007) USA Enrolment 1989–90, follow-up to 1994 Nested case–control study	Cases: 30; see Hunter et al. (1997) Controls: 78 Exposure assessment method: personal monitoring; serum concentrations adjusted for total lipids	NHL	*p,p'*-DDE, median, ng/g lipid			Age	Strengths: large study sample Limitations: blood samples taken after start of follow-up; short follow-up period
			358.0	8	1.0		
			881.4	9	1.2 (0.4–3.7)		
			1788.9	13	2 (0.7–6.1)		
Laden et al. (2010) USA Enrolment 1989–1990; follow-up NR Nested case–control study	Cases: 145; as in Hunter et al. (1997) Controls: 290 Exposure assessment method: personal monitoring; serum concentrations adjusted for total lipids	NHL	*p,p'*-DDE, median, ng/g lipid			Smoking status, region	Strengths: large study Limitations: inconsistent results at different follow-up periods
			343.6	30	1.00		
			779.6	43	1.41 (0.76–2.60)		
			1327.0	27	0.77 (0.39–1.52)		
			2325.2	45	1.56 (0.82–2.97)		
			Trend-test *P* value: 0.33				

Table 2.2 (continued)

Reference, location, enrolment/ follow-up period, study-design	Population size, description, exposure assessment method	Organ site	Exposure category or level	Exposed cases/ deaths	Risk estimate (95% CI)	Covariates	Comments
Engel et al. (2007) Norway Enrolment 1972–1978; follow-up to 1999 Nested case–control study	Cases: 190 from Norwegian cancer registry Controls: 190 Exposure assessment method: personal monitoring; exposure corrected for total lipids	NHL (200 202)	*p,p′*-DDE, median, ng/g lipid			Sex, age, smoking status, county	Strengths: clear diagnosis, long follow-up Limitations:
			2059.1	39	1.0		
			3247.2	50	1.4 (0.7–2.6)		
			4673.2	52	1.4 (0.7–2.6)		
			7513.0	49	1.4 (0.7–2.9)		
			Early follow-up, 2–16 yrs:				
			2059.1	17	1.0		
			3247.2	31	3.1 (1.1–8.6)		
			4673.2	28	2.4 (0.9–6.4)		
			7513.0	26	4.3 (1.2–15)		
			Late follow-up, 17–25 yrs:				
			2059.1	22	1.0		
			3247.2	19	0.8 (0.3–2.0)		
			4673.2	24	1.3 (0.5–3.2)		
			7513.0	23	0.8 (0.3–2.0)		
Bertrand et al. (2010) USA 1982–2003 Nested case–control study	Cases: 205; annual questionnaires, confirmed from medical records Controls: 409; cohort; matched on race, age, fasting status Exposure assessment method: personal monitoring; lipid-adjusted serum concentrations	NHL	ng/g lipid			Alcohol consumption, weight, smoking	Strengths: diagnoses confirmed using medical records; prospective measurement of serum Limitations: may be unmeasured confounding from dietary factors
			43–1045	37	1.00		
			> 1045–1741	37	0.97 (0.55–1.7)		
			> 1741–2523	52	1.4 (0.78–2.50)		
			> 2523–3595	29	0.71 (0.39–1.30)		
			> 3595–1897	50	1.30 (0.74–2.30)		
			Trend-test *P* value: 0.7				

Table 2.2 (continued)

Reference, location, enrolment/ follow-up period, study-design	Population size, description, exposure assessment method	Organ site	Exposure category or level	Exposed cases/ deaths	Risk estimate (95% CI)	Covariates	Comments
Bräuner et al. (2012) Denmark Enrolment 1993–1997; follow-up to 2008 Nested case–control study	Cases: 239 from Danish cancer registry Controls: 245; random sample of the entire cohort Exposure assessment method: personal monitoring; concentrations in adipose tissue (see Raaschou-Nielsen et al., 2005)	NHL	*p,p′*-DDT, µg/kg lipids			Age, sex	Case–cohort analysis Strengths: adipose tissue used rather than blood (preferred indicator because it represents cumulative exposure) Limitations: did not adjust for co-exposure to other pesticides or PCBs; large number of samples were below the LOD
			6–15	29	1.00		
			15–22	32	1.06 (0.52–2.14)		
			22–36	35	1.03 (0.51–2.09)		
			36–49	23	1.21 (0.53–2.75)		
			49–460	18	1.64 (0.68–3.96)		
			68–390	59	1.00		
			390–680	53	0.83 (0.49–1.39)		
			p,p′-DDE, µg/kg lipids				
			680–1100	63	1.04 (0.62–1.73)		
			1100–1700	34	0.79 (0.43–1.46)		
			1700–8000	29	1.10 (0.57–2.14)		

CI, confidence interval; DDE, dichlorodiphenyldichloroethylene; DDT, dichlorodiphenyltrichloroethane; LOD, limit of detection; NHL, non-Hodgkin lymphoma; PCBs, polychlorinated biphenyls

concentration of *p,p′*-DDE (OR, 1.5; 95% CI, 0.7–3.2, adjusting for years of education, current smoking status), which was stronger in the earlier period of 0–12 years of follow-up (Engel et al., 2007). Neither analysis showed clear exposure–response trends.

An analysis of the association between NHL and serum *p,p′*-DDE concentrations in the previously described United States Nurses' Health Study (see Hunter et al., 1997) was also reported in the same paper by Engel et al. (2007). Thirty participants with incident NHL diagnosed between the date of blood collection and May 1994 (median follow-up, 1 year) were included as cases and 78 cohort members selected previously as controls for another study served as controls. A non-significant increased risk of NHL was found for increasing quartile of *p,p′*-DDE, with the odds ratio for the highest quartile being 2.0 (95% CI, 0.7–6.1).

After an extended follow-up of the Nurses' Health Study cohort [follow-up interval not specified] 145 cases of NHL were identified and two controls matched on age, race, month of blood draw, and fasting status were selected for each case (Laden et al., 2010). Median concentrations of *p,p′*-DDE were 996.2 ng/g lipid in cases and 1002.3 ng/g lipid in controls. No consistent pattern of association was observed for quartiles of total serum concentrations of *p,p′*-DDE after adjustment for potential confounders, including smoking and reproductive history. For all NHL combined, the odds ratio for the fourth versus the first quartile was 1.56 (95% CI, 0.82–2.97).

In the JANUS cohort described previously (see Ward et al., 2000), lipid-corrected concentrations of *p,p′*-DDE were available for 190 confirmed cases of NHL ascertained with follow-up to 1999 (Engel et al., 2007). An equal number of controls from the cohort were matched by age, sex, county, and date of examination. In the analysis, further adjustments were made for BMI and smoking status. The odds ratios for the association of NHL with *p,p′*-DDE was the same (1.4) for all quartiles above the first. For the early period of follow-up, 2–16 years, an increase in risk was reported for the second exposure quartile (OR, 3.1; 95% CI, 1.1–8.6) and for the fourth exposure quartile (OR, 4.3; 95% CI, 1.2–15.0) compared with the lowest quartile, with a significant upward trend across quartiles. However, no excess or trend was found for the later follow-up period, 17–25 years.

The Physicians' Health Study began in 1982 in the USA as a randomized trial for the primary prevention of cardiovascular disease and cancer in 22 071 male physicians aged 40–84 years at enrolment. A total of 14 916 participants provided a blood sample in 1982–84 (before randomization) and were followed until 2003 using annual questionnaires confirmed by review of medical records to identify newly diagnosed NHL (Bertrand et al., 2010). After exclusions, 205 cases with available blood samples were included. For each case, two controls matched on baseline by race, age, date of blood collection, and fasting status at blood draw were selected from the cohort. Lipid-corrected concentrations of *p,p′*-DDE in serum were determined for cases and controls. There was no significant association of NHL with *p,p′*-DDE, with the odds ratio in the highest quintile being 1.3 (95% CI, 0.74–2.3) in a multivariable adjusted analysis.

In a further study based on the Danish Diet, Cancer and Health study described previously (see Raaschou-Nielsen et al., 2005), NHL cases were ascertained among 57 053 persons followed until 2008 (Bräuner et al., 2012). Exposures of cases were compared with those of a random sample of the cohort in a case-cohort analysis that included 239 cases and 245 individuals from the cohort with measurements of *p,p′*-DDT and *p,p′*–DDE. The median concentration of *p,p′*-DDT was 24 μg/kg lipid for cases and 21 μg/kg lipid for controls; the median concentration of *p,p′*-DDE μg/kg lipid was 700 μg/kg lipid for cases and 640 μg/kg lipid for controls. There were suggestions of a monotonic dose–response relationship between risk of NHL and *p,p′*-DDT concentration. The incidence rate ratio for the highest versus lowest exposure category was 1.64 (95% CI, 0.68–3.96). In a linear analysis, the incidence rate ratio for a one interquartile range increase in exposure to *p,p′*-DDT was 1.35 (95% CI, 1.10–1.66). There was no clear pattern of increased risk for *p,p′*-DDE. [Although measures were made of 8 pesticides and 10 PCB congeners, no adjustment was made for co-exposure. In addition, concentrations in a large number of samples were below the limit of detection.]

2.1.3 Cancer of the testis

See Table 2.3

McGlynn et al. (2008) reported the results of a case–control study on testicular germ cell tumours conducted among United States military personnel. The cases included in the analysis were 739 men who had donated blood to the Department of Defense Serum Repository between 1987 and 2002, and who were subsequently diagnosed with testicular germ cell tumour between 1988 and 2003. Controls were 915 men with a serum sample available in the repository, matched on birth year, ethnicity, and date of serum sample. Eleven organochlorine compounds, including *p,p′*-DDT and *p,p′*-DDE, were analysed in the serum. Data on other risk factors were collected by telephone interview. Testicular germ cell tumour was statistically significantly associated (*P* for trend, 0.0002) with increasing levels of *p,p′*-DDE (highest versus lowest quartile: OR, 1.71; 95% CI, 1.23–2.38). This association was apparent for seminoma (OR, 1.91; 95% CI, 1.22–2.99) and for non-seminoma (OR, 1.63; 95% CI, 1.10–2.42). *p,p′*-DDT was detected in only 20% of the subjects, and was not significantly associated with testicular germ cell tumour. [This was a well-conducted study with a large number of subjects, and high response rates. The use of prediagnostic serum samples was a major advantage of this study, as serum levels of organochlorine compounds are unlikely to be influenced by the disease.]

In a paper by Purdue et al. (2009), the authors reported the findings of a case–control study on testicular germ cell tumours that was nested within the previously described Janus Serum Bank cohort of Norway (see Ward et al., 2000). Cases were Janus cohort members who were diagnosed with testicular germ cell tumour between 1972 and 1999 through linkage with the Norwegian cancer registry. One male control from the cohort was matched to each case by region, time period, and age at blood draw. The analysis included 49 cases and 51 controls. Concentrations of 11 organochlorine insecticides, including *p,p′*-DDE, *p,p′*-DDT, and *o,p′*-DDT, and of 34 PCBs were measured in serum. In case–control comparisons, the odds ratio for testicular germ cell tumour increased with increasing serum *p,p′*-DDE concentration: the odds ratio for the highest to the lowest exposure tertile was 2.2 (95% CI, 0.7–6.5). A similar association was observed when restricting the analysis to seminoma cases. Elevated odds ratios were also observed for *p,p′*-DDT (OR, 2.1; 95% CI, 0.6–7.2 for the third versus the first quartile), but not for *o,p′*-DDT. [This was a small but well-conducted study. An important strength was the use of serum samples collected before diagnosis, minimizing the possibility that measurements were affected by the disease. Moreover, DDT had only recently been banned at the time of blood sample collection (1972–1978).]

In California, USA, Cohn et al. (2010) examined maternal serum concentrations of DDT-related compounds in relation to sons' risk of testicular cancer as assessed more than 30 years later in the Child Health and Development Studies (described previously, see Cohn et al., 2007). Of study participants who had serum samples, 15 sons were diagnosed with germ cell testicular tumour. The cases were matched to three controls each by race and year of birth; most analyses were not adjusted for other risk factors because

Table 2.3 Cohort studies of cancer of the testis and exposure to DDT and its metabolites

Reference, location enrolment/follow-up period, study design	Population size, description, exposure assessment method	Organ site	Exposure category or level	Exposed cases/deaths	Risk estimate (95% CI)	Covariates	Comments
McGlynn et al. (2008) USA 1988–2003 Nested case–control study	Cases: 739 from Defence Medical Surveillance system Controls: 915 men with a sample in the registry matched on birth year, race/ethnicity, and date of sample Exposure assessment method: personal monitoring; GC-MS frequency for *p,p′*-DDT 20%	Testis (TGCT)	*p,p′*-DDE, µg/g			Age at blood donation, ethnicity, date of serum draw, age at reference date, cryptorchidism, family history of testicular cancer, height, BMI	Strengths: large study size; analysis of pre-diagnostic serum samples; high response rate; histologically confirmed tumours Limitations: some participants could not be contacted due to military deployment; adjustment for self-reported BMI; multiple comparisons
			≤ 0.157	186	1.00		
			0.158–0.250	167	1.01 (0.75–1.36)		
			0.251–0.390	146	1.00 (0.73–1.38)		
			> 0.390	236	1.71 (1.23–2.38)		
			Trend-test *P* value: 0.0002				
		Testis (seminoma)	*p,p′*-DDE, µg/g				
			≤ 0.157	59	1.00		
			0.158–0.250	68	1.17 (0.76–1.78)		
			0.251–0.390	57	0.98 (0.62–1.54)		
			> 0.390	128	1.91 (1.22–2.99)		
			Trend-test *P* value: 0.0008				
		Testis (non-seminoma)	*p,p′*-DDE, µg/g				
			≤ 0.157	127	1.00		
			0.158–0.250	98	0.98 (0.69–1.39)		
			0.251–0.390	89	1.10 (0.76–1.59)		
			> 0.390	108	1.63 (1.10–2.42)		
			Trend-test *P* value: 0.0044				

Table 2.3 (continued)

Reference, location enrolment/follow-up period, study design	Population size, description, exposure assessment method	Organ site	Exposure category or level	Exposed cases/deaths	Risk estimate (95% CI)	Covariates	Comments
McGlynn et al. (2008) USA 1988–2003 Nested case–control study (cont.)		Testis (TGCT)	*p,p′*-DDT, µg/g				
			≤ 0.029	630	1.00		
			0.0210–0.259	27	0.81 (0.49–1.35)		
			0.260–0.397	40	1.27 (0.81–2.01)		
			> 0.397	37	1.13 (0.71–1.82)		
			Trend-test *P* value: 0.5				
		Testis (seminoma)	*p,p′*-DDT, µg/g				
			≤ 0.029	260	1.00		
			0.0210–0.259	11	0.59 (0.29–1.19)		
			0.260–0.397	19	1.20 (0.67–2.14)		
			> 0.397	22	1.30 (0.73–2.30)		
			Trend-test *P* value: 0.4				
		Testis (non-seminoma)	*p,p′*-DDT, µg/g				
			≤ 0.029	369	1.00		
			0.0210–0.259	16	1.02 (0.55–1.9)		
			0.260–0.397	21	1.39 (0.79–2.42)		
			> 0.397	15	0.94 (0.50–1.77)		
			Trend-test *P* value: 0.86				

Table 2.3 (continued)

Reference, location enrolment/follow-up period, study design	Population size, description, exposure assessment method	Organ site	Exposure category or level	Exposed cases/ deaths	Risk estimate (95% CI)	Covariates	Comments
Purdue et al. (2009) Norway enrolment 1972 – 1978; follow-up 1999 Nested case–control study	Cases: 49 from Norwegian cancer registry Controls: 51 men from the Janus cohort matched by region, time period and age at blood draw Exposure assessment method: personal monitoring	Testis	*o,p′*-DDT:				
			Tertile 1	16	1.0	Region, age at blood donation, time period of blood draw, total lipids	Strengths: use of serum samples collected before diagnosis – DDT had only recently been banned at the time of blood sample collection (1972–1978); completeness of the Norway cancer registry Limitations: small study size
			Tertile 2	11	0.8 (0.2–2.6)		
			Tertile 3	21	1.4 (0.4–4.5)		
			p,p′-DDT:				
			Tertile 1	11	1.0		
			Tertile 2	20	2.3 (0.7–7.2)		
			Tertile 3	18	2.1 (0.6–7.2)		
			p,p′-DDE:				
			Tertile 1	11	1.0		
			Tertile 2	18	1.8 (0.6–5.5)		
			Tertile 3	20	2.2 (0.7–6.5)		
		Testis (seminomas)	*p,p′*-DDE:				
			Tertile 1	10	1.0		
			Tertile 2	10	1.0 (0.3–3.8)		
			Tertile 3	14	2.2 (0.5–8.7)		
Cohn et al. (2010) California, USA Enrolment 1959–1967; follow-up to 2000 Nested case–control study	Cases: 15 from california cancer registry Controls: 45 from cohort, matched by race and birth year Exposure assessment method: personal monitoring	Testis	Interquartile range			None	Strengths: assessment of exposure in utero from perinatal blood samples; maternal blood samples taken during time period of DDT use Limitations: very small study size; number lost to follow-up is not known; no adjustment for TGCT risk factors
			p,p′-DDT	15	0.70 (0.26–1.64)		
			p,p′-DDE	15	0.19 (0.04–0.62)		
			o,p′-DDT	15	0.77 (0.37–1.33)		
			p,p′-DDT/ *p,p′*-DDE ratio	15	3.56 (1.34–11.88)		
			Trend-test *P* value: 0.34				

BMI, body-mass index; CI, confidence interval; DDE, dichlorodiphenyldichloroethylene; DDT, dichlorodiphenyltrichloroethane; GC-ECD, gas chromatography with electron capture detector; GC-MS, gas chromatography-mass spectrometry; TGCT, testicular germ cell tumour; yr, year

of small numbers. Concentrations of *p,p′*-DDT, *o,p′*-DDT, and *p,p′*-DDT in mothers' serum were inversely associated with testicular cancer in univariate analyses; however, after adjustment for *p,p′*-DDE, the odds ratio for *p,p′*-DDT was increased to 4.81 (95% CI, 0.92–48.62). The odds ratio for *p,p′*-DDE in the same model was 0.08 (95% CI, 0.01–0.41). Testicular cancer cases had significantly higher DDT/DDE ratios than their matched controls ($P < 0.05$). [Direct measurements of three DDT-related compounds from maternal serum obtained 1–3 days after delivery was a strength of this study; however, the risk estimate for cancer of the testis associated with DDE concentration was extremely low in this population, and adjusting for DDE in this data may have introduced bias in risk estimates for DDT]. Since cancer of the testis is relatively rare and only 15 exposed cases were available for analysis, risks estimates were unstable and excessively sensitive to DDE adjustment. In addition, there was no adjustment for other risk factors. The results must therefore be considered with caution.]

2.1.4 Cancer of the liver

See Table 2.4

A nested case–control study was conducted among the participants in the Nutritional Intervention Trial in Linxian, China, to evaluate the association between DDT and primary cancer of the liver (McGlynn et al., 2006). Compared with the coastal regions of China, which have a high incidence of cancer of the liver attributed in part to aflatoxin exposure, Linxian has a relatively low incidence of cancer of the liver and low exposure to aflatoxin. The trial consisted of 29 584 women and men aged 40–69 years at enrolment (1986–1991). A 10 mL blood sample was collected from each participant at baseline. The cases included 168 individuals who developed cancer of the liver in 2001, and the control group included 385 individuals frequency-matched on age and sex who were alive and had never had cancer of the liver. In multivariable models controlled for hepatitis B surface antigen (HBsAg) status, serum DDE concentration and other covariates, the risk of developing liver cancer increased with increased serum DDT concentration (OR for quintile 5 versus quintile 1, 3.8; 95% CI, 1.7-8.6; *P* for trend, 0.002). The odds ratio for the same comparison without adjustment for DDE was 2.0 (95% CI, 1.1–3.9; *P* for trend, 0.049). In contrast there was no statistically significant association between liver cancer and DDE concentration, regardless of adjustment for DDT. The association between high serum DDT concentration and liver cancer was stronger among individuals with DDE concentrations below the median (OR, 3.55; 95% CI, 1.45–8.74) compared with those with DDE concentrations above the median (OR, 1.70; 95% CI, 0.97–2.98). [Risks may be particularly increased among persons exposed directly to DDT (resulting in a higher ratio of DDT to DDE) or alternatively may be associated with an individuals' ability to metabolize DDT and DDE. The odds ratios were not adjusted for alcohol drinking or exposure to aflatoxin.]

In Haimen City, China, a prospective study to identify environmental and genetic risk factors for hepatocellular carcinoma (HCC) in addition to hepatitis B virus (HBV) infection enrolled 83 794 people between February 1992 and December 1993 (Persson et al., 2012). Pre-diagnostic blood samples were collected at baseline and used to determine lipid-corrected DDT concentrations. Information on multiple risk factors and HBV infection status was determined based on HBsAg. Participants were actively followed until September 2000, and then passively followed via death-certificate determination. Incident cases of HCC were diagnosed by histology and/or liver imagining, α-fetoprotein elevation (> 400 ng/mL), clinical criteria, or by death certificate with post-mortem interviews of family members. HCC cases ($n = 488$) and controls ($n = 492$) were frequency-matched by

Table 2.4 Cohort studies of cancer of the liver and exposure to DDT and its metabolites

Reference, location enrolment/follow-up period, study design	Population size, description, exposure assessment method	Organ site	Exposure category or level	Exposed cases/ deaths	Risk estimate (95% CI)	Covariates	Comments
McGlynn et al. (2006) Linxian, China baseline 1984–85; follow-up 2001 Nested case–control study	Cases: 168; multiple methods Controls: 385; frequency-matched on age and sex Exposure assessment method: personal monitoring; *p,p′*-DDT and *p,p′*-DDE in serum	Liver (HCC)	DDT, ng/g lipid			Age, sex, HBsAg status, commune of residence	Strengths: pre-diagnostic blood samples; high DDT/DDE levels in Chinese population; controlling for other well-known risk factors for liver cancer Limitations: not all liver tumours histologically diagnosed; exposure to aflatoxin (AFB1) not known; no adjustment for alcohol drinking
			< 265 (Q1)	26	1.0		
			265–382 (Q2)	35	1.3 (0.7–2.5)		
			383–521 (Q3)	34	1.4 (0.7–2.6)		
			522–787 (Q4)	33	1.4 (0.7–2.7)		
			> 787 (Q5)	40	2.0 (1.1–3.9)		
			Trend-test *P* value: 0.049				
		Liver (HCC)	DDT, ng/g lipid			Age, sex, HBsAg status, commune of residence, DDE	
			< 265 (Q1)	26	1.0		
			265–382 (Q2)	35	1.5 (0.8–2.7)		
			383–521 (Q3)	34	1.7 (0.9–3.3)		
			522–787 (Q4)	33	2.1 (1.0–4.3)		
			> 787 (Q5)	40	3.8 (1.7–8.6)		
			Trend-test *P* value: 0.0024				
		Liver (HCC)	DDE, ng/g lipid			Age, sex, HBsAg status, commune of residence	
			< 1767 (Q1)	27	1.0		
			1767–2443 (Q2)	27	1.0 (0.5–1.9)		
			2444–3478 (Q3)	48	1.7 (0.9–3.1)		
			3479–5458 (Q4)	47	1.8 (1.0–3.3)		
			> 5458 (Q5)	19	0.7 (0.3–1.5)		
			Trend-test *P* value: 0.75				

Table 2.4 (continued)

Reference, location enrolment/follow-up period, study design	Population size, description, exposure assessment method	Organ site	Exposure category or level	Exposed cases/deaths	Risk estimate (95% CI)	Covariates	Comments
McGlynn et al. (2006) Linxian, China baseline 1984–85; follow-up 2001 Nested case–control study (cont.)		Liver (HCC)	DDE, ng/g lipid			Age, sex, HBsAg status, commune of residence, DDT	
			< 1767 (Q1)	27	1.0		
			1767–2443 (Q2)	27	1.0 (0.5–1.9)		
			2444–3478 (Q3)	48	1.7 (0.9–3.1)		
			3479–5458 (Q4)	47	1.9 (1.0–3.1)		
			> 5458 (Q5)	19	0.8 (0.3–1.7)		
			Trend-test *P* value: 0.75				
Persson et al. (2012) Haimen, China Enrolment 1992–1993; – follow-up to 2000 Nested case–control study	Cases: 473; HCC diagnosis ascertained by histology and/or liver imaging, alpha-fetoprotein elevation, clinical criteria, or by death certificate with post-mortem interviews of family members Controls: 488; frequency-matched by age, sex, residence area Exposure assessment method: personal monitoring	Liver (HCC)	*p,p′*-DDT, ng/g lipid			Age, sex, area of residence, HBsAg, family history of HCC, history of acute hepatitis, smoking, alcohol, occupation, continuous serum level of *p,p′*-DDE	Strengths: prospective design; completeness of follow up; large numbers; prediagnostic serum samples; adjustment for HBV chronic infection Limitations: no assessment of aflatoxin B1 exposure
			≤ 261	112	1.00		
			262–404	99	1.26 (0.65–2.46)		
			404–545	74	0.86 (0.41–1.80)		
			545–810	76	1.29 (0.57–2.92)		
			≥ 810	112	2.96 (1.19–7.40)		
			Trend-test *P* value: 0.04				
		Liver (HCC)	*p,p′*-DDE, ng/g lipid			Age, sex, area of residence, HBsAg, family history of HCC, history of acute hepatitis, smoking, alcohol, occupation, continuous serum level of *p,p′*-DDT	
			≤ 10 000	140	1.00		
			10 000–14 746	81	0.70 (0.36–1.35)		
			14 746–21 579	93	0.73 (0.36–1.46)		

Table 2.4 (continued)

Reference, location enrolment/follow-up period, study design	Population size, description, exposure assessment method	Organ site	Exposure category or level	Exposed cases/ deaths	Risk estimate (95% CI)	Covariates	Comments
Persson et al. (2012) Haimen, China Enrolment 1992–1993; – follow-up to 2000 Nested case–control study (cont.)			21 579–32 222	66	0.77 (0.35–1.70)		
			≥ 32 222	93	0.81 (0.33–2.03)		
			Trend-test *P* value: 0.79				
			DDT tertile 2	7	1.5 (0.5–4.9)		
			DDT tertile 3	7	1.8 (0.5–6.2)		
			Trend-test *P* value: 0.34				

BMI, body-mass index; CI, confidence interval DDE, dichlorodiphenyldichloroethylene; DDT, dichlorodiphenyltrichloroethane; HBsAg, hepatitis B virus surface antigen; HBV, hepatitis B virus; HCC, hepatocellular carcinoma; yr, year

age, sex, and area of residence. In a multivariate analysis adjusted for alcohol drinking, HBsAg, and other risk factors, the highest quintile of serum *p,p′*-DDT concentration was associated with an increased risk of HCC (OR, 2.96; 95% CI, 1.19–7.40) and there was a significant linear trend in the exposure-response relationship (*P* for trend, 0.04). There was no association with *p,p′*-DDE.

2.1.5 Other cancer sites

See Table 2.5

Sawada et al. (2010) reported the findings of a nested case–control analysis of cancer of the prostate nested within the Japan Public Health Center-based Prospective Study, a population-based cohort that included 65 657 men aged 40–69 years at baseline in 1990–1993, of whom 14 203 provided blood samples. During follow-up until 2005, 201 cases were identified from major hospitals, cancer registries, and death certificates. Two controls per case were matched by age, area of residence, date and time of blood sampling, and duration of fasting at blood collection. Concentrations of OCPs and PCBs were measured in plasma. No association between incidence of cancer of the prostate and concentration of *o,p′*-DDT, *p,p′*-DDT, or *p,p′*-DDE was observed (*P* for trend across quartiles of exposure distribution, 0.61, 0.45, and 0.65, respectively).

A prospective follow-up study enrolled 2283 adult residents of Charleston, South Carolina, USA in 1960. Venous blood samples were obtained from 919 subjects in 1974–1975 and analysed for *p,p′*-DDT and *p,p′*-DDE (Austin et al., 1989). In internal analysis using mortality until 1985, involving 209 deaths, a monotonic, but not statistically significant, rise in relative risk by tertile of exposure duration was observed for cancer of the respiratory tract. Adjusted relative risks relative to the lowest tertile of exposure were 1.5 (95% CI, 0.5–4.9) for tertile 2 and 1.8 (95% CI, 0.5–6.2) for tertile 3 (*P* for trend, 0.34). [Although follow-up was nearly complete, the small size of the cohort and the potential for residual confounding from smoking due to a lack of intensity and duration of smoking information limited the conclusions that can be made from the results of this study.]

2.1.6 Occupational cohort studies

See Table 2.6

(a) Non-Hodgkin lymphoma

The association between NHL and exposure to DDT was evaluated in the AHS by Purdue et al. (2007) and subsequently by Alavanja et al. (2014). Based on 523 incident cases of NHL among 54 306 study participants free of cancer at the time of enrolment, 98 participants with incident NHL, including multiple myeloma and chronic lymphocytic leukaemia (CLL), provided detailed data on DDT use before onset of disease (Alavanja et al., 2014). The primary use of DDT in this cohort occurred between the 1950s and the early 1970s. DDT was banned for use on crops in 1972 in the USA. Exposure assessment methods for the AHS are described in Section 1.4.4. Risk estimates were adjusted for age, state of residence, race, and total days of herbicide use. Ever use of DDT was not associated with total NHL (RR, 1.0; 95% CI, 0.8–1.3) or any NHL subtype; however, statistically significant positive exposure-response trends for total NHL were observed with lifetime days of DDT use (*P* for trend, 0.02)]. This positive association with total NHL was attenuated in analyses using an earlier definition of NHL that did not include multiple myeloma and CLL. In subtype analyses, the highest category of lifetime DDT use was associated with small B-cell lymphocytic lymphoma/CLL/mantle cell lymphoma (RR, 2.6; 95% CI, 1.3–4.8). The excess risk in this NHL subtype was, however, not significantly different to that for other NHL subtypes in the polytomous regression analysis.

Table 2.5 Cohort studies of cancer at other organ sites and exposure to DDT and its metabolites

Reference, location enrolment/follow-up period, study design	Population size, description, exposure assessment method	Organ site	Exposure category or level	Exposed cases/ deaths	Risk estimate (95% CI)	Covariates	Comments
Sawada et al. (2010) Japan Enrolment 1990–1994; follow;up to 2005 Nested case–control study	Cases: 201 from hospitals, population-based cancer registries and death certificates Controls: 402 cohort members matched by age, area of residence, date and time of blood sampling, fasting at blood collection Exposure assessment method: personal monitoring; measurements made from blood samples taken at baseline	Prostate	*o,p*′-DDT, ng/g lipid			Smoking status, alcohol consumption, marital status, BMI, intake of green tea, intake of green tea and miso soup	Strengths: pre-diagnosis blood samples Limitations: response rate not known
			< 2.5	43	1.00		
			2.5–4.2	57	1.39 (0.79–2.44)		
			4.3–7.6	54	1.29 (0.71–2.34)		
			≥ 7.7	47	1.04 (0.54–2.03)		
			Trend-test *P* value: 0.61				
			p,p′-DDT, ng/g lipid			Smoking status, alcohol consumption, marital status, BMI, intake of green tea, intake of green tea and miso soup	
			< 24	41	1.00		
			24–40	64	1.51 (0.87–2.63)		
			41–63	50	0.92 (0.50–1.70)		
			≥ 64	46	1.0 (0.52–1.92)		
			Trend-test *P* value: 0.45				
			p,p′-DDE, ng/g lipid			Smoking status, alcohol consumption, marital status, BMI, intake of green tea, intake of green tea and miso soup	
			< 560	49	1.00		
			560–939	52	1.00 (0.60–1.66)		
			940–1599	47	0.89 (0.52–1.53)		
			≥ 1600	53	0.90 (0.52–1.54)		
			Trend-test *P* value: 0.65				

Table 2.5 (continued)

Reference, location enrolment/follow-up period, study design	Population size, description, exposure assessment method	Organ site	Exposure category or level	Exposed cases/ deaths	Risk estimate (95% CI)	Covariates	Comments
Austin et al. (1989) Charleston, South Carolina, USA 1974– 1985	919 residents with available blood sample at baseline (1974) Exposure assessment method: personal monitoring; total DDT calculated as a combination of *p,p'*-DDT and *p,p'*-DDE death certificate and expert committee	Respiratory tract:	DDT tertile 1	5		Age, sex, race, years of schooling, smoking status	Strengths: serum sampling in 1974 (high exposure period) Limitations: small study size; cancer mortality not incidence
			DDT tertile 2	7	1.5 (0.5–4.9)		
			DDT tertile 3	7	1.8 (0.5–6.2)		
			Trend-test *P* value: 0.34				

BMI, body-mass index; CI, confidence interval DDE, dichlorodiphenyldichloroethylene; DDT, dichlorodiphenyltrichloroethane; yr, year

Table 2.6 Occupational cohort studies of exposure to DDT

Reference, location enrolment/follow-up period, study design	Population size, description, exposure assessment method	Organ site	Exposure category or level	Exposed cases/ deaths	Risk estimate (95% CI)	Covariates	Comments
Alavanja et al. (2014b) Iowa and North Carolina, USA 1993–2011	57 310 certified private and commercial male pesticide applicators enrolled in the Agricultural Health Study Exposure assessment method: questionnaires administered 1993–1997 and 1999–2005	NHL	No exposure to DDT	152	1.0	Age, state, race, total herbicide days	Strengths: large prospective cohort study; pesticide applicators knowledgeable about use of specific pesticides; detailed exposure estimates Limitations: field verification of exposure estimates done on only a sample of applicators
			< 8.75 days of use	43	1.3 (0.9–1.8)		
			8.75–56 days of use	28	1.1 (0.7–1.7)		
			> 56–1627.5	27	1.7 (1.1–2.6)		
			Trend-test P value: 0.02				
Garabrant et al. (1992) USA Employment 1948–1971, follow-up dates NR Nested case–control study	Cases: 28; company death records Controls: 112 living cohort members matched by age, sex, & race Exposure assessment method: company records	Pancreas	Ever exposure	6	4.8 (1.3–17.6)	NR	Strengths: DDT manufacturing workers had relatively high exposure to DDT Limitations: small number of cases; exposure ascertainment by department and job title can allow for misclassification
			Trend-test P value: 0.02				
Andreotti et al. (2009) Iowa and North Carolina, USA 1993–2004 Nested case–control study	Cases: 93; state cancer registries Controls: 82 503; cancer free cohort members Exposure assessment method: questionnaire	Pancreas	Ever use	6	0.4 (0.2–0.9)	Age, smoking, diabetes, applicator or spouse of applicator	Strengths: large prospective cohort of knowledgeable pesticide applicators Limitations: few exposed cases available for analysis
Beard et al. (2003) Australia 1935–1996	1999 workers and 1984 non-exposed workers; outdoor pest-control workers Exposure assessment method: company records; national death and insurance registers	Pancreas	Exposure duration			Age and calendar period	Strengths: DDT-exposed workers were compared with a non-exposed group of workers and with the general population Limitations: total exposed population was not large; limited exposure assessment
			< yrs	1	5.27 (1.09–15.4)		
			≥ 3 yrs	2	1.35 (0.37–3.44)		
		Leukaemia	Exposure duration			Age and calendar period	
			< 3 yrs	2	2.57 (0.06–14.29)		
			≥ 3 yrs	1	1.52 (0.31–4.45)		

Table 2.6 (continued)

Reference, location enrolment/follow-up period, study design	Population size, description, exposure assessment method	Organ site	Exposure category or level	Exposed cases/ deaths	Risk estimate (95% CI)	Covariates	Comments
Cocco et al. (2005) Sardinia, Italy 1956–1999	4552 men exposed to DDT during antimalaria operation Exposure assessment method: JEM; algorithm based on the European Predictive Operator Exposure Model (EUROPOEM) database Sardinia and Italian mortality records	All cancers combined	Cumulative DDT exposure (mg)			Age, age at first exposure, ethnic origin	Strengths: documented exclusive exposure to DDT by job type and algorithm estimate Limitations: small numbers for some types of cancer
			Unexposed	228	1.0		
			0.01–21.6	154	1.1 (0.9–1.4)		
			21.7–531.4	133	0.9 (0.7–1.1)		
			531.5–2,755	134	0.9 (0.7–1.1)		
			> 2755	152	1 (0.8–1.2)		
		Stomach	Cumulative DDT exposure (mg)			Age, age at first exposure, ethnic origin	
			Unexposed	11	1.0		
			0.01–21.6	8	1.2 (0.5–3.0)		
			21.7–531.4	8	1.1 (0.5–2.8)		
			531.5–2755	13	1.7 (0.7–3.7)		
			≥ 2755	16	2.0 (0.9–4.4)		
			Trend-test *P* value: > 0.05				
Koutros et al. (2013) Iowa and North Carolina, USA 1993–2007	54 412; Agricultural Health Study Exposure assessment method: expert assessment	Prostate: total	Cumulative lifetime exposure to DDT				See Alavanja et al. (2014b) for details
			Unexposed*	578	1.00	Age, state, smoking, fruit servings, leisure time physical activity in winter, race, family history	Strengths: large cohort study, in agricultural population thus high exposure prevalence, good exposure assessment
			Q1*	96	0.98 (0.78–1.22)		
			Q2*	97	1.27 (1.02–1.58)		
			Q3*	96	1.27 (1.02–1.58)		
			Q4*	65	1.18 (0.95–1.34)		
			*Trend-test *P* value: 0.14				
		Prostate: aggressive	Cumulative lifetime exposure to DDT				
			Unexposed*	267	1.00	Age, state, smoking, fruit servings, leisure time physical activity in winter, race, family history	
			Q1*	47	1.06 (0.76–1.48)		
			Q2*	46	1.17 (0.85–1.61)		
			Q3*	46	1.56 (1.13–2.15)		
			Q4*	46	1.3 (0.94–1.80)		
			*Trend-test *P* value: 0.10				

CI, confidence interval; DDE, dichlorodiphenyldichloroethylene; DDT, dichlorodiphenyltrichloroethane; JEM, job–exposure matrix; NHL, non-Hodgkin lymphoma; NR, not reported; OR, odds ratio; yr, year

(b) Cancer of the prostate

Koutros et al. (2013) studied the association between total and aggressive cancer of the prostate with 48 pesticides, including DDT, in the AHS. In follow-up until 2007, there were 1962 incident cases of cancer of the prostate, of which 919 were classified as aggressive. Relative risks for the association between aggressive cancer of the prostate and cumulative lifetime exposure to DDT were greater than unity, but statistically significant only in the third quartile of exposure (RR, 1.56; 95% CI, 1.13–2.15). The relative risk in the fourth quartile was lower (RR, 1.30; 95% CI, 0.94–1.80) and there was no significant exposure-response trend (*P* for trend, 0.10). Similar findings were reported for total prostate cancer and for prostate cancer stratified by family history.

(c) Cancer of the pancreas

Garabrant et al. (1992) conducted a cohort study of mortality among 5886 chemical-manufacturing workers in the USA, which included a nested case–control study of exposure to DDT and cancer of the pancreas. There were 28 deceased cases of pancreatic cancer and 112 controls identified from the cohort matched on age, sex, and race, who were living at the time of the case's death. Only men were included. Next-of-kin of cases and controls were interviewed by telephone for lifestyle factors, occupational history, and past chemical exposures. Exposure histories were constructed from company records and interviews with co-workers that sought to determine whether a subject would have been exposed to specific chemical given their work location and job title. DDT was associated with pancreatic cancer in an ever versus never comparison (OR, 4.8; 95% CI, 1.3–17.6). Among workers with exposure duration greater than the median [not reported; mean exposure duration, 47 months] the risk was 7.4-times [95% CI not reported] that of those with no exposure. The risk also increased with time since first exposure. In models accounting for potential confounding by occupational and lifestyle factors, the odds ratio was 6.1 [95% CI not reported] ($P = 0.02$). [This study was quite small and included only deceased cases. Exposure assessment by location and job title may allow some misclassification.]

The potential link between the use of pesticides, including DDT, and cancer of the pancreas was evaluated in a nested case–control analysis (Andreotti et al., 2009) based on the AHS (see Alavanja et al., 2014),. Incident cases diagnosed through 2004 were included in this analysis (93 cases: 64 male applicators, 29 female spouses). Information on exposure was obtained from questionnaires administered at enrolment. Ever use of DDT by either applicators or spouses was inversely associated with cancer of the pancreas (OR, 0.4 [95% CI, 0.2–0.9]) when adjusted for age, cigarette smoking, diabetes, and applicator type (farmer, commercial applicator, non-certified applicator [spouse]). These results were not appreciably changed when the associations was limited to only certified applicators. [Although the analysis was able to control for age, applicator type, smoking history, and diabetes, only six DDT-exposed applicators were available for this evaluation, which limits the strength of the conclusion.]

(d) Studies of multiple cancer sites

Beard et al. (2003) studied mortality among 1999 men in New South Wales, Australia, who were part of an insecticide-application programme at some point during the period 1935–1996. Mortality and incidence of cancer was compared with that in a cohort of 1984 male outdoor workers not occupationally exposed to pesticides, and with the Australian population. A small portion of surviving subjects also completed a lifestyle morbidity questionnaire. Exposure to specific pesticides was estimated from records of when each pesticide was used and individual employment dates: 394 of the pest-control workers were employed during the period of DDT use from 1955 to 1962. Results were reported for all cancers, prostate cancer, and leukaemia. Among workers employed ≥ 3 years during the period of DDT use, the standardized mortality ratio (SMR) for all cancers combined was 1.16 (95% CI, 0.92–1.46) relative to the national population. Mortality from cancer of the pancreas was more frequent in subjects exposed to DDT for < 3 years (SMR, 5.27; 95% CI, 1.09–15.40; 1 case); the association was weaker in those working for ≥ 3 years. Mortality from cancer of the pancreas was also compared in exposed workers and the unexposed cohort and gave largely similar results. A similar pattern was observed for leukaemia (SMR, 2.57; 95% CI, 0.06–14.29) among workers employed < 3 years. [The small number of exposed cases and the lack of information on exposure to individual pesticides limited the conclusions that could be drawn from this study.]

Cocco et al. (2005) investigated the association between exposure to DDT and cancer mortality in a cohort of 4552 men exposed to DDT as a part of antimalaria operation in Sardinia, Italy, during 1946–1950. Employment records and information on DDT use during the operation were used to develop individual estimates of average and cumulative exposure. Among applicators, an algorithm based on a European Predictive operator exposure model (EUROPOEM) database was used to estimate dermal and inhalation exposure to DDT. Apart from a short period when chlordane was used in unspecified areas, all the individuals were exposed only to DDT. Mortality of the cohort was analysed in comparison to the general Sardinian population and in internal comparisons to an unexposed subcohort. Relative to the general population, overall cancer mortality was decreased among DDT-exposed workers, mainly due to a low risk of cancer of the lung. The standardized mortality ratio for cancer of the stomach was 1.4 (95% CI, 0.7–2.7; 45 observed deaths) among exposed workers, while the standardized mortality ratios for all of the other cancer sites reported, including liver, pancreas, lympho-haematopoietic system, and leukaemia were unity or below. Similar patterns were observed among DDT applicators. In internal analyses, mortality from cancer of the stomach increased with increasing cumulative exposure to DDT (RR, 2.0; 95% CI, 0.9–4.4, in the highest quartile compared with unexposed workers), but the trend was not significant. Increased relative risks for cancer of the bladder, but no significant trend, were also observed in the higher categories of exposure to DDT (RR, 1.4; 95% CI, 0.7–3.4; for the fourth quartile compared with the unexposed). There was no indication of exposure-response relationships for other cancer sites. [Information on some important potential confounders for cancers of the lung and pancreas, such as tobacco use and alcohol consumption, were not individually available for study participants, and this may have resulted in a bias towards the null, since indirect evidence in the paper suggested that applicators smoked less than the comparison group.]

2.2 Case–control studies

The Working Group reviewed the available meta-analyses with the case–control studies for the cancers concerned, as most of the included studies had that design.

2.2.1 Cancer of the breast

(a) Exposure measured in blood

See Table 2.7

Twenty-two case–control studies of cancer of the breast and exposure to DDE or DDT were identified. Most studies adjusted for some or all of the standard reproductive and demographic risk factors for breast cancer, such as BMI, adult body-weight gain, family history, menopausal status, age at menarche, age at first birth, and lactation history. Some studies reported data for

Table 2.7 Case–control studies on cancer of the breast and exposure to DDT and its metabolites measured in blood or adipose tissue

Reference, location, enrolment/ follow-up period	Population size, description, exposure assessment method	Organ site	Exposure category or level	Exposed cases/ deaths	Risk estimate (95% CI)	Covariate	Comments
Zheng et al. (2000) Connecticut, USA 1995–1997	Cases: 475; hospital pathology department and cancer centre Controls: 502; hospital patients with BBD (*n* = 347, response rate 71%) and 155 population-based by random-digit dialling (response rate, 61%) Exposure assessment method: biomarker; serum DDE, lipid-corrected gravimetric lipids; blood drawn post-diagnosis	Breast	DDE, ng/g lipid			BMI, menarche, lactation, FFTP age, parity, HRT, dietary fat, family income, family history, race, study site	Strengths: relatively large sample size Limitations: mainly hospital cases & controls, control groups combined; blood drawn after diagnosis
			< 295.0	139	1.00		
			295.0–660.0	157	1.05 (0.76–1.47)		
			> 660.0	179	0.96 (0.67–1.36)		
			Trend-test *P* value: 0.58				

Table 2.7 (continued)

Reference, location, enrolment/ follow-up period	Population size, description, exposure assessment method	Organ site	Exposure category or level	Exposed cases/ deaths	Risk estimate (95% CI)	Covariate	Comments
Demers et al. (2000) Quebec, Canada 1994–1997	Cases: 314; hospital Controls: 523; 219 hospital; 305 population Exposure assessment method: personal monitoring	Breast	DDE, ng/g lipid, population controls			Age, region, BMI, lactation, FFTP, fertility years, family history, BBD history	Strengths: relatively large sample size, some population controls Limitations: possible confounding as higher DDE in cases could be to more aggressive cancer; cases had lower BMI than controls and were younger; blood collected after surgery for cases & hospital controls
			184.4– < 282.5	52	0.75 (0.45–1.25)		
			282.5– < 427.8	56	1.06 (0.62–1.79)		
			427.8– < 680.0	67	0.86 (0.52–1.42)		
			≥ 680.0	72	1.00 (0.60–1.67)		
			DDT, ng/g lipid, population controls				
			6.0 – < 7.9	52	0.57 (0.34–0.95)		
			7.9 – < 10.6	50	0.5 (0.30–0.84)		
			10.6 – < 15.0	63	0.71 (0.43–1.19)		
			≥ 15.0	70	0.81 (0.48–1.37)		
			DDE, ng/g lipid, hospital controls				
			184.4 – < 282.5	41	0.85 (0.45–1.59)		
			282.5 – < 427.8	57	0.66 (0.37–1.19)		
			427.8 – < 680.0	36	1.54 (0.81–2.95)		
			≥ 680.0	40	1.36 (0.71–2.63)		
			DDT, ng/g lipid, hospital controls				
			6.0 – < 7.9	44	0.85 (0.45–1.59)		
			7.9 – < 10.6	36	1.06 (0.57–1.98)		
			10.6 – < 15.0	44	1.07 (0.59–1.94)		
			≥ 15.0	45	1.37 (0.73–2.56)		

Table 2.7 (continued)

Reference, location, enrolment/ follow-up period	Population size, description, exposure assessment method	Organ site	Exposure category or level	Exposed cases/ deaths	Risk estimate (95% CI)	Covariate	Comments
Millikan et al. (2000) North Carolina, USA 1993–1996	Cases: 748 from atate cancer registry Controls: 659; driving license and Medicare lists Exposure assessment method: personal monitoring; plasma GC-ECD, lipid correction, *o,p*-DDT internal standardd	Breast	DDE, μg/g lipid			Age, race (white/ black), menopause, BMI, parity, lactation, HRT use, income	292 African-American cases; 456 white cases Strengths: population-based, relatively large sample size Limitations: limited precision in stratified analysis
			All < 0.394	274	1.00		
			All 0.394–< 1.044	231	1.05 (0.79–1.40)		
			≥ 1.044	243	1.09 (0.79–1.51)		
			African-Americans < 0.71	89	1.00		
			African-Americans 0.71–< 1.8	90	1.12 (0.70–1.77)		
			African-Americans ≥ 1.8	113	1.41 (0.87–2.29)		
			Whites < 0.30	176	1.00		
			Whites 0.30–< 0.66	146	0.97 (0.68–1.40)		
			Whites ≥ 0.66	134	0.98 (0.67–1.43)		
Gammon et al. (2002) Long Island, NY, USA 1996–1997	Cases: 646; hospital records Controls: 429; population (random-digit dialling and insurance records) Exposure assessment method: personal monitoring; serum, ECD, lipid-corrected	Breast	DDE, ng/g lipid			Age, race, fertility, BBD history	Strengths: population-based, large sample size Limitations: blood drawn after diagnosis
			< 306.91	122	1.00		
			306.91–515.00	110	0.88 (0.58–1.32)		
			515.01–798.24	127	0.94 (0.63–1.43)		
			798.25–1373.48	123	0.92 (0.60–1.42)		
			1378.49–11 818.78	150	1.20 (0.76–1.9)		
			DDT, ng/g lipid				
			< 44.79	129	1.00		
			44.79–61.43	96	0.69 (0.44–1.07)		
			61.44–81.20	123	1.04 (0.66–1.63)		
			81.21–108.03	134	1.16 (0.75–1.80)		
			108.03–747.29	133	1.15 (0.74–1.79)		
Gatto et al. (2007) Los Angeles, USA 1994–1998	Cases: 355 from cancer registry Controls: 327; random-digit dialling Exposure assessment method: personal monitoring; serum DDE	Breast	Serum DDE, μg/g lipid			Age, BMI, Lactation, strata	African-American women only, ages 35–64 yrs Strengths: population-based, large sample of African-American women Limitations: timing of blood sampling not specified
			≤ 0.44	61	1.00		
			> 0.44–0.73	62	0.98 (0.59–1.62)		
			> 0.73–1.15	76	1.07 (0.65–1.75)		
			> 1.15–1.91	81	1.14 (0.69–1.88)		
			> 1.91	75	1.02 (0.61–1.72)		
			Trend-test *P* value: 0.74				

Table 2.7 (continued)

Reference, location, enrolment/ follow-up period	Population size, description, exposure assessment method	Organ site	Exposure category or level	Exposed cases/ deaths	Risk estimate (95% CI)	Covariate	Comments
Charlier et al. (2004) Liege, Belgium 2001–2002	Cases: 231; hospital Controls: 290; hospital, healthy screening patients Exposure assessment method: personal monitoring; DDE serum GC-MS/EI	Breast:	DDE > 0.5 ppb	176	2.21 (1.41–3.48)	Parity, lactation, menopause, HRT, family history	Strengths: sample size; blood draw before surgery Limitations: no BMI information; hospital-based controls
Itoh et al. (2009) Nagano, Japan 2001–2005	Cases: 403; hospital presurgery Controls: 403; women having medical checkups, matched for age, residence Exposure assessment method: personal monitoring; blood draw before surgery	Breast	Quartile median (ng/g lipid)			BMI, menopause, smoked fish intake, vegetable intake, family history, menarche, breast cancer history, screening, lactation, history of chemotherapy, FFTP, Age	Strengths: large sample, GC-MS method, presurgery ascertainment Limitations: controls may have medical conditions
			p,p′-DDE:				
			160	116	1.00		
			300	89	0.47 (0.24–0.92)		
			490	107	0.99 (0.48–2.02)		
			1100	91	1.02 (0.46–2.26)		
			p,p′-DDT:				
			5.6	136	1.00		
			8.5	79	0.58 (0.27–1.25)		
			12.0	97	0.99 (0.47–2.07)		
			23.0	91	0.58 (0.27–1.25)		
			o,p′-DDT:				
			0.9	103			
			1.3	100	0.57 (0.25–1.29)		
			2.0	122	1.13 (0.53–2.38)		
			4.1	78	0.67 (0.30–1.50)		
López-Carrillo et al. (1997) Mexico City, Mexico 1994–1996	Cases: 141; hospital Controls: 141; hospital, non-gynaecology, non-oncology, matched on age and residence Exposure assessment method: biomarker	Breast	DDE, ng/g lipid			Age, BMI, lactation, menopause, parity, family history, FFTP age	Strengths: blood sampled before treatment; control for major risk factors Limitations: controls were hospital-based
			< 242.11	50	1.00		
			242.11–509.25	42	0.60 (0.31–1.16)		
			> 509.25	49	0.76 (0.41–1.42)		

Table 2.7 (continued)

Reference, location, enrolment/ follow-up period	Population size, description, exposure assessment method	Organ site	Exposure category or level	Exposed cases/ deaths	Risk estimate (95% CI)	Covariate	Comments
Olaya-Contreras et al. (1998) Bogota, Colombia 1995–1996	Cases: 153; hospital; incident primary cancer Controls: 153; non-cancer hospital controls, age-matched Exposure assessment method: biomarker; serum DDE ECD; blood drawn before chemotherapy	Breast	DDE, ng/mL			Family history, BMI, parity, menopause, breast cancer history, lactation	Strengths: menopausal stratification Limitations: not lipid-adjusted; no stage information
			0.10–0.14	39	1.00		
			0.15–1.96	45	1.20 (0.64–2.25)		
			1.97–19.20	69	1.95 (1.10–3.52)		
			Premenopausal:				
			0.10–0.14	15	1.00		
			0.15–2.06	20	1.40 (0.55–3.43)		
			2.07–19.20	25	2.46 (0.96–6.30)		
			Postmenopausal:				
			0.10–0.19	24	1.00		
			0.20–1.90	25	1.14 (0.50–2.75)		
			0.91–19.00	44	1.85 (0.84–4.05)		
Moysich et al. (1998) New York state, USA 1986–1991	Cases: 154; area hospitals, incident primary cancer Controls: 192; population (motor vehicle and health insurance rolls) Exposure assessment method: biomarker; serum, ECD	Breast	DDE, ng/g lipid			Age, education, family history, parity, BMI, lactation, FFTP age, years since pregnancy, fruit/vegetable intake, lipids	Postmenopausal women only Strengths: population-based Limitations: low participation rates; blood collected after surgery
			1st tertile	54	1.00		
			2nd tertile	46	1.01 (0.56–1.86)		
			3rd tertile	54	1.34 (0.71–2.55)		
Mendonça et al. (1999) Rio de Janeiro, Brazil 1995–1996	Cases: 177; hospital, admitted within 6 months of diagnosis Controls: 350; hospital visitors without breast cancer Exposure assessment method: biomarker; serum ECD	Breast	Serum DDE, ng/mL			Age, lactation, education, parity, smoking, family history, breast size	OR in abstract differs from tables Strengths: NR Limitations: no adjustment for BMI or alcohol; not lipid-adjusted
			< 1.3	29	1.00		
			1.3–2.4	32	0.95 (0.49–1.80)		
			2.5–3.9	35	1.34 (0.68–2.60)		
			4.0–7.6	37	1.12 (0.58–2.10)		
			≥ 7.6	29	0.83 (0.4–1.60)		
			Trend-test *P* value: 0.79				

Table 2.7 (continued)

Reference, location, enrolment/ follow-up period	Population size, description, exposure assessment method	Organ site	Exposure category or level	Exposed cases/ deaths	Risk estimate (95% CI)	Covariate	Comments
Dello Iacovo et al. (1999) Naples, Italy 1997–1998 (cases), 1993–1998 (controls)	Cases: 170; hospital, first breast cancer surgery Controls: 195; community controls from ongoing cohort study Exposure assessment method: biomarker; fasting blood draw	Breast	Serum DDE, ng/mL			Age, BMI, lactation, parity, serum lipids, education, smoking, menopause	Strengths: sizeable sample; > 30% having DDE > 10 ng/mL Limitations: cases and controls from different sources and not concurrent; criteria not stated for selection of controls; menopause not controlled
			< 6	51	1.00		
			6–10.2	49	0.84 (0.47–1.51)		
			> 10.2	70	1.24 (0.7–2.2)		
Romieu et al. (2000) Mexico City, Mexico 1990–1995	Cases: 120; public hospital network Controls: 126; age-stratified population sample Exposure assessment method: biomarker; lipid-corrected ECD	Breast	DDE, ng/g lipid			Age, menarche, lactation, BMI, menopause	Strengths: high participation rates Limitations: smaller sample size; age difference in cases and controls could affect DDE/DDT risk; selection criteria for subsample not given
			0.20–1.16	18	1.00		
			1.17–1.96	20	1.06 (0.44–2.55)		
			1.17–3.48	38	1.75 (0.76–4.09)		
			3.49–14.84	44	2.16 (0.85–5.50)		
			Trend-test *P* value: 0.06				
Wolff et al. (2000b) East Harlem, NY, USA 1994–1995	Cases: 175; hospital, presurgery Controls: 355; patients with benign breast disease (181) and patients seeking screening or minor procedures (175) Exposure assessment method: biomarker; serum DDE/DDT/ chlordane/PCB; ECD; most blood pre-surgery	Breast	DDE, μg/g lipid			Age, age-square, menopause, race, strata, lactation, HRT, parity	Control groups pooled for analysis Strengths: stage and receptor information, multiracial Limitations: hospital-based
			0–0.44	56	1.00		
			0.45–1.03	42	0.80 (0.49–1.30)		
			> 1.04–12.90	55	0.93 (0.56–1.50)		
			Trend-test *P* value: 0.499				
			DDT, μg/g lipid				
			0–0.0207	44	1.00		
			0.0208–0.033	50	1.19 (0.73–2.00)		
			0.034–1.3	56	1.34 (0.82–2.20)		
			Trend-test *P* value: 0.241				

Table 2.7 (continued)

Reference, location, enrolment/ follow-up period	Population size, description, exposure assessment method	Organ site	Exposure category or level	Exposed cases/ deaths	Risk estimate (95% CI)	Covariate	Comments
Schecter et al. (1997) Hanoi, Viet Nam 1994	Cases: 21; hospital, histologically-confirmed invasive cancer Controls: 21; hospital; BBD Exposure assessment method: biomarker; serum; blood taken at diagnosis	Breast	DDE, ng/mL			Menarche, parity, lactation, weight	DDE 16.7 ng/mL, DDT 2.4 ng/mL in controls Strengths: NR Limitations: BBD controls; cases likely at an advanced stage
			3rd vs 1st tertile	8	1.14 (0.23–5.68)		
			DDT, ng/mL				
			3rd vs 1st tertile	5	1.21 (0.15–9.65)		
Soliman et al. (2003) Egypt Period NR	Cases: 69; hospital, premenopausal Controls: 53; hospital visitors, age-matched Exposure assessment method: biomarker; serum; ECD; blood sampled pre-treatment	Breast	DDE (ppb)			Age, lactation, residence	Strengths: highly exposed younger women Limitations: small sample size; limited control for potential confounders
			> 4.7	69	1.41 (0.63–3.19)		
Pavuk et al. (2003) Slovakia 1997–1999	Cases: 24; hospital, all had treatment Controls: 88; participants in another study Exposure assessment method: biomarker; serum gravimetric lipids, ECD	Breast	DDE, ng/g lipid			Age, menarche, education, alcohol intake, smoking	Strengths: NR Limitations: blood drawn after treatment and up to 2 yrs since diagnosis
			233–2582	8	1.00		
			2583–4388	2	0.53 (0.08–3.27)		
			4389–19 912	14	3.04 (0.65–14.3)		
			Trend-test *P* value: 0.10				
			DDT, ng/g lipid				
			29–81	8	1		
			82–136	3	0.33 (0.06–1.7)		
			137–562	13	1.19 (0.27–5.23)		
			Trend-test *P* value: 0.68				

Table 2.7 (continued)

Reference, location, enrolment/ follow-up period	Population size, description, exposure assessment method	Organ site	Exposure category or level	Exposed cases/ deaths	Risk estimate (95% CI)	Covariate	Comments
Rubin et al. (2006) Alaska 1981–1987	Cases: 63; from Alaska Native Tumor registry with a prior banked serum sample Controls: 63; pair-matched cancer-free Alaska native women with a sample in the serum bank the same year as a case sample Exposure assessment method: biomarker	Breast	DDE (ppb)			Parity, family history of breast cancer, ethnicity, triglycerides, cholesterol levels	Geometric mean DDE, 7.36 ng/mL in controls; Alaska native population Strengths: blood sampled years before diagnosis; wide range of exposure Limitations: small sample size; limited control for breast-cancer risk factors
			< 6.17	15	1.00		
			6.17–9.61	18	0.57 (0.15–2.19)		
			> 9.62	30	1.43 (0.46–4.47)		
van't Veer et al. (1997) Germany, the Netherlands, Northern Ireland, Switzerland, and Spain 1991–1992	Cases: 265; EURAMIC study, postmenopausal women age 50–74 Controls: 341; hospital, matched by centre and age Exposure assessment method: biomarker; needle aspirates from gluteal adipose tissue	Breast	DDE, µg/g			Age, centre, BMI, age at first birth, alcohol consumption	Strengths: women with substantial weight loss in the past year excluded Limitations: low response rates among controls
			≤ 0.86	73	1.00		
			0.87- 1.89	75	1.14 (0.62–2.12)		
			1.89–3.46	63	0.71 (0.38–1.34)		
			> 3.46	54	0.48 (0.25–0.95)		
			Trend-test *P* value: 0.02				
Liljegren et al. (1998) Sweden 1993–1995	Cases: 43; hospital Controls: 35; hospital, BBD Exposure assessment method: biomarker; frozen adipose tissue from surgery	Breast	DDE, ng/g lipid			Age, parity	Patients operated by one surgeon for malignant or benign breast diseases Strengths: ER status was determined Limitations: very small sample size
			≤ 700	31	1.0		
			> 700	12	0.4 (0.1–1.2)		

Table 2.7 (continued)

Reference, location, enrolment/ follow-up period	Population size, description, exposure assessment method	Organ site	Exposure category or level	Exposed cases/ deaths	Risk estimate (95% CI)	Covariate	Comments
Zheng et al. (1999) Connecticut, USA 1994–1997	Cases: 304; patients with surgery for incident breast cancer and available tissue sample Controls: 186; surgical patients with incident BBD and available tissue sample Exposure assessment method: biomarker; GC; breast adipose tissue for cases and controls	Breast	DDE, ng/g lipid			Age, BMI, lifetime months of lactation, age at menarche, age at FFTP, menopausal status, race, income 10 years before the disease diagnosis or interview	Women aged 40–79 yrs Strengths: relatively large sample size Limitations: based in only one hospital
			< 412.6	65	1		
			412.6–779.2	85	1.3 (0.7–2.2)		
			779.3–1355.9	71	0.9 (0.5–1.6)		
			≥ 1356.0	83	0.9 (0.5–1.5)		
			Trend-test *P* value: 0.4				
Aronson et al. (2000) Ontario, Canada 1995–1997	Cases: 217; hospital: women scheduled for biopsy with subsequent diagnosis of cancer Controls: 213; as for cases but with nonmalignant diagnoses Exposure assessment method: biomarker; biopsy specimens obtained before diagnosis	Breast	DDE, µg/kg lipid			Age, study site, menopausal status, present use of HRT, ethnicity, BMI fat and alcohol intake	Strengths: control for multiple risk factors; tissue sampled before diagnosis Limitations: Based on only two hospitals; DDE analysis only for about 50% of eligible women
			≤ 368	55	1.00		
			369–727	59	0.96 (0.55–1.68)		
			728–1389	54	0.92 (0.51–1.67)		
			> 1390	49	1.62 (0.84–3.11)		

Table 2.7 (continued)

Reference, location, enrolment/ follow-up period	Population size, description, exposure assessment method	Organ site	Exposure category or level	Exposed cases/ deaths	Risk estimate (95% CI)	Covariate	Comments
Ibarluzea et al. (2004) Spain 1996–1998	Cases: 198; hospital: women undergoing surgery for newly diagnosed malignant breast cancer Controls: 260; hospital: women undergoing non-cancer-related surgery (65% gall bladder) Exposure assessment method: biomarker; adipose tissue from breast for cases and abdomen for controls	Breast	DDE, ng/g lipid			Age, reference hospital, number of children, age at FFTP, family history of breast cancer, and alcohol and tobacco consumption	Women aged 36–70 yrs Strengths: relatively good sample size, high response rates
			≤ 201.72	33	1.00		
			201.73–397.67	40	1.04 (0.59–1.84)		
			397.68–675.97	29	1.23 (0.69–2.17)		
			≥ 675.98	19	1.22 (0.68–2.21)		

BBD, benign breast disease; BMI, body-mass index; CI, confidence interval; DDE, dichlorodiphenyldichloroethylene; DDT, dichlorodiphenyltrichloroethane; ER, estrogen receptor; FFTP, first full-term pregnancy; GC-ECD, gas chromatography-electron capture detector; GC-MS/EI, gas chromatography-mass spectrometry- electron ionization; HRT, hormone replacement therapy; NR, not reported; OR, odds ratio; PCB, polychlorinated biphenyl; vs, versus; yr, year

DDT as well as for DDE, but the predominant findings were related to DDE. Most studies used blood serum; a few used plasma; one used whole blood. Unless otherwise indicated, DDE refers to p,p′-DDE and DDT to p,p′-DDT. These studies are described below in order of decreasing size. In addition, a pooled analysis of three case–control studies and two cohort studies also reviewed individually in this volume is described in this section (Laden et al., 2001b). Four studies reported incomplete study details (Charlier et al., 2003; Li et al., 2006a; Chang et al., 2008; Zhang et al., 2013) and were excluded from further review.

Seven studies with more than 200 cases and a similar number of controls were availabile to the Working Group; four were from the USA and the remaining three were from Canada, Belgium, and Japan (Demers et al., 2000; Millikan et al., 2000; Zheng et al., 2000; Gammon et al., 2002; Charlier et al., 2004; Gatto et al., 2007; Itoh et al.,2009). Three were population based, four hospital-based. Data collection occurred in the mid 1990s, except for Itoh et al. (2009), which began in 2001, and blood draw was generally soon after diagnosis and before treatment. The highest levels of DDE in blood serum were observed among black women in two studies in the USA, with an average of approximately 8 ng/mL among controls in both North Carolina (Millikan et al., 2000) and Los Angeles (Gatto et al., 2007).

Zheng et al. (2000) conducted a case–control study in Connecticut, USA. Cases (n = 475) were enrolled at a New Haven hospital and through a cancer centre that captured cases from a nearby county. Controls (n = 502) were frequency-matched on age, comprising 347 hospital patients with benign breast disease and 155 population-based controls sampled by random-digit dialling. Mean serum DDE concentration was 456 ng/g lipid in controls. Associations were null (OR, 0.96; 0.67–1.36; P for trend, 0.58) for the third versus first tertile of DDE concentration (P for trend, 0.58), adjusted for standard risk factors for breast cancer. Results were similar in analyses stratified by parity and breast feeding. [This was a large study, but blood was drawn after diagnosis and most controls were hospital-based and had benign breast disease, which shares risk factors with breast cancer. Data for DDT were not reported.]

Demers et al. (2000) undertook a hospital based case–control study in Québec, Canada. Cases (n = 314) were women with invasive cancer identified in 1994–1997 before treatment. Controls were surgical patients (n = 218) with no gynaecologic conditions and women randomly selected from insurance lists (305). Controls and cases were frequency matched on age and residence. Mean serum DDE was 463 ng/g lipid in controls; DDT was 12 ng/g lipid. Associations were null in analyses with population controls (e.g. OR, 1.00; 95% CI, 0.60–1.67; for DDE fifth versus first quintile); however, in analyses using hospital controls the odds ratio was non-significantly increased in the highest exposure categories (OR, 1.37; 95% CI, 0.71–2.63; for the fifth versus the first quintile). Similar patterns were observed for DDT. In case-case analyses, more aggressive cancer was more strongly associated with higher DDE concentration. [This was a large study, with some population-based controls. The Working Group was concerned that a significant proportion of aggressive cases could have higher organochlorine serum levels because of disease.]

Millikan et al. (2000) conducted a population-based case–control study of cancer of the breast among black and white women in North Carolina, USA. DDT exposures in this population in the southern USA were relatively high: mean plasma DDE concentrations among controls were 1690 ng/g lipid among black women, and 760 ng/g lipid among white women. Data for DDT were not reported. The risk of breast cancer increased with DDE concentration among black women (OR, 1.41; 95% CI, 0.87–2.29; for thirrd versus first tertile of DDE), but not among white women (OR, 0.98; 95% CI, 0.67–1.43), or all women combined (OR, 1.09; 95% CI, 0.79–1.51). The risk was higher but imprecise among black women with BMI < 25 kg/m^2 (OR, 3.84; 95% CI, 0.98–15.08; third versus first tertile of DDE). Analyses across other strata, history of having lived or worked on a farm, parity, and lactation gave imprecise results. [This large population-based study included a large number of black women, with higher exposures than in many other studies. Participation was much lower in controls than cases, and the analyses by BMI strata suggested unresolved confounding from pharmacokinetic factors or chance.]

Gammon et al. (2002) reported on a study of cancer of the breast and exposure to organochlorine compounds in Long Island, New York, USA. Incident cases of breast cancer in 1996 and 1997 were identified from hospital pathology records, and age-matched population controls were sampled by random-digit dialling or from health insurance lists for those aged > 65 years. Data on the concentration of DDE and DDT in serum were available for a subset of cases and controls (643 cases and 427 controls for DDE; 633 cases and 418 controls for DDT). No increase in the risk of cancer of the breast was associated with the concentration of either DDE or DDT. Findings were similar in analyses stratified by BMI and tumour hormone-receptor status. [This large population-based study included many environmental risk factors, and results were stratified by suspected risk modifiers. Participation was lower in controls than cases.]

Gatto et al. (2007) studied the association of cancer of the breast and serum DDE concentration among African-American women in a population-based study in Los Angeles, USA. Organochlorine concentrations were measured in serum of a subset of the larger study (355 cases, 327 controls). The lipid-adjusted concentration of serum DDE was higher than in most studies (1250 ng/g lipid among controls); however, there was no association between risk of breast cancer and DDE concentration (OR, 1.02; 95% CI, 0.61–1.74; for fifth versus first quintile; *P* for trend, 0.74. [The large sample was population-based for cases and controls. Participation was lower in controls than cases. DDT was not reported. It was not clear whether blood was sampled before treatment.]

Charlier et al. (2004) studied the association between cancer of the breast and serum DDE concentration among 231 cases recruited from a hospital surgery unit and 290 age-matched controls seeking cytology screening in Belgium, 2001–2002. Blood was collected pre-surgery, and DDE was measured in serum by mass spectrometry. Both DDE and DDT levels were low compared with other studies (DDE, 310 ng/g lipid; DDT, 20 ng/g lipid among controls; other DDT isomers were not detectable). The odds ratio was 2.21 (95% CI, 1.41–3.48) for DDE concentration above the limit of quantification (0.5 ppb) and 1.24 (95% CI, 1.15–1.34) per ppb DDE in serum. Risk data for DDT were not reported. [Reasonably large sample size and blood draw before surgery were strengths of this study. Limitations included use of hospital controls and absence of BMI information. The Working Group noted that the risk estimates per unit of exposure exceeded those in most other studies; however, this was not readily explained by any aspect of the study design or methods.]

Itoh et al. (2009) recruited 403 consecutive patients with cancer of the breast from four hospitals in Japan, in 2001–2005. Control patients (n = 403) having medical checkups were matched for age and residence. Blood samples, collected before surgery for cases, were analysed for nine OCPs and PCBs. DDE and DDT levels in serum were low (370 and 9.9 ng/g lipid, respectively, in controls). After adjustment for multiple risk factors, no association or significant trend was found with the concentration of DDE, DDT, or *o,p′*-DDT. The odds ratio of for the fourth versus first *p,p′*-DDE quartiles was 1.02 (95% CI, 0.46–2.26). Stratified analyses by ER and menopausal status were also null. [This large study used high-quality laboratory methods and obtained blood before surgery. Controls could have had medical conditions that affected DDE levels.]

[This group of large studies included two population-based studies with large numbers of African-American women who had higher DDT exposures than whites. Nevertheless, the range of exposures was modest; no median DDE levels exceeded 10 ng/mL. The Working Group noted that DDT blood levels in these studies were generally less than about 10% the concentration of DDE, consistent with past, not current, exposure. Several studies measured exposure to other pesticides or to PCBs in addition to DDT, but potential confounding from these exposures was generally not assessed. Blood sampling for cases occurred after diagnosis, but several studies reported that blood samples were obtained before treatment. Not all studies reported excluding metastatic or secondary cancer.]

An additional seven studies each with between 100 and 200 cases of cancer of the breast provided pertinent data (López-Carrillo et al., 1997; Moysich et al., 1998; Olaya-Contreras et al., 1998; Dello Iacovo et al., 1999; Mendonça et al., 1999; Romieu et al., 2000; Wolff et al., 2000b). The data were collected between 1986 and 1995; two studies were from the USA, four were from Latin America, and one was from Italy.

López-Carrillo et al. (1997) conducted a hospital-based case–control study in 1994–1996 in Mexico City, Mexico. Cases (n = 141) aged 20–79 years were recruited from participating hospitals and an equal number of age-matched cancer-free controls were selected from other hospital services. Mean serum concentrations of 505.5 ng/g lipid and 84.5 ng/g lipid were reported for DDE and DDT, respectively. DDE concentration was higher in cases and DDT was higher in controls, with neither difference being statistically significant. DDE level was not associated with the risk of cancer of the breast, including in models adjusted for multiple risk factors for breast cancer; the odds ratio for the third versus the first tertile of DDE was 0.76 (95% CI, 0.41–1.42). Analyses stratified by menopausal status gave similar results. No risk data were reported for DDT.

Olaya-Contreras et al. (1998) recruited women with incident breast cancer from a cancer-speciality hospital and controls from a hospital providing non-cancer care in Bogota, Colombia, during 1995–1996 (153 pairs). Blood samples for the cases were obtained before treatment; mean plasma DDE was 2.5 ng/mL among controls and was higher in cases than controls, regardless of menopausal status. The risk of breast cancer increased with plasma DDE concentration (OR, 1.95; 95% CI, 1.10–3.52; for the thirrd versus first tertile), but the trend was not statistically significant (P for trend, 0.09). DDD and DDT were measured, but results were not reported [DDE was not lipid-adjusted].

Moysich et al. (1998) studied the association between postmenopausal cancer of the breast with serum concentrations of several organochlorine compounds, including DDE, in a case–control study in the state of New York, USA. Women with incident, primary postmenopausal cancer of the breast were identified from hospitals, while the controls were postmenopausal women from the community sampled from motor-vehicle and health-insurance records. Data on DDE in serum and risk factors for cancer of the breast were available for 154 cases and 192 controls; blood samples were obtained after surgery for most cases. Among all women, the risk of breast cancer was increased in the highest category of DDE exposure (OR, 1.34; 95% CI, 0.71–2.55), but there was no significant exposure-response trend (P for trend, 0.25). Stronger associations, but no significant trend, were observed for women who had never lactated.

A case–control study in Rio de Janeiro, Brazil, enrolled women admitted to a national cancer hospital with a diagnosis of breast cancer within

6 months as cases, and female hospital visitors without breast cancer as controls (Mendonça et al., 1999). Cases and controls were interviewed in hospital and blood specimens were obtained (before surgery for most cases) and analysed for DDE and related compounds. DDE concentrations were available for 162 cases and 331 controls. No association between the risk of breast cancer and increasing serum DDE was observed (*P* for trend, 0.79). [The Working Group noted that the odds ratios reported in the abstract and tables did not match; the figures tabulated here are from Table 3 of the paper.]

Dello Iacovo et al. (1999) conducted a case–control study in Naples, Italy, among 170 women undergoing surgery for breast cancer in a cancer-speciality hospital and 190 controls from a cohort study on diet and cancer at the same hospital. Blood samples were analysed for several organochlorine compounds, including DDE and DDT. The odds ratios for breast cancer and serum DDE concentration were 0.84 (95% CI, 0.47–1.51) for 6.0–10.2 ng/mL, and 1.24 (95% CI, 0.70–2 20) for > 10.2 ng/mL, relative to a referent group with < 6.0 ng/mL. Data for DDT were not reported.

A hospital-based study in New York City, USA, included 175 women with incident breast cancer, a control group of 181 women having surgery or biopsies for benign breast disease, and a second control group of 175 women without either disease who were undergoing screening or minor procedures (Wolff et al., 2000b). Concentrations of organochlorines were determined in blood samples, most of which were collected before surgery. DDE concentration was not associated with the risk of breast cancer. However, risks were non-significantly increased for DDT (OR, 1.34; 95% CI, 0.82–2.2; in the highest exposure category. No significant trend was observed for either exposure indicator.

Romieu et al. (2000) measured serum DDE and DDT concentrations for a subsample of 120 cases of cancer of the breast and 126 controls from a larger study initiated in 1990 in Mexico City, Mexico. In the original study, cases were recruited from a network of hospitals affiliated with the government health system and controls were an age-stratified random sample of the general population. Levels of DDE and DDT in controls were 2510 ng/g lipid, and 230 ng/g lipid, respectively. The concentration of DDE was significantly higher in cases than in controls, while the concentration of DDT was non-significantly higher in controls. With adjustment for risk factors for breast cancer, but not DDT, the odds ratio for DDE was 2.16 (95% CI, 0.85–5.50) for the fourth compared with the first quartile (*P* for trend, 0.06). Similar odds ratios were observed in analyses stratified by menopausal status. Risk data were not reported for DDT.

Four smaller studies from several countries had comparatively high average serum or plasma DDE concentrations of up to 17 ng/mL (Schecter et al., 1997; Pavuk et al., 2003; Soliman et al., 2003; Rubin et al., 2006).

In a hospital-based study in Hanoi, Viet Nam, Schecter et al. (1997) measured serum DDE and DDT concentrations in 21 women with invasive cancer of the breast and 21 control women with fibrocystic breast disease. Mean DDE contentration in the controls was 16.7 ng/mL. Neither DDE nor DDT concentration was significantly associated with the occurrence of breast cancer. [While women in this study had relatively high exposures to DDE, the small sample size limited precision.]

Soliman et al. (2003) studied the association of serum organochlorine levels with cancer of the breast among 69 premenopausal women newly diagnosed with cancer of the breast from three centres in Egypt and 69 women hospital visitors selected as controls. Mean DDE concentration in controls was 17 ppb and was higher among rural than urban women. The odds ratio for DDE concentration above the median (4.7 ppb) was 1.41 (95% CI, 0.63–3.19). DDT concentrations were measured, but no risk data were reported.

[Exposures were relatively high in this population, but precision was limited.]

In a study in Slovakia, Pavuk et al. (2003) analysed the association between several organochlorines and cancer of the breast among 24 cases with diagnoses in 1997–1999 identified from a hospital oncology service and 88 controls participating in a cross-sectional study in the same district in 1998. All the cases had been treated before providing blood samples. For DDE, the odds ratio for the third versus first tertile of exposure was 3.04 (95% CI, 0.65–14.3; *P* for trend, 0.10). For DDT, the odds ratio was not notably increased. [The limitations of this study included blood collection up to 2 years after diagnosis and treatment, different sources for cases and controls, limited control for potential confounders, and limited precision.]

Rubin et al. (2006) measured DDE in banked serum samples collected from native women in Alaska, USA, in 1981–1987. Cases (n = 63) were women in the Alaska Native Tumor Registry with incident cancer of the breast cancer diagnosed up until 1995 and an earlier serum sample, Pair-matched controls were women with a sample in the same serum bank collected in the same year as a case's sample. In a multivariable model, the odds ratio for the third tertile of DDE exposure (> 9.62 ppb) was 1.43 (95% CI, 0.46–4.47). [This population had fairly high exposures to DDT and blood was sampled years before diagnosis; however, there was no adjustment for reproductive risk factors or BMI, and precision was limited.]

A pooled analysis of three of the case–control studies (Moysich et al., 1998; Wolff et al., 2000b; Zheng et al., 2000) and of the two cohort studies (Hunter et al., 1997; Helzlsouer et al., 1999) reviewed in this monograph included 1400 cases of cancer of the breast and 1642 controls from five different areas of the USA (Laden et al., 2001b). Median concentrations of DDE in serum or plasma ranged from 2.6 to 11.1 ng/mL across the five studies. In pooled case–control analyses adjusted for key breast cancer risk factors, there was no association between breast cancer risk and lipid-adjusted DDE concentration. [The pooled analysis included a wide range of DDE serum concentrations of > 10 ng/mL; the strengths of the pooled analysis, in addition to improved precision, included unified quality control during the course of the laboratory assays in four separate laboratories.]

(b) Exposure measured in adipose tissue

See Table 2.7

Seven case–control studies of the relationship between DDE or DDT concentration in adipose tissue and risk of cancer of the breast were available to the Working Group (van't Veer et al., 1997; Liljegren et al., 1998; Zheng et al., 1999; Aronson et al., 2000; Bagga et. al., 2000; Woolcott et al., 2001; Ibarluzea et al., 2004).

The European community multicenter study on antioxidants, myocardial infarction, and breast cancer (EURAMIC) study in Germany, the Netherlands, Northern Ireland, Switzerland, and Spain enrolled 265 postmenopausal women with cancer of the breast and 341 controls matched for age and centre in 1991–1992 (van't Veer et al., 1997). The mean DDE concentration in adipose tissue aspirates was 1.51 μg/g among controls and 1.35 μg/g among cases. After adjustment for BMI, age at first birth, and current alcohol drinking, odds ratios for cancer of the breast decreased with increasing DDE levels; women in the highest category of DDE exposure had an odds ratio of 0.48 (95% CI, 0.25–0.95). [This study had low response rates (22–50%) among controls, except for in Spain (91%).]

In a case–control study in Connecticut, USA (Zheng et al., 1999), 304 women with incident breast cancer and 186 control women with benign breast disease were enrolled beween 1994 to 1997. Cases and controls were aged 40–79 years and had had breast-related surgery at a single hospital and an available sample of breast adipose tissue. DDT and DDE concentrations were measured in stored adipose tissue specimens. Age-adjusted geometric mean tissue DDE and DDT concentrations were similar for cases and controls (respectively, 736.5 ppb and 784.1 ppb for DDE; 51.8 ppb and 55.6 ppb for DDE). Analyses adjusted for reproductive and demographic risk factors did not indicate an association between adipose tissue levels of DDE or DDT and risk of breast cancer. [This study had a relatively large sample size, although it was based in a single hospital. The population may have overlapped with that of a study by Zheng et al. (2000) on breast cancer and DDT measured in blood.]

Liljegren et al. (1998) investigated the association between cancer of the breast and exposure to several chlorinated compounds, including DDE, in adipose tissue in a case–control study in surgical patients in Sweden. The population included 43 patients treated for malignant breast lesions and 35 treated for benign breast lesions by a single surgeon in 1993–1995. DDE concentration was measured in adipose tissue samples obtained at surgery. The concentration of DDE was higher among controls than among cases (1026 versus 767 ng/g lipid, respectively). Odds ratios adjusted for age and parity were below unity, but not statistically significant for all women, as well as for postmenopausal women and cases with ER-positive tumours. [Conclusions were limited by incomplete control for established risk factors, and small sample size.]

The relationship between risk of cancer of the breast and the concentration of DDE and other organochlorines was evaluated by Aronson et al. (2000) in a hospital-based case–control study in Ontario, Canada. Women aged < 80 years in 1995–1997 were enrolled at the time of biopsy. Through subsequent review of pathology records, 217 women with breast cancer and 213 control women, most with benign breast lesions, were identified. Concentrations of DDE and DDT were measured in biopsy tissue (geometric mean DDE concentration: cases, 693 μg/kg; controls, 596 μg/kg). For DDE, odds ratios adjusted for established risk factors were about unity except in the highest exposure category (OR, 1.62; 95% CI, 0.84–3.11; for concentrations ≥ 1390 μg/kg). This association was largely confined to premenopausal women. DDT concentrations were not associated with risk of breast cancer. In a further study in this population (Woolcott et al., 2001), breast cancer subtypes were defined by tumour characteristics, including ER status, progesterone receptor (PR) status, tumour size, and grade. Although the odds ratios did not differ significantly by subtype, DDE levels were higher with risk of ER-negative (OR, 2.4; 95% CI, 1.0–5.4) breast cancer than ER-positive breast cancer (OR, 1.1; 95% CI, 0.6–1.9) [This study included adjustments for multiple risk factors for breast cancer, including fat intake. However, participants were enrolled exclusively from two hospitals and only 50% of eligible women had enough adipose tissue available for analysis.]

Bagga et al. (2000) conducted a case–control study on the association betweem cancer of the breast and OCPs in breast adipose tissue among participants in a health plan in California, USA. The cases were 73 women with breast cancer and an equal number of women undergoing breast-reduction surgery were enrolled as controls, respectively. Concentrations of DDT and its metabolites were measured in adipose tissue obtained at biopsy. The concentration of DDE, but not DDE, was significantly higher in cases than in controls (800 versus 709 ng/g; $P = 0.006$). The association between risk of breast cancer and DDE and DDT concentrations was modelled using quadratic terms in logistic regression models adjusting for age. The coefficients were positive for both exposure metrics, but were not statistically significant. [This study was reported in a brief communication and details on the methods used were limited. Analyses did not control for important risk factors, and the modelling methods and results were not readily comparable with other studies.]

In Spain, Ibarluzea et al. (2004) carried out a case–control study with 198 women with cancer of the breast and 260 age- and hospital-matched control women. Cases were undergoing surgery for breast cancer and controls were receiving surgery for other, non-cancer conditions, mostly of the gall bladder. Concentrations of DDT and its metabolites, as well as those of several other chemicals, were measured in samples of breast or abdominal adipose tissue from cases and controls, respectively. Geometric mean DDE concentrations were 326.86 ng/g lipid among cases and 307.34 ng/g lipid among controls. Adjusted odds ratios were greater than unity in the two higher exposure groups (OR, 1.22; 95% CI, 0.68–2.21; for DDE ≥ 675.98 ng/g), but there was no significant trend with increasing exposure. In stratified analyses, increased risk was limited to postmenopausal women. Risk data were not reported for DDT. [This study had a relatively good sample size, and high response rates. There were no controls for lactation.]

(c) Other exposure assessment methods

In the Long Island Breast Cancer Study Project (LIBCSP), White et al. (2013) assessed self-reported acute exposure to a fogger truck (used in the area to spray DDT before it was banned in 1972) as a proxy measure of exposure to DDT in 1508 cases and 1556 controls. Among all women, 33% reported ever seeing a fogger truck at their residence before 1972. For women reporting such exposure, the odds ratio for breast cancer was 1.16 (95% CI, 0.98–1.37). Odds ratios were near unity for reported exposure before age 14 or age 20 years, or after 1972. Compared with other breast cancer subtypes, women with ER+PR+ tumours had an increased odds ratio for ever seeing a pre-1972 fogger truck (OR, 1.44; 95% CI, 1.08–1.93). Self-reported exposure to a fogger truck was not correlated with serum DDT or DDE concentrations. [There were low response rates among controls. Differential recall of exposure may have affected the results.]

(d) Meta-analyses

Four meta-analyses evaluated the association between cancer of the breast and DDT and/or DDE (Adami et al., 1995; López-Cervantes et al., 2004; Ingber et al., 2013; Park et al., 2014).

Forty-six studies on the association between cancer of the breast and exposure to DDT or DDE from 500 published studies screened through June 2012 were included in a meta-analysis by Ingber et al. (2013). The meta-odds ratio for DDE was 1.04 (95% CI, 0.94–1.15) and DDT was 1.02 (95% CI, 0.92–1.13). There was no indication of publication bias (Begg's *P*-value, 0.09; Egger's *P*-value, 0.14). Heterogeneity was moderate for DDE (I^2, 31.72%) and high (I^2, 64.5%) for DDT. Lipid-adjusted differences in mean concentration were significantly higher for DDE concentrations in cases versus controls (difference, 110.30 ng/g lipids; $P = 0.01$), while no differences were found for non-lipid adjusted estimates or for DDT.

A meta-analysis by Park et al. (2014) included 35 case–control studies (8160 cases and 9280 controls) on DDE exposure and cancer of the breast with nested, population-based, and hospital-based designs published in English until August 2012. The summary odds ratio for DDE was 1.03 (95% CI, 0.95–1.12). There was no evidence of publication bias (funnel plots were symmetric; Egger's test *P* value, 0.145) and moderate heterogeneity was indicated (I^2, 40.9; $P = 0.006$) overall. Subgroup meta-analysis indicated no significant association between exposure to DDE and risk of breast cancer by type of design, study years, biological specimens, and geographical region of the study.

Both the Park et al. (2014) and the Ingber et al. (2013) meta-analyses were largely consistent with the 2004 meta-analysis by López-Cervantes et al. (2004) which examined the scientific literature through February 2001 and by Adami et al. (1995), which reviewed the literature until 1993. A general conclusion of these reviews was that currently available studies do not support the view that DDE increases the risk of breast cancer in humans (Fig. 2.1). [Many of the studies included in the available meta-analyses adjusted for common risk factors, such as age, BMI, family history of breast cancer, but few adjusted for breast feeding and diet, both of which have been related to DDT/DDE body burden. The Working Group concluded that adjustment for DDE in assessments of the risk of cancer associated with DDT is generally inappropriate because the practice produces variable effects on the risk estimate, depending on whether the exposure was from DDE in the diet or from DDT directly from application or manufacturing. Moreover, the age at exposure to chemicals such as DDE seems to be an important modifier in explaining the relationship between exposure and the risk of disease, for example, Cohn et al. (2007) reported that DDT was primarily associated with breast cancer in women potentially exposed before age 14 years. Unexamined variations in DDT metabolizing enzymes may also be an important determinant of increased risk of breast cancer.]

2.2.2 Lympho-haematopoietic cancers

See Table 2.8

Case–control studies of exposure to DDT and risk of lymphoma and leukaemia are grouped here by method of exposure assessment according to whether exposure was estimated from measurements in biological samples or by questionnaire or environmental monitoring. It should be noted that interpretation of the published literature was complicated by the change over time in the classification and coding systems for NHL and its subtypes.

(a) Studies based on biological samples

The hypothesis of immune system disturbances in modulating DDT-related risk of NHL, previously examined in relation to concurrent allergic conditions, was explored in three small Swedish case–control studies that measured the Epstein-Barr virus (EBV) early antigen (EBV EA) titre along with the lipid-adjusted plasma concentration of PCBs and organochlorines, including *p,p'*-DDE. The risk of hairy cell leukaemia was 6.6-fold (95% CI, 1.3–41.6) among subjects with above median plasma levels of *p,p'*-DDE and with an EBV EA titre above 40, while no association was observed in the overall study population (OR, 0.6; 95% CI, 0.2–1.5) (Nordström et al., 2000).

Similar results with respect to EBV EA were observed for total NHL in two further small studies by the same group (Hardell et al., 2001, 2009). For B-cell lymphoma subtypes, increased risk was associated with above median *p,p'*-DDE levels, independent of the EBV EA titre, for diffuse large B-cell lymphoma (OR, 2.8; 95% CI, 1.1–6.7), but not follicular lymphoma (Hardell et al., 2009). [While these studies highlighted the possible interaction between DDT body burden and immune factors, their small size and the heterogeneity of the exposure indicator in one study, limited the interpretation of the findings.]

A population survey on pesticide exposure was conducted in the USA by the EPA in 1970–87, during which organochlorine measurements were made in adipose tissue samples from surgical procedures or autopsy in a stratified random sample of the United States population. From the resulting database, 175 people with NHL and pesticide measurements were identified and matched by age, sex, hospital, and race to 481 control individuals in the database with diagnoses of myocardial infarction or accidental injury, but no cancer diagnosis (Quintana et al., 2004). DDT and DDE adipose tissue levels above the fourth quartile were associated with an increase in risk of NHL, which was non-significant for DDT (OR, 1.39; 95% CI, 0.78–2.47) and significant for DDE (OR, 1.99; 95% CI, 1.14–3.47). Although the trends were significant in both instances, risks below the upper quartile were near or below unity. [Organochlorines were measured post mortem or after diagnosis, which limited the interpretation. No further information was available on conditions possibly associated with DDT and DDE body burden, apart from age, sex, geographical region of the patient's hospital, and race.]

De Roos et al. (2005) conducted a case–control study in four areas of the USA, including Iowa, Los Angeles County, and the cities of Chicago and Seattle, in 1998–2000, with 100 cases identified through the Surveillance, Epidemiology, and End results Program (SEER) cancer registries, and 100 population controls. Study subjects donated a blood sample before treatment was initiated and plasma samples were tested for 40 PCB congeners, and 13 OCPs. After adjusting by sex, study site, date of birth, and date of blood draw, risk of NHL did not increase with increasing quartile of plasma concentration of DDT (ORs, 1, 1.12, 1.02, 1.2, respectively; $P = 0.75$) or DDE (ORs, 1, 0.64, 0.33, 0.85, respectively; $P = 0.74$). No analysis was conducted for the individual lymphoma subtypes.

A similar design was applied in a population-based study of 422 NHL cases and 460 controls in British Columbia, Canada (Spinelli et al., 2007). After adjusting by age, sex, study area, education, family history of lympho-haematopoietic cancer, and ethnicity, risk of NHL showed a significant upward trend with *p,p′*-DDE blood concentration (*P* for trend, 0.027), although none of the individual risk estimates was statistically significant. Risk was mostly elevated for follicular lymphoma (OR, 1.8; 95% CI, 0.9–3.3), and a group of "other B-cell lymphomas" excluding follicular or diffuse large B-cell lymphoma (OR, 1.8; 95% CI, 1.0–3.2). No association was observed with DDT blood concentration (OR, 0.91; 95% CI, 0.68–1.20; for > 3.24 ng/g lipid). [The Working Group noted that this was the largest study of DDT and risk of NHL based on biomarkers (blood levels of DDE); subjects with > 10% body-weight loss before diagnosis were excluded from the study. However, the timing of blood draw, whether before or after treatment, might be reasons for caution in interpreting the results.]

A multicentre European study of 174 NHL cases and 203 controls (Cocco et al., 2008) assessed exposure to 17 OCPs, including 6 DDT isomers, and 9 PCBs, in blood plasma. Analyses did not show an association between increasing blood *p,p′*-DDE concentration and risk of all NHL (*P* for trend, 0.48), or the subtypes, diffuse large B-cell lymphoma (*P* for trend, 0.48) and CLL (*P* for trend, 0.92), after adjusting by age, sex, education, and study centre. Analyses limited to subjects whose blood samples were taken before treatment did not modify the risk estimates.

A small biomarker study was conducted in a French area polluted by emissions from an incinerator (Viel et al., 2011). Lipid-adjusted DDT and DDE serum concentrations were measured in 34 NHL cases and 1-to-1 matched blood donors: the results showed a 3% increase in risk of NHL for each 10 ng/g lipid increase in *p,p′*-DDE (95% CI, 0.99–1.08) and a 20% (95% CI, 1.01–1.45) increase for a 10 ng/g increase in *p,p′*-DDT level. [The Working Group noted that while the finding of detectable DDT blood levels in the years when this study was conducted was unexpected, the small study size and the lack of adjustment for the highly chlorinated organochlorines because of incinerator location in this contaminated area were reasons for caution in interpreting the results. The choice of blood donors as controls was a potential source of selection bias should those individuals might not be representative of the source population of the cases.]

(b) Questionnaire-based studies

Early case–control studies of human exposure to DDT took place in the late 1980s. Woods et al. (1987) conducted a case–control study of 576 NHL cases and 694 controls in Washington state, USA. Exposure data were obtained by interview with study participants or proxies, using a detailed questionnaire including a section on pesticides. Exposure assessment was supported by local

Fig. 2.1 Effect of *p,p′*-DDE on breast cancer risk from each study in a meta-analysis, according to *p,p′*-DDE (ng/g) body burden levels

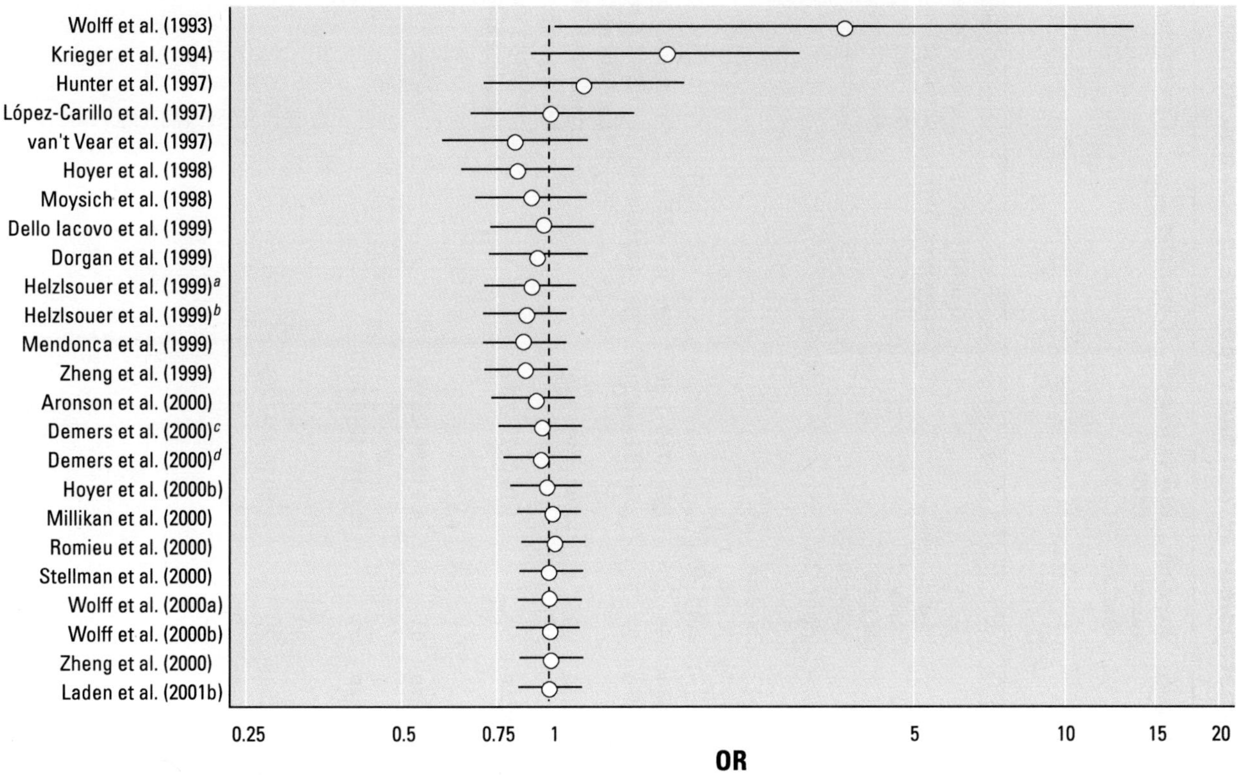

[a] Biological samples taken in 1974.
[b] Biological samples taken in 1989
[c] Controls are population-based
[d] Controls are clinically based
From López-Cervantes et al. (2004)

experts in forestry, wood products, and agricultural industries. After adjusting for age, ever exposure to DDT was associated with increased risk of NHL (OR, 1.82, 95% CI, 1.04–3.20). [No trend was evaluated in this study.]

Occupational exposure to pesticides was also considered in a population-based case–control study of 121 Hodgkin lymphoma (HL) cases and 948 controls in Kansas, USA (Hoar Zahm et al., 1988). Information on use of insecticides, including DDT, was obtained by questionnaire. No association with the risk of HL was found for ever use of insecticides, but no data were reported for DDT, specifically.

Persson et al. (1989) interviewed 106 cases of NHL and 54 cases of HL identified through the register of patients diagnosed at the Department of Oncology at Orebro Medical Centre Hospital, Sweden, in 1964–1986, and still alive when the study was conducted. Cases of T-cell lymphoma and malignant histiocytosis were excluded. Controls were a sample of 275 subjects who had participated in earlier studies. Cases and controls self-reported about a list of occupational exposures, which included DDT, which had lasted a minimum of 1 year and occurred at least 5 years before interview. No case of NHL nor the corresponding controls reported exposure, while three cases of HL and three of their controls

Table 2.8 Case–control studies on cancers of the lympho-haematopoietic system and exposure to DDT and its metabolites

Reference, location follow-up/ enrolment period, study-design	Population size, description, exposure assessment method	Organ site (ICD-O)	Exposure category or level	Exposed cases/ deaths	Risk estimate (95% CI)	Covariates	Comments
Nordström et al. (2000) Sweden 1987–92	Cases: 54; national cancer registry Controls: 54; national population registry, matched for age, sex, & county Exposure assessment method: biomarker; lipid-adjusted concentrations in plasma	NHL (hairy cell leukaemia)	*p,p′*-DDE > median	19	0.6 (0.2–1.5)	Age, other occupational exposures, BMI	Strengths: based on national population; diagnoses validated by the cancer registry Limitations: small sample size
				p,p′-DDE above median & EBV EA < 40	9		
				p,p′-DDE above median & EBV EA ≥ 40	10		
Hardell et al. (2001) Sweden 1994–97	Cases: 82; hospital Controls: 83; hospital and population register Exposure assessment method: biomarker; OC concentrations in abdominal fat or plasma	NHL	*p,p′*-DDE > median	44	1.20 (0.60–2.50)	Age, sex, BMI, type of specimen	Strengths: assessment by EBV EA status Limitations: small study size; heterogeneous control group and tissue sampling material
			p,p-DDE > median & EBV EA ≤ 80	17	2.00 (0.64–6.50)		
			p,p-DDE > median & EBV EA > 80	18	2.90 (0.93–9.70)		
		NHL (B-cell lymphoma), low grade	*p,p*-DDE > median EBV EA ≤ 80	5	1.20 (0.23–7.80)		
			p,p-DDE > median & EBV EA > 80	11	4.4 (0.96–26.0)		
		NHL (B-cell lymphoma): high grade	*p,p*-DDE > median & EBV EA ≤ 80	9	1.60 (0.41–6.70)		
			p,p′-DDE > median & EBV EA > 80	7	1.80 (0.42–7.70)		

Table 2.8 (continued)

Reference, location follow-up/enrolment period, study-design	Population size, description, exposure assessment method	Organ site (ICD-O)	Exposure category or level	Exposed cases/deaths	Risk estimate (95% CI)	Covariates	Comments
Hardell et al. (2009) Sweden 2000–2002	Cases: 99; hospitals in 4 health service regions Controls: 99; population registries in the same regions, matched by age & sex Exposure assessment method: biomarker; lipid-adjusted concentration in plasma	NHL	*p,p′*-DDE > median	53	1.5 (0.8–2.9)	Age, sex, BMI	Strengths: population-based; pathology review Limitations: small study size
			p,p′-DDE > median & EBVEA ≤ 40	14	1.0 (0.4–2.7)		
			p,p′-DDE > median & EBVEA > 40	39	3.3 (1.4–7.7)		
		NHL (follicular)	*p,p′*-DDE > median	10	1.2 (0.4–3.5)		
		NHL (DLBCL)	*p,p′*-DDE > median	24	2.8 (1.1–6.7)		
Quintana et al. (2004) United States 1970–87	Cases: 175; NHL cases from national database of tissue samples with pesticide measurements Controls: 481; ondividuals in the database with diagnosis of accidental injuries or myocardal infarction Exposure assessment method: biomarker; OC concentration in surgical or autopsy fat tissue samples	NHL	DDT (ppm)			Age, sex, race, centre	Strengths: large database; random selection of subjects independent of diagnosis Limitations: exposure measurements post mortem or post diagnosis; limited information on potential confounders
			< 0.55	58	1.00		
			0.55–0.92	34	0.80 (0.47–1.35)		
			0.93–1.56	38	0.97 (0.56–1.70)		
			> 1.56	45	1.39 (0.78–2.47)		
			Trend-test *P* value: 0.04				
		NHL	DDE (ppm)				
			< 2.40	48	1		
			2.40–4.38	24	0.53 (0.29–0.96)		
			4.39–7.21	38	1.12 (0.64–1.98)		
			> 7.21	65	1.99 (1.14–3.47)		
			Trend-test *P* value: 0.002				

Table 2.8 (continued)

Reference, location follow-up/enrolment period, study-design	Population size, description, exposure assessment method	Organ site (ICD-O)	Exposure category or level	Exposed cases/deaths	Risk estimate (95% CI)	Covariates	Comments
De Roos et al. (2005) Four states, USA 1998–2000	Cases: 100; SEER registries Controls: 100; Medicare records, random-digit dialling, matched by age, sex, race Exposure assessment method: biomarker; 36 PCB congeners & 14 OC pesticides or metabolites measured in plasma	NHL	*p,p′*-DDT, ng/g lipids			Sex, study site, birth date, date of blood draw	OCs detected in 30% or more study subjects Strengths: good statistical power; availability of pathology data; information on large number of possible confounders. Limitations: multiple comparisons; biomonitoring at the time of diagnosis
			≤ 3.7	18	1.00		
			> 3.7–5.9	23	1.12 (0.33–3.84)		
			> 5.9–9.9	32	1.02 (0.35–2.99)		
			> 9.9	27	1.20 (0.39–3.70)		
			Trend-test *P* value: 0.75				
		NHL	*p,p′*-DDE, ng/g lipids				
			≤ 254.5	35	1.00		
			> 254.5–450.5	25	0.64 (0.28–1.43)		
			> 450.5–872.5	11	0.33 (0.14–0.80)		
			> 872.5	29	0.85 (0.37–1.94)		
			Trend-test *P* value: 0.74				
Spinelli et al. (2007) British Columbia, Canada 2000–2004	Cases: 422; cancer registry Controls: 460; population (client registry of Ministry of Health) Exposure assessment method: lipid-adjusted plasma of 14 PCBs and 11 OC pesticides	NHL	DDE, ng/g lipid			Age, ethnicity, BMI	Time of blood collection not reported Strengths: large study size; exclusion of subjects with weight loss before diagnosis Limitations: low response among controls
			> 134.41–263.91	84	0.84 (0.56–1.25)		
			> 263.91–512.02	100	1.04 (0.70–1.56)		
			> 512.02–18 898	121	1.42 (0.92–2.19)		
			Highest vs lowest quartile	NR	1.40 (0.90–2.20)		
			Trend-test *P* value: 0.027				
		NHL (follicular)	DDE, highest vs lowest quartile	NR	1.8 (0.9–3.3)	Age, sex, study area, education, family history lymphopoietic cancer, ethnicity, BMI, farming jobs	
			Trend-test *P* value: 0.027				
		NHL (DLBCL)	DDE, highest vs lowest quartile	NR	0.6 (0.2–1.5)		
			Trend-test *P* value: 0.027				

Table 2.8 (continued)

Reference, location follow-up/ enrolment period, study-design	Population size, description, exposure assessment method	Organ site (ICD-O)	Exposure category or level	Exposed cases/ deaths	Risk estimate (95% CI)	Covariates	Comments
Spinelli et al. (2007) British Columbia, Canada 2000–2004 (cont.)		NHL, other histological type	DDE, highest vs lowest quartile	NR	1.8 (1.0–3.2)		
			Trend-test *P* value: 0.027				
		NHL					
			DDT, above vs below LOD	133	0.91 (0.68–1.20)		
			Trend-test *P* value: 0.491				
		NHL (DLBCL)					
			DDT above vs below DL	NR	1.0 (0.6–1.7)		
			Trend-test *P* value: 0.027				
		NHL (T-cell)	DDT				
			Above vs below LoD	NR	1.1 (0.6–2.1)		
			Trend-test *P* value: 0.027				
		NHL	DDT				
			Above vs below LOD	NR	1.0 (0.7–1.5)		
			Trend-test *P* value: 0.027				
		NHL (follicular)	DDE				
			Highest vs lowest quartile	NR	1.8 (0.9–3.3)		
			Trend-test *P* value: 0.027				

Table 2.8 (continued)

Reference, location follow-up/enrolment period, study-design	Population size, description, exposure assessment method	Organ site (ICD-O)	Exposure category or level	Exposed cases/deaths	Risk estimate (95% CI)	Covariates	Comments
Cocco et al. (2008) Spain, France, Germany 1998–2004	Cases: 174; resident in the referral area of the participating centres Controls: 203; population controls in Germany; hospital controls in Spain and France Exposure assessment method: biomarker; no difference in median OC levels by pre/post diagnostic sampling	NHL	ng/g lipid *p,p′*-DDE			Age, sex, education, study site	Epilymph study Strengths: Pathology review. Limitations: Limited sample size; low prevalence of exposure
			395.0–791.02	36	0.8 (0.4–1.5)		
			791.03–1431.07	43	0.9 (0.5–1.7)		
			> 1431.08	56	1.2 (0.7–2.4)		
			Trend-test *P* value: 0.48				
		NHL (DLBCL)	*p,p′*-DDE, ppb				
			395.0–791.02	8	0.7 (0.3–2.0)		
			791.03–1431.07	10	0.9 (0.4–2.6)		
			> 1431.08	14	1.3 (0.5–3.6)		
			Trend-test *P* value: 0.48				
		NHL (SLL/CLL)	*p,p′*-DDE, ppb				
			395.0–791.02	7	0.4 (0.1–1.1)		
			791.03–1431.07	13	0.6 (0.2–1.5)		
			> 1431.08	20	1.0 (0.4–2.5)		
			Trend-test *P* value: 0.92				
Viel et al. (2011) Besancon area, France 2003–05	Cases: 34; hospital Controls: 34; register of blood donors, matched by age, sex and date of blood draw Exposure assessment method: questionnaire; lipid-adjusted serum measurements of OC	NHL	10 ng/g increment			NR	Strengths: incident cases; well-defined area, questionnaire information was available; pathology was available Limitations: small study size; controls may not be representative of the base population; adjustment by PCB not considered
			DDE	NR	1.03 (0.99–1.08)		
			DDT	NR	1.20 (1.01–1.45)		

Table 2.8 (continued)

Reference, location follow-up/ enrolment period, study-design	Population size, description, exposure assessment method	Organ site (ICD-O)	Exposure category or level	Exposed cases/ deaths	Risk estimate (95% CI)	Covariates	Comments
Woods et al. (1987) Washington State, USA 1981–1984	Cases: 576; population-based cancer registry Controls: 694; random-digit dialling and social security records Exposure assessment method: questionnaire	NHL	Ever DDT	NR	1.82 (1.04–3.20)	Age	Strengths: large study size Limitations: multiple co-exposures; no trends evaluated; non consideration of confounders other than age
Persson et al. (1989) Sweden 1964–86	Cases: 160 (106 NHL, 54 HL); hospital Controls: 275; population Exposure assessment method: questionnaire; minimum duration of exposure, 1 yr; minimum latency, 5 yrs	NHL	Ever DDT	0	0	Age, sex, farming, fresh wood	Malignant histiocytosis and T-cell lymphoma were excluded Strengths: pathology was available Limitations: small size; only surviving cases included; reliance on self-reported exposure information
		HL	Ever DDT	3	7.5 (0.8–70.0)		
Flodin et al. (1988) Sweden 1975–84	Cases: 111; 5 hospitals Controls: 431; population Exposure assessment method: postal questionnaire; minimum exposure duration, 1 yr; minimum latency, 5 yrs	NHL (CLL)	Ever contact with DDT	6	6 (1.5–23)	Age, sex, other occupational exposures	Strengths: clinically and cytologically confirmed diagnoses Limitations: small size; only living cases included; reliance on self-report

Table 2.8 (continued)

Reference, location follow-up/ enrolment period, study-design	Population size, description, exposure assessment method	Organ site (ICD-O)	Exposure category or level	Exposed cases/ deaths	Risk estimate (95% CI)	Covariates	Comments
Eriksson & Karlsson (1992) Sweden 1982–86	Cases: 256; national cancer registry Controls: 256; national population registry, or national death registry of the causes of death, matched by age, sex, vital status, and county Exposure assessment method: postal questionnaire with telephone interview for subjects reporting farm work	MM	Days exposed to DDT			Co-exposures	Prevalence of exposure to DDT was unusually high Strengths: cases validated from cancer registry Limitations: overlapping exposures not considered; self-administered questionnaires
			Ever exposed	53	1.75 [1.07–2.86]		
			Ever exposed, farming and forestry occupations only	NR	1.86 [0.92–3.75]		
			≤ 5 days	NR	1.08 [0.35–3.37]		
			6–20 days	NR	1.54 [0.77–3.10]		
			≥ 21 days	NR	1.61 [0.53–4.93]		
Brown et al. (1990) USA (Iowa and Minnesota) 1980–1983	Cases: 578; tumour registry (Iowa) and hospital records (Minnesota) Controls: 1245; population, matched by age, vital status and state Exposure assessment method: detailed questionnaires with supplemental interview on pesticide exposure	Leukaemia	DDT days/yr			Vital status, age, state, tobacco use, family history lymphopoietic cancer, high-risk occupations, high-risk exposures other than farming	Strengths: large size – pathology review. Limitations: overlap with exposure to other pesticides; self-report
			Ever use of DDT in crops	35	1.2 (0.7–1.8)		
			1–4 days/yr	7	0.7 (0.3–1.8)		
			5–9	8	2.4 (0.9–6.4)		
			10+	5	1.0 (0.3–2.8)		
			Ever use of DDT in animal breeding	80	1.3 (1.0–1.8)		
			1–4 days/yr	7	0.6 (0.3–1.4)		
			5–9	7	1.1 (0.4–2.7)		
			10+	21	2.1 (1.1–3.9)		

Table 2.8 (continued)

Reference, location follow-up/ enrolment period, study-design	Population size, description, exposure assessment method	Organ site (ICD-O)	Exposure category or level	Exposed cases/ deaths	Risk estimate (95% CI)	Covariates	Comments
Cantor et al. (1992) USA (Iowa & Minnesota) 1980–83	Cases: 622; Iowa state health registry, Minnesota hospital and pathology records Controls: 1245; population Exposure assessment method: questionnaire; self report based on pesticide list	NHL	Ever use in animals	79	1.2 (0.9–1.7)	Vital status, age, state, cigarette smoking, family history of lympho-haematopoietic cancer, high-risk occupations, high-risk exposures	Strengths: good statistical power; availability of pathological information Limitations: multiple exposures; reliance on self-report
			Ever use in animals before 1965	68	1.3 (0.9–1.8)		
			Ever use in crops	57	1.7 (1.2–2.6)		
			Ever use in crops before 1965	45	1.8 (1.1–2.7)		
Baris et al. (1998) USA (Nebraska, Iowa, Minnesota, Kansas) 1983–1986	Cases: 993; cancer registry (Iowa & Kansas), hospitals and special surveillance (Nebraska & Minnesota) Controls: 2918; population, matched by state, race, sex, age and vital status Exposure assessment method: questionnaire; telephone or in-person interview	NHL	Use of DDT			Age, state of residence, respondent type (proxy/direct)	Pooled analysis of four case–control studies; men only; some states excluded from exposure–response analyses Strengths: large size; pathology review; exposure–response analysis Limitations: self-report; high proportion of proxy respondents, particularly among controls
			Ever	161	1.2 (1.0–1.6)		
			1–4 yrs	36	1.1 (0.7–1.6)		
			5–9 yrs	31	1.4 (0.8–2.2)		
			10–14 yrs	29	1.1 (0.6–1.9)		
			≥ 15 yrs	39	1.5 (0.9–2.3)		
			≤ 5 days/yr	12	1.0 (0.5–2.1)		
			> 5 days/yr	11	2.1 (0.9–4.9)		
			Ever, adjusted for 2,4-D & OPs	161	0.9 (0.4–1.8)		
			1–4 yrs, adjusted for 2,4-D & OPs	36	0.9 (0.4–2.0)		
			5–9 yrs adjusted for 2,4-D & OPs	31	1.0 (0.4–2.5)		
			10–14 yrs adjusted for 2,4-D & OPs	29	0.9 (0.4–2.3)		
			≥ 15 yrs adjusted for 2,4-D & OPs	39	1.2 (0.5–2.8)		

Table 2.8 (continued)

Reference, location follow-up/ enrolment period, study-design	Population size, description, exposure assessment method	Organ site (ICD-O)	Exposure category or level	Exposed cases/ deaths	Risk estimate (95% CI)	Covariates	Comments
De Roos et al. (2003) Midwestern USA 1979–86	Cases: 870; records of health registry, hospitals and pathology laboratories Controls: 2569; population controls from random-digit dialling and Medicare records Exposure assessment method: questionnaire; self report based on a list	NHL (200, 202)	Ever exposure to DDT	98	1.0 (0.7–1.3)	Age, study site, use of any pesticide in a list	Strengths: study size control for confounders assessment of multiple exposures Limitations: multiple comparisons, reliance on self-report, weakness in the exposure assessment
			Ever exposed to DDT only	68	0.9 (0.6–1.3)		
			Ever exposed to DDT and chlordane	30	1.7 (0.7–3.2)		
Assennato et al. (1995) Apulia, Italy 1987–89	Cases: 26; family physicians and pathology registers Controls: 74; other cancers, excluding sites associated with farm work Exposure assessment method: JEM; support from an agronomist	Lymphatic and haematopoetic	Ever exposed to DDT	7	4.18 (1.04–16.76)	Age, sex, respondent type (proxy/direct), smoking	Multiple exposure to different pesticides Strengths: detailed exposure assessment Limitations: extremely small study size
Nanni et al. (1996) Forli, Italy 1987–90	Cases: 61; 11 incident cases of NHL and CLL in the study area Controls: 217; general population matched to cases by sex and age group Exposure assessment method: JEM; questionnaire and crop-exposure matrices	NHL (CLL)	DDT			Sex, age, family history lymphopoietic cancer, farming, herpes zoster, altitude	Strengths: pathological classification; more detailed exposure assessment Limitations: small size; number and type of controls is unclear; outdated classification of lymphoma; no results presented for exposure in crop farming without animal breeding
			Ever used (recall)	27	1.74 (0.93–3.27)		
			Ever used (JEM)	28	1.70 (0.91–3.17)		
			1 kg cumulative dose	5	1.22 (0.95–1.57)		

Table 2.8 (continued)

Reference, location follow-up/enrolment period, study-design	Population size, description, exposure assessment method	Organ site (ICD-O)	Exposure category or level	Exposed cases/deaths	Risk estimate (95% CI)	Covariates	Comments
McDuffie et al. (2001) Canada (six provinces) 1991–1994	Cases: 517; Cancer registries and hospitals Controls: 1506; health insurance and voting records, matched by age & province Exposure assessment method: postal questionnaire with telephone follow-up on pesticides	NHL	Use of DDT			Age, province of residence, family history lymphopoietic cancer, medical history	Strengths: large size; pathology review. Limitations: self-reported exposure; low response rate; multiple comparisons
			10+ hours/yr	32	1.73 (1.08–2.76)		
			1–2 days/yr	18	1.75 (0.96–3.21)		
			3+ days/yr	14	1.50 (0.77–2.91)		
Pahwa et al. (2012) Six provinces, Canada 1991–94	Cases: 513; cancer registries & hospitals Controls: 1506; population Exposure assessment method: questionnaire; see McDuffie et al. (2001)	NHL	Use of DDT			Age, province of residence, respondent type (proxy/direct), diesel oil exposure	Same population as McDuffie et al. (2001) Strengths: analysis by concurrent immunological conditions; large size; pathology available Limitations: unknown proportion of refusals to participate; use of a postal questionnaire; lack of control for exposure to other pesticides
			Ever	33	1.69 (1.07–2.67)		
			Ever, no asthma, allergy or hay fever	18	1.31 (0.73–2.36)		
			Ever, with asthma, allergy or hay fever	15	2.53 (1.17–5.47)		

Table 2.8 (continued)

Reference, location follow-up/ enrolment period, study-design	Population size, description, exposure assessment method	Organ site (ICD-O)	Exposure category or level	Exposed cases/ deaths	Risk estimate (95% CI)	Covariates	Comments
Cocco et al. (2013) Six European centres 1998–2003	Cases: 2348; residents in the referral area of participating centres Controls: 2462; population and hospital controls Exposure assessment method: expert assessment; crop–exposure matrix	NHL (B-cell lymphoma)	Ever exposed to DDT	3	1.2 (0.2–5.9)	Age, sex, education, study site	Strengths: large study size; pathology review; detailed exposure assessment Limitations: low prevalence of exposure to individual chemicals
Colt et al. (2005) USA (Iowa, Los Angeles, Detroit and Seattle) 1998–2000	Cases: 603; cancer registries Controls: 443; population, frequency-matched on age, sex, race, and centre Exposure assessment method: environmental monitoring; vacuum sample of household carpet dust; measurement of DDT and derivatives, PCBs and other OC insecticides	NHL	DDE, 1 ng/g			Age, study site, education, sex	Strengths: study size; objective exposure measurements Limitations: other occupational and environmental (non-ousehold) sources of exposure not considered
			< LOD	304	1.0		
			> LOD	299	1.3 (1.0–1.7)		
			20.8–34.9	94	1.3 (0.8–1.8)		
			35–55.9	83	1.1 (0.7–1.6)		
			56–2450	122	1.6 (1.1–2.2)		
			Trend-test *P* value: 0.02				
			DDT, 1 ng/g				
			< LOD	197	1.0		
			> LOD	406	0.9 (0.7–1.2)		
			20.8–98.7	124	0.8 (0.6–1.2)		
			98.8–248	111	0.8 (0.5–1.1)		
			248.1–24 600	170	1.2 (0.8–1.6)		
			Trend-test *P* value: 0.09				
		NHL (follicular)	DDE > LOD	156	1.3 (0.9–2.0)		
			DDT > LOD	156	1.0 (0.7–1.6)		
		NHL (DLBCL)	DDE > LOD	189	1.3 (0.9–1.9)		
			DDT > LOD	189	0.8 (0.5–1.1)		
		NHL (T-cell)	DDE > LOD	36	2.6 (1.3–5.4)		
			DDT > LOD	36	2.8 (1.1–7.1)		
		Lymphatic and hematopoetic	DDE > LOD	206	1.2 (0.8–1.6)		
			DDT > LOD	206	0.8 (0.5–1.1)		

Table 2.8 (continued)

Reference, location follow-up/ enrolment period, study-design	Population size, description, exposure assessment method	Organ site (ICD-O)	Exposure category or level	Exposed cases/ deaths	Risk estimate (95% CI)	Covariates	Comments
Ward et al. (2009) USA (California) 2001–06	Cases: 184; paediatric clinical centres Controls: 212; birth certificate registries, individually matched on age, sex, race, Hispanic ethnicity, and maternal residence Exposure assessment method: environmental monitoring; vacuum sample of household carpet dust; measurement of DDT and derivatives, PCBs and other OC insecticides	Leukaemia (childhood ALL)	ng/g dust			Age, sex, race, family income, year of enrolment	Strengths: pathology available; comprehensive recruitment of cases; high participation rate; objective measurement of exposure Limitations: other parental, occupational sources of exposure not considered; carpet dust measurements were missing for 10% of cases and 15% of controls
			DDE > LOD detection limit	145	0.87 (0.51–1.50)		
			2.0–9.4	38	0.74 (0.39–1.41)		
			9.4–21.7	59	1.08 (0.58–2.02)		
			21.7–850.4	48	0.83 (0.43–1.59)		

ALL, acute lymphoblastic/lymphocytic leukaemia; BBD, benign breast disease; BMI, body-mass index; CI, confidence interval; CLL, chronic lymphocytic leukaemia; DDE, dichlorodiphenyldichloroethylene; DDT, dichlorodiphenyltrichloroethane; DLBCL, diffuse large B-cell lymphoma; EBV, Epstein-Barr virus; EBV EA, Epstein-Barr virus early antigen; ER, estrogen receptor; FFTP, first full-term pregnancy; GC, gas chromatography; HRT, hormone replacement therapy; HL, Hodgkin lymphoma; JEM, job–exposure matrix; LOD, limit of detection; MM, multiple myeloma; NR, not reported; OC, organochlorine; OPs, organophosphates; PCB, polychlorinated biphenyl; SLL, small lymphocytic lymphoma; vs, versus; yr, year

did, resulting in non-significant although large excess risk in a multivariate analysis (OR, 7.5; 95% CI, 0.8–70).

Another study in Sweden focused on CLL, comparing 111 CLL cases identified in five hospitals in central and southern Sweden with 431 population controls from the same catchment areas. Exposures were assessed by postal questionnaire. Based on six cases, ever exposure to DDT was associated with a sixfold risk of CLL (95% CI, 1.5–23), after adjusting by age, sex, and other occupational exposures (Flodin et al., 1988).

A third study in Sweden, conducted in 1982–1986, focused on occupational and environmental risk factors for multiple myeloma (Eriksson & Karlsson, 1992). Cases were 256 patients with multiple myeloma who were identified through the Swedish cancer registry, individually matched by age, sex, vital status, and county to 256 controls selected from population registries or mortality registries. Exposure to DDT and other occupational exposures was assessed by mailed questionnaire, followed by a second in-person interview for subjects working on a farm or in other occupations potentially involving pesticides. Risk of multiple myeloma increased monotonically up to 1.6-fold with highest category of days of contact with DDT [95% CI, 0.53–4.93], although none of the individual risk estimates was statistically significant. When restricting the analysis to subjects working in agriculture or forestry, the risk was 1.86[(95% CI, 0.92–3.75). [The prevalence of farm work and exposure to DDT in this study population was unexpectedly high (farm work, 44%; ever exposure to DDT, about 12% among controls); a large proportion of interviews were conducted with next of kin; 90% confidence intervals of risk estimates were reported; a list of overlapping exposures, including phenoxy acids and livestock were evaluated individually, but no reciprocal adjustment was sought.]

[The Working Group considered that results from this group of studies were difficult to interpret as they were severely underpowered, referred to different pathological entities, and were further limited by the methodological concerns illustrated in preceding Working Group comments.]

Brown et al. (1990) included 578 cases of leukaemia identified from the Iowa, USA cancer registry and from hospital records in Minnesota between 1980 and 1983, and 1245 population controls matched by age, vital status, and state. All types of leukaemia, whether myeloid or lymphatic, acute or chronic, were included. Exposures were assessed by in-person interview, with a supplemental telephone interview concerning reported pesticide exposures. Risk of all leukaemia combined did not show a consistent trend when use of DDT in crop farming was considered; however, risk associated with the highest category of frequency of use (10+ days/year) was twofold (95% CI, 1.1–3.9) for DDT use in raising livestock. No substantial changes in the risk estimates were observed when a 20-year lag was applied to define exposure; risk was consistently elevated for both CLL or chronic myeloid leukaemia. [Interpretation was complicated by multiple exposures. Information was obtained by self-report or indirectly from next-of-kin in a substantial proportion of cases and controls.]

Another case–control study was conducted in the early 1980s in Iowa and Minnesota, USA, and included 622 NHL cases and 1245 controls (Cantor et al., 1992). This study evaluated the risk of NHL associated with ever exposure to DDT whether in disinfecting animals, or in crops, and whether occurring before 1965 or thereafter. The exposure assessment methods were similar to those used by Brown et al. (1990) with detailed questions on pesticide use. Regardless of whether exposure occurred before 1965 or thereafter, after adjusting by vital status, age, state, cigarette smoking, family history of lympho-haemopoietic cancer, high-risk occupations, and high-risk exposures, ever use of DDT on animals was associated with a weak increase in risk, while there was a stronger association with ever use on crops (OR, 1.7; 95% CI, 1.2–2.6; for any use). [Information was indirectly obtained by next-of-kin for a substantial proportion of cases and controls.]

A pooled analysis of case–control studies on NHL conducted in the 1980s in four states in the USA (Baris et al., 1998) included the study by Cantor et al. (1992), and consisted of 993 histologically confirmed NHL cases and 2918 population controls, in part selected from mortality files, frequency-matched to cases by race, age, sex, and vital status at the time of interview. Next-of-kin interviews accounted for 32% of cases and 40% of the controls. Risk of NHL increased with days/year of use up to 2.1-fold (95% CI, 0.9–4.9) for the highest category (>5 days/year), However, when adjusting by exposure to 2,4-D and organophosphates, no increase in risk was observed for ever exposure (OR, 0.9; 95% CI, 0.4–1.8) and the dose–response curve was flattened (OR, 0.7; 95% CI, 0.0–15.0; for > 5 days/year). [This was the largest and most informative case–control study on NHL and occupational exposure to DDT, but information was indirectly obtained from next-of-kin in a substantial proportion of cases and controls.] A further analysis of the same data set was published in 2003 (De Roos et al., 2003), and included ever use of a list of 47 insecticides and herbicides. Any subject with missing information on any of the 47 pesticides was dropped from the analysis, and for the most frequently used pesticides, adjustments were made for multiple exposure using conventional and hierarchical logistic regression. Ever exposure to DDT was not associated with an increase in risk of NHL (OR, 1.0; 95% CI, 0.7–1.3), nor was reported use of DDT only (OR, 0.9; 95% CI, 0.6–1.3); however, a non-significant increase in risk was observed for the combined use of DDT and chlordane (OR, 1.7; 95% CI, 0.7–3.2; in conventional logistic regression).

In a small study in Italy that included 26 NHL cases and 74 controls, and was based partially on proxy interviews, ever exposure to DDT was attributed in seven cases using a crop–exposure matrix supported by a consultant agronomist (Assennato et al., 1995). The resulting odds ratio for exposure to DDT was increased more than fourfold (95% CI, 1.04–16.76).

In another small study in Italy (Nanni et al., 1996), cases of NHL and CLL combined (n = 61) and controls who were raising livestock were selected. The underlying hypothesis was that using insecticides in livestock might have involved higher exposure levels. Within this category, the authors explored risk associated with ever exposure to DDT using two strategies: self-report or using a job–exposure matrix (JEM). Either way, an approximate excess risk of 70% was observed when all NHL subtypes were combined with CLL (for the JEM: OR, 1.70; 95% CI, 0.91–3.17; for recall: 1.74 (95% CI, 0.93–3.27). When the analysis was limited to low-grade NHL combined with CLL, risks were elevated (based on JEM: OR, 2.16; 95% CI, 0.86–5.43; and based on the subjects' recall: OR, 2.33; 95% CI, 0.93–5.85). However, when calculating lifetime cumulative exposure, the excess risk among subjects with a cumulative exposure to ≥ 1 kg of DDT was 22% in respect to subjects with a lower cumulative exposure (OR, 1.22; 95% CI, 0.95–1.57). [This study used a more detailed exposure assessment, but the number and type of controls was unclear, the classification of lymphomas was outdated, and precision was limited.]

Risk of NHL risk related to exposure to DDT was also explored in a study in Canada that covered six provinces and gathered questionnaire data from 517 cases and 1506 population controls (Mc Duffie et al., 2001). Exposure to pesticides was assessed by questionnaire and detailed telephone interview with subjects who reported any exposure. More than 16% of cases and 12% of controls reported exposure to five or more chemicals, and it was mostly occasional, with the highest exposure frequency set at ≥ 3 days/year and a median of 10 hours/year. DDT exposure above the median was associated with a significant increase in risk of NHL (OR, 1.73; 95% CI, 1.08–2.76), but risk did not increase by categories of exposure frequency. A subsequent analysis of the same data set explored the effect modification by concurrent asthma, allergy, or hay fever on the association between ever exposure to DDT and risk of NHL (Pahwa et al., 2012). The results showed that the excess risk associated with ever exposure to DDT was concentrated among subjects with a concurrent allergic condition (OR, 2.53; 95% CI, 1.17–5.47), while there was no association among those who did not report such conditions. [Major reasons for concern in interpreting findings from this study included reliance on self-reported exposures and a low response rate.]

DDT was evaluated in relation to overall lymphoma (including all B-cell and T-cell subtypes and HL) in a multicentre case–control study in six European countries (Cocco et al., 2013). Exposure to pesticides was estimated using a crop–exposure matrix and occupational histories collected by interview, which included a detailed questionnaire specific for the agricultural work. Data for DDT were reported only with respect to B-cell lymphoma. The prevalence of ever exposure to DDT was extremely low (about 0.01%) and the associated risk for all B-cell lymphomas was not elevated (OR, 1.2; 95% CI, 0.2–5.9).

(c) *Environmental monitoring-based studies*

Two studies in the USA examined environmental exposure to DDT by using a vacuum to sample carpet dust in the household of the study subjects (Colt et al., 2005; Ward et al., 2009). One study examined associations between organochlorines and NHL in adults in four areas covered by the SEER programme of the United States National Cancer Institute in 1998–2000 (Colt et al., 2005). The second study, conducted in northern and central California in 2001–2006, investigated associations between childhood leukaemia and PCBs and OCPs (Ward et al., 2009). Although both studies collected data more than 30 years after use of DDT was discontinued in the USA, the SEER study was characterized by an elevated frequency of detection of DDT in the carpet dust (cases, 67%; controls, 71%), that was higher than its major and most persistent derivative, DDE (cases, 50%; controls, 44%). A significant upward trend in risk of NHL was associated with DDE concentration in the carpet dust (*P* for trend, 0.02). There was no trend for DDT (*P* for trend, 0.09). In the analysis by individual NHL subtype, risk associated with DDE and DDT concentrations above the limit of detection was highest for T-cell lymphoma (ORs: 2.6; 95% 1.3–5.4; and 2.8; 95% CI, 1.1–7.1; respectively), while a non-significant 30% excess risk associated with DDE concentrations above the limit of detection was also observed for diffuse large B-cell lymphoma (OR, 1.3; 95% CI, 0.9–1.9) and follicular lymphoma (OR, 1.3; 95% CI, 0.9–2.0). [The Working Group noted that in this study risks were adjusted by age, sex, study site, and education, but not by occupation or rural/urban residence, which might have helped to discriminate possible alternative sources of the observed excess risk. In addition, it was not clear how DDT concentrations in carpet dust related to individual exposure in the etiologically relevant period.]

In a similar study of childhood leukaemia in California (Ward et al., 2009), which included family income and year of recruitment among the adjusting covariates, and used geographical mapping of rural areas and crops around the residence of study subjects, DDT was detected in 57% of dust samples, and DDE was detected in 82%. However, risk did not increase with increasing level of DDE (*P* for trend, 0.794), or DDT (*P* for trend, 0.709) in the carpet dust.

(d) Meta-analyses

Numerous meta-analyses and pooled analyses have evaluated associations between cancer risk and farmers' or farmworkers' occupational exposures to pesticides, but only one of these (Schinasi & Leon, 2014) examined associations of specific pesticides with NHL.

In that meta-analysis, Schinasi & Leon (2014), 44 papers published after 1980, with information about 21 pesticide chemical groups and 80 active ingredients were evaluated. [The Working Group noted that all of the reported results were from high-income countries.] Meta-risk ratio (meta RR) estimates and 95% confidence intervals using random effect models were computed, allowing between-study heterogeneity to contribute to the variance, and I^2 values, which represent the percentage of the total variance explained by study heterogeneity, and measures of inconsistency between studies were reported. Confidence limit ratios (CLRs, the ratio of the upper to the lower CI limits) were also reported as an indicator of precision. Sensitivity analyses were conducted to evaluate the robustness of results by potential sources of heterogeneity including study design, sex, geographical area, decade of cancer diagnosis, and source of the controls in the case–control studies

Seven papers contributed to the meta-analysis of DDT (Woods et al., 1987; Persson et al., 1993; Baris et al., 1998; Hardell et al., 2002; Purdue et al., 2007; Eriksson et al., 2008; Pahwa et al., 2012). The meta RR estimate for DDT with NHL overall was 1.3 (95% CI, 1.1–1.5; I^2 = 0%). The association with NHL subtypes was 1.4 (95% CI, 1.0–1.5) for B-cell lymphoma; 1.2 (95% CI, 0.9–1.7) for diffuse large B-cell lymphoma; and 1.5 (95% CI, 1.0–2.4) for follicular lymphoma (Schinasi & Leon, 2014).

[The Working Group noted that most of the available studies were on NHL overall and not NHL subtypes. Since NHL subtypes are believed to be etiologically heterogeneous, the lack of subtype-specific analyses in many of the previously published studies may be masking important associations in the available meta-analyses. Since the definition of NHL has changed over time, care must also be exercised in comparing findings from studies using older definitions of NHL with more recent studies based on the current understanding of NHL. The definition of NHL used by the SEER coding scheme is now based on the Pathology Working Group of the International Lymphoma Epidemiology Consortium (ICD-O-3 Interlymph modification) (Morton et al., 2007).]

2.2.3 Cancer of the prostate

See Table 2.9

In two studies on cancer of the prostate, exposure to DDT from agricultural work was assessed by experts or by means of JEM. These studies on occupational exposure are described first.

Settimi et al. (2003) investigated risk of cancer of the prostate in relation to exposure to pesticides in five areas in Italy using data from a hospital-based case–control study carried out between 1990 and 1992. The case group was composed of 124 patients with prostate cancer. The control group included 659 patients with other cancers. Exposure to 217 different pesticides, including DDT, was assessed by a team of agronomists using data obtained from crop-specific forms collected from subjects engaged in agricultural work, national statistics on pesticide use, supplier's records, and personal experience. Men ever exposed to DDT had an odds ratio for prostate cancer of 2.1 (95% CI, 1.2–3.8). Odds ratios in men exposed to DDT for ≤ 15 years or > 15 years were 2.1 (95% CI, 0.9–2.1) and 2.2 (95% CI, 1.1–4.8), respectively. [The Working Group noted the low specificity of exposure assessment, and that selection of controls with other cancers may have biased risk estimates.]

Band et al. (2011) conducted a case–control study on cancer of the prostate in British

Columbia, Canada. Cases were patients with prostate cancer ascertained by the population-based British Columbia cancer registry for the years 1983–1990 (n = 1153). Controls were patients with cancers at all other sites excluding lung and unknown primary (n = 3999), matched to the cases on year of birth and year of diagnosis. Information on lifetime job description, occupation, and industry titles, location, duration and time period of work, as well as alcohol and tobacco consumption was obtained from a self-administered questionnaire. Farmers' exposure to 180 pesticides was assessed using a JEM based on region, crop, task, and job title. Exposures via pesticide application were quantified using estimates derived from the North American Pesticide Handlers Exposure Database. A significant association was reported between prostate cancer and exposure to DDT (OR, 1.68; 95% CI, 1.04–2.70 for high-level exposure with reference to the unexposed group), with a dose–response trend (P for trend, 0.03). [This was a large case–control study. The use of a JEM for assessing exposure may have led to exposure misclassification, but exposure indicators were calculated over the lifetime.]

Several other studies reported on associations between cancer of the prostate and exposure to DDT and its derivatives as measured in biological samples. Ritchie et al. (2003) examined the relationship between serum concentrations of OCPs and prostate cancer in a population-based case–control study in Iowa, USA. Cases were 58 men diagnosed with prostate cancer between May 2000 and May 2001 who were enrolled at a university hospital and a smaller urology clinic. Controls were 99 men who received physical check-ups in the hospital. Concentrations of 48 organochlorinated compounds including *p,p′*-DDE and *p,p′*-DDT were measured in serum samples. Since the detection rate of *p,p′*-DDT was close to 0%, only *p,p′*-DDE (detection rate, > 99%) was investigated in relation to risk of prostate cancer. The odds ratio for *p,p′*-DDE in the highest exposure tertile when compared with the lowest was 1.08 (95% CI, 0.47–2.50). [This study was the first to investigate prostate cancer and biological measurements of organochlorine compounds, and was based on very small numbers of cases and controls.]

In another hospital-based case–control study on cancer of the prostate conducted at Örebro University hospital, Sweden, Hardell et al. (2006a) measured the levels of several chlorinated and brominated pollutants, including *p,p′*-DDE, in adipose tissue biopsy from the abdominal wall taken during surgery for 57 cases diagnosed with prostate cancer in 1997–1999, and 20 controls undergoing transurethral resection for benign hyperplasia. The odds ratio for prostate cancer among men with *p,p′*-DDE levels above the median was 2.30 (95% CI, 0.77–6.85). [The sample size of the study was small and choice of controls may have led to selection bias.]

Aronson et al. (2010) reported the results of a case–control study of cancer of the prostate in patients who had visited any of a group of five urologists in Kingston, Ontario, Canada, between 1997 and 1999. Cases were selected from men diagnosed with incident primary cancer of the prostate at biopsy (n = 79). A group of urological controls (n = 194) included men with non-cancerous urological disease (erectile dysfunction, prostatitis, benign prostate hyperplasia, haematospermia/haematuria, urinary obstruction/pain/ infection etc.), and a group of biopsy controls (n = 135) included men in whom no prostate cancer was detected at biopsy. Concentrations of 14 PCB congeners and 13 pesticides were measured in blood plasma. Odds ratios for *p,p′*-DDE and *p,p′*-DDT comparing the highest exposure tertile to the lowest were 0.73 (95% CI, 0.38–1.40) and 1.05 (0.55–2.00), respectively, indicating no association between risk of cancer and blood exposure levels at diagnosis. Results based on the urology control group only were similar. [This was a small study with limited statistical power to detect associations.]

Table 2.9 Case–control studies on cancer of the prostate and exposure to DDT and its metabolites

Reference, location, enrolment/ follow-up period	Population size, description, exposure assessment method	Organ site	Exposure category or level	Exposed cases/ deaths	Risk estimate (95% CI)	Covariates	Comments
Settimi et al. (2003) Italy (4 areas) 1990–1992	Cases: 124; local and university hospitals Controls: 659; hospital patients with other cancers Exposure assessment method: team of agronomists assessed exposure to pesticides based on interview data	Prostate	DDT:			Age, family history of prostate cancer, interview (direct/ indirect)	Strengths: same study base for cases and controls; adjustment for main confounders Limitations: relatively small study size; low specificity of exposure assessment; cancer sites of controls possibly associated with exposure
			Ever exposed	20	2.1 (1.2–3.8)		
			≤ 15 yrs	16	2.1 (0.9–2.1)		
			> 15 yrs	4	2.2 (1.1–4.8)		
Band et al. (2011) Canada, British Columbia 1983–1990	Cases: 1153; cancer registry Controls: 3999; other cancer patients from the same registry excluding lung cancer and cancer of unknown primary site Exposure assessment method: JEM for 1950–1998, including 45 animal and crops; information on exposure (quantitative or never/ever) to 139 pesticide active ingredients determined for type of work and time; quantification derived from models used for pesticide registration	Prostate	No exposure	1104	1.00	Alcohol consumption, cigarette years, pipe years, education level, respondent type (proxy/direct)	Spearman correlation coefficient between DDT and lindane 0.72 Strengths: large size; histological confirmation-high response rates; lifetime cumulative exposure assessment Limitations: exposure misclassification due to the JEM; potential selection bias due to use of cancer controls; multiple comparisons (142 active chemicals evaluated); strong inter-correlations between exposures; no mutual adjustment
			Low exposure	19	1.24 (0.71–2.16)		
			High exposure	30	1.68 (1.04–2.70)		
			Trend-test P value: 0.03				
			Ever vs never exposed	49	1.47 (1.02–2.12)		
			Trend-test P value: 0.03				

Table 2.9 (continued)

Reference, location, enrolment/ follow-up period	Population size, description, exposure assessment method	Organ site	Exposure category or level	Exposed cases/ deaths	Risk estimate (95% CI)	Covariates	Comments
Ritchie et al. (2003) Iowa, USA 2000–2001	Cases: 58; university hospital and urology clinic Controls: 99; men receiving annual check-ups in the hospital Exposure assessment method: biomarker; detection rates for *p,p′*-DDE, > 99%; detection rates for *p,p′*-DDT, 0% cases; 2% controls (not investigated)	Prostate	*p,p′*-DDE (µg/g) ≤ 0.180	20	1	Age, body mass index, prostatis	Strengths: first study on prostate-cancer risk based on biological measurements of OC Limitations: very small numbers of cases; source population for the controls not well defined
			0.181–0.340	15	0.72 (0.31–1.71)		
			> 0.340	23	1.08 (0.47–2.5)		
Hardell et al. (2006a) Örebro, Sweden 1997–1999	Cases: 58; hospital Controls: 20; hospital (men undergoing transurethral resection for benign hyperplasia) Exposure assessment method: biomarker; chlorinated and brominated compounds measured in abdominal adipose tissue biopsy	Prostate	*p,p′*-DDE, < 291 ng/g lipid	15	1.0	Age, BMI	Strengths: adipose tissue biopsy; high response rate Limitations: small sample size
			p,p′-DDE, > 291 ng/g lipid	42	2.3 (0.77–6.85)		
Aronson et al. (2010) Kingston, Ontario, Canada 1997–1999	Cases: 79; incident prostate cancer diagnosed by biopsy at urology clinics Controls: 329; men without prostate cancer seen at the same clinics (135 with biopsy, 194 without) Exposure assessment method: biomarker; 13 pesticides and 14 PCBs measured in plasma (LOD, 8 µg/kg lipid for DDT; 4 µg/kg lipid for DDE)	Prostate	*p,p′*-DDE, µg/kg lipid			Age, teenage physical activity, alcohol consumption, smoking pack-years	Strengths: PSA and DRE screening in cases and controls Limitations: very small number of cases; total response rates not available; controls with urological diseases possibly related to exposure
			< 270	27	1.00		
			270–548.9	27	0.97 (0.52–1.83)		
			548.9–2362.3	24	0.73 (0.38–1.40)		
			Trend-test *P* value: 0.35				
			p,p′-DDT, µg/kg lipid				
			< 5.3	24	1.00		
			5.3–8.4	28	1.19 (0.63–2.26)		
			8.4–49.1	26	1.05 (0.55–2.00)		
			Trend-test *P* value: 0.9				

Table 2.9 (continued)

Reference, location, enrolment/ follow-up period	Population size, description, exposure assessment method	Organ site	Exposure category or level	Exposed cases/ deaths	Risk estimate (95% CI)	Covariates	Comments
Emeville et al. (2015) Guadeloupe (French Caribbean) 2004–2007	Cases: 576; private and public urology clinics Controls: 655; free health screening programme Exposure assessment method: biomarker; *p,p'*-DDE measured in plasma (*p,p'*-DDT not investigated: detection frequency, 36.2%); LOD, 0.05 µg/L for *p,p'*-DDE and *p,p'*-DDT	Prostate (total)	*p,p'*-DDE, µg/L			Age, waist-to-hip ratio, type 2 diabetes, alcohol, total plasma lipid concentration	Strengths: largest study investigating associations between DDE and prostate cancer based on biological measurements of exposure Limitations: source population for the controls was not well defined; *p',p'*-DDT not investigated; no adjustment for BMI
			< 0.79	106	1.00		
			0.79–1.62	96	0.96 (0.66–1.42)		
			1.63–2.89	111	1.05 (0.71–1.55)		
			2.90–5.18	104	1.02 (0.67–1.53)		
			> 5.19	159	1.53 (1.02–2.30)		
			Trend-test *P* value: 0.01				
		Prostate (aggressive/ advanced)	*p,p'*-DDE, µg/L				
			< 1.37	20	1.00		
			1.37–3.41	34	1.55 (0.85–2.85)		
			> 3.42	47	1.92 (1.04–3.54)		
			Trend-test *P* value: 0.06				
			DDT, 1–20 yrs of exposure	34	1.6 (0.9–4.6)		

BMI, body-mass index; CI, confidence interval; DDE, dichlorodiphenyldichloroethylene; DDT, dichlorodiphenyltrichloroethane; DRE, digital rectal examination; JEM, job-exposure matrix; LOD, limit of detection; NR, not reported; OC, organochlorine; OPs, organophosphates; PCB, polychlorinated biphenyl; PSA, prostate-specific antigen; vs, versus; yr, year

Emeville et al. (2015) reported the results of a population-based case–control study of cancer of the prostate and plasma concentrations of DDE conducted in Guadeloupe (French Caribbean). The case group included 576 men with incident prostate cancer (81% of the cases with a blood sample available) identified from private and public clinics covering the entire territory of Guadeloupe. Controls were men aged 45 years or older selected from a free health screening programme open to the general population, and who had normal findings upon digital rectal examination and normal prostate-specific antigen (PSA) concentrations. Organochlorine compounds including *p,p′*-DDE and *p,p′*-DDT were measured in the blood samples via high-resolution gas chromatography, but only *p,p′*-DDE (detection frequency, 96.2% among controls) was investigated in relation to risk of prostate cancer. The odds ratio for men in the highest quintile of DDE concentration compared with men in the lowest quintile was 1.53 (1.02–2.30). There was an overall statistically significant trend ($P = 0.01$), which was mainly driven by the odds ratio in the highest exposure quintile, as the odds ratios in lower quintiles were close to the null. The odd ratio for cases with high-grade Gleason score was 1.92 (95% CI, 1.04–3.54) (worse prognostic value) for men in the highest tertile relative to men in the lowest tertile, but was not significantly different from the corresponding odds ratio for cases with low-grade Gleason score. [This was the largest study to date investigating associations between DDE and prostate cancer based on biological measurements of exposure; the source population for the controls was not well defined and may have introduced some selection bias.]

Lim et al. (2015) conducted a meta-analysis of studies of the associations between cancer of the prostate and measurements of persistent organic pollutants, including *p,p′*-DDE, in blood or adipose tissue. The analysis included four case–control studies and a nested case–control study reviewed in this *Monograph* and a cross-sectional study of the United States National Health and Nutrition Examination Survey (NHANES) study (Xu et al., 2010). The meta-odds ratio for cancer of the prostate with *p,p′*-DDE comparing high versus low concentrations was 1.41 (95% CI, 1.12–1.78). Based on four of the included studies, the odds ratio per 1 μg/g lipid of *p,p′*-DDE was 1.25 (95% CI, 0.86–1.84).

2.2.4 Cancer of the testis

See Table 2.10

Hardell et al. (2003) conducted a population-based case–control study on cancer of the testis in Sweden. Case patients and their mothers were recruited from 1997 to 2000 from urology or oncology departments of several hospitals in Sweden. Controls and their mothers were drawn from the Swedish population registry, age-matched with the cases (5-year age group) and control mothers were age-matched with case mothers (5-year age group). Persistent organic pollutants, including *p,p′*-DDE, were measured at diagnosis in the blood samples provided by 58 cases and 44 mothers of cases, and by 61 controls and 45 mothers of controls. The odds ratio for exposure to *p,p′*-DDE at above median concentration versus below median for male subjects was 1.7 (95% CI, 0.8–3.7), and 1.3 (95% CI, 0.5–3.0) for mothers of study subjects. Stratification of the case group by seminoma and non-seminoma did not reveal a notably different association with *p,p′*-DDE. In a later publication on the same study (Hardell et al., 2006b), results on *p,p′*-DDE in mothers were also presented separately by age group, showing increased odds ratios only among mothers younger than age 55 years at the time of diagnosis (OR, 2.3, 95% CI, 0.6–9.5). [This was the first study examining testicular cancer in relation to biological measurements of persistent organic pollutants. Selection biases were minimized by use of a population registry to select the controls and their mothers, and participation rates were high, but mothers' blood was sampled only after case diagnosis (on average, at age 31 years) and the results were imprecise.]

Biggs et al. (2008) conducted a population-based case–control study on testicular germ cell tumours in Washington State, USA. The case group included 246 men aged 18–45 years with invasive testicular germ cell carcinoma diagnosed between 1999 and 2008, identified from the files of the cancer surveillance system, a part of the SEER programme of the United States National Cancer Institute. The control group included 630 men frequency-matched to the cases on 5-year age group, and selected using a random-digit dialling procedure. Concentations of 12 OCPs, including *p,p′*-DDT, *o,p*-DDT, *p,p′*-DDE, and 36 PCB congeners were measured from blood samples drawn at the date of enrolment in the study. The odds ratios for men exposed above the 85th percentile of exposure compared with men exposed below the median were 0.61 (95% CI, 0.32–1.14), 1.17 (95% CI, 0.68–2.00), 1.30 (95% CI, 0.67–2.53), and for *p,p′*-DDE, *p,p′*-DDT, and *o,p*-DDT, respectively. Analysis based on the continuous variables did not show any significant association between these compounds and testicular germ cell carcinoma. No interaction between *p,p′*-DDE levels and the androgen receptor genotype was observed. [In this study, the response rate for cases and controls was low; the use of post-diagnostic blood samples in this study may have distorted the results, despite the fact that odds ratios were adjusted for change in BMI between reference date and blood draw.]

In a hospital-based case–control study in Rome, Italy, Giannandrea et al. (2011) examined the association between serum levels of *p,p′*-DDE and cancer of the testis. Cases were 50 patients with testicular cancer recruited between October 2006 and September 2008 at the Laboratory of Seminology-Sperm Bank. The controls included 48 men recruited from the same department among men undergoing examination to ascertain their fertility status, and of approximately the same age and BMI as the cases. Only 26% of cases and 10% of controls had detectable serum *p,p′*-DDE concentrations (limit of detection, 0.2 ng/mL). For men with detectable levels of *p,p′*-DDE, the adjusted odds ratio for testicular cancer was 3.21 (95% CI, 0.77–13.30). [This was a small study. Choosing controls among men consulting for fertility problems could introduce bias. There was a large proportion of subjects under the limit of detection, possibily because the study was conducted long after the use of DDT was discontinued. There was no adjustment for total lipid concentration.]

2.2.5 Other cancer sites

See Table 2.11

(a) Cancer of the pancreas

Fryzek et al. (1997) conducted a case–control study in Michigan, USA, to investigate the relationship between cancer of the pancreas and exposure to pesticides. The case group included 66 patients diagnosed with this cancer in 1994 and 1995 in seven hospitals (response rate, 81%). The controls (n = 131) were frequency-matched on age, sex, ethnicity, and county of residence, and were identified by random-digit dialling (response rate, 27%). Exposure to pesticides and to DDT during leisure time activities and at work were assessed by questionnaire. Ever using DDT was associated with a non-statistically significant increased risk of cancer of the pancreas (OR, 1.6; 95% CI, 0.8–3.1). [Unlike most studies on this cancer, for which the prognosis is poor, this study was based on direct interviews of cases. However, the response rate for cases and controls was low and the use of questionnaire for assessing exposure to DDT may lead to recall bias.]

Hoppin et al. (2000) conducted a population-based case–control study on cancer of the pancreas in the San Francisco bay area, USA. Cases were 108 patients aged 21–85 years with pancreatic cancer, recruited using a rapid case ascertainment between October 1996 and May

Table 2.10 Case–control studies on cancer of the testis and exposure to DDT and its metabolites

Reference, location, enrolment/follow-up period	Population size, description, exposure assessment method	Organ site	Exposure category or level	Exposed cases/deaths	Risk estimate (95% CI)	Covariates	Comments
Hardell et al. (2003) Sweden 1997–2000	Cases: 58 (44 mothers); hospital urology and oncology departments Controls: 61 (45 mothers); population registry; matched by age Exposure assessment method: biomarker	Testis	*p,p′*-DDE > median	34	1.7 (0.8–3.7)	Age, BMI	Strengths: population controls; high participation rates, including mothers Limitations: small size; blood drawn from mothers after case diagnosis
			Seminoma	14	1.5 (0.5–4.5)		
			Non-seminoma	20	1.9 (0.8–3.7)		
			Mothers *p,p′*-DDE > median	22	1.3 (0.5–3.0)		
Biggs et al. (2008) Washington state, USA 1999–2008	Cases: 246; cancer registry Controls: 630; population; frequency-matched on age Exposure assessment method: biomarker; *p,p′*-DDT not detected in 6.7% of subjects, *o,p′*-DDT not detected in 42.3%, *p,p′*-DDE not detected in 0%; poor between-run reliability of the analytic method for several of the analyses	Testis (testicular germ cell tumours)	*p,p′*-DDE, pg/g			Age, ethnicity, change in BMI between reference date and blood draw, assay run number, serum lipids	Strengths: relatively large study size Limitations: use of post-diagnostic blood samples; low response rates of cases and controls
			≤ 1101	130	1.00		
			> 1101–2473	94	1.14 (0.78–1.67)		
			> 2473	21	0.61 (0.32–1.14)		
			Trend-test *P* value: 0.36				
			p,p′-DDT, pg/g				
			≤ 27	110	1.00		
			27–47	94	1.39 (0.96–2.02)		
			> 47	32	1.17 (0.68–2.00)		
			Trend-test *P* value: 0.3				
			o,p-DDT, pg/g				
			≤ 5	104	1.00		
			5–13	83	1.26 (0.83–1.91)		
			> 13	37	1.30 (0.67–2.53)		
			Trend-test *P* value: 0.28				

Table 2.10 (continued)

Reference, location, enrolment/follow-up period	Population size, description, exposure assessment method	Organ site	Exposure category or level	Exposed cases/deaths	Risk estimate (95% CI)	Covariates	Comments
Giannandrea et al. (2011) Rome, Italy 2006–2008	Cases: 50; university laboratory and sperm bank Controls: 48; men with fertility examination in the same department as cases Exposure assessment method: personal monitoring	Testis (testicular cancer)	*p,p′*-DDE, < 0.2 ng/mL	37	1.00	Mother's age at birth, education, parity	The late period of the study (2008) may explain the large proportion of subjects below the detection limit Limitations: small study; large proportion of subjects under LOD; selection of controls among men consulting for fertility problems may not be adequate if fertility is associated with DDT/DDE; no adjustment for total lipid concentration
			p,p′-DDE, ≥ 0.2 ng/mL	13	3.21 (0.77–13.30)		
			DDT, 1–20 yrs of exposure	34	1.6 (0.9–4.6)		

BMI, body mass index; CI, confidence interval; DDE, dichlorodiphenyldichloroethylene; DDT, dichlorodiphenyltrichloroethane; JEM, job–exposure matrix; LOD, limit of detection; NR, not reported; OC, organochlorine; OPs, organophosphates; PCB, polychlorinated biphenyl; vs, versus; yr, year

1998, who were alive and provided a blood sample. Control subjects ($n = 82$) were frequency-matched on sex and age to the cases and identified using random-digit dialling and random samples from the Health Care Financing Administration lists. Detailed in-person interviews were conducted and blood samples were obtained for organochlorine analyses that included DDE, DDT, and 11 PCB congeners. DDE was detected in more than 50% of the samples. The odds ratio for DDE in the highest exposure tertile (≥ 1880 ng/g lipids) compared with DDE in the lowest (< 850 ng/g lipid) was 2.1 (95% CI, 0.9–4.7), and there was some indication of a dose–response trend (*P* for trend, 0.08). However, when adjusting for total PCBs, the odds ratio in the highest tertile decreased to 1.1 (95% CI, 0.4–2.8), whereas the association observed with PCBs was not modified by adjustment for DDE. [This was the first population-based study on pancreatic cancer using serum measurements of organochlorine chemicals.]

In a hospital-based case–control study of cancer of the pancreas in Sweden, Hardell et al. (2007) measured concentrations of organochlorine compounds in adipose tissue from 21 cases diagnosed between 1996 and 1999, and 59 controls undergoing surgery for benign prostate hyperplasia (20 men) or hysterectomy (39 women). For *p,p'*-DDE, the odds ratio for exposure above the median (controls, 261 ng/g lipid) adjusted for BMI at tissue sampling, age, and sex was 2.39 (95% CI, 0.73–7.78). The odds ratio was unchanged after consideration of loss of weight during the preceding year. [This was a small study with measurements of organochlorine compounds in the adipose tissue; the effect of the wasting syndrome characteristic of pancreatic cancer on organochlorine concentrations may have influenced the result.]

In a case–control (case–case) study in Spain evaluated the relation between levels of *p,p'*-DDT, *p,p'*-DDE, and PCBs, and mutations in codon 12 of the *K-ras* gene in patients with exocrine cancer of the pancreas (Porta et al., 1999). Cases of pancreatic cancer with wild-type *K-ras* ($n = 17$) were frequency-matched for age and sex to cases of pancreatic cancer with a *K-ras* mutation ($n = 34$). Serum concentrations of *p,p'*-DDT were significantly higher in pancreatic cancer cases with a *K-ras* mutation than in cases without a mutation (unadjusted odds ratio for upper tertile, 8.7 (95% CI, 1.6–48.5), *P* for trend , 0.005). For *p,p'*-DDE, the corresponding figures were 5.3 (95% CI, 1.1–25.2; *P* for trend, 0.03). These associations remained significant after adjusting for covariates, including smoking. A specific association was observed between glycine-to-valine substitution at codon 12 and both *p,p'*-DDT and *p,p'*-DDE concentrations. [The Working Group noted the small size of this study.]

(b) Cancer of the endometrium

Sturgeon et al. (1998) reported the results of a case–control study on cancer of the endometrium in five geographical areas of the USA. Cases were 90 women diagnosed with the disease in seven hospitals between 1987 and 1990. Controls with intact uterus matched on age, race, and areas of residence were selected from random-digit dialling or from the files of the Health Care Financing Administration. Among the 498 eligible cases and 477 eligible controls, only 90 sets of cases and matched controls had sufficient blood volume for organochlorine analyses. Among other organochlorine compounds, four DDT-related compounds (*o,p'*-DDT, *p,p'*-DDT, *o,p'*-DDE, *p,p* '-DDE), as well as 13 other OCPs and 27 PCB congeners were measured in stored serum samples. Among these, *o,p'*-DDE was detected too infrequently for analysis, and only *p,p'*-DDT was higher in cases than in controls ($P = 0.03$). The odds ratio for women in the highest tertile of *p,p'*-DDT was 1.8 (95% CI, 0.7–4.4) compared with the women in the lowest tertile. Odds ratios were not increased for *p,p'*-DDE or for *o,p'*-DDT. [This was the first study examining the relationship between serum organochlorine

Table 2.11 Case–control studies on other cancers and exposure to DDT and its metabolites

Reference, location, enrolment/ follow-up period	Population size, description, exposure assessment method	Organ site	Exposure category or level	Exposed cases/ deaths	Risk estimate (95% CI)	Covariates	Comments
Fryzek et al. (1997) Michigan, USA 1994–1995	Cases: 66; 7 hospitals Controls: 131; population (random-digit dialling); frequency-matched by age group, sex, ethnicity and county Exposure assessment method: questionnaire	Pancreas	DDT			None	Strengths: rapid identification of pancreas cancer cases allowing direct interviews; strict case definition Limitations: small study size; only living cases; low response rate in controls; possible recall bias when assessing exposure to DDT
			Ever exposed	17	1.6 (0.8–3.1)		
			Low exposure	5	1.1 (0.4–3.3)		
			High exposure	7	1.7 (0.6–4.8)		
Hoppin et al. (2000) San Francisco Bay Area, USA Oct–1998	Cases: 108; rapid case ascertainment Controls: 82; random-digit dialling and health care financing records, frequency-matched by sex and age Exposure assessment method: biomarker; measurements of DDE, DDT, PCBs and other OCs	Pancreas	DDE, ng/g lipid			Age, race, sex	Strengths: direct interviews only; consideration of effects of cachexia on OC serum levels in sensitivity analyses; adjustment for exposure to PCBs Limitations: low response rates
			< 850	30	1.0		
			850–1880	37	1.5 (0.7–3.3)		
			≥ 1880	41	2.1 (0.9–4.7)		
			Trend-test P value: 0.08				
Hardell et al. (2007) Sweden 1996–1999	Cases: 21; hospital Controls: 59; surgery patients (benign prostate hyperplasia, 20 men or hysterectomy, 39 women) Exposure assessment method: biomarker; OCs measured in adipose tissue	Pancreas	*p,p′*-DDE > median	14	2.39 (0.73–7.78)	BMI at tissue sampling, age, sex	Strengths: adipose tissue samples; 100% response rate in cases; consideration of BMI 1 yr and 10 yrs before tissue sampling Limitations: small study size

Table 2.11 (continued)

Reference, location, enrolment/ follow-up period	Population size, description, exposure assessment method	Organ site	Exposure category or level	Exposed cases/ deaths	Risk estimate (95% CI)	Covariates	Comments
Sturgeon et al. (1998) USA 198–1990	Cases: 90; 7 hospitals in 5 USA geographical areas Controls: 90; population, with intact uterus, matched by age, race, residence Exposure assessment method: biomarker; measurements of *o,p′*-DDT; *p,p′*-DDT, *o,p′*-DDE, *p,p′*-DDE; 13 other OC compounds & 27 PCBs	Endometrium	*p,p′*-DDE, ng/g lipids			Age, area of residence, weight, ethnicity	Strengths: first study examining the relation between serum OC concentrations and endometrial-cancer risk Limitations: small study size; low proportion of cases & controls with available blood samples
			256–943	27	1.0		
			954–1357	17	0.5 (0.2–1.2)		
			1359–2276	27	1.0 (0.4–2.5)		
			2391–10 486	19	0.7 (0.2–2.0)		
			o,p′-DDT, ng/g lipid				
			0	43	1.0		
			5–75.8	27	0.9 (0.4–2.1)		
			78.2–386.6	20	0.5 (0.1–1.9)		
			p,p′-DDT, ng/g lipids				
			0	41	1.0		
			31.6–98.2	15	0.6 (0.2–1.6)		
			99.0–278.0	34	1.8 (0.7–4.4)		
Weiderpass et al. (2000) Sweden 1996–1997	Cases: 154; hospital gynaecology and oncology departments in 12 counties Controls: 205; population, matched by age and with intact uterus Exposure assessment method: biomarker; 10 chlorinated pesticides, including DDT, and 10 PCBs measured in serum	Endometrium	*p,p′*-DDT:			Age, BMI	Strengths: population-based design; restriction to women who never used HRT; control for possible confounders- collection of blood samples immediately after diagnosis (no effect of cancer treatment) Limitations: relatively low participation rates
			quartile 1	NR	1.0		
			quartile 2	NR	1.1 (0.6–2.2)		
			quartile 3	NR	0.8 (0.4–1.6)		
			quartile 4	NR	1.1 (0.5–2.1)		
			Trend-test *P*-value: 0.95				
			p,p′-DDE:				
			quartile 1	NR	1.0		
			quartile 2	NR	0.9 (0.5–1.8)		
			quartile 3	NR	1.1 (0.6–2.0)		
			quartile 4	NR	1.0 (0.6–2.0)		
			Trend-test *P* value: 0.78				

Table 2.11 (continued)

Reference, location, enrolment/ follow-up period	Population size, description, exposure assessment method	Organ site	Exposure category or level	Exposed cases/ deaths	Risk estimate (95% CI)	Covariates	Comments
Hardell et al. (2004) Sweden 1997–1998	Cases: 76; hospital, surgery for endometrial cancer Controls: 39; same hospitals, women having hysterectomy for endometrial hyperplasia Exposure assessment method: biomarker; *p,p′*-DDE and other OC compounds measured in adipose tissue collected during surgery	Endometrium	*p,p′*-DDE:			Age, BMI	Strengths: adipose tissue samples; high response rates Limitations: controls with benign disease (endometrial hyperplasia) possibly related to OC exposure; very small numbers
			< median	21	1.0		
			≥ median	55	1.9 (0.8–4.8)		
			50–75th percentile	32	2.4 (0.8–6.8)		
			> 75th percentile	23	1.3 (0.4–4.1)		
Zhao et al. (2012) Xiamen, China 2007–2009	Cases: 345; 3 hospitals Controls: 961; healthy control subjects recruited from the same three hospitals Exposure assessment method: biomarker; analytical methods not reported in detail	Liver (HCC)	*p,p′*-DDT, µg/L			Age, sex, education, alcohol consumption, smoking, aflatoxin, HBV, HCV	DDT/DDE ratio elevated (4.89) indicating recent exposure to DDT Strengths: large numbers; elevated levels of OC exposures; data on other risk factors for liver cancer Limitations: recruitment details for the controls not reported; details of exposure and covariate measurement not reported; DDE measurements not lipid-adjusted
			< 16.11	41	1.0		
			16.11–34.63	53	1.3 (0.81–2.08)		
			34.64–43.08	85	2.08 (1.34–3.22)		
			≥ 43.09	166	4.07 (2.72–6.10)		
			Trend-test *P* value: 0				
			p,p′-DDE, µg/L				
			< 2.62	76	1.00		
			2.62–6.84	60	0.79 (0.54–1.17)		
			6.85–10.55	61	0.81 (0.54–1.20)		
			≥ 10.56	148	1.96 (1.39–2.76)		
			Trend-test *P* value: 0.00 001				

Table 2.11 (continued)

Reference, location, enrolment/follow-up period	Population size, description, exposure assessment method	Organ site	Exposure category or level	Exposed cases/deaths	Risk estimate (95% CI)	Covariates	Comments
Howsam et al. (2004) Barcelona, Spain 1996–1998	Cases: 132; hospital Controls: 76; age-sex matched patients at the same hospital with other diseases Exposure assessment method: biomarker	Colon & rectum	*p,p'*-DDE, ng/g lipids			Age, sex, energy intake, BMI	Strengths: careful design, robust analytical methods, potential confounders examined in detail, study of interaction between OC values and K-ras and p53 mutations Limitations: use of hospital controls may introduce selection bias
			< 2574	38	1.00		
			2574–5565	49	2.17 (1.03–4.54)		
			> 5565	45	1.6 (0.79–3.25)		
			Trend-test *P* value: 0.19				
		Colon & rectum	*p,p'*-DDT, ng/g lipids				
			< 336	44	1.00		
			336–740	52	1.58 (0.74–3.36)		
			> 740	36	0.56 (0.27–1.17)		
			Trend-test *P* value: 0.12				
De Stefani et al. (1996) Uruguay 1993–1994	Cases: 270; 5 hospitals, men only Controls: 383; men with other cancers from the same hospitals Exposure assessment method: questionnaire	Lung	DDT, ever exposed	50	1.7 (1.0–2.8)	Age, residence, education, cigarette smoking, alcohol consumption	Overall response rate (all cancer sites), 97.4% Strengths: inclusion of all patients admitted to major hospital; high response rate; adjustment for major confounders Limitations: other cancers as controls; no indication of adjustment of other occupational lung carcinogens
			DDT, 1–20 yrs of exposure	34	1.6 (0.9–4.6)		

BBD, benign breast disease; BMI, body-mass index; CI, confidence interval; DDE, dichlorodiphenyldichloroethylene; DDT, dichlorodiphenyltrichloroethane; HCC, hepatocellular carcinoma; HBV, hepatitis B virus; HCV, hepatitis C virus; HRT, hormone replacement therapy; JEM, job–exposure matrix; LOD, limit of detection; NR, not reported; OC, organochlorine; OPs, organophosphates; PCB, polychlorinated biphenyl; vs, versus; yr, year

concentrations and risk of endometrial cancer. The Working Group noted concerns about the potential for selection bias related to the small proportion of subjects with available blood samples and about the limited precision.]

Weiderpass et al. (2000) conducted a population-based case–control study on cancer of the endometrium in Sweden. Cases were 154 women with endometrial cancer diagnosed in 1996 and 1997 in gynaecology or gynaecological oncology departments in 12 Swedish counties and who had never used hormone replacement therapy. Controls were 205 age-matched women selected from the population register, who had never used hormone replacement therapy and who had not undergone hysterectomy. Four DDT and DDE compounds were measured in serum. No association was found between endometrial cancer and serum concentration of any of the four compounds. [This was a well-conducted study using a population-based design, in which only women who had never used hormonal treatment were included to examine the hormonal effects of environmental exposures on the endometrium.]

In a small hospital-based case–control study in Sweden, Hardell et al. (2004) compared concentrations of *p,p′*-DDE in adipose tissue from 76 cases with cancer of the endometrium, and 39 controls with endometrial hyperplasia undergoing surgery. The odds ratio for women with *p,p′*-DDE concentrations above the median compared with women with concentrations below the median was 1.9 (95% CI, 0.8–4.8), but the odds ratio decreased to 1.3 (95% CI, 0.4–4.1) for women in the highest exposure group (concentration > 75% percentile). [This was a very small study.]

(c) Cancer of the liver

Zhao et al. (2012) published the findings of a hospital-based case–control study of HCC in Xiamen, China. Cases diagnosed with HCC (n = 345) and healthy controls (n = 961) were recruited between 2007 and 2009 in three hospitals, from subjects who had been living for at least 10 years in Xiamen. Cases and controls were interviewed using a structured questionnaire. Organochlorine pesticides and PCB congeners were measured in blood serum samples taken after diagnosis. Higher concentrations of *p,p′*-DDT and *p,p′*-DDE were observed than in a previous study of cancer of the liver in Linxian (McGlynn et al., 2006; see Section 2.1.4). The odds ratio comparing the highest exposure quartile of *p,p′*-DDT to the lowest was 4.07 (95% CI, 2.72–6.10) with adjustment for risk factors including alcohol drinking, HBV, and hepatitis C virus (HCV) infection, and aflatoxin B1, and the dose–response trend was highly significant ($P < 10^{-5}$). Concentrations of *p,p′*-DDE was also associated with cancer of the liver (corresponding adjusted OR, 1.96; 95% CI, 1.39–2.76; P for trend, 10^{-4}). The authors also reported positive interactions of OCPs with other risk factors for HCC, particularly between *p,p′*-DDT and aflatoxin B1. [This was a large study on HCC in a region of high incidence of HCC, with intensive past use and limited current use of OCPs. The Working Group noted concerns about incomplete reporting of this study, particularly with respect to the selection of controls and measurement of covariates.]

(d) Cancer of the colon

Howsam et al. (2004) conducted a case–control study on organochlorine exposure and risk of cancer of the colorectum within a larger hospital-based study in Barcelona, Spain. Cases were a random sample of 132 patients with a new diagnosis of colorectal adenocarcinoma attending a university hospital between 1996 and 1998, frequency-matched on age, sex, and energy intake, to a sample of 76 controls with new diagnoses of other diseases at the same hospital. Organochlorine compounds were measured in blood samples obtained at diagnosis. Overall, as compared with subjects in the first exposure tertile of exposure distribution of *p,p′*-DDE, odds ratios in the second and third exposure tertiles were 2.17 (95% CI, 1.03–4.54) and 1.30 (95% CI, 0.79–3.25), respectively (*P* for trend, 0.19). Corresponding values for *p,p′*-DDT were 1.58 (95% CI, 0.74–3.36) and 0.56 (95% CI, 0.27–1.17), *P* for trend, 0.12. *p,p′*-DDE significantly interacted with *p53* (*P* for interaction, 0.047) and *K-ras* gene mutations (*P* for interaction, 0.012). [DDT was only analysed in a subset of cases and controls from a larger study. The hospital-based design with other patients as controls was a potential source of selection bias.]

(e) Cancer of the lung

De Stefani et al. (1996) conducted a hospital-based case–control study in Montevideo, Uruguay, on occupational risk factors for cancer of the lung. The study was part of a large multisite case–control study: all incident cases of cancer occurring in men aged 30–75 years admitted in any of five major hospitals in Montevideo were included. The overall response rate was 97.4%. The paper reported results on lung cancer (*n* = 270 cases), using patients with cancer at other sites as the control group (*n* = 383), after excluding cancer sites that shared occupational etiologies with lung cancer. Cancers of the colorectum and prostate were the most common diagnoses among controls. Exposure to DDT was assessed from an occupational questionnaire, along with exposures to other occupational hazards. Ever being exposed to DDT was associated with an odds ratio of 1.7 (95% CI, 1.0–2.8) that increased to 2.0 (95% CI, 0.9–4.7) for men exposed to DDT for more than 20 years. The analysis by histological subtype indicated odds ratios of 1.3 (95% CI, 0.7–2.3), 3.6 (95% CI, 1.5–8.9), and 2.3 (95% CI, 1.2–4.7) for squamous cell cancer, small cell cancer, and adenocarcinoma of the lung, respectively. [Assessing exposure to pesticides from the questionnaire was the major limitation of this study; using a cancer reference group was also a limitation if the cancers used as controls shared common occupational exposures with cases of lung cancer.]

3. Cancer in Experimental Animals

The carcinogenicity of DDT in experimental animals was previously reviewed by the Working Group in 1973 (Some Organochlorine Pesticides; IARC, 1974) and in 1991 (Occupational Exposures in Insecticide Application, and Some Pesticides; IARC, 1991).

The Working Group previously classified DDT as having *sufficient evidence* of carcinogenicity in experimental animals (IARC, 1991). The Working Group for the present monograph reviewed all studies, including those published after 1991, and summarized those judged adequate for an evaluation of carcinogenicity. The findings of the pertinent studies are summarized in Table 3.1.

3.1 Mouse

3.1.1 Oral administration

In a screening study on about 70 compounds, groups of 18 male and 18 female (C57Bl/6 × C3H/Anf) F_1 and (C57Bl/6 × AKR)F_1 mice (age, 7 days) were given daily single doses of *p,p′*-DDT [purity unspecified] at 46.4 mg/kg bw (maximum tolerated dose) by gavage until age 28 days, when the mice were transferred to a diet containing *p,p′*-DDT at a concentration of 140 mg/kg. Groups of 90 mice served as controls. About 30% of females of both strains died during the treatment. The surviving mice were killed at age 81 weeks. The incidence of hepatoma (benign or malignant, combined) was increased significantly in male and female mice of each strain, except in female (C57Bl/6 × AKR)F_1 mice, and the incidence of malignant lymphoma was significantly increased in (C57Bl/6 × AKR)F_1 females (NTIS, 1968; Innes et al., 1969).

In a five-generation study, *p,p′*-DDT-treated and control groups of male and female BALB/c mice from each of the five generations (F_1–F_5) were studied for tumour incidence. Groups of mice from the F_1–F_5 generations (a total of 683 mice, including males and females) received a diet containing *p,p′*-DDT [purity unspecified] at a concentration of 2.8–3 mg/kg for 6 months, and other groups of mice from the F_1–F_5 generations (a total of 406 mice, including males and females) received a control diet. At experimental month 26, the incidence of pulmonary carcinoma was significantly increased in treated mice (all generations combined; 116/683 [$P < 0.001$]) compared with controls (5/406). The incidences of lymphosarcoma (all generations combined; 15/683 versus 1/406 controls [$P < 0.001$]), and leukaemia (all generations combined; control, 10/406; treated, 85/683 (64/683 in females) [$P < 0.001$]) were also significantly increased in treated mice (Tarján & Kemény, 1969). [The Working Group determined that the higher incidence of pulmonary carcinoma, leukaemia, and lymphosarcoma, attained significance in the F_2, F_3, and F_3 generations of treated mice, respectively, and subsequently increased in each succeeding generation.]

In a two-generation dose–response study, groups of 90–127 male and female CF-1 mice (including parent F_0, and offspring F_1) were fed a diet containing technical-grade DDT at concentrations of 0, 2, 10, 50, or 250 mg/kg, starting at age 6–7 weeks for the F_0 generation and continuing in the F_0 and F_1 for life. There was excess mortality from week 60 onwards among mice of the F_0 and F_1 generations that had received DDT at 250 mg/kg diet. The incidence (both generations combined) of hepatoma (benign or malignant, combined) was increased by exposure to DDT. The incidences were in males: 25/113 (controls), 57/124, 52/104, 67/127, and 82/103, respectively; and in females: 4/111 (controls), 4/105, 11/124, 13/104, and 60/90, respectively. The increase in the incidence of hepatoma over that in controls in male and female mice fed DDT at 250 mg/kg diet was significant [$P < 0.01$]. In females, the excess over that of the controls was also significant in the group fed DDT at 50 mg/kg diet [$P < 0.05$] (Tomatis et al., 1972).

In a continuation of the study byTomatis et al. (1972), the effects of the same doses of DDT were studied by Turusov et al. (1973) in six consecutive generations of CF-1 mice [including the first two generations described by Tomatis et al. (1972)]. The experiment involved a total of 2764 exposed and 668 control animals. Exposure to all four levels of technical-grade DDT (2, 10, 50, 250 mg/kg diet) for life significantly increased the incidence (all generations combined) of hepatoma (benign or malignant, combined) in males. In females, the incidence of hepatoma was significantly increased after exposure at 10, 50, or 250 mg/kg, with a significant positive trend [$P < 0.001$]. Hepatoblastoma was observed at a significantly increased incidence in DDT-treated male mice: 3/328 in control males, and 5/354, 14/362, 12/383, and 25/350 in males treated at 2, 10, 50, and 250 ppm, respectively, with a significant positive trend [$P < 0.001$]. DDT did not significantly alter the incidence of tumours at sites other than the liver (Turusov et al., 1973).

In a two-generation study, 515 female and 430 male BALB/c mice were given diets containing technical-grade DDT at a concentration of 0, 2, 20, or 250 mg/kg for life. In females, the survival rates were comparable in all groups; in males, early deaths occurred in all groups as a consequence of fighting and because of toxicity (at the highest dose). In male and female (F_0 and F_1 combined) mice that survived more than 60 weeks, the incidences of liver cell tumours (benign or malignant, combined) were significantly increased in males and females fed diet containing DDT at 250 mg/kg (Terracini et al., 1973a). Confirmatory results were obtained in two subsequent generations of BALB/c female mice (F_2 and F_3) fed diets containing DDT at 250 mg/kg. Mice of the F_1, F_2 and F_3 generations, which were exposed to DDT both in utero and after birth for life, developed more liver cell tumours than did F_0 mice, which were exposed to DDT only after weaning [data presented only by graph, not by table or exact number] (Terracini et al., 1973b).

Table 3.1 Studies of carcinogenicity with DDT and its metabolites in mice, rats, hamsters, and monkeys

Species, strain (sex) Duration Reference	Route, dosing regimen No. of animals/group at start	Results For each target organ: incidence (%) and/or multiplicity of tumours	Significance	Purity Comments
DDT				
Full carcinogenicity				
Mouse, (C57Bl/6 × C3H/Anf)F_1 (M) 80 wk Innes et al. (1969); NTIS (1968)	0, 46.4 mg/kg bw, by gavage (1×/day), age 7–28 days; subsequently in diet at 0, or 140 mg/kg 90, 18 mice	Hepatoma (benign or malignant, combined): 8/79, 11/18	[$P < 0.01$]	*p,p′*-DDT: purity, NR
Mouse, (C57Bl/6 × C3H/Anf)F_1 (F) 80 wk Innes et al. (1969); NTIS (1968)	0, 46.4 mg/kg bw, by gavage (1×/day), age 7–28 days; subsequently at 0, or 140 mg/kg diet 90, 18 mice	Hepatoma (benign or malignant, combined): 0/87, 4/18	[$P < 0.01$]	*p,p′*-DDT: purity, NR
Mouse, (C57Bl/6 × AKR) F_1 (M) 80 wk Innes et al. (1969); NTIS (1968)	0, 46.4 mg/kg bw, by gavage (1×/day), age 7–28 days; subsequently at 0, or 140 mg/kg diet 90, 18 mice	Hepatoma (benign or malignant, combined): 5/90, 7/18	[$P < 0.01$]	*p,p′*-DDT: purity, NR
Mouse, (C57Bl/6 × AKR) F_1 (F) 80 wk Innes et al. (1969); NTIS (1968)	0, 46.4 mg/kg bw, by gavage (1×/day), age 7–28 days; subsequently at 0, or 140 mg/kg diet 90, 18 mice	Hepatoma (benign or malignant, combined): 1/82, 1/18 Lymphoma: 4/82, 6/18	[NS] [$P < 0.01$]	*p,p′*-DDT: purity, NR
Mouse, BALB/c (M+F) 5 generations, for life Tarján & Kemény (1969)	0, 2.8–3 mg/kg diet for 6 mo for all 5 generations 406, 683 mice	Pulmonary carcinoma: 5/406, 116/683 Leukaemia: 10/406, 85/683 Lymphosarcoma: 1/406, 15/683	[$P < 0.001$] [$P < 0.001$] [$P < 0.001$]	*p,p′*-DDT: purity, NR Tumour incidences for each of the 5 generations at experimental month 26 were combined The increases in the incidence of pulmonary carcinoma, leukaemia or lymphosarcoma were consistent across all generations.
Mouse, CF-1 (M) 2-generation for life Tomatis et al. (1972)	0, 2, 10, 50, 250 mg/kg diet for life for both generations 113, 124, 104, 127, 103 mice	Hepatoma (benign or malignant, combined): 25/113, 57/124, 52/104, 67/127, 82/103*	* [$P < 0.01$]	73–78% *p,p′*-DDT, 20% *o,p′*-DDT, 1% *m,p′*-DDT, 0.5–1.5% *p,p′*-TDE and 0.5% *p,p′*-DDE Tumour incidences for both generations were combined

Table 3.1 (continued)

Species, strain (sex) Duration Reference	Route, dosing regimen No. of animals/group at start	Results For each target organ: incidence (%) and/or multiplicity of tumours	Significance	Purity Comments
Mouse, CF-l (F) 2-generation for life Tomatis et al. (1972)	0, 2, 10, 50, 250 mg/kg diet for life for both generations 111, 105, 104, 104, 90 mice	Hepatoma (benign or malignant, combined): 4/111, 4/105, 11/124, 13/104*, 60/90**	* [$P < 0.05$]; ** [$P < 0.01$]	73–78% *p,p'*-DDT, 20% *o,p'*-DDT, 1% *m,p'*-DDT, 0.5–1.5% *p,p'*-TDE and 0.5% *p,p'*-DDE Tumour incidences for both generations were combined
Mouse, CF-l (M) 6-generation for life Turusov et al. (1973)	0, 2, 10, 50, 250 mg/kg diet for life for all generations 328, 354, 362, 383, 350 mice	Hepatoma (benign or malignant, combined): 97/328, 179/354, 181/362, 214/383, 301/350 Hepatoblastoma: 3/328, 5/354, 14/362*, 12/383*, 25/350**	[$P < 0.01$ (all doses)] * [$P < 0.05$]; ** [$P < 0.01$]; [$P < 0.001$ (trend)]	73–78% *p,p'*-DDT, 20% *o,p'*-DDT, 1% *m,p'*-DDT, 0.5–1.5% *p,p'*-TDE and 0.5% *p,p'*-DDE Tumour incidences for all generations were combined
Mouse, CF-l (F) 6-generation for life Turusov et al. (1973)	0, 2, 10, 50, 250 mg/kg diet for life for all generations 340, 339, 355, 328, 293 mice	Hepatoma (benign or malignant, combined): 16/340, 12/339, 32/355*, 43/328**, 192/293**	* [$P < 0.05$]; ** [$P < 0.01$]; [$P < 0.001$ (trend)]	73–78% *p,p'*-DDT, 20% *o,p'*-DDT, 1% *m,p'*-DDT, 0.5–1.5% *p,p'*-TDE and 0.5% *p,p'*-DDE Tumour incidences for all generations were combined
Mouse, BALB/c (M) 2-generation for life Terracini et al. (1973a)	0, 2, 20, 250 mg/kg diet for life for both generations 107, 112, 105, 106 mice	Liver cell tumours (benign or malignant, combined): 1/62, 3/48, 0/48, 14/31*	* [$P < 0.01$]; [$P < 0.001$ (trend)]	70–75% *p,p'*-DDT, 20% *o,p'*-DDT and 0.2–4% *p,p'*-TDE Incidence in mice that died after 60 wk Tumour incidences for both generations were combined
Mouse, BALB/c (F) 2-generation for life Terracini et al. (1973a)	0, 2, 20, 250 mg/kg diet for life for both generations 131, 135, 128, 121 mice	Liver cell tumours (benign or malignant, combined): 0/124, 0/130, 1/126, 71/115*	* [$P < 0.01$; [$P < 0.001$ (trend)]	70–75% *p,p'*-DDT, 20% *o,p'*-DDT and 0.2–4% *p,p'*-TDE Incidence in mice that died after 60 wk Tumour incidences for both generations were combined
Mouse, CF-1 (M) 2 years Walker et al. (1973)	0, 50, 100 mg/kg diet 47, 32, 32 mice	Liver cell tumours [benign or malignant, combined]: 6/47, 12/32*, 17/32**	* [$P < 0.05$]; ** [$P < 0.01$]; [$P < 0.001$ (trend)]	*p,p'*-DDT; purity, > 99.5%
Mouse, CF-1 (F) 2 years Walker et al. (1973)	0, 50, 100 mg/kg diet 47, 30, 32 mice	Liver cell tumours [benign or malignant, combined]: 8/47, 15/30*, 24/32*	* [$P < 0.01$]; [$P < 0.001$ (trend)]	*p,p'*-DDT; purity, > 99.5%
Mouse, CF-1 (M) 110 wk Thorpe & Walker (1973)	0, 100 mg/kg diet 45, 30 mice	Liver cell tumours [benign or malignant, combined]: 11/45, 23/30	$P < 0.01$	*p,p'*-DDT: purity, > 99.5%

Table 3.1 (continued)

Species, strain (sex) Duration Reference	Route, dosing regimen No. of animals/group at start	Results For each target organ: incidence (%) and/or multiplicity of tumours	Significance	Purity Comments
Mouse, CF-1 (F) 110 wk Thorpe & Walker (1973)	0, 100 mg/kg diet 44, 30 mice	Liver cell tumours [benign or malignant, combined]: 10/44, 26/30	$P < 0.01$	*p,p′*-DDT: purity, > 99.5%
Mouse, Swiss (M) 80 wk Kashyap et al. (1977)	0 (untreated control), 100 mg/kg diet or 0.25 mg/animal by gavage 30, 30, 30 mice	Lymphoma: 2/26, 8/27*, 6/24	*[$P < 0.05$]	70.5% *p,p′*-DDT and 21.3% *o,p′*-DDT
Mouse, Swiss (F) 80 wk Kashyap et al. (1977)	0 (untreated control), 100 mg/kg diet or 0.25 mg/animal by gavage 30, 30, 30 mice	Lymphoma: 2/20, 8/22*, 8/24	*[$P < 0.05$]	70.5% *p,p′*-DDT and 21.3% *o,p′*-DDT
Mouse, B6C3F$_1$ (M) 91 wk NCI (1978)	0, 22, 44 mg/kg diet TWA for 78 wk 20, 50, 50 mice	Lymphoma: 0/19, 2/49, 1/50	NS	Technical-grade DDT; purity, about 70%, assumed to be *p,p′*-DDT Survival at 70 wk: 12/20, 20/50, 37/50
Mouse, B6C3F$_1$ (F) 92 wk NCI (1978)	0, 87, 175 mg/kg diet TWA for 78 wk 20, 50, 50 mice	Lymphoma: 0/20, 3/49, 7/46	$P = 0.026$ (trend), Cochran-Armitage test	Technical-grade DDT; purity, about 70%, assumed to be *p,p′*-DDT Survival at 70 wk: 20/20, 45/50, 36/50
Mouse, BALB/c (M) 75 wk Lipsky et al. (1989)	0 (control) or 175 ppm in the diet (0–16 wk), 125 ppm (16–24 wk) then 100 ppm (24–75 wk) 90, 90 mice	Hepatocellular adenoma: 0/10, 1/4 (at 52 wk) 2/36, 4/12 (at 75 wk) HCC: 1/36, 2/12 (at 75 wk) Hepatocellular adenoma or HCC (combined): 3/36, 5/12 (at 75 wk)	 [NS] [$P < 0.03$] [NS] [$P < 0.02$]	DDT: purity, 99%; the Working Group was unable to determine whether this was technical grade or *p,p′*-DDT Interim sacrifices at 2, 4, 8, 16, 24, 36, and 52 wk
Mouse, CF-1 (M) 65 wk	0 (control) or 250 ppm in the diet (15 or 30 wk treatment) 70, 60, 60 mice	Hepatoma [not further classified]: 12/70, 13/60 (15 wk), 38/60* (30 wk)	* [$P < 0.01$]	Technical-grade DDT: purity, NR
95 wk Tomatis et al. (1974a)	0 (control) or 250 ppm in the diet (15 or 30 wk treatment) 83, 60, 60 mice	Hepatoma [not further classified]: 24/83, 25/60 (15 wk), 41/60* (30 wk)	* [$P < 0.01$]	

Table 3.1 (continued)

Species, strain (sex) Duration Reference	Route, dosing regimen No. of animals/group at start	Results For each target organ: incidence (%) and/or multiplicity of tumours	Significance	Purity Comments
120 wk Tomatis et al. (1974a)	0 (control) or 250 ppm in the diet (15 or 30 wk treatment) 98, 60, 60 mice	Hepatoma [not further classified]: 33/98, 25/60 (15 wk), 37/60* (30 wk)	* [$P < 0.01$]	
Mouse, CF-1 (F) 65 wk	0 (control) or 250 ppm in the diet (15 or 30 wk treatment) 69, 60, 54 mice	Hepatoma [not further classified]: 0/69, 3/60 (15 wk), 4/54* (30 wk)	* [$P < 0.05$]	Technical-grade DDT: purity, NR
95 wk	0 (control) or 250 ppm in the diet (15 or 30 wk treatment) 72, 60, 55 mice	Hepatoma [not further classified]: 0/72, 11/60* (15 wk), 11/55* (30 wk)	* [$P < 0.01$]	
120 wk Tomatis et al. (1974a)	0 (control) or 250 ppm in the diet (15 or 30 wk treatment) 90, 60, 54 mice	Hepatoma [not further classified]: 1/90, 5/60* (15 wk), 11/54** (30 wk)	* [$P < 0.05$]; ** [$P < 0.01$]	
Mouse, Swiss (M) 80 wk Kashyap et al. (1977)	Skin application, 0 (untreated controls), 0.25 mg/animal, twice per wk 30, 30 mice	No significant increase	[NS]	70.5% *p,p′*-DDT and 21.3% *o,p′*-DDT
Mouse, Swiss (F) 80 wk Kashyap et al. (1977)	Skin application, 0 (untreated controls) or 0.25 mg/animal, twice per wk 30, 30 mice	No significant increase	[NS]	70.5% *p,p′*-DDT and 21.3% *o,p′*-DDT
Mouse, Swiss (M) 80 wk Kashyap et al. (1977)	Subcutaneous injection, 0 (untreated controls), or 0.25 mg/animal (2×/mo) 30, 30 mice	Liver cell carcinoma: 1/26, 3/28	[NS]	70.5% *p,p′*-DDT and 21.3% *o,p′*-DDT
Mouse, Swiss (F) 80 wk Kashyap et al. (1977)	Subcutaneous injection, 0 (untreated controls), or 0.25 mg/animal (2×/mo) 30, 30 mice	Liver cell carcinoma: 0/20, 7/26	[$P = 0.0123$]	70.5% *p,p′*-DDT and 21.3% *o,p′*-DDT
Co-administration with known carcinogens or modifying factors				
Mouse, dd (F) 8 wk Uchiyama et al. (1974)	0, 100 ppm in the diet 1 wk after start of DDT treatment (for 8 wk), 3-methylcholanthrene was applied for 4 wk in the uterus 39, 16 mice	Cervical epithelial carcinoma: 0/39, [3/16 (about 20%)]	[$P < 0.03$]	DDT, not further specified: purity NR

Table 3.1 (continued)

Species, strain (sex) Duration Reference	Route, dosing regimen No. of animals/group at start	Results For each target organ: incidence (%) and/or multiplicity of tumours	Significance	Purity Comments
Mouse, B6C3F$_1$ (M) 43 wk Williams & Numoto (1984)	0, 50 ppm in the diet Initiated by NDEA at 20 ppm in drinking-water for 14 wk before DDT treatment (after 4 wk) for 25 wk 30, 30 mice	Hepatocellular adenoma or carcinoma (combined): 8/20, 14/21	[NS]	Technical-grade DDT: purity, 97.6%
Full carcinogenicity				
Rat, Osborne-Mendel (M+F) 24 mo Fitzhugh & Nelson (1947)	0, 100, 200, 400, 600, 800 ppm in the diet 36, 12, 36, 36, 48, 60 rats	"Low-grade" HCC: 4/81 (all treated groups)	[NS]	81.8% *p,p′*-DDT and 18.2% *o,p′*-DDT Tumour incidence in rats that survived ≥ 18 mo
		Liver nodular adenomatoid hyperplasias [adenoma]: 11/81 (all treated groups)	[NS]	
		No liver tumours in 20 controls		
Rat, Osborne-Mendel (M) 2 yr Reuber (1978)	0, 200, 400, 600, 800 ppm in the diet 12, 12, 12, 24, 24 rats	HCC: 0/5, 6/21 (all DDT treatment groups)	[NS]	81.8% *p,p′*-DDT and 18.2% *o,p′*-DDT The effective number of rats is the total number of rats that survived 84 wk or longer
		Liver neoplastic nodules [adenomas]: 0/5, 4/21 (all DDT treatment groups)	[NS]	
		Total liver tumours (HCC, neoplastic nodules, Küpffer cell sarcoma): 0/5, 11/21 (all DDT treatment groups)	[$P < 0.05$]	
		Lymphosarcoma: 0/6, 14/50 (all DDT treatment groups)	[NS]	The effective number of rats is the total number of rats that survived 52 wk or longer
Rat, Osborne-Mendel (F) 2 yr Reuber (1978)	0, 200, 400, 600, 800 ppm in the diet 12, 12, 12, 24, 24 rats	HCC: 0/6, 4/12 (all DDT treatment groups combined)	[NS]	81.8% *p,p′*-DDT and 18.2% *o,p′*-DDT The effective number of rats is the total number of rats that survived 84 wk or longer
		Ovary carcinoma: 0/6, 11/12 (all DDT treatment groups combined)	[$P = 0.0004$]	The effective number of rats is the total number of rats that survived ≥ 89 wk
Rat, Osborne-Mendel (M, F) 2 yr Radomski et al. (1965)	0, 80 mg/kg diet 30 M and 30 F/group	Undifferentiated bronchogenic carcinoma (M+F): 2/60, 8/60	[$P < 0.05$]	DDT, not further specified: purity, NR

Table 3.1 (continued)

Species, strain (sex) Duration Reference	Route, dosing regimen No. of animals/group at start	Results For each target organ: incidence (%) and/or multiplicity of tumours	Significance	Purity Comments
Rat, Osborne- Mendel (F) Up to 27 mo Deichmann et al. (1967)	0, 200 mg/kg diet 30 M and 30 F/group	No lung tumours	NS	DDT, not further specified: purity, NR
Rat, Wistar (M) 145 wk Rossi et al. (1977)	0, 500 mg/kg diet 36, 37 rats	Liver cell tumours [benign or malignant, combined]: 0/35, 9/27	[$P = 0.002$]	70–75% *p,p'*-DDT, 20% *o,p'*-DDT and 0.2–4% *p,p'*-TDE
Rat, Wistar (F) 145 wk Rossi et al. (1977)	0, 500 mg/kg diet 35, 35 rats	Liver cell tumours [benign or malignant, combined]: 0/32, 15/28	[$P < 0.0001$]	70–75% *p,p'*-DDT, 20% *o,p'*-DDT and 0.2–4% *p,p'*-TDE
Rat, Wistar (M) 60 wk Angsubhakorn et al. (2002)	0, 500 ppm in the diet (from wk 6–12) 30, 35 rats	Liver neoplastic nodules [benign]: 0/18, 1/19	[NS]	*p, p'*-DDT; purity, 86% Animals of age ≥ 1 year; weight, 400–500 g
Rat, F344 (M) 103 wk Shivapurkar et al. (1986)	0, 0.05% for 72 wk then fed un-supplemented diet 30, 30 rats	Liver neoplastic nodules [benign]: 2/28, 6/28 HCC: 1/28, 0/28	[NS] [NS]	DDT, not further specified: purity, NR Age NR; weight, > 50–60 g
Rat, Osborne-Mendel (M) 111 wk NCI (1978)	0, 321, 642 mg/kg diet TWA for 78 wk 20, 50, 50 rats	No significant increase	NS	Purity, about 70%; assumed to be *p,p'*-DDT Survival not affected
Rat, Osborne-Mendel (F) 111 wk NCI (1978)	0, 210, 420 mg/kg diet TWA for 78 wk 20, 50, 50 rats	Adrenal gland pheochromocytoma: 0/19, 0/38, 3/24 Thyroid follicular cell adenoma or carcinoma (combined): 1/19, 13/45*, 10/43 Thyroid follicular cell adenoma: 1/19, 10/45, 5/43 Thyroid follicular cell carcinoma: 0/19, 4/45, 6/43	$P = 0.031$ (trend), Cochran-Armitage test * $P = 0.032$	Purity, about 70%; assumed to be *p,p'*-DDT Survival not affected
Rat, MRC Porton (M) 144 wk Cabral et al. (1982a)	0, 125, 250, 500 mg/kg diet 38, 30, 30, 38 rats	Liver cell tumours [benign or malignant, combined]: 1/38, 0/30, 1/30, 2/38	[NS]	78.9% *p,p'*-DDT, 16.7% *o,p'*-DDT, 1.6% *p,p'*-DDE, 0.6% *p,p'*-TDE, 0.2% *o,p'*- DDE, 0.1% *o,p'*-TDE and 1.9% unknown
Rat, MRC Porton (F) 144 wk Cabral et al. (1982a)	0, 125, 250, 500 mg/kg diet 38, 30, 30, 38 rats	Liver cell tumours [benign or malignant, combined]: 0/38, 2/30, 4/30, 7/38*	$P < 0.001$ (trend), * [$P < 0.01$]	78.9% *p,p'*-DDT, 16.7% *o,p'*-DDT, 1.6% *p,p'*-DDE, 0.6% *p,p'*-TDE, 0.2% *o,p'*- DDE, 0.1% *o,p'*-TDE and 1.9% unknown

Table 3.1 (continued)

Species, strain (sex) Duration Reference	Route, dosing regimen No. of animals/group at start	Results For each target organ: incidence (%) and/or multiplicity of tumours	Significance	Purity Comments
Rat, F344/DuCrj (M) 78 wk (satellite experiment) Harada et al. (2003)	0, 5, 50, 500 ppm in the diet 8, 8, 8, 7 rats	Hepatocellular adenoma: 0/8, 0/8, 0/8, 6/7*	* $P < 0.01$	*p,p′*-DDT: purity, > 98%
Rat, F344/DuCrj (F) 78 wk (satellite experiment) Harada et al. (2003)	0, 5, 50, 500 ppm in the diet 8, 8, 7, 8 rats	Hepatocellular adenoma: 0/8, 0/8, 0/7, 1/8	NS	*p,p′*-DDT: purity, > 98%
Rat, F344/DuCrj (M) 104 wk Harada et al. (2003)	0, 5, 50, 500 ppm in the diet (0, 0.17, 1.7, 19.1 mg/kg per day) 40, 40, 40, 40 rats	Hepatocellular adenoma: 0/40, 0/40, 5/40*, 22/40** HCC: 0/40, 0/40, 0/40, 14/40*	* $P < 0.05$; ** $P < 0.01$ * $P < 0.01$	*p,p′*-DDT: purity, > 98%
Rat, F344/DuCrj (F) 104 wk Harada et al. (2003)	0, 5, 50, 500 ppm in the diet (0, 0.21, 2.2, 25.2 mg/kg per day) 40, 40, 40, 40 rats	Hepatocellular adenoma: 0/40, 0/40, 0/40, 16/40* HCC: 0/40, 0/40 0/40, 2/40	* $P < 0.01$ NS (see comments)	*p,p′*-DDT: purity, > 98% The Working Group considered that the incidence of hepatocellular adenoma or carcinoma (combined) [16–18/40] was significantly increased in the high-dose group
Co-administration with known carcinogens or modifying factors				
Rat, Buffalo (M) 82 wk Angsubhakorn et al. (2002)	0, 100 ppm in the diet (from wk 1–20) Initiated by a single dose of AFB_1 at 5 mg/kg bw by gavage before DDT treatment for 20 wk 14, 19 rats	Liver neoplastic nodules [benign]: 1/14, 3/19	NS	*p, p′*-DDT: purity, 86%
Rat, Wistar (M) 60 wk Angsubhakorn et al. (2002)	0, 500 ppm in the diet (from wk 6–12) Initiated by AFB_1 at 4 ppm in the diet for 6 wk before DDT treatment 35, 43 rats	Liver neoplastic nodules [benign]: 9/29, 9/28 Malignant hepatic tumours: 8/29, 10/28 HCC: 6/29, 7/28 Cholangiocellular carcinoma: 1/29, 1/28 Hepato-cholangiocellular carcinoma: 1/29, 2/28	[NS] [NS] [NS] [NS] [NS]	*p, p′*-DDT: purity, 86% Animals of age ≥1 year; weight, 400–500 g

Table 3.1 (continued)

Species, strain (sex) Duration Reference	Route, dosing regimen No. of animals/group at start	Results For each target organ: incidence (%) and/or multiplicity of tumours	Significance	Purity Comments
Rat, Wistar (M) 40 or 52 wk Nishizumi (1979)	0, 1.25 mg/rat by gavage, twice per wk for 12 wk Initiated by NDEA at 50 ppm in drinking-water for 2 wk before DDT treatment 1 wk after. The animals were killed at weeks 40 or 52 10, 10, 10, 10 rats	Liver tumours at 40 wk (> 5 mm): 0/8, 1/6 Liver tumours at 52 wk (> 5 mm): 0/8, 3/8	[NS] [NS]	Purity, NR
Rat, Sprague- Dawley (M) 389 days Peraino et al. (1975)	0, 0.05% in the diet Initiated by AAF at the dose of 0.02% in diet for 18 days, then 1 wk later were treated with DDT 120, 120 rats	Hepatocellular adenoma or carcinoma (combined): 31/108, 77/103	[$P < 0.0001$]	Technical-grade: 70% *p,p′*-DDT, 12% *o,p′*-DDT, 3% *p,p′*-TDE and 15% *p,p′*-DDE
Rat, F344 (M) 103 wk Shivapurkar et al. (1986)	0, 0.05% in the diet for 72 wk then fed un-supplemented diet Initiated by single i.p. injection of NDEA at 200 mg/kg bw before DDT treatment 30, 30 rats	Liver neoplastic nodules [benign]: 3/28, 1/28 HCC: 18/28, 28/28 Cholangioma or cholangiocarcinoma (combined): 0/28, 6/28	[NS] [$P < 0.0001$] [$P < 0.02$]	DDT, not further specified: purity, NR Age NR; weight, > 50–60 g
Rat, F344 (M) 43 wk Kushida et al. (2005)	0, 0.005, 0.5, 500 ppm in the diet Initiated by two i.p. injections of NDEA at 100 mg/kg bw with a 1-wk interval before DDT treatment 20, 20, 20, 20 rats	Hepatocellular adenoma: 11/20, 13/20, 10/20, 18/18* Multiplicity: 1.10, 0.95, 1.10, 11.44* tumours/rat HCC: 12/20, 7/20, 13/20, 18/18* Multiplicity: 1.00, 0.50, 0.90, 28.67* tumours/rat Hepatocellular adenoma or HCC (combined): 15/20, 14/20, 17/20, 18/18* Multiplicity: 2.10, 1.45, 2.00, 40.11* tumours/rat	* [$P < 0.002$] * $P < 0.01$, Dunnett's test * [$P < 0.01$] * $P < 0.01$, Dunnett's test * [$P < 0.05$] * $P < 0.01$, Dunnett's test	*p,p′*-DDT: purity, > 98%

Table 3.1 (continued)

Species, strain (sex) Duration Reference	Route, dosing regimen No. of animals/group at start	Results For each target organ: incidence (%) and/or multiplicity of tumours	Significance	Purity Comments
Full carcinogenicity				
Hamster, Syrian golden (M) 120 wk Cabral et al. (1982b)	0, 125, 250, 500 mg/kg diet 40, 30, 31, 40 hamsters	Adrenal cortex tumours (mostly adenomas): 3/40, 4/30, 6/31, 8/39	[$P = 0.04$ (trend)]	78.9% *p,p'*-DDT, 16.7% *o,p'*-DDT, 1.6% *p,p'*-DDE, 0.6% *p,p'*- TDE, 0.2% *o,p'*-DDE, 0.1% *o,p'*-TDE and 1.9% unknown
Hamster, Syrian golden (F) 120 wk Cabral et al. (1982b)	0, 125, 250, 500 mg/kg diet 40, 30, 29, 40 hamsters	Adrenal cortex tumours (mostly adenomas): 0/39, 0/28, 1/28, 3/40	[NS]	78.9% *p,p'*-DDT, 16.7% *o,p'*-DDT, 1.6% *p,p'*-DDE, 0.6% *p,p'*- TDE, 0.2% *o,p'*-DDE, 0.1% *o,p'*-TDE and 1.9% unknown
Hamster, Syrian golden (M) 120 wk Rossi et al. (1983)	0, 1000 mg/kg diet 31, 35 hamsters	Adrenal gland tumours (mainly cortical adenoma): 8/31, 14/35	[NS]	70–75% *p,p'*-DDT, 20% *o,p'*-DDT and 0.2–4% *p,p'*-TDE
Hamster, Syrian golden (F) 120 wk Rossi et al. (1983)	0, 1000 mg/kg diet 42, 36 hamsters	Adrenal gland tumours (mainly cortical adenoma): 2/42, 10/36	[$P < 0.01$]	70–75% *p,p'*-DDT, 20% *o,p'*-DDT and 0.2–4% *p,p'*-TDE
Hamster, strain NR (M) 18 months Graillot et al. (1975)	0, 250, 500, 1000 mg/kg diet 30, 30, 30, 30 hamsters	Lymphosarcoma: 50.0%, 22.7%, 13.0%, 0%	NS (for increase)	Technical-grade DDT: 70% *p,p'*-DDT No other tumour types observed
Hamster, strain NR (F) 18 months Graillot et al. (1975)	0, 250, 500, 1000 mg/kg diet 30, 30, 30, 30 hamsters	Lymphosarcoma: 41.0%, 17.4%, 0%, 0%	NS (for increase)	Technical-grade DDT: 70% *p,p'*-DDT No other tumour types observed
Co-administration with known carcinogens or modifying factors				
Hamster, Syrian golden (M) 31 wk Tanaka et al. (1987)	0, 500 ppm in the diet for 30 wk Initiated by i.p. injection of NDMA at 6 mg/kg bw, then DDT 1 wk after 15, 15 hamsters	Hepatocellular adenoma: 1/15, 1/15	[NS]	Purity, NR

Table 3.1 (continued)

Species, strain (sex) Duration Reference	Route, dosing regimen No. of animals/group at start	Results For each target organ: incidence (%) and/or multiplicity of tumours	Significance	Purity Comments
Full carcinogenicity				
Monkey, cynomolgus and rhesus (M+F) Up to 304 mo Takayama et al. (1999)	0 (control), 20 mg/kg bw in the diet for 130 mo 17, 24 monkeys	One HCC and one prostatic adenocarcinoma in two DDT-treated cynomolgus monkeys; and two leiomyoma of the uterus and one of the oesophagus in three other DDT-treated monkeys No tumours in controls		*p,p′*-DDT: purity, NR Treated: 13 cynomolgus monkeys and 11 rhesus monkeys Controls: 9 cynomolgus monkeys and 8 rhesus monkeys
DDD (TDE)				
Full carcinogenicity				
Mouse, CF-l (M) 123–124 wk Tomatis et al. (1974b)	0, 250 mg/kg diet 100, 60 mice	Hepatomas [benign or malignant, combined]: 33/98, 31/59	[$P < 0.05$]	*p,p′*-TDE: purity, 99%
		Lung tumours (adenoma or adenocarcinoma, combined): 53/98, 51/59	[$P < 0.0001$]	
Mouse, CF-l (F) 123–124 wk Tomatis et al. (1974b)	0, 250 mg/kg diet 90, 60 mice	Hepatomas [benign or malignant, combined]: 1/90, 1/59	[NS]	*p,p′*-TDE: purity, 99%
		Lung tumours (adenoma or adenocarcinoma, combined): 37/90, 43/59	[$P < 0.0001$]	
Mouse, B6C3F$_1$ (M) 90–92 wk NCI (1978)	0, 411, 822 mg/kg diet TWA for 78 wk 20, 50, 50 mice	HCC: 2/18, 12/44, 14/50	NS	TDE (principal component, 60%, assumed to be *p,p′*-TDE; 19 unidentified impurities)
Mouse, B6C3F$_1$ (F) 90–93 wk NCI (1978)	0, 411, 822 mg/kg diet TWA for 78 wk 20, 50, 50 mice	HCC: 0/20, 2/48, 3/47	NS	TDE (principal component, 60%, assumed to be *p,p′*-TDE; 19 unidentified impurities)
Rat, Osborne-Mendel (M) 111–112 wk NCI (1978)	0, 1647, 3294 mg/kg diet TWA for 78 wk 20, 50, 50 rats	Thyroid follicular cell adenoma: 0/19, 11/49*, 9/49*	* [$P < 0.05$], [$P < 0.05$ (trend)]	TDE (principal component, 60%, assumed to be *p,p′*-TDE; 19 unidentified impurities)
		Thyroid follicular cell carcinoma: 1/19, 6/49, 3/49	NS	
		Thyroid follicular cell adenoma or carcinoma (combined): 1/19, 16/49*, 11/49	* $P = 0.016$	

Table 3.1 (continued)

Species, strain (sex) Duration Reference	Route, dosing regimen No. of animals/group at start	Results For each target organ: incidence (%) and/or multiplicity of tumours	Significance	Purity Comments
Rat, Osborne-Mendel (F) 111–113 wk NCI (1978)	0, 850, 1700 mg/kg diet for 78 wk 20, 50, 50 rats	Thyroid follicular cell adenoma: 0/19, 6/48, 5/50 Thyroid follicular cell adenoma or carcinoma (combined): 2/19, 11/48, 6/50	[NS] NS	TDE (principal component, 60%, assumed to be *p,p′*-TDE; 19 unidentified impurities)
DDE				
Full carcinogenicity				
Mouse, CF-1 (M) 123–124 wk Tomatis et al. (1974b)	0, 250 mg/kg diet 100, 60 mice	Hepatoma [benign or malignant, combined]: 33/98, 39/53	[$P < 0.0001$]	*p,p′*-DDE: purity, 99%
Mouse, CF-1 (F) 123–124 wk Tomatis et al. (1974b)	0, 250 mg/kg diet 90, 60 mice	Hepatoma [benign or malignant, combined]: 1/90, 54/55	[$P < 0.0001$]	*p,p′*-DDE: purity, 99%
Mouse, $B6C3F_1$ (M) 92 wk NCI (1978)	0, 148, 261 mg/kg diet for 78–79 wk 20, 50, 50 mice	HCC: 0/19, 7/41, 17/47*	* $P = 0.001$; $P = 0.001$ (trend), Cochran-Armitage test	*p,p′*-DDE: purity, > 95%; 1 minor impurity Survival at 70 wk: 5/20, 35/50, 31/50
Mouse, $B6C3F_1$ (F) 92–93 wk NCI (1978)	0, 148, 261 mg/kg diet for 78 wk 20, 50, 50 mice	HCC: 0/19, 19/47*, 34/48*	* $P < 0.001$; $P < 0.001$ (trend), Cochran-Armitage test	*p,p′*-DDE: purity, > 95%; 1 minor impurity Survival at 75 wk: 19/20, 47/50, 28/50
Rat, Osborne-Mendel (M) 111 wk NCI (1978)	0, 437, 839 mg/kg diet TWA for 74–78 wk 20, 50, 50 rats	Thyroid follicular cell adenoma or carcinoma (combined): 3/20, 12/49, 10/47	NS	*p,p′*-DDE: purity, > 95%; 1 minor impurity
Rat, Osborne-Mendel (F) 111–112 wk NCI (1978)	0, 242, 462 mg/kg diet TWA for 73–78 wk 19, 50, 50 rats	Thyroid follicular cell adenoma or carcinoma (combined): 2/19, 9/48, 12/48	NS	*p,p′*-DDE: purity, > 95%; 1 minor impurity
Hamster, Syrian Golden (M) 120 wk Rossi et al. (1983)	0, 500, 1000 mg/kg diet 40–47 hamsters	Hepatocellular adenoma or HCC (combined): 0/10, 7/15*, 8/24*	* [$P < 0.05$]	*p,p′*-DDE: purity, 99% Hepatocellular tumours were mostly carcinomas
Hamster, Syrian Golden (F) 120 wk Rossi et al. (1983)	0, 500, 1000 mg/kg diet 43–46 hamsters	Hepatocellular adenoma or HCC (combined): 0/31, 4/26*, 5/24*	* [$P < 0.05$]	*p,p′*-DDE: purity, 99% Hepatocellular tumours were mostly carcinomas

AAF, 2-acetylaminofluorene; AFB_1, aflatoxin B_1; DDD, dichlorodiphenyldichloroethane; DDE, dichlorodiphenyldichloroethylene; DDT, dichlorodiphenyltrichloroethane; i.p., intraperitoneal; mo, month; NDEA, *N*-nitrosodiethylamine; NDMA, *N*-nitrosodimethylamine; NR, not reported; NS, not significant; TWA, time-weighted average; wk, week

Groups of 30–32 male and 30–32 female CF-1 mice (age, 4 weeks) were fed diets containing *p,p′*-DDT (purity, > 99.5%) at a concentration of 0, 50, or 100 mg/kg for 2 years. A control group consisted of 47 males and 47 females. A significant increase in the incidence of liver cell tumours (including simple nodular growth of parenchymal cells and areas of papilliform and adenoid growth of tumour cells) [benign or malignant hepatocellular tumours] was observed in all groups of treated males [$P < 0.05$] and females [$P < 0.01$] (Walker et al., 1973).

In a subsequent study, 30 male and 30 female CF-1 mice (age, 4 weeks) were fed a diet containing *p,p′*-DDT (purity, > 99.5%) at a concentration of 100 mg/kg for 110 weeks. Forty five males and 44 females served as controls. The mice were killed when the intra-abdominal masses reached a size that caused the mice to become anorexic or clinically affected. A significant increase ($P < 0.01$) in the incidence of liver tumours (including simple nodular growth of parenchymal cells, and areas of papilliform and adenoid growth of tumour cells) [benign or malignant, combined] (23/30 treated males and 26/30 treated females compared with 11/45 male and 10/44 female controls, respectively) was observed within 26 months (Thorpe & Walker, 1973).

Groups of 30 male and 30 female Swiss inbred mice (age, 6–8 weeks) were given diet containing technical-grade DDT at 100 mg/kg or a daily dose of DDT of 0.25 mg by gavage in olive oil for 80 weeks. Groups of 30 male and 30 female mice served as untreated controls. Survival and body-weight gains were not affected by treatment. The incidence of malignant lymphoma was increased in males (feeding, 8/27 [$P < 0.05$]; gavage, 6/24 [not significant]; controls, 2/26) and females (feeding, 8/22 [$P < 0.05$]; gavage, 8/24 [not significant]; controls, 2/20) (Kashyap et al., 1977). [The Working Group noted the inadequate number of animals used and lack of vehicle controls.]

Groups of 50 male and 50 female B6C3F$_1$ mice (age, 6 weeks) were fed diets containing technical-grade DDT for 78 weeks and were then held for 14 (male) or 15 (female) additional weeks before termination. Groups of 20 mice were fed a control diet for 91 (male) or 92 (female) weeks. Initially, males received diets containing DDT at 10 or 20 mg/kg and females received diets containing DDT at 50 or 100 mg/kg; after 9 weeks, these concentrations were gradually increased up to 25 and 50 mg/kg for males and 100 and 200 mg/kg for females because of the absence of toxicity. The time-weighted average dietary concentrations were 22 and 44 mg/kg for males and 87 and 175 mg/kg of diet for females. Survival in all groups of male mice was poor, possibly due to fighting. Survival of male mice at week 70 was 12/20 among controls, 20/50 at the lower dose, and 37/50 at the higher dose; terminal survival of female mice was 20/20 among controls, 45/50 at the lower dose, and 36/50 at the higher dose. There was no difference in body-weight gain between treated and control mice. The incidence of malignant lymphoma was increased only in females (males: controls, 0/19; lower dose, 2/49; higher dose, 1/50 (not significant); females: controls, 0/20; lower dose, 3/49; higher dose, 7/46 ($P = 0.026$, trend test) (NCI, 1978). [The Working Group noted the small number of controls, and that females received doses that were four times higher than those for males.]

Ninety male BALB/c mice [age, 8–10 weeks] were fed diets containing DDT (purity, 99%; Aldrich [The Working Group was unable to determine whether this was *p,p′*-DDT or technical-grade DDT]). The control group contained 90 male mice. Ten mice were killed after 2, 4, and 8 weeks of exposure to DDT at 175 ppm. The DDT concentration was lowered to 125 ppm and 6 mice were killed at 16 weeks because of toxicity. Due to increasing mortality, the DDT concentration was lowered to 100 ppm and 4 mice were killed at 24, 36, and 52 weeks, and 12 mice at 75 weeks. Ten mice from the control group were killed after 2, 4, 8, 16, 24, 36, and 52 weeks, and 20 mice were killed after 75 weeks. The incidence of hepatocellular adenoma was: controls, 0/10; DDT, 1/4 at 52 weeks; and controls, 2/36; treated, 4/12 [$P < 0.03$] at 75 weeks. The incidence of HCC was: controls, 1/36; treated, 2/12 at 75 weeks. The incidence of hepatocellular adenoma or carcinoma (combined) was: controls, 3/36; treated, 5/12 [$P < 0.02$] at 75 weeks (Lipsky et al., 1989).

Six groups of 60 male and 54–60 female CF-1 mice (age, 9–10 weeks) were fed diets containing technical-grade DDT [purity not reported] at a concentration of 250 ppm for 15 or 30 weeks. Three control groups of 70–98 males and 69–90 females were fed a normal diet. The mice were killed at 65, 95, or 120 weeks. For the control groups, the incidences of hepatoma [not further classified] in males were 12/70 at 65 weeks, 24/83 at 95 weeks, and 33/98 at 120 weeks, and in females were 0/69 at 65 weeks, 0/72 at 95 weeks, and 1/90 at 120 weeks. For the 15-week treatment groups, there was no increase in the incidence of hepatoma in males, and for females the incidences were 3/60 at 65 weeks [not significant], 11/60 at 95 weeks [$P < 0.01$], and 5/60 at 120 weeks [$P < 0.05$]. For the 30-week treatment groups, the incidences of hepatoma for males were 38/60 at 65 weeks [$P < 0.01$], 41/60 at 95 weeks [$P < 0.01$], and 37/60 at 120 weeks [$P < 0.01$], and for females were 4/54 at 65 weeks [$P < 0.05$], 11/55 at 95 weeks [$P < 0.01$], and 11/54 at 120 weeks [$P < 0.01$] (Tomatis et al., 1974a).

3.1.2 Skin application

Groups of 30 male and 30 female Swiss inbred mice (age, 6–8 weeks) were given technical-grade DDT as a dose of 0.25 mg in 0.1 mL olive oil twice per week by skin application for 80 weeks. Groups of 30 males and 30 females served as untreated controls. Survival and body-weight gains were not affected by treatment, and no increase in tumour incidence was observed (Kashyap et al., 1977).

3.1.3 Subcutaneous injection

Groups of 30 male and 30 female Swiss inbred mice (age, 6–8 weeks) received technical-grade DDT at a dose of 0.25 mg by subcutaneous injection in 0.1 mL of olive oil twice per month for 80 weeks. Groups of 30 males and 30 females served as untreated controls. Survival and body-weight gain were not affected by treatment. The incidence of liver cell carcinoma was 7/26 [$P = 0.0123$] in treated females and 0/20 in control females, and 3/28 [not significant] in treated males and 1/26 in control males (Kashyap et al., 1977).

3.1.4 Co-administration with known carcinogens or other modifying factors

Groups of 30 male B6C3F$_1$ mice (age, 8 weeks) were given drinking-water containing *N*-nitrosodiethylamine (NDEA) at a concentration of 20 ppm for 14 weeks. After 4 weeks, the mice were fed diets containing technical-grade DDT (purity, 97.6%) at either 0 or 50 ppm for 25 weeks. At 43 weeks, the number of DDT-treated mice with hepatocellular adenoma or carcinoma (combined) was non-significantly increased compared with controls (NDEA, 8/20; NDEA/DDT, 14/21) (Williams & Numoto, 1984).

Groups of 39 and 16 female dd mice [age not reported; weight, 22–25 g] were fed DDT [not further specified, purity unspecified] at a concentration of 0 (control) or 100 ppm for 8 weeks (termination of the experiment). One week after the DDT treatment was started, a thread impregnated with 3-methylcholanthrene was inserted into the uterus and removed after 4 weeks. The incidence of carcinoma of the cervical epithelium was increased in DDT-treated mice compared with controls (controls, 0/39; DDT, [about 20%; 3/16, estimated; $P < 0.03$]) (Uchiyama et al., 1974).

3.2 Rat

3.2.1 Oral administration

In two long-term studies started at an interval of 1 year, a total of 192 male and female Osborne-Mendel rats (age, 3 weeks) received diets containing technical-grade DDT, as a powder or as a solution in corn oil, at various concentrations from 100 to 800 ppm for 24 months. A total of 36 male and female rats served as controls (corn oil only). Tumour incidences for all treated groups for both studies were pooled. Among the 81 treated rats that survived at least 18 months, four had "low-grade" HCCs (measuring 0.5–1.2 cm), and 11 had nodular adenomatoid hyperplasia (liver nodules measuring up to 0.3 cm [adenomas]). No liver tumours were found in 20 rats in the control group (Fitzhugh & Nelson, 1947). [The Working Group noted the inadequate reporting and that incidences for both studies were pooled, which made the study impossible to interpret.]

In a re-analysis of one of the two studies by Fitzhugh & Nelson (1947) (see above), Reuber (1978) reported that groups of 12 male and 12 female Osborne-Mendel rats (age, 3 weeks) were given diets containing technical-grade DDT (as a powder or as a solution in corn oil) at a concentration of 0, 200, 400, 600, or 800 ppm for 2 years. Tumour incidences for groups of treated rats were pooled. Hepatocellular carcinomas were present in 6 out of 21 male rats, liver neoplastic nodules [adenomas] in 4 out of 21 male rats, and a Küpffer cell sarcoma in 1 out of 21 male rats receiving DDT that survived for 84 weeks or longer. None of the five male rats in the control group had hepatocellular neoplasms. Neoplasms of the liver (all types) were seen in 11 out of 21 exposed male rats (54% [$P < 0.05$]). Half of the 12 liver neoplasms developed in male rats receiving DDT at 800 ppm. Four out of 12 (36%) exposed female rats (DDT concentration, 200–600 ppm) and 0 out of 6 female control rats that survived for 84 weeks or longer developed HCCs. Fourteen out of 50 male rats receiving DDT, and 0 of 6 male control rats that survived for 52 weeks or longer developed lymphosarcoma. Carcinomas of the ovary were seen in 11 out of 12 females (92% [$P = 0.0004$]) receiving DDT for 89 weeks or longer, compared with 0 of 6 controls (Reuber, 1978). [The Working Group noted the inadequate reporting, and that incidences in treated groups were pooled, which made the study difficult to interpret.]

In two studies of similar design reported by the same institute, groups of 30 male and 30 female Osborne-Mendel rats [age not reported] were exposed to diets containing DDT [not further identified, purity unspecified] at a concentration of 0 or 80 ppm from weaning for 2 years in the first study (Radomski et al., 1965), and to DDT [not further identified, purity unspecified] at a concentration of 0 or 200 ppm from weaning for up to 27 months in the second study (Deichmann et al., 1967). In the first study, undifferentiated bronchogenic carcinomas were seen in 2 out of 60 rats in the control group (male and females combined), and in 8 out of 60 rats (males and females combined) fed DDT at 80 ppm [$P < 0.05$]. In the second study, no tumours of the lung were observed.

Four groups of 36 or 37 male and 35 female outbred Wistar rats (age, 7 weeks) were fed diets containing technical-grade DDT at a concentration of 0 or 500 mg/kg of diet until age 152 weeks. Survival was not affected by the treatment. Body-weight gains were decreased [by 10–20%] in the treated groups when compared with the controls. The average dose of DDT was 34.1 mg/kg bw per day in males and 37.0 mg/kg bw per day in females. The incidence of liver cell tumours [benign or malignant, combined] was increased in treated males (9/27; controls, 0/35) [$P = 0.002$] and females (15/28; controls, 0/32) [$P < 0.0001$] (Rossi et al., 1977).

Groups of 50 male and 50 female Osborne-Mendel rats (age, 7 weeks) were fed diets containing technical-grade DDT for 78 weeks and killed at 111 weeks. The initial concentrations of DDT were 420 or 840 mg/kg diet for males and 315 or 630 mg/kg diet for females. In females, these concentrations were decreased (after 26 weeks) to 158 and 315 mg/kg diet for females when signs of toxicity (tremors) appeared. In males, these concentrations were increased (after 12 weeks) to 500 and 1000 mg/kg diet, and then decreased (after 14 weeks) to 250 and 500 mg/kg diet (because of the signs of toxicity observed in females at the same time). The time-weighted average concentrations were 321 and 642 mg/kg diet for males and 210 and 420 mg/kg diet for females. Groups of 20 males and 20 females received a control diet. Compound-related mean body-weight depression was observed in male and female rats at the higher dose. Survival was not affected by the treatment. In females, there was a significant positive trend in the incidence of pheochromocytoma of the adrenal gland (0/19, 0/38, 3/24; $P = 0.031$) and a significant increase in the incidence of thyroid follicular cell adenoma or carcinoma (combined) at the lower dose (1/19, 13/45, 10/43; $P = 0.032$). There was no increase in the incidence of tumours that could be attributed to treatment with DDT in males (NCI, 1978). [The Working Group noted the small number of controls, and that dose levels were changed during the course of the study.]

Groups of 38 male and 38 female MRC Porton rats (age, 6–7 weeks) were fed a control diet or a diet containing technical-grade DDT at 500 mg/kg for 144 weeks. Additional groups of 30 male and 30 female rats were fed diets containing DDT at 125 or 250 mg/kg. Survival and body-weight gains were not significantly different between treated and control groups; survival at 80 weeks was > 70% in all groups except males at the highest dose (61%). The incidence of liver cell tumours [benign or malignant, combined] was significantly increased in female rats at the highest dose (controls, 0/38; lowest dose, 2/30; intermediate dose, 4/30; highest dose, 7/38 [$P < 0.01$]; trend test, $P < 0.001$). There was no significant increase in the incidence of any neoplasm in males (Cabral et al., 1982a).

Groups of 30 male Fischer 344 rats [age not reported; weight, 50–60 g] were fed with chow diet for 5 days, then were fed with chow diet with or without 0.05% DDT (Aldrich Chemical Co., [purity unspecified]) for 72 weeks, then fed un-supplemented diet until week 103. The incidence of liver neoplastic nodules [benign] was 2/28 in the controls, and 6/28 [not significant] in the group receiving DDT; while the incidence of HCC was 1/28 in the controls, and 0/28 in the group receiving DDT (Shivapurkar et al., 1986).

Groups of 30 or 35 male Wistar rats (age, at least 1 year; weight, 400–500 g) were given diets containing *p,p′*-DDT (purity, 86%) at a concentration of 0 or 500 ppm in experimental weeks 6–12, and then normal diet until the end of the study (60 weeks). Only one rat developed liver neoplastic nodules [benign] (control, 0/18; DDT, 1/19) (Angsubhakorn et al., 2002). [The Working Group noted the short duration of the experiment.]

Groups of 40 male and 40 female F344/DuCrj rats (age, 5 weeks) were fed diets containing *p,p′*-DDT (purity, > 98%) at a concentration of 0, 5, 50, or 500 ppm for 2 years. *p,p′*-DDT intake for males was estimated as 0, 0.17, 1.7 or 19.1 mg/kg per day, and for females was 0, 0.21, 2.2, or 25.2 mg/kg per day. Groups of 20 male and 20 female F344DuCrj rats were used for a satellite experiment with 6 males and 6 females for each dose level killed after 26 and 52 weeks of treatment, and with all surviving rats killed after 78 weeks of treatment. After 2 years, males and females at the highest dose (500 ppm) had whole body tremors in the late stages of treatment (weeks 70 to 104); however, there were no significant differences in mortality between treated and control groups. The mortality rates in the groups at 0, 5, 50, or 500 ppm at termination were 5/40, 10/40, 4/40, 7/40 for males, and 7/40, 13/40, 8/40, 7/40 for females, respectively. Mean body weights of males and females at the highest dose were reduced by 12% and 25% during the study compared with controls, but those of other dose groups were similar to those of the controls. The incidences of hepatocellular adenoma at 104 weeks of treatment were significantly increased in males at 50 and 500 ppm (control, 0/40; 50 ppm, 5/40; 500 ppm, 22/40; $P < 0.05$ and $P < 0.01$, respectively) and females at 500 ppm (control, 0/40; 500 ppm, 16/40; $P < 0.01$). The incidence of HCC was also increased in males at 500 ppm (control, 0/40; 500 ppm, 14/40; $P < 0.01$), but not in females (control, 0/40; 500 ppm, 2/40) after 104 weeks of treatment. [The Working Group considered that the incidence of hepatocellular adenoma or carcinoma (combined) was significantly increased in treated female rats at 500 ppm.] In the satellite experiment, hepatocellular adenomas were observed in males and females at 500 ppm after 78 weeks of treatment (males: 6/7 versus 0/8 controls, $P < 0.01$; and females: 1/8) (Harada et al., 2003).

3.2.2 Co-administration with known carcinogens or other modifying factors

Groups of 14 or 19 male Buffalo rats (age, 6 weeks) were given a single oral dose of aflatoxin B_1 (AFB_1) at 5 mg/kg bw by gavage, followed by diet containing DDT (*p,p′*-DDT; purity, 86%) at a concentration of 0 or 100 ppm for 20 weeks, and then normal diet until the end of the study (82 weeks). Neoplastic nodules [benign] were observed in the liver (AFB_1 group, 1/14; AFB_1/DDT group, 3/19 [not significant]) (Angsubhakorn et al., 2002).

Groups of 35 or 43 male Wistar rats (age, at least 1 year; weight, 400–500 g), were given diets containing AFB_1 at a concentration of 4 ppm for 6 weeks, followed by a 6-week exposure to *p,p′*-DDT (purity, 86%) at a dose of 0 or 500 ppm, then normal diet until the end of the study (60 weeks). Neoplastic nodules [benign] were observed in the liver (AFB_1 group, 9/29; AFB_1/DDT group, 9/28 [not significant]). There was a slight [not significant] increase in the incidence of malignant hepatic tumours in the AFB_1/DDT-treated group (8/29 AFB_1 group; 10/28, AFB_1/DDT group) (Angsubhakorn et al., 2002).

A group of 120 male Sprague- Dawley rats (age, 22 days) were fed diet containing 2-acetylaminofluorene (AAF) at 0.02% for 18 days. After a pause of 1 week, the rats were then given diet containing technical-grade DDT at 0.05% for 389 days. A control group of 120 male rats was given AAF only. Average daily intake of DDT was estimated to be 50 mg/kg per day (by the 4th to 6th experimental month) or 20 mg/kg per day (by the 15th month). The incidence of hepatocellular adenoma or carcinoma (combined) was increased by treatment with DDT (AAF group, 31/108; and AAF/DDT group, 77/103, [$P < 0.0001$]) (Peraino et al., 1975).

A group of 30 male Fischer 344 rats [age not reported; weight, 50–60 g] were fed a chow diet for 5 days, then were injected with a single intraperitoneal dose of NDEA (200 mg/kg bw), then fed either the same chow diet as previously or the chow diet supplemented with 0.05% DDT (Aldrich Chemical Co., [purity unspecified]) for 72 weeks, then fed un-supplemented diet until 103 weeks. The incidence of hepatic tumours was increased by treatment with DDT: for liver neoplastic nodules [benign], there were 3/28 in the group receiving NDEA compared with 1/28 in the group receiving NDEA/DDT [not significant]; for HCC, there were 18/28 in the group receiving NDEA compared with 28/28 in the group receiving NDEA/DDT [$P < 0.0001$]; and for cholangioma or cholangiocarcinoma (combined), there were 0/28 in the group receiving NDEA compared with 6/28 in the group receiving NDEA/DDT [$P < 0.02$] (Shivapurkar et al., 1986).

Four groups of 10 male Wistar rats (age, 28 days) received drinking-water containing NDEA at a concentration of 50 ppm for 2 weeks, and (1 week later) were given DDT (Nakarai Chemical Co., Kyoto; [purity unspecified]) at a dose of 0 or 1.25 mg/rat (a solution of 0.1 mL of 1.25%

DDT) by gavage twice per week for 12 weeks. The rats were killed at 40 or 52 weeks. The incidence of liver tumours (> 5 mm) was 0/8 in the group receiving NDEA and 3/8 [not significant] in the group receiving NDEA/DDT at 52 weeks (Nishizumi, 1979).

Groups of 20 male F344 rats (age, 6 weeks) received two intraperitoneal injections of NDEA (100 mg/kg bw) with a 1-week interval, and then were fed diets containing *p,p′*-DDT (purity, > 98%) at a dose of 0 (control), 0.005, 0.5, or 500 ppm for 43 weeks. Final body weights were significantly decreased in the group receiving DDT at 500 ppm for 43 weeks compared with controls. Rats fed DDT at 500 ppm had significantly increased incidences and multiplicities of hepatocellular tumours. Tumour incidences and multiplicities were as follows: incidences of hepatocellular adenoma, 11/20 in NDEA controls, 13/20 in the group receiving DDT at 0.005 ppm, 10/20 at 0.5 ppm, and 18/18 at 500 ppm [$P < 0.002$]; the multiplicities of hepatocellular adenoma were 1.10, 0.95, 1.10, and 11.44 tumours per rat ($P < 0.01$); incidences of HCC were 12/20 in NDEA controls, 7/20 in the group receiving DDT at 0.005 ppm, 13/20 at 0.5 ppm, and 18/18 at 500 ppm [$P < 0.01$]; multiplicities of HCC were 1.00, 0.50, 0.90, and 28.67 tumours per rat ($P < 0.01$); the incidence of hepatocellular adenoma or carcinoma (combined) were 15/20 in NDEA controls, 14/20 in the group receiving DDT at 0.005 ppm, 17/20 at 0.5 ppm, and 18/18 at 500 ppm [$P < 0.05$]; and the multiplicities of of hepatocellular adenoma or carcinoma (combined) were 2.10, 1.45, 2.00, and 40.11 tumours per rat ($P < 0.01$) (Kushida et al., 2005).

3.3 Hamster

3.3.1 *Oral administration*

Groups of 30–40 male and 29–40 female outbred Syrian golden hamsters (age, 5 weeks) were fed diets containing technical-grade DDT at a concentration of 0, 125, 250, or 500 mg/kg. Survival of the treated hamsters at 52 weeks was similar to that of controls. The study was terminated at 120 weeks, when the last survivor was killed. There was no significant difference in tumour incidence between treated groups and controls; however, a significant positive trend was observed for the incidence of tumours of the adrenal cortex (mostly adenomas) in males (controls, 3/40; lowest dose, 4/30; intermediate dose, 6/31; and highest dose, 8/39; [P for trend, 0.04]), but not in females (controls, 0/39; lowest dose, 0/28; intermediate dose, 1/28; and highest dose, 3/40) (Cabral et al., 1982b).

Groups of 45 or 48 male and 46 or 48 female Syrian golden hamsters (age, 8 weeks) were fed diets containing technical-grade DDT at a concentration of 0 or 1000 mg/kg until age 128 weeks. Survival was 60% or greater in all groups at 80 weeks. Tumours of the adrenal gland (mainly cortical adenoma) occurred in 14/35 treated males compared with 8/31 male controls [not significant], and in 10/36 treated females compared with 2/42 female controls [$P < 0.01$] (Rossi et al., 1983).

Groups of 30 male and 30 female hamsters [strain unspecified; age at start, ~1 month] were given diets containing technical-grade DDT at a concentration of 0, 250, 500, or 1000 mg/kg for 18 months. No difference in body-weight gains between groups was observed. Mean survival time ranged, respectively, from 13.0 and 14.9 months in the male and female control groups, to 17.3 and 17.1 months in the groups of males and females at the highest dose. The incidence of lymphosarcoma was reduced from 50% in male controls and 41% in female controls to 0% in groups of males and females at the highest dose. No other tumour types were observed (Graillot et al., 1975).

3.3.2 Co-administration with known carcinogens or other modifying factors

Groups of 15 male Syrian golden hamsters (age, 6 weeks) were fed diets containing DDT at a concentration of 0 or 500 ppm (Nakarai Chemical Co., Osaka; [purity unspecified]) for 30 weeks. One week before DDT treatment, the hamsters were given an intraperitoneal injection of *N*-nitrosodimethylamine (NDMA) at 6 mg/kg bw. Hepatocellular adenomas were observed, but the incidence was not increased in hamsters treated with DDT (NDMA, 1/15; NDMA/DDT, 1/15) (Tanaka et al., 1987).

3.4 Monkey

Oral administration

A group of 13 cynomolgus monkeys and 11 rhesus monkeys (male and female newborns) were given diets containing *p,p′*-DDT (Aldrich Chemical Co., [purity not specified]) at a dose of 20 mg/kg bw for 130 months, followed by a control diet (without DDT). A control group of 9 cynomolgus monkeys and 8 rhesus monkeys received only the control diet. The monkeys were observed for up to 304 months. One HCC and one prostatic adenocarcinoma were reported in two DDT-treated cynomolgus monkeys. Two leiomyoma of the uterus and one of the oesophagus were reported in three other DDT-treated monkeys. No tumours were observed in the controls (Takayama et al., 1999).

3.5 Carcinogenicity of metabolites of DDT

3.5.1 DDD

(a) Mouse

Groups of 60 male and 60 female CF-1 mice (age, 6–7 weeks) were fed a diet containing *p,p′*-TDE [*p,p′*-DDD] (purity, 99%) at a concentration of 250 mg/kg until age 130 weeks; 100 males and 90 females served as controls. The incidence of hepatoma [benign or malignant, combined] was significantly increased in treated males (control, 33/98 (34%); treated, 31/59 (52%); [$P < 0.05$]), and the incidence of lung tumours (adenoma or adenocarcinoma, combined) was significantly increased in males and females compared with controls (male controls, 53/98 (53%); treated, 51/59 (86%) [$P < 0.0001$]; female controls, 37/90 (41%); treated, 43/59 (73%) [$P < 0.0001$] (Tomatis et al., 1974b).

Groups of 50 male and 50 female B6C3F$_1$ mice (age, 6 weeks) were fed diets initially containing technical-grade TDE (principal component, 60%; assumed to be *p,p′*-TDE [*p,p′*-DDD]; 19 unidentified impurities) at a concentration of 0, 315, or 630 mg/kg. The dietary concentrations of TDE were increased to 425 and 850 mg/kg due to lack of toxicity. The mice were fed DDT for 78 weeks and were killed at 90–93 weeks. The time-weighted average dietary concentrations of TDE were 411 and 822 mg/kg of diet. Additional groups of 20 males and 20 females were fed a control diet. Body-weight gain of treated females was reduced (beginning experimental week 30). Survival was not affected by treatment; terminal survival in males was 13/20 in the controls, 30/50 at the lower dose, and 27/50 at the highest dose; and in females was 18/20 in the controls, 41/50 at the lower dose, and 44/50 at the highest dose. There was no significant increase in the incidence of tumours (NCI, 1978). [The Working Group noted the small number of controls, and changes in dosing during the study.]

(b) Rat

Groups of 50 male and 50 female Osborne-Mendel rats (age, 7 weeks) were fed diets containing technical-grade TDE [*p,p′*-] (principal component, 60%; assumed to be *p,p′*-TDE [*p,p′*-DDD]; 19 unidentified impurities) for 78 weeks and were killed at 111–113 weeks. The initial dietary concentrations of TDE for male rats of 1400 or 2800 mg/kg were increased to 1750 or 3500 mg/kg due to lack of toxicity. Females received diets containing TDE at 850 or 1700 mg/kg of diet throughout the study. The time-weighted average concentrations given to males were 1647 or 3294 mg/kg of diet. Additional groups of 20 males and 20 females were fed a control diet. Body-weight gains were substantially reduced in rats at the higher dose and somewhat reduced in rats at the lower dose compared with controls. Survival was not affected by treatment. Increased incidences of follicular cell adenoma of the thyroid gland were seen in males and females (males: controls, 0/19; lower dose, 11/49 [$P < 0.05$]; higher dose, 9/49 [$P < 0.05$]; [trend, $P < 0.05$]; female: controls, 0/19; lower dose, 6/48 [not significant]; higher dose, 5/50 [not significant]), and significance was reached for follicular cell adenoma or carcinoma (combined) only in males at the lower dose (males: controls, 1/19; lower dose, 16/49 ($P = 0.016$); higher dose, 11/49; females: controls, 2/19; lower dose, 11/48; higher dose: 6/50) (NCI, 1978). [The Working Group noted the small number of controls, and the changes in dosing for males during the study.]

3.5.2 DDE

(a) Mouse

A group of 60 male and 60 female CF-1 mice (age, 6–7 weeks) was fed a diet containing *p,p′*-DDE (purity, 99%) at a concentration of 250 mg/kg until age 130 weeks. The control group comprised 100 males and 90 females. An increased incidence of hepatoma [benign or malignant, combined] was found in treated males and treated females compared with controls (male controls, 33/98 (34%); treated males, 39/53 (74%) [$P < 0.0001$]; female controls, 1/90 (1%); treated females, 54/55 (98%) [$P < 0.0001$]) (Tomatis et al., 1974b).

Groups of 50 male and 50 female B6C3F$_1$ mice (age, 6 weeks) were fed diets containing *p,p′*-DDE (purity, > 95%; one minor impurity) for 78–79 weeks and were killed at 92–93 weeks. The initial dietary concentrations of 125 (lower dose) or 250 mg/kg (higher dose) were increased during the study to 150 or 300 mg/kg due to lack of toxicity. When toxicity became apparent, the concentrations in the diet were held constant, but the higher-dose diets were replaced by control diet every fifth week for the duration of the treatment period. The time-weighted average dietary concentrations were 148 and 261 mg/kg of diet for the groups at the lower and higher dose, respectively. Control groups of 20 males and 20 females were fed a control diet. Body-weight gain was reduced in treated females compared with controls. At 70 weeks, survival in males was 5/20 in controls, 35/50 at the lower dose, and 31/50 at the higher dose; at 75 weeks, survival in females was 19/20 in controls, 47/50 at the lower dose, and 28/50 at the higher dose. The incidences of HCC were significantly increased in males (0/19 in controls, 7/41 at the lower dose, and 17/47 at the higher dose; $P = 0.001$ for trend and for the group at the higher dose) and females (0/19 in controls, 19/47 at the lower dose, and 34/48 at the higher dose; $P < 0.001$ for the groups at the lower dose, higher dose, and for trend) (NCI, 1978). [The Working Group noted the small number of controls, the low survival of male controls, and the changes in dosing during the study.]

(b) Rat

Groups of 50 male and 50 female Osborne-Mendel rats (age, 7 weeks) were fed diets containing *p,p′*-DDE (purity, > 95%; one minor impurity) for 73–78 weeks and were killed at 111–112 weeks. The initial dietary concentrations of 675 or 1350 mg/kg for male rats, and of 375 or 750 mg/kg for females were reduced to 338 or 675 mg/kg of diet for males and 187 or 375 for females due to the onset of toxic signs. After 32–36 weeks, the higher-dose diets were replaced by control diet every fifth week for the duration of treatment period. The time-weighted average concentrations were 437 and 839 mg/kg of diet for males, and 242 and 462 mg/kg of diet for females. Control groups of 20 males and 20 females were fed a control diet. Body-weight gains were somewhat reduced in all treated male and high-dose females compared with controls. Survival at 92 weeks was 16/20 in controls, 34/50 at the lower dose, and 26/50 at the higher dose in males, and 20/20 in controls, 42/50 at the lower dose, and 36/50 at the higher dose in females. There was no significant increase in tumour incidence in treated males and females (NCI, 1978). [The Working Group noted the small number of controls and the changes in dosing during the study.]

(c) Hamster

Groups of 40–47 male and 43–46 female Syrian golden hamsters (age, 8 weeks) were fed a control diet or a diet containing *p,p′*-DDE (purity, 99%) at a concentration of 500 or 1000 mg/kg until age 128 weeks. Survival was 50% or greater in all groups at 80 weeks. There were significantly increased incidences [$P < 0.05$] of hepatocellular adenoma or carcinoma (combined) [mostly carcinomas] in both groups of treated males and females: male controls, 0/10; males at the lower dose, 7/15; and males at the higher dose, 8/24; and female controls, 0/31; females at the lower dose, 4/26; and females at the higher dose, 5/24 (Rossi et al., 1983).

4. Mechanistic and Other Relevant Data

4.1 Toxicokinetic data

4.1.1 Absorption, distribution, and excretion

(a) Humans

p,p′-DDT, *o,p′*-DDT, and their metabolites DDE, *p,p′*-DDD, and *o,p′*-DDD are highly lipophilic compounds. They are expected to be readily absorbed in exposed humans. The toxicokinetics of *p,p′*-DDT have been more extensively studied than those of *o,p′*-DDT. Absorption after inhalation exposure has not been directly measured experimentally in humans, but uptake has been estimated to be 100% for the gaseous phase and 44% for the particulate phase (Volckens & Leith, 2003). Uptake from direct dermal exposures to *p,p′*-DDT has been estimated to be around 10%, with about an order of magnitude lower for uptake from contaminated soil (Wester et al., 1990). Human experimental studies have confirmed *p,p′*-DDT absorption via the oral route, although urinary recovery is not complete in the timeframe of the studies because of the long excretion half-lives of the compounds (Morgan & Roan, 1971; Roan et al., 1971).

p,p′-DDT, *o,p′*-DDT, and their metabolites readily distribute through the body via lymphatic and blood circulation, preferentially concentrating in lipids due to their high lipophilicity (Morgan & Roan, 1970, 1971). In an analysis of human autopsy cases, concentrations of *p,p′*-DDT and DDE were highest in adipose tissue, with concentrations in blood, liver, and kidney proportion to their lipid content (Morgan & Roan, 1970). Levels of *p,p′*-DDT and DDE in the brain, however, were about an order of magnitude lower than would be expected based on lipid content alone (Morgan & Roan, 1970). *p,p′*-DDT and its metabolites have also been detected in breast milk and cord blood (Galetin-Smith et al., 1990; Minh et al., 2004), and have been found to be transported across the placenta to the fetus (Sala et al., 2001; Vizcaino et al., 2014). Due to their lipophilicity, *p,p′*-DDT and DDE remain sequestered in adipose tissue, with a long biological half-lives, estimated to be around 5 years for *p,p′*-DDT and almost 9 years for DDE (Smith, 1999; Wolff et al., 2000b). Due to these long half-lives, comprehensive mass-balance studies have not been performed in humans. Available data suggest that excretion of *p,p′*-DDT products after exposure occurs largely via the urine, with *p,p′*-DDA (2,2-bis(4-chlorophenyl)-acetic acid) being the most commonly measured metabolite (Hayes et al., 1971; Roan et al., 1971). However, the excretion rate of *p,p′*-DDA may be more closely related to ongoing exposures than overall body burden (Roan et al., 1971). On the other hand, *o,p′*-DDT is rapidly excreted as urinary metabolites (Morgan & Roan, 1974).

(b) Experimental systems

p,p′-DDT, *o,p′*-DDT, and their metabolites DDE, *p,p′*-DDD, and *o,p′*-DDD are readily absorbed by all experimental animal species tested after dermal application or ingestion. There are many more toxicokinetic data on *p,p′*-DDT than *o,p′*-DDT. No studies were identified in which *p,p′*-DDT, *o,p′*-DDT, or their metabolites were administered via inhalation. Several radiolabel studies with *p,p′*-DDT in rats or in vitro have demonstrated dermal absorption of < 5% after up to 5 days (Shah & Guthrie, 1983; Reifenrath et al., 1991; Toś-Luty et al., 2002). In rhesus monkeys, dermal absorption of *p,p′*-DDT was measured to be 9–31% in an acetone vehicle and 3–4% in soil (Wester et al., 1990). Absorption after ingestion was much greater, with uptake of > 70% measured in rats given *p,p′*-DDT in vegetable oil (Rothe et al., 1957; Keller & Yeary, 1980). Because of its lipophilicity, *p,p′*-DDT is more poorly absorbed if a non-absorbable vehicle such as mineral oil or paraffin is used (Keller & Yeary, 1980; Palin et al., 1982). Most of this *p,p′*-DDT absorption occurs by way of the intestinal lymphatic system, with a smaller amount through portal blood (Rothe et al., 1957; Jandacek et al., 2009). The carrier for *p,p′*-DDT in lymph is predominatly the lipid core of chylomicrons (Pocock & Vost, 1974).

After absorption, *p,p′*-DDT and its metabolites DDE and *p,p′*-DDD are readily distributed to tissues via lymph and blood circulation. In plasma, *p,p′*-DDT is carried by both lipoproteins and albumin (Mohammed et al., 1990). The highest concentrations of *p,p′*-DDT are generally in fat, liver, and brain, consistent with their lipophilicity, based on experiments in mice, rats, and dogs (Finnegan et al., 1949; Woolley & Talens, 1971; Mühlebach et al., 1991; Tomiyama et al., 2003, 2004; Tebourbi et al., 2006). In a study in pregnant rabbits, concentrations of *p,p′*-DDT and metabolites *p,p′*-DDD and DDE were higher in the fetus, placenta, uterus, and ovaries than in maternal plasma, indicating accumulation in fetal and reproductive tissues (Hart et al., 1972). Placental and/or lactational transfer to offspring after maternal exposure to *p,p′*-DDT has also been demonstrated in other species, including rats, dogs, and cows, with milk being a major source of neonatal exposure due to its lipid content (Woodard et al., 1945; Carter & Mann, 1949; Finnegan et al., 1949; Woolley & Talens, 1971).

Excretion of *p,p′*-DDT and *o,p′*-DDT occurs mainly through the urine, and to a lesser extent, milk (discussed previously) and faeces. *p,p′*-DDA is the predominant urinary metabolite after exposure to *p,p′*-DDT or *p,p′*-DDD, and is reported to account for 85–99% of urinary metabolites in hamsters (Gold & Brunk, 1983). *p,p′*-DDT in faeces after oral exposure may represent unabsorbed compound, while *p,p′*-DDT metabolites in faeces are likely due to biliary excretion (Jensen et al., 1957). Moreover, there are data indicating enterohepatic recirculation of *p,p′*-DDA in the rat (Gingell, 1975). Elimination in experimental animals is slow, although not as slow as estimated in humans, with most studies expected to recover only a fraction of the administered dose. In one radiolabel study in rats, excretion half-lives (the time estimated to excrete 50% of the administered dose) were estimated to be 12 days for *p,p′*-DDT, 3 days for *p,p′*-DDD, and 24 days for DDE (Fawcett et al., 1987). In another study in rats treated with DDE, the total body-burden half-life was estimated to be 120 days (Mühlebach et al., 1991). Urinary excretion of metabolites is also the main pathway of excretion of *o,p′*-DDT, based on experiments in rats, but occurs at a faster rate than for *p,p′*-DDT (Feil et al., 1973; Reif & Sinsheimer, 1975).

4.1.2 Metabolism

(a) Humans

Based on experiments in humans given *p,p′*-DDT, DDE, or *p,p′*-DDD via ingestion, *p,p′*-DDT conversion is primarily to *p,p′*-DDD, with a smaller amount being dehydrochlorinated to DDE (Hayes et al., 1956, 1971; Morgan & Roan, 1971; Roan et al., 1971). *p,p′*-DDD readily degrades through several intermediates to form *p,p′*-DDA, which is readily excreted in the urine, while DDE is poorly eliminated and accumulates in lipid-rich tissues (Morgan & Roan, 1971). Only small amounts (near the limit of detection) of DDE have been detected in the urine of individuals given *p,p′*-DDT or DDE by ingestion (concentrations orders of magnitude lower than the urinary concentrations of *p,p′*-DDA) (Roan et al., 1971).

Studies in humans fed *o,p′*-DDT showed rapid metabolism and excretion (Morgan & Roan, 1974). Humans fed *o,p′*-DDD excreted more than half of the daily dose via the urine each day (Reif et al., 1974).

(b) Experimental systems

p,p′-DDT is metabolized to multiple intermediates and metabolites generated through complex reaction pathways. Although the same basic pathways are thought to exist in various species, including humans, there may be quantitative differences across species and tissues that have not yet been fully elucidated. The postulated metabolism scheme for *p,p′*-DDT is shown in Fig. 4.1 (based on a review by Smith, 2010).

Reductive dechlorination of *p,p′*-DDT to *p,p′*-DDD in experimental animals appears to be primarily catalysed by microsomal cytochrome P450s (CYPs), and has been measured directly in vitro as well as inferred indirectly by the ability of phenobarbital and diphenylhydantoin to induce excretion (Alary et al., 1971; Fries et al., 1971; Kitamura et al., 2002). This reaction may also proceed non-enzymatically, but at a much slower rate based on experiments in rat liver microsomes and rat blood (Kitamura et al., 2002). The other fate of *p,p′*-DDT is dehydrochlorination to DDE, but the catalysis of this reaction is unclear. While earlier studies suggested that formation of DDE and *p,p′*-DDD are completely separate pathways (Peterson & Robison, 1964), later studies suggested that *p,p′*-DDD can also be converted to DDE (Gold & Brunk, 1982; Kitamura et al., 2002). However, based on the much smaller amount of urinary DDE produced after administration of *p,p′*-DDD when compared with *p,p′*-DDT, most DDE produced after exposure to *p,p′*-DDT is likely to be direct conversion from *p,p′*-DDT, at least at lower exposures (Gold & Brunk, 1982; Fox et al., 1998).

1-Chloro-2,2-bis(*p*-chlorophenyl)ethylene (*p,p′*-DDMU) is thought to be derived from *p,p′*-DDD, but not DDE (Kitamura et al., 2002). *p,p′*-DDMU and another downstream intermediate, 1-chloro-4-[1-(4-chlorophenyl)ethenyl] benzene (*p,p′*-DDNU), may form reactive epoxide intermediates (Planche et al., 1979; Gold et al., 1981).

The major urinary metabolite after administration of *p,p′*-DDT is *p,p′*-DDA. *p,p′*-DDA is produced from *p,p′*-DDD through multiple potential pathways, principally an acyl chloride intermediate (Cl^- *p,p′*-DDA) (Peterson & Robison, 1964; Gold & Brunk, 1982; Fawcett et al., 1987). Cl^- *p,p′*-DDA may can then either undergo hydrolysis to form *p,p′*-DDA or acylate cellular nucleophiles (Gold & Brunk, 1982).

o,p′-DDT metabolism has been less extensively studied. Based on studies in rats, the primary metabolic pathway appears to be through *o,p′*-DDD to *o,p′*-DDA, with no formation of DDE (Feil et al., 1973; Reif & Sinsheimer, 1975). A postulated metabolism scheme for *o,p′*-DDT is shown in Fig. 4.2, based on a review (Smith, 2010).

Fig. 4.1 Postulated metabolism scheme of *p,p'*-DDT in rats

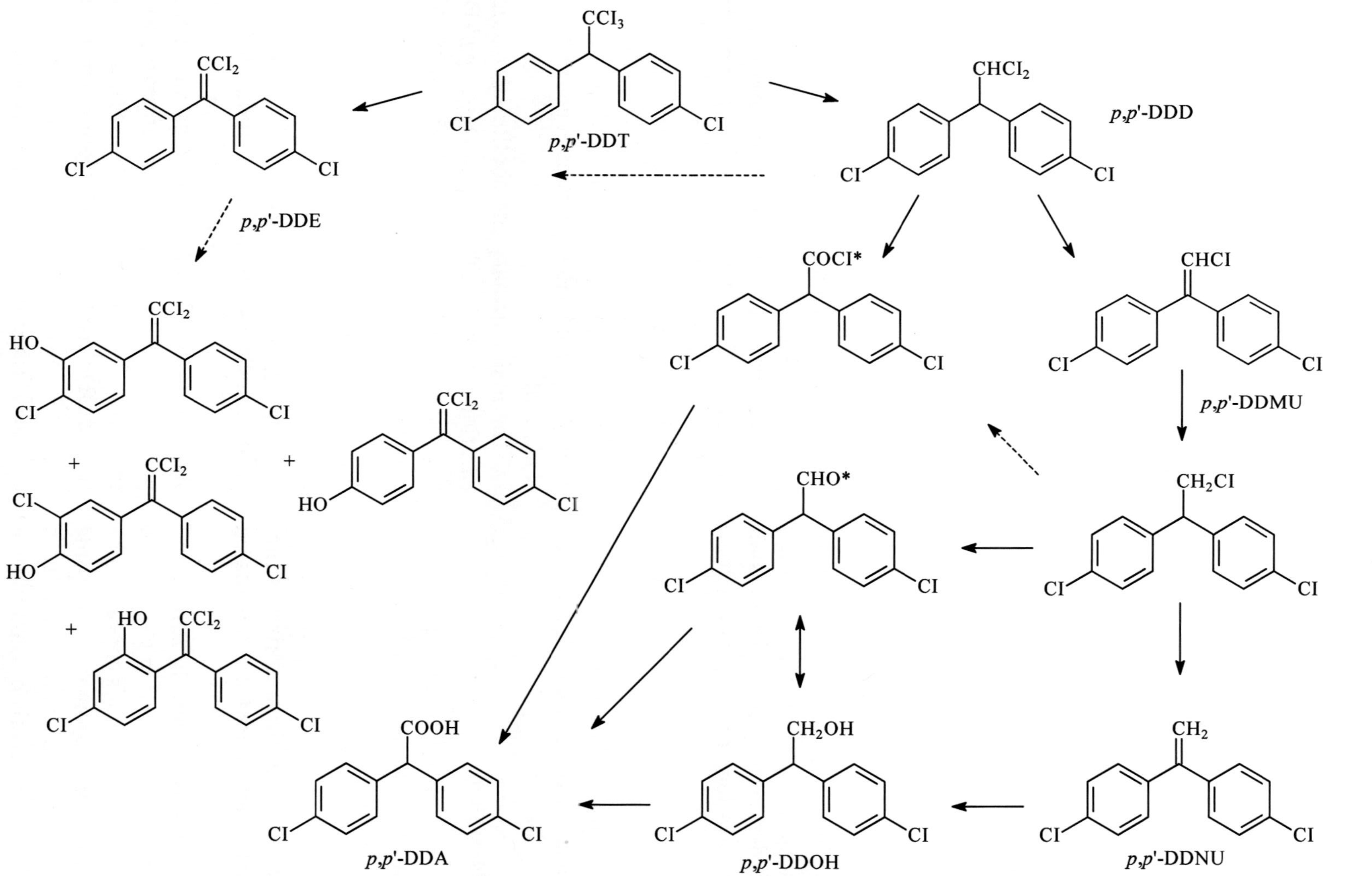

Reprinted from Smith (2010). In: Hayes' Handbook of Pesticide Toxicology, third edition, Robert Krieger, Chapter 93: Toxicology of DDT and some analogues, Pages No. 1975–2032, Copyright (2010), with permission from Elsevier.
The metabolites indicated by an asterisk are likely to be reactive with cellular constituents

Fig. 4.2 Postulated metabolism scheme of *o,p′*-DDT in rats

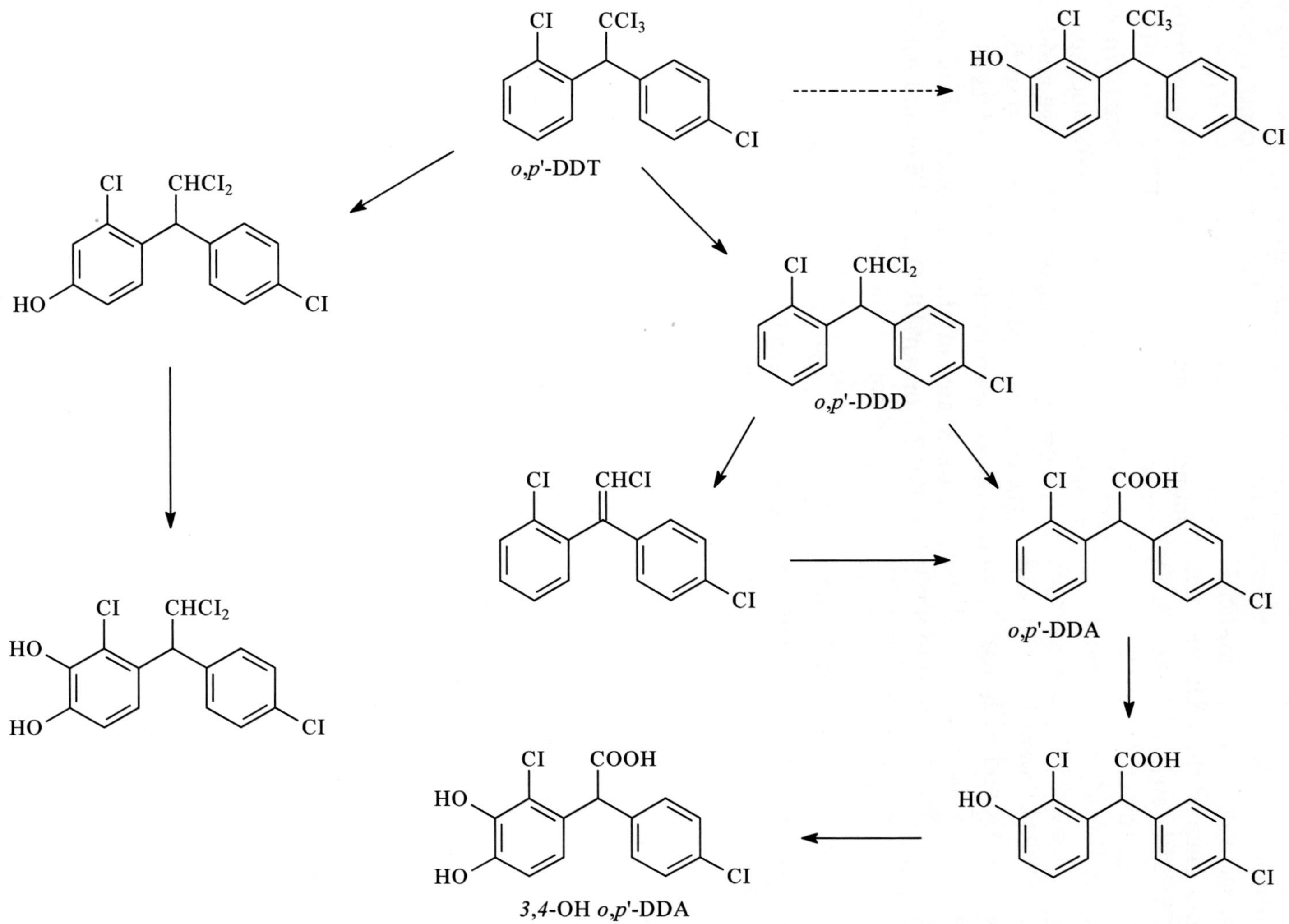

Reprinted from Smith (2010). In: Hayes' Handbook of Pesticide Toxicology, third edition, Robert Krieger, Chapter 93: Toxicology of DDT and some analogues, Pages No. 1975–2032, Copyright (2010), with permission from Elsevier.

4.1.3 Modulation of xenobiotic metabolism enzymes

No studies in exposed humans were available to the Working Group.

p,p′-DDT transactivated the pregnane X receptor (PXR) and induced the PXR-mediated expression of CYP3A4 and CYP2B6 in a hepatoma cell line (Lemaire et al., 2004).

In male Sprague-Dawley rats exposed to *p,p′*-DDE (100 mg/kg for 7 days by oral gavage), liver CYP2B1 and CYP3A1 were induced and were correlated with CAR and PXR activity in a receptor transactivation assays (Wyde et al., 2003; see Section 4.2.1(c)). *p,p′*-DDE (intraperitoneal dose of 0, 20, 60, or 100 mg/kg bw every other day for 10 days) significantly upregulated rat hepatic enzymes CYP1A1, CYP2B1, and uridine diphosphate-glucuronosyltransferase (UDPGTs), although CYP1A2 did not change significantly (Liu et al., 2011, 2014).

4.2 Mechanisms of carcinogenesis

This section summarizes evidence for the six of ten key characteristics of carcinogens (Smith et al., 2016) that had adequate data for evaluation, concerning whether DDT modulates receptor-mediated effects; is immunosuppressive; induces oxidative stress; alters cell proliferation and death; is genotoxic; and induces chronic inflammation.

4.2.1 Receptor-mediated effects

(a) Exposed humans

(i) Serum hormone levels

In studies of occupationally exposed men, there were conflicting results regarding associations between *p,p′*-DDT or *p,p′*-DDE and serum levels of testosterone, sex hormone-binding globulin (SHBG), and estradiol, while there was no association between *p,p′*-DDT or *p,p′*-DDE and luteinizing hormone (LH) or follicle-stimulating hormone (FSH). For instance, in a cross-sectional study of 59 men employed for 15.8 ± 7.8 (mean ± standard deviation, SD) years (range, 4–34 years) as DDT sprayers or working with DDT sprayers in malaria vector control in South Africa, significant positive associations were found by linear regression analysis between serum testosterone and *p,p′*-DDT ($P = 0.014$) and *p,p′*-DDD ($P = 0.050$), and between 17β-estradiol (E_2) and *p,p′*-DDT ($P = 0.001$) and *p,p′*-DDD ($P = 0.003$) after adjustment for age and SHBG (Dalvie et al., 2004b). There were no such significant associations for SHBG, and associations between any of the hormones measured and *o,p′*-DDT or *o,p′*-DDD, *o,p′*-DDE and *p,p′*-DDD were also not significant.

In a study of 107 men in Italy who had been exposed to DDT because of antimalaria spraying when they were young, *p,p′*-DDT and *p,p′*-DDE were detectable and quantifiable in 14 men and 106 men, respectively. There were no significant associations between levels of *p,p′*-DDT and *p,p′*-DDE and serum levels of any of the hormones (LH, FSH, SHBG, E_2, and testosterone) assessed approximately 50 years after exposure (Cocco et al., 2004). *o,p′*-DDT and *o,p′*-DDE were undetectable.

In a group of 137 African-American farmers (aged 33–88 years; mean, 62 years) from North Carolina, USA, who reported using DDT in their work, serum levels of DDE (presumably *p,p′*-DDE) were not correlated with serum levels of testosterone, 5α-dihydrotestosterone (DHT), SHBG, or free testosterone (Martin et al., 2002).

In men and women exposed through habitual consumption of fish, there were no associations between *p,p′*-DDT and/or *p,p′*-DDE and sex steroids, LH and FSH, and thyroid hormones triiodothyronine, thyroxine, and thyroid-stimulating hormone (T3, T4, and TSH, respectively). For example, in a study of 110 men (age range, 23–79 years) from Latvia ($n = 67$; mean age, 48 years) and Sweden ($n = 43$; mean age, 42 years) who regularly consumed fish, there were no significant associations between plasma levels of *p,p′*-DDT or *p,p′*-DDE and plasma levels of free testosterone, LH, FSH, prolactin, TSH, and total and free T3 and T4 (Hagmar et al., 2001). Similarly, in 150 regular consumers of fish from the United States Great Lakes region, DDE was not significantly correlated with testosterone, SHBG, free testosterone, SHBG-bound testosterone, dehydroepiandrosterone, estrone sulfate, LH, and FSH, or the thyroid hormones T3, T4, and T3 uptake and free T4 index (Persky et al., 2001). In 51 women (42 were regular consumers of fish), no significant correlations between DDE and thyroid hormones were reported. A significant inverse correlation was found between DDE and estrone sulfate in the serum in a subset of 56 men from this study, but there was again no correlation between DDE and the other hormones (Turyk et al., 2006).

In men and women generally exposed through the environment, there were conflicting outcomes (both changes in hormone levels and negative results). In a study of 304 men and 300 women from an area in Brazil heavily polluted with OCPs, linear regression analysis found a significant inverse association ($P < 0.05$) between testosterone and *o,p′*-DDT, and a borderline significant inverse association ($P < 0.05$) between testosterone and *p,p′*-DDE in the serum of the men; there was no such association for *p,p′*-DDT or *p,p′*-DDD (Freire et al., 2014). When *p,p′*-DDE levels were divided into quartiles, a significant inverse association with testosterone levels was found (*P* for trend, 0.02). In a study of 257 men (age range, 18–82 years; mean, 42 years) and 436 women (age range, 18–95 years; mean, 42 years) from the native American Mohawk nation in the USA, serum levels of DDE [presumably *p,p′*-DDE] and serum testosterone levels were not significantly associated (Goncharov et al., 2009).

In peri- or postmenopausal women ($n = 77$), there was a significant inverse association between LH and serum levels of *p,p′*-DDT or *p,p′*-DDD, and between FSH and serum levels of *p,p′*-DDD (Freire et al., 2014). There was a significant inverse association between LH and *p,p′*-DDE levels across *p,p′*-DDE quartiles ($P < 0.001$), but no significant associations between serum organochlorine levels and other hormones (E_2, progesterone, and prolactin) in peri- or postmenopausal women.

Several studies of populations exposed through the environment found a positive association between *p,p′*-DDT and/or *p,p′*-DDE and T3 and/or T4, and in some there was an inverse association with TSH. However, there were also several studies that did not find such associations. In a study in Brazil, a significant inverse association was seen in men between *p,p′*-DDT and free T4 levels, but not total T3 or TSH (Freire et al., 2013). In women, in contrast, there was a positive association between *p,p′*-DDT and free T4 levels and total T3, while serum *o,p′*-DDT was also positively associated with free T4 levels.

A large study among participants of the National Health and Nutrition Examination Survey (NHANES) in the USA in 1999–2000 ($n = 986$) and 2001–2002 ($n = 1443$) reported an inverse association between serum *p,p′*-DDE and TSH, but only in women aged 66 years and older in the 1999–2000 period (Turyk et al., 2007).

When serum levels of *p,p′*-DDE and sex steroid hormones were compared in a group of 341 men (age range, 18–51 years) from an infertility clinic in Boston, USA, no significant associations were found for testosterone, SHBG, free testosterone, and E_2 (Ferguson et al., 2012). However, when serum levels of *p,p′*-DDE and thyroid hormones were compared using multivariate linear regression analysis, there were significant positive associations for total T3 and free T4, and a significant inverse association for TSH (Meeker et al., 2007).

In 48 men and 66 women (age range, 55–74 years) from a polluted area in the upper Hudson River area in New York State, USA, there was a significant ($P < 0.05$) association between the sum of *p,p′*-DDT and *p,p′*-DDE and total T4 and total T3 in women, but not in men, after adjustment for several covariates, including other organochlorine exposures (Bloom et al., 2014). TSH and free T4 levels were not significantly associated with the sum of *p,p′*-DDT and *p,p′*-DDE in either sex.

In 834 men and 1212 women (age range, 20–75 years) in a polluted area in east Slovakia, *p,p′*-DDE levels were positively associated with those of T3 ($r = 0.072$; $P < 0.01$), but not associated with free T4 or TSH (Langer et al., 2007). No results were presented for *p,p′*-DDT.

There was no association between prenatal DDT exposure (measured in cord blood) and thyroid hormone status in newborns. In blood samples collected during pregnancy from 147 Canadian women, a significant association between *p,p′*-DDE and total T3 was found, but not for free T4 or TSH, while there were no significant associations for *p,p′*-DDT and thyroid hormones (Takser et al., 2005). In another study of cord blood, significant inverse associations with total T4 were found for *p,p′*-DDT, *p,p′*-DDE, and *o,p′*-DDE, but not for *o,p′*-DDT and *p,p′*-DDD in samples collected from 39 women in Thailand (Asawasinsopon et al., 2006). No associations were found for free T4 or TSH. In a similar study of 247 women in China, no significant associations found between *p,p′*-DDE and free T3, free T4, and TSH (Li et al., 2014b). In 259 children (age, 4 years) in Spain, blood levels of *p,p′*-DDT and *p,p′*-DDE were not associated with serum levels of T3, T4, or TSH (Alvarez-Pedrerol et al., 2008b). No significant correlations have been reported between cord blood levels of *p,p′*-DDT and *p,p′*-DDE and TSH levels in 27 newborns (age, 3 days) from the same area in Spain (Álvarez-Pedrerol et al., 2008a). In a cohort of 453 newborns in the Valencia region of Spain, cord blood levels of *p,p′*-DDT and *p,p′*-DDE were not associated with TSH serum levels. In a study of newborns in another Spanish population, *p,p′*-DDT and *p,p′*-DDE and TSH levels in the cord blood of 453 babies were not significantly associated (Lopez-Espinosa et al., 2010). An earlier report from the same group of 70 newborns also indicated the absence of a significant association between *p,p′*-DDT and *p,p′*-DDE and TSH levels in cord blood (Ribas-Fitó et al., 2003).

(ii) Other endocrine-related effects

In a nested case–control study in Spain, newborns with cryptorchidism and/or hypospadias were compared with boys without malformations and placental levels of DDT were measured (Fernandez et al., 2007). The odds ratio for cryptorchidism and/or hypospadias was 2.17 (95% CI, 0.96–5.00) for *o,p′*-DDT concentration of ≥ LOD, and 2.17 (95% CI, 0.95–5.00) for *p,p′*-DDT concentration of ≥ LOD, but no odds ratios specifically for cryptorchidism were presented. In three nested case–control studies, no such effect was found: *p,p*-DDT, *p,p′*-DDE, *p,p′*-DDD, and *o,p′*-DDT in breast milk was not correlated with cryptorchidism (Damgaard et al., 2006). Further, there was no correlation between cryptorchidism and maternal serum levels of *p,p′*-DDT and *p,p′*-DDE (Bhatia et al., 2005) and DDE (Longnecker et al., 2002). [Overall, there appeared to be no association in humans between maternal DDT exposure and cryptorchidism in male offspring.]

One study found no association between DDT exposure and anogenital distance (Longnecker et al., 2007). Another study reported an effect on anogenital distance with maternal serum DDE levels, but not with DDT levels (Torres-Sanchez et al., 2008).

In a cross-sectional study of young women in China, age at menarche was 1.11 years less ($P < 0.001$) in women in the fourth quartile of total serum DDT concentration compared with the lowest quartile, and a 10 ng/g increase in serum DDT concentrations was significantly ($P < 0.001$) associated with a 0.2 year reduction in age at menarche (Ouyang et al., 2005). In contrast, there was no association between the presence and level of *p,p′*-DDE in serum and adipose tissues and precocious puberty in a study in Turkey of girls with premature breast development (Ozen et al., 2012). [The data indicating an association between DDT and early menarche in these and other studies were thus mixed and potentially confounded by body weight and growth, which can affect DDT congener levels in growing girls, and time between menarche and blood sampling (Wolff et al., 2007).]

Studies on diabetes and obesity are presented in Section 4.5.

(iii) Tumour receptor expression

[The Working Group noted the paucity of data evaluating DDT and receptors in tumours in humans. Available data concerning an association between *o,p′*-DDT and human epidermal growth factor receptor 2 (HER2)-positive breast cancer are presented in Section 4.4.]

(b) Human cells in vitro

(i) Estrogen receptor-mediated effects

[The Working Group noted that *o,p′*-DDT binds to and activates ER in a variety of human cell types, effects of which are blocked by an ER antagonist. *o,p′*-DDT is 100–1000 times less potent in ER binding than estradiol, and *p,p′*-DDT and DDT metabolites bind ER less potently than *o,p′*-DDT. Notably, DDT and DDT metabolite levels in human adipose tissue are 100s to 1000 times higher than in serum (Kanja et al., 1992; Toppari et al., 1996), while nipple aspirate estradiol is only 4–45 times higher than in serum (Petrakis et al., 1987).]

o,p′-DDT binds the human ER less potently than estradiol in a variety of assays: (i) cytosolic binding assays (Klotz et al., 1996; Scippo et al., 2004), binding to both ERα and ERβ (Kuiper et al., 1998), *o,p′*-DDT having greater affinity for human ERα than for human ERβ (Legler et al., 2002); (ii) assays with cells transfected with reporter vectors for ERα and ERβ (Shelby et al., 1996; Lemaire et al., 2006); (iii) yeast-based assays in which 4-hydroxy-tamoxifen abrogated the effect (Klotz et al., 1996; Dhooge et al., 2006); and (iv) assays of binding to recombinant ERα and ERβ (Scippo et al., 2004). In MCF-7 cells, nuclear redistribution of ER was induced by *o,p′*-DDT at 1 μM (Steinmetz et al., 1996). Some studies observed relatively strong ER binding (Andersen et al., 1999) and physiologically relevant ERα and ERβ transactivation, such as via CAT (chloramphenicol acetyltransferase) reporter gene transcription, in MCF-10 and MCF-7 cells (Shekhar et al., 1997; Lemaire et al., 2006). While *o,p′*-DDT and *p,p′*-DDT were not inhibitory to E_2-induced ER transactivation in the chemically activated luciferase expression (CALUX) system using human osteosarcoma cells (Sonneveld et al., 2005), *o,p′*-DDT and *p,p′*-DDT stimulated MCF-7 cells to enter and progress through the cell cycle in an ER-dependent manner that was ablated by ER antagonist ICI182,780 (Dees et al., 1997a; Shekhar et al., 1997).

The human ER, but not the rat ER, also binds *o,p′*-DDD, *o,p′*-DDE and *p,p′*-DDT (Kelce et al., 1995; Li et al., 2008). [The Working Group noted that these data suggest that the human ER is more permissive of binding by DDT-related compounds than the ER of other species.] *p,p′*-DDT is less estrogenic in vitro than *o,p′*-DDT, showing little if any binding to the human ER (Danzo, 1997; Kuiper et al., 1998; Scippo et al., 2004), weak to very weak ER transactivation, and sometimes no effect (Chen et al., 1997; Li et al., 2008;). *p,p′*-DDT also stimulated estrogenic proliferation in these systems (see Section 4.2.6).

o,p′-DDE is also less estrogenic than *o,p′*-DDT, showing weak to very weak binding to the ER (Zava et al., 1997; Scippo et al., 2004), weak ER transactivation in MCF-7 and HeLa cells and yeast systems (Balaguer et al., 1999), and weak stimulation of MCF-7 cell proliferation (Soto et al., 1995); lack of ER binding and transactivation have also been reported in Sf9 and human embryonal kidney 293 cells (Kuiper et al., 1998; Sheeler et al., 2000). *p,p′*-DDE has even less estrogenic effects, showing weak to no ER binding, and mostly no ER transactivation in

MCF-7 cells (Soto et al., 1995; Aubé et al., 2011). *o,p′*-DDD is also estrogenic, but weaker than *o,p′*-DDT showing weak to very weak ER binding, weak ER transactivation, and very weak stimulation of MCF-7 cell proliferation (Klotz et al., 1996; Aubé et al., 2011). *p,p′*-DDD had hardly any estrogenic effect, showing weak ER binding, but no ER transactivation and very weak stimulation of MCF-7 cell proliferation (Tully et al., 2000; Lee et al., 2002; Scippo et al., 2004), although ER transactivation in MCF-7 cells stronger than *o,p′*-DDT has also been reported, which effect was abrogated by 4-hydroxy-tamoxifen (Klotz et al., 1996).

ER antagonist ICI 182,780 blocked the effects of *o,p′*-DDT and *p,p′*-DDE on induction of PR and downregulation of ER; these effects were stronger than interference of Erk activation with an inhibitor of MEK1 (Silva et al., 2010).

There are data, mostly from studies with MCF-7 cells in vitro, that DDT and DDE can also act via non-genomic mechanisms involving plasma membrane-located ERs that can activate signalling pathways resulting in rapid responses (Bratton et al., 2012). In human breast-cancer cells, *o,p′*-DDE has binding affinity for GPR30 (Thomas et al., 2005). Both *o,p′*-DDE and *p,p′*-DDT also bind GPR30 in a transformed human embryonic kidney cell line (Thomas & Dong, 2006).

(ii) Effects on aromatase and steroid-hormone production

Estrogens in humans and other mammals are formed from precursor androgens by the enzyme aromatase. *p,p′*-DDE at concentrations of 50–100 ng/mL caused significant 130–135% induction of aromatase enzyme activity in primary human endometrial stromal cells (Holloway et al., 2005). In contrast, *p,p′*-DDE inhibited aromatase activity in HEK 239 human embryonic kidney cells at a concentration of 20 μM (Benachour et al., 2007). No effect of *p,p′*-DDE was found on aromatase activity in H295R human adrenocortical cancer cells at concentrations of 1–10 μM, while *o,p′*-DDT, *p,p′*-DDT, and *o,p′*-DDE were inhibitory but only at a cytotoxic concentration of 10 μM (Sanderson et al., 2002).

In human JEG-3 choriocarcinoma cells, *o,p′-DDT, p,p′-DDT, p,p′-DDE*, and *o,p′-DDE* at concentrations as low as 1 ng/mL all increased progesterone secretion as well as secretion of human chorionic gonadotropin at concentrations of 10 ng/mL and higher (Wójtowicz et al., 2007b). In human placental explant cultures, *o,p′*-DDT, *p,p′*-DDE, and *o,p′*-DDE, but not *p,p′*-DDT, increased progesterone secretion at concentrations of 100 ng/mL and higher, while conversion of dehydroepiandrosterone to E_2 was decreased by all four compounds at concentrations of 1 ng/mL and higher (10 ng/mL and higher for *o,p′*-DDE) (Wójtowicz et al., 2007c). Effects on aromatase activity were inconsistent, although the DDT and DDE isomers tended to decrease it, and aromatase protein expression [not quantified] was decreased by all four, with *o,p′-DDT, o,p′-DDE* being the most active inhibitors.

(iii) Androgen receptor-mediated effects

[The Working Group noted that DDT and its metabolites antagonize the AR, with *p,p′*-DDE being the most potent, in cells from humans and other species and in non-human experimental systems in vivo.]

In HepG2 cells stably transfected with human AR reporter constructs, transactivation by DHT was most strongly inhibited by *p,p′*-DDE (at concentrations of one to two orders of magnitude above the maximum activating DHT concentration), followed by *p,p′*-DDT, and least strongly by *o,p′*-DDT (Kelce et al., 1995; Maness et al., 1998). Inhibitory activity of *p,p′*-DDE was also observed in AR-negative human PC-3 prostate-cancer cells transfected with human AR and a reporter construct (Schrader & Cooke, 2000). *p,p′*-DDE inhibited androgen-induced human AR transactivation in yeast systems (Gaido et al., 1997; Li et al., 2008) and bound to recombinant human

AR (Scippo et al., 2004). *o,p'*-DDD and *p,p'*-DDD more strongly bound to recombinant human AR, while *o,p'*-DDT and *p,p'*-DDT showed binding comparable to that of *p,p'*-DDE (Scippo et al., 2004; Sonneveld et al., 2005). *o,p'*-DDT also binds and inhibits transactivation of the human AR in recombinant yeast (Chen et al., 1997; Gaido et al., 1997). Inhibitory activity of *p,p'*-DDE was also observed in AR-negative human PC-3 prostate-cancer cells transfected with AR and a reporter construct (Schrader & Cooke, 2000). *o,p'*-DDT and *p,p'*-DDT were equally inhibitory in the CALUX system using human osteosarcoma cells (Sonneveld et al., 2005).

p,p'-DDE induced concentration dependent proliferation of human breast-cancer cell lines expressing AR (e.g. CAMA-1) and MCF-7 cells transfected with AR (Aubé et al., 2008).

Analysis in silico indicated human AR binding of *p,p'*-DDT, but other isoforms and related congeners were not evaluated (Wang et al., 2010). A similar analysis showed that additional AR binding sites, other than that used by DHT, were subject to binding by *p,p'*-DDE (Xu et al., 2013).

(iv) Progesterone receptor-mediated effects

o,p'-DDT induced PR in MCF-7 cells in a dose-dependent manner reaching the same level as induced by 10 nM estradiol at a concentration of 1 µM; *p,p'*-DDT, *o,p'*-DDE, and *o,p'*-DDD had the same effect but to a lesser extent (Chen et al., 1997). Binding to the human PR has been reported in a competitive binding assay in T47D human breast-cancer cells with 1 nM radiolabeled R5020, a synthetic PR agonist, for *o,p'*-DDT, *o,p'*-DDD, *p,p'*-DDD, and *o,p'*-DDE at concentrations of 150 nM and higher, but not for *p,p'*-DDT, *p,p'*-DDE, and *p,p'*-DDA (Klotz et al., 1997). In a yeast system and in T47D cells containing PR reporter constructs, PR was not significantly transactivated by any of these compounds, but all inhibited PR transactivation by progesterone, mostly at concentrations of 100 nM and higher (in cells) or 1 µM and higher (in yeast) (Klotz et al., 1997). Binding to recombinant human PR was also found for *p,p'*-DDT, but not *p,p'*-DDD, while *o,p'*-DDT, *p,p'*-DDE, and *o,p'*-DDE had lower affinity (Scippo et al., 2004).

(v) Effects on HER2 receptors

At a low dose, *o,p'*-DDT (1 nM) enhanced the tyrosine kinase activity of HER2 in human MCF-7 breast-cancer cells irrespective of tamoxifen exposure or estrogen depletion of culture media (Enan & Matsumura, 1998; Hatakeyama & Matsumura, 1999). *o,p'*-DDT also gave rise to increased MCF-7 foci (abnormal concentric piling up of cells in post confluent cultures) (Hatakeyama & Matsumura, 1999). The effects of *o,p'*-DDT on tyrosine kinase activity of HER2 and MCF-7 foci formation were blocked by a mononuclear antibody specific to HER2.

(vi) Other receptor-mediated effects

Induction of cAMP production by TSH in Chinese hamster ovarian cells transfected with recombinant human TSH receptor was slightly inhibited by *p,p'*-DDT at 1 µM and significantly at 10 µM and higher; *p,p'*-DDT interfered with internalization of the TSH receptor induced by TSH, possibly by direct interaction with the receptor (Santini et al., 2003; Picchietti et al., 2009).

There are conflicting reports on whether DDT compounds can interact with the aryl hydrocarbon receptor (AhR). In human placental tissue, *o,p'*-DDT, *p,p'*-DDT, *o,p'*-DDE, and *p,p'*-DDE (100 ng/mL) all decreased the expression of AhR protein (Wójtowicz et al., 2011). In human MCF-7 and LNCaP cells *o,p'*-DDT reduced the mRNA and protein expression of Arnt2, but at a high dose of 10 µM (Qin et al., 2011).

(c) Experimental systems in vivo

(i) Estrogen receptor-mediated effects

[The Working Group noted that estrogenic effects of *o,p′*-DDT were consistently shown across numerous experimental non-human systems.]

Results from *o,p′*-DDT administration in vivo extensively supported the notion that ER activation leads to physiological consequences, e.g. increased lordosis in rats (Brown & Blaustein, 1984) and increased uterine wet weight in birds, mink, rats, and mice (Bitman et al., 1968; Duby et al., 1971; Al-Jamal & Dubin, 2000).

The first reports of estrogenic activity of DDT were published in 1968 and 1969, showing increased oviduct weights in fowl by *o,p′*-DDT, but not *p,p′*-DDT, and high tissue concentrations of both (Bitman et al., 1968). Increased uterine weights occurred in immature rats after administration of *o,p′*-DDT, technical DDT and, to a lesser extent, *p,p′*-DDT and the DDT metabolite *o,p′*-DDD, but not *p,p′*-DDD (1 or 5 mg/kg) (Welch et al., 1969). *o,p′*-DDT, technical DDT, and *p,p′*-DDT, and *o,p′*-DDD also inhibited uterine uptake of radiolabeled E_2 (Welch et al., 1969).

ER transcription-inducing effects of single intraperitoneal injections of *p,p′*-DDT (5–500 µg/kg) have been observed in liver, brain, thymus, and prostate of immature male transgenic mice with estrogen-reporter constructs in all estrogen-responsive tissues (Di Lorenzo et al., 2002). Effects were synergized by co-treatment with E_2 (50 µg/kg), and abolished by intraperitoneal injections with ICI 182,780, once before and once after the DDT injection. However, similar treatment with *o,p′*-DDT had the opposite effects in the liver by itself and antagonized the effects of co-administered E_2.

In ovariectomized DA/Han rats, three daily oral treatments of *o,p′*-DDT (500 mg/kg bw per day) markedly reduced mRNA expression of ER in the uterus (Diel et al., 2000). In prepubertal Sprague-Dawley rats given seven daily intraperitoneal injections of *p,p′*-DDT or *p,p′*-DDE (25 mg/kg), the uterine-weight response to estrone treatment was decreased and uptake of radiolabeled estrone was reduced (Welch et al., 1971). The uterine-weight response to *o,p′*-DDT and *p,p′*-DDT has been reproduced (Bitman & Cecil, 1970; Kanno et al., 2003), as has the lack of this in response to *p,p′*-DDD, *p,p′*-DDE, and *p,p′*-DDA (Bitman & Cecil, 1970). These findings were consistent with increases in uterine and vaginal epithelial thickness caused by a treatment for 7 days with *o,p′*-DDD delivered via silastic implants to ovariectomized young adult CD-1 mice at doses resulting in serum *o,p′*-DDD levels of 18 ng/mL and higher (Ulrich et al., 2000). Of note, the estrogenicity of the active DDT compounds was weak, about three to four orders of magnitude less than that of E_2. The uterotropic effect of *o,p′*-DDT was blocked by co-administration of the antiestrogen raloxifene, which does not have estrogenic effects on the uterus as does tamoxifen, demonstrating that this is an ER-mediated effect (Al-Jamal & Dubin, 2000).

(ii) Androgen receptor-mediated effects

Experimental data in vivo are consistent with *p,p′*-DDE antagonism of AR. Mice exposed to DDT had decreased AR mRNA and protein in their testes (Chaturvedi et al., 2010). Rat puberty is delayed and seminal vesicle and ventral prostate weights are reduced in adult rats exposed to DDT (Kelce et al., 1995). Three daily oral doses (12.5–50 mg/kg bw per day) of technical-grade DDT inhibited uptake of radiolabeled testosterone in the prostate of Swiss-Webster mice and reduced formation of testosterone metabolites in the mouse prostate and liver (Smith et al., 1972; Lloyd et al., 1974), and reduced formation of testosterone metabolites in the rat liver (Sierra-Santoyo et al., 2005). In one study, serum testosterone was decreased in CD rats, but increased in Long Evans rats; in both strains, E_2 levels increased, while DHT was increased only in Long Evans rats and FSH decreased only in CD rats (LH and prolactin were not affected) (O'Connor et al., 1999). In ovariectomized DA/Han rats, three daily oral treatments of *o,p'*-DDT (500 mg/kg bw per day) markedly reduced mRNA expression of AR in the uterus (Diel et al., 2000). The anti-androgenic activity of *p,p'*-DDE in vitro is consistent with delayed preputial separation in pubertal rats and reduced accessory sex gland weights in castrated, testosterone-supplemented adult rats treated with an oral dose of 200 mg/kg bw per day for 4 days (Kelce et al., 1995) or 100 mg/kg bw per day for 7 days (O'Connor et al., 1999), although a lack of such effects has also been reported from a study in intact rats (Krause, 1977).

(iii) Other receptor-mediated effects

Regarding effects on aromatase and steroid-hormone production, oral treatment of adult male Sprague Dawley rats with *p,p'*-DDE (100 mg/kg bw per day) for 7 days caused significant induction of aromatase protein expression and aromatase activity in the liver, and elevation of circulating E_2 levels (You et al., 2001).

The progestin-binding site of the PR can also be bound by *o,p'*-DDT in rats (Brown & Blaustein, 1984). The PR is bound by *o,p'*-DDE, and to a lesser extent by *p,p'*-DDT and *p,p'*-DDE in birds (Lundholm, 1991). The PR is also bound by *o,p'*-DDT in fish (Das & Thomas, 1999; Berg et al., 2005).

When exposure to *p,p'*-DDE was initiated at weaning, the latency of HER2-positive mouse mammary tumours was shortened (Johnson et al., 2012).

Oral treatment of adult SD rats and Long Evans rats with *p,p'*-DDE for 15 days at 100–300 mg/kg bw per day decreased serum T4, but not T3, and increased serum levels of TSH (O'Connor et al., 1999). Treatment of adult Wistar rats with intraperitoneal injections of *p,p'*-DDT for 10 days at 50–100 mg/kg bw per day increased thyroid relative weights, decreased activity of type I, but not type II, 5'-deiodinase in liver, kidney, and in a biphasic manner thyroid, decreased serum T4 and, to a lesser extent,T3 and increased serum levels of TSH (Tebourbi et al., 2010). [The Working Group noted that it was not clear whether these DDT effects involved TSH receptor mediation or other mechanisms.]

o,p'-DDD caused a glucocorticoid deficiency in dogs, decreasing circulating 17-OH-corticosteroids and ablating adrenocorticotropic hormone (ACTH) action (Cueto & Moran, 1968). The attenuation of these effects by glucocorticoid receptor (GCR) agonist prednisolone was suggestive of GR antagonism by *o,p'*-DDD (Cueto & Moran, 1968). The eosinophilic response to ACTH was impaired in beagles after a high exposure to *o,p'*-DDT (50 mg/kg bw per day for 32 days), consistent with adrenocortical inhibitory effects of *o,p'*-DDT (Copeland & Cranmer, 1974). These effects could be secondary to liver toxicity, as elevated microsomal enzymes could have stimulated extra-adrenal metabolism of cortisol. Technical-grade DDT (mostly *p,p'*-DDT) altered the adrenal weight of rats (Foster, 1968), and *p,p'*-DDT exposure elevated corticosterone in sparrows (Scollon et al., 2004).

Several reports indicate effects of *o,p'*-DDT on expression of P450 isoforms and other genes mediated by the constitutive androstane receptor (CAR) and/or the pregnane X receptor (PXR) in rats and mice (Wyde et al., 2003; Medina-Díaz et al., 2007; Kiyosawa et al., 2008).

(d) Experimental systems in vitro

(i) Estrogen receptor-mediated effects

The agonism of ER by *o,p'*-DDT in cells from animals and other experimental systems in vitro has been well characterized (Nelson, 1974). *o,p'*-DDT has been shown to bind ER from rats (Nelson, 1974), rabbits (Andersen et al., 1999), cattle (Tiemann et al., 1996), vultures (ERα) (Naidoo et al., 2008), zebrafish (ERα and ERβ) (Legler et al., 2002), yeast (Chen et al., 1997), as well as humans (Chen et al., 1997), albeit at differing potencies. [The Working Group noted that in most of these studies, it was not specified whether ERα or ERβ was investigated.] In primary porcine ovarian cells (granulosa and theca cells together) *p,p′*-DDE and, to a lesser extent, *o,p′*-DDT and *o,p′*-DDE, stimulated E_2 secretion at concentrations of 0.4 and 4 µg/mL, while progesterone secretion was suppressed by *o,p′*-DDT, *p,p′*-DDT, and *p,p′*-DDE, but not by *o,p′*-DDE (Wójtowicz et al., 2007a). *o,p′*-DDT, *p,p′*-DDT, *p,p′*-DDE, and *o,p′*-DDE all increased the conversion of testosterone to E_2 due to increased aromatase activity, *p,p′*-DDE being most active.

(ii) Androgen receptor-mediated effects

In monkey CV-1 cells, *o,p′*-DDT, *p,p′*-DDT, *o,p′*-DDE, and *p,p′*-DDE all potently inhibited AR transcriptional activity (Chaturvedi et al., 2010). *p,p′*-DDE strongly inhibited binding and transactivation of radiolabeled DHT to the rat prostate androgen receptor (Wakeling & Visek, 1973; Kelce et al., 1995; Danzo, 1997), while *o,p′*-DDT and *p,p′*-DDT and *p,p′*-DDD had a weak inhibitory effect (Kelce et al., 1995; Danzo, 1997). *o,p′*-DDT, *p,p′*-DDT, *p,p′*-DDD also antagonized the rat and bird AR with lower affinity than *p,p′*-DDE (Lundholm, 1991; Kelce et al., 1995). In Chinese hamster ovarian cells, *p,p′*-DDE was also the strongest inhibitor of androgen (R1881) activation of the AR, while *o,p′*-DDT and *o,p′*-DDE were also inhibitory and *o,p′*-DDD and *p,p′*-DDD were inactive (Roy et al., 2004).

(iii) Other receptor-mediated effects

In monkey cells exposed in vitro, there was no change in transcriptional activity or nuclear translocation of the GCR by o,p′-DDT, p,p′-DDT, o,p′-DDE, or p,p′-DDE (Chaturvedi et al., 2010). p,p′-DDE displaces dexamethasone from the GCR of birds (Lundholm, 1991). At higher concentrations, o,p′-DDT, p,p′-DDT and o,p′-DDE also bind the GCR in birds (Lundholm, 1991).

4.2.2 Immunosuppression

[The Working Group noted that there were consistent findings that DDT suppresses the humoral immune response across multiple species.]

(a) Humans

(i) Exposed humans

In a study of 49 patients who worked as farmers or farmhands and had been occupationally exposed to multiple insecticides for at least 6 months, detection of DDT and DDE in blood was associated with elevated levels of interleukin-4 (IL-4), a Th2 cytokine, in plasma. DDE and DDD suppressed the Th1 cytokines IL-2 and interferon-gamma (IFN-γ) (Daniel et al., 2002).

Significantly higher levels of *p,p′*-DDE were detected in the blood of patients with systemic lupus erythematosus than in healthy controls (Dar et al., 2012).

In a study of environmental exposure to DDT in 302 residents living near a Superfund site in Aberdeen, North Carolina, USA or in neighboring communities, few differences in immune markers were associated with residential location. In residents who lived closer to the site, mitogen-induced lymphoproliferative activity was statistically significantly lower than that in residents who lived farther away ($P < 0.05$) (Vine et al., 2000).

(ii) Human cells in vitro

In highly purified natural killer (NK) cells exposed to to 4,4′-DDT (5 µM) for 24 hours, ability to destroy K562 tumour cells was inhibited by 61 ± 13%. The loss of cytotoxic function seen with 4,4′ -DDT increased when exposure was increased to 6 days (Reed et al., 2004). [The Working Group noted that this study shows that DDT decreases the tumour-cell killing (lytic) function of human NK cells.]

In a follow-up study, exposure to DDT (2.5 µM) for 24 hours (which caused > 55% loss of lytic function) decreased NK binding function by about 22%, and decreased CD16 cell-surface protein by 20% (Hurd-Brown et al., 2013). In another study, DDT substantially and persistently decreased NK cell lytic function (by 55% at 2.5 μM) within 24 hours (Udoji et al., 2010). [The Working Group noted that three independent studies showed that DDT impairs human NK cell function.]

Ndebele et al. (2004) showed that DDT significantly suppressed IL-2 production in activated CD4+ Jurkat T-cells, at the transcriptional and translational levels. Comitogenic and immunosuppressant effects of *p,p′*-DDT and its derivatives on phytohaemagglutinin-stimulated human lymphocytes have also been reported (Nikolaeva et al., 1980).

(b) Experimental systems in vivo

(i) Mice

Banerjee and colleagues showed a depression in the primary and secondary humoral immune response in mice after subchronic exposure to DDT (Banerjee et al., 1986), and also demonstrated suppression of the humoral immune response to a T-independent antigen (Banerjee, 1987a). In a subsequent study, restraint and other forms of stress potentiated DDT-induced suppression of the humoral immune response in mice (Banerjee et al., 1997).

DDT injections (once every 4 days) markedly increased the incidence of albuminuria and reduced uterine weight, but were without effect on immunity or mortality in the New Zealand Black/New Zealand White F1 hybrid (B/W) mouse model of systemic lupus erythematosus (Li & McMurray, 2009). A separate study showed that DDT accelerated the natural course of systemic lupus erythematosus in this mouse model of the disease (Sobel et al., 2005). *o,p*-DDT significantly reduced the time to development of renal impairment, the primary clinical indication of lupus in the model system (Sobel et al., 2005). There was no clear correlation between autoimmune effects and estrogenicity, as assessed via measurement of uterine hypertrophy (Sobel et al., 2005).

(ii) Rats

Banerjee (1987b) reported suppression of humoral and cell-mediated immune responses in albino rats after short-term exposure to DDT. This suppression increased in a dose- and time-dependent manner (Banerjee, 1987b). Sensitivity to these effects was increased by protein deficiency. Humoral and cellular immune suppression were induced by DDT (50 and 100 ppm) only in rats fed a diet containing 3% protein. In rats immunized with tetanus toxoid, inhibition of leukocyte and macrophage migration was also diminished (Banerjee et al., 1995). A later study demonstrated that DDT, DDE, and DDD, but not DDA, induced different degrees of humoral and cellular immune suppression, with the potency of effects in the order of DDE > DDD > DDT (Banerjee et al., 1996). In rats given DDT (100 and 200 ppm) by oral administration, the humoral immune response was markedly suppressed as assessed by anti-sheep erythrocyte antibody titres via a mechanism that may involve free radicals (Koner et al., 1998).

(iii) Rabbits

In rabbits, DDT (20 mg/kg) significantly reduced organ weights (of the lung, liver, and spleen), antiovalbumin synthesis in the lung and spleen, and maximum serum antibody titre. Serum protein levels decreased as a result of a reduction in levels of albumin and γ- and β-globulin (Chung et al., 1989).

(v) Marine mammals

A higher incidence of bacterial infections were seen in harbour porpoises from the North and Baltic seas of Germany, in comparison to whales from less polluted Arctic waters (Beineke et al., 2005). In free-ranging bottlenose dolphins (Lahvis et al., 1995), the immune response decreased with increasing concentrations of several contaminants, including DDT, in whole blood; inverse correlations were found between concanavalin A-induced lymphocyte proliferation and concentrations of *p,p′*-DDT (Lahvis et al., 1995).

(iv) Birds

The primary humoral immune response in Japanese quail was not affected by in ovo exposure to either isomer of DDT (Bryan et al., 1989). Total circulating erythrocyte numbers were reduced in females after injection in ovo with *o,p′*-DDT but not *p,p′*-DDT. Exposure in ovo to *o,p′*-DDT, but not to *p,p′*-DDT, had long-term and estrogen-like effects on behaviour and haematology. Reproductive behaviours were attenuated with *o,p′*-DDT, which also increased the total number of eggshell malformations (Bryan et al., 1989).

Antibody-mediated immunity was not affected in chickens treated with DDT (Glick, 1974).

An antagonistic action rather than a synergistic or additive effect on blood parameters was observed when DDT and PCB (Aroclor 1254) were co-administered in White Leghorn cockerels (*Gallus domesticus*) (Iturri et al., 1982).

(vi) Frogs

In northern leopard frogs (*Rana pipiens*), sublethal doses of DDT (923 ng/g wet weightt) markedly suppressed antibody response, whereas DTH reactions were enhanced, and respiratory burst was lower (Gilbertson et al., 2003). No differences in the response were seen when pathogens were administered before DDT. A companion field study found significant differences in immune function with pesticide exposure in frogs from Ontario, Canada (Gilbertson et al., 2003).

(c) Experimental systems in vitro

In mouse J774A.1 macrophages, DDT inhibited functional activation, and reduced the capacity to limit intracellular growth of intracellular pathogens, using *Mycobacterium microti* as a model (Nuñez et al., 2002). Technical-grade DDT (a mixture of *p,p′*-DDT (85%), *o,p′*-DDT (15%), and trace amounts of *o,o′*-DDT) and *p,p′*-DDE were more potent than *p,p′*-DDT (Nuñez et al., 2002).

In a study in rat peritoneal exudate cell suspensions, macrophages treated for 4.5 hours in vitro with lipoprotein-sequestered DDT (2.5 μM) showed significant inhibition of their ability to phagocytize yeast particles (Kaminski et al., 1986). An earlier study reported no statistical difference in phagocytic activity was observed in either in vitro or in vivo in rat leukocytes (Kaliser, 1968).

Kannan and Sharma described defective lymphocyte transformation by DDT and responsiveness of rabbit peripheral blood lymphocytes to phytohaemagglutinin in vitro (Kannan & Sharma, 1979).

In beluga whale peripheral blood leukocytes and splenocytes exposed in vitro, *p,p′*-DDT and *p,p′*-DDE had no marked effect on phagocytosis (De Guise et al., 1998). However, *p,p′*-DDT, but not *p,p′*-DDE, significantly reduced the proliferative response of the splenocytes cultured either with or without phytohaemagglutinin (De Guise et al., 1998).

4.2.3 Oxidative stress

(a) Humans

In a study of 44 breast tumours and 21 benign breast biopsies, there was no correlation between *p,p′*-DDT or *p,p′*-DDE and levels of 8-hydroxy-2′-deoxyguanosine (8-OHdG; a biomarker of oxidative DNA damage) (Charles et al., 2001). [The Working Group noted that post-diagnostic measurements in breast tumours and biopsies are of questionable relevance to the potential for DDT to induce oxidative stress.]

In in vitro studies, *p,p′*-DDT or *p,p′*-DDE increased the production of intracellular ROS and inhibited superoxide dismutase (SOD) activity in human colorectal adenocarcinoma DLD1 cells or SW620 cells (Song et al., 2014a, b); see also xenograft experiments in nude mice, described below. *p,p′*-DDE also activated NADPH oxidase, and reduced catalase activity and glutathione content. NAC, a ROS inhibitor, suppressed the induction of Wnt/β-catenin signalling and cell proliferation by *p,p′*-DDT in DLD1 cells and by *p,p′*-DDE in both cell lines. *p,p′*-DDT also significantly elevated ROS levels in HepG2 human liver-cancer cells (Jin et al., 2014). Gamma-glutamylcysteine synthetase (γ-GCS) and SOD activity were inhibited. Beta-catenin and its downstream target genes c-Myc and cyclin D1 were significantly upregulated; co-treatment of DDT with NAC inhibited this overexpression. Moreover, *p,p′*-DDT-induced proliferation of HepG2 cells was inhibited by co-treatment with NAC or β-catenin siRNA (Jin et al., 2014).

In human peripheral blood mononuclear cells in vitro (Pérez-Maldonado et al., 2005), DDT and DDD increased oxidative stress levels by 19-fold compared with controls; for DDE, the increase was 25-fold. These increases in reactive oxygen species (ROS) preceded induction of apoptosis, which was significantly inhibited by *N*-acetyl-*L*-cysteine (NAC) for all three compounds (Pérez-Maldonado et al., 2005).

(b) Experimental systems in vivo

In xenograft experiments in nude mice, *p,p′*-DDT (5 nmol/kg, ip) increased tumour size, oxidative stress and activation of Wnt/β-catenin signalling in human DLD1 tumours (Song et al., 2014b) or HepG2 tumours (see also Section 4.2.4b). These studies were consistent with the in vitro findings in human colorectal (Song et al., 2014a, b) or HepG2 cells (Jin et al. 2014) by the same research group.

In male and female rats exposed to *p,p′*-DDT (5, 50, or 500 ppm) for 2 years, hepatic levels of 8-OH-dG were elevated throughout the study at 500 ppm in males and females, whereas hepatic lipid peroxidation was increased at 50 and 500 ppm but in males only (Harada et al., 2003; see also Section 3).

In rats exposed orally for 8 weeks, *p,p′*-DDT dose-dependently increased levels of thiobarbituric acid reactive substance (TBARS) in serum and SOD activity in erythrocytes, and suppressed the humoral immune response. Ascorbic acid attenuated the effects of *p,p′*-DDT on lipid peroxidation, SOD activity, and humoral immune suppression (Koner et al., 1998). *p,p′*-DDT (40 mg/kg) caused a time-dependent increase in hepatic mitochondrial and microsomal lipid peroxidation and DNA single-strand breaks in rats (Hassoun et al., 1993). In male F344 rats, a 16-week exposure to *p,p′*-DDT at dietary concentrations of 20 ppm or more increased glutathione *S*-transferase placental form (GST-P)-positive foci (putative preneoplastic lesions) in the liver. At concentrations of DDT of 0.5 ppm or less, oxidative stress in the liver, assessed by measuring 8-OH-dG as a marker of oxidative DNA damage, was decreased; this was associated with a decreased number of GST-P-positive foci (Sukata et al., 2002). [The Working Group noted that oxidative stress levels were not determined at the higher doses, although other studies confirm that oxidative stress is induced at these doses of *p,p′*-DDT.]

4.2.4 Altered cell proliferation or death

(a) Humans

(i) Exposed humans

Blood levels of DDT and DDE were associated with a higher frequency of apoptotic cells in exposed children (the specific isomer responsible was not identified) (Pérez-Maldonado et al., 2004).

Elevated levels of apoptosis have been noted in blood mononuclear cells isolated from children and exposed ex vivo to *p,p′*-DDT or *p,p′*-DDE (Pérez-Maldonado et al., 2005). Simultaneous elevation of ROS was also noted (see Section 4.2.3).

(ii) Human cells in vitro

o,p′-DDT induces proliferation of ER-responsive cells such as MCF-7 cells (Soto et al., 1995; Payne et al., 2000b), acting both singly and synergistically in mixtures (Payne et al., 2001) (see also Section 4.2.1). This isomer can also upregulate genes such as vascular endothelial growth factor A (VEGFA) in estrogen-responsive MCF-7 breast cancer cells in an ER-independent fashion through crosstalk between MAPK signalling pathways and transcriptional coactivators (Bratton et al., 2012). *p,p′*-DDE also stimulates proliferation of ER-positive breast-cancer cells and this induction primarily occurred after exposures at lower concentrations (Aubé et al., 2011). However, *p,p′*-DDE suppressed proliferation of androgen-dependent LNCaP cells, while *o,p′*-DDT induced MAPK phosphorylation and slightly increased proliferation of these cells, but not of androgen-independent PC-3 cells; these effects did not appear to occur via direct interaction with the AR (Tessier & Matsumura, 2001) (see also Section 4.2.1). Consistent with these observations, *o,p′*-DDT exhibited weak cell proliferation stimulatory effects, compared with estradiol, in ER-positive human MCF-7, T-47D, and MVLN breast-cancer cells (Dees et al., 1997a; Silva et al., 2007). In parallel with stimulation of MCF-7 cell proliferation, *o,p′*-DDT suppressed apoptosis at concentrations of 100 nM and higher (Diel et al., 2002).

o,p′-DDT induced the expression of the cell-death ligand TNF-α, and elevated apoptosis via a p38 MAPK-dependent mechanism in endometrial Ishikawa and human embryonic kidney (HEK) 293 cells; the apoptotic pathway involves mitochondrial release of cytochrome c with subsequent effector caspase-3/7 activation (Frigo et al., 2005). DDT exposure in vitro modulates the proliferation and viability of human endometrial endothelial cells (Welch et al., 1969; Bredhult et al., 2007, 2008). Environmental concentrations (range, 0.1–10 nM) of *p,p′*-DDT stimulated the proliferation of human colorectal adenocarcinoma DLD1 cells via a Wnt/β-catenin pathway mediated by oxidative stress (Song et al., 2014b).

Induction of cytotoxicity (e.g. in HepG2, HaCaT and primary hepatocytes) has been reported, albeit at ≈100 μM (Delescluse et al., 1998; Gerić et al., 2012; Jin et al., 2014).

(b) Experimental systems

(i) In vivo

In contrast to in vivo (using nude mice) and in vitro (in HepG2 cells) studies of low -doses of *p,p′*-DDT studies indicating a role of β-catenin (e.g., Jin et al., 2014; see Section 4.2.3b) in anti-apoptotic effects and stimulation of cell-cycle progression in liver, other studies in male C57BL mice have instead pointed to constitutive hepatic CAR- and ERα-mediated gene activation (e.g. Kazantseva et al., 2013; this study used a technical mixture containing 85% *p,p′*-DDT and 15% *o,p′*-DDT).

In rats, sublethal DDT concentrations can uncouple oxidative phosphorylation (Byczkowski, 1976). In combination with a carcinogen, *o,p′*-DDT elevated mammary-gland cell proliferation and chromosomal alterations in a rat cancer model (Uppala et al., 2005). It stimulated AP-1 activity via p38 MAPK (Bratton et al., 2009).

(ii) In vitro

In murine embryos, *o,p′*-DDT significantly reduced development to blastocyst and cell number while increasing apoptosis levels (Greenlee et al., 1999). Bredhult et al. (2007, 2008) showed dose (1–100 μM)-related decreases in *o,p′*-DDT-induced cell proliferation and cell viability (increasing proportions of apoptotic and necrotic cells).

DDT was slightly anti-apoptotic in transforming TGF-β-induced apoptosis in rat hepatoma FTO-2B cells as determined by reductions in DNA fragmentation and CPP32 (caspase-3)-like caspase activity (Buchmann et al., 1999). At nanomolar concentrations, *o,p′*-DDE transiently elevated extracellular-regulated kinase (ERK) phosphorylation in rat pituitary GH_3/B6/F10 cells, which express ERα (Bulayeva & Watson, 2004). In primary rat Sertoli cells, DDT exposure (isomer not specified) decreases the level of follicle stimulating hormone (FSH) binding sites; FSH stimulates Sertoli proliferation perinatally, which conditions spermatogenesis (Bernard et al., 2007).

o,p′-DDT may inhibit the ER-mediated effects of other environmental contaminants, such as PCB-126, on proliferation in co-cultured porcine ovarian theca and granulosa cells (Gregoraszczuk et al., 2008).

4.2.5 Genetic and related effects

DDT, its isomers and its metabolites have been studied in a variety of assays for genotoxic and related potential. Table 4.1, Table 4.2, Table 4.4, and Tables 4.5 and 4.7 (available online at: http://monographs.iarc.fr/ENG/Monographs/vol113/index.php) summarize the studies carried out with DDT, including several isomeric forms in humans in vivo and in vitro, in non-human mammals in vivo, in non-human mammals in vitro, and in non-mammalian systems, respectively. Table 4.3, and Tables 4.6 and 4.8 (available online at: http://monographs.iarc.fr/ENG/Monographs/vol113/index.php) summarize the studies carried out with different metabolites of DDT in humans in vitro, in non-human mammals in vivo, in non-human mammals in vitro, and in non-mammalian systems, respectively.

(a) Exposed humans

See Table 4.1

No association was seen between DNA adducts (8-OH-dG) in primary breast adenocarcinomas and p,p′-DDT and, *p,p′*-DDE tissue levels (Charles et al., 2001).

Induction of DNA strand breaks was demonstrated by the comet assay in lymphocytes from maternal and umbilical cord blood collected from 50 mother–infant pairs exposed to different pesticides, including DDT (with a stronger response in infants than in mothers) (Alvarado-Hernandez et al., 2013); in the lymphocytes of children from different Mexican communities (Pérez-Maldonado et al., 2006; Pérez-Maldonado et al., 2011; Jasso-Pineda et al., 2015) [Causative effect of DDT alone cannot be demonstrated in the latter]; and in lymphocytes from women from Mexican communities affected by malaria and with a history of indoor DDT spraying (Yáñez et al., 2004). No association was found between DNA strand breaks assessed by the comet assay in sperm cells and *p,p′*-DDE, PCB, and hexachlorobenzene in blood samples from subfertile male donors (Hauser et al., 2003).

No association was seen between DNA strand breaks assessed by the comet assay and micronucleus formation and the blood level of *p,p′*-DDE (among other pollutants in the blood and urine) in a study of residents from areas with different types of pollution (De Coster et al., 2008). However, higher p53 protein levels in serum were associated with higher values of serum DDE, HCB and certain PCBs. Furthermore, higher levels of DDE were associated with higher levels of carcinoembryonic antigen after correction for confounding factors (De Coster et al., 2008).

A weak association was found between DNA damage in sperm chromatid and DDT/DDE plasma levels in donors from malaria-endemic areas living in houses sprayed annually with DDT (de Jager et al., 2009). No association was found between DNA damage by the sperm chromatid structure assay in fishermen with low and high consumption of fatty fish (an important source of exposure to persistent organochlorine pollutants) and serum levels of 2,2′,4,4′,5,5′-hexachlorobiphenyl and *p,p′*-DDE (Rignell-Hydbom et al., 2005).

Table 4.1 Genetic and related effects of DDT in exposed humans

Tissue	Cell type (if specified)	End-point	Test	Description of exposed and controls	Response, significance	Comments	Reference
Breast	Tumour cells (carcinoma)	DNA damage	Oxidative DNA damage	44 primary tumours (cancerous) and 21 benign breast biopsy (non-cancerous) tissues from USA residents	–		Charles et al. (2001)
Blood	Lymphocytes	DNA damage	Comet assay	Maternal and umbilical cord blood was collected from 50 mother–infant pairs living in a rural area from Mexico where agriculture is the main economic activity. Determinations of *p,p′*-DDT, *p,p′*-DDE, aldrin, heptachlor epoxide, oxichlordane, chlordane, nonachlor, mirex, endosulfan, α-, β-, and γ-HCH were performed	+	Positive response higher in infants than in mothers [Causative effect of DDT alone cannot be demonstrated]	Alvarado-Hernandez et al. (2013)
Blood	Lymphocytes	DNA damage	Comet assay	276 children from 11 communities in four states of Mexico exposed to a chemical mixture including agrochemicals, i.e. high levels of DDT, polycyclic aromatic hydrocarbons, arsenic, among others. Levels of arsenic and 1-hydroxypyrene in urine and lead and total DDT (*p,p′*-DDE and *p,p′*-DDT) in blood were quantified	+	[Causative effect of DDT alone cannot be demonstrated]	Jasso-Pineda et al. (2015)
Blood	Lymphocytes isolated	DNA damage	Comet assay	61 healthy children in 2003 and during the y 2004, 57 children from the same communities in southern Mexico were assessed. Level of *p,p′*-DDT, *p,p′*-DDD and *p,p′*-DDE was determined	+	Association between DNA damage and *p,p′*-DDT and *p,p′*-DDE levels was found, but not for *p,p′*-DDD	Pérez-Maldonado et al. (2006)
Blood	Lymphocytes isolated	DNA damage	Comet assay	73 children from 4 different Mexican communities. Level of *p,p′*-DDT, *p,p′*-DDD and *p,p′*-DDE was determined	+	Association between DNA damage and *p,p′*-DDT and *p,p′*-DDE levels was found but not for *p,p′*-DDD	Pérez-Maldonado et al. (2011)

Table 4.1 (continued)

Tissue	Cell type (if specified)	End-point	Test	Description of exposed and controls	Response, significance	Comments	Reference
Blood	Lymphocytes isolated	DNA damage	Comet assay	54 healthy women residents of the state of San Luis Potosi, Mexico. One group of women had a history of indoor DDT spraying, and another group had no history of DDT spraying. All women had a similar ethnic and socioeconomic background (low-income Mexican indigenous), and had lived in their community for at least 5 yr before the study. DDT, DDE and DDD level were determined in blood samples	+	Number of donors from each studied group not mentioned. Blood levels of DDT, DDD, and DDE were significantly correlated with DNA damage. This association remained significant after controlling for nutritional status, smoking habits and alcohol ingestion	Yáñez et al. (2004)
Sperm	Sperm cells	DNA damage	Comet assay	212 male partners of a studied subfertile couple and 142 non-study subjects fitting the same inclusion criteria as the men recruited in the study from Massachusetts (USA). PCB, HCB and *p,p′*-DDE levels were determined in blood samples from the donors	(+)	No statistically significant consistent associations between the comet assay parameters and any of the individual PCB congeners, sum of PCB, or *p,p′*-DDE [Causative effect of DDT alone cannot be demonstrated]	Hauser et al. (2003)
Blood	Blood cells	DNA damage	Comet assay	1583 residents aged 50–65 from 9 areas with different types of pollution in Flanders (Belgium). *p,p′*-DDE was found in donors together with several other pollutants compounds including cadmium, lead, HCB, PCBs and dioxins measured in blood and urine	–	No association between *p,p′*-DDE and DNA damage was found	De Coster et al. (2008)

Table 4.1 (continued)

Tissue	Cell type (if specified)	End-point	Test	Description of exposed and controls	Response, significance	Comments	Reference
Sperm	Spermatocytes	DNA damage	Sperm chromatin structure assay	311 male donors from a malaria endemic area (Limpopo Province, South Africa). The housing in these communities consists of traditional mud dwellings with thatched (grass) roofs or brick and cement houses. DDT is sprayed inside onto unpainted brick, cement and mud walls annually, but not on the painted walls. Levels of *p,p′*-DDT and *p,p′*-DDE were determined in blood donor samples	+	Weak association between DDT/DDE plasma levels and chromatin integrity of sperm	de Jager et al. (2009)
Sperm	Spermatocytes	DNA damage	Sperm chromatin structure assay	176 Swedish fishermen (with low and high consumption of fatty fish, a very important exposure source of persistent organochlorine pollutants). Serum levels of 2,2',4,4',5,5'-hexachlorobiphenyl (PCB-153) and *p,p′*-DDE were determined	–	Results indicate a trend to an association DNA fragmentation and *p,p′*-DDE levels	Rignell-Hydbom et al. (2005)
Paraffin-embedded tumour tissue	Pancreatic tumour cells	Mutation	*KRAS* mutation	103 pancreatic ductal adenocarcinoma Spanish patients in whom *KRAS* mutation was determined	+	Cases whose tumours harboured a *KRAS* mutation had higher concentrations of *p,p′*-DDT, *p,p′*- DDE and polychlorinated biphenyls. *p,p′*-DDT, *p,p′*-DDE and PCB were significantly associated with the two most prevalent *KRAS* mutations (Val and Asp) [Causative effect of DDT alone cannot be demonstrated]	Porta et al. (2009)

Table 4.1 (continued)

Tissue	Cell type (if specified)	End-point	Test	Description of exposed and controls	Response, significance	Comments	Reference
Paraffin-embedded tumour tissue	Pancreatic tumour cells	Mutation	*KRAS* mutation	51 patients with pancreatic ductal carcinoma from eastern Spain. Cases of pancreatic cancer with wildtype *KRAS* (n = 17) were frequency matched for age and sex to cases of pancreatic cancer with a *KRAS* mutation (n = 34, case–case study) with positive levels in serum of *p,p′*-DDE, *p,p′*-DDT and PCB congeners	+	Serum concentrations of *p,p′*-DDT and *p,p′*-DDE were significantly higher in pancreatic cancer cases with a *KRAS* mutation than in cases without a mutation. A specific association was observed between a glycine to valine substitution at codon 12 and both *p,p′*-DDT and *p,p′*-DDE concentrations [Causative effect of DDT alone cannot be demonstrated]	Porta et al. (1999)
Paraffin-embedded tumour tissue	Pancreatic tumour cells	Mutation	*KRAS* mutation	61 diagnosed patients identified with pancreatic cancer in the San Francisco Bay area (USA) with positive levels in serum of *p,p′*-DDE, *p,p′*-DDT and 11 PCB congeners	-		Slebos et al. (2000)
Blood	Lymphocytes	Chromosomal damage	Chromosomal aberrations	Exp 1: 50 workers from insecticide plants S. Paulo (Brazil), 25 of them directly exposed to DDT from 2 mo to 10 yr (average weekly exposure of 48 h) DDT levels were determined in plasma donor samples	+		Rabello et al. (1975)

Table 4.1 (continued)

Tissue	Cell type (if specified)	End-point	Test	Description of exposed and controls	Response, significance	Comments	Reference
Blood (cont.)				Exp 2: 8 subjects directly exposed to DDT (São Paulo, Brazil) for at least 20 days up to 2 yr (mean 2 mo; 48 h/wk) and 10 labourers with no history of occupational exposure to DDT. DDT levels were determined in plasma donor samples Exp 3: 15 workers at a pesticide plant in the city of São Paulo, S. Paulo (Brazil). Exposure (including to methylparathion and DDT) varied from 1 wk up to 7 yr, with intermittent periods of non-exposure. 13 unexposed controls			Rabello et al. (1975) (cont.)
Blood	Lymphocytes	Chromosomal damage	Chromosomal aberrations	42 male workers (Idaho) occupationally exposed to DDT, 2,4-D, malathion, ethyl parathion, endosulfan, atrazine, dicamba, among other pesticides; 16 normal healthy donors from the same age group	+	[Causative effect of DDT alone cannot be demonstrated]	Yoder et al. (1973)
Blood	Lymphocytes	Chromosomal damage	Chromosomal aberrations	25 male workers (India) occupationally exposed to DDT, BHC, malathion, parathion, dimethoate, fenitrothion, urea and gromor. 30 normal healthy males from the same age group and socioeconomic class for the control	+	[Causative effect of DDT alone cannot be demonstrated]	Rupa et al. (1988)
Blood	Lymphocytes	Chromosomal damage	Micronucleus formation	Maternal and umbilical cord blood was collected from 50 mother–infant pairs from living in a rural area from Mexico where agriculture is the main economic activity. Determinations of aldrin, heptachlor epoxide, oxichlordane, chlordane, nonachlor, mirex, endosulfan, α,α-,β-,γ-HCH, *p′p′*-DDT,and *p′p′*-DDE were performed	–	No variation between infants and mothers	Alvarado-Hernandez et al. (2013)

Table 4.1 (continued)

Tissue	Cell type (if specified)	End-point	Test	Description of exposed and controls	Response, significance	Comments	Reference
Blood	Lymphocytes	Chromosomal damage	Micronucleus formation	151 residents near a pesticide dump site, 151 controls comparison area residents from Aberdeen, North Carolina, USA; Exposure to complex pesticide mixture (DDT, aldrin, dieldrin, endosulfan, among others), organic solvents and heavy metals. Of 20 organochlorines tested in exposed residents, only DDE was detected in the blood of participants (except for one individual)	–		Vine et al. (2000)
Blood	Lymphocytes	Chromosomal damage	Sister chromatid exchanges	25 male workers (India) occupationally exposed to DDT, BHC, malathion, parathion, dimethoate, fenitrothion, urea and gromor were selected. 30 normal healthy males from the same age group and socioeconomic class for the control	+	[Causative effect of DDT alone cannot be demonstrated]	Rupa et al. (1988)
Blood	Lymphocytes	Chromosomal damage	Sister-chromatid exchanges	61 male pesticide applicators who worked in cotton fields (India) and regularly sprayed pesticides such as DDT, BHC, endosulfan, malathion, methyl parathion, phosphamidon, dimethoate, monocrotophos, quinalphos fenvelrate, and cypermethrin. Median use of pesticides 8 h/day and 9 mo/yr; 45 unexposed men were used as matched control group	+	[Causative effect of DDT alone cannot be demonstrated]	Rupa et al. (1991)
Blood	Lymphocytes	Chromosomal damage	Micronucleus formation	1583 residents aged 50–65 yrs from 9 areas with different types of pollution in Antwerp and Ghent (Belgium). *p,p′*-DDE was found in donors together with several other pollutants compounds including cadmium, lead, HCB, PCBs, and dioxins measured in blood and urine	–	No association between *p,p′*-DDE and chromosomal damage was found	De Coster et al. (2008)

+, positive; –, negative; BHC, β-hexachlorocyclohexane; DDD, dichlorodiphenyldichloroethane; DDE, dichlorodiphenyldichloroethylene; DDT, dichlorodiphenyltrichloroethane; HCB, hexachlorobenzene; HCH, hexachlorocyclohexane; mo, month; PCB, polychlorinated biphenyl; wk, week; yr, year

Regarding mutations, results were inconclusive for associations between *KRAS* mutation in cancer patients and serum levels of DDT or PCB congeners. A positive association was reported in patients with pancreatic ductal adenocarcinoma and higher serum levels of *p,p′*-DDT, *p,p′*-DDE, and PCBs. Specifically, an association was observed between a glycine-to-valine substitution at codon 12 of *KRAS* and concentrations of both *p,p′*-DDT and *p,p′*-DDE (Porta et al., 1999, 2009) [A causative effect of DDT alone could not be demonstrated]. The absence of *KRAS* mutation in pancreatic cancer was associated with higher serum levels of *p,p′*-DDE, but the association lacked statistical significance, and no association with absence or presence of the mutation was found for 11 PCB congeners (Slebos et al., 2000).

Chromosomal aberrations were induced in lymphocytes from insecticide-plant workers directly and continuously exposed to DDT; DDT induced mostly chromatid-type aberrations (Rabello et al., 1975). No induction was seen when workers were chronically but intermittently exposed to low doses of pesticides; thus a long-term exposure to small doses of pesticides in the workplace did not seem to increase the basal level of chromosomal damage (Rabello et al., 1975).

Induction of chromosomal aberrations in lymphocytes was reported in workers occupationally exposed to DDT, among other pesticides (Yoder et al., 1973; Rupa et al., 1988). [A causative effect of DDT alone could not be demonstrated]

No induction of micronucleus formation was seen in lymphocytes from individuals living near a pesticide dump site and exposed to a complex pesticide mixture, including DDT (Vine et al., 2000). No induction of micronucleus formation was reported in circulating lymphocytes from maternal and umbilical cord blood collected from 50 mother–infant pairs exposed to different pesticides including DDT (Alvarado-Hernandez et al., 2013).

Sister-chromatid exchanges were observed in lymphocytes from workers occupationally exposed to DDT, among other pesticides (Rupa et al., 1988, 1991).

(b) Human cells in vitro

See Table 4.2 and Table 4.3

DNA adducts were induced by *p,p′*-DDT in L-02 embryo hepatocyte cells (Shi et al., 2010). DNA strand breaks, assessed by the comet assay, were induced by *p,p′*-DDT in L-02 embryo hepatocyte cells (Shi et al., 2010), and by *p,p′*-DDT, *p,p′*-DDE and *p,p′*-DDD in lymphocytes (Yáñez et al., 2004; Gajski et al., 2007; Gerić et al., 2012). *p,p′*-DDT, *p,p′*-DDE and *p,p′*-DDD did not induce DNA strand breaks by the Fpg-modified comet methodology in lymphocytes (Gerić et al., 2012). DDT was negative in the assay for unscheduled DNA synthesis in lymphocytes (Rocchi et al., 1980), in VA-4 (Ahmed et al., 1977), or in HeLa (Brandt et al., 1972) cells. A positive result for inhibition of the metabolic cooperation was reported in ASS skin fibroblast cells exposed to DDT (Davidson et al., 1985).

Chromosomal aberrations were not induced in lymphocytes by *p,p′*-DDT (Hart et al., 1972), or by a non-specified formulation (*p,p′*-DDT, 63–77%; *o,p′*-DDT, 8–20%; *p,p′*-DDE, 3–5%; and *o,p′*-DDD, 0.2–4%) (Lessa et al., 1976).

Micronuclei were induced in L-02 embryo hepatocyte cells by *p,p′*-DDT (Shi et al., 2010), and in lymphocytes by *p,p′*-DDT, *p,p′*-DDE, and *p,p′*-DDD (Gajski et al., 2007; Garaj-Vrhovac et al., 2008; Gerić et al., 2012). Other nuclear abnormalities (nuclear buds and nucleoplasmic bridges) were induced in lymphocytes by *p,p′*-DDT (Garaj-Vrhovac et al., 2008). Micronuclei were not induced in HepG2 cells by DDT (Wu et al., 2003).

(c) Non-human mammals

See Table 4.4 (and Tables 4.5 and 4.6 online http://monographs.iarc.fr/ENG/Monographs/vol113/index.php).

Table 4.2 Genetic and related effects of DDT in human cells in vitro

Tissue, cell line	End-point	Test	Results		Dose (LED or HID)	Comments	Reference
			Without metabolic activation	With metabolic activation			
L-02 embryo hepatocyte cell line	DNA oxidative damage	DNA adducts (8-OHdG) formation	+	NT	0.001 µmol/L	*p,p′*-DDT	Shi et al. (2010)
L-02 embryo hepatocyte cell line	DNA damage	Comet assay	+	NT	0.01 µmol/L	*p,p′*-DDT	Shi et al. (2010)
Lymphocytes	DNA damage	Comet assay	+	NT	250 µg/mL	*p,p′*-DDT	Gajski et al. (2007)
Lymphocytes	DNA damage	Comet assay	+	NT	0.1 µg/mL	*p,p′*-DDT	Gerić et al. (2012)
Lymphocytes	DNA damage	Comet assay	+	NT	40 µg/mL	*p,p′*-DDT	Yáñez et al. (2004)
Lymphocytes	DNA damage	Fpg-modified comet assay	–	NT	0.1 µg/mL	*p,p′*-DDT	Gerić et al. (2012)
VA-4 cell line	DNA damage	UDS assay	–	-	1000 µM	DDT	Ahmed et al. (1977)
HeLa cell line	DNA damage	UDS assay	–	NT	18 µg/mL	DDT	Brandt et al. (1972)
Lymphocytes	DNA damage	UDS assay	–	NT	500 µg/mL	DDT	Rocchi et al. (1980)
ASS- skin fibroblast cell line	Mutation	Inhibition of metabolic cooperation assay	+	NT	5 µg/mL	DDT	Davidson et al. (1985)
Lymphocytes	Chromosomal damage	Chromosomal aberrations	–	NT	100 µg/mL	*p,p′*-DDT	Hart et al. (1972)
Lymphocytes	Chromosomal damage	Chromosomal aberrations	–	NT	40 µg/mL	Formulated product DDT (63–77% of *p,p′*-DDT; 8–20% of *o,p′*-DDT; 3–5% of *p,p′*-DDE and 0.2–4% of *o,p′*-DDD).	Lessa et al. (1976)
L-02 embryo hepatocyte cell line	Chromosomal damage	Micronucleus formation	+	NT	0.01 µmol/L	*p,p′*-DDT	Shi et al. (2010)
Lymphocytes	Chromosomal damage	Micronucleus formation	+	NT	0.1 µg/mL	*p,p′*-DDT	Gerić et al. (2012)

Table 4.2 (continued)

Tissue, cell line	End-point	Test	Results		Dose (LED or HID)	Comments	Reference
			Without metabolic activation	With metabolic activation			
Lymphocytes	Chromosomal damage	Micronucleus formation	+	NT	250 µg/mL	*p,p′*-DDT	Gajski et al. (2007)
Lymphocytes	Chromosomal damage	Micronucleus formation/Other nuclear abnormalities	+	NT	0.025 µg/mL	*p,p′*-DDT	Garaj-Vrhovac et al. (2008)
HepG2 cell line	Chromosomal damage	Micronucleus formation	–	NT	60 µM	DDT	Wu et al. (2003)

+, positive; –, negative; DDD, dichlorodiphenyldychloroethane; DDE, dichlorodiphenyldichloroethylene; DDT, dichlorodiphenyltrichloroethane; HCB, hexachlorobenzene; HID, highest ineffective dose; LED, lowest effective dose; NT, not tested; 8-OH-dG, 8-oxo-2′-deoxyguanosine; PAHs, polycyclic aromatic hydrocarbons; PCBs, polychlorinated biphenyls; SEQ, sediment equivalent; UDS, unscheduled DNA synthesis

Table 4.3 Genetic and related effects of metabolites of DDT in human cells in vitro

Tissue, cell line	End-point	Test	Results		Dose (LED or HID)	Comments	Reference
			Without metabolic activation	With metabolic activation			
Lymphocytes	DNA damage	Comet assay	+	NT	4.1 µg/mL	*p,p′*-DDE	Gerić et al. (2012)
Lymphocytes	DNA damage	Comet assay	+	NT	40 µg/mL	*p,p′*-DDE	Yáñez et al. (2004)
Lymphocytes	DNA damage	Comet assay	+	NT	3.9 µg/mL	*p,p′*-DDD	Gerić et al. (2012)
Lymphocytes	DNA damage	Comet assay	+	NT	40 µg/mL	*p,p′*-DDD	Yáñez et al. (2004)
Lymphocytes	DNA damage	Fpg-modified comet assay	–	NT	4.1 µg/mL	*p,p′*-DDE	Gerić et al. (2012)
Lymphocytes	DNA damage	Fpg-modified comet assay	–	NT	3.9 µg/mL	*p,p′*-DDD	Gerić et al. (2012)
Lymphocytes	Chromosomal damage	Micronucleus formation	+	NT	4.1 µg/mL	*p,p′*-DDE	Gerić et al. (2012)
Lymphocytes	Chromosomal damage	Micronucleus formation	+	NT	3.9 µg/mL	*p,p′*-DDD	Gerić et al. (2012)

+, positive; –, negative; DDD, dichlorodiphenyldichloroethane; DDE, dichlorodiphenyldichloroethylene; DDT, dichlorodiphenyltrichloroethane; HID, highest ineffective dose; LED, lowest effective dose; NT, not tested

Table 4.4 Genetic and related effects of DDT in non-human mammals in vivo

Species, strain, sex	Tissue	End-point	Test	Results	Dose (LED or HID)	Route, duration, dosing regimen	Comments	Reference
Mammals								
Rat, Sprague-Dawley, female	Hepatic cells	DNA damage	Alkaline elution	+	40 mg/kg	p.o. × 1, sampled after 6, 12, and 24 h	DDT	Hassoun et al. (1993)
Rat, Wistar, female	Hepatic cells	DNA damage	Primary DNA-lesions	+	500 mg/kg	p.o. × 1 by gavage	DDT	Hilpert et al. (1983)
Rat, Wistar, female	Lymphocytes isolated	DNA damage	Comet assay	+	7 mg/m^3	Inh. 8 h/d, 6 d/wk for 5 mo	DDT	Canales-Aguirre et al. (2011)
Rat, Wistar, male	Epithelial mammary cells	DNA damage	Comet assay	+	7 mg/m^3	Inh. 8 h/d, 6 d/wk for 5 mo	DDT	Canales-Aguirre et al. (2011)
Rat, Wistar, male	Stomach, colon, liver, kidney, bladder, lung,	DNA damage	Comet assay	+	75 mg/kg	p.o. × 1, sampled after 3, 8 and 24 h	*p,p'*-DDT	Sekihashi et al. (2002)
Rat, Wistar, male	Brain, bone marrow	DNA damage	Comet assay	–	75 mg/kg	p.o. × 1, sampled after 3, 8 and 24 h	*p,p'*-DDT	Sekihashi et al. (2002)
Mouse, ddY, male	Stomach, colon, liver, kidney, bladder, lung, brain, bone marrow	DNA damage	Comet assay	–	75 mg/kg	p.o. × 1 killed after 3, 8, and 24 h	*p,p'*-DDT	Sekihashi et al. (2002)
Rat, albino, male	Germ cells	Mutation	Dominant lethal assay	+	50 mg/kg	p.o. 1 × /d for 5 days then mated with females immediately after the last injection. Females killed 13 days after the mid-week of mating	*p,p'*-DDT	Palmer et al. (1973)
Rat, albino, male	Germ cells	Mutation	Dominant lethal assay	–	80 mg/kg	p.o. 1 × /d for 5 days then mated with females immediately after the last injection. Females killed 13 days after the mid-week of mating	*p,p'*-DDT	Palmer et al. (1973)

Table 4.4 (continued)

Species, strain, sex	Tissue	End-point	Test	Results	Dose (LED or HID)	Route, duration, dosing regimen	Comments	Reference
Mouse, Swiss ICR/Ha, male	Germ cells	Mutation	Dominant lethal (acute) assay	–	130 mg/kg	i.p. × 1	DDT	Epstein et al. (1972)
Mouse, NMRI, male	Germ cells	Mutation	Dominant lethal (acute) assay	–	1200 mg/kg	i.p. x 1	DDT	Buselmaier et al. (1972)
Mouse, Swiss CD-1, male	Germ cells	Mutation	Dominant lethal (acute) assay	–	105 mg/kg	i.p. x 1	DDT	Epstein & Shafner (1968)
Mouse, CF/1, aabbcc, male	Germ cells	Mutation	Spot test	–	25 mg/kg	p.o. in diet for life span over 5 generations	DDT	Wallace et al. (1976)
Mouse, Swiss albino, male	Germ cells	Mutation	Dominant lethal (acute) assay	+	150 mg/kg	p.o. 1 × /day for 2 days	Formulation (*p,p′*-DDT, 80%; *o,p′*-DDT, 18%; *p,p′*-DDE, 2%)	Clark (1974)
			Dominant lethal (chronic) assay	+	100 mg/kg	p.o. 2 × /wk for 10 wk	Formulation (*p,p′*-DDT, 80%; *o,p′*-DDT, 18%; *p,p′*-DDE, 2%)	
Rat, Sprague-Dawley, female	Mammary gland tissue sections	Chromosomal damage	Chromosomal aberrations	+	50 mg/kg	s.c. on d 21, 23, 25, 27, 29, 31, 32, and 34 postpartum, sampled on day 35	*o,p′*-DDT	Uppala et al. (2005)
Rat, Osborne-Mendel, male	Bone marrow cells	Chromosomal damage	Chromosomal aberrations	–	200 mg/kg	i.p × 1, sampled after 18, 24, and 48 h	DDT	Legator et al. (1973)
				–	200 mg/kg	i.p 1 × /d for 5 days, sampled 6 h after the last injection	DDT	
				–	100 mg/kg	p.o. × 1. sampled after 18, 24, and 48 h	DDT	
				–	100 mg/kg	p.o. 1 × /day for 5 days, sampled 6 h after the last injection	DDT	
Rat, Wistar, female	Oral mucosa cells	Chromosomal damage	Micronucleus induction	+	7 mg/m³	Inh. 8 h/day, 6 d/wk for 5 mo	DDT	Canales-Aguirre et al. (2011)
Rat, Sprague-Dawley, female	Mammary gland tissue sections	Chromosomal damage	Micronucleus formation	–	50 mg/kg	s.c. on day 21, 23, 25, 27, 29, 31, 32 and 34 postpartum, sampled on day 35	*o,p′*-DDT	Uppala et al. (2005)

Table 4.4 (continued)

Species, strain, sex	Tissue	End-point	Test	Results	Dose (LED or HID)	Route, duration, dosing regimen	Comments	Reference
Mouse, BALB/c, sex NR	Bone marrow cells	Chromosomal damage	Chromosomal aberrations	+	100 mg/kg	i.p. × 1	DDT	Johnson & Jalal (1973)
Mouse, BALB/c, sex NR	Bone marrow cells	Chromosomal damage	Chromosomal aberrations	+	50 mg/kg	i.p. × 1, killed 48 h after treatment	DDT	Larsen & Jalal (1974)
Mouse, sex NR	Spleen cells	Chromosomal damage	Chromosomal aberrations	+	5.5 mg/kg	i.p. × 1, killed 6, 24, and 48 h after treatment	DDT	Amer et al. (1996)
Mouse, Swiss albino, male	Spermatocytes	Chromosomal damage	Chromosomal aberrations	+	250 mg/kg	p.o. × 1 × /day for 2 d	Formulation (*p,p'*-DDT, 80%; *o,p'*-DDT, 18%; *p,p'*-DDE, 2%)	Clark (1974)
Rabbit, New Zealand White, female	Lymphocytes	Chromosomal damage	Chromosomal aberrations	–	50 mg/kg	p.o. × 1 at day 7, 8, 9, and 28 of gestation	*p,p'*-DDT	Hart et al. (1972)
Mouse, BALB/c, sex NR	Bone marrow cells	Chromosomal damage	Chromosomal stickiness	+	100 mg/kg	i.p. × 1	DDT	Johnson & Jalal (1973)
Mouse, BALB/c, sex NR	Bone marrow cells	Chromosomal damage	Chromosomal stickiness	–	250 mg/kg	i.p. × 1	DDT	Larsen & Jalal (1974)
Mouse, CD1, male	Hair follicle	Chromosomal damage	Nuclear aberration assay	+	1/4 dermal LD50	Topically × 1, 24 h before analysis	DDT	Schop et al. (1990)
Mouse, CD1, male	Bone marrow cells	Chromosomal damage	Micronucleus formation	–	1/4 dermal LD50	Topically × 1, 24 h before analysis	DDT	Schop et al. (1990)
Mouse hybrid C57BL/6 × C3H/He	Bone marrow cells	Chromosomal damage	Micronucleus formation	–	1/2 LC50	i.p 1 × /day for 5 d, killed 4 h after the last injection	DDT	Bruce & Heddle (1979)

+, positive; –, negative; ±, equivocal (variable response in several experiments within an adequate study); DDE, dichlorodiphenyldichloroethylene; DDT, dichlorodiphenyltrichloroethane; HID, highest ineffective dose; inh., inhalation; i.p., intraperitoneal; LC, lethal concentration; LD, lethal dose; LED, lowest effective dose (units as reported); mo, month; NR, not reported; NT, not tested; NA, not applicable; p.o., oral; s.c., subcutaneous; wk, week

(i) In vivo

The comet assay for DNA strand breaks gave negative results in mouse stomach, colon, liver, kidney, lung, brain, and bone marrow cells after oral exposure to *p,p′*-DDT (Sekihashi et al., 2002). In rats, oral exposure to *p,p′*-DDT induced DNA strand breaks assessed by the comet assay in stomach, colon, liver, kidney, bladder, and lung, but not in brain and bone marrow cells (Sekihashi et al., 2002). DNA strand breaks by alkaline elution were seen in rat hepatocytes after oral exposure to DDT (Hassoun et al., 1993). Inhalation exposure to DDT induced DNA strand breaks in the comet assay in rat lymphocytes and epithelial mammary cells (Canales-Aguirre et al., 2011).

In mice, there was no mutagenic effect of DDT as seen by the dominant lethal assay after intraperitoneal exposure (Epstein & Shafner, 1968; Buselmaier et al., 1972; Epstein et al., 1972), or by the spot test after oral exposure to DDT (Wallace et al., 1976). A positve mutagenic effect was seen in the dominant lethal assay in mice following 2-day or 10-week oral exposure to a non-specified DDT formulation (*p,p′*-DDT, 80%; *o,p′*-DDT, 18%; and *p,p′*-DDE, 2%) (Clark, 1974). In rats, there are inconsistent results for mutagenic effects. Positive results were also seen for the dominant lethal assay after oral exposure to *p,p′*-DDT, but negative results were seen after intraperitoneal exposure (Palmer et al., 1973). Positive results were reported for inhibition of metabolic cooperation in hepatocytes of rats exposed orally to *p,p′*-DDT (Sugie et al., 1987).

Induction of chromosomal aberrations in mouse bone marrow and spleen cells was observed after intraperitoneal exposure to DDT (Johnson & Jalal, 1973; Larsen & Jalal, 1974; Amer et al., 1996), and in spermatocytes after oral exposure to a non-specified DDT formulation (*p,p′*-DDT, 80%; *o,p′*-DDT, 18%; and *p,p′*-DDE, 2%) (Clark, 1974). Induction of chromosomal damage by chromosomal stickiness analysis gave both positive (Johnson & Jalal, 1973) and negative (Larsen & Jalal, 1974) results in mouse bone marrow cells after intraperitoneal exposure to DDT. Chromosomal damage by the hair follicle assay was induced in mice after topical exposure to DDT (Schop et al., 1990). No induction of micronuclei was seen in mouse bone marrow cells after topical exposure to DDT (Schop et al., 1990), or in hybrid C57BL/6 × C3H/He mice after intraperitoneal treatment (Bruce & Heddle, 1979).

Inconclusive results were reported for the induction of chrosomal aberrations by DDT in the rat. Positve results were seen after subcutaneous treatment of rats with *o,p′*-DDT (Uppala et al., 2005). No induction of chromosomal aberrations was reported in rat bone marrow cells after intraperitoneal and oral exposure to DDT (Legator et al., 1973). Positive results for micronucleus formation were found in rat oral mucosa after oral exposure to DDT (Canales-Aguirre et al., 2011), but not in rat mammary gland cells after subcutaneous exposure to *o,p′*-DDT (Uppala et al., 2005). In rabbits, no induction of chromosomal aberrations in lymphocytes was seen in pregnant females treated orally with *p,p′*-DDT on days 7, 8, 9, and 28 of gestation (Hart et al., 1972).

(ii) In vitro

DNA strand breaks were induced by *p,p′*-DDE, but not DDT, in the alkaline elution assay in rat primary hepatocytes (Sina et al., 1983).

Results were negative with DDT, DDE, or *p,p′*-DDE in the assay for unscheduled DNA synthesis in primary hepatocytes of mice, rats, and Chinese hamsters (Klaunig et al., 1984; Maslansky & Williams, 1981; Probst et al., 1981; Williams et al., 1982). No induction of DNA damage was found with *p,p′*-DDT in the alkaline elution assay with and without S9 in Chinese hamster V79 cells (Swenberg et al., 1976; Swenberg, 1981).

In the cell transformation assay, DDT was positive in mouse Balb/c 3T3 fibroblasts with or without S9 (Fitzgerald et al., 1989), but *p,p′*-DDT and *p,p′*-DDE gave negative results in mouse embryo cells (Langenbach & Gingell, 1975). *p,p′*-DDE gave positive results in assays for TK (Clive et al., 1979; McGregor et al., 1988) and *Hprt* mutation (Amacher & Zelljadt, 1984) in L5178Y mouse lymphoma cells. DDT was not mutagenic in the ARL/HGPRT assay in rat liver cells (Telang et al., 1981). *p,p′*-DDT was negative in *Hprt*, 6-thioguanine (6-TO) resistance, and diphtheria toxin *(DT)* resistance assays in Chinese hamster V79 cells (Kelly-Garvert & Legator, 1973; Tsushimoto et al., 1983).

p,p′-DDE, but not *p,p′*-DDT, induced chromosomal aberrations in Chinese hamster V79 cells (Kelly-Garvert & Legator, 1973). In Chinese hamster B14 F28 cells, chromosomal aberrations were induced by *p,p′*-DDT, *p,p′*-DDE, and *p,p′*-DDA, but not by *p,p′*-DDD (Mahr & Miltenburger, 1976). In Chinese hamster ovary cells, DDE did not induce chromosomal aberrations, regardless of the presence or absence of S9 microsomal fraction, but induced a borderline increase in the frequency of sister-chromatid exchange when S9 was present (Galloway et al., 1987).

In rat kangaroo cells, chromosomal aberrations were induced after exposure to *p,p′*-DDT, *o,p′*-DDT, *p,p′*-DDE, *o,p′*-DDE, *p,p′*-DDD, and *o,p′*-DDD, but not after exposure to *p,p′*-DDA (Palmer et al., 1972). In rabbit lymphocytes, *p,p′*-DDT did not induce chromosomal aberrations (Hart et al., 1972). In whale skin fibroblasts, micronucleus formation was induced by *p,p′*-DDT when S9 was present (Gauthier et al., 1999).

(e) Non-mammalian systems

See Tables 4.7 and 4.8 (available online at http://monographs.iarc.fr/ENG/Monographs/vol113/index.php).

(i) Insects

In *Drosophila melanogaster*, positive findings were reported in *Accord* insertion assays for DDT (Catania et al., 2004), as well as in the dominant lethal assay for a non-specified DDT formulation (*p,p′*-DDT, 80%; *o,p′*-DDT, 18%; and *p,p′*-DDE, 2%) (Clark, 1974). Mutagenic effects in the sex-linked recessive lethal assay were inconsistent. A positive mutagenic effect was reported for *p,p′*-DDE (in feeding media but not when injected) (Valencia et al., 1985) and 2,2-bis(*p*-chlorophenyl) acetic acid (DDA) (Vogel, 1972). Negative results were reported in the sex-linked recessive lethal test with *p,p′*-DDT, DDE, DDD, DDOM and a non-specified DDT formulation (*p,p′*-DDT, 80%; *o,p′*-DDT, 18%; and *p,p′*-DDE, 2%) (Pielou, 1952; Vogel, 1972; Clark, 1974). The formulation was positive in the dominant lethal test and induced chromosomal aberrations (Clark, 1974). Chromosomal aberrations were not induced by an unspecified formulation (Woodruff et al., 1983) or by *p,p′*-DDE in the heritable translocation assay (Valencia et al., 1985). In other insects, chromosomal damage was induced in *Anopheles arabiensis* and *Anopheles gambie* survivors of DDT exposure (Nigatu et al., 1995; Brooke et al., 2002).

(ii) Lower eukaryotes

In *Saccharomyces cerevisiae*, results were negative in assays for mitotic gene conversion and colony formation with DDT, DDE, DDD, or DDA (Fahrig, 1974), and in the assay for mitotic chromosomal loss with *p,p′*-DDT (Albertini et al., 1988). Chromosomal damage was induced in the intrachromosomal recombination assay by *p,p′*-DDE in the absence but not in the presence of S9 (Schiestl, 1989; Schiestl et al., 1989).

In *Aspergillus nidulans*, *p,p′*-DDT was negative in assays for forward mutation and for chromosomal aberrations (Crebelli et al., 1986).

In *Neurospora crassa*, a formulation of DDT (*p,p′*-DDT, 80%; *o,p′*-DDT, 18%; *p,p′*-DDE; 2%)

gave negative (in vivo) or equivocal (in vitro) results in the host-mediated assay (Clark, 1974).

(iii) Prokaryotes

In *Salmonella typhimurium*, DDT was negative in the assay for reverse mutation, regardless of the presence or absence of S9, in strains TA92, TA98, TA100, TA1535, TA1536, TA1537, TA1538, TA1978, C3076, D3052, and G46 (Byeon et al., 1976; Marshall et al., 1976; Van Dijck & Van de Voorde, 1976; Probst et al., 1981; Glatt & Oesch, 1987). No mutagenic effects were reported with *p,p′*-DDT or DDE, regardless of the presence or absence of S9, in numerous strains (TA92, TA98, TA100, TA1535, TA1536, TA1537, TA1538, TA1950, and TA1978) (Marshall et al., 1976; Van Dijck & Van de Voorde, 1976; Moriya et al., 1983; Glatt & Oesch, 1987).

A positive result was reported in the assay for reverse mutation with the DDT metabolite 1-chloro-2,2-bis(*p*-chlorophenyl)ethene (DDMU)-epoxide in strain TA100 in the absence of S9 (Gold et al., 1981). No mutagenic effect of the DDT metabolite 2,2-bis(*p*-chlorophenyl)-2-chloroacetaldehyde (αCl-DDCHO) was reported in strain TA100 (Gold et al., 1981), or with DDT metabolite 1,1-bis(*p*-chlorophenyl)-2,2-dichloroethane in strain TA98 (Glatt & Oesch, 1987). However, mutagenic activity was observed for the latter when norharman was added to the S9 (Glatt & Oesch, 1987).

In *Escherichia coli* Q-13 exposed to DDT in the presence of S9, no DNA damage was seen in the DNA-cell-binding assay (Kubinski et al., 1981). In *E. coli*, DDT was not mutagenic in the SOS chromotest assay, with or without S9, in strain PQ37 (Dayan et al., 1987). No mutagenic effects were seen with DDT, DDE, or *p,p′*-DDE, with or without S9, in strain WP2 *uvrA*, or in strain WP2 *hcr* with *p,p′*-DDT or *p,p′*-DDE (Probst et al., 1981; Moriya et al., 1983; Mamber et al., 1984; Glatt & Oesch, 1987). *p,p′*-DDT, *p,p′*-DDE, and *p,p′*-DDD were not mutagenic in the assay for reverse mutation, with or without S9, in strain Pol-A (Fluck et al., 1976). In *Bacillus subtilis*, a negative result in the Rec assay was reported with DDT (Shirasu et al., 1976).

In *Serratia marcescens*, DDD gave positive results in the assay for reverse mutation (Buselmaier et al., 1973). Negative results were reported with DDT, DDE, and DDA in the host-mediated assay (in vivo) (Buselmaier et al., 1973).

4.2.6 Inflammation

(a) Humans

No studies in exposed humans were available to the Working Group.

DDT and its metabolites induced a pro-inflammatory state, with production of pro-inflammatory cytokines and prostaglandins, in multiple types of human cells in vitro. In peripheral blood mononuclear cells isolated from healthy individuals (not exposed to DDE), a proinflammatory state was induced at a low concentration of p,p′-DDE (10 μg/mL), while apoptosis was triggered at the higher concentration of p,p′-DDE (80 μg/mL) (Alegría-Torres et al., 2009; Cárdenas-González et al., 2013). p,p′-DDE enhanced the expression of proinflammatory cytokines (TNF-α, IL-1beta, IL-6) and COX-2 induction at the protein level (Cárdenas-González et al., 2013). In a human trophoblast-derived cell line, exposure to DDE and DDD induced the expression of COX-2 protein, leading to increased production of prostaglandin E(2) (PGE2) (Dominguez-Lopez et al., 2012). In the U937 human macrophage cell line, *p,p′*-DDT upregulated mRNA expression of the pro-inflammatory genes COX-2 and VEGF (Sciullo et al., 2010). Studies by Dutta et al. (2008) showed that DDT significantly enhanced production of tumour necrosis factor-α (TNF-α) and nitric oxide (NO) in macrophages. In human MCF-7 and MDA-MB-231 breast-cancer cell lines, *o,p′*-DDT markedly increased COX-2 protein levels, COX-2 mRNA expression and promoter activity, and production of PGE(2), activating the cyclic-AMP response element (CRE), and raising levels of cAMP and binding of cyclic-AMP response element binding protein (CREB). *o,p′*-DDT induced the activity of aromatase, and was correlated with upregulation of COX-2, and mediated via CRE activation and PKA and PI3-kinase/Akt signalling pathways in breast-cancer cells (Han et al., 2010).

(b) Experimental systems

(i) In vivo

In the intestine of CYP3A4-transgenic mice, *o,p′*-DDT (1 mg/kg) induced increases in CYP3A4 and mouse Cyp3a11 mRNA (Medina-Díaz et al., 2007). The inducibility of CYP3A4 was attenuated at higher doses of *o,p′*-DDT, accompanied by moderate increases in the mRNA of interleukin-6 (a repressor of CYP3A4 transcription) (Medina-Díaz et al., 2007). [The Working Group notes the relevance of this finding to the induction of lymphomas in the gastrointestinal tract and liver tumours.]

In rats exposed to *p,p′*-DDT (0, 5, 50, or 500 ppm) for up to 2 years, microcystic anaemia was induced in a dose-dependent manner (Tomita et al., 2013). In ovarectomized rats, *p,p′*-DDT significantly increased (by threefold) the number of blood eosinophils, and increased their degranulation (Bustos et al., 1995).

(ii) In vitro

Similar to findings in human macrophages, *o,p′*-DDT increased the production of NO and proinflammatory cytokines (IL-1beta, IL-6, TNF-α) in a dose-dependent manner in murine macrophages (Kim et al., 2004). In murine RAW 264.7 macrophages, exposure to *o,p′*-DDT caused a marked increase in the production of PGE2 (a COX-2 metabolite), and a dose-dependent increase in levels of COX-2 protein and mRNA Han et al. (2008).

In bovine epithelial cells and muscle strips of bovine oviducts, DDT and DDE significantly enhanced prostaglandin secretion at concentrations that did not affect cell viability (Wrobel et al., 2012), consistent with findings in cells from humans and other mammalian species.

4.2.7 Other mechanisms

Several studies were identified on epigenetic alterations with DDT exposure. In humans, Huen et al. (2014) reported an association between higher prenatal exposure to DDT and/or DDE and lower Alu methylation at birth, particularly after adjusting for cell type composition ($P = 0.02$ for *o,p′* -DDT) in a birth cohort of Mexican-American children in the Center for the Health Assessment of Mothers and Children of Salinas (CHAMACOS) study. In leukocyte DNA from Japanese women, the global methylation level was significantly decreased by 0.33–0.83% per quartile category for serum *o,p′*-DDT and *p,p′*-DDT, amongst other compounds (Itoh et al., 2014). Similarly, after adjusting for age and cigarette smoking, Rusiecki et al. (2008) reported statistically significant inverse linear relationships in the Alu assay with *p,p′*-DDT, *p,p′*-DDE, and the sum of all persistent organic pollutants, but no multivariate analyses were conducted and the results were not adjusted for multiple comparisons. Studies in rats also reported epigenetic effects with DDT (Shutoh et al., 2009; Skinner et al., 2013; Chanyshev et al., 2014).

Regarding immortalization, the AHS reported that the mean relative telomere length in buccal cells decreased significantly with lifetime intensity-weighted days ($P = 0.04$), but not with lifetime days of DDT use ($P = 0.08$) (Hou et al., 2013).

Regarding DNA repair, Kushida et al. (2005) reported that mRNA levels for 8-oxoguanine glycosylase 1 were inversely correlated with GST-P-positive foci in liver in studies of *N*-diethylnitrosamine initiation and DDT promotion.

4.3 Data relevant to comparisons across agents and end-points

4.3.1 General description of the database

The analysis of the in-vitro bioactivity of the agents reviewed in *IARC Monographs* Volume 113 (i.e. 2,4-D, lindane, and DDT) was informed by data from high-throughput screening assays generated by the Toxicity Testing in the 21st Century (Tox21) and Toxicity Forecaster (ToxCast™) research programmes of the government of the USA (Kavlock et al., 2012; Tice et al., 2013). At its meeting in 2014, the Advisory Group To Recommend Priorities for the *IARC Monographs* programme encouraged inclusion of analysis of high-throughput and high-content data (including from curated government databases) (Straif et al., 2014).

Lindane, DDT (*p,p′*-DDT, *o,p′*-DDT, *p,p′*-DDE, *p,p′*-DDD), and 2,4-D were among the approximately 1000 chemicals tested across the full assay battery of the Tox21 and ToxCast research programmes as of 27 April 2015. This assay battery includes 342 assays, for which data on 821 assay end-points (several assays include multiple end-point readouts) are publicly available on the website of the ToxCast research programme (EPA, 2015a). Detailed information about the chemicals tested, assays used, and associated procedures for data analysis is also publicly available (EPA, 2015b). It should be noted that the metabolic capacity of the cell-based assays is variable, and generally limited.

4.3.2 Aligning in vitro assays to 10 "key characteristics" of known human carcinogens

In order to explore the bioactivity profiles of the agents being evaluated in *IARC Monographs* Volume 113 with respect to their potential impact on mechanisms of carcinogenesis, the 821 available assay end-points in the ToxCast/Tox21 database were first mapped to the 10 key characteristics of known human carcinogens (Smith et al., 2016). Working Group members and *IARC Monographs* staff made independent assignments for each assay type to one or more "key characteristics." The assignment was based on the biological target being probed by each assay. The consensus assignments comprise 265 assay end-points that mapped to 6 of the 10 "key characteristics" as shown below. Within each key characteristic, the assays were further divided by the Working Group into subsets of similar end-points.

1. *Is electrophilic or can undergo metabolic activation (31 end-points)*: no assays directly measure electrophilicity or metabolic activation. However, assay end-points measuring CYP inhibition (29 end-points) and aromatase inhibition (2 end-points) were mapped to this characteristic.
2. *Is genotoxic (0 end-points):* no assay end-points were mapped to this characteristic.
3. *Alters DNA repair or causes genomic instability (0 end-points):* no assay end-points were mapped to this characteristic.
4. *Induces epigenetic alterations (11 end-points):* the assay end-points mapped to this characteristic measure targets associated with DNA binding (e.g. transcription factors) (4 end-points) and transformation catalysts (e.g. histone deacetylase) (7 end-points).
5. *Induces oxidative stress (18 end-points):* the assay end-points mapped to this characteristic measure oxidative stress via cell imaging (7 end-points), markers of oxidative stress (e.g. nuclear factor erythroid 2-related factor, NRF2) (6 end-points), and metalloproteinase (5 end-points).
6. *Induces chronic inflammation (45 end-points):* the assay end-points mapped to this characteristic measure cellular adhesion (14 end-points), cytokines (e.g. IL8) (29

end-points), and nuclear factor kappa-light-chain-enhancer of activated B cells (NF-κB) activity (2 end-points).

7. *Is immunosuppressive (0 end-points):* no assay end-points were mapped to this characteristic.
8. *Modulates receptor-mediated effects (92 end-points):* a large and diverse collection of cell-free and cell-based assay end-points measuring nuclear and other receptor bioactivity, specifically aryl hydrocarbon receptor (AhR) (2 end-points), androgen receptor (11 end-points), ER (18 end-points), farnesoid X receptor (FXR) (7 end-points), peroxisome proliferator-activated receptor (PPAR) (12 end-points), pregnane X receptor_vitamin D receptor (PXR_VDR) (7 end-points), retinoic acid receptor (RAR) (6 end-points), and others (29 end-points), were mapped to this characteristic.
9. *Causes immortalization (0 end-points):* no assay end-points were mapped to this characteristic.
10. *Alters cell proliferation/death or nutrient supply (68 end-points):* the assay end-points mapped to this characteristic measure cytotoxicity (41 end-points), mitochondrial toxicity (7 end-points), cell cycle (16 end-points), and cell proliferation (4 end-points).

By matching assays to key characteristics, additional insights could be obtained on the bioactivity profile for each compound specifically for the purpose of evaluating their potential to interact with or affect mechanisms involved in carcinogenesis. In addition, for each chemical, the results of the in-vitro assays that represent each "key characteristic" can be compared with the results for a larger compendium of substances with similar in-vitro data, so that a particular chemical can be aligned with other chemicals with similar toxicological effects. Nonetheless, the available assays do not cover the full spectrum of targets that may be associated with these mechanisms, and metabolic capacity in many of the assays is limited, which could account for any absence of bioactivity. Conversely, the presence of bioactivity alone does not definitively imply that the agent exhibits that key characteristic, as the assay data are considered along with other information, both in vivo and in vitro.

The Working Group then extracted information from the ToxCast database concerning whether a chemical was "active" or "inactive" for each of the selected assay end-points (Sipes et al., 2013; EPA, 2015b). In the analysis by the Working Group, each "active" was given a value of 1, and each "inactive" was given a value of 0. Thus, by assigning all active compounds a value of 1, the micromolar "potency" estimates from the concentration–response data were not explicitly modelled.

Next, to integrate the data across individual assay end-points into the cumulative score for each "key characteristic," the toxicological prioritization index (ToxPi) approach (Reif et al., 2010) and associated software (Reif et al., 2013; Filer et al., 2014) were used. In the Working Group's analyses, the ToxPi score provides a visual measure of the potential for a chemical to be associated with a "key characteristic" relative to 181 chemicals that have been previously evaluated by the *IARC Monographs* and that have been screened by ToxCast. Assay end-point data were available in ToxCast for these 181 chemicals, and not for other chemicals previously evaluated by IARC. ToxPi is a dimensionless index score that integrates multiple, different, assay results and displays them visually. Within each subset of end-points ("slice"), data are translated into ToxPi slice-wise scores for all compounds as detailed below and in the publications describing the approach and the associated software package (Reif et al., 2013). Within each individual slice for a given chemical, the distance from the origin represents the relative chemical-elicited activity of the component assays (i.e. slices extending farther from the origin were associated with "active" calls on more assays). The overall score for a chemical, visualized as a radial ToxPi profile, is the aggregation of all slice-wise scores.

The list of ToxCast/Tox21 assay end-points included in the analysis by the Working Group, description of the target and/or model system for each end-point (e.g. cell type, species, detection technology, etc.), their mapping to 6 of the 10 "key characteristics" of known human carcinogens, and the decision as to whether each chemical was "active" or "inactive" are available as supplemental material to *Monographs* Volume 113 (IARC, 2017b). The output files generated for each "key characteristic" are also provided in the supplemental material, and can be opened using ToxPi software that is freely available for download without a licence (Reif et al., 2013).

4.3.3 Specific effects across 6 of the 10 "key characteristics" based on data from high-throughput screening data in vitro

The relative effects of DDT were compared with those of 181 chemicals selected from the more than 800 chemicals previously evaluated by the *IARC Monographs* and also screened by the Tox21/ToxCast programmes, and with those of the other compounds evaluated in the present volume of the *IARC Monographs* (Volume 113) and with their metabolites. Of these 181 chemicals previously evaluated by the *IARC Monographs* and screened in the ToxCast/Tox21 programmes, 8 are classified in Group 1 (*carcinogenic to humans*), 18 are in Group 2A (*probably carcinogenic to humans*), 59 are in Group 2B (*possibly carcinogenic to humans*), 95 are in Group 3 (*not classifiable as to its carcinogenicity to humans*), and 1 is in Group 4 (*probably not carcinogenic to humans*). The results are presented in a dot plot as a rank order of all compounds in the analysis arranged in the order of their relative activity). The results are presented in a dot plot as a rank order of all compounds in the analysis arranged in the order of their relative activity. The relative positions of lindane, DDT (*p,p′*-DDT, *o,p′*-DDT, *p,p′*-DDE, *p,p′*-DDD), and 2,4-D in the ranked list are also shown on the *y*-axis. The colour scheme legend (lower left in each plot) annotates each compound according to its previous *IARC Monographs* group classification. The legend key (lower right graphic in each plot) lists components of the ToxPi chart as subcategories that comprise assay end-points in each characteristic, as well as their respective colour-coding (see Section 4.3.2; IARC, 2017b). The ToxPi profile and numeric score are shown for the highest-ranked chemical in each analysis (directly above the legend key) to represent the maximum ToxPi score and and for DDT (upper frames).

Characteristic (1) *Is electrophilic or can undergo metabolic activation*: *p,p′*-DDT was not active for any of the assay end-points tested. *o,p′*-DDT was active for 1 of the assay end-points tested. *p,p′*-DDE was not active for any of the assay end-points tested. *p,p′*-DDD was active for 1 of the assay end-points tested. In comparison, the highest-ranked chemical, malathion (IARC Group 2A; IARC, 2017a), was active for 20 out of 29 assay end-points related to CYP inhibition and in 1 out of 2 related to aromatase inhibition (Fig. 4.3).

Characteristic (4) *Induces epigenetic alterations: p,p′*-DDT was active for two of the assay end-points tested. *o,p′*-DDT was active for 2 of the assay end-points tested. *p,p′*-DDE was active for 2 of the assay end-points tested. *p,p′*-DDD was active for 4 of the assay end-points tested. In comparison, the highest-ranked chemical, captan (IARC Group 3; IARC, 1983) was active for 0 out of 4 DNA binding assay end-points and 5 out of 7 transformation catalyst (e.g. histone modification) assay end-points (Fig. 4.4).

Characteristic (5) *Induces oxidative stress*: *p,p′*-DDT was active for 9 of the assay end-points tested. *o,p′*-DDT was active for 4 of the assay end-points tested. *p,p′*-DDE was active for 5 of the assay end-points tested. *p,p′*-DDD was active

Fig. 4.3 ToxPi ranking for DDT and its metabolites using ToxCast assay end-points mapped to metabolic activation

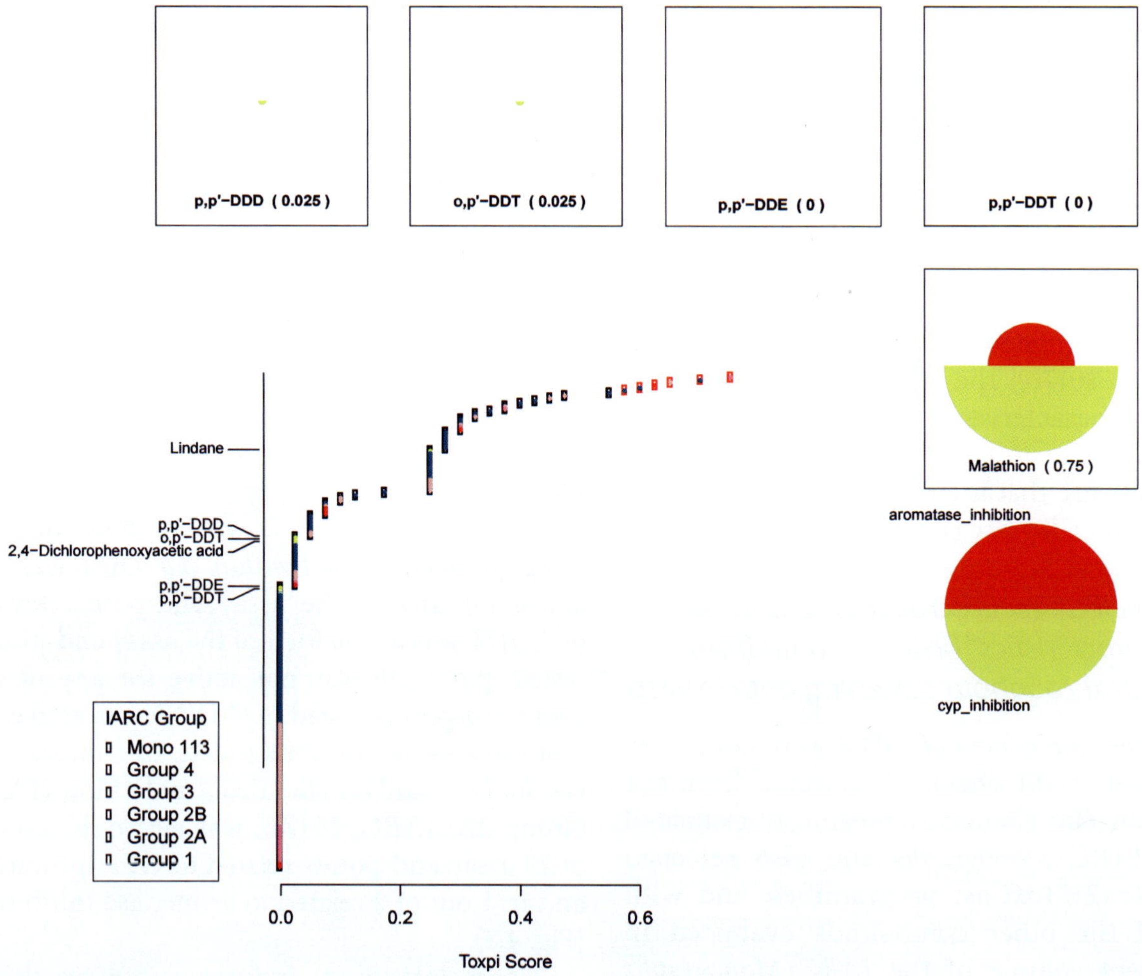

On the left-hand side, the relative rank of DDT, and its metabolites, is shown (*y*-axis) with respect to their ToxPi score (*x*-axis) as compared with the other chemicals evaluated in the present volume (IARC Monograph 113) and the 181chemicals previously evaluated by IARC. The inset in the scatter plot shows subcategories of the ToxPi chart, as well as their respective colour coding. On the right-hand side, the ToxPi charts of the top-ranked chemical (in this case, DDT) and the target chemical(s) (DDT and its metabolites) are shown with their respective ToxPi score in parentheses.

Fig. 4.4 ToxPi ranking for DDT and its metabolites using ToxCast assay end-points mapped to epigenetic alterations

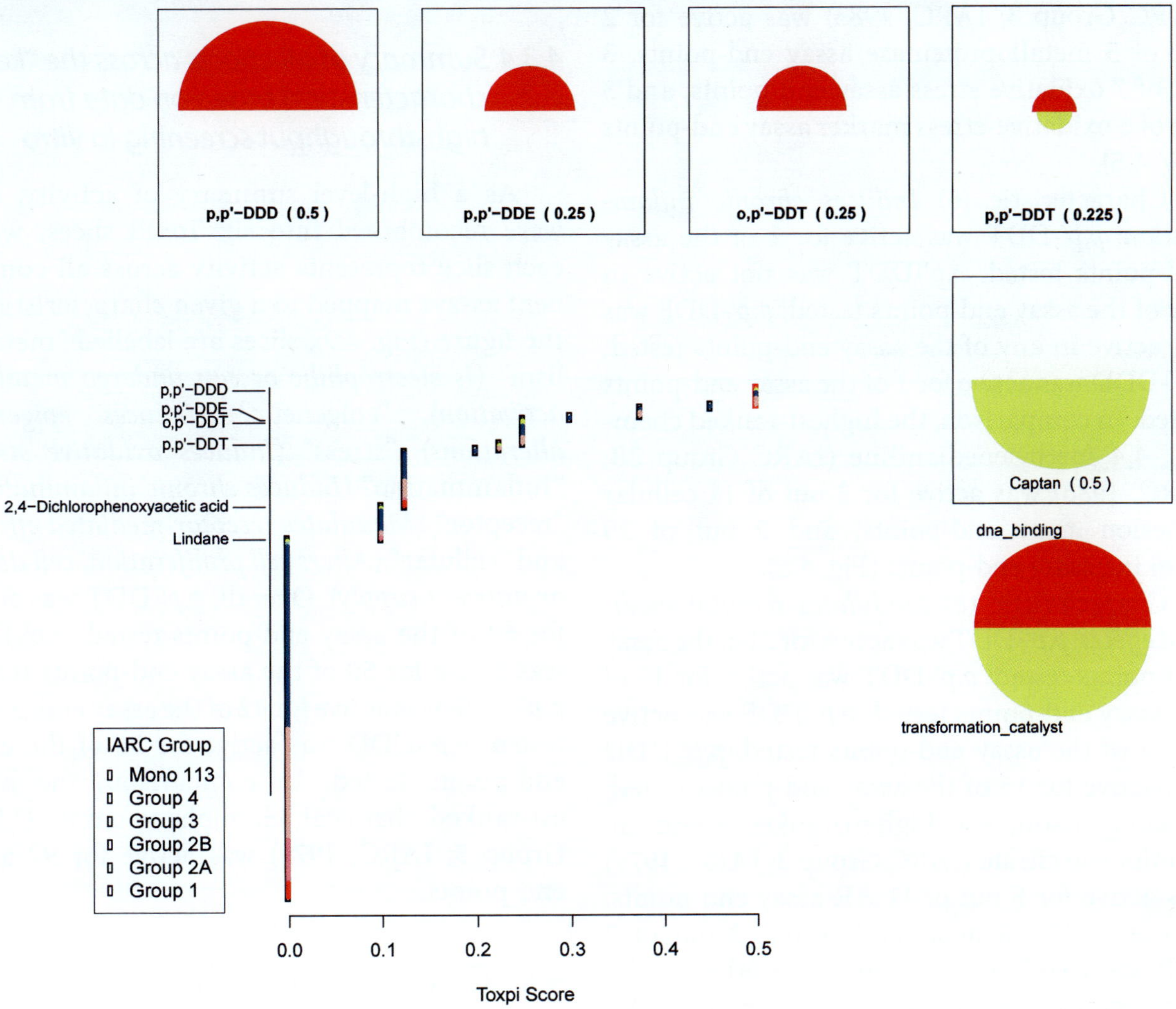

On the left-hand side, the relative rank of DDT, and its metabolites, is shown (*y*-axis) with respect to their ToxPi score (*x*-axis) as compared with the other chemicals evaluated in the present volume (IARC Monograph 113) and the 181 chemicals previously evaluated by IARC. The inset in the scatter plot shows subcategories of the ToxPi chart, as well as their respective colour coding. On the right-hand side, the ToxPi charts of the top-ranked chemical (in this case, captan) and the target chemical(s) (DDT and its metabolites) are shown with their respective ToxPi score in parentheses.

for 8 of the assay end-points tested. In comparison, the highest-ranked chemical, carbaryl (IARC Group 3; IARC, 1983) was active for 2 out of 5 metalloproteinase assay end-points, 3 out of 7 oxidative stress assay end-points, and 3 out of 6 oxidative-stress marker assay end-points (Fig. 4.5).

Characteristic (6) *Induces chronic inflammation*: *p,p′*-DDT was active for 1 of the assay end-points tested. *o,p′*-DDT was not active in any of the assay end-points tested. *p,p′*-DDE was not active in any of the assay end-points tested. *p,p′*-DDD was active for 1 of the assay end-points tested. In comparison, the highest-ranked chemical, 4,4′-methylenedianiline (IARC Group 2B; IARC, 1986) was active for 2 out of 14 cellular adhesion assay end-points, and 2 out of 29 cytokine assay end-points (Fig. 4.6).

Characteristic (8) *Modulates receptor-mediated effects: p,p′*-DDT was active for 21 of the assay end-points tested. *o,p′*-DDT was active for 19 of the assay end-points tested. *p,p′*-DDE was active for 17 of the assay end-points tested. *p,p′*-DDD was active for 15 of the assay end-points tested. In comparison, the highest-ranked chemical, clomiphene citrate (IARC Group 3; IARC, 1979) was active for 5 out of 11 AR assay end-points, 13 out of 18 ER assay end-points, 3 out of 7 FXR assay end-points, 6 out of 29 other nuclear-receptor assay end-points, 2 out of 12 PPAR assay end-points, 5 out of 7 PXR_VDR assay end-points, and 1 out of 6 RAR assay end-points (Fig. 4.7).

Characteristic (10) *Alters cell proliferation, cell death, or nutrient supply:p,p′*-DDT was active for 29 of the assay end-points tested. *o,p′*-DDT was active for 24 of the assay end-points tested. *p,p′*-DDE was active for 18 of the assay end-points tested. *p,p′*-DDD was active for 27 of the assay end-points tested. In comparison, the highest-ranked chemical, ziram (IARC Group 3; Monograph volume 53) was active in 2 out of 16 cell-cycle assay end-points, 33 out of 41 cytotoxicity end-points, and 2 out of 7 mitochondrial-toxicity assay end-points (Fig. 4.8).

4.3.4 Summary of all effects across the "key characteristics" based on data from high-throughput screening in vitro

As a high-level summary of activity, data were recombined into six ToxPi slices, where each slice represents activity across all component assays mapped to a given characteristic. In the figure (Fig. 4.9), slices are labelled "metabolism" (*Is electrophilic or can undergo metabolic activation*), "epigenetic" (*Induces epigenetic alterations*), "stress" (*Induces oxidative stress*), "inflammation" (*Induces chronic inflammation*), "receptor" (*Modulates receptor-mediated effects*), and "cellular" (*Alters cell proliferation, cell death, or nutrient supply*). Overall, *p,p′*-DDT was active for 62 of the assay end-points tested. *o,p′*-DDT was active for 50 of the assay end-points tested. *p,p′*-DDE was active for 42 of the assay end-points tested. *p,p′*-DDD was active for 56 of the assay end-points tested. In comparison, the highest-ranked chemical, clomiphene citrate (IARC Group 3; IARC, 1979) was active for 97 assay end-points.

4.4 Cancer susceptibility data

4.4.1 Inter-individual variability

(a) Genetic susceptibility

The Working Group identified a single study in humans on genetic susceptibility to cancer associated with exposure to DDT or its metabolites. Excessive nucleotide repeats in the AR did not modify the association between *p,p′*-DDE and testicular germ cell carcinoma in a case–control study in a population of men in the USA (Biggs et al., 2008).

Fig. 4.5 ToxPi ranking for DDT and its metabolites using ToxCast assay end-points mapped to oxidative stress markers

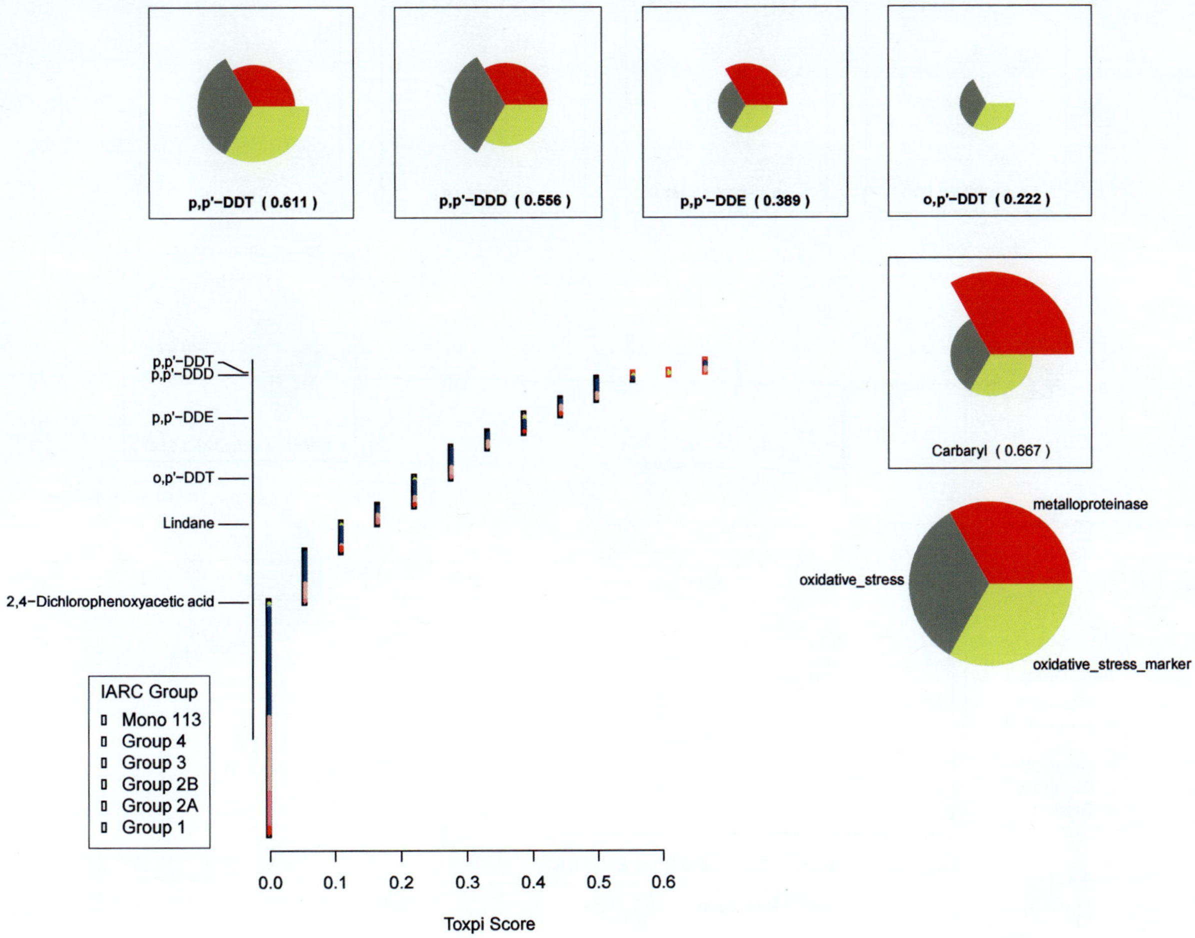

On the left-hand side, the relative rank of DDT, and its metabolites, is shown (*y*-axis) with respect to their ToxPi score (*x*-axis) as compared with the other chemicals evaluated in the present volume (IARC Monograph 113) and the 181 chemicals previously evaluated by IARC. The inset in the scatter plot shows subcategories of the ToxPi chart, as well as their respective colour coding. On the right-hand side, the ToxPi charts of the top-ranked chemical (in this case, carbaryl) and the target chemical(s) (DDT and its metabolites) are shown with their respective ToxPi score in parentheses.

Fig. 4.6 ToxPi ranking for DDT and its metabolites using ToxCast assay end-points mapped to chronic inflammation

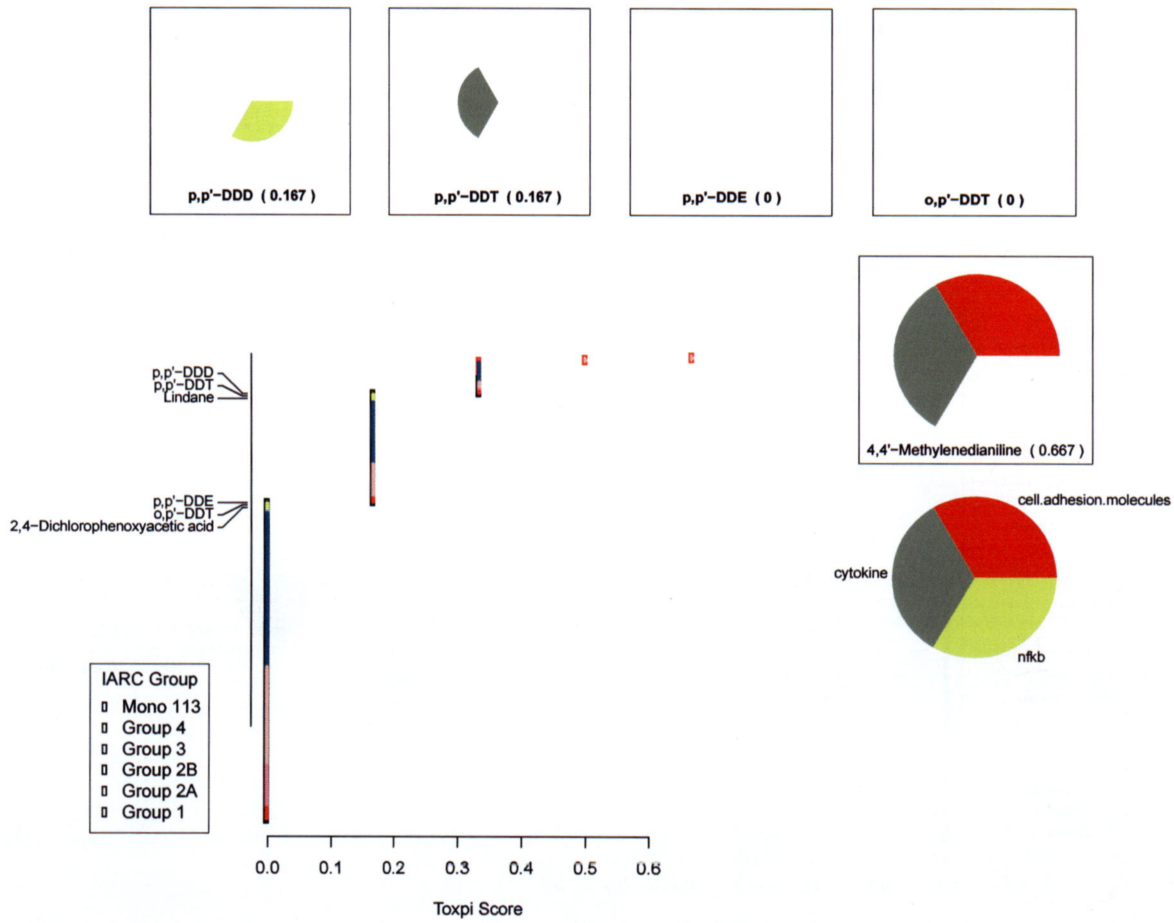

On the left-hand side, the relative rank of DDT, and its metabolites, is shown (y-axis) with respect to their ToxPi score (x-axis) as compared with the other chemicals evaluated in the present volume (IARC Monograph 113) and the 181 chemicals previously evaluated by IARC. The inset in the scatter plot shows subcategories of the ToxPi chart, as well as their respective colour coding. On the right-hand side, the ToxPi charts of the top-ranked chemical (in this case, 4,4'-methylenedianiline) and the target chemical(s) (DDT and its metabolites) are shown with their respective ToxPi score in parentheses.

Fig. 4.7 ToxPi ranking for DDT and its metabolites using ToxCast assay end-points mapped to modulation of receptor-mediated effects

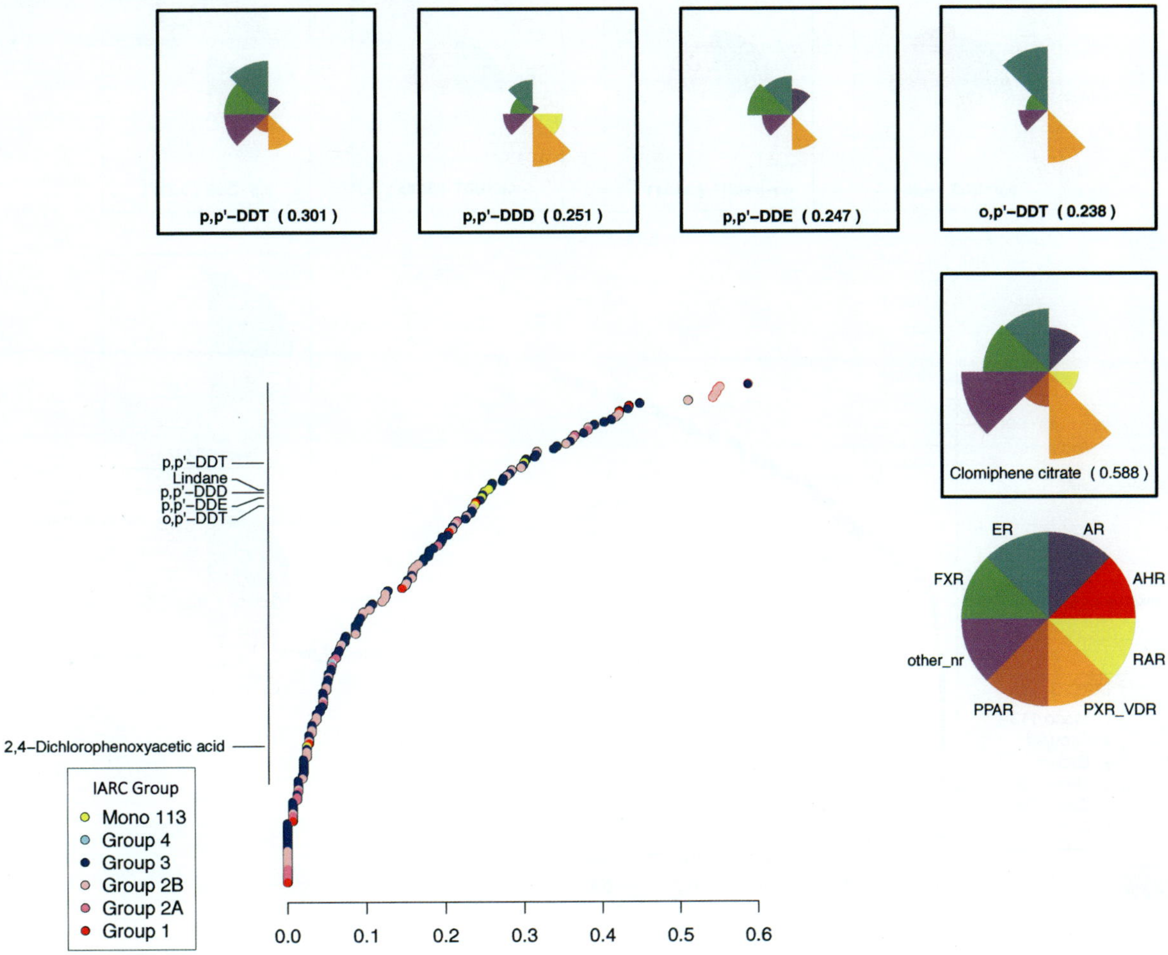

On the left-hand side, the relative rank of DDT, and its metabolites, is shown (*y*-axis) with respect to their ToxPi score (*x*-axis) as compared with the other chemicals evaluated in the present volume (IARC Monograph 113) and the 181 chemicals previously evaluated by IARC. The inset in the scatter plot shows subcategories of the ToxPi chart, as well as their respective colour coding. On the right-hand side, the ToxPi charts of the top-ranked chemical (in this case, clomiphene citrate) and the target chemical(s) (DDT and its metabolites) are shown with their respective ToxPi score in parentheses.

Fig. 4.8 ToxPi ranking for DDT and its metabolites using ToxCast assay end-points mapped to cytotoxicity and proliferation

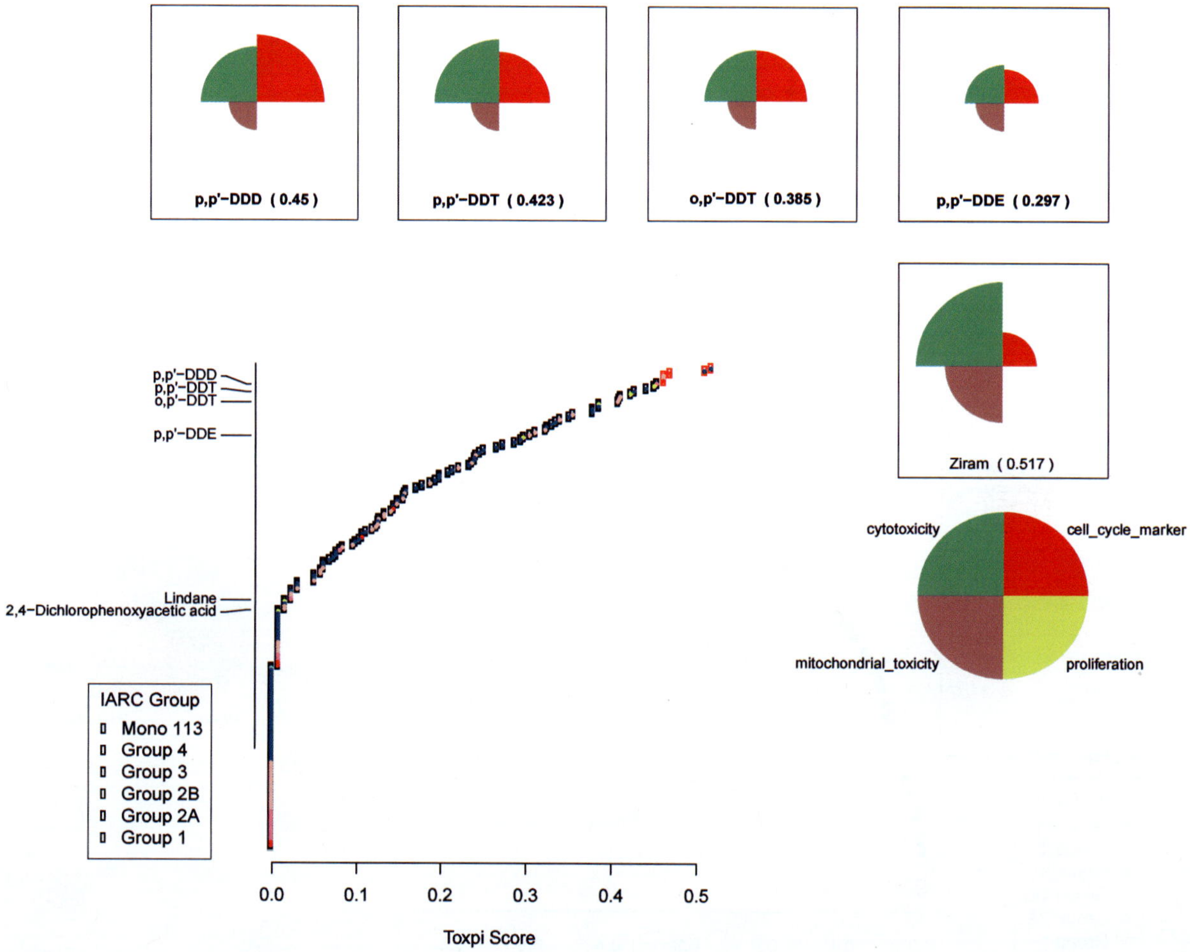

On the left-hand side, the relative rank of DDT, and its metabolites, is shown (*y*-axis) with respect to their ToxPi score (*x*-axis) as compared with the other chemicals evaluated in the present volume (IARC Monograph 113) and the 181 chemicals previously evaluated by IARC. The inset in the scatter plot shows subcategories of the ToxPi chart, as well as their respective colour coding. On the right-hand side, the ToxPi charts of the top-ranked chemical (in this case, ziram) and the target chemical(s) (DDT and its metabolites) are shown with their respective ToxPi score in parentheses.

Fig. 4.9 ToxPi ranking for DDT and its metabolites using ToxCast assay end-points: summary of key characteristics

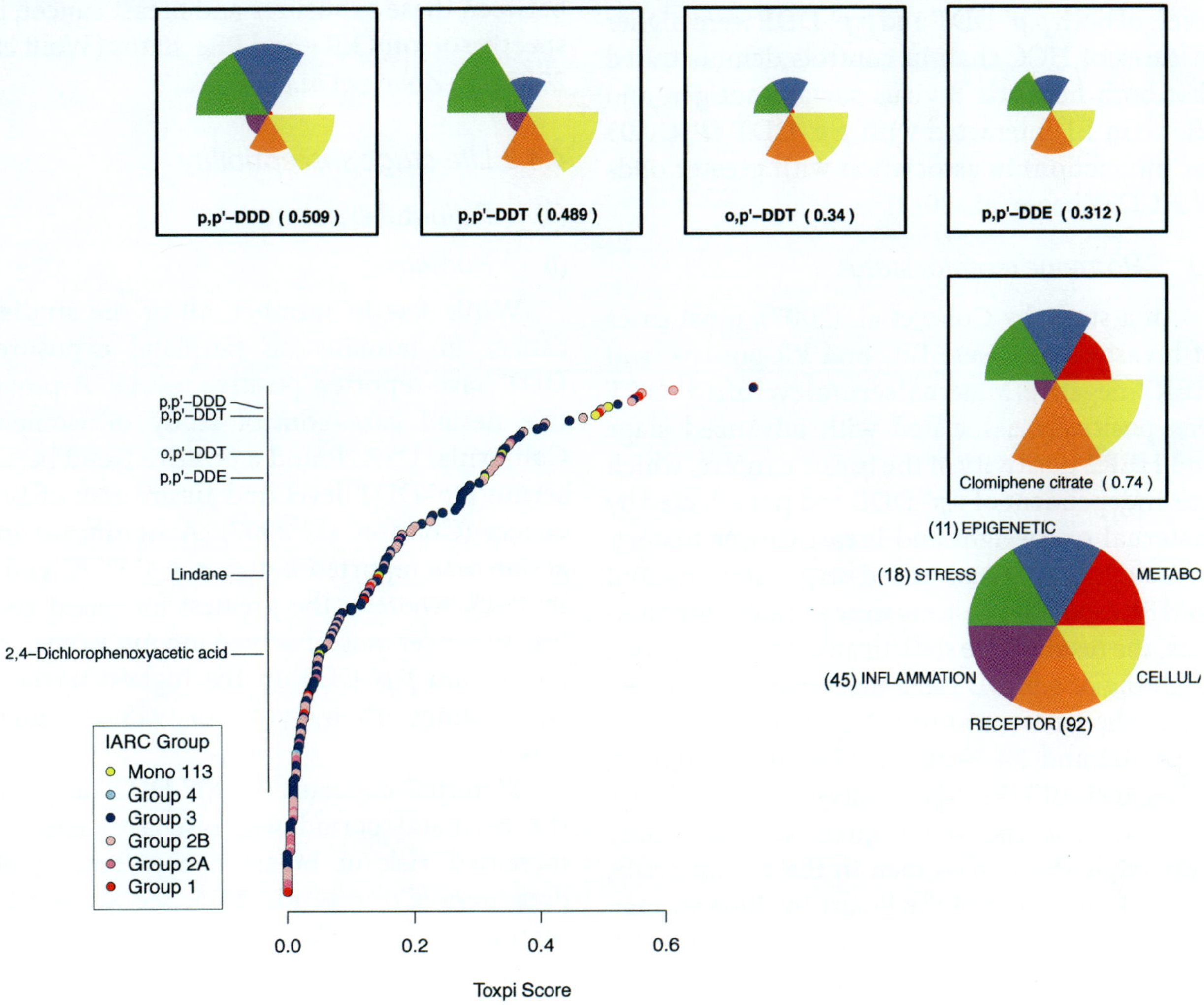

On the left-hand side, the relative rank of DDT, and its metabolites, is shown (*y*-axis) with respect to their ToxPi score (*x*-axis) as compared with the other chemicals evaluated in the present volume (IARC Monograph 113) and the 181 chemicals previously evaluated by IARC. The inset in the scatter plot shows subcategories of the ToxPi chart, as well as their respective colour coding. On the right-hand side, the ToxPi charts of the top-ranked chemical (in this case, clomiphene citrate) and the target chemical(s) (DDT and its metabolites) are shown with their respective ToxPi score in parentheses.

(b) Immune susceptibility

A case–control study in China, in which serum levels of both *p,p′*-DDT and *p,p′*-DDE were higher in cases of HCC than in controls demonstrated that both hepatitis B virus surface antigen and aflatoxin B1 interacted with *p,p′*-DDT ($P < 0.05$ for interaction) in association with greater odds of HCC (Zhao et al., 2012).

(c) Hormone receptor status

In a study by Cohn et al. (2007), most cases of breast cancer were ER- and PR-positive and HER2-negative. Maternal serum level of *o,p′*-DDT was positively associated with advanced stage and HER2 positivity of the breast cancers, which was independent of *p,p′*-DDE and not affected by maternal overweight and breast cancer history. Although only 22 advanced-stage tumours and 16 HER2-positive cancers were available for analyses, the results were statistically significant; for a doubling of *o,p′*-DDT at a diagnosis of advanced stage, the odds ratio was 2.2 (95% CI, 1.1–4.2; $P = 0.02$) and 2.1 (95% CI, 1.0–4.8; $P = 0.05$ at a diagnosis of HER2-positivity). *o,p′*-DDT levels for women in the fourth quartile of the study were triple those of women in the first quartile, and risk of cancer of the breast in these women was quadrupled (for advanced stage disease: OR, 4.6; 95% CI, 1.3–16.5 for 4th quartile compared with the 1st quartile; for HER2-positive cancers: OR, 4.6; 95% CI,1.1–19.7). [These observations suggested a strong effect on breast cancer stage and HER2 status at diagnosis of *o,p′*-DDT exposure in utero in these women.]

The possibility that PR+, HER2+, and triple-negative status modifies the association between DDT or DDE exposures and cancer was indirectly suggested by a case–control study on breast cancer in Canadia, which found a significant trend for the odds ratios associated with *p,p′*-DDE and ER negative breast cancer status, but no association with ER-positive breast cancer status (Woolcott et al., 2001).

Three case–control studies on *p,p′*-DDT and *p,p′*-DDE in the USA did not find an association between these exposures and breast cancer, irrespective of joint ER+ and PR+ status (Wolff et al., 2000a, b; Zheng et al., 2000).

4.4.2 Life-stage susceptibility

(a) Perinatal exposure

(i) Humans

While few in number, all of the studies of cancer in humans on perinatal exposure to DDT have reported positive results. A prospective nested case–control study of women in California, USA, found a positive trend between serum *p,p′*-DDT level and future risk of breast cancer (Cohn et al., 2007). A significant interaction was reported between *p,p′*-DDT and age in 1945, whereby the greatest increased risk of breast cancer was observed among women who had serum *p,p′*-DDT in the highest tertile and were younger than 4 years in 1945 (Cohn et al., 2007).

Maternal exposure to *o,p′*-DDT assessed in the perinatal period was also associated with increased risk of breast cancer among adult daughters (Cohn et al., 2015; see Section 2 for further study details).

In a multi-decade prospective study examining the association between prenatal exposure to DDT compounds detected in maternal sera in the USA, male offspring had an increased risk of testicular cancer associated with higher DDT/DDE ratio and lower *o,p′*-DDT and *p,p′*-DDE levels in their mothers' sera (Cohn et al., 2010). [The Working Group noted that this pattern of associations was consistent with a slower rate of *p,p′*-DDT metabolism in association with elevated testicular cancer risk in offspring.] However, in a case–control study in Sweden, *p,p′*-DDE level measured in maternal sera was not associated with testicular cancer status (Hardell et al., 2006b).

(ii) Experimental systems

Carcinogenesis was enhanced when DDT exposures occured before tissues were fully developed (such as cross-generational, prenatal, preweaning and prepuberty). Mice with prepubertal exposure to *p,p′*-DDE had an early onset of HER2-positive tumours (Johnson et al., 2012). In a study in which mice were exposed to technical DDT (*p,p′*-DDT, 73–78%; *o,p′*-DDT, 20%; *m,p′*-DDT, 1%; *p,p′*-DDE, 0.5% ; and *p,p′*-DDD, 0.5–1.5%) and followed to the F_3 generation (Tomatis et al., 1972), there was an increase in tumour number (see Section 4.3) that was potentiated as generations increased up to F_3 (Terracini et al., 1973b). There was also an apparent increase in body mass in F_0 females and F_1 males and females [no additional follow-up reported] (Tomatis et al., 1972). Body mass and abdominal fat was increased in F_3 rats for which F_0 females had been exposed to *p,p′*-DDT while pregnant (Skinner et al., 2013).

(b) Menopause

Data regarding an association between DDE and cancer of the breast among postmenopausal women are inconsistent. In studies demonstrating a positive association with DDE and breast cancer and that also looked for DDE interaction with menopausal status, DDE was associated with heightened risk in postmenopausal women. The positive association between serum *p,p′*-DDE level and breast cancer was heightened among postmenopausal women compared with pre- or perimenopausal women in hospital-based studies in Belgium and Mexico (Romieu et al., 2000; Charlier et al., 2004).

In case–control studies of women in Spain, Poland, and the USA, in whom *p,p′*-DDE level had no association with cancer of the breast, menopausal status did not modify the null association (Zheng et al., 2000; Ibarluzea et al., 2004; Ociepa-Zawal et al., 2010). Similarly, in a 10-year prospective follow-up of women in Japan, in whom *p,p′*-DDE level had no association with breast cancer, menopausal status did not modify the null association (Iwasaki et al., 2008).

4.5 Other adverse effects

4.5.1 Humans

Prospective studies in Spain and the USA have indicated that exposure to DDE in utero and in the early postnatal period is associated with increased risk of overweight/obesity in toddlers and adult offspring (Karmaus et al., 2009; Warner et al., 2014). A more extensive literature also demonstrates positive associations between *p,p′*-DDE and *p,p′*-DDT exposure with diabetes in humans (reviewed in Taylor et al., 2013). [DDT and metabolites are associated with type 2 diabetes and obesity in humans.] *p,p′*-DDT and *p,p′*-DDE showed a strong association with type 2 diabetes after adjusting for age, sex, BMI, alcohol consumption, and cigarette smoking (Son et al., 2010).

Studies of maternal DDT exposure and adverse birth outcomes (e.g. pre-term birth, small for gestational age, spontaneous abortion, birth weight) have reported mixed results. One large study found a statistically significant trend for pre-term birth and small-for-gestational-age babies (Longnecker et al., 2001). Another smaller study found statistically significant correlations between lower birth weight and DDE in placenta or DDT in breast milk (Dewan et al., 2013). Other, mostly smaller, studies found weak or non-statistically significant changes in outcomes such as preterm birth, small-for-gestational-age birth, birth weight, or gestational age (Gladen et al., 2003; Farhang et al., 2005; Jusko et al., 2006; Khanjani & Sim, 2006; Sagiv et al., 2007; Vafeiadi et al., 2014). One study suggested that DDT and its metabolite DDE have opposite associations with birth weight (Kezios et al., 2013).

A few studies have examined other developmental or reproductive outcomes. A study by Korrick et al. (2001) reported an association between spontaneous abortion and maternal serum DDE levels.

Venners et al. (2005) showed a positive associations between the risk of subsequent early pregnancy loss and the preconception serum total DDT in 388 newly married, female textile workers in China between 1996 and 1998 (Venners et al., 2005). One study reported the rate of birth defects to be associated with occupational exposure to DDT (Salazar-García et al., 2004).

4.5.2 Experimental systems

Reproductive and developmental toxicity of DDT in experimental systems has been reviewed by ATSDR (2002) and Smith (2010). Decreased foetal body weights were reported in rabbits given DDT during gestation (Hart et al., 1971; Fabro et al., 1984). Male reproductive effects have been reported in several studies in rats (Kelce et al., 1995; You et al., 1998; Ben Rhouma et al., 2001). A two-generation study of reproductive toxicity in rats found no reproductive or developmental effects except for decreased pup viability at one time-point at the highest dose and, at the two higher doses, altered hormone levels and delayed male sexual maturation (Hojo et al., 2006). In a study of mouse preimplantation embryos exposed in vitro to DDT, some alterations in development in vitro were observed, but upon transfer to recipient mice, no measureable effects on implantation rates, transfer efficiencies, or multiple other pup characteristics were reported (Greenlee et al., 2005).

5. Summary of Data Reported

5.1 Exposure data

From the discovery of its insecticidal properties in 1939 until its production and use began to be phased out in the early 1970s, 1,1′-(2,2,2-trichloro-ethylidene)bis(4-chlorobenzene) – DDT – was used extensively for insect control in public health and agriculture worldwide. It has been estimated that a total of 1.8 million tonnes of DDT have been produced globally since the 1940s. Apart from its use as a pesticide, DDT is also reported to be used in some countries as an intermediate in the production of the pesticide dicofol and of antifouling paint. While its use in agriculture has been largely prohibited, DDT was, and still is is in some countries, used to control vectors for malaria and a few other diseases of public health importance (e.g. leishmaniasis).

Currently, DDT is manufactured in one country, and its use is officially limited to vector control in several countries in Africa and Asia.. Accordingly, occupational exposure to DDT can still occur among workers in manufacturing and sprayers in vector-control programs, but the number of people affected is small.The population at large is still exposed to DDT, despite the fact that it is no longer used in many countries because of earlier widespread application and the environmental and biological persistence of the compound and its metabolites, DDT and its metabolites have been detected in air, rain, soil, glaciers, water, animal and plant tissues, food, and the work environment. In most countries, exposure of the general population in most countries occurs mainly through the diet. Blood DDT and DDE levels in the general population have dropped at least two to three orders of magnitude over time in most parts of the world, but to a lesser degree where DDT continues to be used.

5.2 Human carcinogenicity data

The risk of cancer associated with exposure to DDT has been evaluated in numerous cohort and case–control studies in several countries. The largest quantity of data is available for cancer of the breast and lymphoma. Cancers of the liver, testis, prostate, endometrium, pancreas, lung, and colon have also been studied. Exposure has been assessed in these studies by biological measurement of markers of exposure to DDT, mostly *p,p*-DDE and *p,p*′-DDT, as well as with questionnaires, sometimes in combination with expert assessment. An important consideration in studies using biological markers is whether the samples were obtained before or after disease onset, when disease progression or treatment may affect the concentration of the marker.

5.2.1 Cancer of the liver

The association of liver cancer with DDT exposure was assessed in three large studies in Linxian, Haimen, and Xiamen, China. Nested case–control studies in Linxian and Haimen reported a strong association with a significant trend between hepatocellular carcinoma and concentrations of *p,p*′-DDT, but not *p,p*′-DDE, in blood samples collected before diagnosis. These studies are consistently positive for *p,p*′-DDT, and inconsistent for *p,p*-DDE; the observed associations for DDT are strong, with dose–response relationships, and the odds ratios were adjusted for important risk factors for hepatocellular carcinoma, including markers for hepatitis (HBV sAg). The population-based case–control study in Xiamen, which had higher exposure to DDT than the study in Linxian, also reported a strong association with *p,p*′-DDT concentration, with a significant dose–response trend in blood samples taken after diagnosis, as well as a weaker association with *p,p*′-DDE. Risk estimates were adjusted for hepatitis and aflatoxin exposure. However, it was unclear how the controls in the Xiamen study were selected. In contrast, no increased risk of cancer of the liver was observed in a cohort of men occupationally exposed to DDT during an antimalarial spraying campaign in Sardinia, Italy.

5.2.2 Cancer of the testis

Six studies have assessed the association between DDT or DDE measurements in blood samples and cancer of the testis. A statistically significant positive association between *p,p*′-DDE and testicular cancer was seen in a large nested case–control study using blood samples taken before diagnosis among United States servicemen. Positive but non -significant associations were found in a smaller nested case–control study in Norway using blood samples taken before diagnosis. These two studies provide the strongest evidence for an association between exposure to DDT and testicular cancer. Positive non-significant associations were found in two small case–control small studies using post-diagnostic blood samples, while no association was detected in a large population-based case–control study using post-diagnostic blood samples. Results were inconsistent in two small studies (one also evaluating levels among cases, mentioned above) that examined DDT and/or DDE levels in mothers of the cases of testicular cancer.

5.2.3 Cancer of the breast

More than 40 epidemiological studies conducted in North America, Latin America, Asia, and Europe since 1993 have assessed the relationship between exposure to DDT and risk of cancer of the breast . Almost all studies used *p,p*′-DDE measurements in blood or adipose tissue as an exposure indicator, and some reported results for *p,p*′-DDT. Biological measurements of exposure were made at diagnosis or several years before. No association overall was found between *p,p′*-DDE or *p,p′*-DDT levels and breast cancer. Stratification by hormone-receptor status of the breast tumour, or menopausal status, did not modify the results. Several meta-analyses on *p,p′*-DDE exposure was that the available studies supported the view that DDE is not associated with an increased risk of breast cancer in humans. However, the potential influence of age at exposure to DDT in relation to risk of breast cancer remains of interest, as suggested by two studies that reported an increased risk of breast cancer in women highly exposed to DDT early in life.

5.2.4 Non-Hodgkin lymphoma

More than 30 studies have evaluated risk of lympho-haematopoietic malignancies in relation to DDT exposure using biomarkers or questionnaires, in a few instances supported by expert assessment of agricultural exposures. The large Agricultural Health Study in the USA observed significant upward trends in risk of non-Hodgkin lymphoma in relation to several indicators of DDT use while controlling for other suspected risk factors. However, a retrospective cohort mortality study of applicators involved in an antimalarial campaign in Sardinia, Italy, with almost exclusive use of DDT, did not identify any association between DDT exposure and lymphoma. Evidence from case–control studies based on self-reported or expert assessments from questionnaire was inconsistent, with no association in several large studies and positive associations in some smaller studies. Evidence was also inconsistent in studies using measurements in biological specimens as biomarkers of exposure to DDT. In studies that adjusted for exposure to other pesticides or persistent organochorines, the associations with DDT were typically weakened. Conflicting results on the association of DDT with the overall group of lymphomas might be related to heterogeneity of association by subtype, but studies of leukaemia and lymphoma subtypes were based on relatively small numbers.

5.2.5 Cancer of the prostate

Several studies have examined the association between exposure to DDT and cancer of the prostate. Positive associations with DDT were found in two population-based case–control studies, in which exposure was assessed by experts or by a job–exposure matrix from the job history. In the United States Agricultural Health Study, there was no clear relationship between lifetime cumulative exposure to DDT and the incidence of total or aggressive cancer of the prostate. The largest study using *p,p′*-DDE measurements, conducted in the French Caribbean, reported a modest but statistically-significant increased risk of prostate cancer associated with *p*,p′-DDE serum concentration measured in blood sampled after diagnosis. Conversely, no significant association with *p,p′*-DDE was observed in four other studies that used serum measurements of *p,p′*-DDE, including a large nested case–control study in Japan.

5.2.6 Other cancer sites

Exposure to DDT or DDE and risk of cancer has also been examined at other cancer sites, including the pancreas, endometrium, colon, and lung. There was no evidence for an association between these cancers and exposure to DDT.

5.3 Animal carcinogenicity data

In mice, 12 out of 13 studies of carcinogenicity with DDT (11 oral administration studies by feeding or gavage, and one subcutaneous injection study) in males and/or females gave positive results (some for multiple sites). One skin application study gave negative results.

In treated mice, DDT consistently increased the incidence of benign and/or malignant tumours of the liver that were classified across the various studies as benign or malignant liver cell tumours, hepatomas (not further classified), benign or malignant hepatomas, or hepatocellular adenoma or carcinoma.

In nine of these positive studies (including the subcutaneous injection study), there was an increase in the incidence of liver cell tumours (benign, malignant, or combined benign or malignant): six studies were positive for males and females, two studies for males only, and one study for females only; in one of these studies there was also an increase in the incidence of hepatoblastoma.

In three of these positive studies, there was an increase in the incidence of malignant lymphoma: one study was positive for males and females, one for males, and one for females. In another of these positive studies, there was an increase in the incidences of malignant lymphoma, leukaemia and pulmonary carcinoma in males and females (combined).

In rats, six of nine carcinogenicity studies by oral administration (feeding) in males and/or females were positive. In four studies, there was an increase in the incidence of liver cell tumours (benign, malignant, or combined benign or malignant): three studies were positive for males and four for females; in one of these studies, there was also an increase in the incidence of ovary carcinoma. The incidences of thyroid follicular cell adenoma or carcinoma (combined) and of adrenal gland pheochromocytoma were also increased in females in the fifth positive study. In the sixth positive study, there was a small but significant increase in the incidence of bronchogenic carcinoma in males and females (combined). In several initiation-promotion studies, DDT promoted benign and/or malignant liver tumours.

In hamsters, two out of three carcinogenicity studies by oral administration (feeding) were positive: there was an increase in the incidence of adrenal cortex adenoma in males in one study and in females in another study.

In one study in monkeys, one prostatic adenocarcinoma and one HCC were reported in two cynomolgus monkeys out of 24 DDT-exposed cynomolgus or rhesus monkeys. No tumours were observed in 17 untreated cynomolgus or rhesus monkeys.

The DDT metabolite DDD was carcinogenic in one of two mouse oral administration (feeding) studies. In this study, DDD caused an increase in the incidence of hepatoma (benign or malignant, combined) in males, and of lung adenoma or adenocarcinoma (combined) in males and females. In one feeding study in rats, DDD caused an increase in the incidence of thyroid follicular cell adenoma or carcinoma (combined) in males.

The DDT metabolite DDE was carcinogenic in mice with an increase in the incidence of hepatocellular tumours (benign and malignant) in two oral administration (feeding) studies in males and females, but not in one study in male and female rats. One feeding study in hamsters showed an increase in the incidence of hepatocellular adenoma or carcinoma (combined) in males and females.

5.4 Mechanistic and other relevant data

p,p′-DDT and *o,p′*-DDT are highly lipophilic and readily absorbed via all routes of exposure. Both DDT isomers are distributed widely in the body by both lymphatic and blood circulation, with a preference for adipose and other lipid-rich tissues. *p,p′*-DDT and *o,p′*-DDT are metabolized to *p,p′*-DDD and *o,p′*-DDD, respectively, which readily degrade to *p,p′*-DDA and *o,p′*-DDA excreted in urine. *p,p′*-DDT is also metabolized to DDE, which is poorly eliminated and more lipophilic than the parent compound. Human half-lives of *p,p′*-DDT and DDE are long, on the order of 5 (for *p,p′*-DDT) to 10 (for DDE) years, whereas *o,p′*-DDT is rapidly metabolized and excreted. *p,p′*-DDT has been reported to induce several P450 enzymes in rats and several PXR-mediated P450s in a human hepatoma cell line. Metabolism in humans and experimental systems are expected to be similar.

With respect to the key characteristics of human carcinogens, adequate data were available to evaluate whether DDT modulates receptor-mediated effects, is immunosuppressive, induces oxidative stress, alters cell proliferation, cell death or nutrient supply, is genotoxic, and induces chronic inflammation.

The evidence is *strong* that DDT modulates receptor-mediated effects that can operate in humans. DDT and its metabolites can modulate thyroid hormones in exposed humans. While evidence is less clear for effects on the sex steroid hormone axis in men and women, estrogenic effects of *o,p′*-DDT and *p,p′*-DDT, such as binding and activation of ER, were consistently seen across numerous experimental systems, including human cells, and were blocked by anti-estrogens in human breast cancer cells and in mice. Evidence that DDT and its metabolites antagonize the AR, with *p,p′*-DDE being the most potent, was consistent across non-human experimental systems in vivo and in cells from a variety of species including humans. DDT and its metabolites can bind and activate progesterone in cells from multiple species including humans. DDT induced PR expression in ER-positive breast cancer cells. It also activated PR in such cells and in a yeast system, and blocked transactivation of PR by progesterone. Some studies also report activation of CAR or PXR by DDT. Some studies suggest a relationship between *o,p′*-DDT or *p,p′*-DDE and breast or mammary cancer that involves HER2.

The evidence is *strong* that DDT is immunosuppressive, and this can operate in humans. In studies in exposed humans, immunological changes are reported to be correlated with *p,p′*-DDT or *p,p′*-DDE levels, though subjects in these studies were also exposed to other contaminants that may be correlated with *p,p′*-DDT or *p,p′*-DDE levels. Additionally, human natural killer cell suppression has been observed in multiple in vitro studies of *p,p′*-DDT, indicative of its potential to decrease immunosurveillance. In mice exposed in vivo to *p,p′*-DDT or *o,p′*-DDT, suppression of B-cell function indicating effects on humoral immune response have been reported, with potentiation via stress that is consistent with data showing modulation of glucocorticoids. Suppression of humoral immune response by *p,p′*-DDT, *o,p′*-DDT, or *p,p′*-DDE has also been reported in rats, rabbits, and marine mammals in vivo, as well as mice and rats in vitro.

The evidence is *strong* that DDT induces oxidative stress, and this can occur in humans. *p,p′*-DDT, *p,p′*-DDD, and/or *p,p′*-DDE activated the Wnt/β-catenin pathway via ROS and stimulated proliferation of human colorectal and liver cancer cells in vitro and in xenografted mice. *p,p′*-DDT, *p,p′*-DDD, and/or *p,p′*-DDE increased ROS levels in human peripheral blood mononuclear cells. Several of these effects were inhibited by anti-oxidant treatment. In liver of exposed rats, *p,p′*-DDT increased lipid peroxidation and levels of 8-hydroxy-2′-deoxyguanosine (8-OHdG).

The evidence is *moderate* that DDT alters cell proliferation or cell death. There is evidence, primarily for *o,p′*-DDT, of ER-dependent induction of cell proliferation in particular cell types *in vitro*. In some cell types *in vitro*, *o,p′*-DDT induces apoptosis although in others, a suppression of this end-point has been noted.

The evidence for genotoxicity of DDT is *moderate*. There is some evidence of DNA damage, chromosome aberrations, and micronuclei in human cells exposed to DDT in vitro. Experimental mammalian in vivo and in vitro data are mixed, for some of these same end-points. Data in non-mammalian experimental systems are predominantly negative.

The evidence that DDT induces inflammation is *moderate*. While the number of studies is small, there are data in human and mammalian cells in vitro that *p,p*′-DDT, *o,p*′-DDT, *p,p*′-DDD, or *p,p*′-DDE can induce a pro-inflammatory state. However, there are no data in exposed humans and very few data in experimental systems in vivo.

In high throughput testing in the Tox21 and ToxCast research programs of the United States government, *p,p*′-DDT, *o,p*′-DDT, *p,p*′-DDE, and *p,p*′-DDD were positive in between 42 and 62 high throughput assay end-points, mostly related to receptor-mediated effects or cell proliferation/cell death/nutrient supply, among the 265 assay end-points relevant to the key characteristics of human carcinogens.

There is some evidence pertaining to cancer susceptibility factors related to infectious agents and perinatal exposures.

Overall, the mechanistic data provide strong support for the carcinogenicity findings of DDT. This includes strong evidence that DDT modulates receptor-mediated effects, is immunosuppressive, and induces oxidative stress, and that these effects can operate in humans.

6. Evaluation

6.1 Cancer in humans

There is *limited evidence* in humans for the carcinogenicity of DDT. Positive associations have been observed between DDT and cancers of the liver and testis, and non-Hodgkin lymphoma.

6.2 Cancer in experimental animals

There is *sufficient evidence* in experimental animals for the carcinogenicity of DDT.

There is *sufficient evidence* in experimental animals for the carcinogenicity of DDE.

There is *sufficient evidence* in experimental animals for the carcinogenicity of DDD.

6.3 Overall evaluation

DDT is *probably carcinogenic to humans (Group 2A)*.

6.4 Rationale

In addition to limited evidence for the carcinogenicity of DDT in humans and sufficient evidence for the carcinogenicity of DDT in experimental animals, there is strong mechanistic evidence for DDT that three key characteristics of known human carcinogens can operate in humans: receptor-mediated effects, immunosuppression, and oxidative stress.

References

Abbassy MS, Ibrahim HZ, Abu El-Amayem MM (1999). Occurrence of pesticides and polychlorinated biphenyls in water of the Nile river at the estuaries of Rosetta and Damiatta branches, north of delta, Egypt. *J Environ Sci Health*, B34(2):255–67. doi:10.1080/03601239909373196

Acquavella JF, Alexander BH, Mandel JS, Burns CJ, Gustin C (2006). Exposure misclassification in studies of agricultural pesticides: insights from biomonitoring. *Epidemiology*, 17(1):69–74. doi:10.1097/01.ede.0000190603.52867.22 PMID:16357597

Adami HO, Lipworth L, Titus-Ernstoff L, Hsieh CC, Hanberg A, Ahlborg U et al. (1995). Organochlorine compounds and estrogen-related cancers in women. *Cancer Causes Control*, 6(6):551–66. doi:10.1007/BF00054165 PMID:8580305

Adetona O, Horton K, Sjodin A, Jones R, Hall DB, Aguillar-Villalobos M et al. (2013). Concentrations of select persistent organic pollutants across pregnancy trimesters in maternal and in cord serum in Trujillo, Peru. *Chemosphere*, 91(10):1426–33. doi:10.1016/j.chemosphere.2013.01.043 PMID:23453434

Ahmed FE, Hart RW, Lewis NJ (1977). Pesticide induced DNA damage and its repair in cultured human cells. *Mutat Res*, 42(2):161–74. doi:10.1016/S0027-5107(77)80020-1 PMID:190533

Ahmed MT, Loutfy N, El Shiekh E (2002). Residue levels of DDE and PCBs in the blood serum of women in the

Port Said region of Egypt. *J Hazard Mater*, 89(1):41–8. doi:10.1016/S0304-3894(01)00283-7 PMID:11734345

Al-Jamal JH, Dubin NH (2000). The effect of raloxifene on the uterine weight response in immature mice exposed to 17beta-estradiol, 1,1,1-trichloro-2, 2-bis(p-chlorophenyl)ethane, and methoxychlor. *Am J Obstet Gynecol*, 182(5):1099–102. doi:10.1067/mob.2000.105407 PMID:10819839

Al-Mughrabi KI, Qrunfleh IM (2002). Pesticide residues in soil from the Jordan Valley, Jordan. *Bull Environ Contam Toxicol*, 68(1):86–96. doi:10.1007/s00128-001-0223-7 PMID:11731836

Al-Othman AA, Abd-Alrahman SH, Al-Daghri NM (2015). DDT and its metabolites are linked to increased risk of type 2 diabetes among Saudi adults: a cross-sectional study. *Environ Sci Pollut Res Int*, 22(1):379–86. doi:10.1007/s11356-014-3371-0 PMID:25077657

Al-Saleh I, Al-Doush I, Alsabbaheen A, Mohamed GD, Rabbah A (2012). Levels of DDT and its metabolites in placenta, maternal and cord blood and their potential influence on neonatal anthropometric measures. *Sci Total Environ*, 416:62–74. doi:10.1016/j.scitotenv.2011.11.020 PMID:22192892

Alary JG, Guay P, Brodeur J (1971). Effect of phenobarbital pretreatment on the metabolism of DDT in the rat and the bovine. *Toxicol Appl Pharmacol*, 18(2):457–68. doi:10.1016/0041-008X(71)90138-4 PMID:5569383

Alavanja MC, Hofmann JN, Lynch CF, Hines CJ, Barry KH, Barker J et al. (2014). Non-hodgkin lymphoma risk and insecticide, fungicide and fumigant use in the Agricultural Health Study. *PLoS ONE*, 9(10):e109332. doi:10.1371/journal.pone.0109332 PMID:25337994

Alawi MA, Tamimi S, Jaghabir M (1999). Storage of organochlorine pesticides in human adipose tissues of Jordanian males and females. *Chemosphere*, 38(12):2865–73. doi:10.1016/S0045-6535(98)00488-3 PMID:10214717

Albertini S, Friederich U, Würgler FE (1988). Induction of mitotic chromosome loss in the diploid yeast Saccharomyces cerevisiae D61.M by genotoxic carcinogens and tumor promoters. *Environ Mol Mutagen*, 11(4):497–508. doi:10.1002/em.2850110410 PMID:3286249

Alegría-Torres JA, Díaz-Barriga F, Gandolfi AJ, Pérez-Maldonado IN (2009). Mechanisms of *p,p*′-DDE-induced apoptosis in human peripheral blood mononuclear cells. *Toxicol In Vitro*, 23(6):1000–6. doi:10.1016/j.tiv.2009.06.021 PMID:19545618

Aliyeva G, Halsall C, Alasgarova K, Avazova M, Ibrahimov Y, Aghayeva R (2013). The legacy of persistent organic pollutants in Azerbaijan: an assessment of past use and current contamination. *Environ Sci Pollut Res Int*, 20(4):1993–2008. doi:10.1007/s11356-012-1076-9 PMID:22825638

Alvarado-Hernandez DL, Montero-Montoya R, Serrano-García L, Arellano-Aguilar O, Jasso-Pineda Y, Yáñez-Estrada L (2013). Assessment of exposure to organochlorine pesticides and levels of DNA damage in mother-infant pairs of an agrarian community. *Environ Mol Mutagen*, 54(2):99–111. doi:10.1002/em.21753 PMID:23355095

Álvarez-Pedrerol M, Ribas-Fitó N, Torrent M, Carrizo D, Garcia-Esteban R, Grimalt JO et al. (2008a). Thyroid disruption at birth due to prenatal exposure to beta-hexachlorocyclohexane. *Environ Int*, 34(6):737–40. doi:10.1016/j.envint.2007.12.001 PMID:18207242

Alvarez-Pedrerol M, Ribas-Fitó N, Torrent M, Carrizo D, Grimalt JO, Sunyer J (2008b). Effects of PCBs, *p,p*′-DDT, *p,p*′-DDE, HCB and beta-HCH on thyroid function in preschool children. *Occup Environ Med*, 65(7):452–7. doi:10.1136/oem.2007.032763 PMID:17933884

Amacher DE, Zelljadt I (1984). Mutagenic activity of some clastogenic chemicals at the hypoxanthine guanine phosphoribosyl transferase locus of Chinese hamster ovary cells. *Mutat Res*, 136(2):137–45. doi:10.1016/0165-1218(84)90156-3 PMID:6717480

Amdany R, Chimuka L, Cukrowska E, Kukučka P, Kohoutek J, Tölgyessy P et al. (2014). Assessment of bioavailable fraction of POPS in surface water bodies in Johannesburg City, South Africa, using passive samplers: an initial assessment. *Environ Monit Assess*, 186(9):5639–53. doi:10.1007/s10661-014-3809-3 PMID:24869948

Amer SM, Fahmy MA, Donya SM (1996). Cytogenetic effect of some insecticides in mouse spleen. *J Appl Toxicol*, 16(1):1–3. doi:10.1002/(SICI)1099-1263(199601)16:1<1::AID-JAT294>3.0.CO;2-A PMID:8821669

Andersen HR, Andersson A-M, Arnold SF, Autrup H, Barfoed M, Beresford NA et al. (1999). Comparison of short-term estrogenicity tests for identification of hormone-disrupting chemicals. *Environ Health Perspect*, 107:Suppl 1:89–108. doi:10.1289/ehp.99107s189 PMID:10229711

Andreotti G, Freeman LE, Hou L, Coble J, Rusiecki J, Hoppin JA et al. (2009). Agricultural pesticide use and pancreatic cancer risk in the Agricultural Health Study Cohort. *Int J Cancer*, 124(10):2495–500. doi:10.1002/ijc.24185 PMID:19142867

Angsubhakorn S, Pradermwong A, Phanwichien K, Nguansangiam S (2002). Promotion of aflatoxin B1-induced hepatocarcinogenesis by dichlorodiphenyl trichloroethane (DDT). *Southeast Asian J Trop Med Public Health*, 33(3):613–23. PMID:12693600

Aronson KJ, Miller AB, Woolcott CG, Sterns EE, McCready DR, Lickley LA et al. (2000). Breast adipose tissue concentrations of polychlorinated biphenyls and other organochlorines and breast cancer risk. *Cancer Epidemiol Biomarkers Prev*, 9(1):55–63. PMID:10667464

Aronson KJ, Wilson JW, Hamel M, Diarsvitri W, Fan W, Woolcott C et al. (2010). Plasma organochlorine levels and prostate cancer risk. *J Expo Sci Environ Epidemiol*, 20(5):434–45. doi:10.1038/jes.2009.33 PMID:19513097

Asawasinsopon R, Prapamontol T, Prakobvitayakit O, Vaneesorn Y, Mangklabruks A, Hock B (2006). The association between organochlorine and thyroid hormone levels in cord serum: a study from northern Thailand. *Environ Int*, 32(4):554–9. doi:10.1016/j.envint.2006.01.001 PMID:16492389

Asawasinsopon R, Prapamontol T, Prakobvitayakit O, Vaneesorn Y, Mangklabruks A, Hock B (2006a). The association between organochlorine and thyroid hormone levels in cord serum: a study from northern Thailand. *Environ Int*, 32(4):554–9. doi:10.1016/j.envint.2006.01.001 PMID:16492389

Asawasinsopon R, Prapamontol T, Prakobvitayakit O, Vaneesorn Y, Mangklabruks A, Hock B (2006b). Plasma levels of DDT and their association with reproductive hormones in adult men from northern Thailand. *Sci Total Environ*, 355(1–3):98–105. doi:10.1016/j.scitotenv.2005.03.004 PMID:15964612

Assennato G, Ferri GM, Tria G, Porro A, Macinagrossa L, Ruggieri M (1995). [Tumors of the hemolymphopoietic tract and employment in agriculture: a case-control study carried out in an epidemiologic area in southern Italy.] *G Ital Med Lav*, 17(1–6):91–7. PMID:8991832

ATSDR (2002). Toxicological profile for DDT, DDE, and DDD. Atlanta (GA), USA: U.S. Department of Health and Human Services, Agency for Toxic Substances and Disease Registry (ATSDR). Available from: http://www.atsdr.cdc.gov, accessed March 2015.

Aubé M, Larochelle C, Ayotte P (2008). 1,1-dichloro-2,2-bis(p-chlorophenyl)ethylene (*p,p′*-DDE) disrupts the estrogen-androgen balance regulating the growth of hormone-dependent breast cancer cells. *Breast Cancer Res*, 10(1):R16. doi:10.1186/bcr1862 PMID:18275596

Aubé M, Larochelle C, Ayotte P (2011). Differential effects of a complex organochlorine mixture on the proliferation of breast cancer cell lines. *Environ Res*, 111(3):337–47. doi:10.1016/j.envres.2011.01.010 PMID:21295777

Austin H, Keil JE, Cole P (1989). A prospective follow-up study of cancer mortality in relation to serum DDT. *Am J Public Health*, 79(1):43–6. doi:10.2105/AJPH.79.1.43 PMID:2909181

Axmon A, Thulstrup AM, Rignell-Hydbom A, Pedersen HS, Zvyezday V, Ludwicki JK et al.; INUENDO (2006). Time to pregnancy as a function of male and female serum concentrations of 2,2′4,4′5,5′-hexachlorobiphenyl (CB-153) and 1,1-dichloro-2,2-bis (p-chlorophenyl)-ethylene (p,p′-DDE). *Hum Reprod*, 21(3):657–65. doi:10.1093/humrep/dei397 PMID:16361295

Aydin ME, Ozcan S, Beduk F, Tor A (2013). Levels of organochlorine pesticides and heavy metals in surface waters of Konya closed basin, Turkey. *ScientificWorldJournal*, 2013:849716. doi:10.1155/2013/849716 PMID:23533363

Azandjeme CS, Delisle H, Fayomi B, Ayotte P, Djrolo F, Houinato D et al. (2014). High serum organochlorine pesticide concentrations in diabetics of a cotton producing area of the Benin Republic (West Africa). *Environ Int*, 69:1–8. doi:10.1016/j.envint.2014.04.002 PMID:24769438

Bachelet D, Truong T, Verner MA, Arveux P, Kerbrat P, Charlier C et al. (2011). Determinants of serum concentrations of 1,1-dichloro-2,2-bis(p-chlorophenyl) ethylene and polychlorinated biphenyls among French women in the CECILE study. *Environ Res*, 111(6):861–70. doi:10.1016/j.envres.2011.06.001 PMID:21684540

Badach H, Nazimek T, Kamińska IA (2007). Pesticide content in drinking water samples collected from orchard areas in central Poland. *Ann Agric Environ Med*, 14(1):109–14. PMID:17655187

Bagga D, Anders KH, Wang HJ, Roberts E, Glaspy JA (2000). Organochlorine pesticide content of breast adipose tissue from women with breast cancer and control subjects. *J Natl Cancer Inst*, 92(9):750–3. doi:10.1093/jnci/92.9.750 PMID:10793112

Bakke B, De Roos AJ, Barr DB, Stewart PA, Blair A, Freeman LB et al. (2009). Exposure to atrazine and selected non-persistent pesticides among corn farmers during a growing season. *J Expo Sci Environ Epidemiol*, 19(6):544–54. doi:10.1038/jes.2008.53 PMID:19052531

Balaguer P, François F, Comunale F, Fenet H, Boussioux AM, Pons M et al. (1999). Reporter cell lines to study the estrogenic effects of xenoestrogens. *Sci Total Environ*, 233(1–3):47–56. doi:10.1016/S0048-9697(99)00178-3 PMID:10492897

Band PR, Abanto Z, Bert J, Lang B, Fang R, Gallagher RP et al. (2011). Prostate cancer risk and exposure to pesticides in British Columbia farmers. *Prostate*, 71(2):168–83. doi:10.1002/pros.21232 PMID:20799287

Banerjee BD (1987a). Sub-chronic effect of DDT on humoral immune response to a thymus-independent antigen (bacterial lipopolysaccharide) in mice. *Bull Environ Contam Toxicol*, 39(5):822–6. doi:10.1007/BF01855861 PMID:3318960

Banerjee BD (1987b). Effects of sub-chronic DDT exposure on humoral and cell-mediated immune responses in albino rats. *Bull Environ Contam Toxicol*, 39(5):827–34. doi:10.1007/BF01855862 PMID:3690008

Banerjee BD, Koner BC, Ray A (1997). Influence of stress on DDT-induced humoral immune responsiveness in mice. *Environ Res*, 74(1):43–7. doi:10.1006/enrs.1997.3749 PMID:9339213

Banerjee BD, Ramachandran M, Hussain QZ (1986). Sub-chronic effect of DDT on humoral immune response in mice. *Bull Environ Contam Toxicol*, 37:433–40. doi:10.1007/BF01607785 PMID:3527308

Banerjee BD, Ray A, Pasha ST (1996). A comparative evaluation of immunotoxicity of DDT and its metabolites in rats. *Indian J Exp Biol*, 34(6):517–22. PMID:8792639

Banerjee BD, Saha S, Mohapatra TK, Ray A (1995). Influence of dietary protein on DDT-induced immune responsiveness in rats. *Indian J Exp Biol*, 33(10):739–44. PMID:8575804

Baris D, Kwak LW, Rothman N, Wilson W, Manns A, Tarone RE et al. (2000). Blood levels of organochlorines before and after chemotherapy among non-Hodgkin's lymphoma patients. *Cancer Epidemiol Biomarkers Prev*, 9(2):193–7. PMID:10698481

Baris D, Zahm SH, Cantor KP, Blair A (1998). Agricultural use of DDT and risk of non-Hodgkin's lymphoma: pooled analysis of three case-control studies in the United States. *Occup Environ Med*, 55(8):522–7. doi:10.1136/oem.55.8.522 PMID:9849538

Barr JR, Maggio VL, Barr DB, Turner WE, Sjödin A, Sandau CD et al. (2003). New high-resolution mass spectrometric approach for the measurement of polychlorinated biphenyls and organochlorine pesticides in human serum. *J Chromatogr B Analyt Technol Biomed Life Sci*, 794(1):137–48. doi:10.1016/S1570-0232(03)00451-3 PMID:12888206

Bartsch H, Malaveille C, Camus AM, Martel-Planche G, Brun G, Hautefeuille A et al. (1980). Validation and comparative studies on 180 chemicals with S. typhimurium strains and V79 Chinese hamster cells in the presence of various metabolizing systems. *Mutat Res*, 76(1):1–50. doi:10.1016/0165-1110(80)90002-0 PMID:6993936

Bates MN, Buckland SJ, Garrett N, Ellis H, Needham LL, Patterson DG Jr et al. (2004). Persistent organochlorines in the serum of the non-occupationally exposed New Zealand population. *Chemosphere*, 54(10):1431–43. doi:10.1016/j.chemosphere.2003.09.040 PMID:14659945

Beard J, Sladden T, Morgan G, Berry G, Brooks L, McMichael A (2003). Health impacts of pesticide exposure in a cohort of outdoor workers. *Environ Health Perspect*, 111(5):724–30. doi:10.1289/ehp.5885 PMID:12727601

Behfar A, Nazari Z, Rabiee MH, Raeesi G, Oveisi MR, Sadeghi N et al. (2013). The organochlorine pesticides residue levels in Karun river water. *Jundishapur J Nat Pharm Prod*, 8(1):41–6. doi:10.5812/jjnpp.6783 PMID:24624185

Behrooz RD, Sari AE, Bahramifar N, Ghasempouri SM (2009). Organochlorine pesticide and polychlorinated biphenyl residues in human milk from the Southern Coast of Caspian Sea, Iran. *Chemosphere*, 74(7):931–7. doi:10.1016/j.chemosphere.2008.10.014 PMID:19042005

Beineke A, Siebert U, McLachlan M, Bruhn R, Thron K, Failing K et al. (2005). Investigations of the potential influence of environmental contaminants on the thymus and spleen of harbor porpoises (Phocoena phocoena). *Environ Sci Technol*, 39(11):3933–8. doi:10.1021/es048709j PMID:15984767

Ben Hassine S, Ameur WB, Gandoura N, Driss MR (2012). Determination of chlorinated pesticides, polychlorinated biphenyls, and polybrominated diphenyl ethers in human milk from Bizerte (Tunisia) in 2010. *Chemosphere*, 89(4):369–77. doi:10.1016/j.chemosphere.2012.05.035 PMID:22743186

Ben Hassine S, Hammami B, Ben Ameur W, El Megdiche Y, Barhoumi B, El Abidi R et al. (2014). Concentrations of organochlorine pesticides and polychlorinated biphenyls in human serum and their relation with age, gender, and BMI for the general population of Bizerte, Tunisia. *Environ Sci Pollut Res Int*, 21(10):6303–13. doi:10.1007/s11356-013-1480-9 PMID:23338993

Ben Rhouma K, Tébourbi O, Krichah R, Sakly M (2001). Reproductive toxicity of DDT in adult male rats. *Hum Exp Toxicol*, 20(8):393–7. doi:10.1191/096032701682692946 PMID:11727789

Benachour N, Moslemi S, Sipahutar H, Seralini GE (2007). Cytotoxic effects and aromatase inhibition by xenobiotic endocrine disrupters alone and in combination. *Toxicol Appl Pharmacol*, 222(2):129–40. doi:10.1016/j.taap.2007.03.033 PMID:17599374

Berg AH, Thomas P, Olsson PE (2005). Biochemical characterization of the Arctic char (Salvelinus alpinus) ovarian progestin membrane receptor. *Reprod Biol Endocrinol*, 3(1):64. doi:10.1186/1477-7827-3-64 PMID:16281974

Bergonzi R, Specchia C, Dinolfo M, Tomasi C, De Palma G, Frusca T et al. (2009). Distribution of persistent organochlorine pollutants in maternal and foetal tissues: data from an Italian polluted urban area. *Chemosphere*, 76(6):747–54. doi:10.1016/j.chemosphere.2009.05.026 PMID:19539348

Bernard L, Martinat N, Lécureuil C, Crépieux P, Reiter E, Tilloy-Ellul A et al. (2007). Dichlorodiphenyltrichloroethane impairs follicle-stimulating hormone receptor-mediated signalling in rat Sertoli cells. *Reprod Toxicol*, 23(2):158–64. doi:10.1016/j.reprotox.2006.11.002 PMID:17157474

Bertrand KA, Spiegelman D, Aster JC, Altshul LM, Korrick SA, Rodig SJ et al. (2010). Plasma organochlorine levels and risk of non-Hodgkin lymphoma in a cohort of men. *Epidemiology*, 21(2):172–80. doi:10.1097/EDE.0b013e3181cb610b PMID:20087190

Bhatia R, Shiau R, Petreas M, Weintraub JM, Farhang L, Eskenazi B (2005). Organochlorine pesticides and male genital anomalies in the child health and development studies. *Environ Health Perspect*, 113(2):220–4. doi:10.1289/ehp.7382 PMID:15687061

Bhatnagar VK, Kashyap R, Zaidi SS, Kulkarni PK, Saiyed HN (2004). Levels of DDT, HCH, and HCB residues in human blood in Ahmedabad, India. *Bull Environ Contam Toxicol*, 72(2):261–5. doi:10.1007/s00128-003-9049-9 PMID:15106760

Bi X, Thomas GO, Jones KC, Qu W, Sheng G, Martin FL et al. (2007). Exposure of electronics dismantling workers to polybrominated diphenyl ethers, polychlorinated biphenyls, and organochlorine pesticides in South China. *Environ Sci Technol*, 41(16):5647–53. doi:10.1021/es070346a PMID:17874768

Biggs ML, Davis MD, Eaton DL, Weiss NS, Barr DB, Doody DR et al. (2008). Serum organochlorine pesticide residues and risk of testicular germ cell carcinoma: a population-based case-control study. *Cancer Epidemiol Biomarkers Prev*, 17(8):2012–8. doi:10.1158/1055-9965. EPI-08-0032 PMID:18708392

Bimenya GS, Harabulema M, Okot JP, Francis O, Lugemwa M, Okwi AL (2010). Plasma levels of DDT/DDE and liver function in malaria control personnel 6 months after indoor residual spraying with DDT in northern Uganda, 2008. *S Afr Med J*, 100(2):118–21. PMID:20459917

Binelli A, Riva C, Cogni D, Provini A (2008a). Assessment of the genotoxic potential of benzo(a)pyrene and pp′-dichlorodiphenyldichloroethylene in Zebra mussel (Dreissena polymorpha). *Mutat Res*, 649(1–2):135–45. doi:10.1016/j.mrgentox.2007.08.011 PMID:17997130

Binelli A, Riva C, Cogni D, Provini A (2008b). Genotoxic effects of *p,p′*-DDT (1,1,1-trichloro-2,2-bis-(chlorophenyl)ethane) and its metabolites in Zebra mussel (D. polymorpha) by SCGE assay and micronucleus test. *Environ Mol Mutagen*, 49(5):406–15. doi:10.1002/em.20400 PMID:18418866

Bitman J, Cecil HC (1970). Estrogenic activity of DDT analogs and polychlorinated biphenyls. *J Agric Food Chem*, 18(6):1108–12. doi:10.1021/jf60172a019 PMID:5483049

Bitman J, Cecil HC, Harris SJ, Fries GF (1968). Estrogenic activity of *o,p′*-DDT in the mammalian uterus and avian oviduct. *Science*, 162(3851):371–2. doi:10.1126/science.162.3851.371 PMID:5677532

Blackwood A, Wolff M, Rundle A, Estabrook A, Schnabel F, Mooney LA et al. (1998). Organochlorine compounds (DDE and PCB) in plasma and breast cyst fluid of women with benign breast disease. *Cancer Epidemiol Biomarkers Prev*, 7(7):579–83. PMID:9681525

Blair A, Tarone R, Sandler D, Lynch C, Rowland A, Wintersteen W et al. (2000). Reliability of reporting on lifestyle and agricultural factors by a sample of participants in the agricultural health study from iowa. *Ann Epidemiol*, 10(7):478. doi:10.1016/S1047-2797(00)00113-7 PMID:11018423

Bloom MS, Jansing RL, Kannan K, Rej R, Fitzgerald EF (2014). Thyroid hormones are associated with exposure to persistent organic pollutants in aging residents of upper Hudson River communities. *Int J Hyg Environ Health*, 217(4–5):473–82. doi:10.1016/j.ijheh.2013.09.003 PMID:24138783

Bouwman H, Becker PJ, Schutte CHJ (1994). Malaria control and longitudinal changes in levels of DDT and its metabolites in human serum from KwaZulu. *Bull World Health Organ*, 72(6):921–30. PMID:7867138

Bouwman H, Bornman R, van den Berg H, Kylin H (2013). DDT: fifty years since Silent Spring. Chapter 11. In: Late lessons from early warnings: science, precaution, innovation. Brussels, Belgium: European Environment Agency; pp. 272–91.

Bouwman H, Kylin H, Sereda B, Bornman R (2012). High levels of DDT in breast milk: intake, risk, lactation duration, and involvement of gender. *Environ Pollut*, 170:63–70. doi:10.1016/j.envpol.2012.06.009 PMID:22766005

Bouwman H, Sereda B, Meinhardt HM (2006). Simultaneous presence of DDT and pyrethroid residues in human breast milk from a malaria endemic area in South Africa. *Environ Pollut*, 144(3):902–17. doi:10.1016/j.envpol.2006.02.002 PMID:16564119

Brams A, Buchet JP, Crutzen-Fayt MC, De Meester C, Lauwerys R, Léonard A (1987). A comparative study, with 40 chemicals, of the efficiency of the Salmonella assay and the SOS chromotest (kit procedure). *Toxicol Lett*, 38(1–2):123–33. doi:10.1016/0378-4274(87)90120-2 PMID:3307023

Brandt WN, Flamm WG, Bernheim NJ (1972). The value of hydroxyurea in assessing repair synthesis of DNA in HeLa cells. *Chem Biol Interact*, 5(5):327–39. doi:10.1016/0009-2797(72)90072-5 PMID:4345371

Bratton MR, Frigo DE, Segar HC, Nephew KP, McLachlan JA, Wiese TE et al. (2012). The organochlorine *o,p′*-DDT plays a role in coactivator-mediated MAPK crosstalk in MCF-7 breast cancer cells. *Environ Health Perspect*, 120(9):1291–6. doi:10.1289/ehp.1104296 PMID:22609851

Bratton MR, Frigo DE, Vigh-Conrad KA, Fan D, Wadsworth S, McLachlan JA et al. (2009). Organochlorine-mediated potentiation of the general coactivator p300 through p38 mitogen-activated protein kinase. *Carcinogenesis*, 30(1):106–13. doi:10.1093/carcin/bgn213 PMID:18791200

Bräuner EV, Sørensen M, Gaudreau E, LeBlanc A, Eriksen KT, Tjønneland A et al. (2012). A prospective study of organochlorines in adipose tissue and risk of non-Hodgkin lymphoma. *Environ Health Perspect*, 120(1):105–11. doi:10.1289/ehp.1103573 PMID:22328999

Bredhult C, Bäcklin BM, Olovsson M (2007). Effects of some endocrine disruptors on the proliferation and viability of human endometrial endothelial cells in vitro. *Reprod Toxicol*, 23(4):550–9. doi:10.1016/j.reprotox.2007.03.006 PMID:17493787

Bredhult C, Sahlin L, Olovsson M (2008). Gene expression analysis of human endometrial endothelial cells exposed to op′-DDT. *Mol Hum Reprod*, 14(2):97–106. doi:10.1093/molehr/gam091 PMID:18204070

Brilhante OM, Franco R (2006). Exposure pathways to HCH and DDT in Cidade dos Meninos and its surrounding districts of Amapa, Figueiras and Pilar, metropolitan regions of Rio de Janeiro, Brazil. *Int J Environ Health Res*, 16(3):205–17. doi:10.1080/09603120500538291 PMID:16611565

Brooke BD, Hunt RH, Chandre F, Carnevale P, Coetzee M (2002). Stable chromosomal inversion polymorphisms and insecticide resistance in the malaria vector mosquito Anopheles gambiae (Diptera: Culicidae). *J Med Entomol*, 39(4):568–73. doi:10.1603/0022-2585-39.4.568 PMID:12144286

Brooks GT (1974). Chlorinated insecticides of the DDT Group. Vol 1. Technology and Application. Cleveland (OH), USA: CRC press; 7-83.

Brown LM, Blair A, Gibson R, Everett GD, Cantor KP, Schuman LM et al. (1990). Pesticide exposures and other agricultural risk factors for leukemia among men in Iowa and Minnesota. *Cancer Res*, 50(20):6585–91. PMID:2208120

Brown TJ, Blaustein JD (1984). 1-(o-Chlorophenyl)-1 (p-chlorophenyl)2,2,2-trichloroethane induces functional progestin receptors in the rat hypothalamus and pituitary gland. *Endocrinology*, 115(6):2052–8. doi:10.1210/endo-115-6-2052 PMID:6499760

Bruce WR, Heddle JA (1979). The mutagenic activity of 61 agents as determined by the micronucleus, Salmonella, and sperm abnormality assays. *Can J Genet Cytol*, 21(3):319–33. doi:10.1139/g79-036 PMID:393369

Bryan TE, Gildersleeve RP, Wiard RP (1989). Exposure of Japanese quail embryos to *o,p*′-DDT has long-term effects on reproductive behaviors, hematology, and feather morphology. *Teratology*, 39(6):525–35. doi:10.1002/tera.1420390603 PMID:2475919

Buchmann A, Willy C, Buenemann CL, Stroh C, Schmiechen A, Schwarz M (1999). Inhibition of transforming growth factor beta1-induced hepatoma cell apoptosis by liver tumor promoters: characterization of primary signalling events and effects on CPP32-like caspase activity. *Cell Death Differ*, 6(2):190–200. doi:10.1038/sj.cdd.4400475 PMID:10200566

Bulayeva NN, Watson CS (2004). Xenoestrogen-induced ERK-1 and ERK-2 activation via multiple membrane-initiated signalling pathways. *Environ Health Perspect*, 112(15):1481–7. doi:10.1289/ehp.7175 PMID:15531431

Burns JS, Williams PL, Sergeyev O, Korrick SA, Lee MM, Revich B et al. (2012). Serum concentrations of organochlorine pesticides and growth among Russian boys. *Environ Health Perspect*, 120(2):303–8. doi:10.1289/ehp.1103743 PMID:21984531

Buselmaier W, Roehrborn G, Propping P (1973). Comparative investigations on the mutagenicity of pesticides in mammalian test systems. *Mutat Res*, 21(1):25–6. doi:10.1016/0165-7992(73)90034-1

Buselmaier W, Röhrborn G, Propping P (1972). Mutagenicity investigations with pesticides in the host-mediated assay and the dominant lethal test in mice. *Biol Zbl*, 91:310–25.

Bussolaro D, Filipak Neto F, Glinski A, Roche H, Guiloski IC, Mela M et al. (2012). Bioaccumulation and related effects of PCBs and organochlorinated pesticides in freshwater fish Hypostomus commersoni. *J Environ Monit*, 14(8):2154–63. doi:10.1039/c2em10863a PMID:22772567

Bustos S, Soto J, Bruzzone N, Vásquez V, Tchernitchin AN (1995). Effect of *p,p*′-DDT and estrogen on the presence in the circulation and degranulation of blood eosinophil leukocytes. *Bull Environ Contam Toxicol*, 55(2):309–15. doi:10.1007/BF00203026 PMID:7579940

Byczkowski JZ (1976). The mode of action of *p,p*′-DDT on mammalian mitochondria. *Toxicology*, 6(3):309–14. doi:10.1016/0300-483X(76)90034-2 PMID:136771

Byeon W-H, Hyun HH, Lee SY (1976). Mutagenicity of pesticides in the Salmonella/microsome system. *Korean J Microbiol.*, 14:128–34.

Cabral JRP, Hall RK, Rossi L, Bronczyk SA, Shubik P (1982a). Effects of long-term intake of DDT on rats. *Tumori*, 68(1):11–7. PMID:6280347

Cabral JRP, Hall RK, Rossi L, Bronczyk SA, Shubik P (1982b). Lack of carcinogenicity of DDT in hamsters. *Tumori*, 68(1):5–10. PMID:7071944

Cai QY, Mo CH, Wu QT, Katsoyiannis A, Zeng QY (2008). The status of soil contamination by semivolatile organic chemicals (SVOCs) in China: a review. *Sci Total Environ*, 389(2–3):209–24. doi:10.1016/j.scitotenv.2007.08.026 PMID:17936334

Campos A, Lino CM, Cardoso SM, Silveira MI (2005). Organochlorine pesticide residues in European sardine, horse mackerel and Atlantic mackerel from Portugal. *Food Addit Contam*, 22(7):642–6. doi:10.1080/02652030500136969 PMID:16019839

Canales-Aguirre A, Padilla-Camberos E, Gómez-Pinedo U, Salado-Ponce H, Feria-Velasco A, De Celis R (2011). Genotoxic effect of chronic exposure to DDT on lymphocytes, oral mucosa and breast cells of female rats. *Int J Environ Res Public Health*, 8(2):540–53. doi:10.3390/ijerph8020540 PMID:21556202

Cantor KP, Blair A, Everett G, Gibson R, Burmeister LF, Brown LM et al. (1992). Pesticides and other agricultural risk factors for non-Hodgkin's lymphoma among men in Iowa and Minnesota. *Cancer Res*, 52(9):2447–55. PMID:1568215

Cárdenas-González M, Gaspar-Ramírez O, Pérez-Vázquez FJ, Alegría-Torres JA, González-Amaro R, Pérez-Maldonado IN (2013). *p,p*′-DDE, a DDT metabolite, induces proinflammatory molecules in human peripheral blood mononuclear cells "in vitro". *Exp Toxicol Pathol*, 65(5):661–5. doi:10.1016/j.etp.2012.07.006 PMID:22901987

Carter RH, Mann HD (1949). The DDT content of milk from a cow sprayed with DDT. *J Econ Entomol*, 42(4):708. doi:10.1093/jee/42.4.708 PMID:18138208

Casida JE, Quistad GB (1998). Golden age of insecticide research: past, present, or future? *Annu Rev Entomol*, 43(1):1–16. doi:10.1146/annurev.ento.43.1.1 PMID:9444749

Catania F, Kauer MO, Daborn PJ, Yen JL, Ffrench-Constant RH, Schlötterer C (2004). World-wide survey of an Accord insertion and its association with DDT resistance in Drosophila melanogaster. *Mol Ecol*, 13(8):2491–504. doi:10.1111/j.1365-294X.2004.02263.x PMID:15245421

CDC (2015). NIOSH Pocket Guide to Chemical Hazards: DDT. Atlanta (GA), USA: Centers for Disease Control and Prevention. Available from: http://www.cdc.gov/niosh/npg/nengapdxa.html.

Cerná M, Krsková A, Cejchanová M, Spěváčková V (2012). Human biomonitoring in the Czech Republic: an overview. *Int J Hyg Environ Health*, 215(2):109–19. doi:10.1016/j.ijheh.2011.09.007 PMID:22014893

Chang Y, Feng L, Miao W (2011). Toxicogenomic investigation of Tetrahymena thermophila exposed to dichlorodiphenyltrichloroethane (DDT), tributyltin (TBT), and 2,3,7,8-tetrachlorodibenzo-p-dioxin (TCDD). *Sci China Life Sci*, 54(7):617–25. doi:10.1007/s11427-011-4194-6 PMID:21748585

Chang YL, Li J, Yao SQ, Hu WN, Jiang SF, Guo Z et al. (2008). [A case-control study on serum organochlorines residues, genetic polymorphisms of glutathione S-transferase T1 and the risks of breast cancer.] *Zhonghua Liu Xing Bing Xue Za Zhi*, 29(8):763–6. [Chinese] PMID:19103108

Channa K, Röllin HB, Nøst TH, Odland JO, Sandanger TM (2012). Prenatal exposure to DDT in malaria endemic region following indoor residual spraying and in non-malaria coastal regions of South Africa. *Sci Total Environ*, 429:183–90. doi:10.1016/j.scitotenv.2012.03.073 PMID:22578843

Chanyshev MD, Kosorotikov NI, Titov SE, Kolesnikov NN, Gulyaeva LF (2014). Expression of microRNAs, CYP1A1 and CYP2B1 in the livers and ovaries of female rats treated with DDT and PAHs. *Life Sci*, 103(2):95–100. doi:10.1016/j.lfs.2014.03.031 PMID:24727239

Charles MJ, Schell MJ, Willman E, Gross HB, Lin Y, Sonnenberg S et al. (2001). Organochlorines and 8-hydroxy-2′-deoxyguanosine (8-OHdG) in cancerous and noncancerous breast tissue: do the data support the hypothesis that oxidative DNA damage caused by organochlorines affects breast cancer? *Arch Environ Contam Toxicol*, 41(3):386–95. doi:10.1007/s002440010264 PMID:11503078

Charlier C, Albert A, Herman P, Hamoir E, Gaspard U, Meurisse M et al. (2003). Breast cancer and serum organochlorine residues. *Occup Environ Med*, 60(5):348–51. doi:10.1136/oem.60.5.348 PMID:12709520

Charlier C, Foidart JM, Pitance F, Herman P, Gaspard U, Meurisse M et al. (2004). Environmental dichlorodiphenyltrichlorethane or hexachlorobenzene exposure and breast cancer: is there a risk? *Clin Chem Lab Med*, 42(2):222–7. doi:10.1515/CCLM.2004.040 PMID:15061365

Chaturvedi NK, Kumar S, Negi S, Tyagi RK (2010). Endocrine disruptors provoke differential modulatory responses on androgen receptor and pregnane and xenobiotic receptor: potential implications in metabolic disorders. *Mol Cell Biochem*, 345(1–2):291–308. doi:10.1007/s11010-010-0583-6 PMID:20830510

ChemIDplus (2015). TOXNET Toxicology data network. Bethesda (MD), USA: United States National Library of Medicine. Available from: http://chem2.sis.nlm.nih.gov/chemidplus/rn/startswith/50-29-3.

Chen CW, Hurd C, Vorojeikina DP, Arnold SF, Notides AC (1997). Transcriptional activation of the human estrogen receptor by DDT isomers and metabolites in yeast and MCF-7 cells. *Biochem Pharmacol*, 53(8):1161–72. doi:10.1016/S0006-2952(97)00097-X PMID:9175721

Chikuni O, Nhachi CF, Nyazema NZ, Polder A, Nafstad I, Skaare JU (1997). Assessment of environmental pollution by PCBs, DDT and its metabolites using human milk of mothers in Zimbabwe. *Sci Total Environ*, 199(1–2):183–90. doi:10.1016/S0048-9697(97)05494-6 PMID:9200862

Chung RA, Williams CS, Naidu YM, Thakore VR (1989). Influence of DDT and PCBs in rabbits and goats as related to nucleic acid, protein and lipid metabolism. *J Environ Pathol Toxicol Oncol*, 9(3):283–302. PMID:2509680

Chung SW, Kwong KP, Yau JC (2008). Dietary exposure to DDT of secondary school students in Hong Kong. *Chemosphere*, 73(1):65–9. doi:10.1016/j.chemosphere.2008.05.049 PMID:18599106

Clark JM (1974). Mutagenicity of DDT in mice, Drosophila melanogaster and Neurospora crassa. *Aust J Biol Sci*, 27(4):427–40. PMID:4279650

Clive D, Johnson KO, Spector JFS, Batson AG, Brown MMM (1979). Validation and characterization of the L5178Y/TK+/− mouse lymphoma mutagen assay system. *Mutat Res*, 59(1):61–108. doi:10.1016/0027-5107(79)90195-7 PMID:372791

Coble J, Arbuckle T, Lee W, Alavanja M, Dosemeci M (2005). The validation of a pesticide exposure algorithm using biological monitoring results. *J Occup Environ Hyg*, 2(3):194–201. doi:10.1080/15459620590923343 PMID:15764542

Cocco P, Brennan P, Ibba A, de Sanjosé Llongueras S, Maynadié M, Nieters A et al. (2008). Plasma polychlorobiphenyl and organochlorine pesticide level and risk of major lymphoma subtypes. *Occup Environ Med*, 65(2):132–40. doi:10.1136/oem.2007.033548 PMID:17699548

Cocco P, Fadda D, Billai B, D'Atri M, Melis M, Blair A (2005). Cancer mortality among men occupationally exposed to dichlorodiphenyltrichloroethane. *Cancer Res*, 65(20):9588–94. doi:10.1158/0008-5472.CAN-05-1487 PMID:16230425

Cocco P, Loviselli A, Fadda D, Ibba A, Melis M, Oppo A et al. (2004). Serum sex hormones in men occupationally exposed to dichloro-diphenyl-trichloro ethane (DDT) as young adults. *J Endocrinol*, 182(3):391–7. doi:10.1677/joe.0.1820391 PMID:15350181

Cocco P, Satta G, Dubois S, Pili C, Pilleri M, Zucca M et al. (2013). Lymphoma risk and occupational exposure to pesticides: results of the Epilymph study. *Occup Environ Med*, 70(2):91–8. doi:10.1136/oemed-2012-100845 PMID:23117219

Cohen JM, Smith DL, Cotter C, Ward A, Yamey G, Sabot OJ et al. (2012). Malaria resurgence: a systematic review and assessment of its causes. *Malar J*, 11(1):122. doi:10.1186/1475-2875-11-122 PMID:22531245

Cohn BA, Cirillo PM, Christianson RE (2010). Prenatal DDT exposure and testicular cancer: a nested case-control study. *Arch Environ Occup Health*, 65(3):127–34. doi:10.1080/19338241003730887 PMID:20705572

Cohn BA, La Merrill M, Krigbaum NY, Yeh G, Park JS, Zimmermann L et al. (2015). DDT exposure in utero and breast cancer. *J Clin Endocrinol Metab*, 100(8):2865–72. doi:10.1210/jc.2015-1841 PMID:26079774

Cohn BA, Wolff MS, Cirillo PM, Sholtz RI (2007). DDT and breast cancer in young women: new data on the significance of age at exposure. *Environ Health Perspect*, 115(10):1406–14. PMID:17938728

Colt JS, Lubin J, Camann D, Davis S, Cerhan J, Severson RK et al. (2004). Comparison of pesticide levels in carpet dust and self-reported pest treatment practices in four US sites. *J Expo Anal Environ Epidemiol*, 14(1):74–83. doi:10.1038/sj.jea.7500307 PMID:14726946

Colt JS, Severson RK, Lubin J, Rothman N, Camann D, Davis S et al. (2005). Organochlorines in carpet dust and non-Hodgkin lymphoma. *Epidemiology*, 16(4):516–25. doi:10.1097/01.ede.0000164811.25760.f1 PMID:15951670

Cooper GS, Martin SA, Longnecker MP, Sandler DP, Germolec DR (2004). Associations between plasma DDE levels and immunologic measures in African-American farmers in North Carolina. *Environ Health Perspect*, 112(10):1080–4. doi:10.1289/ehp.6892 PMID:15238281

Copeland MF, Cranmer MF (1974). Effects of *o,p*′-DDT on the adrenal gland and hepatic microsomal enzyme system in the beagle dog. *Toxicol Appl Pharmacol*, 27(1):1–10. doi:10.1016/0041-008X(74)90168-9 PMID:4851764

Crebelli R, Bellincampi D, Conti G, Conti L, Morpurgo G, Carere A (1986). A comparative study on selected chemical carcinogens for chromosome malsegregation, mitotic crossing-over and forward mutation induction in Aspergillus nidulans. *Mutat Res*, 172(2):139–49. doi:10.1016/0165-1218(86)90070-4 PMID:3531838

Cueto C Jr, Moran NC (1968). The circulatory effects of catecholamines and ouabain in glucocorticoid-deficient animals. *J Pharmacol Exp Ther*, 164(1):31–44. PMID:4301845

Cupul-Uicab LA, Hernández-Avila M, Terrazas-Medina EA, Pennell ML, Longnecker MP (2010). Prenatal exposure to the major DDT metabolite 1,1-dichloro-2,2-bis(p-chlorophenyl)ethylene (DDE) and growth in boys from Mexico. *Environ Res*, 110(6):595–603. doi:10.1016/j.envres.2010.06.001 PMID:20566194

Curtis CF (2002). Should the use of DDT be revived for malaria vector control? *Biomedica*, 22(4):455–61. doi:10.7705/biomedica.v22i4.1171 PMID:12596442

Dahmardeh Behrooz R, Barghi M, Bahramifar N, Esmaili-Sari A (2012). Organochlorine contaminants in the hair of Iranian pregnant women. *Chemosphere*, 86(3):235–41. doi:10.1016/j.chemosphere.2011.09.031 PMID:22047617

Dalvie MA, Myers JE, Lou Thompson M, Dyer S, Robins TG, Omar S et al. (2004b). The hormonal effects of long-term DDT exposure on malaria vector-control workers in Limpopo Province, South Africa. *Environ Res*, 96(1):9–19. doi:10.1016/j.envres.2003.09.003 PMID:15261779

Dalvie MA, Myers JE, Thompson ML, Robins TG, Omar S, Riebow J (2004a). Exploration of different methods for measuring DDT exposure among malaria vector-control workers in Limpopo Province, South Africa. *Environ Res*, 96(1):20–7. doi:10.1016/j.envres.2003.09.004 PMID:15261780

Damgaard IN, Skakkebaek NE, Toppari J, Virtanen HE, Shen H, Schramm KW et al.; Nordic Cryptorchidism Study Group (2006). Persistent pesticides in human breast milk and cryptorchidism. *Environ Health Perspect*, 114(7):1133–8. doi:10.1289/ehp.8741 PMID:16835070

Daniel V, Huber W, Bauer K, Suesal C, Conradt C, Opelz G (2002). Associations of dichlorodiphenyltrichloroethane (DDT) 4.4 and dichlorodiphenyldichloroethylene (DDE) 4.4 blood levels with plasma IL-4. *Arch Environ Health*, 57(6):541–7. doi:10.1080/00039890209602086 PMID:12696651

Danzo BJ (1997). Environmental xenobiotics may disrupt normal endocrine function by interfering with the binding of physiological ligands to steroid receptors and binding proteins. *Environ Health Perspect*, 105(3):294–301. doi:10.1289/ehp.97105294 PMID:9171990

Dar SA, Das S, Ramachandran VG, Bhattacharya SN, Mustafa MD, Banerjee BD et al. (2012). Alterations in T-lymphocyte sub-set profiles and cytokine secretion by PBMC of systemic lupus erythematosus patients upon in vitro exposure to organochlorine pesticides. *J Immunotoxicol*, 9(1):85–95. doi:10.3109/1547691X.2011.642103 PMID:22214240

Darnerud PO, Aune M, Larsson L, Lignell S, Mutshatshi T, Okonkwo J et al. (2011). Levels of brominated flame retardants and other pesistent organic pollutants in breast milk samples from Limpopo Province, South Africa. *Sci Total Environ*, 409(19):4048–53. doi:10.1016/j.scitotenv.2011.05.054 PMID:21708397

Das S, Thomas P (1999). Pesticides interfere with the nongenomic action of a progestogen on meiotic maturation by binding to its plasma membrane receptor on fish oocytes. *Endocrinology*, 140(4):1953–6. doi:10.1210/endo.140.4.6781 PMID:10098537

Davidson JS, Baumgarten I, Harley EH (1985). Use of a new citrulline incorporation assay to investigate inhibition of intercellular communication by 1,1,1-trichloro-2,2-bis(p-chlorophenyl)ethane in human fibroblasts. *Cancer Res*, 45(2):515–9. PMID:3967226

Davodi M, Esmaili-Sari A, Bahramifarr N (2011). Concentration of polychlorinated biphenyls and organochlorine pesticides in some edible fish species from the Shadegan Marshes (Iran). *Ecotoxicol Environ Saf*, 74(3):294–300. doi:10.1016/j.ecoenv.2010.07.045 PMID:21168210

Dayan J, Deguingand S, Truzman C, Chevron M (1987). Application of the SOS chromotest to 10 pharmaceutical agents. *Mutat Res*, 187(2):55–66. doi:10.1016/0165-1218(87)90118-2 PMID:2433580

de Azevedo e Silva CE, Azeredo A, Lailson-Brito J, Torres JP, Malm O (2007). Polychlorinated biphenyls and DDT in swordfish (Xiphias gladius) and blue shark (Prionace glauca) from Brazilian coast. *Chemosphere*, 67(9):S48–53. doi:10.1016/j.chemosphere.2006.05.089 PMID:17223179

De Coster S, Koppen G, Bracke M, Schroijen C, Den Hond E, Nelen V et al. (2008). Pollutant effects on genotoxic parameters and tumor-associated protein levels in adults: a cross sectional study. *Environ Health*, 7(1):26. doi:10.1186/1476-069X-7-26 PMID:18522717

De Flora S (1981). Study of 106 organic and inorganic compounds in the Salmonella/microsome test. *Carcinogenesis*, 2(4):283–98. doi:10.1093/carcin/2.4.283 PMID:7023727

De Flora S, Camoirano A, Izzotti A, D'Agostini F, Bennicelli C (1989). Photoactivation of mutagens. *Carcinogenesis*, 10(6):1089–97. doi:10.1093/carcin/10.6.1089 PMID:2655963

De Guise S, Martineau D, Béland P, Fournier M (1998). Effects of in vitro exposure of beluga whale leukocytes to selected organochlorines. *J Toxicol Environ Health A*, 55(7):479–93. doi:10.1080/009841098158287 PMID:9860322

de Jager C, Aneck-Hahn NH, Bornman MS, Farias P, Leter G, Eleuteri P et al. (2009). Sperm chromatin integrity in DDT-exposed young men living in a malaria area in the Limpopo Province, South Africa. *Hum Reprod*, 24(10):2429–38. doi:10.1093/humrep/dep249 PMID:19608568

de Jager C, Aneck-Hahn NH, Bornman MS, Farias P, Spanò M (2012). DDT exposure levels and semen quality of young men from a malaria area in South Africa. *Malar J*, 11:Suppl 1: 21. Available from: http://www.ncbi.nlm.nih.gov/entrez/query.fcgi?cmd=Retrieve&db=PubMed&list_uids=22243516&dopt=Abstract doi:10.1186/1475-2875-11-S1-P21 PMID:22243516

de Mora S, Fowler SW, Tolosa I, Villeneuve JP, Cattini C (2005). Chlorinated hydrocarbons in marine biota and coastal sediments from the Gulf and Gulf of Oman. *Mar Pollut Bull*, 50(8):835–49. doi:10.1016/j.marpolbul.2005.02.022 PMID:16115501

De Roos AJ, Hartge P, Lubin JH, Colt JS, Davis S, Cerhan JR et al. (2005). Persistent organochlorine chemicals in plasma and risk of non-Hodgkin's lymphoma. *Cancer Res*, 65(23):11214–26. doi:10.1158/0008-5472.CAN-05-1755 PMID:16322272

De Roos AJ, Zahm SH, Cantor KP, Weisenburger DD, Holmes FF, Burmeister LF et al. (2003). Integrative assessment of multiple pesticides as risk factors for non-Hodgkin's lymphoma among men. *Occup Environ Med*, 60(9):E11. doi:10.1136/oem.60.9.e11 PMID:12937207

De Stefani E, Kogevinas M, Boffetta P, Ronco A, Mendilaharsu M (1996). Occupation and the risk of lung cancer in Uruguay. *Scand J Work Environ Health*, 22(5):346–52. doi:10.5271/sjweh.152 PMID:8923607

Dees C, Askari M, Foster JS, Ahamed S, Wimalasena J (1997a). DDT mimicks estradiol stimulation of breast cancer cells to enter the cell cycle. *Mol Carcinog*, 18(2):107–14. doi:10.1002/(SICI)1098-2744(199702)18:2<107::AID-MC6>3.0.CO;2-D PMID:9049186

Deichmann WB, Keplinger M, Sala F, Glass E (1967). Synergism among oral carcinogens. IV. The simultaneous feeding of four tumorigens to rats. *Toxicol Appl Pharmacol*, 11(1):88–103. doi:10.1016/0041-008X(67)90030-0 PMID:6056159

Delescluse C, Ledirac N, de Sousa G, Pralavorio M, Lesca P, Rahmani R (1998). Cytotoxic effects and induction of cytochromes P450 1A1/2 by insecticides, in hepatic or epidermal cells: binding capability to the Ah receptor. *Toxicol Lett*, 96–97(97):33–9. doi:10.1016/S0378-4274(98)00047-2 PMID:9820645

Dello Iacovo R, Celentano E, Strollo AM, Iazzetta G, Capasso I, Randazzo G (1999). Organochlorines and breast cancer. A study on Neapolitan women. *Adv Exp Med Biol*, 472:57–66. PMID:10736616

Demers A, Ayotte P, Brisson J, Dodin S, Robert J, Dewailly E (2000). Risk and aggressiveness of breast cancer in relation to plasma organochlorine concentrations. *Cancer Epidemiol Biomarkers Prev*, 9(2):161–6. PMID:10698476

Den Hond E, Dhooge W, Bruckers L, Schoeters G, Nelen V, van de Mieroop E et al. (2011). Internal exposure to pollutants and sexual maturation in Flemish adolescents. *J Expo Sci Environ Epidemiol*, 21(3):224–33. doi:10.1038/jes.2010.2 PMID:20197795

Deribe E, Rosseland BO, Borgstrøm R, Salbu B, Gebremariam Z, Dadebo E et al. (2013). Biomagnification of DDT and its metabolites in four fish species of a tropical lake. *Ecotoxicol Environ Saf*, 95:10–8. doi:10.1016/j.ecoenv.2013.03.020 PMID:23790590

Dewan P, Jain V, Gupta P, Banerjee BD (2013). Organochlorine pesticide residues in maternal blood, cord blood, placenta, and breastmilk and their relation to birth size. *Chemosphere*, 90(5):1704–10. doi:10.1016/j.chemosphere.2012.09.083 PMID:23141556

Dhananjayan V, Ravichandran B, Rajmohan HR (2012). Organochlorine pesticide residues in blood samples of agriculture and sheep wool workers in Bangalore (rural), India. *Bull Environ Contam Toxicol*, 88(4):497–500. doi:10.1007/s00128-012-0546-6 PMID:22323047

Dhooge W, Arijs K, D'Haese I, Stuyvaert S, Versonnen B, Janssen C et al. (2006). Experimental parameters affecting sensitivity and specificity of a yeast assay for estrogenic compounds: results of an interlaboratory validation exercise. *Anal Bioanal Chem*, 386(5):1419–28. doi:10.1007/s00216-006-0669-x PMID:16896612

Di Lorenzo D, Villa R, Biasiotto G, Belloli S, Ruggeri G, Albertini A et al. (2002). Isomer-specific activity of dichlorodyphenyltrichloroethane with estrogen receptor in adult and suckling estrogen reporter mice. *Endocrinology*, 143(12):4544–51. doi:10.1210/en.2002-220448 PMID:12446581

Díaz-Barriga Martinez F, Trejo-Acevedo A, Betanzos AF, Espinosa-Reyes G, Alegría-Torres JA, Maldonado IN (2012). Assessment of DDT and DDE levels in soil, dust, and blood samples from Chihuahua, Mexico. *Arch Environ Contam Toxicol*, 62(2):351–8. doi:10.1007/s00244-011-9700-0 PMID:21822982

Diel P, Olff S, Schmidt S, Michna H (2002). Effects of the environmental estrogens bisphenol A, *o,p′*-DDT, p-tert-octylphenol and coumestrol on apoptosis induction, cell proliferation and the expression of estrogen sensitive molecular parameters in the human breast cancer cell line MCF-7. *J Steroid Biochem Mol Biol*, 80(1):61–70. doi:10.1016/S0960-0760(01)00173-X PMID:11867264

Diel P, Schulz T, Smolnikar K, Strunck E, Vollmer G, Michna H (2000). Ability of xeno- and phytoestrogens to modulate expression of estrogen-sensitive genes in rat uterus: estrogenicity profiles and uterotropic activity. *J Steroid Biochem Mol Biol*, 73(1–2):1–10. doi:10.1016/S0960-0760(00)00051-0 PMID:10822019

Dominguez-Lopez P, Diaz-Cueto L, Olivares A, Ulloa-Aguirre A, Arechavaleta-Velasco F (2012). Differential effect of DDT, DDE, and DDD on COX-2 expression in the human trophoblast derived HTR-8/SVneo cells. *J Biochem Mol Toxicol*, 26(11):454–60. doi:10.1002/jbt.21444 PMID:23132776

Dorea JG, Cruz-Granja AC, Lacayo-Romero ML, Cuadra-Leal J (2001). Perinatal metabolism of dichlorodiphenyldichloroethylene in Nicaraguan mothers. *Environ Res*, 86(3):229–37. doi:10.1006/enrs.2001.4277 PMID:11453673

Dores EF, Carbo L, de Abreu AB (2003). Serum DDT in malaria vector control sprayers in Mato Grosso State, Brazil. *Cad Saude Publica*, 19(2):429–37. doi:10.1590/S0102-311X2003000200009 PMID:12764458

Dorgan JF, Brock JW, Rothman N, Needham LL, Miller R, Stephenson HE Jr et al. (1999). Serum organochlorine pesticides and PCBs and breast cancer risk: results from a prospective analysis (USA). *Cancer Causes Control*, 10(1):1–11. doi:10.1023/A:1008824131727 PMID:10334636

Dosemeci M, Alavanja MC, Rowland AS, Mage D, Zahm SH, Rothman N et al. (2002). A quantitative approach for estimating exposure to pesticides in the Agricultural Health Study. *Ann Occup Hyg*, 46(2):245–60. doi:10.1093/annhyg/mef011 PMID:12074034

Duby RT, Travis HF, Terrill CE (1971). Uterotropic activity of DDT in rats and mink and its influence on reproduction in the rat. *Toxicol Appl Pharmacol*, 18(2):348–55. doi:10.1016/0041-008X(71)90127-X PMID:5569378

Dutta R, Mondal AM, Arora V, Nag TC, Das N (2008). Immunomodulatory effect of DDT (bis[4-chlorophenyl]-1,1,1-trichloroethane) on complement system and macrophages. *Toxicology*, 252(1–3):78–85. doi:10.1016/j.tox.2008.07.063 PMID:18755234

Emeville E, Giusti A, Coumoul X, Thomé JP, Blanchet P, Multigner L (2015). Associations of plasma concentrations of dichlorodiphenyldichloroethylene and polychlorinated biphenyls with prostate cancer: a case-control study in Guadeloupe (French West Indies). *Environ Health Perspect*, 123(4):317–23. PMID:25493337

Enan E, Matsumura F (1998). Activation of c-Neu tyrosine kinase by *o,p′*-DDT and beta-HCH in cell-free and intact cell preparations from MCF-7 human breast cancer cells. *J Biochem Mol Toxicol*, 12(2):83–92. doi:10.1002/(SICI)1099-0461(1998)12:2<83::AID-JBT3>3.0.CO;2-K PMID:9443065

Engel LS, Laden F, Andersen A, Strickland PT, Blair A, Needham LL et al. (2007). Polychlorinated biphenyl levels in peripheral blood and non-Hodgkin's lymphoma: a report from three cohorts. *Cancer Res*, 67(11):5545–52. doi:10.1158/0008-5472.CAN-06-3906 PMID:17545638

Ennaceur S, Gandoura N, Driss MR (2008). Distribution of polychlorinated biphenyls and organochlorine pesticides in human breast milk from various locations in Tunisia: levels of contamination, influencing factors, and infant risk assessment. *Environ Res*, 108(1):86–93. doi:10.1016/j.envres.2008.05.005 PMID:18614165

EPA (1975a). *DDT;* a review of scientific and economic aspects of the decision to ban its use as a pesticide. Washington (DC), USA: Office of Pesticide Programs, United States Environmental Protection Agency.

EPA (1975b). DDT regulatory history: a brief survey (to 1975). Washington (DC), USA: United States Environmental Protection Agency. Available from: http://www2.epa.gov/aboutepa/ddt-regulatory-historybrief-survey-1975.

EPA (2008). Health Effects Support Document for 1,1-Dichloro-2,2-bis(p-chlorophenyl)ethylene (DDE). Document No. EPA-822-R-08-003. Washington (DC), USA: United States Environmental Protection Agency.

Available from: http://www.epa.gov/safewater/ccl/pdf/DDE.pdf, accessed March 2015.

EPA (2015a). Pesticide Analytical Methods an Environmental Chemistry Methods Index. Washington (DC), USA: United States Environmental Protection Agency. Available from: http://www2.epa.gov/pesticide-analytical-methods.

EPA (2015b). DDT - A brief history and status. Washington (DC), USA: United States Environmental Protection Agency. Available from: http://www2.epa.gov/ingredients-used-pesticide-products/ddt-brief-history-and-status.

Epstein SS, Arnold E, Andrea J, Bass W, Bishop Y (1972). Detection of chemical mutagens by the dominant lethal assay in the mouse. *Toxicol Appl Pharmacol*, 23(2):288–325. doi:10.1016/0041-008X(72)90192-5 PMID:5074577

Epstein SS, Shafner H (1968). Chemical mutagens in the human environment. *Nature*, 219(5152):385–7. doi:10.1038/219385a0 PMID:5662155

Eriksson M, Hardell L, Carlberg M, Akerman M (2008). Pesticide exposure as risk factor for non-Hodgkin lymphoma including histopathological subgroup analysis. *Int J Cancer*, 123(7):1657–63. doi:10.1002/ijc.23589 PMID:18623080

Eriksson M, Karlsson M (1992). Occupational and other environmental factors and multiple myeloma: a population based case-control study. *Br J Ind Med*, 49(2):95–103. PMID:1536825

European Commission (2015). EU - pesticides database [online database]. Brussels, Belgium: European Commission. Available from: http://ec.europa.eu/food/plant/pesticides/index_en.htm.

European Food Safety Authority (2015). The 2013 European Union report on pesticide residues in food. *EFSA Journal*, 13(3):4038.

Fabro S, McLachlan JA, Dames NM (1984). Chemical exposure of embryos during the preimplantation stages of pregnancy: mortality rate and intrauterine development. *Am J Obstet Gynecol*, 148(7):929–38. doi:10.1016/0002-9378(84)90535-0 PMID:6711631

Fahrig R (1974). Comparative mutagenicity studies with pesticides. *IARC Sci Publ*, 10:161–81.

Fang Y, Nie Z, Yang Y, Die Q, Liu F, He J et al. (2015). Human health risk assessment of pesticide residues in market-sold vegetables and fish in a northern metropolis of China. *Environ Sci Pollut Res Int*, 22(8):6135–43. doi:10.1007/s11356-014-3822-7 PMID:25395327

Farhang L, Weintraub JM, Petreas M, Eskenazi B, Bhatia R (2005). Association of DDT and DDE with birth weight and length of gestation in the Child Health and Development Studies, 1959–1967. *Am J Epidemiol*, 162(8):717–25. doi:10.1093/aje/kwi276 PMID:16120698

Fawcett SC, King LJ, Bunyan PJ, Stanley PI (1987). The metabolism of 14C-DDT, 14C-DDD, 14C-DDE and 14C-DDMU in rats and Japanese quail. *Xenobiotica*, 17(5):525–38. doi:10.3109/00498258709043960 PMID:3604258

Feil VJ, Lamoureux CJ, Styrvoky E, Zaylskie RG, Thacker EJ, Holman GM (1973). Metabolism of *o,p′*-DDT in rats. *J Agric Food Chem*, 21(6):1072–8. doi:10.1021/jf60190a013 PMID:4755830

Ferguson KK, Hauser R, Altshul L, Meeker JD (2012). Serum concentrations of p, p′-DDE, HCB, PCBs and reproductive hormones among men of reproductive age. *Reprod Toxicol*, 34(3):429–35. doi:10.1016/j.reprotox.2012.04.006 PMID:22564984

Fernandez MF, Olmos B, Granada A, López-Espinosa MJ, Molina-Molina JM, Fernandez JM et al. (2007). Human exposure to endocrine-disrupting chemicals and prenatal risk factors for cryptorchidism and hypospadias: a nested case-control study. *Environ Health Perspect*, 115(S-1):Suppl 1:8–14. doi:10.1289/ehp.9351 PMID:18174944

Ferreira CP, De-Oliveira AC, Paumgartten FJ (2011). Serum concentrations of DDT and DDE among malaria control workers in the Amazon region. *J Occup Health*, 53(2):115–22. doi:10.1539/joh.O10026 PMID:21233591

Filer D, Patisaul HB, Schug T, Reif D, Thayer K (2014). Test driving ToxCast: endocrine profiling for 1858 chemicals included in phase II. *Curr Opin Pharmacol*, 19:145–52. doi:10.1016/j.coph.2014.09.021 PMID:25460227

Finnegan JK, Haag HB, Larson PS (1949). Tissue distribution and elimination of DDD and DDT following oral administration to dogs and rats. *Proc Soc Exp Biol Med*, 72(2):357–60. doi:10.3181/00379727-72-17431 PMID:15398310

Fitzgerald DJ, Piccoli C, Yamasaki H (1989). Detection of non-genotoxic carcinogens in the BALB/c 3T3 cell transformation/mutation assay system. *Mutagenesis*, 4(4):286–91. doi:10.1093/mutage/4.4.286 PMID:2674607

Fitzhugh OG, Nelson AA (1947). The chronic oral toxicity of DDT (2,2-bis(p-chlorophenyl-1,1,1-trichloroethane). *J Pharmacol Exp Ther*, 89(1):18–30. PMID:20281336

Flodin U, Fredriksson M, Persson B, Axelson O (1988). Chronic lymphatic leukaemia and engine exhausts, fresh wood, and DDT: a case-referent study. *Br J Ind Med*, 45(1):33–8. PMID:2449239

Fluck ER, Poirier LA, Ruelius HW (1976). Evaluation of a DNA polymerase-deficient mutant of E. coli for the rapid detection of carcinogens. *Chem Biol Interact*, 15(3):219–31. doi:10.1016/0009-2797(76)90148-4 PMID:793736

Foster TS (1968). Effect of some pesticides on the adrenal glands in the rat. *Can J Biochem*, 46(9):1115–20. doi:10.1139/o68-166 PMID:5687639

Fox SD, Roman JM, Issaq HJ, Nims RW (1998). Metabolic conversion of 1,1-dichloro-2,2-bis(p-chlorophenyl) ethane (DDD) to 1,1-dichloro-2,2-bis(p-chlorophenyl) ethylene (DDE) in the male F344/NCr rat. *Arch*

Environ Contam Toxicol, 35(1):104–8. doi:10.1007/s002449900356 PMID:9601927

Freire C, Koifman RJ, Sarcinelli PN, Rosa AC, Clapauch R, Koifman S (2014). Association between serum levels of organochlorine pesticides and sex hormones in adults living in a heavily contaminated area in Brazil. *Int J Hyg Environ Health*, 217(2–3):370–8. doi:10.1016/j.ijheh.2013.07.012 PMID:23972672

Freire C, Koifman RJ, Sarcinelli PN, Simões Rosa AC, Clapauch R, Koifman S (2013). Long-term exposure to organochlorine pesticides and thyroid status in adults in a heavily contaminated area in Brazil. *Environ Res*, 127:7–15. doi:10.1016/j.envres.2013.09.001 PMID:24183346

Fries GF, Marrow GS Jr, Lester JW, Gordon CH (1971). Effect of microsomal enzyme inducing drugs on DDT and dieldrin elimination from cows. *J Dairy Sci*, 54(3):364–8. doi:10.3168/jds.S0022-0302(71)85845-9 PMID:5096116

Frigo DE, Vigh KA, Struckhoff AP, Elliott S, Beckman BS, Burow ME et al. (2005). Xenobiotic-induced TNF-α expression and apoptosis through the p38 MAPK signalling pathway. *Toxicol Lett*, 155(2):227–38. doi:10.1016/j.toxlet.2004.09.008 PMID:15603917

Fryzek JP, Garabrant DH, Harlow SD, Severson RK, Gillespie BW, Schenk M et al. (1997). A case-control study of self-reported exposures to pesticides and pancreas cancer in southeastern Michigan. *Int J Cancer*, 72(1):62–7. doi:10.1002/(SICI)1097-0215(19970703)72:1<62::AID-IJC9>3.0.CO;2-2 PMID:9212224

FSANZ (2003). Food standards. The 20th Australian Total Diet Survey. A total diet survey of pesticide residues and contaminants. Canberra, Australia: Food Standards Australia New Zealand. Available from: http://www.foodstandards.gov.au/publications/documents/Final_20th_Total_Diet_Survey.pdf.

Furberg AS, Sandanger T, Thune I, Burkow IC, Lun E (2002). Fish consumption and plasma levels of organochlorines in a female population in northern Norway. *J Environ Monit*, 4(1):175–81. doi:10.1039/b106207g PMID:11871702

Gaido KW, Leonard LS, Lovell S, Gould JC, Babaï D, Portier CJ et al. (1997). Evaluation of chemicals with endocrine modulating activity in a yeast-based steroid hormone receptor gene transcription assay. *Toxicol Appl Pharmacol*, 143(1):205–12. doi:10.1006/taap.1996.8069 PMID:9073609

Gajski G, Ravlić S, Capuder Z, Garaj-Vrhovac V (2007). Use of sensitive methods for detection of DNA damage on human lymphocytes exposed to *p,p'*-DDT: Comet assay and new criteria for scoring micronucleus test. *J Environ Sci Health B*, 42(6):607–13. doi:10.1080/03601230701465445 PMID:17701695

Galetin-Smith R, Pavkov S, Roncevic N (1990). DDT and PCBs in human milk: implication for breast feeding infants. *Bull Environ Contam Toxicol*, 45(6):811–8. doi:10.1007/BF01701076 PMID:2126216

Galindo Reyes JG, Leyva NR, Millan OA, Lazcano GA (2002). Effects of pesticides on DNA and protein of shrimp larvae Litopenaeus stylirostris of the California Gulf. *Ecotoxicol Environ Saf*, 53(2):191–5. doi:10.1006/eesa.2002.2156 PMID:12568452

Galloway SM, Armstrong MJ, Reuben C, Colman S, Brown B, Cannon C et al. (1987). Chromosome aberrations and sister chromatid exchanges in Chinese hamster ovary cells: evaluations of 108 chemicals. *Environ Mol Mutagen*, 10(S10):Suppl 10:1–35. doi:10.1002/em.2850100502 PMID:3319609

Gammon MD, Wolff MS, Neugut AI, Eng SM, Teitelbaum SL, Britton JA et al. (2002). Environmental toxins and breast cancer on Long Island. II. Organochlorine compound levels in blood. *Cancer Epidemiol Biomarkers Prev*, 11(8):686–97. PMID:12163320

Garabrant DH, Held J, Langholz B, Peters JM, Mack TM (1992). DDT and related compounds and risk of pancreatic cancer. *J Natl Cancer Inst*, 84(10):764–71. doi:10.1093/jnci/84.10.764 PMID:1573662

Garaj-Vrhovac V, Gajski G, Ravlić S (2008). Efficacy of HUMN criteria for scoring the micronucleus assay in human lymphocytes exposed to a low concentration of *p,p'*-DDT. *Braz J Med Biol Res*, 41(6):473–6. PMID:18592119

García de la Parra LM, Cervantes-Mojica LJ, González-Valdivia C, Martínez-Cordero FJ, Aguilar-Zárate G, Bastidas-Bastidas P et al. (2012). Distribution of pesticides and PCBs in sediments of agricultural drains in the Culiacan Valley, Sinaloa, Mexico. *Arch Environ Contam Toxicol*, 63(3):323–36. doi:10.1007/s00244-012-9780-5 PMID:22811371

Gascon M, Vrijheid M, Garí M, Fort M, Grimalt JO, Martinez D et al. (2015). Temporal trends in concentrations and total serum burdens of organochlorine compounds from birth until adolescence and the role of breastfeeding. *Environ Int*, 74:144–51. doi:10.1016/j.envint.2014.10.010 PMID:25454231

Gatto NM, Longnecker MP, Press MF, Sullivan-Halley J, McKean-Cowdin R, Bernstein L (2007). Serum organochlorines and breast cancer: a case-control study among African-American women. *Cancer Causes Control*, 18(1):29–39. doi:10.1007/s10552-006-0070-2 PMID:17186420

Gauthier JM, Dubeau H, Rassart E (1999). Induction of micronuclei in vitro by organochlorine compounds in beluga whale skin fibroblasts. *Mutat Res*, 439(1):87–95. doi:10.1016/S1383-5718(98)00178-8 PMID:10029683

Gebremichael S, Birhanu T, Tessema DA (2013). Analysis of organochlorine pesticide residues in human and cow's milk in the towns of Asendabo, Serbo and Jimma in South-Western Ethiopia. *Chemosphere*, 90(5):1652–7. doi:10.1016/j.chemosphere.2012.09.008 PMID:23062941

Gerić M, Ceraj-Cerić N, Gajski G, Vasilić Ž, Capuder Ž, Garaj-Vrhovac V (2012). Cytogenetic status of human lymphocytes after exposure to low concentrations of *p,p′*-DDT, and its metabolites (*p,p′*-DDE, and *p,p′*-DDD) in vitro. *Chemosphere*, 87(11):1288–94. doi:10.1016/j.chemosphere.2012.01.037 PMID:22354074

Giannandrea F, Gandini L, Paoli D, Turci R, Figà-Talamanca I (2011). Pesticide exposure and serum organochlorine residuals among testicular cancer patients and healthy controls. *J Environ Sci Health B*, 46(8):780–7. PMID:21902556

Gilbertson MK, Haffner GD, Drouillard KG, Albert A, Dixon B (2003). Immunosuppression in the northern leopard frog (Rana pipiens) induced by pesticide exposure. *Environ Toxicol Chem*, 22(1):101–10. doi:10.1897/1551-5028(2003)022<0101:IITN-LF>2.0.CO;2 PMID:12503752

Gillespie MJ, Lythgo CM, Plumb AD, Wilkins JPG (1994). A survey comparing the chemical compostion of dicofol formulations sold in the UK before and after the introduction of EC 'Prohibition Directive 79/117/EEC'. *Pestic Sci*, 42(4):305–14. doi:10.1002/ps.2780420408

Gingell R (1975). Enterohepatic circulation of bis(p-chlorophenyl)acetic acid in the rat. *Drug Metab Dispos*, 3(1):42–6. PMID:234833

Gladen BC, Shkiryak-Nyzhnyk ZA, Chyslovska N, Zadorozhnaja TD, Little RE (2003). Persistent organochlorine compounds and birth weight. *Ann Epidemiol*, 13(3):151–7. doi:10.1016/S1047-2797(02)00268-5 PMID:12604157

Glatt HR, Oesch F (1987). Species differences in enzymes controlling reactive epoxides. *Arch Toxicol Suppl*, 10:Suppl 10: 111–24. doi:10.1007/978-3-642-71617-1_9 PMID:2437883

Glick B (1974). Antibody-mediated immunity in the presence of Mirex and DDT. *Poult Sci*, 53(4):1476–85. doi:10.3382/ps.0531476 PMID:4850595

Global Environment Facility (2006). Improvement of production technology of dicofol from DDT and introduction of alternative technology including IPM technology for leaf mite control in China. Geneva, Switzerland: United Nations Development Programme.

Gold B, Brunk G (1982). Metabolism of 1,1,1-trichloro-2,2-bis(p-chlorophenyl)-ethane and 1,1-dichloro-2,2-bis(p-chlorophenyl)ethane in the mouse. *Chem Biol Interact*, 41(3):327–39. doi:10.1016/0009-2797(82)90109-0 PMID:7105253

Gold B, Brunk G (1983). Metabolism of 1,1,1-trichloro-2,2-bis(p-chlorophenyl)ethane (DDT), 1,1-dichloro-2,2-bis(p-chlorophenyl)ethane, and 1-chloro-2,2-bis(p-chlorophenyl)ethene in the hamster. *Cancer Res*, 43(6):2644–7. PMID:6850581

Gold B, Leuschen T, Brunk G, Gingell R (1981). Metabolism of a DDT metabolite via a chloroepoxide. *Chem Biol Interact*, 35(2):159–76. doi:10.1016/0009-2797(81)90140-X PMID:7214600

Goncharov A, Rej R, Negoita S, Schymura M, Santiago-Rivera A, Morse G et al.; Akwesasne Task Force on the Environment (2009). Lower serum testosterone associated with elevated polychlorinated biphenyl concentrations in Native American men. *Environ Health Perspect*, 117(9):1454–60. doi:10.1289/ehp.0800134 PMID:19750113

Graillot C, Gak JC, Lancret C, Truhaut R (1975). On the modes and mechanisms of toxic activity of organochlorine insecticides. II. Study in hamster of long-term toxicity of DDT (Fr.). *Eur J Toxicol*, 8:353–9.

Greenlee AR, Ellis TM, Berg RL, Mercieca MD (2005). Pregnancy outcomes for mouse preimplantation embryos exposed in vitro to the estrogenic pesticide *o,p′*-DDT. *Reprod Toxicol*, 20(2):229–38. doi:10.1016/j.reprotox.2005.02.004 PMID:15907658

Greenlee AR, Quail CA, Berg RL (1999). Developmental alterations in murine embryos exposed in vitro to an estrogenic pesticide, *o,p′*-DDT. *Reprod Toxicol*, 13(6):555–65. doi:10.1016/S0890-6238(99)00051-9 PMID:10613404

Gregoraszczuk EL, Ptak A, Karniewska M, Ropstad E (2008). Action of defined mixtures of PCBs, *p,p′*-DDT and its metabolite *p,p′*-DDE, on co-culture of porcine theca and granulosa cells: steroid secretion, cell proliferation and apoptosis. *Reprod Toxicol*, 26(2):170–4. doi:10.1016/j.reprotox.2008.07.003 PMID:18692563

Griffin DE 3rd, Hill WE (1978). In vitro breakage of plasmid DNA by mutagens and pesticides. *Mutat Res*, 52(2):161–9. doi:10.1016/0027-5107(78)90138-0 PMID:368611

Gunasekaran K, Sahu SS, Jambulingam P, Das PK (2005). DDT indoor residual spray, still an effective tool to control Anopheles fluviatilis-transmitted Plasmodium falciparum malaria in India. *Trop Med Int Health*, 10(2):160–8. doi:10.1111/j.1365-3156.2004.01369.x PMID:15679559

Gunderson EL (1995). FDA Total Diet Study, July 1986-April 1991, dietary intakes of pesticides, selected elements, and other chemicals. *J AOAC Int*, 78(6):1353–63. PMID:8664570

Gutiérrez R, Ortiz R, Vega S, Schettino B, Ramirez ML, Perez JJ (2013). Residues levels of organochlorine pesticide in cow's milk from industrial farms in Hidalgo, Mexico. *J Environ Sci Health B*, 48(11):935–40. doi:10.1080/03601234.2013.816592 PMID:23998305

Gutiérrez R, Ruíz JL, Ortiz R, Vega S, Schettino B, Yamazaki A et al. (2012). Organochlorine pesticide residues in bovine milk from organic farms in Chiapas, Mexico. *Bull Environ Contam Toxicol*, 89(4):882–7. doi:10.1007/s00128-012-0764-y PMID:22864634

Hagmar L, Björk J, Sjödin A, Bergman A, Erfurth EM (2001). Plasma levels of persistent organohalogens and hormone levels in adult male humans. *Arch Environ Health*, 56(2):138–43. doi:10.1080/00039890109604065 PMID:11339677

Han EH, Kim HG, Hwang YP, Choi JH, Im JH, Park B et al. (2010). The role of cyclooxygenase-2-dependent signalling via cyclic AMP response element activation on aromatase up-regulation by *o,p′*-DDT in human breast cancer cells. *Toxicol Lett*, 198(3):331–41. doi:10.1016/j.toxlet.2010.07.015 PMID:20678559

Han EH, Kim JY, Kim HK, Hwang YP, Jeong HG (2008). *o,p′*-DDT induces cyclooxygenase-2 gene expression in murine macrophages: role of AP-1 and CRE promoter elements and PI3-kinase/Akt/MAPK signalling pathways. *Toxicol Appl Pharmacol*, 233(2):333–42. doi:10.1016/j.taap.2008.09.003 PMID:18840457

Hanaoka T, Takahashi Y, Kobayashi M, Sasaki S, Usuda M, Okubo S et al. (2002). Residuals of beta-hexachlorocyclohexane, dichlorodiphenyltrichloroethane, and hexachlorobenzene in serum, and relations with consumption of dietary components in rural residents in Japan. *Sci Total Environ*, 286(1–3):119–27. doi:10.1016/S0048-9697(01)00969-X PMID:11886087

Harada T, Yamaguchi S, Ohtsuka R, Takeda M, Fujisawa H, Yoshida T et al. (2003). Mechanisms of promotion and progression of preneoplastic lesions in hepatocarcinogenesis by DDT in F344 rats. *Toxicol Pathol*, 31(1):87–98. doi:10.1080/01926230390173941 PMID:12597452

Hardell E, Carlberg M, Nordström M, van Bavel B (2010b). Time trends of persistent organic pollutants in Sweden during 1993–2007 and relation to age, gender, body mass index, breast-feeding and parity. *Sci Total Environ*, 408(20):4412–9. doi:10.1016/j.scitotenv.2010.06.029 PMID:20643475

Hardell E, Eriksson M, Lindström G, Van Bavel B, Linde A, Carlberg M et al. (2001). Case-control study on concentrations of organohalogen compounds and titers of antibodies to Epstein-Barr virus antigens in the etiology of non-Hodgkin lymphoma. *Leuk Lymphoma*, 42(4):619–29. doi:10.3109/10428190109099322 PMID:11697490

Hardell K, Carlberg M, Hardell L, Björnfoth H, Ericson Jogsten I, Eriksson M et al. (2009). Concentrations of organohalogen compounds and titres of antibodies to Epstein-Barr virus antigens and the risk for non-Hodgkin lymphoma. *Oncol Rep*, 21(6):1567–76. doi:10.3892/or_00000389 PMID:19424638

Hardell L, Andersson SO, Carlberg M, Bohr L, van Bavel B, Lindström G et al. (2006a). Adipose tissue concentrations of persistent organic pollutants and the risk of prostate cancer. *J Occup Environ Med*, 48(7):700–7. doi:10.1097/01.jom.0000205989.46603.43 PMID:16832227

Hardell L, Bavel B, Lindström G, Eriksson M, Carlberg M (2006b). In utero exposure to persistent organic pollutants in relation to testicular cancer risk. *Int J Androl*, 29(1):228–34. doi:10.1111/j.1365-2605.2005.00622.x PMID:16371110

Hardell L, Carlberg M, Hardell K, Björnfoth H, Wickbom G, Ionescu M et al. (2007). Decreased survival in pancreatic cancer patients with high concentrations of organochlorines in adipose tissue. *Biomed Pharmacother*, 61(10):659–64. doi:10.1016/j.biopha.2007.04.006 PMID:17560068

Hardell L, Eriksson M, Nordstrom M (2002). Exposure to pesticides as risk factor for non-Hodgkin's lymphoma and hairy cell leukemia: pooled analysis of two Swedish case-control studies. *Leuk Lymphoma*, 43(5):1043–9. doi:10.1080/10428190290021560 PMID:12148884

Hardell L, van Bavel B, Lindström G, Björnfoth H, Orgum P, Carlberg M et al. (2004). Adipose tissue concentrations of p,p′-DDE and the risk for endometrial cancer. *Gynecol Oncol*, 95(3):706–11. doi:10.1016/j.ygyno.2004.08.022 PMID:15581986

Hardell L, van Bavel B, Lindström G, Carlberg M, Dreifaldt AC, Wijkström H et al. (2003). Increased concentrations of polychlorinated biphenyls, hexachlorobenzene, and chlordanes in mothers of men with testicular cancer. *Environ Health Perspect*, 111(7):930–4. doi:10.1289/ehp.5816 PMID:12782494

Hardell S, Tilander H, Welfinger-Smith G, Burger J, Carpenter DO (2010a). Levels of polychlorinated biphenyls (PCBs) and three organochlorine pesticides in fish from the Aleutian Islands of Alaska. *PLoS ONE*, 5(8):e12396. doi:10.1371/journal.pone.0012396 PMID:20811633

Hart MM, Adamson RH, Fabro S (1971). Prematurity and intrauterine growth retardation induced by DDT in the rabbit. *Arch Int Pharmacodyn Ther*, 192(2):286–90. PMID:5093202

Hart MM, Whang-Peng J, Sieber SM, Fabro S, Adamson RH (1972). Distribution and effects of DDT in the pregnant rabbit. *Xenobiotica*, 2(6):567–74. doi:10.3109/00498257209111084 PMID:4662547

Hassoun E, Bagchi M, Bagchi D, Stohs SJ (1993). Comparative studies on lipid peroxidation and DNA-single strand breaks induced by lindane, DDT, chlordane and endrin in rats. *Comp Biochem Physiol C*, 104(3):427–31. doi:10.1016/0742-8413(93)90013-B PMID:7689940

Hatakeyama M, Matsumura F (1999). Correlation between the activation of Neu tyrosine kinase and promotion of foci formation induced by selected organochlorine compounds in the MCF-7 model system. *J Biochem Mol Toxicol*, 13(6):296–302. doi:10.1002/(SICI)1099-0461(1999)13:6<296::AID-JBT2>3.0.CO;2-S PMID:10487416

Hauser R, Singh NP, Chen Z, Pothier L, Altshul L (2003). Lack of an association between environmental exposure to polychlorinated biphenyls and *p,p′*-DDE and DNA damage in human sperm measured using the neutral comet assay. *Hum Reprod*, 18(12):2525–33. doi:10.1093/humrep/deg508 PMID:14645167

Hayes WJ Jr, Dale WE, Pirkle CI (1971). Evidence of safety of long-term, high, oral doses of DDT for man. *Arch Environ Health*, 22(1):119–35. doi:10.1080/00039896.1971.10665822 PMID:4992918

Hayes WJ Jr, Durham WF, Cueto C Jr (1956). The effect of known repeated oral doses of chlorophenothane (DDT) in man. *J Am Med Assoc*, 162(9):890–7. doi:10.1001/jama.1956.72970260008012 PMID:13366680

Helzlsouer KJ, Alberg AJ, Huang HY, Hoffman SC, Strickland PT, Brock JW et al. (1999). Serum concentrations of organochlorine compounds and the subsequent development of breast cancer. *Cancer Epidemiol Biomarkers Prev*, 8(6):525–32. PMID:10385143

Herrero-Mercado M, Waliszewski SM, Valencia-Quintana R, Caba M, Hernández-Chalate F, García-Aguilar E et al. (2010). Organochlorine pesticide levels in adipose tissue of pregnant women in Veracruz, Mexico. *Bull Environ Contam Toxicol*, 84(6):652–6. doi:10.1007/s00128-010-0028-7 PMID:20449723

Hilpert D, Romen W, Neumann HG (1983). The role of partial hepatectomy and of promoters in the formation of tumors in non-target tissues of trans-4-acetylaminostilbene in rats. *Carcinogenesis*, 4(12):1519–25. doi:10.1093/carcin/4.12.1519 PMID:6317216

Hoar Zahm S, Blair A, Holmes FF, Boysen CD, Robel RJ (1988). A case-referent study of soft-tissue sarcoma and Hodgkin's disease. Farming and insecticide use. *Scand J Work Environ Health*, 14(4):224–30. doi:10.5271/sjweh.1928 PMID:3175554

Hojo H, Aoyama H, Takahashi KL, Shimizu N, Araki M, Takizawa Y et al. (2006). Two-generation reproduction toxicity study in rats with 1,1,1-trichloro-2,2-bis(4-chlorophenyl)ethane (*p,p′*-DDT). *Congenit Anom (Kyoto)*, 46(2):105–14. doi:10.1111/j.1741-4520.2006.00110.x PMID:16732769

Holloway AC, Stys KA, Foster WG (2005). DDE-induced changes in aromatase activity in endometrial stromal cells in culture. *Endocrine*, 27(1):45–50. doi:10.1385/ENDO:27:1:045 PMID:16077170

Hoppin JA, Tolbert PE, Holly EA, Brock JW, Korrick SA, Altshul LM et al. (2000). Pancreatic cancer and serum organochlorine levels. *Cancer Epidemiol Biomarkers Prev*, 9(2):199–205. PMID:10698482

Hoppin JA, Yucel F, Dosemeci M, Sandler DP (2002). Accuracy of self-reported pesticide use duration information from licensed pesticide applicators in the Agricultural Health Study. *J Expo Anal Environ Epidemiol*, 12(5):313–8. doi:10.1038/sj.jea.7500232 PMID:12198579

Hou L, Andreotti G, Baccarelli AA, Savage S, Hoppin JA, Sandler DP et al. (2013). Lifetime pesticide use and telomere shortening among male pesticide applicators in the Agricultural Health Study. *Environ Health Perspect*, 121(8):919–24. doi:10.1289/ehp.1206432 PMID:23774483

Howsam M, Grimalt JO, Guinó E, Navarro M, Martí-Ragué J, Peinado MA et al.; Bellvitge Colorectal Cancer Group (2004). Organochlorine exposure and colorectal cancer risk. *Environ Health Perspect*, 112(15):1460–6. doi:10.1289/ehp.7143 PMID:15531428

Høyer AP, Gerdes AM, Jørgensen T, Rank F, Hartvig HB (2002). Organochlorines, p53 mutations in relation to breast cancer risk and survival. A Danish cohort-nested case-controls study. *Breast Cancer Res Treat*, 71(1):59–65. doi:10.1023/A:1013340327099 PMID:11859874

Høyer AP, Grandjean P, Jørgensen T, Brock JW, Hartvig HB (1998). Organochlorine exposure and risk of breast cancer. *Lancet*, 352(9143):1816–20. doi:10.1016/S0140-6736(98)04504-8 PMID:9851382

Høyer AP, Jørgensen T, Grandjean P, Hartvig HB (2000). Repeated measurements of organochlorine exposure and breast cancer risk (Denmark). *Cancer Causes Control*, 11(2):177–84. doi:10.1023/A:1008926219539 PMID:10710203

Høyer AP, Jørgensen T, Rank F, Grandjean P (2001). Organochlorine exposures influence on breast cancer risk and survival according to estrogen receptor status: a Danish cohort-nested case-control study. *BMC Cancer*, 1(1):8. doi:10.1186/1471-2407-1-8 PMID:11518544

Huen K, Yousefi P, Bradman A, Yan L, Harley KG, Kogut K et al. (2014). Effects of age, sex, and persistent organic pollutants on DNA methylation in children. *Environ Mol Mutagen*, 55(3):209–22. doi:10.1002/em.21845 PMID:24375655

Humphries MS (2013). DDT residue contamination in sediments from Lake Sibaya in northern KwaZulu-Natal, South Africa: implications for conservation in a World Heritage Site. *Chemosphere*, 93(8):1494–9. doi:10.1016/j.chemosphere.2013.07.047 PMID:23972730

Hunter DJ, Hankinson SE, a F, Colditz GA, Manson JE, Willett WC et al. (1997). Plasma organochlorine levels and the risk of breast cancer. *N Engl J Med*, 337(18):1253–8. doi:10.1056/NEJM199710303371801 PMID:9345073

Hurd-Brown T, Udoji F, Martin T, Whalen MM (2013). Effects of DDT and triclosan on tumor-cell binding capacity and cell-surface protein expression of human natural killer cells. *J Appl Toxicol*, 33(6):495–502. doi:10.1002/jat.2767 PMID:22729613

IARC (1974). Some organochlorine pesticides. *IARC Monogr Eval Carcinog Risk Chem Man*, 5:83–124. Available from: http://monographs.iarc.fr/ENG/Monographs/vol1-42/mono5.pdf.

IARC (1979). Sex hormones (II). *IARC Monogr Eval Carcinog Risk Chem Hum*, 21:1–583. Available from: http://monographs.iarc.fr/ENG/Monographs/vol1-42/mono21.pdf.

IARC (1983). Miscellaneous pesticides. *IARC Monogr Eval Carcinog Risk Chem Hum*, 30:1–424. Available from: http://monographs.iarc.fr/ENG/Monographs/vol1-42/mono30.pdf PMID:6578175

IARC (1986). Some chemicals used in plastics and elastomers. *IARC Monogr Eval Carcinog Risk Chem Hum*, 39:1–403. Available from: http://monographs.iarc.fr/ENG/Monographs/vol1-42/mono39.pdf.

IARC (1991). Occupational exposures in insecticide application, and some pesticides. *IARC Monogr Eval Carcinog Risks Hum*, 53:1–612. Available from: http://monographs.iarc.fr/ENG/Monographs/vol53/index.php.

IARC (2017a). Some organophosphate insecticides and herbicides. *IARC Monogr Eval Carcinog Risks Hum*, 112:1–452. Available from: http://monographs.iarc.fr/ENG/Monographs/vol112/index.php.

IARC (2017b). List of ToxCast/Tox21 assay end-points. In: Supplemental Material to *IARC Monographs* Volume 113. Lyon, France: International Agency for Research on Cancer. Available from: http://monographs.iarc.fr/ENG/Monographs/vol113/index.php.

Ibarluzea JM, Fernández MF, Santa-Marina L, Olea-Serrano MF, Rivas AM, Aurrekoetxea JJ et al. (2004). Breast cancer risk and the combined effect of environmental estrogens. *Cancer Causes Control*, 15(6):591–600. doi:10.1023/B:CACO.0000036167.51236.86 PMID:15280638

Ingber SZ, Buser MC, Pohl HR, Abadin HG, Murray HE, Scinicariello F (2013). DDT/DDE and breast cancer: a meta-analysis. *Regul Toxicol Pharmacol*, 67(3):421–33. doi:10.1016/j.yrtph.2013.08.021 PMID:24021539

Innes JRM, Ulland BM, Valerio MG, Petrucelli L, Fishbein L, Hart ER et al. (1969). Bioassay of pesticides and industrial chemicals for tumorigenicity in mice: a preliminary note. *J Natl Cancer Inst*, 42(6):1101–14. PMID:5793189

Itoh H, Iwasaki M, Hanaoka T, Kasuga Y, Yokoyama S, Onuma H et al. (2009). Serum organochlorines and breast cancer risk in Japanese women: a case-control study. *Cancer Causes Control*, 20(5):567–80. doi:10.1007/s10552-008-9265-z PMID:19031103

Itoh H, Iwasaki M, Kasuga Y, Yokoyama S, Onuma H, Nishimura H et al. (2014). Association between serum organochlorines and global methylation level of leukocyte DNA among Japanese women: a cross-sectional study. *Sci Total Environ*, 490:603–9. doi:10.1016/j.scitotenv.2014.05.035 PMID:24880549

Iturri SJ, Gaggero A, Lucero L, Goldzveig M, Giroz MT (1982). The combined effect of DDT and polychlorinated biphenyls on the blood and heart rate of the white leghorn cockerel (Gallus domesticus). *Comp Biochem Physiol C*, 72(1):27–32. doi:10.1016/0306-4492(82)90200-3 PMID:6125333

Iwasaki M, Inoue M, Sasazuki S, Kurahashi N, Itoh H, Usuda M et al.; Japan Public Health Center-based Prospective Study Group (2008). Plasma organochlorine levels and subsequent risk of breast cancer among Japanese women: a nested case-control study. *Sci Total Environ*, 402(2–3):176–83. doi:10.1016/j.scitotenv.2008.05.009 PMID:18555519

Iwata H, Tanabe S, Sakai N, Nishimura A, Tatsukawa R (1994). Geographical distribution of persistent organochlorines in air, water and sediments from Asia and Oceania, and their implications for global redistribution from lower latitudes. *Environ Pollut*, 85(1):15–33. doi:10.1016/0269-7491(94)90234-8 PMID:15091681

Jakszyn P, Goñi F, Etxeandia A, Vives A, Millán E, López R et al. (2009). Serum levels of organochlorine pesticides in healthy adults from five regions of Spain. *Chemosphere*, 76(11):1518–24. doi:10.1016/j.chemosphere.2009.05.048 PMID:19586652

Jandacek RJ, Rider T, Yang Q, Woollett LA, Tso P (2009). Lymphatic and portal vein absorption of organochlorine compounds in rats. *Am J Physiol Gastrointest Liver Physiol*, 296(2):G226–34. doi:10.1152/ajpgi.90517.2008 PMID:19056760

Jarman WM, Ballschmiter K (2012). From coal to DDT: the history of the development of the pesticide DDT from synthetic dyes till Silent Spring. *Endeavour*, 36(4):131–42. doi:10.1016/j.endeavour.2012.10.003 PMID:23177325

Jasso-Pineda Y, Díaz-Barriga F, Yáñez-Estrada L, Pérez-Vázquez FJ, Pérez-Maldonado IN (2015). DNA damage in Mexican children living in high-risk contaminated scenarios. *Sci Total Environ*, 518–519:38–48. doi:10.1016/j.scitotenv.2015.02.073 PMID:25747362

Jensen JA, Cueto C, Dale WE, Rothe CF, Pearce GW, Mattson AM (1957). Metabolism of insecticides: DDT metabolites in feces and bile of rats. *J Agric Food Chem*, 5(12):919–25. doi:10.1021/jf60082a002

Jin XT, Song L, Zhao JY, Li ZY, Zhao MR, Liu WP (2014). Dichlorodiphenyltrichloroethane exposure induces the growth of hepatocellular carcinoma via Wnt/β-catenin pathway. *Toxicol Lett*, 225(1):158–66. doi:10.1016/j.toxlet.2013.12.006 PMID:24355586

Johnson GA, Jalal SM (1973). DDT-induced chromosomal damage in mice. *J Hered*, 64(1):7–8. PMID:4698916

Johnson NA, Ho A, Cline JM, Hughes CL, Foster WG, Davis VL (2012). Accelerated mammary tumor onset in a HER2/Neu mouse model exposed to DDT metabolites locally delivered to the mammary gland. *Environ Health Perspect*, 120(8):1170–6. doi:10.1289/ehp.1104327 PMID:22514210

Jusko TA, Koepsell TD, Baker RJ, Greenfield TA, Willman EJ, Charles MJ et al. (2006). Maternal DDT exposures in relation to fetal and 5-year growth. *Epidemiology*, 17(6):692–700. doi:10.1097/01.ede.0000232226.06807.90 PMID:17003683

Kaliser LA (1968). An in vitro and in vivo study of the effect of DDT on the phagocytic acitivty of rat white blood cells. *Toxicol Appl Pharmacol*, 13(3):353–7. doi:10.1016/0041-008X(68)90110-5 PMID:5726664

Kaminski NE, Wells DS, Dauterman WC, Roberts JF, Guthrie FE (1986). Macrophage uptake of a lipoprotein-sequestered toxicant: a potential route of immunotoxicity. *Toxicol Appl Pharmacol*, 82(3):474–80. doi:10.1016/0041-008X(86)90282-6 PMID:3952730

Kanja LW, Skaare JU, Ojwang SB, Maitai CK (1992). A comparison of organochlorine pesticide residues in maternal adipose tissue, maternal blood, cord blood, and human milk from mother/infant pairs. *Arch Environ Contam Toxicol*, 22(1):21–4. doi:10.1007/BF00213297 PMID:1554250

Kannan K, Sharma JD (1979). Defective lymphocyte transformation by DDT: in vitro responsiveness of rabbit peripheral blood lymphocytes to PHA. *Indian J Exp Biol*, 17(8):805–6. PMID:544455

Kanno J, Onyon L, Peddada S, Ashby J, Jacob E, Owens W (2003). The OECD program to validate the rat uterotrophic bioassay. Phase 2: dose-response studies. *Environ Health Perspect*, 111(12):1530–49. doi:10.1289/ehp.5780 PMID:12948896

Karmaus W, Osuch JR, Eneli I, Mudd LM, Zhang J, Mikucki D et al. (2009). Maternal levels of dichlorodiphenyl-dichloroethylene (DDE) may increase weight and body mass index in adult female offspring. *Occup Environ Med*, 66(3):143–9. doi:10.1136/oem.2008.041921 PMID:19060027

Kashyap SK, Nigam SK, Karnik AB, Gupta RC, Chatterjee SK (1977). Carcinogenicity of DDT (dichlorodiphenyl trichloroethane) in pure inbred Swiss mice. *Int J Cancer*, 19(5):725–9. doi:10.1002/ijc.2910190519 PMID:863549

Kaushik CP, Sharma HR, Gulati D, Kaushik A (2011). Changing patterns of organochlorine pesticide residues in raw bovine milk from Haryana, India. *Environ Monit Assess*, 182(1–4):467–75. doi:10.1007/s10661-011-1890-4 PMID:21331758

Kavlock R, Chandler K, Houck K, Hunter S, Judson R, Kleinstreuer N et al. (2012). Update on EPA's ToxCast program: providing high throughput decision support tools for chemical risk management. *Chem Res Toxicol*, 25(7):1287–302. doi:10.1021/tx3000939 PMID:22519603

Kazantseva YA, Yarushkin AA, Pustylnyak VO (2013). Dichlorodiphenyltrichloroethane technical mixture regulates cell cycle and apoptosis genes through the activation of CAR and ERα in mouse livers. *Toxicol Appl Pharmacol*, 271(2):137–43. doi:10.1016/j.taap.2013.05.008 PMID:23684557

Kelce WR, Stone CR, Laws SC, Gray LE, Kemppainen JA, Wilson EM (1995). Persistent DDT metabolite *p,p'*-DDE is a potent androgen receptor antagonist. *Nature*, 375(6532):581–5. doi:10.1038/375581a0 PMID:7791873

Keller WC, Yeary RA (1980). A comparison of the effects of mineral oil, vegetable oil, and sodium sulfate on the intestinal absorption of DDT in rodents. *Clin Toxicol*, 16(2):223–31. doi:10.3109/15563658008989941 PMID:7398212

Kelly-Garvert F, Legator MS (1973). Cytogenetic and mutagenic effects of DDT and DDE in a Chinese hamster cell line. *Mutat Res*, 17(2):223–9. doi:10.1016/0027-5107(73)90170-X PMID:4346226

Kezios KL, Liu X, Cirillo PM, Cohn BA, Kalantzi OI, Wang Y et al. (2013). Dichlorodiphenyltrichloroethane (DDT), DDT metabolites and pregnancy outcomes. *Reprod Toxicol*, 35:156–64. doi:10.1016/j.reprotox.2012.10.013 PMID:23142753

Khanjani N, Sim MR (2006). Maternal contamination with dichlorodiphenyltrichloroethane and reproductive outcomes in an Australian population. *Environ Res*, 101(3):373–9. doi:10.1016/j.envres.2005.10.003 PMID:16360142

Kim JY, Choi CY, Lee KJ, Shin DW, Jung KS, Chung YC et al. (2004). Induction of inducible nitric oxide synthase and proinflammatory cytokines expression by *o,p'*-DDT in macrophages. *Toxicol Lett*, 147(3):261–9. doi:10.1016/j.toxlet.2003.12.001 PMID:15104118

Kim S, Park J, Kim HJ, Lee JJ, Choi G, Choi S et al. (2013a). Association between several persistent organic pollutants and thyroid hormone levels in serum among the pregnant women of Korea. *Environ Int*, 59:442–8. doi:10.1016/j.envint.2013.07.009 PMID:23928038

Kitamura S, Shimizu Y, Shiraga Y, Yoshida M, Sugihara K, Ohta S (2002). Reductive metabolism of *p,p'*-DDT and *o,p'*-DDT by rat liver cytochrome P450. *Drug Metab Dispos*, 30(2):113–8. doi:10.1124/dmd.30.2.113 PMID:11792678

Kiyosawa N, Kwekel JC, Burgoon LD, Williams KJ, Tashiro C, Chittim B et al. (2008). *o,p'*-DDT elicits PXR/CAR-, not ER-, mediated responses in the immature ovariectomized rat liver. *Toxicol Sci*, 101(2):350–63. doi:10.1093/toxsci/kfm275 PMID:17984292

Klaunig JE, Goldblatt PJ, Hinton DE, Lipsky MM, Trump BF (1984). Carcinogen induced unscheduled DNA synthesis in mouse hepatocytes. *Toxicol Pathol*, 12(2):119–25. doi:10.1177/019262338401200202 PMID:11478312

Klotz DM, Beckman BS, Hill SM, McLachlan JA, Walters MR, Arnold SF (1996). Identification of environmental chemicals with estrogenic activity using a combination of in vitro assays. *Environ Health Perspect*, 104(10):1084–9. doi:10.1289/ehp.961041084 PMID:8930550

Klotz DM, Ladlie BL, Vonier PM, McLachlan JA, Arnold SF (1997). *o,p'*-DDT and its metabolites inhibit progesterone-dependent responses in yeast and human cells. *Mol Cell Endocrinol*, 129(1):63–71. doi:10.1016/S0303-7207(96)04041-5 PMID:9175630

Koner BC, Banerjee BD, Ray A (1998). Organochlorine pesticide-induced oxidative stress and immune suppression in rats. *Indian J Exp Biol*, 36(4):395–8. PMID:9717451

Konishi Y, Kuwabara K, Hori S (2001). Continuous surveillance of organochlorine compounds in human breast milk from 1972 to 1998 in Osaka, Japan. *Arch Environ Contam Toxicol*, 40(4):571–8. doi:10.1007/s002440010212 PMID:11525502

Korrick SA, Chen C, Damokosh AI, Ni J, Liu X, Cho SI et al. (2001). Association of DDT with spontaneous abortion: a case-control study. *Ann Epidemiol*, 11(7):491–6. doi:10.1016/S1047-2797(01)00239-3 PMID:11557181

Koutros S, Beane Freeman LE, Lubin JH, Heltshe SL, Andreotti G, Barry KH et al. (2013). Risk of total and aggressive prostate cancer and pesticide use in the Agricultural Health Study. *Am J Epidemiol*, 177(1):59–74. doi:10.1093/aje/kws225 PMID:23171882

Krause W (1977). Influence of DDT, DDVP and malathion on FSH, LH and testosterone serum levels and testosterone concentration in testis. *Bull Environ Contam Toxicol*, 18(2):231–42. doi:10.1007/BF01686072 PMID:890160

Krebs B, Maasfeld W, Schrader J et al. (2000). Uniform principles for safeguarding the health of workers re-entering crop growing areas after application of plant protection products. In: Honeycutt RC, Day EW Jr editors. *Worker exposure to agrochemicals: methods for monitoring and assessment*. Boca Raton (FL), USA: Lewis Publishers; pp. 107–18.

Krieger N, Wolff MS, Hiatt RA, Rivera M, Vogelman J, Orentreich N (1994). Breast cancer and serum organochlorines: a prospective study among white, black, and Asian women. *J Natl Cancer Inst*, 86(8):589–99. doi:10.1093/jnci/86.8.589 PMID:8145274

Kubinski H, Gutzke GE, Kubinski ZO (1981). DNA-cell-binding (DCB) assay for suspected carcinogens and mutagens. *Mutat Res*, 89(2):95–136. doi:10.1016/0165-1218(81)90118-X PMID:7290095

Kuiper GG, Lemmen JG, Carlsson B, Corton JC, Safe SH, van der Saag PT et al. (1998). Interaction of estrogenic chemicals and phytoestrogens with estrogen receptor beta. *Endocrinology*, 139(10):4252–63. PMID:9751507

Kunisue T, Muraoka M, Ohtake M, Sudaryanto A, Minh NH, Ueno D et al. (2006). Contamination status of persistent organochlorines in human breast milk from Japan: recent levels and temporal trend. *Chemosphere*, 64(9):1601–8. doi:10.1016/j.chemosphere.2005.11.010 PMID:16386779

Kunisue T, Someya M, Kayama F, Jin Y, Tanabe S (2004a). Persistent organochlorines in human breast milk collected from primiparae in Dalian and Shenyang, China. *Environ Pollut*, 131(3):381–92. doi:10.1016/j.envpol.2004.03.008 PMID:15261401

Kunisue T, Someya M, Monirith I, Watanabe M, Tana TS, Tanabe S (2004b). Occurrence of PCBs, organochlorine insecticides, tris(4-chlorophenyl)methane, and tris(4-chlorophenyl)methanol in human breast milk collected from Cambodia. *Arch Environ Contam Toxicol*, 46(3):405–12. doi:10.1007/s00244-003-3008-7 PMID:15195813

Kurata M, Hirose K, Umeda M (1982). Inhibition of metabolic cooperation in Chinese hamster cells by organochlorine pesticides. *Gan*, 73(2):217–21. PMID:7117752

Kushida M, Sukata T, Uwagawa S, Ozaki K, Kinoshita A, Wanibuchi H et al. (2005). Low dose DDT inhibition of hepatocarcinogenesis initiated by diethylnitrosamine in male rats: possible mechanisms. *Toxicol Appl Pharmacol*, 208(3):285–94. doi:10.1016/j.taap.2005.03.018 PMID:15885732

Laden F, Bertrand KA, Altshul L, Aster JC, Korrick SA, Sagiv SK (2010). Plasma organochlorine levels and risk of non-Hodgkin lymphoma in the Nurses' Health Study. *Cancer Epidemiol Biomarkers Prev*, 19(5):1381–4. doi:10.1158/1055-9965.EPI-10-0125 PMID:20406963

Laden F, Collman G, Iwamoto K, Alberg AJ, Berkowitz GS, Freudenheim JL et al. (2001b). 1,1-Dichloro-2,2-bis(p-chlorophenyl)ethylene and polychlorinated biphenyls and breast cancer: combined analysis of five U.S. studies. *J Natl Cancer Inst*, 93(10):768–76. doi:10.1093/jnci/93.10.768 PMID:11353787

Laden F, Hankinson SE, Wolff MS, Colditz GA, Willett WC, Speizer FE et al. (2001a). Plasma organochlorine levels and the risk of breast cancer: an extended follow-up in the Nurses' Health Study. *Int J Cancer*, 91(4):568–74. doi:10.1002/1097-0215(200002)9999:9999<::AID-IJC1081>3.0.CO;2-W PMID:11251983

Lahvis GP, Wells RS, Kuehl DW, Stewart JL, Rhinehart HL, Via CS (1995). Decreased lymphocyte responses in free-ranging bottlenose dolphins (Tursiops truncatus) are associated with increased concentrations of PCBs and DDT in peripheral blood. *Environ Health Perspect*, 103:Suppl 4: 67–72. doi:10.1289/ehp.95103s467 PMID:7556026

Langenbach R, Gingell R (1975). Cytotoxic and oncogenic activities of 1,1,1-trichloro-2,2-bis(p-chlorophenyl)-ethane and metabolites to mouse embryo cells in culture. *J Natl Cancer Inst*, 54(4):981–3. PMID:1127727

Langer P, Kocan A, Tajtáková M, Rádiková Z, Petrík J, Koska J et al. (2007). Possible effects of persistent organochlorinated pollutants cocktail on thyroid hormone levels and pituitary-thyroid interrelations. *Chemosphere*, 70(1):110–8. doi:10.1016/j.chemosphere.2007.06.046 PMID:17692893

Larsen KD, Jalal SM (1974). DDT induced chromosome mutations in mice–further testing. *Can J Genet Cytol*, 16(3):491–7. doi:10.1139/g74-055 PMID:4378424

Lee HS, Miyauchi K, Nagata Y, Fukuda R, Sasagawa S, Endoh H et al. (2002). Employment of the human estrogen receptor beta ligand-binding domain and co-activator SRC1 nuclear receptor-binding domain for the construction of a yeast two-hybrid detection system for endocrine disrupters. *J Biochem*, 131(3):399–405. doi:10.1093/oxfordjournals.jbchem.a003115 PMID:11872169

Lee S, Kim S, Lee HK, Lee IS, Park J, Kim HJ et al. (2013). Contamination of polychlorinated biphenyls and organochlorine pesticides in breast milk in Korea: time-course variation, influencing factors, and exposure assessment. *Chemosphere*, 93(8):1578–85. doi:10.1016/j.chemosphere.2013.08.011 PMID:24112654

Legator MS, Palmer KA, Adler ID (1973). A collaborative study of in vivo cytogenetic analysis. I. Interpretation of slide preparations. *Toxicol Appl Pharmacol*,

24(3):337–50. doi:10.1016/0041-008X(73)90040-9 PMID:4704808

Legler J, Zeinstra LM, Schuitemaker F, Lanser PH, Bogerd J, Brouwer A et al. (2002). Comparison of in vivo and in vitro reporter gene assays for short-term screening of estrogenic activity. *Environ Sci Technol*, 36(20):4410–5. doi:10.1021/es010323a PMID:12387416

Lemaire G, de Sousa G, Rahmani R (2004). A PXR reporter gene assay in a stable cell culture system: CYP3A4 and CYP2B6 induction by pesticides. *Biochem Pharmacol*, 68(12):2347–58. doi:10.1016/j.bcp.2004.07.041 PMID:15548381

Lemaire G, Mnif W, Mauvais P, Balaguer P, Rahmani R (2006). Activation of alpha- and beta-estrogen receptors by persistent pesticides in reporter cell lines. *Life Sci*, 79(12):1160–9. doi:10.1016/j.lfs.2006.03.023 PMID:16626760

Lessa JMM, Beçak W, Nazareth Rabello M, Pereira CAB, Ungaro MT (1976). Cytogenetic study of DDT on human lymphocytes in vitro. *Mutat Res*, 40(2):131–8. doi:10.1016/0165-1218(76)90008-2 PMID:934176

Li C, Cheng Y, Tang Q, Lin S, Li Y, Hu X et al. (2014b). The association between prenatal exposure to organochlorine pesticides and thyroid hormone levels in newborns in Yancheng, China. *Environ Res*, 129:47–51. doi:10.1016/j.envres.2013.12.009 PMID:24529002

Li J, Li N, Ma M, Giesy JP, Wang Z (2008). In vitro profiling of the endocrine disrupting potency of organochlorine pesticides. *Toxicol Lett*, 183(1–3):65–71. PMID:18992306

Li J, Lu Y, Shi Y, Wang T, Wang G, Luo W et al. (2011). Environmental pollution by persistent toxic substances and health risk in an industrial area of China. *J Environ Sci (China)*, 23(8):1359–67. doi:10.1016/S1001-0742(10)60554-2 PMID:22128544

Li J, McMurray RW (2009). Effects of chronic exposure to DDT and TCDD on disease activity in murine systemic lupus erythematosus. *Lupus*, 18(11):941–9. doi:10.1177/0961203309104431 PMID:19762394

Li JY, Li H, Tao P, Lei FM (2006a). [Serum organochlorines pesticides level of non-occupational exposure women and risk of breast cancer:a case-control study.] *Wei Sheng Yan Jiu*, 35(4):391–4. [Chinese] PMID:16986505

Li S, Tian Y, Ding Q, Liu W (2014a). The release of persistent organic pollutants from a closed system dicofol production process. *Chemosphere*, 94:164–8. doi:10.1016/j.chemosphere.2013.09.090 PMID:24161578

Li YF, Zhulidov AV, Robarts RD, Korotova LG, Zhulidov DA, Gurtovaya TY et al. (2006). Dichlorodiphenyltrichloroethane usage in the former Soviet Union. *Sci Total Environ*, 357(1–3):138–45. doi:10.1016/j.scitotenv.2005.06.009 PMID:16125753

Liljegren G, Hardell L, Lindström G, Dahl P, Magnuson A (1998). Case-control study on breast cancer and adipose tissue concentrations of congener specific polychlorinated biphenyls, DDE and hexachlorobenzene. *Eur J Cancer Prev*, 7(2):135–40. PMID:9818775

Lim JE, Park SH, Jee SH, Park H (2015). Body concentrations of persistent organic pollutants and prostate cancer: a meta-analysis. *Environ Sci Pollut Res Int*, 22(15):11275–84. doi:10.1007/s11356-015-4315-z PMID:25797015

Lin T, Hu Z, Zhang G, Li X, Xu W, Tang J et al. (2009). Levels and mass burden of DDTs in sediments from fishing harbors: the importance of DDT-containing antifouling paint to the coastal environment of China. *Environ Sci Technol*, 43(21):8033–8. doi:10.1021/es901827b PMID:19924919

Lipsky MM, Trump BF, Hinton DE (1989). Histogenesis of dieldrin and DDT-induced hepatocellular carcinoma in Balb/c mice. *J Environ Pathol Toxicol Oncol*, 9(1):79–93. PMID:2564434

Liu C, Ha M, Li L, Yang K (2014). PCB153 and *p,p′*-DDE disorder thyroid hormones via thyroglobulin, deiodinase 2, transthyretin, hepatic enzymes and receptors. *Environ Sci Pollut Res Int*, 21(19):11361–9. doi:10.1007/s11356-014-3093-3 PMID:24878560

Liu C, Shi Y, Li H, Wang Y, Yang K (2011). *p,p′*-DDE disturbs the homeostasis of thyroid hormones via thyroid hormone receptors, transthyretin, and hepatic enzymes. *Horm Metab Res*, 43(6):391–6. doi:10.1055/s-0031-1277135 PMID:21512963

Lloyd JW, Thomas JA, Mawhinney MG (1974). Prostatic and hepatic metabolism of (1,2–3H) testosterone as affected by DDT pretreatment in the mouse. *Toxicol Appl Pharmacol*, 28(2):248–52. doi:10.1016/0041-008X(74)90011-8 PMID:4853918

Longnecker MP, Gladen BC, Cupul-Uicab LA, Romano-Riquer SP, Weber JP, Chapin RE et al. (2007). In utero exposure to the antiandrogen 1,1-dichloro-2,2-bis(p-chlorophenyl)ethylene (DDE) in relation to anogenital distance in male newborns from Chiapas, México. *Am J Epidemiol*, 165(9):1015–22. doi:10.1093/aje/kwk109 PMID:17272288

Longnecker MP, Klebanoff MA, Brock JW, Zhou H, Gray KA, Needham LL et al. (2002). Maternal serum level of 1,1-dichloro-2,2-bis(p-chlorophenyl)ethylene and risk of cryptorchidism, hypospadias, and polythelia among male offspring. *Am J Epidemiol*, 155(4):313–22. doi:10.1093/aje/155.4.313 PMID:11836195

Longnecker MP, Klebanoff MA, Zhou H, Brock JW (2001). Association between maternal serum concentration of the DDT metabolite DDE and preterm and small-for-gestational-age babies at birth. *Lancet*, 358(9276):110–4. doi:10.1016/S0140-6736(01)05329-6 PMID:11463412

López-Carrillo L, Blair A, López-Cervantes M, Cebrián M, Rueda C, Reyes R et al. (1997). Dichlorodiphenyltrichloroethane serum levels and breast cancer risk: a case-control study from Mexico. *Cancer Res*, 57(17):3728–32. PMID:9288780

López-Carrillo L, Torres-Sánchez L, Moline J, Ireland K, Wolff MS (2001). Breast-feeding and serum p,p′-DDT levels among Mexican women of childbearing age: a pilot study. *Environ Res*, 87(3):131–5. doi:10.1006/enrs.2001.4296 PMID:11771926

López-Cervantes M, Torres-Sánchez L, Tobías A, López-Carrillo L (2004). Dichlorodiphenyldichloroethane burden and breast cancer risk: a meta-analysis of the epidemiologic evidence. *Environ Health Perspect*, 112(2):207–14. doi:10.1289/ehp.6492 PMID:14754575

Lopez-Espinosa MJ, Vizcaino E, Murcia M, Fuentes V, Garcia AM, Rebagliato M et al. (2010). Prenatal exposure to organochlorine compounds and neonatal thyroid stimulating hormone levels. *J Expo Sci Environ Epidemiol*, 20(7):579–88. doi:10.1038/jes.2009.47 PMID:19707252

Lu D, Wang D, Ni R, Lin Y, Feng C, Xu Q et al. (2015). Organochlorine pesticides and their metabolites in human breast milk from Shanghai, China. *Environ Sci Pollut Res Int*, 22(12):9293–306. doi:10.1007/s11356-015-4072-z PMID:25595932

Lundholm CE (1991). Influence of chlorinated hydrocarbons, Hg2+ and methyl-Hg+ on steroid hormone receptors from eggshell gland mucosa of domestic fowls and ducks. *Arch Toxicol*, 65(3):220–7. doi:10.1007/BF02307312 PMID:2053848

Luo F, Song J, Chen MF, Wei J, Pan YY, Yu HB (2014). Risk assessment of manufacturing equipment surfaces contaminated with DDTs and dicofol. *Sci Total Environ*, 468–469:176–85. doi:10.1016/j.scitotenv.2013.08.043 PMID:24029690

Mahmood A, Malik RN, Li J, Zhang G (2014). Levels, distribution pattern and ecological risk assessment of organochlorines pesticides (OCPs) in water and sediments from two tributaries of the Chenab River, Pakistan. *Ecotoxicology*, 23(9):1713–21. doi:10.1007/s10646-014-1332-5 PMID:25204814

Mahr U, Miltenburger HG (1976). The effect of insecticides on Chinese hamster cell cultures. *Mutat Res*, 40(2):107–18. doi:10.1016/0165-1218(76)90005-7 PMID:934173

Maitani T (2004). Evaluation of exposure to chemical substances through foods – exposure to pesticides, heavy metals, dioxins, acrylamide and food additives in Japan. *J Health Sci*, 50(3):205–9. doi:10.1248/jhs.50.205

Mamber SW, Bryson V, Katz SE (1984). Evaluation of the Escherichia coli K12 inductest for detection of potential chemical carcinogens. *Mutat Res*, 130(3):141–51. doi:10.1016/0165-1161(84)90116-X PMID:6374442

Maness SC, McDonnell DP, Gaido KW (1998). Inhibition of androgen receptor-dependent transcriptional activity by DDT isomers and methoxychlor in HepG2 human hepatoma cells. *Toxicol Appl Pharmacol*, 151(1):135–42. doi:10.1006/taap.1998.8431 PMID:9705896

Marshall TC, Dorough HW, Swim HE (1976). Screening of pesticides for mutagenic potential using Salmonella typhimurium mutants. *J Agric Food Chem*, 24(3):560–3. doi:10.1021/jf60205a013 PMID:818141

Martin SA Jr, Harlow SD, Sowers MF, Longnecker MP, Garabrant D, Shore DL et al. (2002). DDT metabolite and androgens in African-American farmers. *Epidemiology*, 13(4):454–8. doi:10.1097/00001648-200207000-00014 PMID:12094101

Maslansky CJ, Williams GM (1981). Evidence for an epigenetic mode of action in organochlorine pesticide hepatocarcinogenicity: a lack of genotoxicity in rat, mouse, and hamster hepatocytes. *J Toxicol Environ Health*, 8(1–2):121–30. doi:10.1080/15287398109530056 PMID:7328699

Matthews G, Zaim M, Yadav RS, Soares A, Hii J, Ameneshewa B et al. (2011). Status of legislation and regulatory control of public health pesticides in countries endemic with or at risk of major vector-borne diseases. *Environ Health Perspect*, 119(11):1517–22. doi:10.1289/ehp.1103637 PMID:21742577

McCann J, Choi E, Yamasaki E, Ames BN (1975). Detection of carcinogens as mutagens in the Salmonella/microsome test: assay of 300 chemicals. *Proc Natl Acad Sci USA*, 72(12):5135–9. doi:10.1073/pnas.72.12.5135 PMID:1061098

McDuffie HH, Pahwa P, McLaughlin JR, Spinelli JJ, Fincham S, Dosman JA et al. (2001). Non-Hodgkin's lymphoma and specific pesticide exposures in men: cross-Canada study of pesticides and health. *Cancer Epidemiol Biomarkers Prev*, 10(11):1155–63. PMID:11700263

McGlynn KA, Abnet CC, Zhang M, Sun XD, Fan JH, O'Brien TR et al. (2006). Serum concentrations of 1,1,1-trichloro-2,2-bis(p-chlorophenyl)ethane (DDT) and 1,1-dichloro-2,2-bis(p-chlorophenyl)ethylene (DDE) and risk of primary liver cancer. *J Natl Cancer Inst*, 98(14):1005–10. doi:10.1093/jnci/djj266 PMID:16849683

McGlynn KA, Quraishi SM, Graubard BI, Weber JP, Rubertone MV, Erickson RL (2008). Persistent organochlorine pesticides and risk of testicular germ cell tumors. *J Natl Cancer Inst*, 100(9):663–71. doi:10.1093/jnci/djn101 PMID:18445826

McGregor DB, Brown A, Cattanach P, Edwards I, McBride D, Riach C et al. (1988). Responses of the L5178Y tk+/tk- mouse lymphoma cell forward mutation assay: III. 72 coded chemicals. *Environ Mol Mutagen*, 12(1):85–154. doi:10.1002/em.2860120111 PMID:3383842

Medina-Díaz IM, Arteaga-Illán G, de León MB, Cisneros B, Sierra-Santoyo A, Vega L et al. (2007). Pregnane X receptor-dependent induction of the CYP3A4 gene by *o,p′*-1,1,1,-trichloro-2,2-bis (p-chlorophenyl) ethane. *Drug Metab Dispos*, 35(1):95–102. doi:10.1124/dmd.106.011759 PMID:17035600

Meeker JD, Altshul L, Hauser R (2007). Serum PCBs, *p,p′*-DDE and HCB predict thyroid hormone levels in men. *Environ Res*, 104(2):296–304. doi:10.1016/j.envres.2006.11.007 PMID:17189629

Mendonça GA, Eluf-Neto J, Andrada-Serpa MJ, Carmo PA, Barreto HH, Inomata ON et al. (1999). Organochlorines and breast cancer: a case-control study in Brazil. *Int J Cancer*, 83(5):596–600. doi:10.1002/(SICI)1097-0215(19991126)83:5<596::AID-IJC4>3.0.CO;2-P PMID:10521792

Mercado LA, Freille SM, Vaca-Pereira JS, Cuellar M, Flores L, Mutch E et al. (2013). Serum concentrations of p,p′-dichlorodiphenyltrichloroethane (p,p′-DDE) in a sample of agricultural workers from Bolivia. *Chemosphere*, 91(10):1381–5. doi:10.1016/j.chemosphere.2012.12.023 PMID:23399304

Metcalf RL (1955). Organic insecticides, their chemistry and mode of action. New York (NY), USA: Interscience.

Meza-Montenegro MM, Valenzuela-Quintanar AI, Balderas-Cortés JJ, Yañez-Estrada L, Gutiérrez-Coronado ML, Cuevas-Robles A et al. (2013). Exposure assessment of organochlorine pesticides, arsenic, and lead in children from the major agricultural areas in Sonora, Mexico. *Arch Environ Contam Toxicol*, 64(3):519–27. doi:10.1007/s00244-012-9846-4 PMID:23254566

Miersma NA, Pepper CB, Anderson TA (2003). Organochlorine pesticides in elementary school yards along the Texas-Mexico border. *Environ Pollut*, 126(1):65–71. doi:10.1016/S0269-7491(03)00126-X PMID:12860103

Millikan R, DeVoto E, Duell EJ, Tse CK, Savitz DA, Beach J et al. (2000). Dichlorodiphenyldichloroethene, polychlorinated biphenyls, and breast cancer among African-American and white women in North Carolina. *Cancer Epidemiol Biomarkers Prev*, 9(11):1233–40. PMID:11097232

Minh NH, Someya M, Minh TB, Kunisue T, Iwata H, Watanabe M et al. (2004). Persistent organochlorine residues in human breast milk from Hanoi and Hochiminh City, Vietnam: contamination, accumulation kinetics and risk assessment for infants. *Environ Pollut*, 129(3):431–41. doi:10.1016/j.envpol.2003.11.012 PMID:15016464

Ministerie van VROM (2004). Information dossier on DDT used for the production of Dicofol. Final report 4L0002. A1. Nijmegen, The Netherlands: Ministry of Housing, Spatial Planning and the Environment. Available from: http://www.unece.org/fileadmin/DAM/env/lrtap/TaskForce/popsxg/2004/Dossier_DDT%20for%20dicofol%20production.pdf, accessed October 2015.

Mishra K, Sharma RC, Kumar S (2011). Organochlorine pollutants in human blood and their relation with age, gender and habitat from North-east India. *Chemosphere*, 85(3):454–64. doi:10.1016/j.chemosphere.2011.07.074 PMID:21925698

Mohammed A, Eklund A, Ostlund-Lindqvist AM, Slanina P (1990). Distribution of toxaphene, DDT, and PCB among lipoprotein fractions in rat and human plasma. *Arch Toxicol*, 64(7):567–71. doi:10.1007/BF01971836 PMID:2127352

Morgan DP, Roan CC (1970). Chlorinated hydrocarbon pesticide residue in human tissues. *Arch Environ Health*, 20(4):452–7. doi:10.1080/00039896.1970.10665621 PMID:5429987

Morgan DP, Roan CC (1971). Absorption, storage, and metabolic conversion of ingested DDT and DDT metabolites in man. *Arch Environ Health*, 22(3):301–8. doi:10.1080/00039896.1971.10665848 PMID:5541487

Morgan DP, Roan CC (1974). Chapter 5: The metabolism of DDT in man. In: Hayes WH editor. *Essays in toxicology*. New York (NY), USA: Academic Press; pp. 39–97.

Moriya M, Ohta T, Watanabe K, Miyazawa T, Kato K, Shirasu Y (1983). Further mutagenicity studies on pesticides in bacterial reversion assay systems. *Mutat Res*, 116(3–4):185–216. doi:10.1016/0165-1218(83)90059-9 PMID:6339892

Mortelmans K, Haworth S, Lawlor T, Speck W, Tainer B, Zeiger E (1986). Salmonella mutagenicity tests: II. Results from the testing of 270 chemicals. *Environ Mutagen*, 8:Suppl 7:1–55. doi:10.1002/em.2860080802 PMID:3516675

Morton LM, Turner JJ, Cerhan JR, Linet MS, Treseler PA, Clarke CA et al. (2007). Proposed classification of lymphoid neoplasms for epidemiologic research from the Pathology Working Group of the International Lymphoma Epidemiology Consortium (InterLymph). *Blood*, 110(2):695–708. doi:10.1182/blood-2006-11-051672 PMID:17389762

Moysich KB, Ambrosone CB, Vena JE, Shields PG, Mendola P, Kostyniak P et al. (1998). Environmental organochlorine exposure and postmenopausal breast cancer risk. *Cancer Epidemiol Biomarkers Prev*, 7(3):181–8. PMID:9521429

Mueller JF, Harden F, Toms LM, Symons R, Fürst P (2008). Persistent organochlorine pesticides in human milk samples from Australia. *Chemosphere*, 70(4):712–20. doi:10.1016/j.chemosphere.2007.06.037 PMID:17675211

Mühlebach S, Moor MJ, Wyss PA, Bickel MH (1991). Kinetics of distribution and elimination of DDE in rats. *Xenobiotica*, 21(1):111–20. doi:10.3109/00498259109039455 PMID:2003362

Naidoo V, Katsu Y, Iguchi T (2008). The influence of non-toxic concentrations of DDT and DDE on the old world vulture estrogen receptor alpha. *Gen Comp Endocrinol*, 159(2–3):188–95. doi:10.1016/j.ygcen.2008.08.010 PMID:18801367

Nakagawa R, Hirakawa H, Hori T (1995). Estimation of 1992–1993 dietary intake of organochlorine and organophosphorus pesticides in Fukuoka, Japan. *J AOAC Int*, 78(4):921–9. PMID:7580330

Nanni O, Amadori D, Lugaresi C, Falcini F, Scarpi E, Saragoni A et al. (1996). Chronic lymphocytic leukaemias and non-Hodgkin's lymphomas by histological type in farming-animal breeding workers: a population case-control study based on a priori exposure matrices. *Occup Environ Med*, 53(10):652–7. doi:10.1136/oem.53.10.652 PMID:8943828

NCBI (2015). Clofenotane. Compound summary for CID 3036. PubChem Open Chemistry Database. Bethesda (MD), USA: National Center for Biotechnology Information. Available from: https://pubchem.ncbi.nlm.nih.gov/compound/3036, accessed 5 May 2015.

NCI (1978). Bioassays of DDT, TDE, and p,p′-DDE for possible carcinogenicity. *Natl Cancer Inst Carcinog Tech Rep Ser*, 131:1–251. PMID:12799661

Ndebele K, Tchounwou PB, McMurray RW (2004). Coumestrol, bisphenol-A, DDT, and TCDD modulation of interleukin-2 expression in activated CD+4 Jurkat T cells. *Int J Environ Res Public Health*, 1(1):3–11. doi:10.3390/ijerph2004010003 PMID:16696175

Nelson JA (1974). Effects of dichlorodiphenyltrichloroethane (DDT) analogs and polychlorinated biphenyl (PCB) mixtures on 17beta-(3H)estradiol binding to rat uterine receptor. *Biochem Pharmacol*, 23(2):447–51. doi:10.1016/0006-2952(74)90436-5 PMID:4360348

Nigatu W, Curtis CF, Lulu M (1995). Test for association of DDT resistance with inversion polymorphism in Anopheles arabiensis from Ethiopia. *J Am Mosq Control Assoc*, 11(2 Pt 1):238–40. PMID:7595453

NIH (2015). DDT. ChemIDPlus. TOXNET (Toxicology Data Network). Bethesda (MD), USA: National Institutes of Health. Available from: http://chem.sis.nlm.nih.gov/chemidplus/rn/50-29-3, accessed 5 May 2015.

Nikolaeva IS, Golovleva LA, Astashkin EI, Pertsova RN, Kovalev IE (1980). [Comitogenic and immunosuppressant action of p,p-DDT and its derivatives on phytohemagglutinin-stimulated human lymphocytes.] *Dokl Akad Nauk SSSR*, 253(2):503–5. PMID:7428595

Nishimura N, Nishimura H, Oshima H (1982). Survey on mutagenicity of pesticides by the *Salmonella*-microsome test. *J Aichi med. Univ. Assoc.*, 10:305–12.

Nishizumi M (1979). Effect of phenobarbital, dichlorodiphenyltrichloroethane, and polychlorinated biphenyls on diethylnitrosamine-induced hepatocarcinogenesis. *Gan*, 70(6):835–7. PMID:119661

Nordström M, Hardell L, Lindström G, Wingfors H, Hardell K, Linde A (2000). Concentrations of organochlorines related to titers to Epstein-Barr virus early antigen IgG as risk factors for hairy cell leukemia. *Environ Health Perspect*, 108(5):441–5. doi:10.1289/ehp.00108441 PMID:10811571

NTIS; National Technical Information Service (1968). Evaluation of carcinogenic, teratogenic and mutagenic activities of selected pesticides and industrial chemicals. Vol. 1, Carcinogenic Study. Washington (DC), USA: United States National Technical Information Service.

Nuñez GMA, Estrada I, Calderon-Aranda ES (2002). DDT inhibits the functional activation of murine macrophages and decreases resistance to infection by Mycobacterium microti. *Toxicology*, 174(3):201–10. doi:10.1016/S0300-483X(02)00078-1 PMID:12007859

NZFSA (2005). 2003/04 New Zealand Total Diet Survey. Agricultural compound residues, selected contaminants and nutrients. Wellington, New Zealand: New Zealand Food Safety Authority. Available from: http://www.foodsafety.govt.nz/elibrary/industry/2003_04-Analyses_Environmental.pdf.

O'Connor JC, Frame SR, Davis LG, Cook JC (1999). Detection of the environmental antiandrogen p,p-DDE in CD and Long-Evans rats using a tier I screening battery and a Hershberger assay. *Toxicol Sci*, 51(1):44–53. doi:10.1093/toxsci/51.1.44 PMID:10496676

Ociepa-Zawal M, Rubis B, Wawrzynczak D, Wachowiak R, Trzeciak WH (2010). Accumulation of environmental estrogens in adipose tissue of breast cancer patients. *J Environ Sci Health A Tox Hazard Subst Environ Eng*, 45(3):305–12. doi:10.1080/10934520903468038 PMID:20390871

Okonkwo JO, Mutshatshi TN, Botha B, Agyei N (2008). DDT, DDE and DDD in human milk from South Africa. *Bull Environ Contam Toxicol*, 81(4):348–54. doi:10.1007/s00128-008-9495-5 PMID:18663399

Olaya-Contreras P, Rodríguez-Villamil J, Posso-Valencia HJ, Cortez JE (1998). Organochlorine exposure and breast cancer risk in Colombian women. *Cad Saude Publica*, 14:Suppl 3: 125–32. doi:10.1590/S0102-311X1998000700013 PMID:9819471

Ouyang F, Perry MJ, Venners SA, Chen C, Wang B, Yang F et al. (2005). Serum DDT, age at menarche, and abnormal menstrual cycle length. *Occup Environ Med*, 62(12):878–84. doi:10.1136/oem.2005.020248 PMID:16299097

Ozen S, Darcan S, Bayindir P, Karasulu E, Simsek DG, Gurler T (2012). Effects of pesticides used in agriculture on the development of precocious puberty. *Environ Monit Assess*, 184(7):4223–32. doi:10.1007/s10661-011-2257-6 PMID:21805074

Pahwa M, Harris SA, Hohenadel K, McLaughlin JR, Spinelli JJ, Pahwa P et al. (2012). Pesticide use, immunologic conditions, and risk of non-Hodgkin lymphoma in Canadian men in six provinces. *Int J Cancer*, 131(11):2650–9. doi:10.1002/ijc.27522 PMID:22396152

Palin KJ, Wilson CG, Davis SS, Phillips AJ (1982). The effect of oils on the lymphatic absorption of DDT. *J Pharm Pharmacol*, 34(11):707–10. doi:10.1111/j.2042-7158.1982.tb06204.x PMID:6129299

Palmer KA, Green S, Legator MS (1972). Cytogenetic effects of DDT and derivatives of DDT in a cultured mammalian cell line. *Toxicol Appl Pharmacol*, 22(3):355–64. doi:10.1016/0041-008X(72)90241-4 PMID:5064916

Palmer KA, Green S, Legator MS (1973). Dominant lethal study of *p,p'*-DDT in rats. *Food Cosmet Toxicol*, 11(1):53–62. doi:10.1016/0015-6264(73)90061-8 PMID:4716130

Pardío V, Martínez D, Flores A, Romero D, Suárez V, López K et al. (2012). Human health risk of dietary intake of organochlorine pesticide residues in bovine meat and tissues from Veracruz, México. *Food Chem*, 135(3):1873–93. doi:10.1016/j.foodchem.2012.06.079 PMID:22953937

Park JH, Cha ES, Ko Y, Hwang MS, Hong JH, Lee WJ (2014). Exposure to dichlorodiphenyltrichloroethane and the risk of breast cancer: a systematic review and meta-analysis. *Osong Public Health Res Perspect*, 5(2):77–84. doi:10.1016/j.phrp.2014.02.001 PMID:24955316

Pavuk M, Cerhan JR, Lynch CF, Kocan A, Petrik J, Chovancova J (2003). Case-control study of PCBs, other organochlorines and breast cancer in Eastern Slovakia. *J Expo Anal Environ Epidemiol*, 13(4):267–75. doi:10.1038/sj.jea.7500277 PMID:12923553

Payne J, Jones C, Lakhani S, Kortenkamp A (2000b). Improving the reproducibility of the MCF-7 cell proliferation assay for the detection of xenoestrogens. *Sci Total Environ*, 248(1):51–62. doi:10.1016/S0048-9697(99)00479-9 PMID:10807042

Payne J, Scholze M, Kortenkamp A (2001). Mixtures of four organochlorines enhance human breast cancer cell proliferation. *Environ Health Perspect*, 109(4):391–7. doi:10.1289/ehp.01109391 PMID:11335188

Peraino C, Fry RJM, Staffeldt E, Christopher JP (1975). Comparative enhancing effects of phenobarbital, amobarbital, diphenylhydantoin, and dichlorodiphenyltrichloroethane on 2-acetylaminofluorene-induced hepatic tumorigenesis in the rat. *Cancer Res*, 35(10):2884–90. PMID:1157054

Pérez-Maldonado IN, Athanasiadou M, Yáñez L, González-Amaro R, Bergman A, Díaz-Barriga F (2006). DDE-induced apoptosis in children exposed to the DDT metabolite. *Sci Total Environ*, 370(2–3):343–51. doi:10.1016/j.scitotenv.2006.06.026 PMID:16904735

Pérez-Maldonado IN, Díaz-Barriga F, de la Fuente H, González-Amaro R, Calderón J, Yáñez L (2004). DDT induces apoptosis in human mononuclear cells in vitro and is associated with increased apoptosis in exposed children. *Environ Res*, 94(1):38–46. doi:10.1016/S0013-9351(03)00112-9 PMID:14643285

Pérez-Maldonado IN, Herrera C, Batres LE, González-Amaro R, Díaz-Barriga F, Yáñez L (2005). DDT-induced oxidative damage in human blood mononuclear cells. *Environ Res*, 98(2):177–84. doi:10.1016/j.envres.2004.11.001 PMID:15820723

Pérez-Maldonado IN, Pérez-Vázquez FJ, Gaspar-Ramírez O, González-Amaro R, Díaz-Barriga F (2011). Variability in DDT-induced apoptosis in Mexican indigenous populations. *Toxicol Mech Methods*, 21(9):675–80. doi:10.3109/15376516.2011.601354 PMID:22003925

Pérez-Maldonado IN, Trejo A, Ruepert C, Jovel RC, Méndez MP, Ferrari M et al. (2010). Assessment of DDT levels in selected environmental media and biological samples from Mexico and Central America. *Chemosphere*, 78(10):1244–9. doi:10.1016/j.chemosphere.2009.12.040 PMID:20092871

Perry MJ, Ouyang F, Korrick S, Venners SA, Altshul L, Xu X et al. (2005). Body mass index and serum 1,1,1-trichloro-2,2-bis(p-chlorophenyl)ethane in nulliparous Chinese women. *Cancer Epidemiol Biomarkers Prev*, 14(10):2433–8. doi:10.1158/1055-9965.EPI-05-0174 PMID:16214928

Persky V, Turyk M, Anderson HA, Hanrahan LP, Falk C, Steenport DN et al. ; Great Lakes Consortium(2001). The effects of PCB exposure and fish consumption on endogenous hormones. *Environ Health Perspect*, 109(12):1275–83. doi:10.1289/ehp.011091275 PMID:11748036

Persson B, Dahlander AM, Fredriksson M, Brage HN, Ohlson CG, Axelson O (1989). Malignant lymphomas and occupational exposures. *Br J Ind Med*, 46(8):516–20. PMID:2775671

Persson B, Fredriksson M, Olsen K, Boeryd B, Axelson O (1993). Some occupational exposures as risk factors for malignant lymphomas. *Cancer*, 72(5):1773–8. doi:10.1002/1097-0142(19930901)72:5<1773::AID-CNCR2820720542>3.0.CO;2-6 PMID:8348507

Persson EC, Graubard BI, Evans AA, London WT, Weber JP, LeBlanc A et al. (2012). Dichlorodiphenyltrichloroethane and risk of hepatocellular carcinoma. *Int J Cancer*, 131(9):2078–84. doi:10.1002/ijc.27459 PMID:22290210

Peterson JE, Robison WH (1964). Metabolic products of *p,p'*-DDT in the rat. *Toxicol Appl Pharmacol*, 6(3):321–7. doi:10.1016/0041-008X(64)90073-0 PMID:14194764

Petrakis NL, Wrensch MR, Ernster VL, Miike R, Murai J, Simberg N et al. (1987). Influence of pregnancy and lactation on serum and breast fluid estrogen levels: implications for breast cancer risk. *Int J Cancer*, 40(5):587–91. doi:10.1002/ijc.2910400502 PMID:3679587

Picchietti S, Belardinelli M, Taddei AR, Fausto AM, Pellegrino M, Maggio R et al. (2009). Thyroid disruptor 1,1,1-trichloro-2,2-bis(p-chlorophenyl)ethane (DDT) prevents internalization of TSH receptor. *Cell Tissue Res*, 336(1):31–40. doi:10.1007/s00441-008-0749-7 PMID:19214582

Pielou DP (1952). The nonmutagenic action of *p-p'*-DDT and γ-hexachlorocyclohexane in *Drosophila melanogaster* Meig. (Diptera: Drosophilidae). *Can J Zool*, 30(6):375–7. doi:10.1139/z52-033

Planche G, Croisy A, Malaveille C, Tomatis L, Bartsch H (1979). Metabolic and mutagenicity studies on DDT and 15 derivatives. Detection of 1,1-bis(p-chlorophenyl)-2,2-dichloroethane and 1,1-bis(p-chlorophenyl)-2,2,2-trichloroethyl acetate (kelthane acetate) as mutagens in Salmonella typhimurium and of 1,1-bis(p-chlorophenyl) ethylene oxide, a likely metabolite, as an alkylating agent. *Chem Biol Interact*, 25(2–3):157–75. doi:10.1016/0009-2797(79)90043-7 PMID:380827

Pocock DE, Vost A (1974). DDT absorption and chylomicron transport in rat. *Lipids*, 9(6):374–81. doi:10.1007/BF02532054 PMID:4365967

Poolpak T, Pokethitiyook P, Kruatrachue M, Arjarasirikoon U, Thanwaniwat N (2008). Residue analysis of organochlorine pesticides in the Mae Klong River of Central Thailand. *J Hazard Mater*, 156(1–3):230–9. doi:10.1016/j.jhazmat.2007.12.078 PMID:18258355

Porta M, López T, Gasull M, Rodríguez-Sanz M, Garí M, Pumarega J et al. (2012). Distribution of blood concentrations of persistent organic pollutants in a representative sample of the population of Barcelona in 2006, and comparison with levels in 2002. *Sci Total Environ*, 423:151–61. doi:10.1016/j.scitotenv.2012.02.001 PMID:22397902

Porta M, López T, Pumarega J, Jariod M, Crous-Bou M, Marco E et al.; PANKRAS II Study Group (2009). In pancreatic ductal adenocarcinoma blood concentrations of some organochlorine compounds and coffee intake are independently associated with KRAS mutations. *Mutagenesis*, 24(6):513–21. doi:10.1093/mutage/gep037 PMID:19797353

Porta M, Malats N, Jariod M, Grimalt JO, Rifà J, Carrato A et al.; PANKRAS II Study Group (1999). Serum concentrations of organochlorine compounds and K-ras mutations in exocrine pancreatic cancer. *Lancet*, 354(9196):2125–9. doi:10.1016/S0140-6736(99)04232-4 PMID:10609819

Pozo K, Harner T, Lee SC, Wania F, Muir DC, Jones KC (2009). Seasonally resolved concentrations of persistent organic pollutants in the global atmosphere from the first year of the GAPS study. *Environ Sci Technol*, 43(3):796–803. doi:10.1021/es802106a PMID:19245019

Probst GS, McMahon RE, Hill LE, Thompson CZ, Epp JK, Neal SB (1981). Chemically-induced unscheduled DNA synthesis in primary rat hepatocyte cultures: a comparison with bacterial mutagenicity using 218 compounds. *Environ Mutagen*, 3(1):11–32. doi:10.1002/em.2860030103 PMID:7021142

Purdue MP, Engel LS, Langseth H, Needham LL, Andersen A, Barr DB et al. (2009). Prediagnostic serum concentrations of organochlorine compounds and risk of testicular germ cell tumors. *Environ Health Perspect*, 117(10):1514–9. doi:10.1289/ehp.0800359 PMID:20019899

Purdue MP, Hoppin JA, Blair A, Dosemeci M, Alavanja MC (2007). Occupational exposure to organochlorine insecticides and cancer incidence in the Agricultural Health Study. *Int J Cancer*, 120(3):642–9. doi:10.1002/ijc.22258 PMID:17096337

Qin XY, Zaha H, Nagano R, Yoshinaga J, Yonemoto J, Sone H (2011). Xenoestrogens down-regulate aryl-hydrocarbon receptor nuclear translocator 2 mRNA expression in human breast cancer cells via an estrogen receptor alpha-dependent mechanism. *Toxicol Lett*, 206(2):152–7. doi:10.1016/j.toxlet.2011.07.007 PMID:21771643

Qiu X, Zhu T, Yao B, Hu J, Hu S (2005). Contribution of dicofol to the current DDT pollution in China. *Environ Sci Technol*, 39(12):4385–90. doi:10.1021/es050342a PMID:16047771

Quintana PJ, Delfino RJ, Korrick S, Ziogas A, Kutz FW, Jones EL et al. (2004). Adipose tissue levels of organochlorine pesticides and polychlorinated biphenyls and risk of non-Hodgkin's lymphoma. *Environ Health Perspect*, 112(8):854–61. doi:10.1289/ehp.6726 PMID:15175172

Raaschou-Nielsen O, Pavuk M, Leblanc A, Dumas P, Philippe Weber J, Olsen A et al. (2005). Adipose organochlorine concentrations and risk of breast cancer among postmenopausal Danish women. *Cancer Epidemiol Biomarkers Prev*, 14(1):67–74. PMID:15668478

Rabello MN, Beçak W, De Almeida WF, Pigati P, Ungaro MT, Murata T et al. (1975). Cytogenetic study on individuals occupationally exposed to DDT. *Mutat Res*, 28(3):449–54. doi:10.1016/0027-5107(75)90238-9 PMID:1134514

Radomski JL, Deichmann WB, MacDonald WE, Glass EM (1965). Synergism among oral carcinogens. I. Results of the simultaneous feeding of four tumorigens to rats. *Toxicol Appl Pharmacol*, 7(5):652–6. doi:10.1016/0041-008X(65)90120-1 PMID:5866802

Rajaei F, Bahramifar N, Esmaili Sari A, Ghasempouri SM, Savabieasfahani M (2010). PCBs and organochlorine pesticides in ducks of Fereydoon-Kenar wildlife refuge in Iran. *Bull Environ Contam Toxicol*, 84(5):577–81. doi:10.1007/s00128-010-9988-x PMID:20419290

Ramachandran M, Banerjee BD, Gulati M, Grover A, Zaidi SS, Hussain QZ (1984). DDT & HCH residues in the body fat & blood samples from some Delhi hospitals. *Indian J Med Res*, 80:590–3. PMID:6085069

Rawn DFK, Cao XL, Doucet J, Davies DJ, Sun WF, Dabeka RW et al. (2004). Canadian Total Diet Study in 1998: pesticide levels in foods from Whitehorse, Yukon, Canada, and corresponding dietary intake estimates. *Food Addit Contam*, 21(3):232–50. doi:10.1080/02652030310001655470 PMID:15195471

Reed A, Dzon L, Loganathan BG, Whalen MM (2004). Immunomodulation of human natural killer cell cytotoxic function by organochlorine pesticides. *Hum Exp Toxicol*, 23(10):463–71. doi:10.1191/0960327104ht477oa PMID:15553171

Reif DM, Martin MT, Tan SW, Houck KA, Judson RS, Richard AM et al. (2010). Endocrine profiling and prioritization of environmental chemicals using ToxCast data. *Environ Health Perspect*, 118(12):1714–20. doi:10.1289/ehp.1002180 PMID:20826373

Reif DM, Sypa M, Lock EF, Wright FA, Wilson A, Cathey T et al. (2013). ToxPi GUI: an interactive visualization tool for transparent integration of data from diverse sources of evidence. *Bioinformatics*, 29(3):402–3. doi:10.1093/bioinformatics/bts686 PMID:23202747

Reif VD, Sinsheimer JE (1975). Metabolism of 1-(0-chlorophenyl)-1-(p-chlorophenyl)-2,2-dichloroethane (*o,p′*-DDD) in rats. *Drug Metab Dispos*, 3(1):15–25. PMID:234830

Reif VD, Sinsheimer JE, Ward JC, Schteingart DE (1974). Aromatic hydroxylation and alkyl oxidation in metabolism of mitotane (*o,p′*-DDD) in humans. *J Pharm Sci*, 63(11):1730–6. doi:10.1002/jps.2600631113 PMID:4427232

Reifenrath WG, Hawkins GS, Kurtz MS (1991). Percutaneous penetration and skin retention of topically applied compounds: an in vitro-in vivo study. *J Pharm Sci*, 80(6):526–32. doi:10.1002/jps.2600800605 PMID:1941541

Ren A, Qiu X, Jin L, Ma J, Li Z, Zhang L et al. (2011). Association of selected persistent organic pollutants in the placenta with the risk of neural tube defects. *Proc Natl Acad Sci USA*, 108(31):12770–5. doi:10.1073/pnas.1105209108 PMID:21768370

Reuber MD (1978). Carcinomas of the liver in Osborne-Mendel rats ingesting DDT. *Tumori*, 64(6):571–7. PMID:581710

Ribas-Fitó N, Sala M, Cardo E, Mazón C, De Muga ME, Verdú A et al. (2003). Organochlorine compounds and concentrations of thyroid stimulating hormone in newborns. *Occup Environ Med*, 60(4):301–3. doi:10.1136/oem.60.4.301 PMID:12660379

Rignell-Hydbom A, Rylander L, Giwercman A, Jönsson BAG, Lindh C, Eleuteri P et al. (2005). Exposure to PCBs and *p,p′*-DDE and human sperm chromatin integrity. *Environ Health Perspect*, 113(2):175–9. doi:10.1289/ehp.7252 PMID:15687046

Risebrough RW, Jarman WM, Springer AM, Walker W 2nd, Hunt WG (1986). A metabolic derivation of DDE from kelthane®. *Environ Toxicol Chem*, 5(1):13–9. doi:10.1002/etc.5620050103

Rissato SR, Galhiane MS, Ximenes VF, de Andrade RM, Talamoni JL, Libânio M et al. (2006). Organochlorine pesticides and polychlorinated biphenyls in soil and water samples in the Northeastern part of São Paulo State, Brazil. *Chemosphere*, 65(11):1949–58. doi:10.1016/j.chemosphere.2006.07.011 PMID:16919310

Ritchie JM, Vial SL, Fuortes LJ, Guo H, Reedy VE, Smith EM (2003). Organochlorines and risk of prostate cancer. *J Occup Environ Med*, 45(7):692–702. doi:10.1097/01.jom.0000071510.96740.0b PMID:12855910

Ritter R, Scheringer M, MacLeod M, Hungerbühler K (2011). Assessment of nonoccupational exposure to DDT in the tropics and the north: relevance of uptake via inhalation from indoor residual spraying. *Environ Health Perspect*, 119(5):707–12. doi:10.1289/ehp.1002542 PMID:21536537

Rivero-Rodriguez L, Borja-Aburto VH, Santos-Burgoa C, Waliszewskiy S, Rios C, Cruz V (1997). Exposure assessment for workers applying DDT to control malaria in Veracruz, Mexico. *Environ Health Perspect*, 105(1):98–101. doi:10.1289/ehp.9710598 PMID:9074888

Roan C, Morgan D, Paschal EH (1971). Urinary excretion of DDA following ingestion of DDT and DDT metabolites in man. *Arch Environ Health*, 22(3):309–15. doi:10.1080/00039896.1971.10665849 PMID:5100105

Rocchi P, Perocco P, Alberghini W, Fini A, Prodi G (1980). Effect of pesticides on scheduled and unscheduled DNA synthesis of rat thymocytes and human lymphocytes. *Arch Toxicol*, 45(2):101–8. doi:10.1007/BF01270907 PMID:7469786

Rojas-Squella X, Santos L, Baumann W, Landaeta D, Jaimes A, Correa JC et al. (2013). Presence of organochlorine pesticides in breast milk samples from Colombian women. *Chemosphere*, 91(6):733–9. doi:10.1016/j.chemosphere.2013.02.026 PMID:23499217

Röllin HB, Sandanger TM, Hansen L, Channa K, Odland JØ (2009). Concentration of selected persistent organic pollutants in blood from delivering women in South Africa. *Sci Total Environ*, 408(1):146–52. doi:10.1016/j.scitotenv.2009.08.049 PMID:19800104

Romieu I, Hernandez-Avila M, Lazcano-Ponce E, Weber JP, Dewailly E (2000). Breast cancer, lactation history, and serum organochlorines. *Am J Epidemiol*, 152(4):363–70. doi:10.1093/aje/152.4.363 PMID:10968381

Rossi L, Barbieri O, Sanguineti M, Cabral JRP, Bruzzi P, Santi L (1983). Carcinogenicity study with technical-grade dichlorodiphenyltrichloroethane and 1,1-dichloro-2,2-bis(p-chlorophenyl)ethylene in hamsters. *Cancer Res*, 43(2):776–81. PMID:6848191

Rossi L, Ravera M, Repetti G, Santi L (1977). Long-term administration of DDT or phenobarbital-Na in Wistar rats. *Int J Cancer*, 19(2):179–85. doi:10.1002/ijc.2910190207 PMID:838519

Rothe CF, Mattson AM, Nueslein RM, Hayes WJ Jr (1957). Metabolism of cholorophenothane (DDT); intestinal lymphatic absorption. *AMA Arch Ind Health*, 16(1):82–6. PMID:13434502

Rothman N, Cantor KP, Blair A, Bush D, Brock JW, Helzlsouer K et al. (1997). A nested case-control study of non-Hodgkin lymphoma and serum organochlorine residues. *Lancet*, 350(9073):240–4. doi:10.1016/S0140-6736(97)02088-6 PMID:9242800

Roy P, Salminen H, Koskimies P, Simola J, Smeds A, Saukko P et al. (2004). Screening of some anti-androgenic endocrine disruptors using a recombinant cell-based in vitro bioassay. *J Steroid Biochem Mol Biol*, 88(2):157–66. doi:10.1016/j.jsbmb.2003.11.005 PMID:15084347

Rubin CH, Lanier A, Kieszak S, Brock JW, Koller KR, Strosnider H et al. (2006). Breast cancer among Alaska Native women potentially exposed to environmental organochlorine chemicals. *Int J Circumpolar*

Health, 65(1):18–27. doi:10.3402/ijch.v65i1.17885 PMID:16544644

Rudge CV, Sandanger T, Röllin HB, Calderon IM, Volpato G, Silva JL et al. (2012). Levels of selected persistent organic pollutants in blood from delivering women in seven selected areas of São Paulo State, Brazil. *Environ Int*, 40:162–9. doi:10.1016/j.envint.2011.07.006 PMID:21820740

Ruiz-Suárez LE, Castro-Chan RA, Rivero-Pérez NE, Trejo-Acevedo A, Guillén-Navarro GK, Geissen V et al. (2014). Levels of organochlorine pesticides in blood plasma from residents of malaria-endemic communities in Chiapas, Mexico. *Int J Environ Res Public Health*, 11(10):10444–60. doi:10.3390/ijerph111010444 PMID:25310541

Rupa DS, Reddy PP, Sreemannarayana K, Reddi OS, Galloway SM (1991). Frequency of sister chromatid exchange in peripheral lymphocytes of male pesticide applicators. *Environ Mol Mutagen*, 18(2):136–8. doi:10.1002/em.2850180209 PMID:1879405

Rupa DS, Rita P, Reddy PP, Reddi OS (1988). Screening of chromosomal aberrations and sister chromatid exchanges in peripheral lymphocytes of vegetable garden workers. *Hum Toxicol*, 7(4):333–6. doi:10.1177/096032718800700406 PMID:3410481

Rusiecki JA, Baccarelli A, Bollati V, Tarantini L, Moore LE, Bonefeld-Jorgensen EC (2008). Global DNA hypomethylation is associated with high serum-persistent organic pollutants in Greenlandic Inuit. *Environ Health Perspect*, 116(11):1547–52. doi:10.1289/ehp.11338 PMID:19057709

Sagiv SK, Tolbert PE, Altshul LM, Korrick SA (2007). Organochlorine exposures during pregnancy and infant size at birth. *Epidemiology*, 18(1):120–9. doi:10.1097/01.ede.0000249769.15001.7c PMID:17179760

Sala M, Ribas-Fitó N, Cardo E, de Muga ME, Marco E, Mazón C et al. (2001). Levels of hexachlorobenzene and other organochlorine compounds in cord blood: exposure across placenta. *Chemosphere*, 43(4–7):895–901. doi:10.1016/S0045-6535(00)00450-1 PMID:11372882

Salazar-García F, Gallardo-Díaz E, Cerón-Mireles P, Loomis D, Borja-Aburto VH (2004). Reproductive effects of occupational DDT exposure among male malaria control workers. *Environ Health Perspect*, 112(5):542–7. doi:10.1289/ehp.6759 PMID:15064158

Salem NM, Ahmad R, Estaitieh H (2009). Organochlorine pesticide residues in dairy products in Jordan. *Chemosphere*, 77(5):673–8. doi:10.1016/j.chemosphere.2009.07.045 PMID:19695668

Sanderson JT, Boerma J, Lansbergen GW, van den Berg M (2002). Induction and inhibition of aromatase (CYP19) activity by various classes of pesticides in H295R human adrenocortical carcinoma cells. *Toxicol Appl Pharmacol*, 182(1):44–54. doi:10.1006/taap.2002.9420 PMID:12127262

Santini F, Vitti P, Ceccarini G, Mammoli C, Rosellini V, Pelosini C et al. (2003). In vitro assay of thyroid disruptors affecting TSH-stimulated adenylate cyclase activity. *J Endocrinol Invest*, 26(10):950–5. doi:10.1007/BF03348190 PMID:14759065

Saoudi A, Fréry N, Zeghnoun A, Bidondo ML, Deschamps V, Göen T et al. (2014). Serum levels of organochlorine pesticides in the French adult population: the French National Nutrition and Health Study (ENNS), 2006–2007. *Sci Total Environ*, 472:1089–99. doi:10.1016/j.scitotenv.2013.11.044 PMID:24361744

Sapbamrer R, Prapamontol T, Prakobvitayakit O, Vaneesorn Y, Mangklabruks A, Hock B (2008). Placental transfer of DDT in mother-infant pairs from Northern Thailand. *J Environ Sci Health B*, 43(6):484–9. doi:10.1080/03601230802174615 PMID:18665984

Sawada N, Iwasaki M, Inoue M, Itoh H, Sasazuki S, Yamaji T et al.; Japan Public Health Center Based Prospective (JPHC) Study Group (2010). Plasma organochlorines and subsequent risk of prostate cancer in Japanese men: a nested case-control study. *Environ Health Perspect*, 118(5):659–65. doi:10.1289/ehp.0901214 PMID:20435560

Schecter A, Toniolo P, Dai LC, Thuy LT, Wolff MS (1997). Blood levels of DDT and breast cancer risk among women living in the north of Vietnam. *Arch Environ Contam Toxicol*, 33(4):453–6. doi:10.1007/s002449900276 PMID:9419265

Schell LM, Gallo MV, Deane GD, Nelder KR, DeCaprio AP, Jacobs A; Akwesasne Task Force on the Environment (2014). Relationships of polychlorinated biphenyls and dichlorodiphenyldichloroethylene (p,p′-DDE) with testosterone levels in adolescent males. *Environ Health Perspect*, 122(3):304–9. PMID:24398050

Schiestl RH (1989). Nonmutagenic carcinogens induce intrachromosomal recombination in yeast. *Nature*, 337(6204):285–8. doi:10.1038/337285a0 PMID:2643057

Schiestl RH, Gietz RD, Mehta RD, Hastings PJ (1989). Carcinogens induce intrachromosomal recombination in yeast. *Carcinogenesis*, 10(8):1445–55. doi:10.1093/carcin/10.8.1445 PMID:2665967

Schinasi L, Leon ME (2014). Non-Hodgkin lymphoma and occupational exposure to agricultural pesticide chemical groups and active ingredients: a systematic review and meta-analysis. *Int J Environ Res Public Health*, 11(4):4449–527. doi:10.3390/ijerph110404449 PMID:24762670

Schop RN, Hardy MH, Goldberg MT (1990). Comparison of the activity of topically applied pesticides and the herbicide 2,4-D in two short-term in vivo assays of genotoxicity in the mouse. *Fundam Appl Toxicol*, 15(4):666–75. doi:10.1016/0272-0590(90)90183-K PMID:2086312

Schrader TJ, Cooke GM (2000). Examination of selected food additives and organochlorine food contaminants for androgenic activity in vitro. *Toxicol Sci*, 53(2):278–88. doi:10.1093/toxsci/53.2.278 PMID:10696776

Scippo ML, Argiris C, Van De Weerdt C, Muller M, Willemsen P, Martial J et al. (2004). Recombinant human estrogen, androgen and progesterone receptors for detection of potential endocrine disruptors. *Anal Bioanal Chem*, 378(3):664–9. doi:10.1007/s00216-003-2251-0 PMID:14579009

Sciullo EM, Vogel CF, Wu D, Murakami A, Ohigashi H, Matsumura F (2010). Effects of selected food phytochemicals in reducing the toxic actions of TCDD and *p,p′*-DDT in U937 macrophages. *Arch Toxicol*, 84(12):957–66. doi:10.1007/s00204-010-0592-y PMID:20865247

Scollon EJ, Carr JA, Cobb GP (2004). The effect of flight, fasting and *p,p′*-DDT on thyroid hormones and corticosterone in Gambel's white-crowned sparrow, Zonotrichia leucophrys gambelli. *Comp Biochem Physiol C Toxicol Pharmacol*, 137(2):179–89. doi:10.1016/j.cca.2004.01.004 PMID:15050929

Sekihashi K, Yamamoto A, Matsumura Y, Ueno S, Watanabe-Akanuma M, Kassie F et al. (2002). Comparative investigation of multiple organs of mice and rats in the comet assay. *Mutat Res*, 517(1–2):53–75. doi:10.1016/S1383-5718(02)00034-7 PMID:12034309

Settimi L, Masina A, Andrion A, Axelson O (2003). Prostate cancer and exposure to pesticides in agricultural settings. *Int J Cancer*, 104(4):458–61. doi:10.1002/ijc.10955 PMID:12584743

Shah PV, Guthrie FE (1983). Percutaneous penetration of three insecticides in rats: a comparison of two methods for in vivo determination. *J Invest Dermatol*, 80(4):291–3. doi:10.1111/1523-1747.ep12534663 PMID:6403629

Sharma BM, Bharat GK, Tayal S, Nizzetto L, Cupr P, Larssen T (2014). Environment and human exposure to persistent organic pollutants (POPs) in India: a systematic review of recent and historical data. *Environ Int*, 66:48–64. doi:10.1016/j.envint.2014.01.022 PMID:24525153

Sharma RS, Sharma GK, Dhillon GPS (1996). Economics of malaria control programme. In: *Epidemiology and control of malaria in India*. New Delhi, India: National Malaria Eradication Programme, Directorate General of Health Services; pp. 1–23.

Sheeler CQ, Dudley MW, Khan SA (2000). Environmental estrogens induce transcriptionally active estrogen receptor dimers in yeast: activity potentiated by the coactivator RIP140. *Environ Health Perspect*, 108(2):97–103. doi:10.1289/ehp.0010897 PMID:10656848

Shekhar PV, Werdell J, Basrur VS (1997). Environmental estrogen stimulation of growth and estrogen receptor function in preneoplastic and cancerous human breast cell lines. *J Natl Cancer Inst*, 89(23):1774–82. doi:10.1093/jnci/89.23.1774 PMID:9392618

Shelby MD, Newbold RR, Tully DB, Chae K, Davis VL (1996). Assessing environmental chemicals for estrogenicity using a combination of in vitro and in vivo assays. *Environ Health Perspect*, 104(12):1296–300. doi:10.1289/ehp.961041296 PMID:9118870

Shen L, Wania F, Lei YD, Teixeira C, Muir DCG, Bidleman TF (2005). Atmospheric distribution and long-range transport behavior of organochlorine pesticides in North America. *Environ Sci Technol*, 39(2):409–20. doi:10.1021/es049489c PMID:15707039

Shi Y, Zhang JH, Jiang M, Zhu LH, Tan HQ, Lu B (2010). Synergistic genotoxicity caused by low concentration of titanium dioxide nanoparticles and *p,p′*-DDT in human hepatocytes. *Environ Mol Mutagen*, 51(3):192–204. PMID:19708068

Shirasu Y, Moriya M, Kato K, Furuhashi A, Kada T (1976). Mutagenicity screening of pesticides in the microbial system. *Mutat Res*, 40(1):19–30. doi:10.1016/0165-1218(76)90018-5 PMID:814455

Shivapurkar N, Hoover KL, Poirier LA (1986). Effect of methionine and choline on liver tumor promotion by phenobarbital and DDT in diethylnitrosamine-initiated rats. *Carcinogenesis*, 7(4):547–50. doi:10.1093/carcin/7.4.547 PMID:3698186

Shutoh Y, Takeda M, Ohtsuka R, Haishima A, Yamaguchi S, Fujie H et al. (2009). Low dose effects of dichlorodiphenyltrichloroethane (DDT) on gene transcription and DNA methylation in the hypothalamus of young male rats: implication of hormesis-like effects. *J Toxicol Sci*, 34(5):469–82. doi:10.2131/jts.34.469 PMID:19797855

Sierra-Santoyo A, Hernández M, Albores A, Cebrián ME (2005). DDT increases hepatic testosterone metabolism in rats. *Arch Toxicol*, 79(1):7–12. doi:10.1007/s00204-004-0603-y PMID:15372139

Silva E, Kabil A, Kortenkamp A (2010). Cross-talk between non-genomic and genomic signalling pathways–distinct effect profiles of environmental estrogens. *Toxicol Appl Pharmacol*, 245(2):160–70. doi:10.1016/j.taap.2010.02.015 PMID:20206645

Silva E, Scholze M, Kortenkamp A (2007). Activity of xenoestrogens at nanomolar concentrations in the E-Screen assay. *Environ Health Perspect*, 115(S-1):Suppl 1:91–7. doi:10.1289/ehp.9363 PMID:18174956

Simmon VE, Kauhanen K, Tardiff RG (1977). Mutagenic activity of chemicals identified in drinking water. *Dev Toxicol Environ Sci*, 2:249–58.

Sina JF, Bean CL, Dysart GR, Taylor VI, Bradley MO (1983). Evaluation of the alkaline elution/rat hepatocyte assay as a predictor of carcinogenic/mutagenic potential. *Mutat Res*, 113(5):357–91. doi:10.1016/0165-1161(83)90228-5 PMID:6877265

Sipes NS, Martin MT, Kothiya P, Reif DM, Judson RS, Richard AM et al. (2013). Profiling 976 ToxCast chemicals across 331 enzymatic and receptor signalling assays. *Chem Res Toxicol*, 26(6):878–95. doi:10.1021/tx400021f PMID:23611293

Skinner MK, Manikkam M, Tracey R, Guerrero-Bosagna C, Haque M, Nilsson EE (2013). Ancestral dichlorodiphenyltrichloroethane (DDT) exposure promotes epigenetic transgenerational inheritance of obesity. *BMC Med*, 11(1):228. doi:10.1186/1741-7015-11-228 PMID:24228800

Slebos RJC, Hoppin JA, Tolbert PE, Holly EA, Brock JW, Zhang RH et al. (2000). K-ras and p53 in pancreatic cancer: association with medical history, histopathology, and environmental exposures in a population-based study. *Cancer Epidemiol Biomarkers Prev*, 9(11):1223–32. PMID:11097231

Smith AG (2010). Toxicology of DDT and some analogues. Chapter 93. In: Krieger R editor. *Hayes' handbook of pesticide toxicology*. 3rd ed. New York (NY), USA: Academic Press; pp. 1975–2032.

Smith D (1999). Worldwide trends in DDT levels in human breast milk. *Int J Epidemiol*, 28(2):179–88. doi:10.1093/ije/28.2.179 PMID:10342677

Smith MT, Guyton KZ, Gibbons CF, Fritz JM, Portier CJ, Rusyn I et al. (2016). Key characteristics of carcinogens as a basis for organizing data on mechanisms of carcinogenesis. *Environ Health Perspect*, 124(6):713–21. PMID:26600562

Smith MT, Thomas JA, Smith CG, Mawhinney MG, Lloyd JW (1972). Effects of DDT on radioactive uptake from testosterone-1,2- 3 H by mouse prostate glands. *Toxicol Appl Pharmacol*, 23(1):159–64. doi:10.1016/0041-008X(72)90215-3 PMID:5071039

Sobel ES, Gianini J, Butfiloski EJ, Croker BP, Schiffenbauer J, Roberts SM (2005). Acceleration of autoimmunity by organochlorine pesticides in (NZB x NZW)F1 mice. *Environ Health Perspect*, 113(3):323–8. doi:10.1289/ehp.7347 PMID:15743722

Soliman AS, Wang X, DiGiovanni J, Eissa S, Morad M, Vulimiri S et al. (2003). Serum organochlorine levels and history of lactation in Egypt. *Environ Res*, 92(2):110–7. doi:10.1016/S0013-9351(02)00056-7 PMID:12854690

Son HK, Kim SA, Kang JH, Chang YS, Park SK, Lee SK et al. (2010). Strong associations between low-dose organochlorine pesticides and type 2 diabetes in Korea. *Environ Int*, 36(5):410–4. doi:10.1016/j.envint.2010.02.012 PMID:20381150

Song L, Liu J, Jin X, Li Z, Zhao M, Liu W (2014a). p,p′-Dichlorodiphenyldichloroethylene induces colorectal adenocarcinoma cell proliferation through oxidative stress. *PLoS ONE*, 9(11):e112700. doi:10.1371/journal.pone.0112700 PMID:25386960

Song L, Zhao J, Jin X, Li Z, Newton IP, Liu W et al. (2014b). The organochlorine *p,p′*-dichlorodiphenyltrichloroethane induces colorectal cancer growth through Wnt/β-catenin signalling. *Toxicol Lett*, 229(1):284–91. doi:10.1016/j.toxlet.2014.06.003 PMID:24968063

Sonneveld E, Jansen HJ, Riteco JA, Brouwer A, van der Burg B (2005). Development of androgen- and estrogen-responsive bioassays, members of a panel of human cell line-based highly selective steroid-responsive bioassays. *Toxicol Sci*, 83(1):136–48. doi:10.1093/toxsci/kfi005 PMID:15483189

Soto AM, Sonnenschein C, Chung KL, Fernandez MF, Olea N, Serrano FO (1995). The E-SCREEN assay as a tool to identify estrogens: an update on estrogenic environmental pollutants. *Environ Health Perspect*, 103:Suppl 7:113–22. doi:10.1289/ehp.95103s7113 PMID:8593856

Spinelli JJ, Ng CH, Weber JP, Connors JM, Gascoyne RD, Lai AS et al. (2007). Organochlorines and risk of non-Hodgkin lymphoma. *Int J Cancer*, 121(12):2767–75. doi:10.1002/ijc.23005 PMID:17722095

Steinmetz R, Young PC, Caperell-Grant A, Gize EA, Madhukar BV, Ben-Jonathan N et al. (1996). Novel estrogenic action of the pesticide residue beta-hexachlorocyclohexane in human breast cancer cells. *Cancer Res*, 56(23):5403–9. PMID:8968093

Straif K, Loomis D, Guyton K, Grosse Y, Lauby-Secretan B, El Ghissassi F et al. (2014). Future priorities for the IARC Monographs. *Lancet Oncol*, 15(7):683–4. doi:10.1016/S1470-2045(14)70168-8

Stuetz W, Prapamontol T, Erhardt JG, Classen HG (2001). Organochlorine pesticide residues in human milk of a Hmong hill tribe living in Northern Thailand. *Sci Total Environ*, 273(1–3):53–60. doi:10.1016/S0048-9697(00)00842-1 PMID:11419602

Sturgeon SR, Brock JW, Potischman N, Needham LL, Rothman N, Brinton LA et al. (1998). Serum concentrations of organochlorine compounds and endometrial cancer risk (United States). *Cancer Causes Control*, 9(4):417–24. doi:10.1023/A:1008823802393 PMID:9794174

Sudaryanto A, Kunisue T, Kajiwara N, Iwata H, Adibroto TA, Hartono P et al. (2006). Specific accumulation of organochlorines in human breast milk from Indonesia: levels, distribution, accumulation kinetics and infant health risk. *Environ Pollut*, 139(1):107–17. doi:10.1016/j.envpol.2005.04.028 PMID:15992976

Sugie S, Mori H, Takahashi M (1987). Effect of in vivo exposure to the liver tumor promoters phenobarbital or DDT on the gap junctions of rat hepatocytes: a quantitative freeze-fracture analysis. *Carcinogenesis*, 8(1):45–51. doi:10.1093/carcin/8.1.45 PMID:3802394

Sukata T, Uwagawa S, Ozaki K, Ogawa M, Nishikawa T, Iwai S et al. (2002). Detailed low-dose study of 1,1-bis(p-chlorophenyl)-2,2,2- trichloroethane carcinogenesis suggests the possibility of a hormetic effect. *Int J Cancer*, 99(1):112–8. doi:10.1002/ijc.10312 PMID:11948501

Swenberg JA (1981). Utilization of the alkaline elution assay as a short-term test for chemical carcinogens. In: Stich HF, San RHC editors. *Short-term tests for chemical carcinogens*. New York (NY), USA: Springer-Verlag; pp. 48–58. doi:10.1007/978-1-4612-5847-6_5

Swenberg JA, Petzold GL, Harbach PR (1976). In vitro DNA damage/alkaline elution assay for predicting carcinogenic potential. *Biochem Biophys Res Commun*, 72(2):732–8. doi:10.1016/S0006-291X(76)80100-3 PMID:985507

Takayama S, Sieber SM, Dalgard DW, Thorgeirsson UP, Adamson RH (1999). Effects of long-term oral administration of DDT on nonhuman primates. *J Cancer Res Clin Oncol*, 125(3–4):219–25. doi:10.1007/s004320050266 PMID:10235477

Takser L, Mergler D, Baldwin M, de Grosbois S, Smargiassi A, Lafond J (2005). Thyroid hormones in pregnancy in relation to environmental exposure to organochlorine compounds and mercury. *Environ Health Perspect*, 113(8):1039–45. doi:10.1289/ehp.7685 PMID:16079076

Tan J, Cheng SM, Loganath A, Chong YS, Obbard JP (2007). Selected organochlorine pesticide and polychlorinated biphenyl residues in house dust in Singapore. *Chemosphere*, 68(9):1675–82. doi:10.1016/j.chemosphere.2007.03.051 PMID:17490710

Tanaka T, Mori H, Williams GM (1987). Enhancement of dimethylnitrosamine-initiated hepatocarcinogenesis in hamsters by subsequent administration of carbon tetrachloride but not phenobarbital or p,p′-dichlorodiphenyltrichloroethane. *Carcinogenesis*, 8(9):1171–8. doi:10.1093/carcin/8.9.1171 PMID:3304690

Tao S, Yu Y, Liu W, Wang X, Cao J, Li B et al. (2008). Validation of dietary intake of dichlorodiphenyltrichloroethane and metabolites in two populations from Beijing and Shenyang, China based on the residuals in human milk. *Environ Sci Technol*, 42(20):7709–14. doi:10.1021/es801219v PMID:18983097

Tarján R, Kemény T (1969). Multigeneration studies on DDT in mice. *Food Cosmet Toxicol*, 7(3):215–22. doi:10.1016/S0015-6264(69)80325-1 PMID:5804863

Taylor KW, Novak RF, Anderson HA, Birnbaum LS, Blystone C, Devito M et al. (2013). Evaluation of the association between persistent organic pollutants (POPs) and diabetes in epidemiological studies: a National Toxicology Program workshop review. *Environ Health Perspect*, 121(7):774–83. doi:10.1289/ehp.1205502 PMID:23651634

Tebourbi O, Driss MR, Sakly M, Rhouma KB (2006). Metabolism of DDT in different tissues of young rats. *J Environ Sci Health B*, 41(2):167–76. doi:10.1080/03601230500364674 PMID:16393904

Tebourbi O, Hallègue D, Yacoubi MT, Sakly M, Rhouma KB (2010). Subacute toxicity of *p,p′*-DDT on rat thyroid: hormonal and histopathological changes. *Environ Toxicol Pharmacol*, 29(3):271–9. doi:10.1016/j.etap.2010.03.002 PMID:21787613

Teeyapant P, Ramchiun S, Polputpisatkul D, Uttawichai C, Parnmen S (2014). Serum concentrations of organochlorine pesticides p,p′-DDE in adult Thai residents with background levels of exposure. *J Toxicol Sci*, 39(1):121–7. doi:10.2131/jts.39.121 PMID:24418716

Telang S, Tong C, Williams G (1981). Induction of mutagenesis by carcinogenic polycyclic aromatic hydrocarbons but not by organochlorine pesticides in the ARL/HGPRT mutagenesis assay. *Environ Mutagen*, 3:359.

Terracini B, Cabral RJ, Testa MC (1973b). A multigeneration study on the effect of continuous administration of DDT to BALB/c mice. In: Deichmann WB editor. *Proceedings of the 8th Inter-American Conference on Toxicology: Pesticides and the environment, a continuing controversy.* Miami (FL), USA: Intercontinental Medical Book Corporation; pp. 77–85.

Terracini B, Testa MC, Carbral JR, Day N (1973a). The effects of long-term feeding of DDT to BALB-c mice. *Int J Cancer*, 11(3):747–64. doi:10.1002/ijc.2910110326 PMID:4364722

Tessier DM, Matsumura F (2001). Increased ErbB-2 tyrosine kinase activity, MAPK phosphorylation, and cell proliferation in the prostate cancer cell line LNCaP following treatment by select pesticides. *Toxicol Sci*, 60(1):38–43. doi:10.1093/toxsci/60.1.38 PMID:11222871

Thomas KW, Dosemeci M, Coble JB, Hoppin JA, Sheldon LS, Chapa G et al. (2010). Assessment of a pesticide exposure intensity algorithm in the Agricultural Health Study. *J Expo Sci Environ Epidemiol*, 20(6):559–69. doi:10.1038/jes.2009.54 PMID:19888312

Thomas P, Dong J (2006). Binding and activation of the seven-transmembrane estrogen receptor GPR30 by environmental estrogens: a potential novel mechanism of endocrine disruption. *J Steroid Biochem Mol Biol*, 102(1–5):175–9. doi:10.1016/j.jsbmb.2006.09.017 PMID:17088055

Thomas P, Pang Y, Filardo EJ, Dong J (2005). Identity of an estrogen membrane receptor coupled to a G protein in human breast cancer cells. *Endocrinology*, 146(2):624–32. doi:10.1210/en.2004-1064 PMID:15539556

Thorpe E, Walker AIT (1973). The toxicology of dieldrin (HEOD). II. Comparative long-term oral toxicity studies in mice with dieldrin, DDT, phenobarbitone, β-BHC and γ-BHC. *Food Cosmet Toxicol*, 11(3):433–42. doi:10.1016/0015-6264(73)90008-4 PMID:4125578

Tice RR, Austin CP, Kavlock RJ, Bucher JR (2013). Improving the human hazard characterization of chemicals: a Tox21 update. *Environ Health Perspect*, 121(7):756–65. doi:10.1289/ehp.1205784 PMID:23603828

Tiemann U, Schneider F, Tuchscherer A (1996). Effects of organochlorine pesticides on DNA synthesis of cultured oviductal and uterine cells and on estrogen receptor of uterine tissue from heifers. *Arch Toxicol*, 70(8):490–6. doi:10.1007/s002040050303 PMID:8783812

Tomatis L, Turusov V, Charles RT, Boicchi M (1974b). Effect of long-term exposure to 1,1-dichloro-2,2-bis(p-chlorophenyl)ethylene, to 1,1-dichloro-2,2-bis(p-chlorophenyl)ethane, and to the two chemicals combined on CF-1 mice. *J Natl Cancer Inst*, 52(3):883–91. PMID:4826570

Tomatis L, Turusov V, Charles RT, Boiocchi M, Gati E (1974a). Liver tumours in CF-1 mice exposed for limited periods to technical DDT. *Z Krebsforsch Klin Onkol Cancer Res Clin Oncol*, 82(1):25–35. doi:10.1007/BF00304381 PMID:4373976

Tomatis L, Turusov V, Day N, Charles RT (1972). The effect of long-term exposure to DDT on CF-1 mice. *Int J Cancer*, 10(3):489–506. doi:10.1002/ijc.2910100308 PMID:4358037

Tomita M, Yoshida T, Fukumori J, Yamaguchi S, Kojima S, Fukuyama T et al. (2013). *p,p′*-DDT induces microcytic anemia in rats. *J Toxicol Sci*, 38(5):775–82. doi:10.2131/jts.38.775 PMID:24067725

Tomiyama N, Takeda M, Watanabe M, Kobayashi H, Harada T (2004). A further study on the reliability of toxicokinetic parameters for predicting hepatotoxicity in rats receiving a 28-day repeated administration of DDT. *J Toxicol Sci*, 29(5):505–16. doi:10.2131/jts.29.505 PMID:15729006

Tomiyama N, Watanabe M, Takeda M, Harada T, Kobayashi H (2003). A comparative study on the reliablility of toxicokinetic parameters for predicting hepatotoxicity of DDT in rats receiving a single or repeated administration. *J Toxicol Sci*, 28(5):403–13. doi:10.2131/jts.28.403 PMID:14746344

Tong C, Fazio M, Williams GM (1981). Rat hepatocyte-mediated mutagenesis of human cells by carcinogenic polycyclic aromatic hydrocarbons but not organochlorine pesticides. *Proc Soc Exp Biol Med*, 167(4):572–5. doi:10.3181/00379727-167-41217 PMID:6269115

Toppari J, Larsen JC, Christiansen P, Giwercman A, Grandjean P, Guillette LJ Jr et al. (1996). Male reproductive health and environmental xenoestrogens. *Environ Health Perspect*, 104:Suppl 4: 741–803. doi:10.1289/ehp.96104s4741 PMID:8880001

Torres-Sanchez L, Zepeda M, Cebrián ME, Belkind-Gerson J, Garcia-Hernandez RM, Belkind-Valdovinos U et al. (2008). Dichlorodiphenyldichloroethylene exposure during the first trimester of pregnancy alters the anal position in male infants. *Ann N Y Acad Sci*, 1140(1):155–62. doi:10.1196/annals.1454.004 PMID:18991914

Toś-Luty S, Tokarska-Rodak M, Latuszyńska J, Przebirowska D (2002). Distribution of dermally absorbed 14c ddt in the organs of Wistar rats. Annals of agricultural and environmental medicine. *Ann Agric Environ Med*, 9:215–23. PMID:12498591

Trejo-Acevedo A, Rivero-Pérez NE, Flores-Ramírez R, Díaz-Barriga F, Ochoa Angeles AC, Pérez-Maldonado IN (2013). Assessment of persistent organic pollutants levels in blood samples from Quintana Roo, Mexico. *Int J Hyg Environ Health*, 216(3):284–9. doi:10.1016/j.ijheh.2012.09.004 PMID:23098827

Tsushimoto G, Chang CC, Trosko JE, Matsumura F (1983). Cytotoxic, mutagenic, and cell-cell communication inhibitory properties of DDT, lindane, and chlordane on Chinese hamster cells in vitro. *Arch Environ Contam Toxicol*, 12(6):721–9. doi:10.1007/BF01060757 PMID:6197034

Tully DB, Cox VT, Mumtaz MM, Davis VL, Chapin RE (2000). Six high-priority organochlorine pesticides, either singly or in combination, are nonestrogenic in transfected HeLa cells. *Reprod Toxicol*, 14(2):95–102. doi:10.1016/S0890-6238(00)00060-5 PMID:10825672

Turusov V, Rakitsky V, Tomatis L (2002). Dichlorodiphenyltrichloroethane (DDT): ubiquity, persistence, and risks. *Environ Health Perspect*, 110(2):125–8. doi:10.1289/ehp.02110125 PMID:11836138

Turusov VS, Day NE, Tomatis L, Gati E, Charles RT (1973). Tumors in CF-1 mice exposed for six consecutive generations to DDT. *J Natl Cancer Inst*, 51(3):983–97. PMID:4355226

Turyk ME, Anderson HA, Freels S, Chatterton R Jr, Needham LL, Patterson DG Jr et al.; Great Lakes Consortium (2006). Associations of organochlorines with endogenous hormones in male Great Lakes fish consumers and nonconsumers. *Environ Res*, 102(3):299–307. doi:10.1016/j.envres.2006.01.009 PMID:16563369

Turyk ME, Anderson HA, Persky VW (2007). Relationships of thyroid hormones with polychlorinated biphenyls, dioxins, furans, and DDE in adults. *Environ Health Perspect*, 115(8):1197–203. doi:10.1289/ehp.10179 PMID:17687447

Uchiyama M, Chiba T, Noda K (1974). Co-carcinogenic effect of DDT and PCB feeding on methylcholanthrene-induced chemical carcinogenesis. *Bull Environ Contam Toxicol*, 12(6):687–93. doi:10.1007/BF01685915 PMID:4218121

Udoji F, Martin T, Etherton R, Whalen MM (2010). Immunosuppressive effects of triclosan, nonylphenol, and DDT on human natural killer cells in vitro. *J Immunotoxicol*, 7(3):205–12. doi:10.3109/15476911003667470 PMID:20297919

Ulrich EM, Caperell-Grant A, Jung SH, Hites RA, Bigsby RM (2000). Environmentally relevant xenoestrogen tissue concentrations correlated to biological responses in mice. *Environ Health Perspect*, 108(10):973–7. doi:10.1289/ehp.00108973 PMID:11049819

UNEP (2002b). The Stockholm Convention on Persistent Organic Pollutants. Geneva, Switzerland: United Nations Environment Programme. Available from: www.pops.int.

UNEP (2010). Report of the Expert Group on the assessment of the production and use of DDT and its alternatives for disease vector control. Third meeting, Geneva, November 2010. Geneva, Switzerland: United Nations Environment Programme.

UNEP/WHO (2013). Report of the Conference of the Parties to the Stockholm Convention on Persistent Organic Pollutants on the work of its sixth meeting.

Geneva, 28 April–10 May 2013. No. UNEP/POPS/COP.6/INF/33.

UNEP/WHO (2014). Report of the DDT expert group on the assessment of the production and use of DDT and its alternatives for disease vector control. Stockholm Convention on Persistent Organic Pollutants. Geneva, Switzerland: United Nations Environment Programme and the World Health Organization; pp. 1–26.

UNEP/WHO (2015). Report of the Conference of the Parties to The Stockholm Convention On Persistent Organic Pollutants on the work of its seventh meeting. Geneva, 4–15 May 2015. No. UNEP/POPS/COP.7/36. Geneva, Switzerland: United Nations Environment Programme and the World Health Organization.

Uppala PT, Roy SK, Tousson A, Barnes S, Uppala GR, Eastmond DA (2005). Induction of cell proliferation, micronuclei and hyperdiploidy/polyploidy in the mammary cells of DDT- and DMBA-treated pubertal rats. *Environ Mol Mutagen*, 46(1):43–52. doi:10.1002/em.20131 PMID:15880734

USGS (2006). The quality of our nation's waters. Pesticides in the nation's streams and ground water, 1992–2001. USGS Circular 1291, Appendix 7A. Statistical summaries of pesticide compounds in stream water, 1992–2001. Reston (VA), USA: United States Geological Survey. Available from: http://water.usgs.gov/nawqa/pnsp/pubs/circ1291/appendix7/7a.html, accessed 2 June 2015.

Vaclavik E, Tjonneland A, Stripp C, Overvad K, Philippe Weber J, Raaschou-Nielsen O (2006). Organochlorines in Danish women: predictors of adipose tissue concentrations. *Environ Res*, 100(3):362–70. doi:10.1016/j.envres.2005.06.006 PMID:16125695

Vafeiadi M, Vrijheid M, Fthenou E, Chalkiadaki G, Rantakokko P, Kiviranta H et al. (2014). Persistent organic pollutants exposure during pregnancy, maternal gestational weight gain, and birth outcomes in the mother-child cohort in Crete, Greece (RHEA study). *Environ Int*, 64:116–23. doi:10.1016/j.envint.2013.12.015 PMID:24389008

Valencia R, Mason JM, Woodruff RC, Zimmering S (1985). Chemical mutagenesis testing in Drosophila. III. Results of 48 coded compounds tested for the National Toxicology Program. *Environ Mutagen*, 7(3):325–48. doi:10.1002/em.2860070309 PMID:3930234

van den Berg H (2009). Global status of DDT and its alternatives for use in vector control to prevent disease. *Environ Health Perspect*, 117(11):1656–63. doi:10.1289/ehp.0900785 PMID:20049114

van den Berg H, Zaim M, Yadav RS, Soares A, Ameneshewa B, Mnzava A et al. (2012). Global trends in the use of insecticides to control vector-borne diseases. *Environ Health Perspect*, 120(4):577–82. doi:10.1289/ehp.1104340 PMID:22251458

Van Dijck P, Van de Voorde H (1976). Mutagenicity versus carcinogenicity of organochlorine insecticides. *Med. Fac. Landbouau. Rijksuniv. Gent*, 41:1491–8.

Van Dyk JC, Bouwman H, Barnhoorn IE, Bornman MS (2010). DDT contamination from indoor residual spraying for malaria control. *Sci Total Environ*, 408(13):2745–52. doi:10.1016/j.scitotenv.2010.03.002 PMID:20381127

van Hemmen JJ (2001). EUROPOEM, a predictive occupational exposure database for registration purposes of pesticides. *Appl Occup Environ Hyg*, 16(2):246–50. doi:10.1080/104732201460406 PMID:11217718

van't Veer P, Lobbezoo IE, Martín-Moreno JM, Guallar E, Gómez-Aracena J, Kardinaal AF et al. (1997). DDT (dicophane) and postmenopausal breast cancer in Europe: case-control study. *BMJ*, 315(7100):81–5. doi:10.1136/bmj.315.7100.81 PMID:9240045

Venners SA, Korrick S, Xu X, Chen C, Guang W, Huang A et al. (2005). Preconception serum DDT and pregnancy loss: a prospective study using a biomarker of pregnancy. *Am J Epidemiol*, 162(8):709–16. doi:10.1093/aje/kwi275 PMID:16120699

Viel JF, Floret N, Deconinck E, Focant JF, De Pauw E, Cahn JY (2011). Increased risk of non-Hodgkin lymphoma and serum organochlorine concentrations among neighbors of a municipal solid waste incinerator. *Environ Int*, 37(2):449–53. doi:10.1016/j.envint.2010.11.009 PMID:21167603

Vine MF, Stein L, Weigle K, Schroeder J, Degnan D, Tse CKJ et al. (2000). Effects on the immune system associated with living near a pesticide dump site. *Environ Health Perspect*, 108(12):1113–24. doi:10.1289/ehp.001081113 PMID:11133390

Vizcaino E, Grimalt JO, Fernández-Somoano A, Tardon A (2014). Transport of persistent organic pollutants across the human placenta. *Environ Int*, 65:107–15. doi:10.1016/j.envint.2014.01.004 PMID:24486968

Vo TT, Gladen BC, Cooper GS, Baird DD, Daniels JL, Gammon MD et al. (2008). Dichlorodiphenyldichloroethane and polychlorinated biphenyls: intraindividual changes, correlations, and predictors in healthy women from the southeastern United States. *Cancer Epidemiol Biomarkers Prev*, 17(10):2729–36. doi:10.1158/1055-9965.EPI-08-0379 PMID:18843016

Vogel E (1972). [Investigations on the mutagenicity of DDT and the DDT-metabolites DDE, DDD, DDOM and DDA in Drosophila melanogaster.] *Mutat Res*, 16(2):157–64. doi:10.1016/0027-5107(72)90176-5 PMID:4627533

Volckens J, Leith D (2003). Partitioning theory for respiratory deposition of semivolatile aerosols. *Ann Occup Hyg*, 47(2):157–64. doi:10.1093/annhyg/meg015 PMID:12582000

Vongbuddhapitak A, Atisook K, Thoophom G, Sungwaranond B, Lertreungdej Y, Suntudrob J et al.

(2002). Dietary exposure of Thais to pesticides during 1989–1996. *J AOAC Int*, 85(1):134–40. PMID:11878592

Vukavić T, Vojinović Miloradov M, Mihajlović I, Ristivojević A (2013). Human milk POPs and neonatal risk trend from 1982 to 2009 in the same geographic region in Serbia. *Environ Int*, 54:45–9. doi:10.1016/j.envint.2013.01.008 PMID:23403145

Wakeling AE, Visek WJ (1973). Insecticide inhibition of 5alpha-dihydrotestosterone binding in the rat ventral prostate. *Science*, 181(4100):659–61. doi:10.1126/science.181.4100.659 PMID:4353358

Walker AIT, Thorpe E, Stevenson DE (1973). The toxicology of dieldrin (HEOD). I. Long-term oral toxicity studies in mice. *Food Cosmet Toxicol*, 11(3):415–32. doi:10.1016/0015-6264(73)90007-2 PMID:4353861

Wallace ME, Knights P, Dye AO (1976). Pilot study of the mutagenicity of DDT in mice. *Environ Pollut*, 11(3):217–22. doi:10.1016/0013-9327(76)90086-0

Wang H, Li J, Gao Y, Xu Y, Pan Y, Tsuji I et al. (2010). Xeno-oestrogens and phyto-oestrogens are alternative ligands for the androgen receptor. *Asian J Androl*, 12(4):535–47. doi:10.1038/aja.2010.14 PMID:20436506

Ward EM, Schulte P, Grajewski B, Andersen A, Patterson DG Jr, Turner W et al. (2000). Serum organochlorine levels and breast cancer: a nested case-control study of Norwegian women. *Cancer Epidemiol Biomarkers Prev*, 9(12):1357–67. PMID:11142422

Ward MH, Colt JS, Metayer C, Gunier RB, Lubin J, Crouse V et al. (2009). Residential exposure to polychlorinated biphenyls and organochlorine pesticides and risk of childhood leukemia. *Environ Health Perspect*, 117(6):1007–13. doi:10.1289/ehp.0900583 PMID:19590698

Warner M, Wesselink A, Harley KG, Bradman A, Kogut K, Eskenazi B (2014). Prenatal exposure to dichlorodiphenyltrichloroethane and obesity at 9 years of age in the CHAMACOS study cohort. *Am J Epidemiol*, 179(11):1312–22. doi:10.1093/aje/kwu046 PMID:24722999

Weiderpass E, Adami HO, Baron JA, Wicklund-Glynn A, Aune M, Atuma S et al. (2000). Organochlorines and endometrial cancer risk. *Cancer Epidemiol Biomarkers Prev*, 9(5):487–93. PMID:10815693

Welch RM, Levin W, Conney AH (1969). Estrogenic action of DDT and its analogs. *Toxicol Appl Pharmacol*, 14(2):358–67. doi:10.1016/0041-008X(69)90117-3 PMID:5772860

Welch RM, Levin W, Kuntzman R, Jacobson M, Conney AH (1971). Effect of halogenated hydrocarbon insecticides on the metabolism and uterotropic action of estrogens in rats and mice. *Toxicol Appl Pharmacol*, 19(2):234–46. doi:10.1016/0041-008X(71)90109-8 PMID:4105824

Wester RC, Maibach HI, Bucks DA, Sedik L, Melendres J, Liao C et al. (1990). Percutaneous absorption of [14C]DDT and [14C]benzo[a]pyrene from soil. *Fundam Appl Toxicol*, 15(3):510–6. doi:10.1016/0272-0590(90)90037-K PMID:2258015

White AJ, Teitelbaum SL, Wolff MS, Stellman SD, Neugut AI, Gammon MD (2013). Exposure to fogger trucks and breast cancer incidence in the Long Island Breast Cancer Study Project: a case-control study. *Environ Health*, 12(1):24. doi:10.1186/1476-069X-12-24 PMID:23497110

Whitworth KW, Bornman RM, Archer JI, Kudumu MO, Travlos GS, Wilson RE et al. (2014). Predictors of plasma DDT and DDE concentrations among women exposed to indoor residual spraying for malaria control in the South African Study of Women and Babies (SOWB). *Environ Health Perspect*, 122(6):545–52. PMID:24577839

WHO (1989). DDT and its derivatives - environmental aspects. Environ Health Criteria. No. 83. Geneva, Switzerland: World Health Organization. Available from: http://www.inchem.org/documents/ehc/ehc/ehc83.htm

WHO (2009). Specifications and evaluations for public health pesticides. DDT 1,1,1-trichloro-2,2-bis(chlorophenyl)ethane. Geneva, Switzerland: World Health Organization. Available from: http://www.who.int/whopes/quality/en/DDT_Aug_09.pdf?ua=1

WHO (2011a). The use of DDT in malaria vector control. WHO position statement. Geneva, Switzerland: Global Malaria Programme, World Health Organization. Available from: http://whqlibdoc.who.int/hq/2011/WHO_HTM_GMP_2011_eng.pdf

WHO (2011b). *Global insecticide use for vector-borne disease control. A 10-year assessment (2000–2009)*. 5th ed. Geneva, Switzerland: World Health Organization; pp. 1–33.

Williams GM, Laspia MF, Dunkel VC (1982). Reliability of the hepatocyte primary culture/DNA repair test in testing of coded carcinogens and noncarcinogens. *Mutat Res*, 97(5):359–70. doi:10.1016/0165-1161(82)90003-6 PMID:7144803

Williams GM, Numoto S (1984). Promotion of mouse liver neoplasms by the organochlorine pesticides chlordane and heptachlor in comparison to dichlorodiphenyltrichloroethane. *Carcinogenesis*, 5(12):1689–96. doi:10.1093/carcin/5.12.1689 PMID:6499120

Wójtowicz AK, Augustowska K, Gregoraszczuk EL (2007b). The short- and long-term effects of two isomers of DDT and their metabolite DDE on hormone secretion and survival of human choriocarcinoma JEG-3 cells. *Pharmacol Rep*, 59(2):224–32. PMID:17556801

Wójtowicz AK, Honkisz E, Zięba-Przybylska D, Milewicz T, Kajta M (2011). Effects of two isomers of DDT and their metabolite DDE on CYP1A1 and AhR function in human placental cells. *Pharmacol Rep*, 63(6):1460–8. doi:10.1016/S1734-1140(11)70710-1 PMID:22358094

Wójtowicz AK, Kajta M, Gregoraszczuk EL (2007a). DDT- and DDE-induced disruption of ovarian steroidogenesis in prepubertal porcine ovarian follicles: a possible interaction with the main steroidogenic enzymes and estrogen receptor beta. *J Physiol Pharmacol*, 58(4):873–85. PMID:18195494

Wójtowicz AK, Milewicz T, Gregoraszczuk EL (2007c). DDT and its metabolite DDE alter steroid hormone secretion in human term placental explants by regulation of aromatase activity. *Toxicol Lett*, 173(1):24–30. doi:10.1016/j.toxlet.2007.06.005 PMID:17681675

Wolff MS, Anderson HA, Britton JA, Rothman N (2007). Pharmacokinetic variability and modern epidemiology–the example of dichlorodiphenyltrichloroethane, body mass index, and birth cohort. *Cancer Epidemiol Biomarkers Prev*, 16(10):1925–30. doi:10.1158/1055-9965.EPI-07-0394 PMID:17932339

Wolff MS, Berkowitz GS, Brower S, Senie R, Bleiweiss IJ, Tartter P et al. (2000a). Organochlorine exposures and breast cancer risk in New York City women. *Environ Res*, 84(2):151–61. doi:10.1006/enrs.2000.4075 PMID:11068929

Wolff MS, Britton JA, Teitelbaum SL, Eng S, Deych E, Ireland K et al. (2005). Improving organochlorine biomarker models for cancer research. *Cancer Epidemiol Biomarkers Prev*, 14(9):2224–36. doi:10.1158/1055-9965.EPI-05-0173 PMID:16172236

Wolff MS, Toniolo PG, Lee EW, Rivera M, Dubin N (1993). Blood levels of organochlorine residues and risk of breast cancer. *J Natl Cancer Inst*, 85(8):648–52. doi:10.1093/jnci/85.8.648 PMID:8468722

Wolff MS, Zeleniuch-Jacquotte A, Dubin N, Toniolo P (2000b). Risk of breast cancer and organochlorine exposure. *Cancer Epidemiol Biomarkers Prev*, 9(3):271–7. PMID:10750665

Wong CK, Leung KM, Poon BH, Lan CY, Wong MH (2002). Organochlorine hydrocarbons in human breast milk collected in Hong Kong and Guangzhou. *Arch Environ Contam Toxicol*, 43(3):364–72. doi:10.1007/s00244-002-1210-7 PMID:12202934

Wong F, Alegria HA, Bidleman TF, Alvarado V, Angeles F, Galarza AA et al. (2009a). Passive air sampling of organochlorine pesticides in Mexico. *Environ Sci Technol*, 43(3):704–10. doi:10.1021/es802385j PMID:19245005

Wong F, Robson M, Diamond ML, Harrad S, Truong J (2009b). Concentrations and chiral signatures of POPs in soils and sediments: a comparative urban versus rural study in Canada and UK. *Chemosphere*, 74(3):404–11. doi:10.1016/j.chemosphere.2008.09.051 PMID:19022474

Wong MH, Leung AO, Chan JK, Choi MP (2005). A review on the usage of POP pesticides in China, with emphasis on DDT loadings in human milk. *Chemosphere*, 60(6):740–52. doi:10.1016/j.chemosphere.2005.04.028 PMID:15949838

Woodard G, Ofner RR, Montgomery CM (1945). Accumulation of DDT in the body fat and its appearance in the milk of dogs. *Science*, 102(2642):177–8. doi:10.1126/science.102.2642.177-a PMID:17844226

Woodruff RC, Phillips JP, Irwin D (1983). Pesticide-induced complete and partial chromosome loss in screens with repair-defective females of Drosophila melanogaster. *Environ Mutagen*, 5(6):835–46. doi:10.1002/em.2860050608 PMID:6418539

Woods JS, Polissar L, Severson RK, Heuser LS, Kulander BG (1987). Soft tissue sarcoma and non-Hodgkin's lymphoma in relation to phenoxyherbicide and chlorinated phenol exposure in western Washington. *J Natl Cancer Inst*, 78(5):899–910. PMID:3471999

Woolcott CG, Aronson KJ, Hanna WM, SenGupta SK, McCready DR, Sterns EE et al. (2001). Organochlorines and breast cancer risk by receptor status, tumor size, and grade (Canada). *Cancer Causes Control*, 12(5):395–404. doi:10.1023/A:1011289905751 PMID:11545454

Woolley DE, Talens GM (1971). Distribution of DDT, DDD, and DDE in tissues of neonatal rats and in milk and other tissues of mother rats chronically exposed to DDT. *Toxicol Appl Pharmacol*, 18(4):907–16. doi:10.1016/0041-008X(71)90238-9 PMID:5570242

Worthing CR, Walker SR editors. (1987). *The pesticide manual: a world compendium*. 8th ed. Thornton Heath, United Kingdom: British Crop Protection Council; pp. 231–2.

Wrobel MH, Mlynarczuk J, Kotwica J (2012). The effect of DDT and its metabolite (DDE) on prostaglandin secretion from epithelial cells and on contractions of the smooth muscle of the bovine oviduct in vitro. *Toxicol Appl Pharmacol*, 259(2):152–9. doi:10.1016/j.taap.2011.12.019 PMID:22230338

Wu XJ, Lu WQ, Mersch-Sundermann V (2003). Benzo(a)pyrene induced micronucleus formation was modulated by persistent organic pollutants (POPs) in metabolically competent human HepG2 cells. *Toxicol Lett*, 144(2):143–50. doi:10.1016/S0378-4274(03)00198-X PMID:12927358

Wyde ME, Bartolucci E, Ueda A, Zhang H, Yan B, Negishi M et al. (2003). The environmental pollutant 1,1-dichloro-2,2-bis (p-chlorophenyl)ethylene induces rat hepatic cytochrome P450 2B and 3A expression through the constitutive androstane receptor and pregnane X receptor. *Mol Pharmacol*, 64(2):474–81. doi:10.1124/mol.64.2.474 PMID:12869653

Xin J, Liu X, Liu W, Jiang L, Wang J, Niu J (2011). Production and use of DDT containing antifouling paint resulted in high DDTs residue in three paint factory sites and two shipyard sites, China. *Chemosphere*, 84(3):342–7. doi:10.1016/j.chemosphere.2011.04.005 PMID:21550629

Xu B, Xiao X, Webber RH, Lines JD (1998). Comparison of the effect of insecticide-treated bed nets and DDT residual spraying on the prevalence of malaria transmitted by Anopheles anthropophagus in China. *Trans*

R Soc Trop Med Hyg, 92(2):135–6. doi:10.1016/S0035-9203(98)90719-2 PMID:9764314

Xu X, Dailey AB, Talbott EO, Ilacqua VA, Kearney G, Asal NR (2010). Associations of serum concentrations of organochlorine pesticides with breast cancer and prostate cancer in U.S. adults. *Environ Health Perspect*, 118(1):60–6. PMID:20056587

Xu XJ, Su JG, Bizzarri AR, Cannistraro S, Liu M, Zeng Y et al. (2013). Detection of persistent organic pollutants binding modes with androgen receptor ligand binding domain by docking and molecular dynamics. *BMC Struct Biol*, 13(1):16 doi:10.1186/1472-6807-13-16 PMID:24053684

Yalçin SS, Orun E, Yalcin S, Aykut O (2014). Organochlorine pesticide residues in breast milk and maternal psychopathologies and infant growth from suburban area of Ankara, Turkey. *Int J Environ Health Res*, 25(4):364–72. PMID:25155352

Yáñez L, Borja-Aburto VH, Rojas E, de la Fuente H, González-Amaro R, Gómez H et al. (2004). DDT induces DNA damage in blood cells. Studies in vitro and in women chronically exposed to this insecticide. *Environ Res*, 94(1):18–24. doi:10.1016/S0013-9351(03)00047-1 PMID:14643282

Yoder J, Watson M, Benson WW (1973). Lymphocyte chromosome analysis of agricultural workers during extensive occupational exposure to pesticides. *Mutat Res*, 21(6):335–40. doi:10.1016/0165-1161(73)90057-5 PMID:4779319

You L, Casanova M, Archibeque-Engle S, Sar M, Fan LQ, Heck HA (1998). Impaired male sexual development in perinatal Sprague-Dawley and Long-Evans hooded rats exposed in utero and lactationally to *p,p′*-DDE. *Toxicol Sci*, 45(2):162–73. PMID:9848123

You L, Sar M, Bartolucci E, Ploch S, Whitt M (2001). Induction of hepatic aromatase by *p,p′*-DDE in adult male rats. *Mol Cell Endocrinol*, 178(1–2):207–14. doi:10.1016/S0303-7207(01)00445-2 PMID:11403911

Yu Y, Li C, Zhang X, Zhang X, Pang Y, Zhang S et al. (2012). Route-specific daily uptake of organochlorine pesticides in food, dust, and air by Shanghai residents, China. *Environ Int*, 50:31–7. doi:10.1016/j.envint.2012.09.007 PMID:23063733

Yu Y, Wang B, Wang X, Liu W, Cao J, Wong M et al. (2013). Temporal trends in daily dietary intakes of DDTs and HCHs in urban populations from Beijing and Shenyang, China. *Chemosphere*, 91(10):1395–400. doi:10.1016/j.chemosphere.2012.12.073 PMID:23427859

Zava DT, Blen M, Duwe G (1997). Estrogenic activity of natural and synthetic estrogens in human breast cancer cells in culture. *Environ Health Perspect*, 105:Suppl 3: 637–45. doi:10.1289/ehp.97105s3637 PMID:9168008

Zhang H, Liu L, Zhang P, Zhao Y, Wu X, Ni W (2013). [A case-control study on the relationship between organochlorine and female breast cancer.] *Wei Sheng Yan Jiu*, 42(1):44–8. [Chinese] PMID:23596706

Zhao B, Shen H, Liu F, Liu S, Niu J, Guo F et al. (2012). Exposure to organochlorine pesticides is an independent risk factor of hepatocellular carcinoma: a case-control study. *J Expo Sci Environ Epidemiol*, 22(6):541–8. doi:10.1038/jes.2011.29 PMID:21915153

Zhao Z, Zhang L, Wu J, Fan C (2009). Distribution and bioaccumulation of organochlorine pesticides in surface sediments and benthic organisms from Taihu Lake, China. *Chemosphere*, 77(9):1191–8. doi:10.1016/j.chemosphere.2009.09.022 PMID:19819519

Zheng T, Holford TR, Mayne ST, Owens PH, Ward B, Carter D et al. (1999). Beta-benzene hexachloride in breast adipose tissue and risk of breast carcinoma. *Cancer*, 85(10):2212–8. doi:10.1002/(SICI)1097-0142(19990515)85:10<2212::AID-CNCR16>3.0.CO;2-F PMID:10326700

Zheng T, Holford TR, Mayne ST, Tessari J, Ward B, Carter D et al. (2000). Risk of female breast cancer associated with serum polychlorinated biphenyls and 1,1-dichloro-2,2′-bis(p-chlorophenyl)ethylene. *Cancer Epidemiol Biomarkers Prev*, 9(2):167–74. PMID:10698477

Zhou P, Zhao Y, Li J, Wu G, Zhang L, Liu Q et al. (2012). Dietary exposure to persistent organochlorine pesticides in 2007 Chinese total diet study. *Environ Int*, 42:152–9. doi:10.1016/j.envint.2011.05.018 PMID:21715010

Zimmermann E, Pedersen JO, Saraubon K, Tjell JC, Prapamontol T (2005). DDT in human milk from Chiang Mai mothers: a public health perspective on infants' exposure. *Bull Environ Contam Toxicol*, 74(2):407–14. doi:10.1007/s00128-004-0599-2 PMID:15841985

LINDANE

1. Exposure Data

1.1 Identification of the agent

The carcinogenicity of halogenated aromatic hydrocarbon pesticides such as lindane has been evaluated previously in Volume 5, and again in Supplement 7 and Volumes 20 and 53 (IARC, 1974, 1979b, 1987, 1991). New data on lindane have since become available, and have been incorporated into the present monograph and taken into consideration in the evaluation.

1.1.1 Nomenclature

Chem. Abstr. Serv. Reg. No.: 58-89-9

Chem. Abstr. Serv. Name:
lα,2α,3β,4α,5α,6β-Hexachlorocyclohexane

Preferred IUPAC Name:
1,2,3,4,5,6-Hexachlorocyclohexane

Synonyms: γ-Benzene hexachloride; γ-hexachlorocyclohexane; lindane

[The Working Group noted that the terms "γ-BHC" and "γ-lindane" have been used historically, but are incorrect.]

Trade Names: Lindane has been used in formulations for many commercial products. The trade names listed here are presented as examples and are not intended to represent an exhaustive list, or to focus on any particular manufacturer or user: Aalindan; Agrocide; Aparasin; Aphtiria; Ben-Hex; Bexol; Celanex; Chlorosene; Gammalin; Gamene Gammexane; Gexane; Hexide: Hortex; Lindator; Lindex; Nexit; Pflanzol; Quellada.

Additional trade names are available in the PubChem Compound Database (PubChem Compound Identifier 727; NCBI, 2015).

1.1.2 Structural and molecular formulae, and relative molecular mass

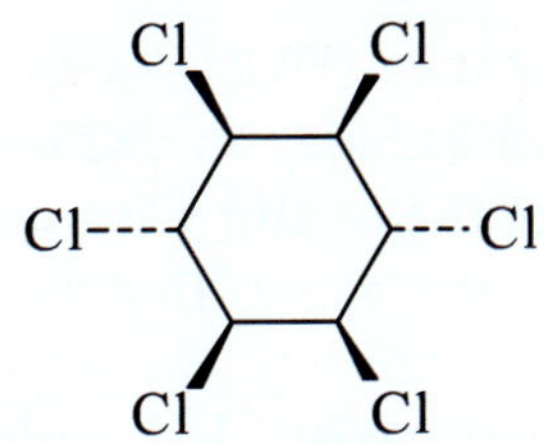

From RSC (2015)

Molecular formula: $C_6H_6Cl_6$ or $ClCH(CHCl)_4CHCl$

Isomers differ in the spatial positions of the chlorine atoms; dashed and solid bridges represent positions below and above the plane, respectively (RSC, 2015)

Relative molecular mass: 290.81

Additional chemical structure information is available in the PubChem Compound Database (NCBI, 2015) and the Merck Index Online (RSC, 2015).

1.1.3 Chemical and physical properties of the pure substance

Description: White crystalline powder (IPCS/ILO, 2009)

Solubility: Volatile in air and poorly soluble in water (g/100 mL at 20 °C, 0.0007) (IPCS, ILO-ICSC 0053). Soluble in ethyl alcohol and ethyl ether, benzene, chloroform (NCBI, 2015)

Octanol/water partition coefficient: log P_{ow}, 3.61–3.72 (IPCS, ILO/ICSC 0053; IPCS/ILO, 2009)

Conversion factor in air (at 25 °C): 1 ppm = 11.89 mg/m^3, assuming normal temperature (25 °C) and pressure (101 kPa) (EPA, 2015a).

1.2 Production and use

Hexachlorocyclohexane (HCH) was first synthesized by Michael Faraday in 1825. After the discovery in 1912 of the δ- and γ-isomers by Teunis Van der Linden, the name "lindane" was given to the γ-isomer.

Many HCH isomers exist, but only six isomers are relatively stable, including α-, β-, γ-, δ-, and ε-isomers. Only γ-HCH has insecticidal properties (Brooks, 1977). Both technical-grade lindane that contains more than 90% γ-HCH (IRPTC, 1983) and technical-grade HCH that contains approximately 60–70% α-HCH, 5–12% β-HCH, 10–40% γ-HCH, 6–10% δ-HCH, and 3–4% ε-HCH (Kutz et al., 1991), have been used worldwide for the insecticidal properties of the γ-isomer. The International Organization for Standardization (ISO) common name and the WHO specification use "lindane" to refer to material containing > 99% γ-HCH.

Table 1.1 Production of lindane

Production	Tonnes per year
World production, 1986[a]	Approx. 38 000
World production, 1988–1993[b]	4 400
World production, 1990–1995	3 222
Production in western Europe, 1990s	Approx. 2 055
Production in the European Economic Community, 1991[c]	1 000–5 000

[a] International Uniform Chemical Information Database (UCLID, 1994)
[b] Detzel et al. (1998)
[c] Rippen (1990/2000)
From UNECE (2004); data from Hazard Substances Committee (OSPAR Commission, 2002)

1.2.1 Production

Technical-grade HCH is produced as a mixture of isomers by photochlorination of benzene, a reaction that can be started by free-radical initiators such as visible or ultraviolet light, X-rays, or γ-rays (ATSDR, 2005). The active γ-HCH (lindane) can be concentrated by treatment with methanol or acetic acid, followed by fractional crystallization, to produce technical-grade lindane containing 99.9% of the γ-isomer.

Historical sites for the production of technical-grade HCH and/or lindane can be found in many countries in Europe, including Bulgaria, Czechia, France, Germany, Italy, Poland, Romania, Spain, Turkey, and the United Kingdom. Production in countries that are members of the United Nations Economic Commission for Europe (UNECE) took mainly place from 1950 or earlier until 1970, and stopped from 1970 onward. Only rough estimates on years of production and on produced volumes were available to the Working Group (Table 1.1; UNECE, 2004).

Commercial production of lindane in the United States of America (USA) began in 1945 and peaked in the 1950s, when 17.6 million pounds [7983 tonnes] was manufactured (IARC, 1974). In 1978, the United States Environmental

Protection Agency (EPA) banned production of technical HCH (UNECE, 2004).

Lindane is produced by 13 manufacturers worldwide, including 7 in India and 4 in China (SRI, 2009), and is available from 42 suppliers, including 19 suppliers in the USA (ChemSources, 2009).

1.2.2 Use

Lindane was extensively used in the past few decades. In 1992, global use of technical-grade HCH was estimated at 0.55 million tonnes, while global use of lindane was estimated at 0.72 million tonnes (Volder & Li, 1995). Technical-grade HCH and lindane were primarily used as insecticides to treat wood and wooden structures, seed, and livestock. Major uses today are as insecticide for fruit and vegetable crops, and in baits and seed treatments for rodent control. Lindane is still used as a human health pharmaceutical as a second-line treatment for control of head lice and scabies (mites) ; it is available in 1% preparations as a lotion, cream, or shampoo (ATSDR, 2005; HSDB, 2009; UNEP/WHO, 2015).

1.3 Measurement and analysis

Methods for the analysis of lindane and other organochlorine pesticides have been developed during the past few decades. Historically, lindane and other HCH isomers have been analysed using gas chromatography and electron capture detection (Prapamontol & Stevenson, 1991; López et al., 2001). In most of the methods currently available, lindane is part of multiple component analytical methods for the measurement of organochlorines and other pesticides in ambient air (Borrás et al., 2011), dust (Regueiro et al., 2007), water (McManus et al., 2013; Regueiro et al., 2008), sediments (Concha-Graña et al., 2010), crops (Walorczyk et al., 2013), human serum (López et al., 2001; Moreno Frías et al., 2001), and urine (Cazorla-Reyes et al., 2011). The analysis of lindane and other HCH isomers involves extraction, clean-up, and gas chromatography or gas chromatography-mass spectrometry-based instrumental analysis (representative analytical methods for lindane in various matrices are listed in Table 1.2).

Extraction methods for organochlorine compounds including lindane vary widely according to the matrix of interest and the sampling method. For example, water samples can be extracted on a rotating disk sorptive extraction technique (Giordano et al., 2011). For air samples collected with a polyurethane foam sampler, various sorbents have been used for solid-phase extraction (Martínez Vidal et al., 1997). Gaseous and particulate phases of lindane in the atmosphere can be extracted on XAD-2 and XAD-4 cartridges and glass fibre filters, respectively (Borrás et al., 2011). In vegetables, fruits, and other plant samples, lindane can be extracted by matrix solid-phase dispersion (Abhilash et al., 2007). Lindane in human serum samples can be extracted with organic solvents, clean-up of the organic extract using acid treatment with sulfuric acid, and elution of the cleaned-up extract by liquid column chromatography (Moreno Frías et al., 2001).

A concern with gas chromatography and electron-capture detection methods is the potential for interference from non-target chemicals, leading to misidentification or incorrect quantitation. More recently, gas chromatography-mass spectrometry methods are being used, which include single quadruple mass-spectrometry detectors running in electron ionization mode with target analyses monitored by selective ion monitoring and gas chromatography coupled with high-resolution mass spectrometry (Barr et al., 2003). These detection methods increase confidence in confirmative analysis by decreasing matrix interferences, improving selectivity.

Table 1.2 Representative methods for the analysis of lindane

Sample matrix	Assay procedure	Limit of detection	Reference
Ambient air	GC-MS		Borrás et al. (2011)
- Gas phase		25.79 pg/m^3	
- Particulate		0.052 pg/m^3	
Indoor dust	GC-μECD	0.22 ng/g	Regueiro et al. (2007)
Water	GC-MS	15 ng/L (LOQ)	McManus et al. (2013)
	GC-MS	21 ng/L	Regueiro et al. (2008)
Sediments	GC-MS	6.10 ng/g	Concha-Graña et al. (2010)
Crops	GC-QqQ-MS/MS	10 ng/g	Walorczyk et al. (2013)
Human serum	GC-ECD	3 ng/mL	López et al. (2001)
		10 ng/mL (β-HCH)	
	GC-ECD	0.006 ng/mL	Moreno Frías et al. (2001)
Urine	GC-IT-MS/MS	0.136 ng/mL	Cazorla-Reyes et al. (2011)

ECD, electron-capture detection; GC, gas chromatography; HCH, hexachlorocyclohexane; IT-MS/MS, ion trap mass spectrometry; LOQ, limit of quantitation; MS, mass spectrometry; QqQ-MS/MS, tandem quadrupole mass spectrometry

1.4 Occurrence and exposure

See Table 1.3, Table 1.4, and Table 1.5

Exposure to lindane is predominantly an exposure to the γ-HCH isomer, which has a very short half-life. Pure and technical forms of lindane exist, with almost pure lindane being γ-HCH, while technical-grade HCH consists of 10–40% γ-HCH plus various other isomers, including β-HCH.

In the present monograph, the terms given to lindane and its isomers are reported as specified by the authors of the referenced paper (e.g. authors may refer to "total HCH" when reporting the summed concentrations of various HCH isomers, or simply to "HCH").

1.4.1 Occupational exposure

Occupational exposure can occur during the manufacture and formulation of lindane, as well as in the treatment of wood and wooden structures, seed grains and in the agricultural application of lindane as a pesticide on livestock and crops (ATSDR, 2005). Exposure in occupational settings is principally through inhalation or dermal contact, although ingestion of lindane due to poor hygiene practices can occur. Hygiene measurement of lindane in air and hand wipes, together with biological measurement of lindane in serum and hair in workers from a range of occupations in the USA, Europe, and Asia have been reported (Table 1.3).

Studies of occupational exposure during the manufacture and formulation of lindane have reported detectable serum concentrations of γ-HCH and β-HCH (Baumann et al., 1980; Kashyap, 1986; Nigam et al., 1986). Baumann et al. reported mean concentrations of γ-HCH and β-HCH of 0.037 mg/L and 0.190 mg/L respectively among workers producing lindane in Germany (Baumann et al., 1980). Mean concentrations of γ-HCH and β-HCH in exposed workers involved in pesticide formulation in India were 0.06 mg/L and 0.413 mg/L respectively (Kashyap, 1986). The mean concentration of lindane in workers directly involved in handling the product in a manufacturing plant in India was 0.057 ppm (Nigam et al., 1986). Farmers in Nigeria were reported to have mean serum concentrations of lindane of 0.08 mg/kg (Sosan et al., 2008).

In commercial seed-conditioning plants in Montana, USA, in 1981–82, lindane levels in hand-wash samples and respiratory pads were

Table 1.3 Occupational exposure to lindane: environmental and biological measurements

Industry Location, year	Job/process	Sampling matrix	Exposure		Comments	References
			Mean	Range		
Lindane production Country and year, NR	NR	Serum	0.37 mg/L (arithmetic mean, γ-HCH)	0.005–0.19 mg/L	α-HCH: mean, 0.07; range, 0.01–0.27 mg/L β-HCH: mean, 0.19; range, 0.02–0.76 mg/L	Baumann et al. (1980)
Seed conditioning Montana, USA, 1981–1982	NR	Handshake sample; respiratory pads	NR	NR	Dermal exposure: site 9: 81.42 mg/hour; site 10: 54.80 mg/hour Respiratory exposure: site 9: 0.36 mg/hour; site 10: 0.54 mg/hour	Grey et al. (1983)
Pesticide manufacture Country and year, NR	Insecticide handlers	Serum	0.06 ppm (arithmetic mean, γ-HCH)	0.01–0.17 ppm	α-HCH: mean, 0.10; range, 0.02–0.18 ppm β-HCH: mean, 0.41; range, 0.16–0.72 ppm	Nigam et al. (1986)
Pesticide manufacture and formulation Country and year, NR	Employees involved in pesticide manufacture of organophosphates, HCH/BHC	Serum	Arithmetic mean (γ-HCH): Formulators: 0.06 mg/L; manufacturers: 0.02 mg/L; controls: 0.001 mg/L	Range (γ-HCH): Formulators: 0.01–0.17 mg/L; manufacturers: 0.0–0.04 mg/L; controls: 0.0–0.01 mg/L	Formulators mean; range (mg/L) α-HCH: mean, 0.10; range, 0.02–0.18 mg/L; β-HCH; mean, 0.41; range, 0.16–0.72 mg/L; Total HCH: mean, 0.60; range, 0.19–1.15 mg/L Biological levels also reported for manufacturing staff, and controls for other isomers and total HCH	Kashyap (1986)
Forestry workers Country NR, 1986		Plasma	NR	NR	Workers monitored for 20 weeks. Detected group mean of γ-HCH (nmol/L): 0 (April–June); 40 (mid July); 16 (August)	Drummond et al. (1988)
Agricultural spraymen and controls Allahabad, India, year NR	NR	Serum	Arithmetic mean (mg/L): Sprayers: γ-HCH: 0.03; β-HCH: 0.22; total HCH: 0.29; Controls: γ-HCH: 0.02; β-HCH: 0.11; total HCH: 0.15	Sprayers (mg/L) γ-HCH: 0.00–0.09; β-HCH: 0.02–0.95; total-HCH: 0.04–1.04 Controls (mg/L) γ-HCH: 0.00–0.71; β-HCH: 0.01–0.69; total HCH: 0.03–0.74	125 exposed sprayers and 47 controls	Joshi et al. (1996)

Table 1.3 (continued)

Industry Location, year	Job/process	Sampling matrix	Exposure		Comments	References
			Mean	Range		
Greenhouses, florists, veterinary departments Paris and suburbs, France, 2002		Air	14.32 ng/m^3 (arithmetic mean, lindane)	0.2–75 ng/m^3	Mean and range calculated from 10 measurements of indoor air	Bouvier et al. (2006)
		Hand wipes	22.76 ng/hand (arithmetic mean, lindane)	0–156.7 ng/hand	Mean and range calculated from 15 measurements	
Greenhouse workers, animal breeders, open cultivation workers Messara and Sitia districts, Crete, Greece, year NR		Hair	70.2 pg/mg (median, lindane)	48.2–95.0 pg/mg	Maximum level was 174.7 pg/mg	Tsatsakis et al. (2008)
Farmers Osun and Ondo states of south-western Nigeria, year NR	Cacao farmers; pesticide application	Serum	0.08 mg/kg (arithmetic mean, lindane)	NR	44 out of 76 farmers had lindane residue measurable in serum samples	Sosan et al. (2008)

HCH, hexachlorocyclohexane; NR, not reported

measured as 81.42 or 54.80 mg/hour and 0.36 or 0.54 mg/hour, respectively (Grey et al., 1983).

In France in 2002, mean lindane levels in indoor-air and hand-wipe samples from greenhouse workers, florists and veterinary-department workers were calculated as 14.32 ng/m^3 and 22.76 ng/hand, respectively (Bouvier et al., 2006). Hair samples from greenhouse workers and farmers in Greece contained lindane at concentrations in the range of 48.2–95.0 pg/mg (Tsatsakis et al., 2008).

Biological levels of lindane have been shown to decrease with reduced occupational exposure. Plasma concentrations of lindane in forestry workers monitored for 20 weeks starting in April 1986 rose from a group mean of zero to 40 nmol/L after 8 weeks, and dropped to 16 nmol/L after 16 weeks (Drummond et al., 1988). The difference in mean serum concentrations between non-exposed and exposed Indian agricultural sprayers was 0.01 mg/L for γ-HCH and 0.12 mg/L for β-HCH (Joshi et al., 1996).

1.4.2 Environmental occurrence

See Table 1.4

Lindane does not occur naturally in the environment. Occurrence may be due to pesticide application processes, release from manufacturing sites or landfills, or to precipitation. Biodegradation is the dominant degradative process for γ-HCH in aquatic systems and soil (ATSDR, 2005).

(a) Air

Historically, air contamination with lindane would have resulted from pesticide application and release from production plants (ATSDR, 2005). Weekly air samples in 1972, 1973, and 1974 taken in Stoneville, Mississippi, USA, contained a maximum lindane concentration of 9.3 ng/m^3 (Arthur et al., 1976). Around 1990, background levels of lindane in the range of 0.01–0.7 ng/m^3 were found in "unpolluted" remote areas, whereas levels in urban and agricultural areas range from 0.1 to 2 ng/m^3 (WHO, 2004). Air monitoring in Ontario, Canada, between 1988 and 1989 showed annual mean levels of β-HCH and γ-HCH of 1.8 and 60 pg/m^3, respectively (Hoff et al., 1992a).

Measurement of background levels of persistent organic pollutants at 71 sites in Europe in 2006 reported mean concentrations of γ-HCH and β-HCH of 35 and 2 pg/m^3, respectively. The presence of γ-HCH was attributed to either technical-grade HCH or lindane (Halse et al., 2011). Passive air sampling in 22 European countries in 2002 reported γ-HCH at concentrations in the range of 1.1–65 pg/m^3 with thehighest concentrations recorded in southern and eastern Europe, in particular, Spain, parts of France, Italy, and the Balkans region (Jaward et al., 2004).

Average annual concentrations in air for γ-HCH, β-HCH, α-HCH, and δ-HCH were reported between 1999 and 2001 in Niigata, Japan, as 32, 23, 92, and 3 × 10^{-3} ng/m^3, respectively (Murayama et al., 2003). A review of various studies on persistent organic pollutants in South China reported mean HCH concentrations of 666 × 10^{-3} ng/m^3 in 2003–2004, decreasing to a mean of 75 × 10^{-3} ng/m^3 in 2006–2007 (Zhang et al., 2013). In Shanghai, China, in 2008–2009, the mean concentration of HCH in air samples was measured as 6.93 × 10^{-3} ng/m^3 (Yu et al., 2012).

The Global Atmospheric Passive Sampling (GAPS) study report that between 2005 and 2008, distinct spatial and temporal patterns show that pesticides such as γ-HCH tend to be more prevalent in developing countries, especially in Asia. Samples taken over Delhi, India, had the highest levels of γ-HCH. In Europe, the levels of γ-HCH are not uniformly distributed, with samples from Paris having the highest levels. Levels of γ-HCH are not very high in North America and South Ameria, which may reflect decreased usage (Fig. 1.1; Shunthirasingham et al., 2010).

Table 1.4 Environmental exposure to lindane

Region, country, city Year	Sampling matrix	Exposure		Comments	Reference
		Mean	Range		
USA Mississippi, 1972–74	Air (weekly sampling)	NR	NR	Maximum level: lindane, 9.3 ng/m³ β-HCH, 49.4 ng/m³	Arthur et al. (1976)
Canada Ontario, 1988–1989	Air	Annual mean: γ-HCH, 60 pg/m³ Arithmetic mean: γ-HCH, 0.47 pg/m³	γ-HCH, 4.0–820 pg/m³	Annual mean (pg/m³): α-HCH, 145 β-HCH, 1.8 Arithmetic mean (pg/m³): α-HCH, 1.0 β-HCH, 0.34 Range (pg/m³): α-HCH, 10–540 β-HCH, MDC–28	Hoff et al. (1992a)
Japan Nigata, 1999–2001	Air	Annual mean (pg/m³): γ-HCH, 32	Annual ranges reported for five sampling points	All POPs decreased 41–80% during 2000 to 2001 except α-HCH and γ-HCH Annual mean (pg/m³): α-HCH, 92 β-HCH, 23 δ-HCH, 3	Murayama et al. (2003)
Europe 2006	Air	Average (pg/m³): γ-HCH, 35 Total HCH, 64	SD: γ-HCH, 38 Total HCH, 59	Average (pg/m³): α-HCH, 26 β-HCH, 2 SD: α-HCH, 24 β-HCH, 7	Halse et al. (2011)
China South China, [review paper]	Air	Mean range, HCH: 75–666 pg/m³	NR	Mean HCH levels from various studies, 2004–2007	Zhang et al. (2013)
Germany 1970–71	Water	238.14 ng/L (unfiltered water)	0–7100 ng/L (unfiltered water)	25 sites sampled in May 1971; 7 sites sampled monthly from April 1970 to June 1971	Herzel (1972)
Israel 1973	Water	NR	γ-HCH, ND–14.9 ng/L	Range: α-HCH, ND–4.1 ng/L	Lahav & Kahanovitch (1974)
USA Georgetown, South Carolina, Year, NR	Water	1.19 ppt	ND–2.21 ppt		Achari et al. (1975)

Table 1.4 (continued)

Region, country, city Year	Sampling matrix	Exposure		Comments	Reference
		Mean	Range		
USA New Orleans, 1970	Water	NR	Lindane, 1.3–2.9 ng/L	Estimated from graph	Brodtmann (1976)
Egypt Nile Delta, 1995–1997	Water	NR	Lindane, 0.286–0.352 μg/L		Abbassy et al. (1999)
Islamic Republic of Iran Karun River, Khuzestan, August 2008–March 2009	Water	Arithmetic mean (μg/L): γ-HCH, 1.58 Total HCH, 4.93	Range (μg/L): γ-HCH, 0.22–4.25 Total HCH, 0.73–11.12	Arithmetic mean (μg/L): α-HCH, 0.08 β-HCH, 1.81 δ-HCH, 1.45 Range (μg/L): α-HCH, 0.01–0.23 β-HCH, 0.08–6.07 δ-HCH, 0.29–3.26	Behfar et al. (2013)
China South China, [review paper]	Water	Mean range: 1.43–285 pg/m^3	NR	Mean HCH levels in water for various studies in 1999–2009	Zhang et al. (2013)
Turkey Konya Basin, 2012	Surface water	NR	March (μg/L): γ-HCH, 0.005–0.010 Total HCH, 0.015–0.065 August (μg/L): γ-HCH, ND–0.005 Total HCH, ND–0.025	March (μg/L): α-HCH, 0.005–0.010 β-HCH, 0.005–0.020 δ-HCH, ND–0.015 August (μg/L): α-HCH, ND–0.005 β-HCH, ND–0.005 δ-HCH, ND–0.025	Aydin et al. (2013)
Pakistan River Chenab, Punjab, January–March 2013	Water	3.3 ng/L (arithmetic mean)	0.33–11.9 ng/L		Mahmood et al. (2014)
Jordan Humrat Al-Sahn, 1998	Soil	NR	NR	Arithmetic mean, α-HCH (ppm): Ghor: 0.14; Wadi Um-Rishrash: 0.02; Wadi Al-Dafali: 0.02 Mean was of five replicates	Al-Mughrabi & Qrunfleh (2002)

Table 1.4 (continued)

Region, country, city Year	Sampling matrix	Exposure		Comments	Reference
		Mean	Range		
China Hong Kong Special Administrative Region Year, NR	Soil	Arithmetic mean (µg/kg): Organic farm, 3.65 Agricultural, 10.50 Abandoned agricultural, 6.01 E-waste storage, 9.71 Container storage, 3.76 Metal recycling workshop, 1.39 Constructions waste, 3.81 Petrol station, 7.94 Open burning site, 2.89 E-waste dismantling workshop, 9.88 E-waste open burning site: 1.14 Car dismantling, 26.80	SD (µg/kg): Organic farm, 2.75 Agricultural, 36.90 Abandoned agricultural, 5.42 E-waste storage, 2.00 Container storage, 1.76 Metal recycling workshop, 0.89 Constructions waste, 2.39 Petrol station, 8.71 Open burning site, 3.43 E-waste dismantling workshop, 28.40 E-waste open burning site, 0.11 Car dismantling, 31.40		Man et al. (2011)
USA 1986–1991	Food	Arithmetic mean (µg/kg bw per day): Age, 6–11 months: γ-HCH, 0.0008; Age, 60–65 yrs: γ-HCH, 0.0006	NR	Mean levels for β-HCH; γ-HCH and α-HCH reported for age groups from 6–11 months to 60–65 yrs by sex Age, 6–11 months: β-HCH, < 0.0001 Age, 60–65 yrs: β-HCH, < 0.0001	Gunderson (1995)
China Beijing and Shenyang, 1970–2007	Food	Arithmetic mean, HCH (µg/kg bw per day): Age, 41–65 yrs: Beijing: 1970: 549; 1992: 26.8; 2005/2007: 1.66	Age 41–65 years: 1.66–549 µg/kg bw per day	Mean levels were reported for Beijing and Shenyang by time period and for five age groups	Yu et al. (2013)

Table 1.4 (continued)

Region, country, city Year	Sampling matrix	Exposure		Comments	Reference
		Mean	Range		
Sweden 2005	Food	Arithmetic mean, total HCH (ng/day): 2005, 70.5 1999, 81.0	NR	Levels reported for each of the five food groups in the article	Törnkvist et al. (2011)
Singapore City and year, NR	Ambient indoor air	Arithmetic mean (ng/g dust): γ-HCH, 2.9 Total HCH, 11	Range (ng/g dust): γ-HCH, < LOD–74 Total HCH, < LOD–240	α-HCH levels were highly correlated with γ-HCH Arithmetic mean (ng/g dust): α-HCH, 0.7 β-HCH, 2.2 δ-HCH, 5.5 Range (ng/g dust): α-HCH, < LOD–8.7 β-HCH, < LOD–57 δ-HCH, < LOD–170	Tan et al. (2007)

HCH, hexachlorocyclohexane; LOD, limit of detection; MDC, minimum detectable concentration; ND, not detected; NR, not reported; POPs, persistent organic pollutants

Fig. 1.1 Air concentrations of organochlorine pesticides (left) and current use pesticides (right) worldwide

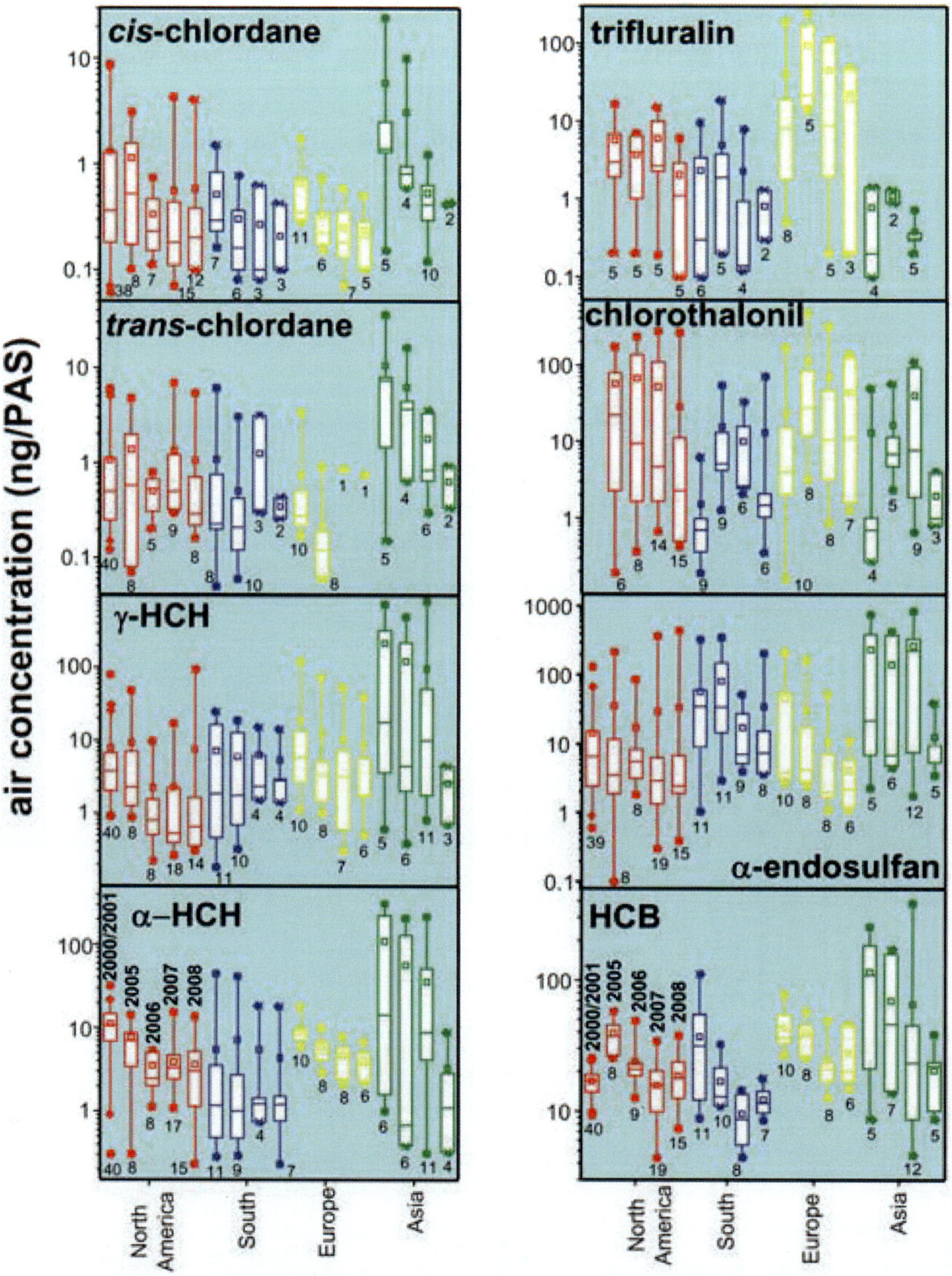

Reproduced from Shunthirasingham et al. (2010). Spatial and temporal pattern of pesticides in the global atmosphere. J Environ Monit. 12(9):1650–7, with permission of The Royal Society of Chemistry

HCB, hexachlorobenzene; HCH, hexachlorocyclohexane; PAS, passive air sampler

(b) Water

Lindane enters water from use in agriculture and forestry, from precipitation and, to a lesser extent, from occasional contamination of wastewater from manufacturing plants (WHO, 2004).

Early studies in the USA reported lindane levels in water. In Georgetown County, South Carolina, a mean lindane concentration of 1.19 ppt was reported in ground water (Achari et al., 1975). In 1970, potable water in Mississippi, USA, contained lindane at levels ranging between 1.3 and 2.9 ng/L (Brodtmann, 1976). In 1982, γ-HCH was detected in urban stormwater samples from Denver, Colorado, and Washington, DC, at 0.0027–0.1 μg/L and 0.052–0.1 μg/L in 20% and 11%, respectively, of the 86 samples collected (Cole et al., 1984; ATSDR, 2005). In the USA, α-, β-, γ-, and δ-HCH have been detected in surface water at 34, 18, 33, and 12, respectively, of the 1662 current or former National Priority Lists sites of the United States Environmental Protection Agency (EPA) (ATSDR, 2005). In the USA in 2001, γ-HCH was detectable in 1.6% and 4.4% of the samples from 83 sites in agricultural areas and from 30 urban sites, respectively (USGS, 2006).

Detectable γ-HCH levels in water have also been reported in the Middle East. In Israel, Lahav & Kahanovitch (1974) reported that γ-HCH levels ranged from non-detectable to 14.9 ng/L in well-water samples collected between 1972 and 1973. Lindane levels in water from the Nile Delta ranged from 0.286 to 0.352 μg/L between 1995 and 1997 (Abbassy et al., 1999). Surface water sampled in 2012 from the Konya Basin in Turkey contained total HCH at levels ranging from non-detectable to 0.065 μg/L, and γ-HCH at levels ranging from non-detectable to 0.010 μg/L (Aydin et al., 2013). Water samples taken monthly between 2009 and 2011 from the Karun river in Khuzestan Province, in the Islamic Republic of Iran, contained total HCH and γ-HCH at mean concentrations of 4.93 μg/L and 1.59 μg/L, respectively (Behfar et al., 2013).

In India, a review of historical and recent data indicated that surface water concentrations decreased three- to fourfold after the banning of production of technical HCH in 1997 (Sharma et al., 2014). In neighbouring Pakistan, the mean concentration of total HCH was 3.31 ng/L for water collected from the Chenab river between January and March 2013 (Mahmood et al., 2014). In southern China, mean HCH levels in surface water decreased from 285 to 1.43 ng/L between 1999 and 2009 (Zhang et al., 2013).

(c) Soil and dust

Lindane can be released into the soil from direct pesticide application, formulation processing, or release from hazardous landfill sites (ATSDR, 2005). In soil, lindane can be degraded under aerobic conditions; the half-life ranges from 88 to 1146 days. In the USA α-, β-, γ-, and δ-HCH have been detected in soil at 63, 78, 90, and 58, respectively, of the 1662 current or former National Priority List sites of the EPA (ATSDR, 2005).

Soil sampled in 1998 from Humrat Al-Sahn in Jordan contained α-HCH at mean levels ranging from 0.02 to 0.14 ppm (Al-Mughrabi & Qrunfleh, 2002). In Hong Kong Special Administrative Region, China, mean HCH levels ranged from 1.14 to 26.8 μg/kg in samples of soil from twelve different types of land use (Man et al., 2011). Elsewhere in China, mean levels of HCH in soil reported between 1999 and 2005 ranged from 3.65 to 5.92 ng/g dry weight (Zhang et al., 2013).

The presence of γ-HCH in household dust may contribute to human exposure; in a study in a household in the USA where γ-HCH in dust was measured at 5.85 ppb, a pesticide formulator and his wife had elevated serum concentrations of γ-HCH (ATSDR, 2005). In Singapore, indoor ambient dust samples from 31 homes contained γ-HCH and β-HCH at mean levels of 2.9 ng/g and 2.23 ng/g, respectively (Tan et al., 2007).

1.4.3 Exposure in the general population

(a) Diet

See Table 1.4

Isomers of HCH are found in dairy products, meat, fish, poultry, garden fruits, oils and fats, leafy and root vegetables, and sugar.

In adult diets in the USA in 1981–1982, γ-HCH was reported to be 8 ng/kg body weight (bw) (ATSDR, 1989). Data on the daily dietary intake of γ-HCH and β-HCH between 1986 and 1991 obtained through the Total Diet Study of the Food and Drug Administration (FDA) in the USA indicated very low levels. Mean intake of γ-HCH ranged from 0.8 to 0.6 ng/kg bw per day for the age groups 6–11 months and 60–65 years, respectively. Mean β-HCH levels were less than 0.1 ng/kg bw per day in all age groups (Gunderson, 1995).

A market basket survey in Sweden reported a dietary estimate of total HCH of 70.5 ng/day in 2005, a slight decrease from 81.0 ng/day in 1999 (Törnkvist et al., 2011).

Temporal trends suggest that dietary intake of HCH appears to be decreasing in China, where dietary intake of HCH has been reported from the 1970s until 2005/2007 across five age groups. The mean levels have shown a decrease for example in those aged 41–65 years in Beijing, from 549 μg/kg per day in the 1970s to 1.66 μg/kg per day in 2005/2007 (Yu et al., 2013).

(b) Biological measurements

See Table 1.5

Biological measurements of lindane in serum, human milk, adipose tissue, placental cord blood, and hair have been reported. In countries such as the USA, the percentage of samples with levels in the low detectable range has been increasing. Globally, the World Health Organization (WHO) has reported low mean levels of γ-HCH in breast milk samples from certain regions in Europe (WHO, 2015; Fig. 1.2).

(i) Europe

The detection of γ-HCH in samples from Europe has been decreasing since the 1970s. The median serum β-HCH concentration for a population of women participating in the Copenhagen City Heart Study, Denmark, decreased from 119.0 to 60.0 ng/g lipid between 1976–1978 and 1981–1983 (Høyer et al., 2000). In Norway, mean serum concentrations of β-HCH measured in women progressively decreased from 81.3 to 19.3 ng/g lipid from 1973–1975 to 1985–1990 (Fig. 1.3; Ward et al., 2000). The German Environmental Health Survey of environmental pollutants in blood quantified γ-HCH and β-HCH in 5.2% and 34% of blood samples in 1998 (Becker et al., 2002). In south-western Germany, the concentration of γ-HCH in whole blood samples from children was 0.07 μg/L in 1996/1997, while mean levels of β-HCH were 0.2, 0.07, and 0.04 μg/L in 1996/1997, 1998/1999, and 2000/2001 respectively. Reporting of γ-HCH and β-HCH was stopped in 1998/199 and 2000/2001, respectively, because levels were predominantly below the limit of detection (Link et al., 2005). In southern Spain, mean lindane concentrations in serum among women and young male volunteers were 1.53 ng/mL and 1.84 ng/mL, respectively (Botella et al., 2004; Carreño et al., 2007). Mean β-HCH concentrations were reported as 167.4 ng/g lipid (range, 155.8–179.9 ng/g lipid) from the European Prospective Investigation into Cancer and Nutrition (EPIC) Spanish cohort between 1992 and 1996 (Jakszyn et al., 2009). In the French National Nutrition and Health Study 2006–2007, γ-HCH was below detectable levels in serum, while the mean serum concentration of β-HCH was 27.0 ng/g lipid (Saoudi et al., 2014).

In Germany, two studies have shown a steady decrease in median β-HCH levels in human milk samples. Between 1986 and 1996, median β-HCH levels decreased from 0.19 to 0.03 mg/kg, while between 1999 and 2006 median β-HCH levels decreased by 0.04 to 0.012 mg/kg (Schade &

Table 1.5 Biological measurements of exposure to lindane (and other HCH isomers) in humans

Region, country, city Year	Sampling matrix	Mean	Range	Comments/ additional data	Reference
USA Wisconsin, Ohio and Michigan, 1993	Serum	Arithmetic mean: γ-HCH, ND β-HCH, 0.05 ppb	γ-HCH, ND β-HCH, 0.04–1.2 ppb		Anderson et al. (1998)
Mexico Mexico City, March 1994 and April 1996	Serum	Median β-HCH: Cases, 104.16 ng/g Controls, 92.98 ng/g	Cases, 53.29–418.54 ng/g Controls, 53.29–270.77 ng/g	95 cases of breast cancer and 95 controls	López-Carrillo et al. (2002)
USA Maryland, 1975–1994	Serum	Median β-HCH: Cases (NHL), 139 ng/g of lipid Controls, 138 ng/g of lipid	Cases, 71.1–286.5 ng/g lipid Controls, 56.9–219.3 ng/g lipid		Cantor et al. (2003)
New Zealand 1996–1997	Serum	Arithmetic mean: γ-HCH: NR β-HCH: 19.7 μg/kg lipid weight	γ-HCH: < 5–91.1 μg/kg lipid weight β-HCH: < 7–73.1	The mean value for γ-HCH was not calculated	Bates et al. (2004)
India Ahmedabad, year NR	Serum	Arithmetic mean (μg/L) γ-HCH: 1.69; α-HCH: 4.49; β-HCH: 35.06;	γ-HCH: 0.72–3.09 μg/L; α-HCH: 1.0–9.16; β-HCH: 20.11–82.09	Median: γ-HCH: 1.54; α-HCH: 3.62; β-HCH: 30.25	Bhatnagar et al. (2004)
Spain, southern, year NR	Serum	Arithmetic mean, lindane: 1.84 μg/L	SD: 2.27 μg/L	Median: 1.47 μg/L; maximum: 17.72 μg/L	Carreño et al. (2007)
Spain 1992–1996	Serum	Geometric mean, β-HCH (ng/g lipid) 167	155.8–179.9 ng/g lipid		Jakszyn et al. (2009)
Gran Canaria Island, Spain 1999–2001	Serum	Arithmetic mean, lindane (ng/g lipid): breast cancer women: 53.2; healthy women: 24.7	Breast cancer: 0.0–111.4; healthy: 0.0–220.0 ng/g lipid		Boada et al. (2012)
Slovakia, eastern, 2002–2004	Serum	Arithmetic mean (ng/mL) Maternal: β-HCH: 0.012; γ-HCH: 0.02; Infant cord blood: β-HCH:0.03; γ-HCH:0.01	SD (ng/mL) Maternal: β-HCH: 10.5; γ-HCH: 1.87; Infant cord blood: β-HCH: 10.4; γ-HCH: 4.67	The concentrations for most samples were higher than the detection limit except for γ-HCH	Patayová et al. (2013)
Benin Borgou, 2011	Serum	Arithmetic mean β-HCH (ng/g): 10.0	SD (ng/g): 20.4		Azandjeme et al. (2014)
USA 1999–2000	Serum	Arithmetic mean, γ-HCH (ng/g lipid): 0.06	< LOD–0.07 ng/g lipids	In subsequent studies of NHANES the levels of γ-HCH were undetectable	CDC (2009)

Table 1.5 (continued)

Region, country, city Year	Sampling matrix	Mean	Range	Comments/ additional data	Reference
France NR, 2006–2007	Serum	Geometric mean (ng/g lipid): γ-HCH: ND; α-HCH: 0.66; β-HCH: 30.4;	P50–P95: α-HCH: 0.74–1.77; β-HCH: 27.0–193.6; γ-HCH: < LOD–3.6 ng/g of lipid		Saoudi et al. (2014)
China Four cities (Beijing, Lanzhou, Taiyun, Xiamne), June–August 2010	Maternal and infant serum	Geometric mean β-HCH (ng/g lipid): Maternal, 67.67 Infant, 33.39	β-HCH (ng/g lipid): Maternal: < LOD–348.03 Infant; < LOD–261.29	Maternal: range, α-HCH: < LOD–8.83; γ-HCH, < LOD–4.24 Levels in infants cord blood and for other isomers were not measurable	Guo et al. (2014)
China Beijing, June 2006-July 2007	Placental and cord sera	Arithmetic mean (ng/g fat): Placenta: α-HCH, 0.85; β-HCH, 71.8; γ-HCH, 5.75; δ-HCH, 2.07; Umbilical cord blood: β-HCH, 97.0	Placenta: α-HCH: 031–1.57; β-HCH: 6.71–193; γ-HCH: ND–15.8; δ-HCH: ND–7.93; Umbilical cord blood: β–HCH: 9.12–336.0 ng/g fat	α-HCH, γ-HCH, and δ-HCH were not detectable in umbilical cord blood samples	Yu et al. (2013)
Spain Granada and Almeria Provinces, year NR	Serum and adipose tissue	Arithmetic mean, lindane serum: 1.56 (ng/mL) adipose tissue: 17.44 (ng/g lipid)	SD: serum, 17.84 ng/mL; adipose tissue, 2.26 ng/g lipid	Maximum levels reported for serum and adipose tissue were 12.77 ng/mL and 113.31 ng/g lipid respectively	Botella et al. (2004)
South Africa year NR	Plasma	Geometric mean range (ng/g lipid): β-HCH: 2.4–10.6; γ-HCH: 1.4–1081	β-HCH: 1.6–44.3; γ-HCH: 150–896 ng/g lipid		Röllin et al. (2009)
South Africa KwaZulu-Natal, year NR	Plasma	Geometric mean, γ-HCH (ng/g lipid): 956	13–164 ng/g lipid	Mean, median, range values reported for three sites for γ-HCH; α -HCH, β-HCH	Channa et al. (2012)
Germany Baden-Wuerttemberg, 1996–2001	Whole blood	Arithmetic mean (μg/L): 1996/1997: β-HCH: 0.2; γ-HCH: 0.07; 1998/1999: β-HCH: 0.07; 2000/2001: β-HCH: 0.04	1996/1997: β-HCH: < 0.02–4.75; γ-HCH: < 0.02–1.38; 1998/1999: β-HCH: 0.02–0.51; 2000/2001: β-HCH: < 0.02–0.62 μg/L	Concentrations of γ-HCH were predominantly less than the detection limit of 0.02 μg/L and stopped in 1998/1999 Concentrations of β-HCH dropped to the detection limit in 2000/2001	Link et al. (2005)

Table 1.5 (continued)

Region, country, city Year	Sampling matrix	Mean	Range	Comments/ additional data	Reference
Germany 1998	Whole blood	Arithmetic mean (µg/L) α-HCH: < 0.1; β-HCH: 0.16; γ-HCH: < 0.1	P50–Max: α-HCH: < 0.1–0.4; β-HCH: < 0.1–7.8; γ-HCH: < 0.1–4.7 µg/L	The percentage of β-HCH concentrations above the limit of quantification (0.1 µg/L) increased with age from 8.6% to 72%	Becker et al. (2002)
China year, NR	Placenta	Median (ng/g lipid): Cases: α-HCH: 1.5, β-HCH: 39, γ-HCH: 1.6, total HCH: 44; Controls: α-HCH: 0.96, β-HCH: 33, γ-HCH 0.99, total HCH: 36	Cases (ng/g lipid): α-HCH: 0.99–2.2, β-HCH: 13–65, γ-HCH: 0.7–2.7, total HCH: 18–71; Controls (ng/g lipid): α-HCH: 0.57–1.4, β-HCH: 22–52, γ-HCH: 0.55–1.7, total HCH: 24–56		Ren et al. (2011)
USA Connecticut, year NR	Breast adipose tissue	Median, β-HCH: Cases: 27.1 ppm; controls: 25.2 ppm	Cases: 16.1–41.6 ppm; controls: 16.3–41.2 ppm	Age-adjusted geometric mean: cases, 27.1; controls, 26.3	Zheng et al. (1999)
USA Connecticut, 1994–1997	Breast adipose tissue	Geometric mean, β-HCH: Cases: 27.1 ppm; controls: 26.3 ppm	NR	Median: Cases: 27.1 (16.1–41.6) ppm; controls, 25.2 (16.3–41.2) ppm	Zheng et al. (1999)
Germany Schleswig-Holstein, 1986–1997	Breast milk	Median, β-HCH: 1986: 0.19 mg/kg; 1997: 0.03 mg/kg	NR	NR	Schade & Heinzow (1998)
Tunisia 2003–2005	Breast milk	Arithmetic mean, γ-HCH: 31 ng/g lipid	SD (ng/g lipid): 17.5; range (ng/g lipid): ND–125.7		Ennaceur et al. (2008)
Germany Lower Saxony, January 1999-December 2006	Breast milk	Arithmetic mean (mg/kg lipid): γ-HCH: 0.001 β-HCH: 0.02	SD (mg/kg lipid): γ-HCH: 0.003 β-HCH: 0.05	Median β-HCH levels decreased 47.1% from 1999 to 2006	Zietz et al. (2008)
Japan Kyushu Island, May 2007-March 2008	Breast milk	Median lindane: 28.3 ng/g lipid	4.5–253 ng/g lipid		Miyake et al. (2011)
Islamic Republic of Iran November 2007-January 2008	Hair	Arithmetic mean, HCH: 14 ng/g	ND–67 ng/g	α-HCH, β-HCH, γ-HCH levels reported for three different sites	Dahmardeh Behrooz et al. (2012)
Tunisia Bizerte, 2010	Breast milk	Arithmetic mean, lindane: 36.5 ng/g lipid wt	ND–125.7 ng/g lipid wt	Median, 27.2 ng/g lipid wt	Ben Hassine et al. (2012)

Table 1.5 (continued)

Region, country, city Year	Sampling matrix	Mean	Range	Comments/ additional data	Reference
Turkey Ankara, year NR	Breast milk	Arithmetic mean (ng/g lipid wt): γ-HCH: 3.1; α-HCH: 7.3; β-HCH: 76.2	γ-HCH: < LOD–42.3 ng/g lipid wt; α-HCH: < LOD–88.7; β-HCH: < LOD–427.6	SD: γ-HCH, 9.0; α-HCH, 16.7; β-HCH, 96.7	Yalçin et al. (2014)
Denmark Copenhagen, 1976–1983	Serum	Median, β-HCH (ng/g lipid): 1976–1978: 119.0; 1981–1983: 60.0	NR	In 353 women, concentrations of β-HCH decreased between 1976–1978 and 1981–1983; in 61 women the β-HCH concentration increased for this period	Høyer et al. (2000)
Norway 1973–1990	Serum	Arithmetic mean, β-HCH (ng/g lipid) Cases, 63.4; controls: 60.0	NR	300 study subjects recruited, but 144 samples had > 90 pesticide compounds above the LOD and were used for analysis	Ward et al. (2000)
Guinea-Bissau 1990–2007	Serum	NR	γ-HCH: 240–< LOD; β-HCH: 320–< LOD ng/g fat	Samples were divided into five age groups and β-HCH; γ-HCH concentrations were reported for each group per time period	Linderholm et al. (2010)

HCH, hexachlorocyclohexane; LOD, limit of detection; NR, not reported

Fig. 1.2 Concentrations of γ-HCH in human milk, by region

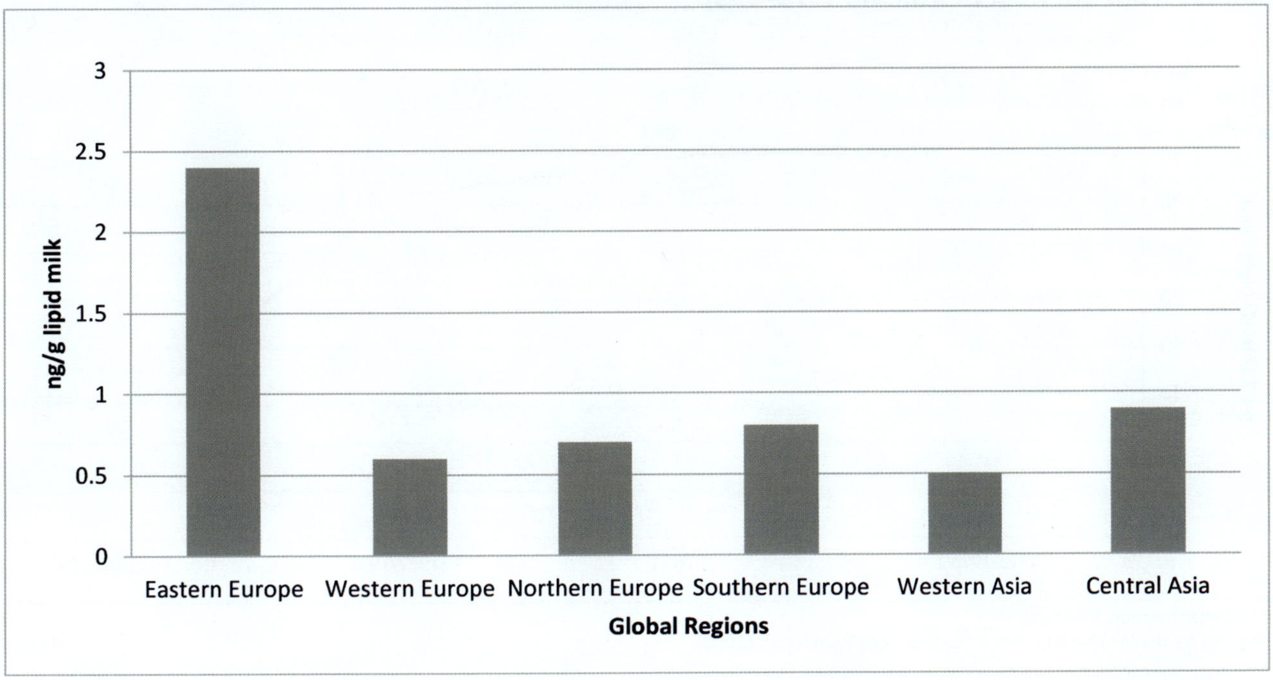

Arithmetic mean concentrations (ng/g lipid) of γ-HCH in human milk in the following regions: eastern Europe: Bulgaria, Czech Republic, Hungary, Republic of Moldova, Russian Federation, Slovakia, Ukraine; western Europe: Belgium, Germany, Luxembourg, Switzerland; northern Europe: Ireland, Lithuania, Norway, Sweden; southern Europe: Italy, Spain; western Asia: Cyprus, Georgia, Israel; central Asia: Tajikistan
HCH, hexachlorocyclohexane
Compiled by the Working Group with data from WHO (2015)

Heinzow, 1998; Zietz et al., 2008). In breast milk samples from 75 mothers in Ankara, Turkey [year not reported], α-HCH, β-HCH, and γ-HCH were reported at mean concentrations of 7.3, 76.2, and 3.1 ng/g lipid weight, respectively (Yalçin et al., 2014).

In Slovakia, between 2002 and 2004, mean levels of β-HCH and γ-HCH in maternal serum and cord-blood samples from infants were 0.12, 0.03, and 0.02, 0.01, respectively. In most samples, γ-HCH was below the limit of detection (LOD) (Patayová et al., 2013).

(ii) The Americas

In Mexico City, Mexico, between March 1994 and April 1996, serum levels of β-HCH levels ranged from 53.3 to 418.5 ng/g (López-Carrillo et al., 2002). In the USA, levels of γ-HCH in the general population have decreased since the 1970s. In a cohort started in 1974 in Maryland, USA, the levels of serum β-HCH among participants in the CLUE I study ranged from 56.9 ng/g lipid in controls to 286.5 ng/g lipid in cases (Cantor et al., 2003). In 1993, a study of the profiles of the Great Lakes critical pollutants reported non-detectable levels of γ-HCH and a median level of 0.05 ppb (range, 0.04–1.2 ppb) for β-HCH in serum from a convenience sample of 32 participants (Anderson et al., 1998). Between 1994 and 1996 in Long Island, New York, the mean concentrations of β-HCH in adipose tissue and serum were reported as 22.21 ng/g lipid and 0.824 ng/mL, respectively (Stellman et al., 1998). Between 1999 and 2000, the mean β-HCH serum level reported from the NHANES study in the USA was 0.058 ng/g, while γ-HCH was below detection limits from 1999 onwards (CDC, 2009).

Fig. 1.3 Temporal trends in organochlorine levels (as reflected in β-HCH concentrations) in serum samples from Norwegian women, 1973–1990

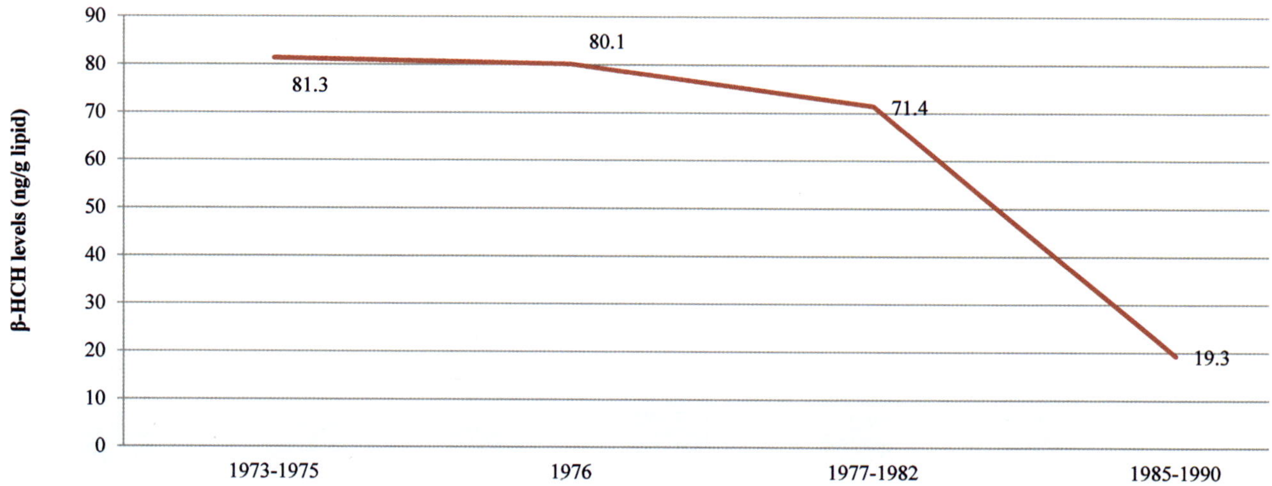

HCH, hexachlorocyclohexane
Compiled by the Working Group with data from Ward et al. (2000)

(iii) Asia and Oceania

Studies in Asia and Oceania have reported levels of lindane, γ-HCH, and β-HCH in samples of serum, human milk, and placental cord blood and tissue.

A mean serum γ-HCH level of 1.69 µg/L was reported among volunteers in Ahmedabad, India, in 2004 (Bhatnagar et al., 2004). Between December 1996 and January 1997, serum analysis from the National Nutrition Survey in New Zealand reported a weighted mean concentration of 19.7 µg/kg lipid weight basis for β-HCH in adults (Bates et al., 2004).

Levels of β-HCH in human milk in a population in Japan studied between 2007 and 2008 ranged from 18.3 ng/g to 45.3 ng/g (Miyake et al., 2011).

In samples of placental tissue in China, lindane levels ranged from 18 to 56 ng/g (Ren et al., 2011). In Beijing between June and July 2006, analysis of placental samples showed mean levels of α-HCH, β-HCH, and γ-HCH of 0.85, 71.8, and 5.75 ng/g fat, while in cord blood there was a mean β-HCH level of 97.0 ng/g fat and all other isomers were undetectable (Yu et al., 2013). Studies in four Chinese cities in 2010 reported levels of γ-HCH in most samples as below detectable limits in maternal blood and none in the cord blood samples while mean β-HCH levels in maternal and cord blood samples were 67.7 ng/g lipid and 33.4 ng/g lipid, respectively (Guo et al., 2014).

(iv) Africa and the Middle East

In various sites in South Africa, mean levels of γ-HCH in plasma ranged from 150 to 896 ng/g lipid (Röllin et al., 2009). In 2011 in north-eastern Benin, the mean serum level of β-HCH was reported as 10.0 ng/g total serum lipids (Azandjeme et al., 2014). In KwaZulu-Natal, South Africa, in 2008 the mean plasma level of γ-HCH was 956 ng/g (Channa et al., 2012).

In Guinea-Bissau, a cohort study conducted in 1990–2007 among men in the general population, with sampling at five different periods, showed a distinct decrease in concentrations of γ-HCH (240 ng/g fat to < LOD) and β-HCH

Fig. 1.4 Concentrations of γ-HCH and β-HCH in three pooled serum samples from Guinea-Bissau, sampled on five occasions between 1990 and 2007

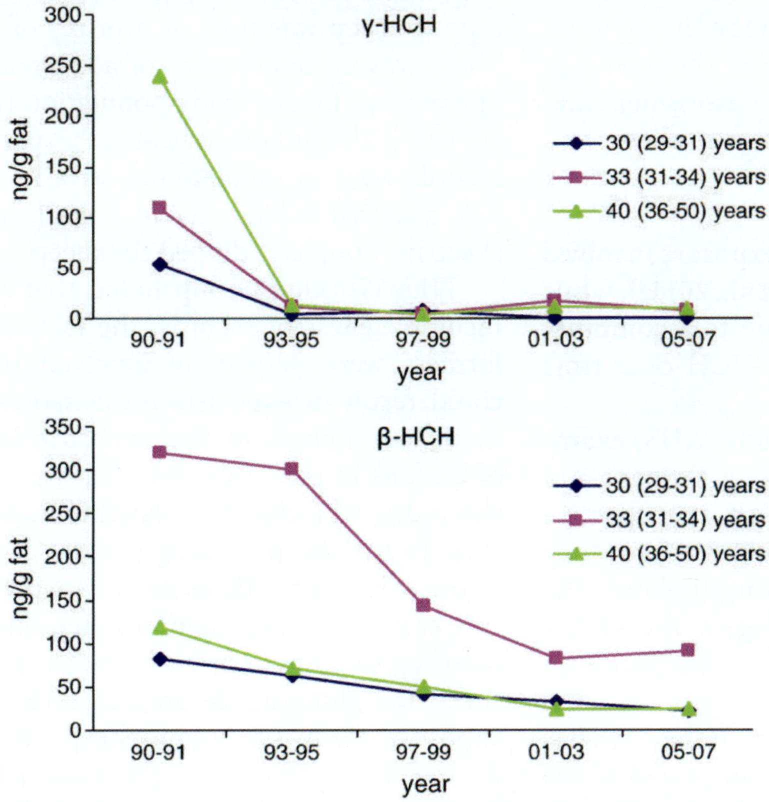

The median and range of age corresponds to the first sampling occasion
HCH, hexachlorocyclohexane
Reprinted from *Environment International*, Volume 36, issue 7, Linderholm et al. (2010). Human exposure to persistent organic pollutants in West Africa — A temporal trend study from Guinea-Bissau, Pages No. 675–682, Copyright (2010), with permission from Elsevier

(320 ng/g fat to < LOD). There was a clear decrease between the first sampling period (1990–1991) and the second sampling period (1993–1995) (Fig. 1.4; Linderholm et al., 2010).

In breast milk samples from Tunisia, the mean concentration of γ-HCH was reported as 31 ng/g lipid in samples from 237 women between 2003 and 2005 (Ennaceur et al., 2008), while a mean lindane level of 36.5 ng/g lipid was reported in a study in 2010 (Ben Hassine et al., 2012).

In the Islamic Republic of Iran between November 2007 and January 2008, hair samples from women contained γ-HCH at a mean concentration of 14 ng/g (Dahmardeh Behrooz et al., 2012).

1.4.4 Exposure assessment in epidemiological studies on lindane

Eight epidemiological studies were evaluated as providing evidence for carcinogenicity of lindane to humans. These studies can be categorized as studies of occupational exposure involving farmers and/or commercial applicators, and studies based upon the general population with exposure assessed using questionnaires or biological assessments. Among these studies, three were based on β-HCH as a marker for exposure to lindane, three on lindane exposure, and two were based on exposure to a commercial product that consisted of technical HCH or γ-HCH, depending on the time period. We here provide an assessment of the strengths and weaknesses of the exposure assessment and assignment methods used in these studies.

(a) Occupational exposure

One study of occupational exposure involved lindane specifically (Alavanja et al., 2014a), while the other two assessed exposure to a combination of technical HCH and γ-HCH over time (Rafnsson, 2006a, b).

The Agricultural Health Study (AHS) examined exposure to lindane among farmers and commercial applicators based on retrospective reporting of ever use and duration and frequency of use of 50 pesticides, including lindane. The exposure assessment methodology of the AHS is described in more detail in the *Monograph* on DDT in the present volume.

[The Working Group noted that the AHS has collected detailed information on pesticide use and practices and validation studies have shown this data to be appropriate for estimating historical exposure to pesticides. However, the validity studies are based on information reported at the time of the exposure surveys and would not necessarily reflect the recall of information reported for all aspects, in particular frequency and duration of use. The assessment of lindane exposure is based on the baseline questionnaire (1993–1997) and relies on recollection. The validity of such recall is unknown, but is not necessarily better or worse than that concerning other pesticides.]

In the other two studies of occupational exposure, exposure to technical-grade HCH and γ-HCH in sheep farmers in Iceland was estimated using records from the Icelandic Veterinary Services, in which the time of the sheep dipping using the pesticide, and the number of sheep owned by the farmer were recorded. The farmer was responsible for mixing the dip, dipping the sheep, and scrubbing the inside of the sheep shed. No advice was given to farmers on the use of personal protective equipment. The number of dipped sheep was used as a proxy of exposure. There was no assessment of additional sources of exposure in this study population (Rafnsson, 2006a, b). The authors indicated that parts of some records were incomplete for certain communities, and that farmers having small numbers of sheep may not have dipped the sheep themselves.

[The Working Group noted that due to the incompleteness of records, the fact that not all farmers were personally involved in dipping could result in exposure misclassification. The main assumption in the exposure assessment of lindane in this study was that the number of sheep dipped is directly related to lindane exposure. In the sheep-dipping process as described in other locations, three events involving exposure can be distinguished: mixing, dipping, and cleaning. Dipping consists of bodily grasping the sheep and plunging the animal in the dip. In an exposure study on organophosphate exposure during sheep dipping in the United Kingdom, Buchanan et al. (2001) identified contact with the pesticide concentrate that occurs during mixing as the most important exposure determinant. Furthermore, people involved in plunging the sheep were more highly exposed than those responsible for throwing or guiding the sheep into the dip. Such exposure to the concentrate is not directly related to the number of sheep, but more directly related to the number of dipping events. The papers did not indicate whether farmers had help in the dipping tasks or performed the dipping themselves. Consequently, the assumption that only the farmer was involved might be questioned.

The two papers under consideration indicated that dipping infested sheep required a longer time (3 minutes) than dipping non-infested sheep (1 minute), but the number of infested versus non-infested sheep per individual farmer was not recorded. Reliance upon the number of sheep dipped as an indicator of exposure assumes an average similar infestation rate of sheep for all farmers and over calendar time. An additional complication is that technical-grade HCH was used for dipping between 1962 and 1970, and was substituted by lindane between 1970 and 1980. Taking these issues together, there is uncertainty regarding the level of lindane exposure, potentially leading to substantial exposure misclassification. Considering that the median of the exposure categories based on the number of sheep owned varied only by a factor of ~8, this could have resulted in substantial misclassification across the exposure categories.]

(b) General population

Among five studies assessing exposure among the general population, two were based on lindane and three were based on β-HCH.

(i) Questionnaire-based approaches

In a study by Blair et al. in the midwest USA, three different questionnaires were used, and these varied in terms of information collected on pesticide use (Blair et al., 1998). [The Working Group noted that studies based on self-reporting are prone to failures of recall and resulting exposure misclassification. Due to the absence of separate evaluations of recall and exposure assignment, the validity of the exposure determination used in this study is not known.]

(ii) Biomonitoring approaches

All of the biomonitoring studies relied upon measurement of β-HCH levels. In one of the studies (Ward et al., 2000), occupational category was also documented, but this did not provide any additional information on exposure. In the other studies, occupational information was not collected.

[The Working Group noted that the validity of using β-HCH as a proxy of lindane depends on the extent to which the two exposures are actually related and a simple equivalence cannot be assumed. In fact, apart from lindane, exposure to β-HCH can occur through the diet, through contact with other environment media, and through other products containing β-HCH (i.e. technical-grade HCH). Accordingly, the extent to which the β-HCH is indicative of lindane will depend on the specific exposure scenario and calendar time the biological sample was collected.]

1.5 Regulation

Most readily available information concerning the regulation of lindane was concerned with action to limit or eliminate usage of this pesticide in various jurisdictions (Weinhold, 2001). Current regulatory controls concerning lindane and other pesticides are briefly listed below.

1.5.1 International treaties and agreements

A summary of earlier international treaties and like measures concerning lindane has been prepared by the Commission for Environmental Cooperation (CEC, 2006), an authority based on collaboration between the countries of North America.

Lindane is listed in Annex III of the Rotterdam Convention on the Prior Informed Consent Procedure for Certain Hazardous Chemicals and Pesticides in International Trade, adopted in February 2004 and legally-binding for Parties. Among the Parties to the Convention, 34 countries have banned all import of lindane, and 38 have restricted or severely restricted the conditions under which it may be imported (Rotterdam Convention, 2004).

Lindane is further regulated under the Stockholm Convention on Persistent Organic Pollutants (for further information about the Stockholm Convention, see Section 1.5 of the *Monograph* on DDT in this volume). As of December 2014, the Stockholm Convention

Secretariat had received three registrations of specific exemptions for allowable uses of lindane applicable in Canada, and Hong Kong and Macao, Special Administrative Regions, China, as described in Annexes A and B of the Convention (UNEP, 2002). Acceptable purposes for use of lindane are restricted to that of a human health pharmaceutical, adjuvant therapeutic drug, for control of head lice and scabies as second-line treatment (Vijgen et al., 2011).

WHO has specified a guideline value for water contamination by lindane of 2 μg/L (WHO, 2004).

1.5.2 Transnational and national regulations

Lindane is banned for use in 52 countries, restricted or severely restricted in 33 countries, not registered in 10 countries, and registered in 17 countries (CEC, 2006, 2015). In the USA, lindane was registered as an agricultural insecticide in the 1940s, and as a pharmaceutical in 1951. It was regulated as a pesticide by the EPA (EPA, 2015a, b), while lindane medications are regulated by the FDA (FDA, 2015). The EPA gradually began restricting its agricultural use in the 1970s. By 2002, use of lindane was limited to seed treatments for six crops, and finally banned in 2007. Under FDA regulation, a 1% γ-HCH lotion is available for the treatment of scabies, and a 1% shampoo is available for the treatment of head lice (CEC, 2006).

In Europe, the European Commission supported withdrawal of approval of lindane as an active substance (European Commission, 2000a, b) for plant protection products. In 2004, the European Parliament adopted Regulation 850/2004 that bans production and use of 13 persistent organic pollutants (European Commission, 2004). For lindane, the regulation allowed member states a phase-out period until December 2007. As assessed in 2006, several countries in Europe still allowed restricted use of lindane (CEC, 2006).

Maximum residue limits for lindane in various foods, particularly crops, are specified under many national and other authorities. As a persistent organic pollutant, lindane is recognized as a water contaminant and subject to regulation by transnational, national, and state or local authorities.

The United States Occupational Safety and Health Administration specifies permissible exposure limits for lindane of 0.5 mg/m^3 (time-weighted average) for general, construction and maritime industries (OSHA, 2017).

2. Cancer in Humans

This section reviews cohort and case–control studies that assessed exposure to lindane or β-HCH and cancer. Although lindane is the gamma isomer of hexachlorocyclohexane (γ-HCH), commercial lindane products may include other isomers, including the β form, which has a longer half-life; in addition, technical-grade HCH containing the γ-isomer and several other isomers has reportedly been used as an insecticide (see Section 1). Studies of cancer using biological measurements of β-HCH as an indicator of exposure to lindane are therefore included in this review, reflecting the Working Group's judgment that the level of exposure to β-HCH could be a proxy indicator of such exposure.

2.1 Cohort studies

See Table 2.1

Using the IARC International Register of Workers Exposed to Phenoxy Herbicides and Their Contaminants, Kogevinas et al. (1995) observed that exposure to lindane was associated with an excess risk of non-Hodgkin lymphoma (NHL), with an odds ratio (OR) of 1.6 (95% confidence interval [CI], 0.3–8.8).

Table 2.1 Cohort studies of cancer and exposure to lindane

Reference, location, enrolment/follow-up period, study design	Population size, description, exposure assessment method	Organ site	Exposure category or level	Exposed cases/deaths	Risk estimate (95% CI)	Covariates controlled	Comments
Kogevinas et al. (1995) Australia, Canada, Europe, New Zealand, 1955 to 1991 Nested case–control study	Cases: 32 NHL, 11 sarcoma; cohort data linked to death certificates and cancer registration records Controls: 158 (NHL controls), 55 (sarcoma controls); selected from cohort, matched to each case by age, sex, and country of residence at time of employment Exposure assessment method: company records; exposure information on qualitative scale, based on departments worked	NHL	Lindane	NR	1.6 (0.3–8.8)	Age, sex, country of residence when employed	Strengths: examined cancer incidence Limitations: no quantitative exposure information
Cantor et al. (2003) Washington County, ML, USA; CLUE study 1974 enrolment/follow-up to 1994 Nested case-control	Cases: 74 incident cases identified from Washington County Cancer Registry for participants from CLUE I or II cohort Controls: 147 controls, 2 per case matched on race, sex, date of birth (within 1 yr), participation in CLUE I or II or private census in 1963–75, date of blood sample (within 15 days), location of stored serum Exposure assessment method: total lipid corrected serum values	NHL (ICD-8, 200 202)	β-HCH (ng/g lipid):			Ever smoked or currently smoking cigarettes, years of education (< 12, ≥ 12), EBV-EA seropositivity, quartile of PCB concentration	Strengths: most cases pathologically confirmed; serum collected pre-diagnosis.; laboratory analysis blind to case-control status Limitations: evidence of possible exposure measurement error
			0.0–84.9	12	1.0		
			85.6–138.0	25	3.0 (1.1–8.4)		
			138.7–177.3	12	1.0 (0.3–3.2)		
			179.4–302.0	25	1.5 (0.5–4.3)		
			Trend-test *P*-value, 0.96				

Table 2.1 (continued)

Reference, location, enrolment/ follow-up period, study design	Population size, description, exposure assessment method	Organ site	Exposure category or level	Exposed cases/ deaths	Risk estimate (95% CI)	Covariates controlled	Comments
Rafnsson (2006a) Iceland Follow-up, 1955–2003	7882 men (213 685 person-yrs), 429 women (10 439 person-yrs); sheep farmers in Iceland Exposure assessment method: records of sheep dipping is kept by the Icelandic Veterinary Services; population-based cancer registry	Lip	Men	41	1.50 (1.08–2.04)		Strengths: complete paper records for 1962–1980; population based cancer registry Limitations: comparison of cancer incidence among sheep farmers with the general population controlled for age only
			Women	2	9.09 (1.02–32.82)		
		Liver and gallbladder (155)	Men	26	0.71 (0.46–1.04)		
			Women	1	0.48 (0.01–2.66)		
		Trachea, bronchus and lung (162)	Men	137	0.54 (0.46–0.64)		
			Women	8	0.77 (0.33–1.52)		
		NHL (200–202)	Men	45	0.90 (0.66–1.21)		
			Women	2	1.04 (0.12–3.76)		
		All cancers combined	Men	1818	0.79 (0.76–0.83)		
			Women	77	0.72 (0.57–0.90)		
		Prostate (all cases)	Men	541	0.92 (0.85–1.00)		
Rafnsson (2006b) Iceland 1962–2003 Nested case–control	Cases: 45; Iceland population-based cancer registry Controls: 221; sample of registered sheep farmers in Iceland Exposure assessment method: paper records on sheep dipping from the Icelandic Veterinary Service	NHL	No. of sheep:			Age, period of birth (< 1910; 1910–1919; 1920 and later)	Strengths: longstanding programme of sheep dipping with lindane; precise estimates of numbers of sheep Limitations: occupational exposures other than lindane were not ascertained; personal habits and medical histories were not available
			3–99	8	1.00		
			> 100	37	3.86 (1.59–8.53)		
			100–199	22	3.83 (1.58–9.31)		
			200–683	15	3.44 (1.31–9.04)		

Table 2.1 (continued)

Reference, location, enrolment/ follow-up period, study design	Population size, description, exposure assessment method	Organ site	Exposure category or level	Exposed cases/ deaths	Risk estimate (95% CI)	Covariates controlled	Comments
Alavanja et al. (2014b) Iowa, North Carolina, USA 1993–2010	54 306 licensed pesticide applicators Exposure assessment method: questionnaire; population-based cancer registries	NHL	Lindane use:			Age, state, race, total days of herbicide use	Strengths: large prospective study, ascertainment of exposure before disease;, large number of lindane exposed participants
			None	396	1.0		
			Ever	85	1.0 (0.8–1.2)		
			Total days of lindane exposure:				
			None	205	1.0		
			Low [≤ 8.75]	18	1.2 (0.7–1.9)		
			Medium [> 8.75–56]	13	1.0 (0.6–1.7)		
			High [> 56–457.25]	14	2.5 (1.4–4.4)		
			Trend-test *P*-value, 0.004				
Mills & Yang (2003) California, USA 1988–1999 Nested case–control	Cases: 222 prostate cancers, resulting from a linkage between UFW union rosters and the population-based California cancer registry Controls: 1110 non-cancer members selected from the UFW union roster Exposure assessment method: California pesticide application records	Prostate	Lindane exposure			Age, date of first union membership, and duration of union affiliation	Strengths: workers with relatively high exposure to pesticides; cases and controls from the same population Limitations: econological exposure assessment method with potential for bias towards the null
			1 (low)	93	1.00		
			High	129	1.32 (0.88–1.96)		
			Level 2	49	1.14 (0.45–1.77)		
			Level 3	47	1.86 (1.10–3.17)		
			Level 4 (high)	33	2.37 (1.12–4.61)		
			Trend-test *P*-value, 0.003				

Table 2.1 (continued)

Reference, location, enrolment/follow-up period, study design	Population size, description, exposure assessment method	Organ site	Exposure category or level	Exposed cases/deaths	Risk estimate (95% CI)	Covariates controlled	Comments
Sawada et al. (2010) Japan enrolment, 1990-1994/follow up to 2005 Nested case–control	Cases: 201 incident cases identified from local hospitals, population-based cancer registries and death certificates, from the cohort of residents of public-health centre areas in Japan Controls: 402, matched by age, area of residence, date time of day, and duration of fasting before blood collection Exposure assessment method: personal monitoring; blood samples taken at baseline	Prostate	β-HCH (ng/g lipid)			Smoking status, alcohol consumption, marital status, BMI, intake of green tea, intake of green tea and miso soup	Strengths: pre-diagnosis blood samples Limitations: response rate not known
			< 200	52	1.00		
			200–319	50	0.89 (0.52–1.50)		
			320–519	56	0.85 (0.50–1.46)		
			≥ 520	43	0.56 (0.30–1.01)		
			Trend-test *P*-value, 0.05				
Koutros et al. (2013) Iowa and North Carolina, USA Enrolment, 1993–1997/follow-up to 31 December 2007	54 412 male pesticide applicators Exposure assessment method: questionnaire	Prostate (all cases)	Lindane exposure:			Age, state, race, family history of prostate cancer, smoking, fruit servings, and leisure-time physical activity in the winter	Agricultural Health Study Strengths: large cohort study, in agricultural population with high exposure prevalence; detailed exposure assessment
			No exposure	840	1.00		
			Q1	43	0.88 (0.63–1.23)		
			Q2	36	1.06 (0.76–1.49)		
			Q3	39	1.06 (0.76–1.48)		
			Q4	39	1.16 (0.84–1.60)		
			Trend-test *P*-value, 0.33				

Table 2.1 (continued)

Reference, location, enrolment/ follow-up period, study design	Population size, description, exposure assessment method	Organ site	Exposure category or level	Exposed cases/ deaths	Risk estimate (95% CI)	Covariates controlled	Comments
McGlynn et al. (2008) USA 1988–2003 Nested case–control	Cases: 739 cases identified among men who donated blood between 1 January 1987 and 31 December 2002 to the DoDSR Controls: 915 controls matched on birth year, race/ethnicity, and date of available serum sample (within 30 days) Exposure assessment method: personal monitoring	Testis (testicular germ cell tumours)	β-HCH (μg/g lipid)			Age at blood donation, ethnicity, date of serum draw, age at reference date, cryptorchidism, family history of testicular cancer, height, BMI	Strengths: large study size; analysis of pre-diagnostic serum samples; high response rate; histologically confirmed tumours Limitations: some participants could not be contacted due to military deployment; adjustment for self-reported BMI; multiple comparisons
			≤ 0.00582	306	1.00		
			0.00583–0.00804	160	1.05 (0.80–1.40)		
			0.00805–0.0115	125	0.82 (0.60–1.11)		
			> 0.0115	143	0.90 (0.65–1.24)		
			Trend-test *P*-value, 0.4				
Purdue et al. (2009) Norway Enrolment, 1972/follow up, 1978–1999 Nested case–control	Cases: 49 members of the Janus serum bank cohort identified through linkage with the Norwegian cancer registry Controls: 51 matched to by region, period of and age at blood draw Exposure assessment method: personal monitoring; median (range) among controls: β-HCH, 129.7 (59.7–295.9) ng/g lipid; γ-HCH, 6.4 (0.1–42.4) ng/g lipid	Testis	β-HCH tertile 1	15	1.0	None (conditional logistic regression; matching criteria region/ age at blood draw/period of blood draw)	Strengths: use of serum samples collected before diagnosis; completeness of the Norway Cancer Registry Limitations: small study size
			β-HCH tertile 2	13	1.0 (0.4–2.7)		
			β-HCH tertile 3	21	1.8 (0.5–6.1)		
		Testis	γ-HCH tertile 1	17	1.0		
			γ-HCH tertile 2	15	1.0 (0.3–3.1)		
			γ-HCH tertile 3	17	1.1 (0.2–5.0)		

BMI, body mass index; CLUE, Campaign Against Cancer and Stroke; DoDSR, Department of Defense Serum Repository; DDT, dichlorodiphenyltrichloroethane; HCH, hexachlorocyclohexane; IQR, interquartile range; EBV-EA, Epstein-Barr virus early antigen; NHL, non-Hodgkin lymphoma; NR, not reported; PCB, polychlorinated biphenyl ; Q, quartile; UFW, United Farmer Workers

In the Clue I and II studies (Cantor et al., 2003), γ-HCH and β-HCH, were measured in serum samples for 74 cases of NHL and 147 matched controls. Data for γ-HCH were not reported as this isomer was detected in < 5% of samples. The concentration of β-HCH was significantly elevated among cases compared with controls. There was no clear pattern of risk with increasing quartile of lipid-corrected and recovery-adjusted level of exposure, with the highest odds ratio being in the second quartile for both unadjusted and adjusted (OR, 3.0; 95% CI, 1.1–8.4) analyses. [In this study most cases were confirmed from pathology information and serum was collected pre-diagnosis. The levels obtained for some compounds such as polychlorinated biphenyls and lindane were higher than expected, which may imply that there was some measurement error.]

A cohort (Rafnsson, 2006a) and nested case–control study (Rafnsson, 2006b) of sheep owners in Iceland provided additional, albeit indirect, evidence to evaluate the association between lindane and NHL. Sheep dipping with lindane formulations was compulsory in Iceland after 1959 to prevent ectoparasites, mainly sheep scab mites. Initially, technical-grade HCH (a mixture of isomers) was used. In the 1970s, technical-grade HCH was replaced by γ-HCH (lindane). A paper record of compliance with this law was available from 1962 to 1980 through the Icelandic Veterinary Services and was used as a surrogate measure of exposure to lindane.

The cohort (Rafnsson, 2006a) comprised 7882 men (213 685 person-years of follow-up) and 429 women (10 439 person-years). Observed and expected cancer cases were compared in an analysis of standardized incidence ratio (SIR), adjusted for age and calendar period. The standardized incidence ratio for all cancers was significantly less than expected for men (SIR, 0.79; 95% CI, 0.76–0.83) and women (SIR, 0.72; 95% CI, 0.57–0.90). Cancer of the lip was the only cancer found at a significantly greater rate than expected; the standardized incidence ratio was 1.5 (95% CI, 1.08–2.04) for men, and 9.09 (95% CI, 1.02–32.82) for women. All other standardized incidence ratios were not statistically significant. The standardized incidence ratio for NHL in men was 0.90 (95% CI, 0.66–1.21; 45 cases) and in women 1.92 (95% CI, 0.12–3.76; 2 cases).

In the nested case–control study (Rafnsson, 2006b), 45 histologically confirmed cases of NHL (International Classification of Disease seventh revision), in men diagnosed in 1962–2003 were recorded in the national cancer registry. A total of 221 cancer-free men sampled at random from the cohort served as controls. The age-adjusted odds ratio for NHL was 3.86 (95% CI, 1.59–8.53) for individuals who had 100 sheep or more compared with those who had less than 100 sheep. No analysis of risk in relation to the period of exposure was conducted. [The overall decreased cancer risk of sheep farmers in Iceland was consistent with findings reported previously among farmers in Iceland and in other countries. The excess risk of cancer of the lip found in this study has been observed in many previous studies of farmers and is usually attributed to long periods of unprotected occupational exposure to solar radiation. The disparity between the results of internal and external analysis may be attributed to more complete controlling for disparities in cancer determinants among farmers. The Working Group noted that the exposure metric used in this study, specifically the number of sheep, could lead to misclassification of the level of lindane exposure, but would likely not have materially affected the exposure rank order.]

In the USA, the AHS considered incident cases of NHL among more than 54 000 study participants who were free of cancer at the time of enrolment (Alavanja et al., 2014a). A history of lindane use was reported for 85 cases of NHL before onset of disease. Lindane was first registered for use in the USA in 1947, but its use was restricted to certified applicators in 1983, and further restricted to specific crops in 2002. Pesticide exposure information was ascertained from two phases of questionnaire administration in this analysis. For participants who did not complete the follow-up questionnaire, a data-driven multiple imputation procedure with logistic regression and stratified sampling was used to impute specific pesticide exposure.

Relative risks were calculated for no, low, or high use of lindane according to histological subtype of NHL; relative risks were elevated for those with high exposure to lindane, although statistically significant increases in risk were seen only for follicular B-cell lymphoma (*P* for trend, 0.04). Statistically significant upward exposure–response trends for NHL were observed with total days of lindane use [1.0 (reference), 1.2 (95% CI, 0.7–1.9), 1.0 (95% CI, 0.6–1.7), 2.5 (95% CI, 1.4–4.4) (*P* for trend, 0.004)], and with intensity-weighted days of exposure [1.0 (reference), 1.3 (95% CI, 0.8–2.2), 1.1 (95% CI, 0.7–1.8), 1.8 (95% CI, 1.0–3.2) (*P* for trend, 0.04)] (Alavanja et al., 2014a). The risk estimates were adjusted for age, state of residence, race, and total days of pesticide use. Lindane was not associated with the use of other insecticides, fungicides, or fumigants. Smoking, other lifestyle factors, and other occupational exposures were not observed to confound the results reported. Similar results were observed in an earlier analysis by Purdue (Purdue et al., 2007).

2.2 Case–control studies nested within cohorts

Several case–control studies nested within cohorts assessed the association between exposure to lindane or γ-HCH and risk of cancer of the prostate or testis. Results for other isomeric forms (β-HCH) are also presented here. The major strength of most of these studies was the use of pre-diagnostic blood samples for measuring HCH.

2.2.1 *Cancer of the prostate*

A nested case–control study of cancer of the prostate was conducted within a large cohort of predominately Hispanic members of the United Farm Workers of America labour union in California, USA (Mills & Yang, 2007). Through electronic linkage between a roster of union members and the California cancer registry for the years 1988–1999, newly diagnosed cases of cancer of the prostate were identified from the union. Age-matched controls were randomly selected from the remained of the cancer-free cohort. Risk for cancer of the prostate was examined by the type of crops and commodities cultivated by the farm workers, as well as by the date of first union activity and duration of union membership. In addition, the risk of cancer of the prostate was evaluated in association with the use of several pesticides recorded by the California Department of Pesticide Regulation by place, time period and crop, rather than at the level of individual workers. Between 1988 and 1999, 222 newly diagnosed cases of cancer of the prostate were identified for analysis and 1110 age-matched controls were selected. The risk of prostate cancer was not associated with patterns of employment in any crop/commodity. However, risk was observed to increase monotonically with an increase in estimated use of lindane; the odds ratio was 2.37 (95% CI, 1.12–4.61) for the highest quartile of use compared with the lowest quartile, adjusted for age, date of first union membership, and duration of union membership. [The Working Group noted that the semi-ecological exposure assessment may have led to measurement error at the level of the individual, but had the advantage that it did not rely on self-report (eliminating the potential for recall bias) as exposure was obtained through record linkage. The Working Group also noted that these methods enabled estimation of whether a cohort member worked in an area with high use of pesticides; however, level of exposure was based on the county, crop, and period when the person worked, and there was no information on job tasks collected from the participants, resulting in possible exposure misclassification.]

One nested case–control study reported results on risk of cancer of the prostate in relation to β-HCH (Sawada et al., 2010). This case–control study was nested within the Japan Public Health Center-based Prospective Study, a population-based cohort of 65 657 men aged 40–69 years at baseline in 1990–1993, of whom 14 203 provided blood samples. During follow up until December 2005, 201 cases of cancer of the prostate were identified from major hospitals, cancer registries, and death certificates. Two controls per case were matched by age, area of residence, date of blood sampling, time of day of blood sampling, and duration of fasting at blood collection. Organochlorine pesticides and polychlorinated biphenyls were measured in the blood samples taken before diagnosis. The incidence of cancer of the prostate tended to be inversely associated with exposure to β-HCH in this study (odds ratio for exposure in the highest exposure category as compared with the lowest: OR, 0.56; 95% CI, 0.30–1.01; *P* for trend, 0.05). [The Working Group noted that measurements of β-HCH in biological samples do not necessarily indicate exposure to lindane in the absence of information about the sources of exposure to HCH compounds.]

In the AHS, 1962 incident cases of cancer of the prostate among 54 412 licensed pesticide applicators were evaluated for exposure to 48 pesticides of widespread use in the cohort (Koutros et al., 2013). Increased use of lindane (lifetime days of use) was not associated with risk of total prostate cancer, nor was it associated with aggressive prostate cancer (i.e. Gleason score > 7). The rate ratio of total prostate cancer comparing the highest quartile of lindane use to those who did not use lindane was 1.16 (96% CI, 0.84–1.60; *P* for trend, 0.33) adjusted for age, state, race, family history of prostate cancer, smoking, fruit servings per day, and leisure-time physical activity in winter. [The Working Group noted that this was a large prospective study with controls for many potential confounders.]

2.2.2 Cancer of the testes

No studies of cancer of the testes and occupational exposure to lindane were available to the Working Group.

McGlynn et al. (2008) reported the results of a case–control study on testicular germ cell tumours among military servicemen in the USA, which included separate analyses for seminoma and non-seminoma, and data on β-HCH but not γ-HCH (lindane). The cases included in the analysis were 739 men who had donated blood to the Department of Defense Serum Repository (DoDSR) between 1987 and 2002, and who were subsequently diagnosed with testicular germ cell tumour. Controls were 915 men with a serum sample available in the DoDSR matched on birth year, ethnicity, and date of available serum sample (within 30 days). Each participant was given a computer-assisted telephone interview to obtain information on height, weight, medical conditions, and family history of cancer. Eleven organochlorine compounds were analysed in the serum. Total testicular germ cell tumours were not associated with β-HCH (highest versus lowest quartile: OR, 0.90; 95% CI, 0.65–1.24; *P* for trend, 0.40). Null results were reported for seminomas (OR, 0.97; 95% CI, 0.63–1.49; 78 exposed cases; *P* for trend, 0.83) and non-seminomas (OR, 0.85; 95% CI, 0.57–1.26; 65 exposed cases; *P* for trend, 0.24). [This was a well-conducted study with a large number of subjects, and high response rates. The use of pre-diagnostic serum samples in this study maybe an advantage; however, β-HCH measurements in biological samples do not necessarily indicate exposure to γ-HCH.]

Purdue et al. (2009) reported the findings of a case–control study on testicular germ cell tumours that was nested within the Janus Serum Bank cohort of Norway. Cases were Janus cohort members with baseline blood collection between 1972 and 1978, who were identified with a diagnosis of testicular germ cell tumour between baseline and 31 December 1999 through linkage with the Norwegian cancer registry. One male control from the Janus cohort was matched to each case by region, time of blood draw, and age at blood draw. The analysis was conducted in 49 cases (80%) and 51 controls (81%) with adequate exposure data. Concentrations of 11 organochlorine pesticides, including β-HCH and γ-HCH (lindane), and of 34 polychlorinated biphenyls were measured in the serum samples. The odds ratios for highest versus lowest exposure tertile were 1.8 (95% CI, 0.5–6.1) and 1.1 (95% CI, 0.2–5.0) for β-HCH and γ-HCH, respectively. [This was a small but well-conducted study. An important strength was the use of serum samples collected before diagnosis; however, as noted previously, β-HCH measurements in biological samples do not necessarily indicate exposure to lindane.]

2.3 Case–control studies

2.3.1 Cancer of the breast

See Table 2.2

Exposure to lindane in relation to cancer of the breast has been assessed in three case–control studies, two in Spain and one in India.

A hospital-based case–control study was conducted in the three largest public hospitals in southern Spain (Ibarluzea et al., 2004). Cases were recruited from women aged 35-70 years undergoing surgery for newly diagnosed malignant carcinoma of the breast. Levels of lindane in adipose tissue samples were higher, but not statistically significantly, in 198 cases of cancer of the breast compared with 260 age-matched controls. After adjusting for potential confounders, a significant odds ratio of 1.76 (95% CI, 1.04–2.98) emerged for postmenopausal women with lindane concentrations greater than the LOD versus those with lindane concentrations less than or equal to the LOD. [More than half of the women had lindane values under the LOD.]

In the Canary Islands, Spain, Boada et al. (2012) found that serum levels of lindane were not associated with cancer of the breast in 121 cases compared with 103 healthy controls. [This study was not matched by age; cases were significantly older than controls, and few women were exposed.]

In a pilot study in India, Siddiqui et al. (2005) found significantly higher levels of lindane in blood samples from 25 cases of cancer of the breast compared with 25 controls with benign disease. No exposure–disease associations were reported. [The limitations of this study included the small sample size, the recruitment of cases from a single hospital, and the lack of controls for confounders.]

2.3.2 Lympho-haematopoietic cancers

See Table 2.3

This section describes case-control studies of risk of NHL or leukaemia and exposure to lindane. It should be noted that interpretation of the published literature was complicated by the change over time in the classification and coding systems for NHL and its subtypes (American Cancer Society, 2016).

Data from an in-person interview study of 622 white men with newly diagnosed NHL and 1245 population-based controls in Iowa and Minnesota, USA, were used to measure the risk associated with farming occupation and specific agricultural exposures (Cantor et al., 1992). Detailed information was collected on farming and pesticide use, including over 100 insecticides used on animals or crops, herbicides, and fungicides. Significantly elevated risks were found for ever handling lindane as a crop insecticide (OR, 2.0; 95% CI, 1.0–3.7) and for handling it before 1965 (OR, 2.2; 95% CI, 1.0–4.7) after adjusting for vital status, age, state, smoking,

Table 2.2 Case-control studies of cancer of the breast and exposure to lindane

Reference, location follow-up/enrolment period	Population size, description, exposure assessment method	Organ site	Exposure category or level	Exposed cases/ deaths	Risk estimate (95% CI)	Covariates controlled	Comments
Boada et al. (2012) Gran Canaria Island, Spain April 1999-June 2001	Cases: 121, from two university hospitals of Gran Canary Island Controls: 103, selected from representative population sample obtained in the Canary Islands Nutrition Survey Exposure assessment method: personal monitoring; serum, GC-ECD	Breast (ICD-O 2nd C50.0-C50.9)	Lindane (ng/g lipid)	105	1.097 (0.420–28.412)	Age, BMI, menopausal status, lactation, and tobacco	Limitations: not matched by age; cases were significantly older than controls, and few women were exposed
			Trend-test *P*-value, 0.988				
Ibarluzea et al. (2004) Granada and Almeria, Spain April 1996-June1998	Cases: 198 women aged 35-70 yrs undergoing surgery for newly diagnosed malignant breast carcinoma (invasive or in situ) without previous history of cancer Controls: 260, age-matched (± 3 yrs) and hospital; exclusion criteria for controls were the presence of gynaecological or endocrine disease, including diabetes, and history of cancer Exposure assessment method: personal monitoring; adipose tissue samples; GC-ECD	Breast (ICD-O 2nd C50.0-C50.9)	LOD (lindane ng/g lipid)			Age, reference hospital, BMI, number of children, age at first full-term pregnancy, family history of breast cancer, alcohol and tobacco	Strengths: control for confounders Limitations: many women had lindane values under the LOD
			< LOD	67	1.00		
			> LOD (all)	54	1.40 (0.92–2.13)		
			> LOD (premenopausal)	33	1.10 (0.50–2.37)		
			> LOD (postmenopausal)	53	1.76 (1.04–2.98)		

BMI, body mass index; GC-ECD, gas chromatography with electron-capture detection; LOD, limit of detection; yr, year

Table 2.3 Case-control studies of lympho-haematopoietic cancers and exposure to lindane

Reference, location follow-up/enrolment period	Population size, description, exposure assessment method	Organ site	Exposure category or level	Exposed cases/ deaths	Risk estimate (95% CI)	Covariates controlled	Comments
Cantor et al. (1992) Iowa and Minnesota, USA Enrolment, 1980–1982	Cases: 622; Iowa Health Registry records and Minnesota hospital and pathology records Controls: 1245; population-based; no cancer of the lympho-haematopoietic system; frequency-matched to cases by age (5-yr group), vital status, state; random-digit dialling (age < 65 yrs); Medicare records (age ≥ 65 yrs); state death certificate files (deceased subjects) Exposure assessment method: questionnaire; in-person interview	NHL	Ever handled lindane:			Age, vital status, state, smoking status, family history of lympho-haematopoietic cancer, high-risk occupations, high-risk exposures	Data subsequently pooled in De Roos et al. (2003) Strengths: large population-based study in farming areas. Minimal evidence for confounding by exposure to other pesticides Limitations: not controlled for exposure to other pesticides; small numbers of cases in some analyses; multiple comparisons; use of proxy respondents; white men only
			As an animal insecticide	55	1.4 (1.0–2.1)		
			before 1965	40	1.7 (1.1–2.7)		
			without PPE	45	1.6 (1.0–2.4)		
			As a crop insecticide	21	2.0 (1.0–3.7)		
			before 1965	14	2.2 (1.0–4.7)		
			without PPE	16	2.6 (1.2–5.5)		
			before 1965 (Iowa residents)	9	1.4 (0.6–3.5)		
			before 1965 (Minnesota residents)	5	6.5 (1.2–35.0)		
Blair et al. (1998) Iowa, Minnesota, Nebraska, Kansas, USA 1980s	Cases: 987 white men Controls: 2895; frequency matched on age, race, state of residence; random-digit dialling for living cases aged < 65 yrs and from the Health Care Financing Administration for those aged ≥ 65 yrs; controls for deceased cases from deaths records in each state, matched for age and year of death Exposure assessment method: telephone interviews with subjects or next of kin in Kansas and Nebraska, and in person in Iowa and Minnesota	NHL	Lindane use:			Age, proxy/direct interview, state of residence	Strengths: pooled analysis with large numbers of cases and controls Limitations: use of proxy respondents
			Non-farmer	243	1.0		
			Farmer (used lindane, no adjustment for other pesticides)	93	1.5 (1.1–2.0)		
		NHL	First lindane use:				
			≥ 20 yrs ago	59	1.7 (1.1–2.5)		
			< 20 yrs ago	18	1.3 (0.7–2.3)		
			Days/year lindane use:				
			≤ 4	8	1.6 (0.6–4.0)		
			≤ 5	5	2.0 (0.6–6.4)		
			Protective equipment:				
			Used	25	1.4 (0.8–2.3)		
			Not used	63	1.5 (1.0–2.2)		

Table 2.3 (continued)

Reference, location follow-up/enrolment period	Population size, description, exposure assessment method	Organ site	Exposure category or level	Exposed cases/deaths	Risk estimate (95% CI)	Covariates controlled	Comments
Blair et al. (1998) Iowa, Minnesota, Nebraska, Kansas, USA 1980s (cont.)			Histological type:				
			Follicular	36	1.6 (1.0–2.5)		
			Diffuse	28	1.5 (0.9–2.5)		
			Small lymphocytic	14	1.9 (0.9–4.0)		
			Other types	15	1.1 (0.6–2.1)		
Schroeder et al. (2001) Iowa and Minnesota, USA 1980–1983	Cases: 622; identified through the State Health Registry of Iowa and active surveillance of hospital and pathology laboratory records in Minnesota Controls: 1245 white males without lympho-haematopoietic cancer, frequency matched to cases on age (± 5 yrs), state, and vital status. identified using random-digit dialling (age < 65 yrs), Health Care Financing Administration Medicare files (age ≥ 65 yrs), and state death certificate files (deceased controls) Exposure assessment method: questionnaire	NHL	Ever exposed to lindane			Age, state	Same study population as Cantor et al. (1992) Limitations: 30% of the study participants were represented by next-of-kin respondents; some bias towards the null may have occurred due to t(14;18) breakpoints outside the range of PCR primers; unable to classify > 60% of NHL cases
			t(14;18)-positive NHL vs controls	14	2.3 (1.3–3.9)		
			t(14;18)-negative NHL vs controls	12	1.0 (0.5–1.7)		
			t(14;18)-positive vs -negative NHL	12	2.1 (0.9–5.1)		

Table 2.3 (continued)

Reference, location follow-up/enrolment period	Population size, description, exposure assessment method	Organ site	Exposure category or level	Exposed cases/ deaths	Risk estimate (95% CI)	Covariates controlled	Comments
De Roos et al. (2003) Nebraska, Iowa, Minnesota, Kansas, USA 1979–1986	Cases: 650, from three previous case-control studies (Hoar et al., 1986; Zahm et al., 1990; Cantor et al., 1992) Controls: 1933, from three previous case-control studies (as for cases) Exposure assessment method: questionnaire; 47 pesticides in the pooled analysis (insecticides and herbicides), for which ≥ 20 persons were exposed; analysis restricted to "potentially carcinogenic pesticides"	NHL	Lindane ever use:			Other pesticides, age, location	Strengths: large number of exposed subjects; analysis of combined pesticide exposure and the number of pesticides used; evaluation of potential superadditivity of pesticides effects; use of hierarchical models Limitations: no quantification of exposure; no information on the timing of pesticide use; use of proxy respondents; many exclusions due to missing data
			Hierarchical regression	59	1.2 (0.8–1.9)		
			Logistic regression	59	1.2 (0.7–2.0)		
Lee et al. (2004) Iowa, Minnesota, Nebraska, USA 1980–83 for Iowa and Minnesota, 1983–86 for Nebraska	Cases: 872 from Iowa State Health Registry, surveillance system of Minnesota hospitals, Nebraska Lymphoma Study Group Controls: 2336, frequency-matched on age, race, state of residence; random-digit dialling for living cases (age < 65 yrs) and from the Health Care Financing Administration (age ≥ 65 yrs; controls for deceased cases from death records in each state, matched for age and year of death Exposure assessment method: questionnaire; telephone interviews with subjects or next-of-kin in Nebraska, and in person in Iowa and Minnesota	NHL	Lindane exposure:			Age, vital status, state of residence	Studies in midwest USA Strengths: pooled study so larger numbers Limitations: use of proxy respondents; no adjustment for co-exposures
			Non-farmers, non-asthmatics	259	1.0		
			Nonasthmatics	84	1.3 (0.97–1.8)		
			Asthmatics	11	2.4 (1.0–5.7)		

Table 2.3 (continued)

Reference, location follow-up/enrolment period	Population size, description, exposure assessment method	Organ site	Exposure category or level	Exposed cases/ deaths	Risk estimate (95% CI)	Covariates controlled	Comments
McDuffie et al. (2001) Alberta, Saskatchewan, Manitoba, Quebec, Ontario, British Columbia, Canada 1991–1994	Cases: 517 newly diagnosed men, age ≥ 19 yrs, enrolled from cancer registries Controls: 1506 (48%) men, age ≥ 19 yrs, enrolled from a random sample of health insurance and voting records; frequency-matched on province and age Exposure assessment method: postal questionnaire, followed by telephone interview for subjects with ≥ 10 h/yr of pesticide exposure and 15% random sample of the remainder; a list of chemical and brand names was mailed to these participants before interview; exposure defined as used at work, in home garden, or as hobby	NHL (200, 202)	Exposure to lindane	15	2.05 (1.01–4.16)	Age, province of residence	Strengths: large study; detailed exposure assessment through telephone interview; deceased were ineligible, reducing the number of surrogate responders; some modelling of multiple pesticide exposures Limitations: potential for recall bias; poor response rates; risk estimates were not adjusted for other pesticides
			Exposure to lindane	15	2.06 (1.01–4.22)	Age, province of residence, medical variables	
Miligi et al. (2003) Italy 1990–1993	Cases: 1145 NHL (ICD-9, 200, 202), 430 leukaemia (ICD-9, 204–208); all incident cases diagnosed in residents of the study area; age 20–74 yrs Controls: 1232 general-population controls, frequency-matched by sex and 5-yr age group, age 20–74 yrs Exposure assessment method: questionnaires reviewed by agronomists to assign pesticide-exposure histories	NHL (ICD-9, 200 & 202) and CLL (ICD-9, 204.1)	Ever occupationally exposed to lindane:			Age, area	NHL and CLL were analysed as a combined category due to biological similarities Strengths: expert assessment of exposure
			Men	9	2.0 (0.6–7.7)		

Table 2.3 (continued)

Reference, location follow-up/enrolment period	Population size, description, exposure assessment method	Organ site	Exposure category or level	Exposed cases/ deaths	Risk estimate (95% CI)	Covariates controlled	Comments
De Roos et al. (2005) Iowa, Los Angeles County, metropolitan areas of Detroit and Seattle, USA July 1998-June 2000	Cases: 100 with blood sample; 1321 incident cases of NHL without HIV infection; age 20–74 yrs; identified from four SEER registry areas Controls: 100 with blood sample; 1057 population controls identified by random-digit dialling (age < 65 yrs) and from Medicare eligibility files ≥ 65 yrs); frequency-matched to cases by age, sex, and race Exposure assessment method: blood sample	NHL	β-HCH (quartiles ng/g lipid)			Sex, study site, birth date, date of blood draw	1% samples had γ-HCH of > LOD, and 82% had β-HCH of > LOD
			≤ 9.1	23	1.00		
			> 9.1–15.0	26	1.19 (0.49–2.89)		
			> 15.0–26.0	28	1.21 (0.50–2.90)		
			> 26.0	23	1.05 (0.42–2.64)		
			All (continuous variable)	100	1.08 (0.94–1.25)		
			Trend-test *P*-value, 0.94				
Cocco et al. (2008) France, Germany, Spain (Epilymph multicentre study) Study period, NR	Cases: 174 incident cases recruited at hospital Controls: 203 from population and hospital Exposure assessment method: lipid-adjusted concentrations	NHL	β-HCH (ppb)			Age, sex, education, centre	Strengths: designed to give sufficient sample size Limitations: biological exposure assessed at diagnosis
			≤ 15	19	1.0		
			15.1–130.73	46	0.7 (0.3–1.6)		
			130.74–338.96	57	0.9 (0.4–1.9)		
			≥ 338.97	52	0.7 (0.3–1.5)		
			Trend-test *P*-value, 0.52				
		DLBCL	β-HCH (ppb)				
			≤ 15	5	1.0		
			15.1–130.73	14	0.8 (0.2–2.5)		
			130.74–338.96	13	0.8 (0.2–2.6)		
			≥ 338.97	12	0.7 (0.3–2.4)		
			Trend-test *P*-value, 0.66				
		SLL/CLL	β-HCH (ppb)				
			≤ 15	8	1.0		
			15.1–130.73	9	0.3 (0.1–1.1)		
			130.74–338.96	20	0.7 (0.2–1.9)		
			≥ 338.97	18	0.5 (0.2–1.5)		
			Trend-test *P*-value, 0.52				

Table 2.3 (continued)

Reference, location follow-up/enrolment period	Population size, description, exposure assessment method	Organ site	Exposure category or level	Exposed cases/ deaths	Risk estimate (95% CI)	Covariates controlled	Comments
Viel et al. (2011) Three electoral wards containing or surrounding a municipal solid waste incinerator, Besancon, France 1 January 2003-31 December 2005	Cases: 34 newly diagnosed at the department of haematology of the university hospital Controls: 34 randomly selected from donor registry of regional blood bank living in the area, matched on sex, age (± 5 yrs), blood draw (± 1 yr) Exposure assessment method: serum from a fasting blood sample; total lipid-adjusted concentration	NHL	ng/g lipid			NR	Strengths: exposure assessment does not rely on participant recall; detailed information on lifestyle factors, diet, occupation Limitations: substantial correlation between several classes of pesticides; the disease process or chemotherapy may have distorted the results
			β-HCH	NR	1.05 (1.00–1.12)		
			γ-HCH	NR	1.16 (0.93–1.49)		
Leukaemia							
Brown et al. (1990) Iowa and Minnesota, USA 1980–1983	Cases: 578 from tumour registry (Iowa) and hospital records (Minnesota) ; review by pathologists Controls: 1245 population-based (random-digit dialling, Medicare records, state death certificates); frequency-matched by 5-yr age group, vital status and state of residence Exposure assessment method: detailed questionnaires: number of animals and crops, use of 24 animal insecticides, 34 crop insecticides, 38 herbicides, 16 fungicides; first and last year of use; tasks (mixing, applying);. days per year for each pesticide	Leukaemia (including myelodys-plasias)	Lindane use on animals:			Vital status, age, state, tobacco use, family history lymphopoietic cancer, high risk occupations, high risk exposures	Studies in midwest USA Strengths: in-person interviews; detailed questionnaires including quantification; information on other potential risk factors; reviewed diagnosis Limitations: self-report of pesticide use
			Ever	38	1.1 (0.7–1.7)		
			Handled > 20 yrs ago	28	1.4 (0.8–2.3)		
			1–4 days/yr	15	1.1 (0.5–2.0)		
			5–9 days/yr	3	1.1 (0.3–4.1)		
			> 10 days/yr	10	1.6 (0.7–3.7)		
		Leukaemia (including myelodys-plasias)	Lindane use on crops:			Vital status, age, state, tobacco use, family history lymphopoietic cancer, high-risk occupations, high-risk exposures	
			Ever	14	1.6 (0.8–3.2)		
			1–4 days/yr	6	3.5 (0.9–12.6)		
			5–9 days/yr	2	1.2 (0.2–6.9)		
			> 10 days/yr	3	1.3 (0.3–5.3)		

CLL, chronic lymphocytic leukaemia; DLBCL, Diffuse large B-cell lymphoma; HCH, hexachlorocyclohexane; HD, Hodgkin disease; LOD, limit of detection; MM, multiple myeloma; NHL, non-Hodgkin lymphoma; NR, not reported; PCR, polymerase chain reaction; PPE, personal protective equipment; SEER, Surveillance, Epidemiology and End Results Program; SLL, small lymphocytic leukaemia; STS, soft tissue sarcoma; vs, versus; yr, year

family history, high-risk occupations, high-risk exposures). [Although exposure to one or more other substances was associated with NHL in this study, information on a large number of exposures was collected presenting potential problems of interpreting risk associated with a particular chemical as multiple comparisons increase the chances of false-positive findings.]

Data from the study by Cantor et al. (1992) were pooled with data from two other population-based case–control studies of NHL and exposure to pesticides in Kansas (Hoar et al., 1986) and Nebraska (Zahm et al., 1990), USA, in which exposure to lindane had not been analysed previously, to evaluate the risk associated with exposure to lindane (Blair et al., 1998). There were 987 cases and 2895 population-based controls. Information was obtained by telephone or in-person interviews, which included detailed questions on farm practices and agricultural use of chemicals. The risk of NHL (adjusted for age, state of residence, type of respondent - index or proxy) was significantly elevated among subjects reporting agricultural use of lindane (OR, 1.5; 95% CI, 1.1–2.0), and was greater among persons who first used the pesticide 20 years before diagnosis, among those who reported more frequent use of ≥ 5 days per year versus < 5 days per year, and from use on crops (OR, 1.9 95% CI, 1.1–1.3), rather than from use on animals (OR, 1.3; 95% CI, 0.9–1.8). However, odds ratios were lower for direct interviews (OR, 1.3; 95% CI, 0.9–1.8) than for proxy respondents (OR, 2.1; 95% CI, 1.0–4.4). Adjustments by use of chemical class of pesticides did not result in large changes in the odds ratios, with slight increases after adjusting for phenoxyacetic acids, triazines, amides, and benzoics, and slight decreases for organophosphates and natural products. Adjustment for reported use of two individual pesticides (dichlorodiphenyltrichloroethane, DDT; and fonofos) had no impact on the odds ratio for use of lindane, while adjustment for use of 2,4-D and diazinon reduced the odds ratios associated with use of lindane from 1.5 to 1.2 (95% CI, 0.5–3.2) and 1.3 (95% CI, 0.9–1.9), respectively.

The data from the study by Cantor et al. (1992) were also used to investigate the hypothesis that some risk factors might act specifically along t(14;18)-dependent pathways - the t(14;18) translocation is a common somatic mutation in NHL - leading to stronger associations with t(14;18)-positive than t(14;18)-negative non-Hodgkin lymphoma (Schroeder et al., 2001). Of 182 cases of NHL, 68 (37%) were t(14;18)-positive. Cases with t(14;18)-positive NHL were associated with several individual pesticides, including lindane (OR, 2.3; 95% CI, 1.3–3.9) in contrast to null or negative associations for the same self-reported exposures and t(14;18)-negative NHL. [No adjustment was made for shared agricultural exposures. This study had several limitations: 30% of the study participants were represented by next-of-kin respondents with consequent increased likelihood of inaccurate or missing data; some bias towards the null may have occurred due to t(14;18) breakpoints outside the range of the polymerase chain reaction (PCR) primers; the authors were unable to classify more than 60% of cases of NHL.]

In a further analysis of the a subset of the data from the same group of studies in the pooled analysis by Blair et al. (1998), De Roos et al. (2003) evaluated multiple exposures, including ever-use of 47 insecticides and herbicides, including lindane. The larger sample size provided adequate numbers of exposed subjects to analyse the set of pesticide exposures simultaneously, using hierarchical regression to adjust estimates based on prior distributions for the pesticide effects. The risk associated with exposures to several pesticides combined, defined as two pesticides used by the same person, but not necessarily at the same time, was also explored. Subjects with any missing information on exposure were omitted, and adjustments were made for the most frequently used pesticides. Ever exposure to lindane was not associated with an increase in risk of NHL (OR, 1.2; 95% CI, 0.8–1.9). In a further analysis of a subset of 100 cases of NHL and 100 matched controls from this pooled data, De Roos et al. (2005) investigated plasma organochlorine levels and risk of NHL. No association was observed for increasing quartiles of β-HCH concentration (*P* for trend, 0.94).

A study of leukaemia in Iowa and Minnesota, USA, which included 578 cases and 1245 population controls, reported a weak non-significant association with ever use of lindane as an insecticide on crops (OR, 1.6; 95% CI, 0.8–3.2, based on 14 cases) or on animals 20 years before the study (OR, 1.4; 95% CI, 0.8–2.3), after adjusting by risk factors including high-risk occupations, and high-risk exposures. Risk associated with exposure to lindane used as a crop insecticide was highest in the lowest category of days per year of use, while a linear, though non-significant, increase in risk was observed for use as an animal insecticide (Brown et al., 1990). [No tests for trend were reported in this paper.]

NHL is known to be associated with a compromised immune status. Asthma is a condition that is also associated with moderate alterations in immune function. For this reason, the data from Iowa, Minnesota, and Nebraska were pooled to investigate whether asthma modified the risk of NHL associated with pesticide exposures; 872 cases and 2336 frequency-matched controls were included (Lee et al., 2004). History of asthma was collected during the interviews. 177 (45 cases, 132 controls) reported having been told by their doctor that they had asthma. Subjects with an asthma history had a lower risk of NHL than non-asthmatics (OR, 0.6; 95% CI, 0.3–1.4). However, asthmatics tended to have larger odds ratios associated with exposure to lindane than did non-asthmatics (OR, 2.4; 95% CI, 1.0–5.7 and 1.3; 95% CI, 0.97–1.8, respectively). The reference category was non-asthmatic farmers. Similar patterns of association were found for many other pesticides. [The use of proxy respondents in this study may have led to misclassification. In addition exposure to a large number of pesticides was assessed but there was no attempt to account for co-exposure in the analyses. The study had limited power to investigate effect modification.]

Concentrations of 17 organochlorine pesticides, including four lindane isomers (α-, β-, γ-, and δ-HCH), were measured in plasma samples from 174 cases of NHL and 203 controls from France, Germany, and Spain, who participated in the Epilymph multicentre European population-based case–control study (Cocco et al., 2008). There was no increased risk of NHL associated with increasing quartiles of β-HCH, the only isomer present at detectable levels in enough subjects to permit analysis, nor for the NHL subtypes investigated (diffuse large B-cell lymphoma and chronic/small lymphocyctic leukaemia). [Although about half of the Spanish patients underwent blood withdrawal after initiating chemotherapy, results did not change after these patients were excluded. As noted previously, measurements of β-HCH in biological samples do not necessarily reflect exposure to lindane.]

A large, Canadian multicentre population-based incident case–control study (517 cases, 1506 controls) was carried out among men in a range of occupations using a postal questionnaire followed by a telephone interview for those reporting pesticide exposure of 10 hours/year or more, and a 15% random sample of the remainder (McDuffie et al., 2001). Exposure to lindane was significantly associated with an increased risk of NHL (OR, 2.05; 95% CI, 1.01–4.16), calculated with stratification by age and province, and was similar when additionally adjusted for significant medical variables such as measles, mumps, previous cancer, allergy desensitization shots, and a positive family history of cancer. [Although this was a large study, the response rate was relatively poor, especially in the controls, for whom it was less than 50%; however, a comparison of non-respondents with respondents using postal code as an indicator of rural residence did not find any indication of rural bias among respondents.]

In an investigation of exposure to organochlorines and the risk of NHL in the neighbourhood of a municipal solid-waste incinerator with high levels of dioxin emissions in Besançon, France, serum concentrations of pesticides, dioxins, furans, and polychlorinated biphenyls (PCBs) were measured for 34 cases of NHL (newly diagnosed in 2003–2005) and 34 controls randomly selected from the donor registry of the regional blood bank (matched for age, sex, and date of blood draw) (Viel et al., 2011). The mean serum concentration of β-HCH was 98.61 ng/g lipid for cases and 48.08 ng/g lipid for controls. An increased risk of NHL was found for β-HCH (OR, 1.05; 95% CI, 1.00–1.12, per 10 ng/g lipid). There was a high correlation with these isomers with several other exposures, including PCBs and lindane. The study excluded eight cases with rapid death or poor prognosis (who were not well enough to be interviewed). [The serum concentrations in this study may have been influenced by the disease process as serum samples were taken after diagnosis. In addition, measurements of β-HCH in biological samples do not necessarily reflect exposure to lindane.]

Miligi et al. (2003) assessed exposure to lindane in a case-control study of NHL and leukaemia that was carried out in Italy in 1990–1993. The analyses included 1145 cases of NHL, 430 cases of leukaemia, and 1232 age- and sex-matched controls from the general population. In-person interviews were conducted by trained interviewers and a questionnaire recorded information on many variables, including lifetime occupational history for all jobs held for more than 6 months and occupational exposure to solvents and pesticides. More specific and detailed data on specific jobs were collected through a job-specific questionnaire developed by industrial hygenists and agronomists. For agricultural exposures, expert agronomists reviewed the data collected and translated it into pesticide-exposure histories that included use of specific active ingredients (including lindane). Cases of NHL (ICD-10, 200, 202) and chronic lymphocytic leukaemia (ICD-10, 204.1) were analysed as a combined category due to biological similarities. Men exposed to lindane had an odds ratio of 2.0 (95% CI, 0.6–7.7; 9 exposed cases) for NHL and chronic lymphocytic leukaemia after adjustment for area and age.

2.3.3 Other cancers

See Table 2.4

(a) Cancer of the prostate

In a case–control study of urology patients in Ontario, Canada, β-HCH was measured in serum from 79 men with incident cancer of the prostate and from 329 age-matched controls (Aronson et al., 2010). No association was observed between concentration of β-HCH and risk of cancer of the prostate (*P* for trend, 0.81). [As cases and controls underwent the same diagnostic procedures and were screened by prostate-specific antigen (PSA) test and digital rectal examination, selection bias was unlikely in this study.]

In a case-control study of cancer of the prostate in British Columbia, Canada, Band et al. (2011) found an elevated risk of cancer of the prostate associated with ever being exposed to lindane (OR, 1.47; 95% CI, 0.94–2.29) and high exposure to lindane (OR, 2.02; 95% CI, 1.15–3.55), with a significant trend in the exposure-response relationship (*P* for trend, 0.03).

(b) Cancer of the testis

Biggs et al. (2008) conducted a population-based case-control study of testicular germ cell carcinoma in three counties of Washington state, USA, in 1999–2003. Generally, concentration of β-HCH was not associated with testicular germ cell carcinoma; however, an odds ratio of 5.54 (95% CI, 1.65–18.56) was associated with an increase of 10 pg/g serum in concentrations of γ-HCH.

Table 2.4 Case-control studies of cancers of the prostate or testis and exposure to lindane

Reference, location, enrolment/follow-up period	Population size, description, exposure assessment method	Organ site	Exposure category or level	Exposed cases/ deaths	Risk estimate (95% CI)	Covariates controlled	Comments
Biggs et al. (2008) Washington state, USA January 1999-December 2008	Cases: 246 men (age, 18–44 yrs) diagnosed with invasive TGCT and resident of King, Pierce, Snohomish counties, Washington State, identified from SEER Controls: 630 men with no history of TGCT, frequency-matched to cases on age (± 5 yrs) and residing in the same three counties as the cases; identified using random-digit dialling Exposure assessment method: personal monitoring; median β-HCH, 4.32 ng/g lipid; median γ-HCH, 1.37 ng/g lipid	Testis (TGCT)	β-HCH (pg/g)			Age, ethnicity, change in BMI between reference date and blood draw, assay run number, serum lipids	Strengths: relatively large study size Limitations: poor between-run reliability of the analytic method for several of the analytes; use of post-diagnostic blood samples; low response rates of cases and controls
			≤ 29	113	1.00		
			> 29–65	83	1.26 (0.86–1.85)		
			> 65	25	0.92 (0.51–1.64)		
			Per 10 pg/g	221	0.97 (0.82–1.13)		
			Trend-test *P*-value, 0.83				
		Testis (TGCT)	γ-HCH (pg/g)				
			≤ 9	130	1.00		
			9–20	73	0.80 (0.53–1.20)		
			> 20	38	1.36 (0.75–2.46)		
			Per 10 pg/g	241	5.54 (1.65–18.56)		
			Trend-test *P*-value, 0.69				
Aronson et al. (2010) Kingston, Ontario, Canada 1997–1999	Cases: 79 cases identified from 1288 men who visited a group of five urologists Controls: 329 controls identified from 1288 men who visited a group of five urologists (same source as the cases); subdivided into 194 urology controls (non-cancerous prostate lesions) and 135 biopsy controls (no prostate cancer detected at biopsy) Exposure assessment method: personal monitoring	Prostate	β-HCH (μg/g lipid)			Age, teenage physical activity, alcohol consumption, smoking pack-years	Strengths: PSA and DRE screening in cases and controls Limitations: very small number of cases; total response rates, NR; controls with urological diseases possibly related to exposure
			< 13.5	22	1.00		
			13.5–21.8	27	1.03 (0.54–1.99)		
			> 21.8 to 230.9	29	1.08 (0.57–2.06)		
			Trend-test *P*-value, 0.81				

Table 2.4 (continued)

Reference, location, enrolment/follow-up period	Population size, description, exposure assessment method	Organ site	Exposure category or level	Exposed cases/deaths	Risk estimate (95% CI)	Covariates controlled	Comments
Band et al. (2011) British Columbia, Canada, 1983–1990	Cases: 1153 histologically confirmed cases from British Columbia cancer registry Controls: 3999 age-matched cancer patients from the same registry: other sites, excluding lung and cancer of unknown primary site Exposure assessment method: lifetime occupational history obtained through a self-administered questionnaire plus JEM to estimate the participants' lifetime cumulative exposure to approximately 180 active compounds in pesticides	Prostate	Lindane exposure:			Alcohol consumption, cigarette yrs, education level, pipe years, respondent type (proxy/direct)	Correlation between specific pesticides as assessed through JEM is reported (e.g. r = 0.72 between lindane and DDT) Strengths: large study size; histological confirmation; high response rates; use of JEM limiting differential exposure misclassification; lifetime cumulative exposure assessment Limitations: multiple comparisons (142 active chemicals evaluated); high correlations between specific pesticides; no mutual adjustment
			Not exposed	1120	1.00		
			Ever	33	1.47 (0.94–2.29)		
			Low	10	0.91 (0.44–1.89)		
			High	23	2.02 (1.15–3.55)		
			Trend-test *P*-value, 0.03				

BMI, body mass index; CI, confidence interval; DRE, digital rectal examination; HCH, hexachlorocyclohexane; JEM, job-exposure matrix; NR, not reported; PSA, prostate-specific antigen; SEER, Surveillance, Epidemiology and End Results Program; TGCT, testicular germ cell tumours; yr, year

2.4 Meta-analysis

In a meta-analysis of NHL and occupational exposures to agricultural pesticides (described in section 2.3.2 of the *Monograph* on DDT in the present volume), the meta relative risk (meta-RR) estimate for NHL and exposure to lindane was significant and relatively precise (meta RR, 1.6; 95% CI, 1.2–2.2; $I^2 = 26\%$) (Schinasi & Leon, 2014). An evaluation of risk for NHL overall and NHL subtype associated with occupational exposure to lindane in the Agricultural Health Study (Alavanja et al., 2014b) was not available at the time at which the meta-analysis by Schinasi & Leon (2014) was accepted for publication. [Inclusion of the paper by Alavanja et al. (2014b) (relative risk [RR] for ever-exposure to lindane, 1.0; 95% CI, 0.8–1.2) would not have changed the meta-estimate substantially, as the results were similar to those of Purdue (RR for lindane exposure, 1.3; 95% CI, 0.8–2.1).] The authors explained that a formal meta-analysis of the exposure-response relationship could not be conducted for lindane because of several limitations of the literature, including variability in the definition of NHL among studies, the small numbers of exposed cases, and because of the differences in cut-off points in the several published studies.

3. Cancer in Experimental Animals

The Working Group has previously reviewed and evaluated the carcinogenicity of hexachlorocyclohexane (HCH) and some of its isomers, including lindane (γ-HCH), in experimental animals (IARC, 1979b; 1987). The α-isomer and technical-grade HCH were classified as having *sufficient evidence* of carcinogenicity in experimental animals, and the β- and γ- (lindane) isomers as having *limited evidence* of carcinogenicity in experimental animals. None of the studies performed and reviewed in the early and mid-seventies followed the current widely accepted study designs for carcinogenicity evaluations in experimental animals.

This *Monograph* revisits the studies that were evaluated previously by the Working Group, describes any updates, and provides details on new studies published subsequently (see also Table 3.1).

3.1 Mouse

Oral administration

Groups of 20 male ICR-JCL mice (age, 5 weeks) were fed diets containing lindane at a concentration of 0 (control), 300, or 600 ppm for 26 weeks. Early mortality was noted in the group at 600 ppm, with five animals dying during the course of the study. This group also displayed lower body weight increases than the controls. At the end of the study, 10 animals per group were killed for histological examination of the liver, kidney, and heart. The incidence of tumours in the liver was 5/10 in the group at 600 ppm (5/10 versus 0/10, [$P < 0.02$]). The tumours were defined as benign hepatomas. No tumours were reported in mice at the lowest dose or in the control group (Goto et al. 1972) [The Working Group noted the short duration of the study, the small number of animals, and that the maximum tolerated dose may have been exceeded (overt toxicity in the group at the highest dose).]

Groups of 10–11 male and female dd strain mice (age, 6 weeks), were fed diets containing lindane [purity not reported] at a concentration of 100, 300, or 600 ppm for 32 weeks, and the surviving animals were killed between week 37 and 38 of the experiment. The control group contained 21 males and 20 females (Hanada et al., 1973). The incidence of hepatoma [not otherwise specified] was 0/14 (controls), 0/10, 0/9, 3/4 [$P < 0.005$] in males; and 0/15 (controls), 0/8, 0/7, and 1/3 in females. [The Working Group noted the short duration of the study, the small number of animals, and that the maximum tolerated dose

Table 3.1 Studies of carcinogenicity in experimental animals fed diets containing lindane

Species, strain (sex); age at start Duration Reference	Dosing regimen No. of animals/group at start	Results For each target organ: incidence (%) and/or multiplicity of tumours	Significance	Comments
Mouse, ICR-JCL (M) ; age, 5 wk 26 wk Goto et al. (1972)	Lindane (purity, NR) at a dietary concentration of 0, 300, or 600 ppm 20/group	*Liver* Benign hepatomas: 0/10, 0/10, 5/10*	*[$P < 0.02$, Fisher 1-tail]	Limitations: small number of animals; short duration; one sex only; MTD may have been exceeded. At the end of the study, 10 animals per group were killed for histological examination of the liver, kidney, and heart Survival, 20, 20, 15
Mouse, CF-1 (M); age, 4 wk 110 wk Thorpe & Walker (1973)	Lindane (purity, 99.5%) at a dietary concentration of 0, or 400 ppm 45, 29/group	*Liver* Liver cell tumours (benign or malignant, combined): 11/45 (24%), 27/29 (93%)*	*$P < 0.01$, Finney 2 × 2 tables	Strengths: covered most of the lifespan Limitations: excessive mortality early in the study; single dose; MTD may have been exceeded Survival, 20, 5 (10% of mice died in first 3 mo of the study)
Mouse, CF-1 (F); age, 4 wk 110 wk Thorpe & Walker (1973)	Lindane (purity, 99.5%) at a dietary concentration of 0, or 400 ppm 45, 30/group	*Liver* Liver cell tumours (benign or malignant, combined): 10/44 (23%), 20/29 (69%)*	*$P < 0.01$, Finney 2 × 2 tables	Strengths: covered most of the lifespan Limitations: excessive mortality early in the study; single dose; MTD may have been exceeded Survival, 14, 1 (20% of female animals died in first 3 mo of the study)
Mouse, dd (M); age, 6 wk 37–38 wk Hanada et al. (1973)	Lindane (purity, NR) at a dietary concentration of 0, 100, 300, or 600 ppm for 32 wk followed by with basal diet for 5–6 wk 21 (control), 10–11 (treated)	*Liver* Hepatoma [NOS]: 0/14, 0/10, 0/9, 3/4*	[$P < 0.005$, Fisher 1-tail]	Limitations: short duration; small number of animals; MTD may have been exceeded Survival, 14, 10, 9, 4 (high mortality in the group at the highest dose)
Mouse, dd (F); age, 6 wk 37–38 wk Hanada et al. (1973)	Lindane (purity, NR) at a dietary concentration of 0, 100, 300, or 600 ppm for 32 wk followed by with basal diet for 5–6 wk 20 (control), 10–11 (treated)	*Liver* Hepatoma [NOS]: 0/15, 0/8, 0/7, 1/3	[NS]	Limitations: short duration; small number of animals; MTD may have been exceeded Survival, 15, 8, 7, 3 (high mortality in the group at the highest dose)
Mouse, NMRI (M); age, 5 wk 80 wk Herbst et al. (1975), Weisse & Herbst (1977)	Pulverized diet containing lindane (purity, NR) at a dietary concentration of 0, 12.5, 25, or 50 ppm 100, 50, 50, 50/group	*Liver* Adenoma: 4/97, 1/49, 0/48, 2/49	NS	Strengths: strain with 2% background incidence of liver tumours Limitations: dose-selection criteria not given; the high dose was below the MTD The studies looked at six sites for tumours: liver, lung, skin, ovary, lympho-haematopoietic system, and uterus; none of the sites had increased incidences of tumours compared with controls Survival, NR

Table 3.1 (continued)

Species, strain (sex); age at start Duration Reference	Dosing regimen No. of animals/group at start	Results For each target organ: incidence (%) and/or multiplicity of tumours	Significance	Comments
Mouse, NMRI (F); age, 5 wk 80 wk Herbst et al. (1975), Weisse & Herbst (1977)	Pulverized diet containing lindane (purity, NR) at a dietary concentration of 0, 12.5, 25, or 50 ppm 100, 50, 50, 50/group	*Liver* Adenoma: 1/98, 1/49, 0/49, 0/48	NS	Strengths: used animal strain with 2% background incidence of liver tumours Limitations: dose selection criteria not given; the high dose was below the MTD The studies looked at six sites for tumours: liver, lung, skin, ovary, lympho-haematopoietic system, and uterus; none of the sites had increased incidences of tumours compared with controls Survival, NR
Mouse, $B6C3F_1$ (M); age, 5 wk 90–91 wk NTP (1977)	Lindane (purity, > 99.9%) at a dietary concentration of 0 (matched controls), 0 (pooled controls), 80, or 160 ppm for 80 wk, followed by basal diet for 10 wk 10, 50, 50, 50/group	*Liver* Hepatocellular carcinoma: 2/10, 5/49, 19/49*, 9/46 *Liver* Neoplastic nodules or hepatocellular carcinoma (combined): 3/10, 8/49, 19/49*, 10/46	 *$P = 0.001$, Fisher exact test vs pooled controls *$P = 0.010$, Fisher exact test vs pooled controls	Strengths: covered most of the lifespan Limitations: number of matched controls was small Other comments: survival for treated and control groups within each sex was similar. Controls were pooled from four other contemporary studies to a total of 50 mice for statistical analysis of the data; the study was judged inadequate for the evaluation No. of survivors, NR
Mouse, $B6C3F_1$ (F); age, 5 wk 90–91 wk NTP (1977)	Lindane (purity, > 99.9%) at a dietary concentration of 0 (matched controls), 0 (pooled controls), 80, or 160 ppm for 80 wk, followed by basal diet for 10–11 wk 10, 50, 50, 50/group	*Liver* Hepatocellular carcinoma: 0/10, 2/47, 2/47, 3/46 *Liver* Neoplastic nodules or hepatocellular carcinoma (combined): 1/10, 3/47, 4/47, 3/46	 NS compared with matched or pooled controls NS compared with matched or pooled controls	Strengths: covered most of the lifespan Limitations: number of matched controls was small Other comments: survival for treated and control groups within each sex was similar. Controls were pooled from other four contemporary studies to a total of 50 mice for statistical analysis of the data; the study was judged inadequate for the evaluation No. of survivors, NR

Table 3.1 (continued)

Species, strain (sex); age at start Duration Reference	Dosing regimen No. of animals/group at start	Results For each target organ: incidence (%) and/or multiplicity of tumours	Significance	Comments
Mouse, Obese mottled yellow *A*^{vy}/*a* (YS × VY) F_1 hybrid (F); age, 4 wk 24 mo Wolff et al. (1987)	Lindane (purity, NR) at a dietary concentration of 0, or 160 ppm 96/group	*Liver* Hepatocellular adenoma or carcinoma (combined): 20/93, 49/94* Hepatocellular adenoma: 8/93, 33/94* Hepatocellular carcinoma 12/93, 16/94 *Lung* Bronchiolo-alveolar tumours: 4/95, 18/95*	 *[$P < 0.0001$, Fisher exact test] *[$P < 0.0001$, Fisher exact test] [NS] *[$P < 0.002$, Fisher exact test]	Strengths: covered most of the lifespan Limitations: use of a single dose and one sex only In a concurrent 18 mo-experiment with 36 mice per group, hepatocellular adenomas developed in 12/36 treated mice vs 0/34 controls [$P < 0.0001$, Fisher exact test] Survival, NR
Mouse, Lean pseudoagouti *A*^{vy}/*a* (YS × VY) F_1 hybrid (F); age, 4 wk 24 mo Wolff et al. (1987)	Lindane (purity, NR) at a dietary concentration 0, or 160 ppm 96/group	*Liver* Hepatocellular adenoma or carcinoma (combined): 7/95, 16/95* Hepatocellular adenoma: 5/95, 11/95 Hepatocellular carcinoma: 2/95, 5/95 *Lung* Bronchiolo-alveolar tumours: 6/95, 13/94	 *[$P < 0.05$, Fisher exact test] [NS] [NS] [NS]	Strengths: covered most of the lifespan Limitations: use of a single dose and one sex only Survival, NR
Mouse, Lean black *a*/*a* (YS × VY) F_1 hybrid mice (F); age, 4 wk 24 mo Wolff et al. (1987)	Lindane (purity, NR) at a dietary concentration of 0, or 160 ppm 96/group	*Liver* Hepatocellular adenoma or carcinoma (combined): 9/96, 4/96 Hepatocellular adenoma: 6/96, 3/96	 [NS] [NS]	Strengths: covered most of the lifespan Limitations: use of a single dose, and one sex only Survival, NR

Table 3.1 (continued)

Species, strain (sex); age at start Duration Reference	Dosing regimen No. of animals/group at start	Results For each target organ: incidence (%) and/or multiplicity of tumours	Significance	Comments
Mouse, Lean black *a/a* (YS × VY) F_1 hybrid mice (F); age, 4 wk 24 mo Wolff et al. (1987) (cont.)		Hepatocellular carcinoma: 3/96, 1/96 *Lung* Bronchiolo-alveolar tumours: 2/96, 3/96	 [NS] [NS]	
Mouse, Crl:CD-1(ICR) BR (M); age, 38–44 days 78 wk EPA (2001a, b)	Lindane (purity, 99.78%) at a dietary concentration of 0 (control), 10, 40, or 160 ppm 50/group	*Lung* Bronchiolo-alveolar adenoma: 16/49, 15/48, 11/49, 8/48 Bronchiolo-alveolar carcinoma: 0/49, 1/48, 3/49, 0/48 Bronchiolo-alveolar adenoma or carcinoma:(combined): 16/49, 16/48, 14/49, 8/48	 Significant negative trend NS (Significant negative trend)	Strengths: adequate duration, GLP study Incidences' denominator excluded mice that died before wk 44 Survival, 41, 34, 35, 38
Mouse, Crl:CD-1(ICR) BR (F); age, 38–44 days 78 wk EPA (2001a, b)	Lindane (purity, 99.78%) at a dietary concentration of 0 (control), 10, 40, or 160 ppm 50/group	*Lung* Bronchiolo-alveolar adenoma: 3/48, 7/46, 7/47, 11/48* Bronchiolo-alveolar carcinoma: 1/48, 2/46, 2/47, 1/48 Bronchiolo-alveolar adenoma or carcinoma (combined): 4/48, 8/46, 9/47, 12/48*	 *P* = 0.0274, exact trend test; **P* = 0.0200, Fisher exact test NS *P* = 0.0389, exact trend test; **P* = 0.0264, Fisher exact test	Strengths: adequate duration; GLP study Incidences' denominator excludes mice that died before wk 44; results of resectioning of lungs did not significantly change initial findings and conclusion Survival, 37, 31, 33, 40

Table 3.1 (continued)

Species, strain (sex); age at start Duration Reference	Dosing regimen No. of animals/group at start	Results For each target organ: incidence (%) and/or multiplicity of tumours	Significance	Comments
Rat, Osborne-Mendel (M); age, 5 wk 108–110 wk NTP (1977)	Lindane (purity,100%) at a dietary concentration of 0 (matched controls), 0 (pooled controls), 236, or 472 ppm TWA for 80 wk, and followed by basal diet for 28–30 wk 10, 55, 50, 50/group	*Spleen* Haemangioma: 0/8, 0/52, 0/44, 3/44	 $P = 0.030$, Cochran-Armitage trend test vs pooled controls	Strengths: covered most of the lifespan Limitations: reduction in dose levels during the course of studies due to death among treated rats; small number of matched controls Other comments: controls were pooled from four other contemporary cancer studies to a total of 55 rats for statistical analysis; the study was judged inadequate for the evaluation Survival, NR
		Thyroid C-cell adenoma: 1/6, 2/42, 3/37, 1/37	 NS	
		Liver Liver neoplastic nodules: 0/10, 0/49, 3/45, 2/45	 NS	
Rat, Osborne-Mendel (F); age, 5 wk 108–110 wk NTP (1977)	Lindane (purity, 100%) at a dietary concentration of 0 (matched controls), 0 (pooled controls), 135, or 270 ppm TWA for 80 wk, and followed by basal diet for 28–30 wk 10, 55, 50, 50/group	*Thyroid* C-cell adenoma: 0/8, 0/48, 4/44*, 3/42	 $^*P = 0.049$, Fisher exact test vs pooled controls	Strengths: covered most of the lifespan Limitations: reduction in dose levels during the course of studies due to death among treated rats; small number of matched controls Other comments: controls were pooled from four other contemporary carcinogenicity studies to a total of 55 rats for statistical analysis; the study was judged inadequate for the evaluation Survival, NR
		Pituitary gland Chromophobe adenoma: 3/7, 6/46, 14/45*, 8/45	 $^*P = 0.033$, Fisher exact test vs pooled controls	
		Liver Neoplastic nodules: 0/10, 1/49, 4/48, 2/45	 NS	

Table 3.1 (continued)

Species, strain (sex); age at start Duration Reference	Dosing regimen No. of animals/group at start	Results For each target organ: incidence (%) and/or multiplicity of tumours	Significance	Comments
Rat, Wistar (M); age, NR 24 mo EPA (2001b)	Lindane (purity, 99.78%) at a dietary concentration of 0 (control), 1, 10, 100, or 400 ppm 50/group	*Adrenal gland* Pheochromocytoma, benign:		Strengths: adequate duration Limitations: limited study details No significant increase in tumour incidence in treated groups in the experiment with female rats with the same study design (EPA, 2001b) Survival, NR
		14%, 16%, 16%, 6%, 24%	NS (statistical tests, NR)	
		Pheochromocytoma, malignant:		
		0%, 0%, 6%, 8%, 2%	NS (statistical tests, NR)	
		Pheochromocytoma, benign or malignant (combined):		
		14%, 16%, 18%, 14%, 26%	NS (statistical tests, NR)	

F, female; GLP, good laboratory practice; M, male; mo, month; MTD, maximum tolerated dose; NR, not reported; NS, not significant; ppm, parts per million; TWA, time-weighted average; vs, versus; wk, week

may have been exceeded (overt toxicity and high mortality in the group at the highest dose).]

In a 2-year comparative study of oral toxicity and carcinogenicity, groups of 29 male and 30 female CF1 mice (age, 4 weeks) were fed diets containing lindane (purity, 99.5%) at 400 ppm for 110 weeks (Thorpe & Walker, 1973). The control group of 45 males and 45 females received basal diet only. During the first 3 months of the experiment, 10% of males and 20% of females died in the treated group. At necropsy, the incidences of benign or malignant (combined) tumours of the liver were 93% in treated males ($P < 0.01$) and 69% in treated females ($P < 0.01$), compared with 24% and 23% in controls, respectively. Metastasis of liver tumours to the lungs was observed in some of the treated animals. [The Working Group noted that the single dose tested may have exceeded the maximum tolerated dose.]

Groups of 50 male and 50 female NMRI mice (age, 5 weeks) were fed diets containing lindane at a concentration of 12.5, 25, or 50 ppm for 80 weeks. The control group contained 100 males and 100 females. No changes in body weight, feed consumption, or mortality were observed in the treated animals. There was no increase in the incidence of neoplasms of the liver, lung, skin, ovary, uterus, or lympho-haematopoietic system in the treated groups (Herbst et al., 1975; Weisse & Herbst, 1977). [The Working Group noted that lindane was tested below the maximum tolerated dose.]

Groups of 50 male and 50 female B6C3F$_1$ mice (age, 5 weeks) were fed diets containing lindane (purity, 100%) at a concentration of 80 or 160 ppm for 80 weeks, followed by an observation period of 10–11 weeks (NTP, 1977). The respective control groups had only 10 matching animals. For the statistical analysis, 10 control animals each were used from four other contemporaneous studies conducted over a period of 1 year at the same laboratory; this "pooled" control group consisted of a total of 50 animals. Survival of males and females in the treated and control groups was similar. Hepatocellular carcinomas and neoplastic nodules of the liver were the most frequent lesions observed in the controls and in the treated groups. The incidence of hepatocellular carcinoma in males at the higher dose (9/46) was not significantly different from that of the pooled control group (5/49). However, the incidence of hepatocellular carcinoma in males at the lower dose was significantly higher than that in the pooled controls (19/49, $P = 0.001$). There was no significant increase in the incidence of any tumour type in females. [The Working Group noted that, because of the small number of matched controls, pooled controls were used for statistical analyses which limited the interpretation of the study. The study was judged inadequate for evaluation.]

Lindane was studied in obese mottled yellow A^{vy}/a, lean pseudoagouti A^{vy}/a and lean black *a/a* (YS × VY) F$_1$ hybrid female mice (Wolff et al., 1987). F$_1$ hybrid mice were produced by mating *a/a* YS females with A^{vy}/a VY male mice. The dominant mutation at the A^{vy} locus in mice results in two phenotypic groups, obese mottled yellow and lean pseudoagouti, that are genetically identical but physiologically different. The hypothesis of the study was that these two phenotypes represent different degrees of expression of this mutation. To prove this hypothesis, the tumourigenic response to lindane of obese mottled yellow A^{vy}/a, lean pseudoagouti A^{vy}/a, and the relatively neoplasia-resistant black *a/a* (YS × VY) F$_1$ hybrid female mice were studied. The mice were given diets containing lindane [purity unspecified] at a concentration of 0 (control) or 160 ppm for 18 or 24 months before termination of the experiment. Each of the three phenotypes had their respective controls and treated groups. The 18-month and 24-month groups had 36 and 96 mice (age, 4 weeks) per group, respectively. At 18 months, only treated obese yellow mice had a significant increase [$P < 0.0001$] in the incidence of hepatocellular adenoma (controls, 0/34; treated, 12/36). At 24 months, the incidences of hepatocellular tumours and lung bronchiolo-alveolar tumours differed quantitatively among the three phenotypes. The incidences of hepatocellular adenoma or carcinoma (combined) in the controls and treated groups were, respectively: 20/93, 49/94 [$P < 0.0001$] in the obese yellow mice; and 7/95, 16/95 [$P < 0.05$] in the lean pseudoagouti mice. In addition, lindane increased the incidence of bronchiolo-alveolar tumours in obese yellow (4/95, 18/95 [$P < 0.002$]) and lean pseudoagouti (6/95, 13/94 [not significant]) mice. Bronchiolo-alveolar tumours observed in this study were classified as either papillary or solid [and not further classified]. The black mouse was resistant to the induction of both hepatocellular and bronchiolo-alveolar tumours. [The Working Group noted the use of a single dose in a study on females only.]

In a study submitted to the EPA (2001a, b), which was carried out according to good laboratory practice, groups of 50 male and 50 female CD-1 mice (age, 38–44 days) were fed diets containing lindane (purity, 99.78%) at a concentration of 0, 10, 40, or 160 ppm for 78 weeks before termination of the experiment. There were no significant differences in body weights and survival between treated animals and controls. In females at the highest dose, there were significant decreases in weights of the uterus plus cervix. In female mice, there was a significant positive trend in the incidence of bronchiolo-alveolar adenoma ($P = 0.0274$), and a significant increase in incidence in the group at the highest dose ($P = 0.0200$) compared with controls. There was also a significant positive trend in the incidence ($P = 0.0389$), and a significant increase in the incidence of bronchiolo-alveolar adenoma or carcinoma (combined) at the highest dose group ($P = 0.0264$) compared with controls. Results of resectioning of lungs did not significantly change the initial findings and conclusion. Lindane did not increase the incidence of any tumour type in treated males.

3.2 Rat

Oral administration

Groups of 50 male and 50 female Osborne-Mendel rats (age, 5 weeks) were fed diets containing lindane (purity, 100%) at time-weighted average doses of 236 ppm (lower dose) or 472 ppm (higher dose) in males, and 135 ppm (lower dose) or 270 ppm (higher dose) in females for 80 weeks (NTP, 1977). The matched controls comprised 10 males and 10 females. The data from 45 untreated males and 45 untreated females from four other similar ongoing cancer bioassays in the same laboratory were pooled for the statistical analysis. Throughout the study, doses were lowered for all groups of treated animals due to mortality. All surviving animals were killed at 108–110 weeks. The incidences of thyroid C-cell adenoma in males were 1/6 (matched controls), 2/42 (pooled controls), 3/37 (lower dose), and 1/37 (higher dose); and in females were 0/8, 0/48, 4/44 ($P = 0.049$, versus pooled controls) and 3/42. The incidences of spleen haemangioma in males were 0/8, 0/52, 0/44, and 3/44, with a significant positive trend ($P = 0.030$, compared with pooled controls). The incidence of chromophobe adenoma of the pituitary gland in females was 3/7, 6/46, 14/45 ($P = 0.033$, versus pooled controls), and 8/45. The incidence of neoplastic nodules of the liver was 0/10, 0/49, 3/45, and 2/45 in males, and 0/10, 1/49, 4/48, and 2/45 in females. [The Working Group noted that dose levels were changed during the course of the study, and that because of the small number of matched controls, pooled controls were used for statistical analyses, which limited the interpretation of the study. The study was judged inadequate for evaluation.]

In a study submitted to the EPA (2001b), groups of 50 male and 50 female Wistar rats [age not reported] were fed diets containing lindane (purity, 99.78%) at a concentration of 0 (control), 1, 10, 100, or 400 ppm for 2 years. Final body weights of males at the highest dose were significantly ($P < 0.05$) less than those of the controls. Body weights and body-weight gains for treated females were similar to those of the controls throughout the study. The percentages of males with tumours of the adrenal gland were: 14%, 16%, 16%, 6%, and 24% for benign tumours; and 0%, 0%, 6%, 8%, and 2% for malignant tumours, respectively. These incidences were not significantly increased compared with controls [statistical tests not reported]. There was no significant increase in tumour incidence in any group of treated females compared with controls. [The Working Group noted the limited reporting.]

4. Mechanistic and Other Relevant Data

4.1 Toxicokinetics

4.1.1 Humans

(a) Absorption, distribution, and excretion

Lindane is a lipophilic compound that is expected to be readily absorbed after exposure. Absorption after inhalation has not been directly measured experimentally in humans, but has been inferred from body-burden measurements of lindane from occupational exposures (e.g. Baumann et al., 1980). While no experimental studies of oral intake of lindane in humans were available to the Working Group, uptake via ingestion has been inferred from accidental or intentional poisoning cases (e.g. Paul et al., 2013; Ramabhatta et al., 2014). Dermal absorption has also been demonstrated, but is dependent on the vehicle in which the chemical is administered. About 9% of the radiolabel was excreted in the urine after administration of lindane dissolved in acetone onto the forearm of healthy adult volunteers (Feldmann & Maibach, 1974). In a separate study, absorption into the systemic circulation after 6 hours was reported to be about 5% when using acetone as the vehicle, and about 60% when using white spirit (Dick et al., 1997). Dermal absorption of lindane from contaminated soil was measured to be in the range of 0.45–2.35% after 24 hours, depending on organic carbon content and soil loading (Duff & Kissel, 1996).

Lindane readily distributes throughout the body via the systemic circulation, with a preference for lipid-rich tissues such as adipose, or the brain (Baumann et al., 1980; Davies et al., 1983). Lindane has also been found in breast milk and umbilical cord blood, indicating lactational and placental transport (Siddiqui et al., 1981). Blood peak concentrations of lindane occurred approximately 4–6 hours after dermal administration (Ginsburg et al., 1977; Lange et al., 1981). Lindane is excreted mainly as metabolites in the urine, with very little excreted unchanged. Excretion terminal half-lives of 18–26 hours have been reported in several studies in adults and/or children (Feldmann & Maibach, 1974; Ginsburg et al., 1977; Aks et al., 1995). A more recent study of an ingestion overdose case estimated a longer half-life of 163 hours (Wiles et al., 2015). One study found that the elimination half-life, like the absorption rate, depended on the vehicle, with a shorter half-life of 25–58 hours with white spirit as vehicle, and a longer half-life of 50–111 hours with acetone (Dick et al., 1997). [The Working Group noted that the half-life of β-HCH is much longer, in the order of 7 years.]

(b) Metabolism

No studies on the metabolism of lindane in vivo in exposed humans were available to the Working Group. Based on biomonitoring of occupational exposures, numerous chlorophenols were identified in the urine of workers exposed to lindane and other HCH isomers (Engst et al., 1976a, b; Angerer et al., 1983). Specifically, 2-monochlorophenol, 3-monochlorophenol, and 4-monochlorophenol; 2,3-dichlorophenol, 2,4-dichlorophenol, 2,6-dichlorophenol, and

3,4-dichlorophenol; 2,4,6-trichlorophenol, 2,4,5-trichlorophenol, 2,3,4-trichlorophenol, 2,3,6-trichlorophenol, and 2,3,5-trichlorophenol; and 2,3,4,6-tetrachlorophenol and 2,3,4,5-tetrachlorophenol were identified. The metabolites 2,4-dichlorophenol and 2,4,6-trichlorophenol, 2,3,5-trichlorophenol, and 2,4,5-trichlorophenol were the most abundant metabolites in the urine, accounting for > 70% of urinary chlorophenols (Angerer et al., 1983). However, co-exposure to dichlorobenzene may have contributed to some of the dichlorophenol metabolites. One study in workers exposed to HCH isomers reported that several of the urinary metabolites were glucuronidated (Engst et al., 1976b). In a study in children given lindane for the treatment of lice, concentrations of 2,4,5- and 2,4,6-trichlorophenol and pentachlorophenol in the urine were elevated, but not statistically significantly so, in comparison to unexposed children, suggesting significant background exposures (Naeher et al., 2009).

In an experiment assessing the metabolism of lindane with human liver microsomes in vitro, six metabolites were reported: γ-1,2,3,4,5,6-hexachlorocyclohex-1-ene (3,6/4,5-HCCH), γ-1,3,4,5,6-pentachlorocyclohex-1-ene (3,6/4,5-PCCH), β-1,3,4,5,6-pentachlorocyclohex-1-ene (3,4,6/5-PCCH), 2,4,6-trichlorophenol 2,3,4,6-tetrachlorophenol, and pentachlorobenzene, the latter two of which were secondary metabolites of 3,6/4,5-HCCH (Fitzloff et al., 1982). The major pathways were through hexachlorocyclohex-1-ene and pentachlorocyclohex-1-ene, with trichlorophenol accounting for a smaller pathway. A separate experiment in human liver microsomes reported 3,4,6/5-PCCH oxide as the major product of 3,4,6/5-PCCH, with 17% of the original 3,4,6/5-PCCH converted to 3,4,6/5-PCCH oxide in 30 minutes (Fitzloff & Pan, 1984). However, this epoxide was found to be stable and not a substrate of epoxide hydrolase (Fitzloff & Pan, 1984). Pentachlorophenol was reported as one of the metabolites of lindane in some studies (Engst et al., 1976a, b), but not others (Angerer et al., 1983).

4.1.2 Experimental systems

(a) Absorption, distribution, and excretion

Lindane is readily absorbed by all species of experimental animal tested. Although no studies were available on the measurement of systemic absorption of lindane after inhalation, based on its lipophilicity, lindane would be expected to be absorbed by this route. Radiolabel studies in vivo with lindane in rats, rhesus monkeys, guinea-pigs, mice, pigs, and dogs have demonstrated dermal absorption (Reifenrath et al., 1984; Moody & Ritter, 1989; Franz et al., 1996), as have studies in vitro (Chang et al., 1994). Lindane is rapidly absorbed after ingestion, with between 30% and 62% reported to be absorbed after 30 minutes in mice and rats (Turner & Shanks, 1980; Ahdaya et al., 1981). In rats, the main route of entry into systemic circulation was blood, with only a small amount entering via the lymphatic system (Turner & Shanks, 1980). In a mouse model, cholestyramine reduced absorption of a single oral dose of lindane that was otherwise acutely toxic (Kassner et al., 1993).

After absorption, lindane and its metabolites are readily distributed to tissues via blood circulation. Due to its lipophilicity, lindane is preferentially stored in lipid-rich tissues such as adipose, but has been detected in a wide range of tissues, including the liver, brain, kidney, adrenals, heart, lungs, spleen, and testis (Eichler et al., 1983; Dalsenter et al., 1996; Siddiqui et al., 1996; Khanna et al., 2002). In all cases, tissues concentrations were higher than in the blood. In one study in the brain, it was reported that lindane preferentially concentrated in white matter as opposed to grey matter, with higher concentrations in the thalamus, mid-brain, and pons-medulla (Sanfeliu et al., 1988). When administered during or after pregnancy in rabbits and rats, lindane transfer via the placenta to the fetus and via lactation to the neonate has been reported (Khanna et al., 1991; Pompa et al., 1994; Seiler et al., 1994; Dalsenter et al., 1997a).

Excretion of lindane occurs mainly through metabolites in the urine, with very little excreted unchanged or in the faeces (Chadwick et al., 1977, 1985; Ahdaya et al., 1981). Excretion appears to be limited by metabolic transformation, as increased body burden and reduced excretion occurred in mice exposed to higher doses that saturated biotransformation (Chadwick et al., 1987).

(b) Metabolism

The metabolism of lindane is complex and involves multiple intermediates and metabolites. Initial metabolism appears to be catalysed by cytochrome P450 (CYP), as shown by activity with rat liver microsomes (Fitzloff et al., 1982; Yamamoto et al., 1983). More than 70 metabolites have been identified in mammalian systems (Macholz & Kujawa, 1985). The ultimate fate of these metabolites is formation of mercapturic acid, glucuronide, or sulfate conjugates excreted in the urine (ATSDR, 2005).

[The Working Group noted that many of the metabolites of lindane, including trichlorophenols, may be biologically active (e.g. exhibit genotoxicity). However, there is a lack of adequate quantitative data to be able to attribute any of the observed effects of lindane to specific metabolites. Therefore, subsequent sections of the present monograph did not include evaluation of any data on metabolites of lindane.]

As with human liver microsomes, rat liver microsomes converted the lindane metabolite 3,4,6/5-PCCH to an epoxide, but this compound was found to be stable, with minimal further metabolism by rat liver microsomes or by purified epoxide hydrolase (Fitzloff & Pan, 1984).

[The Working Group noted that the formation of an epoxide, both in human and experimental systems (albeit in vitro), provides evidence that lindane may be metabolically activated to an electrophile. Additionally, the stability of the epoxide suggests the possibility that it may be systemically available beyond the site of formation (presumably the liver).]

4.1.3 Modulation of metabolic enzymes

(a) Humans

No studies in exposed humans were available to the Working Group.

In a study in vitro, lindane (concentration, 25, 50, and 75 μM; exposure, 10 minutes to 18 hours) aromatase activity was increased after 10 minutes to 6 hours, but inhibited after 18 hours in human placental JEG-3 and transfected kidney E293 cells. There were no effects on *CYP19* mRNA, the gene transcript coding for aromatase (Nativelle-Serpentini et al., 2003). Lindane induced the mRNA expression of CYP2B6 (sevenfold), but inhibited that of CYP2D6 (0.5-fold) and CYP2E1 (0.2-fold) in freshly isolated human hepatocytes (Ellero et al., 2010).

(b) Experimental systems

In CD-1 mice exposed to lindane in utero (dams were dosed with 25 mg/kg bw on days 9–16 of gestation), CYP-dependent steroid hormone metabolism was impaired in male offspring (Di Consiglio et al., 2009). In the adult F_1 mice, CYP-mediated testosterone metabolism was dramatically affected at postnatal days 65–69, in the absence of systemic toxicity. During this period, testosterone 6β- and 2α-hydroxylation and dehydrogenation activities were strongly reduced, suggesting the CYP3A and CYP2C families as the major target of lindane-induced effects. Most changes had almost reversed by postnatal day 100. No effects on aromatase (CYP19) activity were seen. [These findings suggested an impairment of steroid hormone homeostasis, due to CYP-mediated disruption of testosterone catabolism (Di Consiglio et al., 2009).]

Oral administration of lindane (2.5, 5, 10, or 15 mg/kg bw) for 5 days caused a dose-dependent increase in the activity of CYP-dependent 7-ethoxyresorufin-*O*-deethylase (EROD), 7-pentoxyresorufin-*O*-dealkylase (PROD), and *N*-nitrosodimethylamine demethylase (NDMA-*d*) in rat brain and liver (Parmar et al., 2003). Hepatic and brain CYP activity was also increased when the lowest dose (2.5 mg/kg) of lindane was given for a longer duration (15 or 21 days). CYP induction was greater in liver than in brain. Expression of CYP 1A1/1A2, 2B1/2B2 and 2E1 isoenzymes was increased by lindane. [The Working Group noted that each of these enzymes could affect the metabolism of other chemicals.]

Gastrointestinal nitroreductase activity was increased in the small intestine in weanling F344 rats given lindane at a dose of 20 mg/kg bw daily by gavage for 5 weeks (Chadwick et al., 1990). Lindane had no effect on either nitroreductase or dechlorinase enzyme activity in the caecum. [The Working Group noted that increased nitroreductase may account for the previously reported interaction between lindane and parathion.]

4.2 Mechanisms of carcinogenesis

4.2.1 Immunosuppression

(a) Humans

(i) Exposed humans

In comparison to an external control group of 20 clerks, 60 male workers in a factory producing lindane had significantly elevated levels of polymorphonuclear leukocytes and reticulocytes (Brassow et al., 1981). Significantly lower lymphocyte counts, prothrombin (Quick) test, and blood creatinine and uric acid concentrations were also seen in the factory workers. No other significant differences were identified, including in total erythrocytes and leukocytes, platelets, or haemoglobin content, or from case history, physical examination, neurological status, or electrocardiography. [The Working Group noted that the design of this study was weak because of the size and choice of control group.]

Nigam et al. studied 365 individuals exposed to HCH (80% β-HCH) during its manufacture and compared them with 146 controls (Nigam et al., 1993). Beta-globulins significantly increased as total HCH increased. In representative samples, circulating immune complexes were also detected. High concentrations of HCH were reported in the serum of all exposed workers. [The Working Group noted an apparent effect on antibody production in exposed humans.]

Aplastic anaemia was reported in multiple cases of lindane exposure, confirmed through serum blood measurements (Rauch et al., 1990; Rugman & Cosstick, 1990).

In addition to the studies suggesting modest effects of HCH on the immune system of exposed humans, two additional studies reported on mixed exposures to HCHs and other polychlorinated compounds. A study of 146 workers exposed primarily to PCBs for more than 6 months reported only weak associations between immunological abnormalities and concentrations of [α]-HCH, β-HCH, and [γ]-HCH (Daniel et al., 2001). In a separate study, pyruvate metabolism in peripheral blood lymphocytes was different relative to controls for 36 workers occupationally exposed to polychlorinated pesticides (Gammexane, DDT) (Hrycek et al., 1984). [The Working Group noted that these changes were of limited importance.]

Regarding related compounds, Dar et al. reported significantly higher blood β-HCH levels in patients with systemic lupus erythematosus than in healthy controls (Dar et al., 2012). In the patients, HCH concentrations correlated with marked increases in CD3(+)CD4(+) T-lymphocytes and decreases in CD4(+)CD25(+) T-lymphocytes.

(ii) Human cells in vitro

In early studies, Fisher and Mueller reported that γ-HCH inhibited the stimulation of lymphocyte growth by phytohaemagglutinin (Fisher & Mueller, 1971). Roux et al. confirmed the finding that lindane inhibits lymphocyte activation in studies in human peripheral blood mononuclear cells (Roux et al., 1979). Lindane (10^{-4} M) inhibited macromolecular biosynthesis in unstimulated lymphocytes, phytohaemagglutinin-activated lymphocytes, and dividing blast cells (Roux et al., 1979). Dar et al. (2012) corroborated this finding, reporting an inhibitory effect of HCH after treatment in vitro of peripheral blood mononuclear cells from patients with systemic lupus erythematosus. Interleukin-2 (IL-2) and interferon gamma (IFNγ) levels were decreased by HCH, while no effect was seen on IL-4 levels in the patients (Dar et al., 2012).

[The Working Group noted that together these studies show that lindane blocks lymphocyte activation in vitro, which is an immunosuppressive effect.]

Lindane was cytotoxic to human haematopoietic progenitor cells in vitro at concentrations similar to serum concentrations in acute poisonings (Parent-Massin et al., 1994).

(b) Experimental systems

Studies in multiple species show immunosuppressive effects with lindane and HCHs generally, and accumulation of HCHs in lymphoid organs.

(i) Non-human mammals in vivo

Mouse

Dose-dependent immunesuppressive effects with lindane and HCHs were seen in mice. In exposed albino mice, lindane suppressed both primary and secondary humoral immune responses in a time- and dose-dependent manner (Banerjee et al., 1996). With shorter durations of exposure, the secondary antibody response to sheep erythrocytes was more markedly suppressed than the primary response. A biphasic effect on cell-mediated and humoral immune responses was seen in mice exposed for 24 weeks to subtoxic doses of γ-HCH (0.012, 0.12, and 1.2 mg/kg) and evaluated 1 month later (Meera et al., 1992). Initial stimulation was followed by dose-dependent immunosuppression, accompanied by histological changes in lymphoid organs. No effect was seen on peritoneal macrophage function. A second study showed that uptake of ^{45}Ca increased during the initial immunostimulation, and then decreased concomitantly with immunosuppression in mice exposed to γ-HCH (0.012, 0.12, and 1.2 mg/kg) for 4, 12, and 24 weeks (Meera et al., 1993). Verapamil (a calcium-channel blocker) and trifluoperazine (a calmodulin inhibitor) inhibited lymphocyte proliferation during both phases of immunomodulation. Das et al. (1990) showed that HCH (10 and 100 mg/kg bw) can modulate the developing immune system in Swiss albino mice. HCH (α, β, and γ isomers) residues in pups were higher in the lymphoid organs than in the liver, and increased with dose. The delayed hypersensitivity response to sheep erythrocytes was significantly higher at the lower dose, but significantly impaired at the higher dose, compared with controls. The lower dose elevated both the mitogenic responsiveness of the spleen cells and the antibody response to sheep erythrocytes, while no effect on either measure was seen at the higher dose.

Rat

In rats, lindane and technical HCH suppressed the humoral immune response and were haematotoxic. Koner et al. demonstrated that subchronic lindane exposure in rats suppressed the humoral immune response, increased lipid peroxidation, and decreased antioxidant enzymes (Koner et al., 1998). Lindane (40 and 80 ppm in the diet, for 8 weeks) markedly reduced anti-sheep erythrocyte antibody titres, an effect attenuated by daily treatment with ascorbic acid (100 mg/kg, intragastric). A suppressive effect of lindane on humoral immune responses was also seen in weanling rats exposed for 5 weeks (Dewan et al., 1980). The antibody titres attained in response to typhoid vaccine in untreated controls were significantly higher than those in treated animals. Administration of γ-HCH (20 mg/kg per day) for 30 days to ovariectomized rats significant changed the numbers of erythrocytes, neutrophils, and lymphocytes, as well as level of haemoglobin (see also Section 4.2.4) (Raizada et al., 1980). Joseph et al. (1992) showed that the haematotoxicity of dietary HCH (1000 ppm) in male albino rats is enhanced by vitamin A (2000 or 10^5 international units/kg). When vitamin A was absent from the diet, HCH induced severe haematotoxicity, as demonstrated by significantly reduced total leukocyte count, clotting time, and prothrombin time. On the other hand, the only indication of HCH-induced haematotoxicity in rats that received vitamin A supplements was a slight, significant decrease in total leukocyte count.

Rabbit

Kopeć-Szlezak et al. (1990) reported functional changes in granulocytes and structural changes in lymphocytes in the peripheral blood of rabbits exposed for 30 days to lindane (daily doses of 0.1 LD_{50}, or 7 mg/kg). Lindane significantly decreased phagocytic activity, and increased the number of non-phagocytizing granulocytes. Quantitative and qualitative changes were evident in the nucleoli and lysosomes of lymphocytes (Kopeć-Szlezak et al., 1990). Grabarczyk et al. (1990) reported reduced phagocytic activity of neutrophils, and increased number of lymphocytes with inactive nucleoli.

(ii) Non-human mammalian cells in vitro

Lindane caused concentration- and time-dependent cytotoxicity in C57BL/6 mouse splenocytes (Battaglia et al., 2010), and induced apoptotic and necrotic cell death in C57BL/6 mouse thymocytes (Olgun et al., 2004). Mixtures of lindane with either malathion or permethrin demonstrated a significantly greater than additive interaction in apoptotic and necrotic cells. Lindane was cytotoxic to rat haematopoietic progenitor cells, but at concentrations 1000-fold those in human progenitors (Parent-Massin et al., 1994).

(iii) Non-mammalian systems

Fish

At sublethal concentrations, lindane altered haematological parameters in the fish *Etroplus maculatus* (Bijoy Nandan & Nimila, 2012). Significant reductions were seen in erythrocyte count, haemoglobin, and haematocrit (erythrocyte volume fraction) with corresponding changes in mean corpuscular haemoglobin, mean corpuscular volume, and mean corpuscular haemoglobin concentration. In contrast, the leukocyte count) was significantly increased (Bijoy Nandan & Nimila, 2012).

In tilapia, lindane (20 or 40 mg/kg, intraperitoneal administration, for five consecutive days) decreased spleen and pronephros total leukocyte counts (Hart et al., 1997). Hypocellularity of lymphoid regions in the spleen and pronephros was also evident in fish exposed to lindane.

In rainbow trout, lindane (10, 50, or 100 mg/kg, intraperitoneal administration, for 5 days) affected antibody-secreting cells in a dose-dependent fashion. As a result, antibody production in sera, as demonstrated by agglutination, was suppressed (Dunier & Siwicki, 1994). A second study reported effects on immune function in rainbow trout exposed to lindane by oral (daily body dose of 1 mg/kg for 30 days) or intraperitoneal (10, 50, or 100 mg/kg) routes (Dunier et al., 1994).

In a study in leukocytes of gilthead seabream in vitro, lindane had no effect on cell viability, and only slightly altered immunological parameters (e.g. phagocytosis); however, lindane upregulated genes related to the immune system (Cuesta et al., 2008).

Birds

Haematological changes were induced by lindane (5 mg/kg bw, twice in 1 week) in six bird species: house sparrow, baya weaver bird, common myna, rose-ringed parakeet, blue rock pigeon, and domestic duck (Mandal et al., 1986). Lindane induced anaemia, as demonstrated by reductions in erythrocyte count, haematocrit (erythrocyte volume fraction), and haemoglobin content, and in mean corpuscular volume, mean corpuscular haemoglobin, and mean corpuscular haemoglobin concentration. Prolonged bleeding and clotting times were observed. In most exposed birds, lindane also decreased splenic cell counts with minimal increases in splenic weight. Total leukocyte counts were increased. The differential leukocyte count revealed pronounced heterophilia and eosinophilia, with a decline in monocyte, lymphocyte and basophil numbers.

4.2.2 Oxidative stress

(a) Humans

(i) Exposed humans

Markers of oxidative stress were increased in human blood samples obtained from lindane poisoning cases admitted to the Guru Teg Bahadur Hospital, Delhi, India. Lipid peroxidation (thiobarbituric acid-reactive substances) and the activities of superoxide dismutase, catalase, glutathione peroxidase, glutathione-*S*-transferase and γ-glutamyltransferase were increased, while glutathione levels decreased (Banerjee et al., 1999).

In 30 cases of lindane poisoning, γ-glutamyltransferase activity in plasma and glutathione levels in blood were significantly different from controls, but neither was altered in lymphocytes (Seth et al., 2001).

Maternal and cord blood levels of lindane correlated with both intrauterine growth retardation and oxidative stress (Pathak et al., 2011). Specifically, significant correlations were seen with oxidized DNA (8-hydroxy-2′-deoxyguanosine), malondialehyde, and glutathione, as well as protein carbonyl and the ferric reducing ability of plasma. However, as total HCH levels were also correlated with these effects, and considering that the HCH levels are also inter-correlated, the association could not be assigned specifically to lindane. [The Working Group noted the limitations of this study for evaluating these effects.]

(ii) Human cells in vitro

In human peripheral blood lymphocytes, lindane increased the formation of reactive oxygen species and decreased the mitochondrial transmembrane potential, effects that are likely to be responsible for caspase-3 activation (Michałowicz et al., 2013).

At non-cytotoxic concentrations, lindane synergistically increased hydrogen peroxide-induced DNA damage in human fibroblasts (Lueken et al., 2004). In contrast, antagonism was found when measuring DNA breakage in isolated PM2 DNA.

In human HaCaT keratinocytes, lindane increased the production of reactive oxygen species as assessed with dichlorodihydrofluorescein diacetate. Pre-treatment with *N*-acetyl cysteine markedly decreased lindane-induced ERK1/2 phosphorylation, but did not affect Raf or MEK1/2 activation by lindane (Ledirac et al., 2005).

(b) Experimental systems

Hepatic oxidative stress was induced by lindane (30 mg/kg, oral administration) in female Sprague-Dawley rats (Hassoun et al., 1993). Lindane (300 mg/kg, intraperitoneal) also induced oxidative stress in the liver of male Wistar rats as assessed by assay for malondialdehyde, glutathione peroxidase, glutathione reductase, glucose-6-phosphate dehydrogenase, glutathione-*S*-transferase, γ-glutamyltransferase, catalase, and superoxide dismutase (Anilakumar et al., 2009). In the rat heart, oral administration of lindane (1.5 and 7 mg/kg per day for 21 days) induced lipid peroxidation (as measured by thiobarbituric acid-reactive substances), increased superoxide dismutase and catalase activities, and decreased glutathione levels (Ananya et al., 2005).

4.2.3 Receptor-mediated effects

(a) Humans

(i) Exposed humans

In 54 men exposed occupationally to HCH isomers during lindane production, testosterone levels were lower than in 20 unexposed control subjects (clerks of approximately the same age, not otherwise specified), but this was not statistically significant at the $P < 0.05$ level (6.8 ± 2.2 versus 8.0 ± 2.9 ng/mL, mean ± standard deviation) (Tomczak et al., 1981). Luteinizing hormone (LH) levels were significantly higher ($P < 0.01$) in exposed men than in controls (9.6 ± 4.2 versus 6.1 ± 2.1 mIU/mL), consistent with reduced circulating testosterone levels. Follicle-stimulating hormone levels were not different. Hormone levels were measured by a radioimmunoassay with excellent inter-assay variation values. Levels of HCH isomers in the serum measured using an unspecified method were 63.5 ± 45.1, 185 ± 150, and 36.6 ± 40.6 µg/L for α-, β-, and γ-HCH, respectively.

In a study of 304 men and 300 women from an area in Brazil that was heavily polluted with organochlorine pesticides, linear regression analysis found a borderline significant inverse association between testosterone and α-HCH and β-HCH in the serum of the men; there was no such association for γ-HCH (Freire et al., 2014). When β-HCH levels were divided into quartiles, a highly significant inverse association with testosterone levels was found with a P for trend of < 0.001. No significant associations were found between serum sex hormone and HCH levels in premenopausal women ($n = 210$). In peri-/postmenopausal women ($n = 77$), there was a borderline significant association between levels of LH and β-HCH, and a highly significant inverse association between levels of LH and β-HCH across β-HCH quartiles ($P = 0.008$). There were no significant associations between levels of serum γ-HCH and LH.

(ii) Human cells in vitro

Estrogen receptor-mediated effects

γ-HCH does not bind to the estrogen receptor (ER) in cytosolic binding assays (Danzo, 1997) or to ERα and ERβ in cells transfected with reporter vectors (Lemaire et al., 2006). Yeast-based assays yielded mixed results (Lee et al., 2002; Dhooge et al., 2006), while no binding was seen with recombinant ERs (Scippo et al., 2004). Consistent with these observations, γ-HCH does not have estrogenic effects on ER-positive human MCF-7 breast cancer cells (Soto et al., 1995; Briz et al., 2011). [In vivo, the effects of γ-HCH may be complex and display non-linear dose–response relationships based on studies in rats, but no data were available for humans or human cells.]

Estrogen is formed in humans and other mammals by the enzyme aromatase. γ-HCH at concentrations of 25–75 µM inhibited aromatase enzyme activity, but not mRNA expression in human placental JEG-3 cells and human embryonal kidney E293 cells stably transfected with the aromatase gene; proliferation of these cells in the 3-(4,5-dimethylthiazol-2-yl)-2,5-diphenyltetrazolium bromide (MTT) assay was not affected by γ-HCH (Nativelle-Serpentini et al., 2003).

Androgen receptor-mediated effects

γ-HCH (and δ-HCH) inhibited binding of radiolabelled 5α-dihydrotestosterone (DHT) to the androgen receptor and human sex hormone-binding globulin in vitro (Danzo, 1997).

Consistent with this finding, γ-HCH inhibited the growth of LNCaP human prostate cancer cells expressing androgen receptor by 20–30% over 72 hours at concentrations of 1 μM and higher, which may be associated with the presence of ERβ (Maranghi et al., 2007). However, γ-HCH did not activate or inhibit human androgen receptor transfected into PC-3 human prostate cancer cells or Chinese hamster ovarian cells with a reporter construct under transcriptional control of an androgen-responsive promoter (Schrader & Cooke, 2000; Roy et al., 2004). β-HCH did not have effects in this system, but α-HCH had clear anti-androgenic effects (Roy et al., 2004; Pavlíková et al., 2012).

Other receptor-mediated effects

γ-HCH has negligible binding affinity to the human progesterone receptor (Scippo et al., 2004), but inhibited progesterone-receptor transactivation by progesterone in a yeast system in a dose-dependent fashion, being significant at a concentration of 1 μM and above (Jin et al., 1997). γ-HCH at a concentration of 10 μM transactivated the human pregnane X receptor (PXR) and induced the protein expression of CYP3A4 and CAYP2B6, both of which are regulated by PXR (Lemaire et al., 2004).

To examine human adrenocortical NCI-H295R cells as a possible system in vitro for the assessment of adrenal disruption using molecular end-points, lindane was used as positive control. γ-HCH reduced the secretion of cortisol, in addition to downregulating the expression of several steroidogenic enzymes, and blocking the activation of the steroidogenic acute regulatory protein (StAR) gene promoter (Oskarsson et al., 2006).

(b) Experimental systems

(i) Non-human mammals in vivo

Estrogen receptor-mediated effects

Several rodent experiments using substantial γ-HCH doses reported effects in vivo on ER-related end-points, such as the development of the mouse female genital tract. Differences in outcome were evident across species, and other aspects of experimental design. A daily gavage dose of γ-HCH (10 mg/kg) increased circulating 17β-estradiol (E2) in young adult female F344 rats, whereas there was no such effect was seen at 20 mg/kg, and E2 was decreased at 40 mg/kg. LH levels as well as uterine weight were decreased at both higher doses (Cooper et al., 1989). In prepubertal Sprague-Dawley rats given seven daily intraperitoneal injections of γ-HCH at a dose of 15 mg/kg, the uterine-weight response to estrone treatment was decreased, and uptake of radiolabelled estrone was reduced (Welch et al., 1971). Treatment of prepubertal female F344 rats with γ-HCH at 30 mg/kg per day for 7 days also decreased the uterine-weight response to E2 and sharply reduced the E2-induced increase in serum LH (Cooper et al., 1989). Administration of γ-HCH (20 mg/kg per day) for 30 days to ovariectomized rats slightly reduced the estrogen-induced weight increase in tissues of the female genital tract, and caused significant changes in numbers of erythrocytes, neutrophils, and lymphocytes, as well as levels of haemoglobin (see also Section 4.2.4) (Raizada et al., 1980). Blunting of the uterine-weight response to E2 was also found in prepubertal Long Evans rats treated with γ-HCH at a daily dose of 40 mg/kg, but induction of the uterine progesterone receptor and nuclear redistribution of ER induced by E2 were not affected by γ-HCH (Laws et al., 1994). Similarly, in ovariectomized young adult Long Evans rats, progesterone-receptor induction in the uterus and pituitary gland was not affected by γ-HCH at doses up to 40 mg/kg per day (Laws et al., 1994).

In contrast, prenatal treatment of pregnant CD-1 mice with γ-HCH at a daily oral dose of 15 mg/kg given on days 9–16 of gestation, age at vaginal opening of the female offspring was reduced by 2 days and uterine weight at postnatal day 22 was increased by about 15%, an effect that was no longer present at postnatal day 22 (Maranghi et al., 2007).

Aromatase activity in hepatic microsomes of female mice that had been exposed to γ-HCH (15 mg/kg per day) during pregnancy or in utero was not altered 22 days after parturition or birth (Maranghi et al., 2007).

Androgen receptor-mediated effects

In males, lindane caused an impairment of steroid hormone homeostasis, due to CYP-mediated disruption of testosterone catabolism. In particular, γ-HCH inhibited cytosolic 5α-dihydrotestosterone-androgen receptor (DHT-AR) complex formation in the prostate of F344 rats treated for 7 days at a dose of 60 mg/kg (Simić et al., 1991). There are also reports of inhibition in vivo by lindane (4–8 mg/kg per day by gavage for 30–45 days) of the activity of the androgen biosynthetic enzymes 3-hydroxysteroid dehydrogenase and 17-hydroxysteroid dehydrogenase in adult rat testes, and reduction of weights of testes, ventral prostate, and seminal vesicles, and of serum testosterone (by 29–44%) (Chowdhury & Gautam, 1994; Sujatha et al., 2001). Prenatal exposure of pregnant female Wistar rats to γ-HCH (30 mg/kg on day 15 of gestation) resulted in a 43% reduction in serum testosterone in male offspring aged 7 months (Dalsenter et al., 1997b). Prenatal exposure of pregnant female CD-1 mice to γ-HCH (25 mg/kg per day on days 9–16 of gestation) resulted in the impairment of androgen catabolism in pubertal and young adult male offspring (Di Consiglio et al., 2009). The impact of treatment with γ-HCH on the activity of hepatic CYP-mediated testosterone hydroxylase, aromatase, and other testicular parameters was tested at multiple time-points considered critical for the sexual maturation of CD-1 mice (postnatal days 50, 65–69, and 100). On postnatal days 65–69, significant changes to testis weight and spermatid number as well as CYP-mediated changes in testosterone metabolism were observed in the adult F_1 mice without evidence of systemic toxicity. Activities of testosterone 6β- and 2α-hydroxylation and dehydrogenation were most strongly reduced during this period, suggesting the CYP3A and CYP2C families as the major target of lindane-induced effects. Most changes had almost recovered by postnatal day 100. No effects on aromatase activity were seen.

(ii) Non-human mammalian cells in vitro

In rat pituitary tumour cells (MtT/E-2) that are estrogen responsive, estrogenic effects of lindane (γ-HCH) (growth stimulation and ER binding and transcriptional activation) have been reported, but only at concentrations of 100 μM, not at 10 μM and below (Maruyama et al., 1999). Anti-estrogenic effects of γ-HCH (10–30 μM) have been reported for several molecular end-points (activation of Akt and ERK 1/2) and downregulation of ERα, but not ERβ, in primary neuronal cells derived from NMRI mice (Briz et al., 2011).

Using endothelial cell proliferation and thymidine incorporation, wound healing, ascites formation and secretion, chorio-allantoic membrane formation, and an assay for neovascularization in vivo in male mice, lindane was shown to be a potent angiogenesis stimulator (Clere et al., 2012; Bharathi et al., 2013), and neovascularization in male Swiss mice was prevented by silencing of ERα expression (Clere et al., 2012).

γ-HCH (δ-HCH) inhibited binding of radiolabelled DHT in vitro to the rat prostate androgen receptor, but not to rat epididymal androgen binding protein (Danzo, 1997). γ-HCH stimulated testosterone production and proliferation of rat Leydig cells in vitro at concentrations up to 10 μg/mL, effects that disappeared at higher concentrations (Ronco et al., 2001). It strongly counteracted the stimulatory effect of human chorionic gonadotropin on testosterone production by these cells at concentrations of 10 μg/mL and higher, probably via a decrease in cAMP production (Ronco et al., 2001). Inhibition of progesterone biosynthesis by α-, δ-, and γ-HCH at high (50 μM) concentrations was found in a mouse Leydig tumour cell line (Walsh & Stocco, 2000).

4.2.4 Genotoxicity and related effects

Lindane has been studied in a variety of assays for genotoxic and related potential. Tables 4.1–4.4 summarize the studies carried out in humans in vivo and in vitro, in experimental animals in vivo, in mammals in vitro, and in non-mammalian systems both in vitro and in vivo, respectively.

(a) Humans

(i) Exposed humans

No data in exposed humans were available to the Working Group. The relationship between Yq microdeletion in patients with normal karyotype and level of total HCH and its isomers α-HCH, β-HCH and γ-HCH in semen was studied (Khan et al., 2010). No effect was observed with lindane.

(ii) Human cells in vitro

See Table 4.1

No induction of DNA damage measured by DNA-adduct detection was observed after treatment of HepG2 hepatocarcinoma cells line with γ-HCH in vitro (Dubois et al., 1997). Effects were seen in rat cells, as described below. Similarly, negative results were found by the unscheduled DNA synthesis (UDS) assay in the VA-4 cell line after exposure to γ-HCH (Ahmed et al., 1977).

Regarding DNA strand breaks and other types of DNA damage, positive results were found in many but not all cell types. Induction of DNA strand breaks was detected by radioactive labelling of the 5′ broken ends of exposed total leukocytes to lindane (analytical grade) (Sreekumaran Nair et al., 2002). Induction of DNA damage as measured by the comet assay was observed after treatment with γ-HCH in MCL-5 metabolically competent cell line (Martin et al., 1999). Furthermore, a higher level of DNA damage was achieved in the presence of inhibitors of DNA repair. Results indicate that γ-HCH-induced DNA lesions are repaired by nucleotide excision repair (Martin et al., 1999). An increase in the frequency of DNA breaks as determined by the comet assay after lindane treatment was observed in nasal mucosal cells (Pool-Zobel et al., 1994); however, negative results were found in isolated lymphocytes (Pool-Zobel et al., 1993) and in gastric mucosa cells (Pool-Zobel et al., 1994).

Induction of chromosomal damage estimated by the frequency of micronucleus formation was observed after treatment with γ-HCH in MCF-7 cells (Kalantzi et al., 2004; Hewitt et al., 2007).

An increase in the frequency of chromosomal aberration (gap not included) and sister-chromatid exchange was observed after 72 hours of treatment with γ-HCH in human lymphocytes cultured without S9 microsomal fraction (Rupa et al., 1989).

(b) Experimental systems

(i) Non-human mammals in vivo

See Table 4.2

In HPB black mice analysed after a single intraperitoneal treatment, binding of γ-HCH to DNA occurred only at very low levels (Iverson et al., 1984). An increase in the frequency of micronucleus formation was observed in the bone marrow of Park male mice treated with an intraperitoneal dose of γ-HCH (Yaduvanshi et al., 2012). On the other hand, no induction of micronucleus formation was observed in bone marrow

Table 4.1 Genetic and related effects of lindane (γ-HCH) in human cells in vitro

Tissue, cell line	End-point	Test	Results[a]		Concentration (LED or HID)	Comments	Reference
			Without metabolic activation	With metabolic activation			
HepG2 hepatocarcinoma cell line	DNA damage	DNA-adduct ^{32}P-postlabelling	–	NT	50 µM		Dubois et al. (1997)
Leukocytes	DNA damage	DNA strand breaks radioactive labelling assay	+	NT	20 µg/mL	Analytical grade lindane	Sreekumaran Nair et al. (2002)
MCL-5 metabolically competent lymphoblastoid cell line	DNA damage	DNA strand break Comet assay	+	NT	1.56 mM		Martin et al. (1999)
Nasal mucosa cells	DNA damage	DNA strand break Comet assay	+	NT	0.03 mM		Pool-Zobel et al. (1994)
Isolated lymphocytes	DNA damage	DNA strand break Comet assay	–	NT	0.1 mM		Pool-Zobel et al. (1993)
Gastric mucosa cells	DNA damage	DNA strand break Comet assay	–	NT	1 mM		Pool-Zobel et al. (1994)
VA-4 cell line	DNA damage	UDS assay	–	–	1000 µM		Ahmed et al. (1977)
Lymphocytes	Chromosomal damage	Chromosomal aberrations	+	NT	0.05 µg/mL		Rupa et al. (1989)
MCF-7 mammary carcinoma cell line	Chromosomal damage	Micronucleus induction	+	NT	1×10^{-12} M		Hewitt et al. (2007)
MCF-7 mammary carcinoma cell line	Chromosomal damage	Micronucleus induction	+	NT	1×10^{-12} M		Kalantzi et al. (2004)
Lymphocytes	Chromosomal damage	Sister-chromatid exchanges	+	NT	0.1 µg/mL		Rupa et al. (1989)

[a] +, positive result; –, negative result

HID, highest ineffective dose; LED, lowest effective dose; NT, not tested; UDS, unscheduled DNA synthesis

Table 4.2 Genetic and related effects of lindane (γ-HCH) in non-human mammals in vivo

Species, strain, sex	Tissue	End-point	Test	Results[a]	Dose (LED or HID)	Route, duration, dosing regimen	Comments	Reference
Mouse, HPB black	Hepatic cells	DNA damage	DNA binding	+	25 mg/kg	i.p. × 1		Iverson et al. (1984)
Mouse, Swiss albino, male	Germ cells	Chromosomal damage	Dominant lethal mutation	+	500 ppm	p.o. continuously × 4–8 mo	Formulation (not specified)	Lakkad et al. (1982)
Mouse, Swiss albino	Bone marrow	Chromosomal damage	Chromosomal aberration	+	1.6 mg/kg	gastric × 1 × 7 days	Formulation (not specified; 20% γ-HCH)	Kumar et al. (1995)
Mouse, Park, male	Bone marrow	Chromosomal damage	Micronucleus induction	+	35 mg/kg	i.p. × 1		Yaduvanshi et al. (2012)
Mouse, NMRI	Bone marrow	Chromosomal damage	Micronucleus induction	–	70 mg/kg	p.o. × 1		Pool-Zobel et al. (1993)
Mouse, CD-1	Testicle cells	Fertility	Chromatin abnormalities DNA content	+	25 mg/kg	p.o. × 1 on days 9–16 of gestation	Exposure in utero	Traina et al. (2003)
Rat, Sprague-Dawley	Hepatic cells	DNA damage	DNA strand break Alkaline elution	+	30 mg/kg	p.o. × 1		Hassoun et al. (1993)
Rat, Sprague-Dawley	Nasal mucosa cells	DNA damage	DNA strand break Comet assay	+	200 µg/kg	p.o. × 1		Pool-Zobel et al. (1993)
Rat, Sprague-Dawley	Gastric cells	DNA damage	DNA strand break Comet assay	(–)	60 mg/kg	p.o. × 1	[two animals not clearly reported]	Pool-Zobel et al. (1993)
Rat, Sprague-Dawley	Colonic mucous membrane cells	DNA damage	DNA strand break Comet assay	(+)	60 mg/kg	p.o. × 1	[two animals not clearly reported]	Pool-Zobel et al. (1993)
Rat , Sprague-Dawley	Isolated lymphocytes	DNA damage	DNA strand break Comet assay	(–)	60 mg/kg	p.o. × 1	[two animals not clearly reported]	Pool-Zobel et al. (1993)
Rat	Bone marrow	Chromosomal damage	Chromosomal aberration	–	15 mg/kg	p.o. × 12 wk		Gencik (1977)

Table 4.2 (continued)

Species, strain, sex	Tissue	End-point	Test	Results[a]	Dose (LED or HID)	Route, duration, dosing regimen	Comments	Reference
Rat , Sprague-Dawley	Bone marrow	Chromosomal damage	Micronucleus induction	–	60 mg/kg	p.o. × 1		Pool-Zobel et al. (1993)
Rat, Wistar male	Bone marrow	Chromosomal damage	Micronucleus induction	–	100 mg/kg	p.o. × 4 wk		Etim et al. (2006)
Rat, Wistar male	Bone marrow	Chromosomal damage	Micronucleus induction	+	300 mg/kg	i.p. × 1		Anilakumar et al. (2009)
Hamster, Chinese	Bone marrow	Chromosomal damage	Micronucleus induction	–	120 mg/kg	p.o. × 1		Pool-Zobel et al. (1993)
Hamster, Chinese	Bone marrow	Chromosomal damage	Sister-chromatid exchange	–	120 mg/kg	p.o. × 1		Pool-Zobel et al. (1993)

[a] +, positive result; –, negative result; +/–, equivocal (variable response in several experiments within an adequate study); (+) or (–), positive/negative in a study of limited quality
HCH, hexachlorocyclohexane; HID, highest ineffective dose; i.p., intraperitoneal; LED, lowest effective dose; mo, month; NT, not tested; p.o., oral administration; wk, week

cells of NMRI mice after a single oral treatment with γ-HCH (Pool-Zobel et al., 1993). In the male offspring of female CD-1 mice treated in utero with γ-HCH from day 9 to day 16 of gestation, alterations in the DNA content (possibly attributable to DNA strand breaks) of testicle cells was observed (Traina et al., 2003).

In Sprague-Dawley rats treated orally with γ-HCH, DNA damage evaluated by the alkaline elution assay gave positive results in hepatocytes (Hassoun et al., 1993). When DNA damage was evaluated by comet assay, positive results were found in the nasal mucosa of Sprague-Dawley rats treated orally (Pool-Zobel et al., 1993), while results were inconclusive for gastric and colonic mucous membrane cells as well in isolated lymphocytes in the same treated animals (Pool-Zobel et al., 1993). No induction of chromosomal aberration was observed in the bone marrow cells of rats exposed orally to γ-HCH for 12 weeks (Gencik, 1977). While induction of micronucleus formation was reported in bone marrow cells of Wistar male rats after a single intraperitoneal treatment with γ-HCH (Anilakumar et al., 2009), no such effect was reported after oral exposure to γ-HCH, regardless of the treatment period (Pool-Zobel et al., 1993; Etim et al., 2006).

Neither induction of micronucleus formation nor sister-chromatid exchange was observed in bone marrow cells from Chinese hamsters treated orally with γ-HCH (Pool-Zobel et al., 1993).

In several studies, a causative effect of lindane alone could not be demonstrated because the exposure was to a mixture. Induction of DNA damage estimated by the increased frequency of chromosomal aberration was observed in the bone marrow cells of Swiss albino mice after gastric treatment with technical-grade HCH (γ-HCH, 20%) for up to 7 days (Kumar et al., 1995). The dominant-lethal assay gave positive results for DNA damage in the germ cells of Swiss albino males continuously exposed orally to technical-grade HCH for 4–8 months (Lakkad et al., 1982).

(ii) Non-human mammalian cells in vitro

See Table 4.3

γ-HCH gave negative results in two studies in Chinese hamster ovary and V79 cells. In the hypoxanthine-guanine phosphoribosyl-transferase (Hgprt) or sister-chromatid exchange assays, negative results were reported for γ-HCH either in the presence or absence of a S9 microsomal fraction (Pool-Zobel et al., 1993). Interaction of lindane with the DNA of hepatic cells in HPB black mice was analysed in vitro after treatment with γ-HCH. Positive results were observed for γ-HCH either in the presence or in the absence of S9 metabolic fraction (Iverson et al., 1984).

In fetal rat hepatocytes, an increase in the frequency of DNA adducts was observed using ^{32}P post-labelling after treatment in vitro with γ-HCH (Dubois et al., 1997). The comet assay also gave positive results in gastric mucosa and nasal mucosa cells of Sprague-Dawley rats exposed to γ-HCH in vitro (Pool-Zobel et al., 1994). On the other hand, negative results were previously reported for the same end-point and by the same research group for γ-HCH-exposed cells (Pool-Zobel et al., 1993). The comet assay gave negative results in Sprague-Dawley rat primary hepatocytes exposed to γ-HCH (Pool-Zobel et al., 1993).

(iii) Non-mammalian systems

See Table 4.4

The frequency of DNA adduct formation was increased by treatment with γ-HCH in embryonic cells from the quail *Coturnix coturnix* (Dubois et al., 1997).

Treatment with γ-HCH caused DNA damage in the haemocytes of Pacific oyster *Crassostrea gigas*, as determined by the comet assay (Anguiano et al., 2007). Lindane did not induce micronucleus formation in haemocytes from the Mediterranean mussel *Mytilus galloprovincialis* after 15 days of exposure (Raftopoulou et al., 2006).

Table 4.3 Genetic and related effects of lindane (γHCH) in non-human mammalian cells in vitro

Species, strain	Tissue, cell line	End-point	Test	Results[a]		Concentration (LEC or HIC)	Comments	Reference
				Without metabolic activation	With metabolic activation			
Mouse, HPB black	Hepatic cells	DNA damage	DNA binding	+	+	1 µM		Iverson et al. (1984)
Rat	Fetal hepatocytes	DNA damage	DNA adduct-^{32}P-postlabelling	+	NT	50 µM		Dubois et al. (1997)
Rat, Sprague-Dawley	Gastric mucosa cells	DNA damage	DNA strand breaks, comet assay	+	NT	0.125 mM	Positive results for 3 out of the 4 rats tested	Pool-Zobel et al. (1994)
Rat, Sprague-Dawley	Nasal mucosa cells	DNA damage	DNA strand breaks, comet assay	+	NT	0.5 mM		Pool-Zobel et al. (1994)
Rat, Sprague-Dawley	Gastric mucosa cells	DNA damage	DNA strand breaks, comet assay	–	NT	0.1 mM		Pool-Zobel et al. (1993)
Rat, Sprague-Dawley	Primary hepatocytes	DNA damage	DNA strand breaks, comet assay	–	NT	0.1 mM		Pool-Zobel et al. (1993)
Hamster, Chinese	CHO cells	Mutation	*Hprt* assay	–	–	300 mM		Pool-Zobel et al. (1993)
Hamster, Chinese	CHO cells	Chromosomal damage	Sister-chromatid exchange	–	–	300 mM		Pool-Zobel et al. (1993)

[a] +, positive result; –, negative result; ±, equivocal (variable response in several experiments within an adequate study); (+) or (–), positive/negative result in a study of limited quality
CHO, Chinese hamster ovary; HCH, HCH, hexachlorocyclohexane; HIC, highest ineffective concentration; LEC, lowest effective concentration, NT, not tested

Table 4.4 Genetic and related effects of lindane (γ-HCH) in non-mammalian systems

Phylogenetic class	Species, strain, tissue	End-point	Test	Results[a]		Concentration (LEC or HIC)	Comments	Reference
				Without metabolic activation	With metabolic activation			
Bird	Quail, *Coturnix coturnix japonica*, embryonic cells	DNA damage	DNA-adduct ^{32}P-postlabelling	+	NA	50 μM		Dubois et al. (1997)
Mollusc	Pacific oyster, *Crassostrea gigas*	DNA damage	Comet assay	+	NA	0.7 mg/L	Haemocytes exposed to γ-HCH	Anguiano et al. (2007)
	Mediterranean mussel, *Mytilus galloprovincialis*	Chromosomal damage	Micronucleus induction	–	NA	0.03 mg/L	Exposure for 15 days with change of xenobiotic every 2 days	Raftopoulou et al. (2006)
Insect	*Drosophila melanogaster*, Oregon-R strain	Mutation	X-chromosome-linked recessive lethal	+	NA	5.0 μg/L in feeding solution	Formulation (not specified; 20% γ-HCH)	Kumar et al. (1995)
	Drosophila melanogaster, Oregon-R strain	Mutation	Lethal mutation expressed as larval hatchability	+	NA	20.0 μg/L in feeding solution	Formulation (not specified; 20% γ-HCH)	Kumar et al. (1995)
Plant systems	Onion, *Allium cepa*	Chromosomal damage	Chromosomal aberrations	+	NA	Saturated water solution		Hervás (1976)
	Onion, *Allium cepa*	Chromosomal damage	Chromosomal aberrations	+	NA	9.0 mg/L	Formulation (not specified; 20% γ-HCH)	Kumar et al. (1995)
Lower eukaryote (yeast)	*Saccharomyces cerevisiae*	Mutation	Mitotic gene conversion	–	NT	NR		Fahrig (1974)
	Saccharomyces cerevisiae, D61.M	Mutation	Mitotic recombination and mutation	–	NT	0.170 mM		Albertini et al. (1988)
	Saccharomyces cerevisiae, D61.M	Chromosomal damage	Aneuploidy, chromosomal loss assay	–	NT	0.170 mM		Albertini et al. (1988)

Table 4.4 (continued)

Phylogenetic class	Species, strain, tissue	End-point	Test	Results[a]		Concentration (LEC or HIC)	Comments	Reference
				Without metabolic activation	With metabolic activation			
Prokaryote (bacteria)	*Salmonella typhimurium*, TA98, TA100, and TA102,	Mutation	Reverse mutation	+	+	5 µg/plate		Yaduvanshi et al. (2012)
	Salmonella typhimurium, TA98	Mutation	Reverse mutation	–	(+)	50 µg/plate	Positive results in two out of three experiments	Gopalaswamy & Aiyar (1986)
	Bacillus subtilis, M45 Rec$^-$, H17 Rec$^+$	Mutation	Rec assay, differential toxicity	–	NT	NR	Mixture of α-, β-, γ-HCH	Shirasu et al. (1976)

[a] +, positive result; –, negative result; ±, equivocal (variable response in several experiments within an adequate study); (+) or (–), positive/negative result in a study of limited quality
HCH, hexachlorocyclohexane; HIC, highest ineffective concentration; LEC, lowest effective concentration; NA, not applicable; NR, not reported; NT, not tested

In *Drosophila melanogaster*, mutagenicity was observed using the X-chromosome-linked recessive-lethal assay after exposure to a formulation containing 20% γ-HCH (Kumar et al., 1995).

γ-HCH was not mutagenic in *Saccharomyces cerevisiae* (strain not specified) using a mitotic gene-conversion assay in the absence of a microsomal S9 fraction (Fahrig, 1974) as well as in *S. cerevisiae* D61.M in the presence and in the absence of metabolic activation (Albertini et al., 1988). Negative results were also demonstrated for *S. cerevisiae* D61.M exposed to γ-HCH in the chromosomal loss assay, both in the presence and in the absence of metabolic activation (Albertini et al., 1988).

In plants, an increase in the frequency of chromosomal aberration was reported in root meristematic cells of *Allium cepa* after treatment with γ-HCH, with an induction of viable multinucleate cells with aneuploid nuclei after cytokinesis inhibition by caffeine (Hervás, 1976) and by a non-specified formulation containing 20% γ-HCH (Kumar et al., 1995).

In *Salmonella typhimurium*, γ-HCH produced statistically significant mutagenic effects in strains TA98, TA100, and TA102 with or without S9 microsomal fraction (Yaduvanshi et al., 2012). [Although the authors reported a positive finding, the Working Group noted that the reported effect sizes were minimal to achieve statistical significance, i.e. the mutagenic rate was not doubled.] Negative or inconclusive results were reported when *S. typhimurium* strain TA98 was exposed to γ-HCH in the absence or presence of metabolic activation, respectively (Gopalaswamy & Aiyar, 1986).

4.2.5 Altered cell proliferation or death

(a) Humans

No data in exposed humans were available to the Working Group.

In studies in vitro, lindane demonstrated cell type-specific induction of cell proliferation or effects on cell proliferation through receptor-mediated mechanisms in ERα-positive cells, such as MCF-7 human breast cancer cells (Briz et al., 2011). However, lindane may inhibit cell proliferation, in a dose-dependent manner (1–100 μM) through receptor-mediated mechanisms in ERβ- and androgen receptor-positive LNCaP cells (Maranghi et al., 2007). [In comparison with 17β-estradiol, these effects were not marked.]

Through the intracellular release of Ca^{2+}, lindane induces apoptosis in HL-60 cells (Kang et al., 1998), an effect that may contribute to immunotoxicity (Betoulle et al., 2000). While other studies have not observed marked decreases in cell viability, altered levels of apoptosis-related factors such as Bcl-2 (increases in MCF-7 cells after exposure at up to 0.1 nM) or Bax (increases in MCF-7 cells after exposure at 100 μM; increases in PC-3 cells after exposure at 0.01 nM) have been noted (Kalantzi et al., 2004). A consistent observation with lindane is the presence of a biphasic dose–response relationship with distinct effects at low (< 1 μM) and high (> 1 μM to < 100 μM) doses (Llabjani et al., 2011). Kalantzi et al. (2004) reported a distinction between lindane-induced alterations in the Bcl-2:Bax ratio at low concentrations (nM) and cytotoxicity at high doses (μM). Cell type may be critical; for instance, biologically relevant levels of lindane induce apoptosis in human lymphocytes, together with an associated increase in reactive oxygen species and a reduction in mitochondrial transmembrane potential (Michałowicz et al., 2013). Whether lindane-induced modulation of Bcl-2:Bax ratios promotes apoptosis or facilitates cell survival remains to be established, and may depend on a range of other confounding factors (Hewitt et al., 2007).

(b) Experimental systems

Exposure of pregnant mice to lindane in vivo increased apoptosis in primordial germ cells, an effect associated with Akt modulation (La Sala et al., 2009). In male Wistar rats, a single dose of lindane induced testicular apoptosis associated with the nuclear translocation of nuclear factor kappa β (NF-κβ) and the increased expression of a range of pro-apoptotic factors (e.g. cytochrome *c*, caspase-3 and -9, Fas, and FasL) (Saradha et al., 2009). Lindane also increased oxidative stress, triggering NF-κβ translocation and upregulation of target genes such as TNF-α and IL-1α (Videla et al., 2004). Sentinel organisms such as Pacific oysters, *Crassostrea gigas*, exhibit an in-vivo susceptibility to the cytotoxic effects of lindane, with genotoxicity being induced in isolated haemocytes (Anguiano et al., 2007). Finally, in the male offspring of CD-1 females treated in utero with γ-HCH from day 9 to day 16 of gestation, alterations in the DNA content of testicle cells, possibly attributable to DNA strand breaks, were observed (Traina et al., 2003) (see Section 4.2.4).

The results above are consistent with lindane causing a slight elevation in levels of apoptosis in murine splenocytes in vitro (Battaglia et al., 2010). An immunotoxic effect associated with apoptotic and necrotic cell death in murine thymocytes in vitro is linked with the induction of drug metabolizing mixed function oxidase enzymes (Olgun et al., 2004). Despite the generation of reactive oxygen species and the depletion of glutathione, lindane primarily induces apoptosis as opposed to necrosis in exposed Madin-Darby canine kidney cells (Piskac-Collier & Smith, 2009). Its transforming activity in BALB/c 3T3 cells is also associated with cell proliferation (Perocco et al., 1995). This is in contrast to lindane-induced toxicity in primary rat hepatocytes, which could be jointly attributed to the disruption of the autophagic process, the inhibition of apoptotic cell death, and the induction of necrosis (Zucchini-Pascal et al., 2009). Slight increases in levels of Bad mRNA after exposure were noted in PC12 rat pheochromocytoma cells (Aoki et al., 2008). The observation of cytotoxicity at high doses (μM) is consistent with the observation that lindane inhibits phytohaemagglutinin-P-induced stimulation of the mitogenic response in bovine lymphocytes (Kensler & Mueller, 1978).

4.2.6 Inflammation

(a) Humans

No data were available to the Working Group.

(b) Experimental systems

Meade et al. first reported in 1984 that lindane (γ-HCH) induced arachidonic acid release from mouse macrophage phospholipids and also strongly stimulated leukotriene C4 production (Meade et al., 1984). However, in comparison with zymosan, lindane exerts only a modest effect on prostaglandin production. Lindane may therefore affect phosphatidylinositol metabolism (Meade et al., 1984).

The similarity between lindane and inositol 1,4,5-triphosphate (IP3) may explain why lindane releases Ca^{++} from IP3-sensitive intracellular stores in macrophages, myometrial cells as well as cat kidney cells (reviewed in Sauviat & Pages, 2002). Furthermore, lindane was found to influence the metabolic function of hepatic mitochondria, as well as affect the synthesis of inositol phosphate in neuronal cells. Lindane is not a competitive agonist of the IP3 receptor. In mouse peritoneal macrophage cells, lindane can mobilize Ca2+ stores. Lindane also decreases the concentration of phosphatidylinositol (PI), phosphatidylinositol 4-phosphate (PIP), and phosphatidylinositol 4,5-bisphosphate (PIP2) in the membrane of erythrocytes and cerebral cells of rats exposed for 3 or 6 months. Lindane can also promote oxidative stress by modifying the activity of scavenger enzymes, an effect that may also be involved in the inhibition of intercellular gap junctions. [Thus, lindane produces multiple effects on many cell types, rather than being a sole inducer of inflammation.]

In a study in vitro, the influence of γ-HCH (lindane) on the metabolism of arachidonic acid and production of oxygen metabolites was investigated in mouse peritoneal macrophages by Forgue et al. (1990). Lindane stimulated production of prostacyclins (6KPGF1α), prostaglandin E2 (PGE2), leukotriene C4 (LTC4), leukotriene B4 (LTB4), and hydroxyeicosatetraenoic acids (HETEs), and increased luminol-dependent chemiluminescence. Lindane produced a synergistic effect on prostaglandin-leukotriene (PG-LT) and chemiluminescence production when combined with phorbol ester. The calcium ionophore A23187 similarly stimulated chemiluminescence and PG-LT production, suggesting that lindane acts through mobilization of calcium stores (Forgue et al., 1990).

In perfused rat liver, low (5–20 mg/kg), but not high (60 mg/kg) doses of lindane stimulated Kupffer cell activity and led to enhanced liver injury (Videla et al., 1997). At lower exposures, lindane caused an elevation in carbon uptake and in carbon-induced oxygen consumption that was abrogated by the Kupffer cell-inactivator, gadolinium chloride ($GdCl_3$). $GdCl_3$ had no effect in animals given a higher dose of lindane (60 mg/kg), which significantly increased both the rate of oxygen consumption, as well as the sinusoidal efflux of lactate dehydrogenase. Thus, toxicity at higher doses (60 mg/kg) appears to be independent of Kupffer cell activity, and instead related to oxidative stress mechanisms at the parenchymal cell level (Videla et al., 1997).

Paracrine mechanisms leading to enhanced production of prostaglandins (which have been implicated in tumour promotion) were investigated by Kroll et al. (1999). In male Wistar rats, phenobarbital (0.75 g/L in drinking-water) or lindane (350 mg/kg diet) significantly increased the levels of *COX-2* mRNA and protein from isolated Kupffer cells evaluated after 2, 5, or 56 days of exposure. Additionally, treatment in vitro of primary Kupffer cell cultures with lindane (for 1 hour) increased COX-2 protein expression, markedly increased levels of PGE2 and prostaglandin D2 (PGD2) (by 50-fold), and also elevated prostaglandin F2α (PGF2α) (by more than threefold). Lee and Edwards, however, in 2001 challenged the idea that prostaglandins were responsible for the tumour-promoting effects of lindane and phenobarbital in rat liver (Lee & Edwards, 2001). They demonstrated a concentration-dependent increase by PGE2, PGF2α, and PGD2 in the level of DNA synthesis by hepatocytes. Arachidonic acid alone, however, had no effect on DNA synthesis. PGE2 and PGF2α required dexamethasone to mediate their effects and did not further enhance the stimulatory effect of epidermal growth factor (EGF). On the other hand, PGD2 was capable of stimulating DNA synthesis regardless of whether insulin, dexamethasone, or EGF were present or absent. The phorbol ester 12-O-tetradecanoyl phorbol-13-acetate (TPA) significantly increased [(3)*H*]arachidonic acid release, as well as PGE2 formation in hepatocytes, in contrast to α-HCH and other compounds tested, including phenobarbital and DDT. Inhibitors of arachidonic acid metabolism did not selectively block the ability of lindane (or other compounds tested) to stimulate DNA synthesis (Lee & Edwards, 2001).

4.3 Data relevant to comparisons across agents and end-points

4.3.1 General description of the database

The analysis of the in-vitro bioactivity of the agents reviewed in *IARC Monographs* Volume 113 (i.e. 2,4-D, lindane, and DDT) was informed by data from high-throughput screening assays generated by the Toxicity Testing in the 21st Century (Tox21) and Toxicity Forecaster (ToxCast™) research programmes of the government of the USA (Kavlock et al., 2012; Tice et al., 2013). At its meeting in 2014, the Advisory Group To Recommend Priorities for the *IARC Monographs* programme encouraged inclusion of analysis of high-throughput and high-content data (including from curated government databases) (Straif et al., 2014).

Lindane, DDT (*p,p′*-DDT, *o,p′*-DDT, *p,p′*-DDE, *p,p′*-DDD), and 2,4-D were among the approximately 1000 chemicals tested across the full assay battery of the Tox21 and ToxCast research programmes as of 27 April 2015. This assay battery includes 342 assays, for which data on 821 assay end-points (several assays include multiple end-point readouts) are publicly available on the website of the ToxCast research programme (EPA, 2015a). Detailed information about the chemicals tested, assays used, and associated procedures for data analysis is also publicly available (EPA, 2015b). It should be noted that the metabolic capacity of the cell-based assays is variable, and generally limited.

4.3.2 Aligning in-vitro assays to 10 "key characteristics" of known human carcinogens

In order to explore the bioactivity profiles of the agents being evaluated in *IARC Monographs* Volume 113 with respect to their potential impact on mechanisms of carcinogenesis, the 821 available assay end-points in the ToxCast/Tox21 database were first mapped to the 10 key characteristics of known human carcinogens (Smith et al., 2016). Working Group members and *IARC Monographs* staff made independent assignments for each assay type to one or more "key characteristics." The assignment was based on the biological target being probed by each assay. The consensus assignments comprise 265 assay end-points that mapped to 6 of the 10 "key characteristics" as shown below. Within each key characteristic, the assays were further divided by the Working Group into subsets of similar end-points.

1. *Is electrophilic or can undergo metabolic activation (31 end-points)*: no assays directly measure electrophilicity or metabolic activation. However, assay end-points measuring CYP inhibition (29 end-points) and aromatase inhibition (2 end-points) were mapped to this characteristic.
2. *Is genotoxic (0 end-points):* no assay end-points were mapped to this characteristic.
3. *Alters DNA repair or causes genomic instability (0 end-points):* no assay end-points were mapped to this characteristic.
4. *Induces epigenetic alterations (11 end-points):* the assay end-points mapped to this characteristic measure targets associated with DNA binding (e.g. transcription factors) (4 end-points) and transformation catalysts (e.g. histone deacetylase) (7 end-points).
5. *Induces oxidative stress (18 end-points):* the assay end-points mapped to this characteristic measure oxidative stress via cell imaging (7 end-points), markers of oxidative stress (e.g. nuclear factor erythroid 2-related factor, NRF2) (6 end-points), and metalloproteinase (5 end-points).
6. *Induces chronic inflammation (45 end-points):* the assay end-points mapped to this characteristic measure cellular adhesion (14 end-points), cytokines (e.g. IL8) (29 end-points), and nuclear factor kappa-light-chain-enhancer of activated B cells (NF-κB) activity (2 end-points).
7. *Is immunosuppressive (0 end-points):* no assay end-points were mapped to this characteristic.
8. *Modulates receptor-mediated effects (92 end-points):* a large and diverse collection of cell-free and cell-based assay end-points measuring nuclear and other receptor bioactivity, specifically aryl hydrocarbon receptor (AhR) (2 end-points), androgen receptor (11 end-points), ER (18 end-points), farnesoid X receptor (FXR) (7 end-points), peroxisome proliferator-activated receptor (PPAR) (12 end-points), pregnane X receptor_vitamin D receptor (PXR_VDR) (7 end-points), retinoic acid receptor (RAR) (6 end-points), and

others (29 end-points), were mapped to this characteristic.

9. *Causes immortalization (0 end-points):* no assay end-points were mapped to this characteristic.

10. *Alters cell proliferation/death or nutrient supply (68 end-points):* the assay end-points mapped to this characteristic measure cytotoxicity (41 end-points), mitochondrial toxicity (7 end-points), cell cycle (16 end-points), and cell proliferation (4 end-points).

By matching assays to key characteristics, additional insights could be obtained on the bioactivity profile for each compound specifically for the purpose of evaluating their potential to interact with or affect mechanisms involved in carcinogenesis. In addition, for each chemical, the results of the in-vitro assays that represent each "key characteristic" can be compared to the results for a larger compendium of substances with similar in-vitro data, so that a particular chemical can be aligned with other chemicals with similar toxicological effects. Nonetheless, the available assays do not cover the full spectrum of targets that may be associated with these mechanisms, and metabolic capacity in many of the assays is limited, which could account for any absence of bioactivity. Conversely, the presence of bioactivity alone does not definitively imply that the agent exhibits that key characteristic, as the assay data are considered along with other information, both in vivo and in vitro.

The Working Group then extracted information from the ToxCast database concerning whether a chemical was "active" or "inactive" for each of the selected assay end-points (Sipes et al., 2013; EPA, 2015b). In the analysis by the Working Group, each "active" was given a value of 1, and each "inactive" was given a value of 0. Thus, by assigning all active compounds a value of 1, the micromolar "potency" estimates from the concentration–response data were not explicitly modelled.

Next, to integrate the data across individual assay end-points into the cumulative score for each "key characteristic," the toxicological prioritization index (ToxPi) approach (Reif et al., 2010) and associated software (Reif et al., 2013; Filer et al., 2014) were used. In the Working Group's analyses, the ToxPi score provides a visual measure of the potential for a chemical to be associated with a "key characteristic" relative to 181 chemicals that have been previously evaluated by the *IARC Monographs* and that have been screened by ToxCast. Assay end-point data were available in ToxCast for these 181 chemicals, but not for other chemicals previously evaluated by IARC. ToxPi is a dimensionless index score that integrates multiple, different, assay results and displays them visually. Within each subset of end-points ("slice"), data are translated into ToxPi slice-wise scores for all compounds as detailed below and in the publications describing the approach and the associated software package (Reif et al., 2013). Within each individual slice for a given chemical, the distance from the origin represents the relative chemical-elicited activity of the component assays (i.e. slices extending farther from the origin were associated with "active" calls on more assays). The overall score for a chemical, visualized as a radial ToxPi profile, is the aggregation of all slice-wise scores.

The list of ToxCast/Tox21 assay end-points included in the analysis by the Working Group, description of the target and/or model system for each end-point (e.g. cell type, species, detection technology, etc.), their mapping to 6 of the 10 "key characteristics" of known human carcinogens, and the decision as to whether each chemical was "active" or "inactive" are available as supplemental material to *Monographs* Volume 113 (IARC, 2017b). The output files generated for each "key characteristic" are also provided in the supplemental material, and can be opened using ToxPi software that is freely available for download without a licence (Reif et al., 2013).

4.3.3 *Specific effects across 6 of the 10 "key characteristics" based on data from high-throughput screening in vitro*

The relative effects of lindane were compared with those of 181 chemicals selected from the more than 800 chemicals previously evaluated by the *IARC Monographs* and also screened by the Tox21/ToxCast programmes, and with those of the other compounds evaluated in the present volume of the *IARC Monographs* (Volume 113) and with their metabolites. Of these 181 chemicals previously evaluated by the *IARC Monographs* and screened in the ToxCast/Tox21 programmes, 8 are classified in Group 1 (*carcinogenic to humans*), 18 are in Group 2A (*probably carcinogenic to humans*), 59 are in Group 2B (*possibly carcinogenic to humans*), 95 are in Group 3 (*not classifiable as to its carcinogenicity to humans*), and 1 is in Group 4 (*probably not carcinogenic to humans*). The results are presented in a dot plot as a rank order of all compounds in the analysis arranged in the order of their relative activity. The relative positions of lindane, DDT (*p,p*′-DDT, *o,p*′-DDT, *p,p*′-DDE, *p,p*′-DDD), and 2,4-D in the ranked list are also shown on the *y*-axis. The colour scheme legend (lower left in each plot) annotates each compound according to its previous *IARC Monographs* group classification. The legend key (lower right graphic in each plot) lists components of the ToxPi chart as subcategories that comprise assay end-points in each characteristic, as well as their respective colour-coding (see Section 4.3.2; IARC, 2017b). The ToxPi profile and numeric score are shown for the highest-ranked chemical in each analysis (directly above the legend key) to represent the maximum ToxPi score and for lindane (upper frame).

Characteristic (1) *Is electrophilic or can undergo metabolic activation*: Lindane was tested for all 31 assay end-points mapped to this key characteristic. Lindane was active in none of the 29 assay end-points related to CYP inhibition, and in 1 of the 2 assay end-points related to aromatase inhibition. In comparison, the highest-ranked chemical, malathion (IARC Group 2A; IARC, 2017a), was active in 20 out of 29 assay end-points related to CYP inhibition and in 1 out of 2 related to aromatase inhibition (Fig. 4.1).

Characteristic (4) *Induces epigenetic alterations:* Lindane was tested for all 11 assay end-points mapped to this characteristic. Lindane was active in none of the assays. In comparison, the highest-ranked chemical, captan (IARC Group 3; IARC, 1983) was active for 0 out of 4 DNA-binding assay end-points, and 5 out of 7 transformation-catalyst (e.g. histone modification) assay end-points (Fig. 4.2).

Characteristic (5) *Induces oxidative stress*: lindane was tested for all 18 assay end-points mapped to this characteristic. The 18 assay end-points that were mapped to this characteristic are in subcategories of metalloproteinase (5 end-points), oxidative stress (7 end-points), and oxidative stress marker (6 end-points). Lindane was active for 2 of the assay end-points measuring oxidative-stress markers, but none of the other assay end-points. In comparison, the highest-ranked chemical, carbaryl (IARC Group 3; IARC, 1976) was active for 2 out of 5 metalloproteinase assay end-points, 3 out of 7 oxidative stress assay end-points, and 3 out of 6 oxidative-stress marker assay end-points (Fig. 4.3).

Characteristic (6) *Induces chronic inflammation*: lindane was tested for all 45 assay end-points mapped to this characteristic, and was active for 1 assay end-point related to cytokine levels. In comparison, the highest-ranked chemical, 4,4′-methylenedianiline (IARC Group 2B; IARC, 1986) was active for 2 out of 14 cellular-adhesion assay end-points, and 2 out of 29 cytokine-assay end-points (Fig. 4.4).

Fig. 4.1 ToxPi ranking for lindane using ToxCast assay end-points mapped to metabolic activation

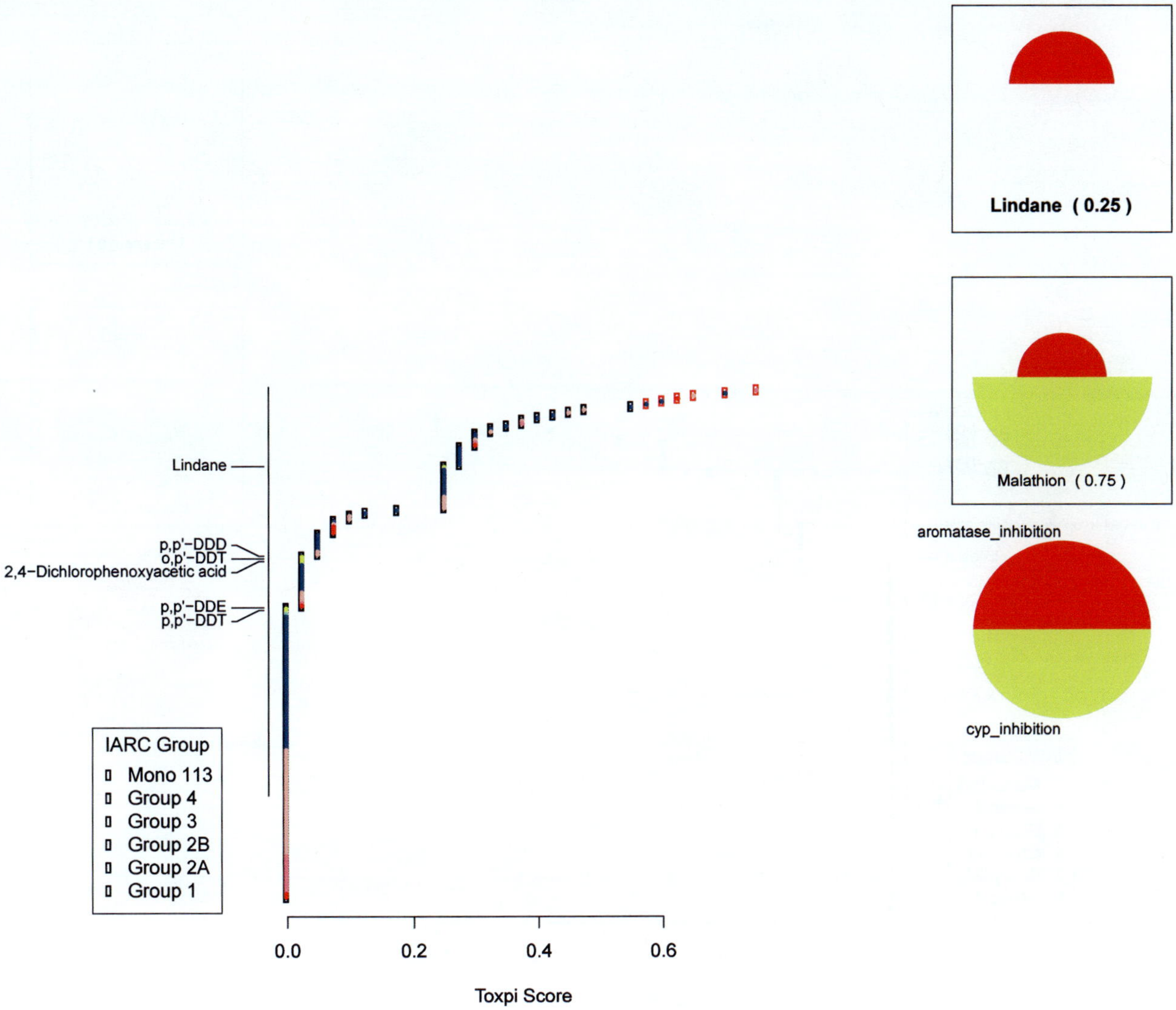

On the left-hand side, the relative rank of lindane is shown (*y*-axis) with respect to its ToxPi score (*x*-axis) as compared with the other chemicals evaluated in the present volume (Volume 113) and the 181 chemicals previously evaluated by IARC. The inset in the scatter plot shows subcategories of the ToxPi chart, as well as their respective colour coding. On the right-hand side, the ToxPi charts of the top-ranked chemical (in this case, malathion) and the target chemical(s) (lindane) are shown with their respective ToxPi score in parentheses.

Fig. 4.2 ToxPi ranking for lindane using ToxCast assay end-points mapped to epigenetic alterations

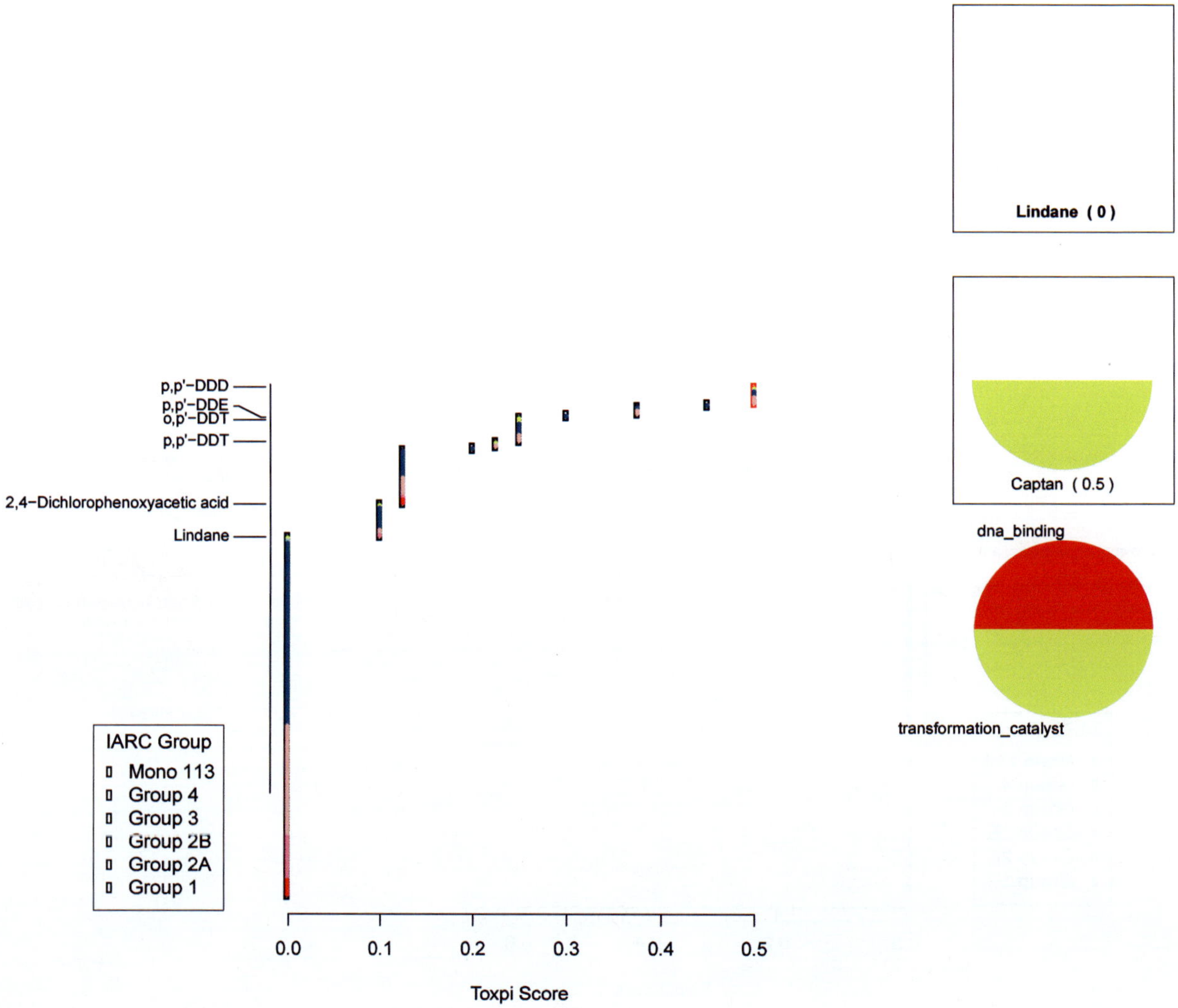

On the left-hand side, the relative rank of lindane is shown (*y*-axis) with respect to its ToxPi score (*x*-axis) as compared with the other chemicals evaluated in the present volume (Volume 113) and the 178 chemicals previously evaluated by IARC. The inset in the scatter plot shows subcategories of the ToxPi chart, as well as their respective colour coding. On the right-hand side, the ToxPi charts of the top-ranked chemical (in this case, captan) and the target chemical(s) (lindane) are shown with their respective ToxPi score in parentheses.

Fig. 4.3 ToxPi ranking for lindane using ToxCast assay end-points mapped to oxidative stress markers

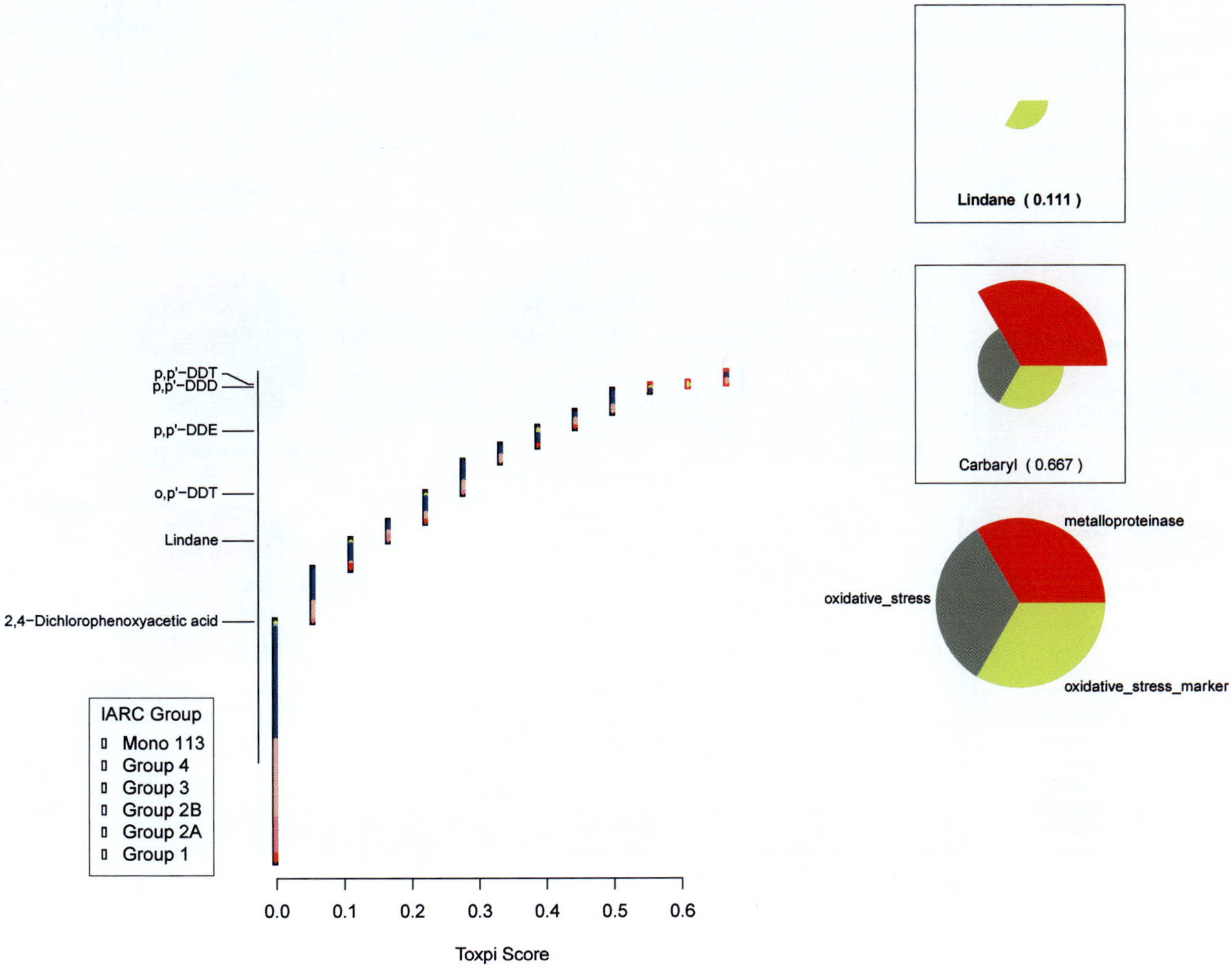

On the left-hand side, the relative rank of lindane is shown (*y*-axis) with respect to its ToxPi score (*x*-axis) as compared with the other chemicals evaluated in the present volume (Volume 113) and the 178 chemicals previously evaluated by IARC. The inset in the scatter plot shows subcategories of the ToxPi chart, as well as their respective colour coding. On the right-hand side, the ToxPi charts of the top-ranked chemical (in this case, carbaryl) and the target chemical(s) (lindane) are shown with their respective ToxPi score in parentheses.

Fig. 4.4 ToxPi ranking for lindane using ToxCast assay end-points mapped to induction of chronic inflammation

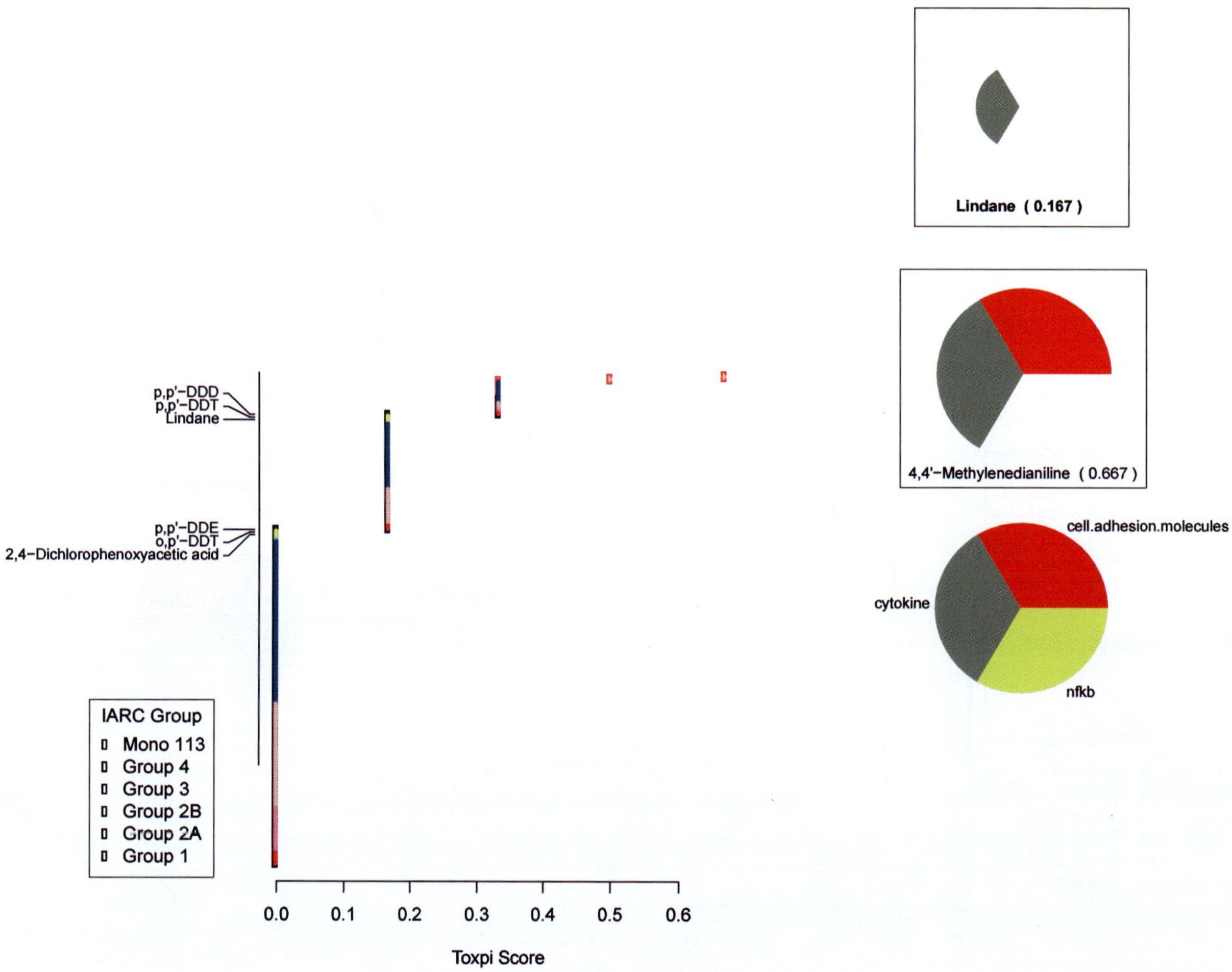

On the left-hand side, the relative rank of lindane is shown (*y*-axis) with respect to its ToxPi score (*x*-axis) as compared with the other chemicals evaluated in the present volume (Volume 113) and the 178 chemicals previously evaluated by IARC. The inset in the scatter plot shows subcategories of the ToxPi chart, as well as their respective colour coding. On the right-hand side, the ToxPi charts of the top-ranked chemical (in this case, 4,4′-methylenedianiline) and the target chemical(s) (lindane) are shown with their respective ToxPi score in parentheses.

Characteristic (8) *Modulates receptor-mediated effects:* lindane was tested for all 92 assay end-points mapped to this characteristic, and was active in 3 out of 18 ER assay end-points, 1 out of 7 FXR assay end-points, 1 out of 29 other nuclear-receptor assay end-points, 1 out of 12 PPAR assay end-points, 3 out of 7 PXR_VDR assay end-points, and 2 out of 6 RAR assay end-points. In comparison, the highest-ranked chemical, clomiphene citrate (IARC Group 3; IARC, 1979a) was active for 5 out of 11 androgen-receptor assay end-points, 13 out of 18 ER assay end-points, 3 out of 7 FXR assay end-points, 6 out of 29 other nuclear-receptor assay end-points, 2 out of 12 PPAR assay end-points, 5 out of 7 PXR_VDR assay end-points, and 1 out of 6 RAR assay end-points (Fig. 4.5).

Characteristic (10) *Alters cell proliferation, cell death, or nutrient supply:* lindane was tested for 67 out of 68 assay end-points mapped to this characteristic and was active for 2 of the 41 assay end-points related to cytotoxicity. In comparison, the highest-ranked chemical, ziram (IARC Group 3; IARC, 1991) was active for 2 out of 16 cell-cycle assay end-points, 33 out of 41 cytotoxicity end-points, and 2 out of 7 mitochondrial-toxicity assay end-points (Fig. 4.6).

4.3.4 Summary of all effects across the "key characteristics" based on data from high-throughput screening in vitro

As a high-level summary of activity, data were recombined into six ToxPi slices, where each slice represents activity across all component assays mapped to a given characteristic. In the figure (Fig. 4.7), slices are labelled "metabolism" (*Is electrophilic or can undergo metabolic activation*), "epigenetic" (*Induces epigenetic alterations*), "stress" (*Induces oxidative stress*), "inflammation" (*Induces chronic inflammation*), "receptor" (*Modulates receptor-mediated effects*), and "cellular" (*Alters cell proliferation, cell death, or nutrient supply*). Lindane was tested for 264 of 265 assay end-points mapped to any characteristic. Overall, lindane was active for 17 of the assay end-points mapped to any characteristic. In comparison, the highest-ranked chemical, clomiphene citrate (IARC Group 3; IARC, 1979a) was active for 97 assay end-points.

4.4 Cancer susceptibility data

There is a paucity of studies that examine susceptibility to cancer associated with exposure to lindane in humans. Two case–control studies of similar size found opposing evidence for postmenopausal status as a susceptibility factor for cancer of the breast (Zheng et al., 1999; Ibarluzea et al., 2004). A history of asthma did not modify a non-significant association between non-Hodgkin lymphoma and exposure to lindane (Lee et al., 2004). No studies of carcinogenicity with lindane in experimental animals have examined susceptibility. [The Working Group noted that few studies evaluated life-stage, and genetic and disease susceptibility. There is no compelling evidence for factors enhancing susceptibility to cancer in association with exposure to lindane.]

4.5 Other adverse effects

Other adverse effects not addressed in Sections 4.1–4.4 that may be relevant to cancer hazard identification for lindane include toxicity in the liver, kidney, haematopoietic system, and testis. These effects have also been reviewed by the Agency for Toxic Substances and Disease Registry (ATSDR, 2005).

Fig. 4.5 ToxPi ranking for lindane using ToxCast assay end-points mapped to modulation of receptor-mediated effects

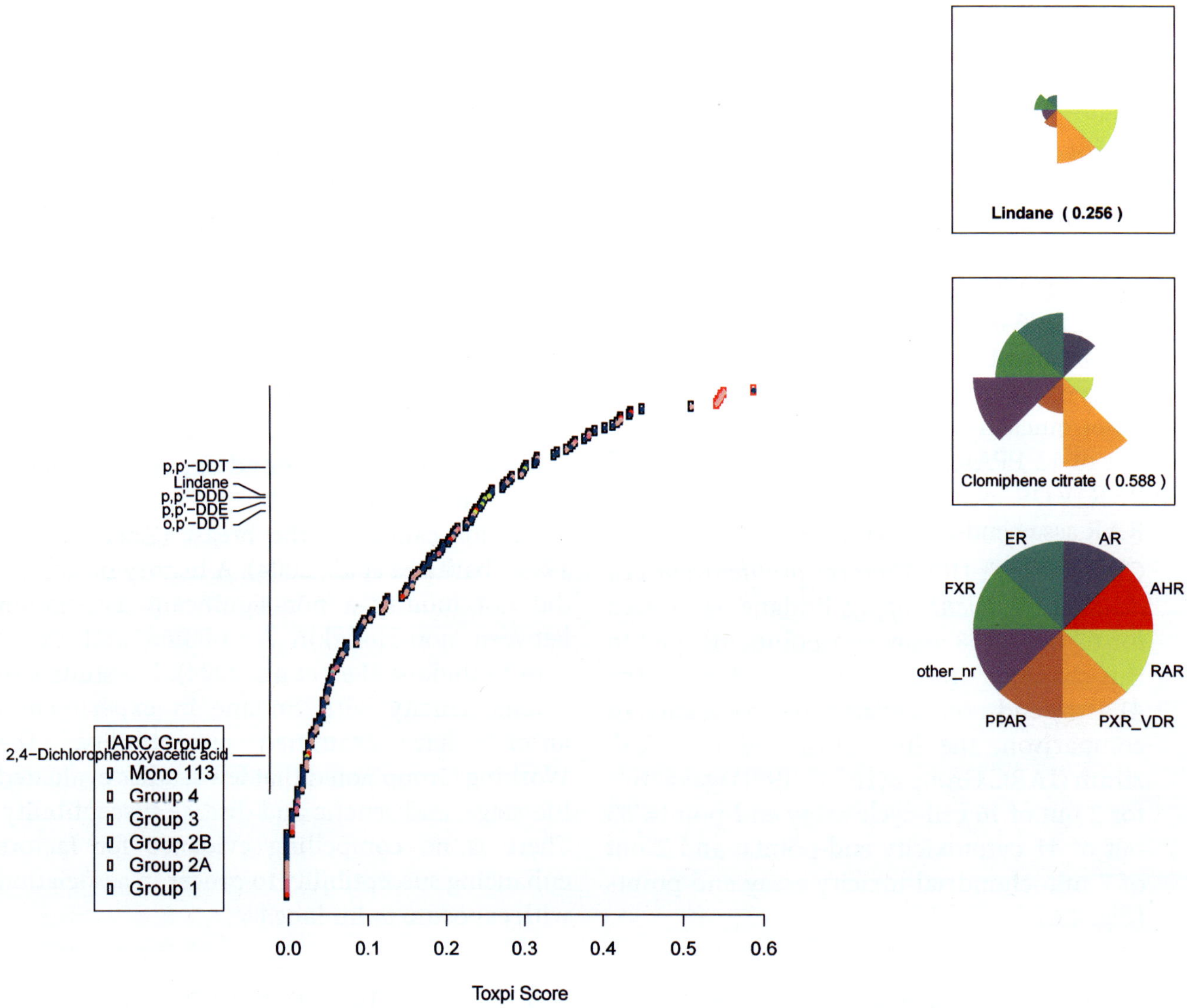

On the left-hand side, the relative rank of lindane is shown (*y*-axis) with respect to its ToxPi score (*x*-axis) as compared with the other chemicals evaluated in the present volume (Volume 113) and the 178 chemicals previously evaluated by IARC. The inset in the scatter plot shows subcategories of the ToxPi chart, as well as their respective colour coding. On the right-hand side, the ToxPi charts of the top-ranked chemical (in this case, clomiphene citrate) and the target chemical(s) (lindane) are shown with their respective ToxPi score in parentheses.

Fig. 4.6 ToxPi ranking for lindane using ToxCast assay end-points mapped to alteration of cell proliferation, cell death, and nutrient supply

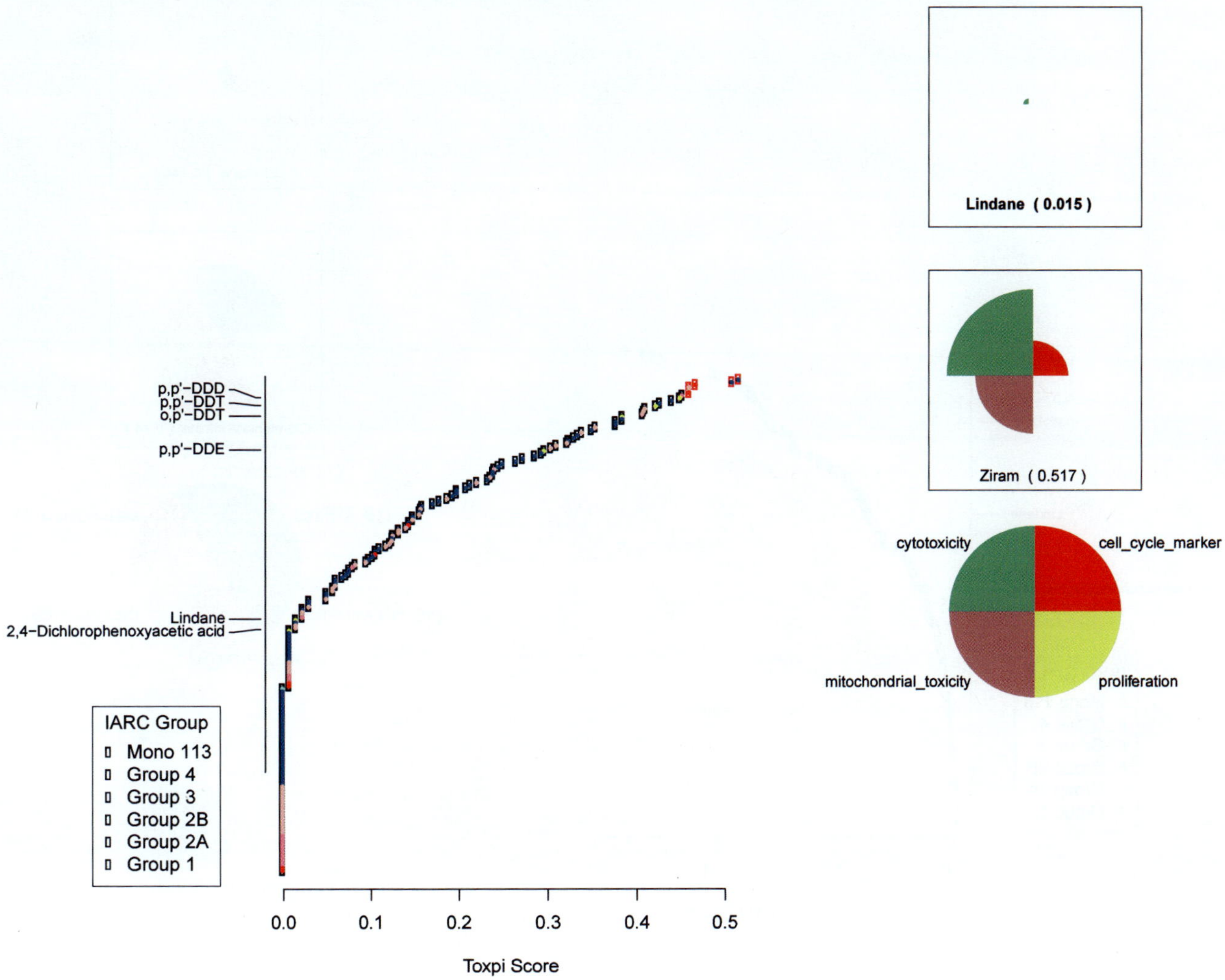

On the left-hand side, the relative rank of lindane is shown (*y*-axis) with respect to its ToxPi score (*x*-axis) as compared with the other chemicals evaluated in the present volume (Volume 113) and the 178 chemicals previously evaluated by IARC. The inset in the scatter plot shows subcategories of the ToxPi chart, as well as their respective colour coding. On the right-hand side, the ToxPi charts of the top-ranked chemical (in this case, ziram) and the target chemical(s) (lindane) are shown with their respective ToxPi score in parentheses.

Fig. 4.7 ToxPi ranking for lindane using ToxCast assay end-points: summary of key characteristics

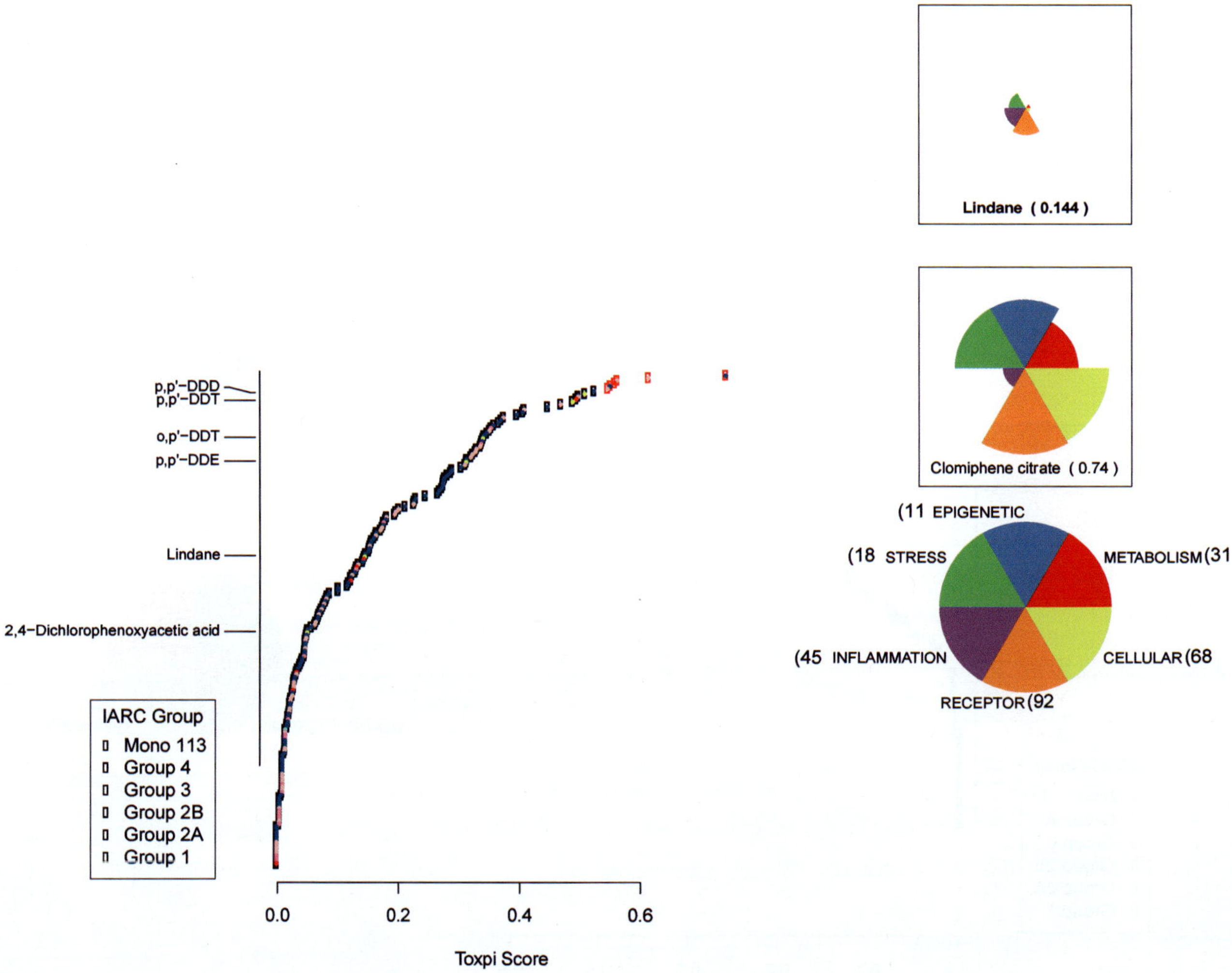

On the left-hand side, the relative rank of lindane is shown (*y*-axis) with respect to its ToxPi score (*x*-axis) as compared with the other chemicals evaluated in the present volume (Volume 113) and the 178 chemicals previously evaluated by IARC. The inset in the scatter plot shows subcategories of the ToxPi chart, as well as their respective colour coding. On the right-hand side, the ToxPi charts of the top-ranked chemical (in this case, clomiphene citrate) and the target chemical(s) (lindane) are shown with their respective ToxPi score in parentheses.

4.5.1 Humans

Several studies have reported increases in liver enzyme levels associated with occupational exposure to technical-grade HCH (Kashyap, 1986; Nigam et al., 1993). Renal failure was reported in multiple cases of acute poisoning with lindane (Munk & Nantel, 1977; Sunder Ram Rao et al., 1988). In one case, simultaneous renal and hepatic injury was reported (Paul et al., 2013).

No studies of associations between exposure to lindane and testicular toxicity in humans were available to the Working Group.

4.5.2 Experimental systems

Experimental studies in rats and rabbits have reported liver toxicity after administration of lindane, as demonstrated by alterations in clinical markers, such as serum aspartate transaminase, alanine transaminase, and lactate dehydrogenase, and by histopathological lesions, such as cytotoxicity, necrosis, and focal degeneration (Schulte Hermann et al., 1971; Grabarczyk et al., 1990; Junqueira et al., 1991; Raizada et al., 2001; Videla et al., 2004; Matsuura et al., 2005; Anilakumar et al., 2007; Vijaya Padma et al., 2011).

Experimental studies in rats and rabbits have reported kidney toxicity after administration of lindane, as demonstrated by clinical markers, such as elevated serum urea, creatinine, and uric acid, and by histopathological lesions, such as focal degeneration (Grabarczyk et al., 1990; Vijaya Padma et al., 2011).

Haematological effects, including reduced phagocytic activity of neutrophils, increased number of lymphocytes with inactive nucleoli, and reduced total leukocyte count have also been reported in rats and rabbits (Grabarczyk et al., 1990; Joseph et al., 1992).

Histological and functional changes in the testes have been reported after administration of lindane in rats. Reported effects include organ-weight changes, alterations in steroidogenesis, decreased spermatogenesis, reduction in antioxidant defence, alterations in testicular morphology, and degeneration of Leydig cells (Samanta et al., 1999a, b; Suwalsky et al., 2000; Pagès et al., 2002; Saradha et al., 2008). Effects in the male reproductive tract in adulthood have also been reported when rats are exposed during lactation (Dalsenter et al., 1997a). In a two-generation study of reproductive toxicity in rats, there were no changes in the testes in the F_0 generation, and decreases in absolute, but not relative, testicular weights in the F_1 and F_2 generations only at the highest dietary concentration of 300 ppm (Matsuura et al., 2005). Srinivasan et al. (1991) did not report any effects on the testes in pups exposed in utero or via lactation through the maternal diet.

5. Summary of data reported

5.1 Exposure data

Lindane is the γ-isomer of hexachlorocyclohexane (HCH). While there are several isomers of HCH, among which the β-, γ-, δ-, and ε-isomers are relatively stable, only γ-HCH has insecticidal properties. Technical-grade lindane containing 90% γ-HCH, and technical-grade HCH containing 10–40% γ-HCH, have both been used worldwide as pesticides.

Lindane has been extensively manufactured and used in the past, primarily as an insecticide to treat wood and wooden structures, seed, crops, and livestock. Occupational exposure to lindane can occur, or has occurred, in the course of its manufacture and formulation, and during agricultural application. Use of lindane decreased significantly from the 1970s to the 1990s due to restrictions and bans on its use in agriculture. Lindane continues to be used for public health purposes, although with decreasing numbers of prescriptions, for the second-line treatment of scabies (mites) and lice in humans (as a specific exemption under the Stockholm Convention on Persistent Organic Pollutants).

Lindane is mobile in the environment and, as a result of long-range atmospheric transport, has been deposited worldwide. Lindane has been measured in food, air, surface water, groundwater, sediment, soil, fish, wildlife, and humans. Current exposure of the general population in countries worldwide occurs mainly through the diet. In most regions of the world, the proportion of human biological samples containing lindane at detectable levels is decreasing.

5.2 Human carcinogenicity data

Cancer risks associated with exposure to lindane have been evaluated in cohort and case–control studies in several countries. The largest body of data concerns non-Hodgkin lymphoma (NHL).

Cohort and case–control studies of mostly occupational exposure to pesticides provide consistent evidence of an association between NHL and lindane. A large prospective cohort study of farmers in the USA estimated exposure through a detailed assessment, and observed significant upward trends in risk of NHL in relation to several indicators of occupational exposure to lindane, while controlling for other risk factors. Another cohort study investigating the mortality of Icelandic sheep farmers reported that the risk of NHL increased with the number of sheep owned. The number of sheep owned is an indirect measure of exposure to lindane because treatment of sheep to control ectoparasites was a legal requirement, with a technical-grade HCH mixture used for this purpose before the 1970s, and lindane alone used afterwards; however, this metric could lead to misclassification of the level of exposure.

A pooled analysis of three population-based case–control studies in the midwestern USA reported a 50% increase in the risk of NHL associated with any use of lindane. The association was stronger with several indicators of higher exposure and remained positive after individual adjustments for other pesticides and pesticide classes. The association between NHL and exposure to lindane was reduced to a 20% excess in a subsequent analysis of a subset of the population with simultaneous adjustment for multiple pesticides. A population-based case–control study in Canada, in which pesticide exposures were assessed using questionnaires, also found moderately increased risks for NHL associated with exposure to lindane. A meta-analysis of four studies examining ever-exposure to lindane in agricultural settings, including the original pooled analysis described above, found a 60% increased risk of NHL with exposure to lindane.

A cohort study in the USA and three casecontrol studies in Europe investigated the association between NHL and β-HCH in the general population using measurements in serum. The cohort study and two case–control studies reported some positive associations between β-HCH and risk, while the other case–control study did not. However, the interpretation of general-population studies using levels of other HCH isomers measured in biological samples is uncertain since such measurements do not necessarily reflect exposure to lindane.

Associations between cancers of the breast, prostate, or testis with lindane or HCH isomers measured in serum have been evaluated in several studies, but the results were not consistent and the findings for HCH isomers may not reflect exposure to lindane, as noted above.

5.3 Animal carcinogenicity data

Lindane was tested for carcinogenicity by oral administration (feeding) in seven studies in mice and two studies in rats. These studies had several limitations, such as short duration of exposure, use of one sex only, inadequate dose selection, lack of rationale for dose selection, and limited reporting.

In treated mice, lindane consistently increased the incidence of benign and/or malignant tumours of the liver, which were classified across the various studies as benign or malignant liver cell tumours, hepatomas (not further classified), benign or malignant hepatomas, or hepatocellular adenoma or carcinoma. In one study in males, there was an increase in the incidence of benign hepatomas. In a second study in males and females, there was an increase in the incidence of benign or malignant (combined) liver cell tumours in both sexes. In a third study in males and females, there was an increase in the incidence of liver hepatoma (not further classified) in males. In a fourth study in females of three different strains of mice, there was an increase in the incidence of hepatocellular adenoma or carcinoma (combined) in two strains, and of bronchiolo-alveolar tumours of the lung (not further classified) in one of these two strains. In a fifth study in males and females, there was a positive trend in the incidence of bronchiolo-alveolar adenoma or carcinoma (combined) in females. One study in males and females gave negative results, and another one study in males and females was inadequate for the evaluation.

In rats, one study in males and females gave negative results, and another study was inadequate for the evaluation.

5.4 Mechanistic and other relevant data

Lindane is highly lipophilic, readily absorbed via all routes of exposure, and distributes widely in the body, with a preference for adipose and other lipid-rich tissues. It is extensively metabolized via cytochrome P450s to a multitude of metabolites and excreted as various conjugates, including mercapturic acids. Terminal half-lives in humans have been estimated to be around 1 day, with a few studies reporting half-lives of up to approximately 1 week. Lindane induces several cytochrome P450 enzymes in rats, and induces CYP2B6 and inhibits CYP2D6 and CYP2E1 in human hepatocytes.

With respect to the key characteristics of human carcinogens, adequate data were available to evaluate whether lindane is immunosuppressive, induces oxidative stress, modulates receptor-mediated effects, is genotoxic, alters cell proliferation, cell death or nutrient supply, and induces chronic inflammation.

The evidence is *strong* that lindane is immunosuppressive, and this characteristic can operate in humans. Lindane causes dose-dependent immunosuppressive effects in vivo in several species, including mice, rats, birds and fish. Lindane suppresses the humoral immune response in mice and rats. In vitro, lindane suppresses the activation of human lymphocytes and is toxic to murine lymphocytes. Immunological studies in exposed humans are inconclusive with respect to immunosuppression, but haematotoxic effects have been observed in humans and rodents.

The evidence is *strong* that lindane induces oxidative stress. Markers of oxidative stress in blood lymphocytes were increased in cases of acute poisoning with lindane in humans. Additional data in experimental systems, including in human lymphocytes in vitro and rats in vivo, provide consistent evidence of increases in reactive oxygen species or markers of oxidative stress.

The evidence is *moderate* that lindane modulates receptor-mediated effects. No consistent associations between lindane and serum hormone levels were found in a few studies in humans. Lindane binds and blocks the androgen receptor in human cells and experimental systems. It did not activate transfected human androgen receptor in vitro. Studies in experimental systems indicate that lindane has anti-estrogenic and anti-androgenic activity.

The evidence is *moderate* that lindane is genotoxic. In human cells in vitro, chromosomal aberration, sister-chromatid exchange, and increased micronucleus formation were reported after treatment with lindane. Negative or mixed results were reported in assays for DNA-adduct formation and other types of DNA damage. In rats, mice, and Chinese hamsters, findings were mixed across experimental systems, including chromosomal and other DNA damage end-points. Negative or statistically significant but very small effects were reported in bacteria and lower eukaryotes. No reliable genotoxicity data were available in exposed humans.

The evidence is *weak* that lindane alters cell proliferation or death, or induces chronic inflammation. No data were available in exposed humans. Studies of apoptosis and proliferation in human cancer cells gave mixed results. Induction of apoptosis was reported in mice in vivo and in several mouse cell types in vitro, but inhibition of apoptosis was reported in rat hepatocytes. A few studies in rats in vivo and in vitro have reported Kupffer cell activation in the liver.

In high-throughput testing in the Toxicity Testing in the 21st Century (Tox21) and Toxicity Forecaster (ToxCast) research programmes of the government of the USA, lindane gave positive results for 17 assay end-points, mostly related to receptor-mediated effects, among the 265 assay end-points relevant to the key characteristics of human carcinogens.

No cancer susceptibility factors have been identified in humans.

Adverse effects of lindane in the liver, kidney, and haematopoietic systems have been reported in exposed humans, as well as in experimental systems. Additionally, testicular toxicity has been reported in rats exposed to lindane.

Overall, the mechanistic data provide strong support for the carcinogenicity of lindane. This includes strong evidence that lindane is immunosuppressive and induces oxidative stress, and that effects can operate in humans.

6. Evaluation

6.1 Cancer in humans

There is *sufficient evidence* in humans for the carcinogenicity of lindane. Lindane causes non-Hodgkin lymphoma.

6.2 Cancer in experimental animals

There is *sufficient evidence* in experimental animals for the carcinogenicity of lindane.

6.3 Overall evaluation

Lindane is *carcinogenic to humans* (*Group 1*).

References

Abbassy MS, Ibrahim HZ, el-Amayem MM (1999). Occurrence of pesticides and polychlorinated biphenyls in water of the Nile river at the estuaries of Rosetta and Damiatta branches, north of Delta, Egypt. *J Environ Sci Health B*, 34(2):255–67. doi:10.1080/03601239909373196 PMID:10192956

Abhilash PC, Jamil S, Singh N (2007). Matrix solid-phase dispersion extraction versus solid-phase extraction in the analysis of combined residues of hexachlorocyclohexane isomers in plant matrices. *J Chromatogr A*, 1176(1-2):43–7. doi:10.1016/j.chroma.2007.11.005 PMID:18035358

Achari RG, Sandhu SS, Warren WJ (1975). Chlorinated hydrocarbon residues in ground water. *Bull Environ Contam Toxicol*, 13(1):94–6. doi:10.1007/BF01684870 PMID:48391

Ahdaya SM, Monroe RJ, Guthrie FE (1981). Absorption and distribution of intubated insecticides in fasted mice. *Pestic Biochem Physiol*, 16(1):38–46. doi:10.1016/0048-3575(81)90070-5

Ahmed FE, Hart RW, Lewis NJ (1977). Pesticide induced DNA damage and its repair in cultured human cells. *Mutat Res*, 42(2):161–74. doi:10.1016/S0027-5107(77)80020-1 PMID:190533

Aks SE, Krantz A, Hryhorczuk DO, Wagner S, Mock J (1995). Acute accidental lindane ingestion in toddlers. *Ann Emerg Med*, 26(5):647–51. doi:10.1016/S0196-0644(95)70020-X PMID:7486377

Al-Mughrabi KI, Qrunfleh IM (2002). Pesticide residues in soil from the Jordan Valley, Jordan. *Bull Environ Contam Toxicol*, 68(1):86–96. doi:10.1007/s00128-001-0223-7 PMID:11731836

Alavanja M, Hofmann J, Lynch C, Hines C, Barry K, Barker J et al. (2014a). Occupational use of insecticides, fungicides and fumigants and risk of non-Hodgkin lymphoma and multiple myeloma in the Agricultural Health Study. *Occup Environ Med*, 71:Suppl 1: A36 doi:10.1136/oemed-2014-102362.111

Alavanja MC, Hofmann JN, Lynch CF, Hines CJ, Barry KH, Barker J et al. (2014b). Non-hodgkin lymphoma risk and insecticide, fungicide and fumigant use in the agricultural health study. *PLoS ONE*, 9(10):e109332 doi:10.1371/journal.pone.0109332 PMID:25337994

Albertini S, Friederich U, Wurgler FE (1988). Induction of mitotic chromosome loss in the diploid yeast Saccharomyces cerevisiae D61.M by genotoxic carcinogens and tumor promoters. *Environ Mol Mutagen*, 11(4):497–508. doi:10.1002/em.2850110410

American Cancer Society (2016). Types of non-Hodgkin lymphoma. Available from: http://www.cancer.org/cancer/non-hodgkinlymphoma/detailedguide/non-hodgkin-lymphoma-types-of-non-hodgkin-lymphoma, accessed 18 April 2016.

Ananya R, Subeena S, Kumar DA, Kumar DT, Kumar MS (2005). Oxidative stress and histopathological changes in the heart following oral lindane (gamma hexachlorohexane) administration in rats. *Med Sci Monit*, 11(9):BR325–9. PMID:16127354

Anderson HA, Falk C, Hanrahan L, Olson J, Burse VW, Needham L et al. ; The Great Lakes Consortium (1998). Profiles of Great Lakes critical pollutants: a sentinel analysis of human blood and urine. *Environ Health Perspect*, 106(5):279–89. PMID:9560354

Angerer J, Maass R, Heinrich R (1983). Occupational exposure to hexachlorocyclohexane. VI. Metabolism of gamma-hexachlorocyclohexane in man. *Int Arch Occup Environ Health*, 52(1):59–67. doi:10.1007/BF00380608 PMID:6192092

Anguiano G, Llera-Herrera R, Rojas E, Vazquez-Boucard C (2007). Subchronic organismal toxicity, cytotoxicity, genotoxicity, and feeding response of Pacific oyster (*Crassostrea gigas*) to lindane (γ-HCH) exposure under experimental conditions. *Environ Toxicol Chem*, 26(10):2192–7. doi:10.1897/06-377R3.1 PMID:17867868

Anilakumar KR, Nagaraj NS, Santhanam K (2007). Reduction of hexachlorocyclohexane-induced oxidative stress and cytotoxicity in rat liver by Emblica officinalis gaertn. *Indian J Exp Biol*, 45(5):450–4. PMID:17569287

Anilakumar KR, Saritha V, Khanum F, Bawa AS (2009). Ameliorative effect of ajwain extract on hexachlorocyclohexane-induced lipid peroxidation in rat liver. *Chem Toxicol*, 47(2):279–82. doi:10.1016/j.fct.2008.09.061

Aoki K, Egawa M, Saito T, Hosokawa T, Kurasaki M (2008). Effects of gamma-hexachlorocyclohexane on apoptosis induced by serum deprivation in PC12 cells. *J Environ Sci Health B*, 43(6):471–5. doi:10.1080/03601230802174573 PMID:18665982

Aronson KJ, Wilson JW, Hamel M, Diarsvitri W, Fan W, Woolcott C et al. (2010). Plasma organochlorine levels and prostate cancer risk. *J Expo Sci Environ Epidemiol*, 20(5):434–45. doi:10.1038/jes.2009.33 PMID:19513097

Arthur RD, Cain JD, Barrentine BF (1976). Atmospheric levels of pesticides in the Mississippi Delta. *Bull Environ Contam Toxicol*, 15(2):129–34. doi:10.1007/BF01685150 PMID:1252627

ATSDR (1989). Hexachlorocyclohexane. Atlanta (GA), USA: Agency for Toxic Substances and Disease Registry. Available from: http://www.atsdr.cdc.gov/toxprofiles/tp43-c4.pdf.

ATSDR (2005). Toxicological profile for alpha-, beta-, gamma-, and delta-hexachlorocyclohexane. Atlanta (GA), USA: Agency for Toxic Substances and Disease Registry. Available from: http://www.atsdr.cdc.gov/toxprofiles/tp43.pdf.

Aydin ME, Ozcan S, Beduk F, Tor A (2013). Levels of organochlorine pesticides and heavy metals in surface waters of Konya closed basin, Turkey. *Scientific WorldJournal*, 2013:849716 doi:10.1155/2013/849716 PMID:23533363

Azandjeme CS, Delisle H, Fayomi B, Ayotte P, Djrolo F, Houinato D et al. (2014). High serum organochlorine pesticide concentrations in diabetics of a cotton producing area of the Benin Republic (West Africa). *Environ Int*, 69:1–8. doi:10.1016/j.envint.2014.04.002 PMID:24769438

Band PR, Abanto Z, Bert J, Lang B, Fang R, Gallagher RP et al. (2011). Prostate cancer risk and exposure to pesticides in British Columbia farmers. *Prostate*, 71(2):168–83. doi:10.1002/pros.21232 PMID:20799287

Banerjee BD, Koner BC, Ray A, Pasha ST (1996). Influence of subchronic exposure to lindane on humoral immunity in mice. *Indian J Exp Biol*, 34(11):1109–13. PMID:9055633

Banerjee BD, Seth V, Bhattacharya A, Pasha ST, Chakraborty AK (1999). Biochemical effects of some pesticides on lipid peroxidation and free-radical scavengers. *Toxicol Lett*, 107(1-3):33–47. doi:10.1016/S0378-4274(99)00029-6 PMID:10414779

Barr JR, Maggio VL, Barr DB, Turner WE, Sjödin A, Sandau CD et al. (2003). New high-resolution mass spectrometric approach for the measurement of polychlorinated biphenyls and organochlorine pesticides in human serum. *J Chromatogr B Analyt Technol Biomed Life Sci*, 794(1):137–48. doi:10.1016/S1570-0232(03)00451-3 PMID:12888206

Bates MN, Buckland SJ, Garrett N, Ellis H, Needham LL, Patterson DG Jr et al. (2004). Persistent organochlorines in the serum of the non-occupationally exposed New Zealand population. *Chemosphere*,

54(10):1431–43. doi:10.1016/j.chemosphere.2003.09.040 PMID:14659945

Battaglia CLR, Gogal RM Jr, Zimmerman K, Misra HP (2010). Malathion, lindane, and piperonyl butoxide, individually or in combined mixtures, induce immunotoxicity via apoptosis in murine splenocytes in vitro. *Int J Toxicol*, 29(2):209–20. doi:10.1177/1091581809357954 PMID:20075186

Baumann K, Angerer J, Heinrich R, Lehnert G (1980). Occupational exposure to hexachlorocyclohexane. I. Body burden of HCH-isomers. *Int Arch Occup Environ Health*, 47(2):119–27. doi:10.1007/BF00716371 PMID:6160111

Becker K, Kaus S, Krause C, Lepom P, Schulz C, Seiwert M et al. (2002). German Environmental Survey 1998 (GerES III): environmental pollutants in blood of the German population. *Int J Hyg Environ Health*, 205(4):297–308. doi:10.1078/1438-4639-00155 PMID:12068749

Behfar A, Nazari Z, Rabiee MH, Raeesi G, Oveisi MR, Sadeghi N et al. (2013). The organochlorine pesticides residue levels in Karun river water. *Jundishapur J Nat Pharm Prod*, 8(1):41–6. doi:10.5812/jjnpp.6783 PMID:24624185

Ben Hassine SB, Ameur WB, Gandoura N, Driss MR (2012). Determination of chlorinated pesticides, polychlorinated biphenyls, and polybrominated diphenyl ethers in human milk from Bizerte (Tunisia) in 2010. *Chemosphere*, 89(4):369–77. doi:10.1016/j.chemosphere.2012.05.035 PMID:22743186

Betoulle S, Duchiron C, Deschaux P (2000). Lindane differently modulates intracellular calcium levels in two populations of rainbow trout (Oncorhynchus mykiss) immune cells: head kidney phagocytes and peripheral blood leucocytes. *Toxicology*, 145(2-3):203–15. doi:10.1016/S0300-483X(99)00226-7 PMID:10771144

Bharathi SP, Raj HM, Jain S, Banerjee BD, Ahmed T, Arora VK (2013). Role of pesticides in the induction of tumor angiogenesis. *Anticancer Res*, 33(1):231–40. PMID:23267150

Bhatnagar VK, Kashyap R, Zaidi SS, Kulkarni PK, Saiyed HN (2004). Levels of DDT, HCH, and HCB residues in human blood in Ahmedabad, India. *Bull Environ Contam Toxicol*, 72(2):261–5. doi:10.1007/s00128-003-9049-9 PMID:15106760

Biggs ML, Davis MD, Eaton DL, Weiss NS, Barr DB, Doody DR et al. (2008). Serum organochlorine pesticide residues and risk of testicular germ cell carcinoma: a population-based case-control study. *Cancer Epidemiol Biomarkers Prev*, 17(8):2012–8. doi:10.1158/1055-9965.EPI-08-0032 PMID:18708392

Bijoy Nandan S, Nimila PJ (2012). Lindane toxicity: histopathological, behavioural and biochemical changes in Etroplus maculatus (Bloch, 1795). *Mar Environ Res*, 76:63–70. doi:10.1016/j.marenvres.2011.10.011 PMID:22115920

Blair A, Cantor KP, Zahm SH (1998). Non-hodgkin's lymphoma and agricultural use of the insecticide lindane. *Am J Ind Med*, 33(1):82–7. doi:10.1002/(SICI)1097-0274(199801)33:1<82::AID-AJIM9>3.0.CO;2-Y PMID:9408531

Boada LD, Zumbado M, Henríquez-Hernández LA, Almeida-González M, Alvarez-León EE, Serra-Majem L et al. (2012). Complex organochlorine pesticide mixtures as determinant factor for breast cancer risk: a population-based case-control study in the Canary Islands (Spain). *Environ Health*, 11(1):28 doi:10.1186/1476-069X-11-28 PMID:22534004

Borrás E, Sánchez P, Muñoz A, Tortajada-Genaro LA (2011). Development of a gas chromatography-mass spectrometry method for the determination of pesticides in gaseous and particulate phases in the atmosphere. *Anal Chim Acta*, 699(1):57–65. doi:10.1016/j.aca.2011.05.009 PMID:21704758

Botella B, Crespo J, Rivas A, Cerrillo I, Olea-Serrano MF, Olea N (2004). Exposure of women to organochlorine pesticides in southern Spain. *Environ Res*, 96(1):34–40. doi:10.1016/j.envres.2003.10.001 PMID:15261782

Bouvier G, Blanchard O, Momas I, Seta N (2006). Pesticide exposure of non-occupationally exposed subjects compared to some occupational exposure: a French pilot study. *Sci Total Environ*, 366(1):74–91. doi:10.1016/j.scitotenv.2005.08.016 PMID:16181660

Brassow HL, Baumann K, Lehnert G (1981). Occupational exposure to hexachlorocyclohexane. II. Health conditions of chronically exposed workers. *Int Arch Occup Environ Health*, 48(1):81–7. doi:10.1007/BF00405934 PMID:6163731

Briz V, Molina-Molina JM, Sánchez-Redondo S, Fernández MF, Grimalt JO, Olea N et al. (2011). Differential estrogenic effects of the persistent organochlorine pesticides dieldrin, endosulfan, and lindane in primary neuronal cultures. *Toxicol Sci*, 120(2):413–27. doi:10.1093/toxsci/kfr019 PMID:21278053

Brodtmann NV Jr (1976). Continuous analysis of chlorinated hydrocarbon pesticides in the lower Mississippi River. *Bull Environ Contam Toxicol*, 15(1):33–9. doi:10.1007/BF01686193 PMID:1276488

Brooks GC (1977). Chlorinated insecticides: retrospect and prospect. In: Plimmer JR, editor. Pesticide chemistry in the 20th century. ACS Symposium Series. Washington (DC), USA: American Chemical Society; pp. 1–19.

Brown LM, Blair A, Gibson R, Everett GD, Cantor KP, Schuman LM et al. (1990). Pesticide exposures and other agricultural risk factors for leukemia among men in Iowa and Minnesota. *Cancer Res*, 50(20):6585–91. PMID:2208120

Buchanan D, Pilkington A, Sewell C, Tannahill SN, Kidd MW, Cherrie B et al. (2001). Estimation of cumulative exposure to organophosphate sheep dips in a study of chronic neurological health effects among

United Kingdom sheep dippers. *Occup Environ Med*, 58(11):694–701. PMID:11600724

Cantor KP, Blair A, Everett G, Gibson R, Burmeister LF, Brown LM et al. (1992). Pesticides and other agricultural risk factors for non-Hodgkin's lymphoma among men in Iowa and Minnesota. *Cancer Res*, 52(9):2447–55. PMID:1568215

Cantor KP, Strickland PT, Brock JW, Bush D, Helzlsouer K, Needham LL et al. (2003). Risk of non-Hodgkin's lymphoma and prediagnostic serum organochlorines: β-hexachlorocyclohexane, chlordane/heptachlor-related compounds, dieldrin, and hexachlorobenzene. *Environ Health Perspect*, 111(2):179–83. doi:10.1289/ehp.4347 PMID:12573902

Carreño J, Rivas A, Granada A, Jose Lopez-Espinosa M, Mariscal M, Olea N et al. (2007). Exposure of young men to organochlorine pesticides in Southern Spain. *Environ Res*, 103(1):55–61. doi:10.1016/j.envres.2006.06.007 PMID:16889768

Cazorla-Reyes R, Fernández-Moreno JL, Romero-González R, Frenich AG, Vidal JL (2011). Single solid phase extraction method for the simultaneous analysis of polar and non-polar pesticides in urine samples by gas chromatography and ultra high pressure liquid chromatography coupled to tandem mass spectrometry. *Talanta*, 85(1):183–96. doi:10.1016/j.talanta.2011.03.048 PMID:21645688

CDC (2009). Fourth national report on human exposure to environmental chemicals. Atlanta (GA), USA: Centers for Disease Control and Prevention. Available from: http://www.cdc.gov/exposurereport/pdf/fourthreport.pdf

CEC (2006). The North American regional action plan (NARAP) on lindane and other hexachlorocyclohexane (HCH) isomers. Montreal, Canada: Commission for Environmental Cooperation. Available from: http://www.cec.org/Storage/36/2797_NARAP_on_Lindane_and_Other_Hexachlorocyclohexane_Isomers.pdf.

CEC (2015). Lindane. Pollutants. Montreal, Canada: Commission for Environmental Cooperation. Available from: http://www.cec.org/lindane, accessed 3 June 2015.

Chadwick RW, Chadwick CJ, Freal JJ, Bryden CC (1977). Comparative enzyme induction and lindane metabolsim in rats pre-treated with various organochlorine pesticides. *Xenobiotica*, 7(4):235–46. doi:10.3109/00498257709035781 PMID:68630

Chadwick RW, Chang JJ, Gilligan PH, Forehand LR, Long JE, Duffy MC (1990). Effect of lindane on nitroreductase and dechlorinase enzyme activity in the gastrointestinal tract. *Toxicol Lett*, 50(2-3):299–308. doi:10.1016/0378-4274(90)90023-F PMID:1689881

Chadwick RW, Copeland MF, Wolff GL, Cooke N, Whitehouse DA, Mole ML (1985). Effects of age and obesity on the metabolism of lindane by black *a/a*, yellow A^{vy}/a, and pseudoagouti A^{vy}/a phenotypes of (YS x VY) F1 hybrid mice. *J Toxicol Environ Health*, 16(6):771–96. doi:10.1080/15287398509530788 PMID:2419580

Chadwick RW, Copeland MF, Wolff GL, Stead AG, Mole ML, Whitehouse DA (1987). Saturation of lindane metabolism in chronically treated (YS x VY) F1 hybrid mice. *J Toxicol Environ Health*, 20(4):411–34. doi:10.1080/15287398709530994 PMID:2435921

Chang SK, Williams PL, Dauterman WC, Riviere JE (1994). Percutaneous absorption, dermatopharmacokinetics and related bio-transformation studies of carbaryl, lindane, malathion, and parathion in isolated perfused porcine skin. *Toxicology*, 91(3):269–80. doi:10.1016/0300-483X(94)90014-0 PMID:7521545

Channa KR, Röllin HB, Wilson KS, Nøst TH, Odland JØ, Naik I et al. (2012). Regional variation in pesticide concentrations in plasma of delivering women residing in rural Indian Ocean coastal regions of South Africa. *J Environ Monit*, 14(11):2952–60. doi:10.1039/c2em30264k PMID:23047303

ChemSources (2009). Chem Sources Online [online search engine]. Clemson (SC), USA: Chemical Sources International. Available from: http://www.chemsources.com/chemonline.html, accessed 3 June 2015.

Chowdhury AR, Gautam AK (1994). Steroidogenic impairment after lindane treatment in male rats. *J UOEH*, 16(2):145–52. PMID:7517064

Clere N, Lauret E, Malthiery Y, Andriantsitohaina R, Faure S (2012). Estrogen receptor alpha as a key target of organochlorines to promote angiogenesis. *Angiogenesis*, 15(4):745–60. doi:10.1007/s10456-012-9288-7 PMID:22829064

Cocco P, Brennan P, Ibba A, de Sanjosé Llongueras S, Maynadié M, Nieters A et al. (2008). Plasma polychlorobiphenyl and organochlorine pesticide level and risk of major lymphoma subtypes. *Occup Environ Med*, 65(2):132–40. doi:10.1136/oem.2007.033548 PMID:17699548

Cole R, Frederick R, Healy R (1984). Preliminary findings of the priority pollutant monitoring project of the nationwide urban runoff program. *J Water Pollut Control Fed*, 56(898):908.

Concha-Graña E, Fernández-González V, Grueiro-Noche G, Muniategui-Lorenzo S, López-Mahía P, Fernández-Fernández E et al. (2010). Development of an environmental friendly method for the analysis of organochlorine pesticides in sediments. *Chemosphere*, 79(7):698–705. doi:10.1016/j.chemosphere.2010.02.052 PMID:20299072

Cooper RL, Chadwick RW, Rehnberg GL, Goldman JM, Booth KC, Hein JF et al. (1989). Effect of lindane on hormonal control of reproductive function in the female rat. *Toxicol Appl Pharmacol*, 99(3):384–94. doi:10.1016/0041-008X(89)90148-8 PMID:2473543

Cuesta A, Meseguer J, Angeles Esteban M (2008). Effects of the organochlorines p,p'-DDE and lindane on gilthead seabream leucocyte immune parameters and gene expression. *Fish Shellfish Immunol*, 25(5):682–8. doi:10.1016/j.fsi.2008.02.006 PMID:18757214

Dahmardeh Behrooz R, Barghi M, Bahramifar N, Esmaili-Sari A (2012). Organochlorine contaminants in the hair of Iranian pregnant women. *Chemosphere*, 86(3):235–41. doi:10.1016/j.chemosphere.2011.09.031 PMID:22047617

Dalsenter PR, Faqi AS, Chahoud I (1997b). Serum testosterone and sexual behavior in rats after prenatal exposure to lindane. *Bull Environ Contam Toxicol*, 59(3):360–6. doi:10.1007/s001289900486 PMID:9256387

Dalsenter PR, Faqi AS, Webb J, Merker HJ, Chahoud I (1996). Reproductive toxicity and tissue concentrations of lindane in adult male rats. *Hum Exp Toxicol*, 15(5):406–10. doi:10.1177/096032719601500508 PMID:8735465

Dalsenter PR, Faqi AS, Webb J, Merker HJ, Chahoud I (1997a). Reproductive toxicity and toxicokinetics of lindane in the male offspring of rats exposed during lactation. *Hum Exp Toxicol*, 16(3):146–53. doi:10.1177/096032719701600303 PMID:9088967

Daniel V, Huber W, Bauer K, Suesal C, Conradt C, Opelz G (2001). Associations of blood levels of PCB, HCHS, and HCB with numbers of lymphocyte subpopulations, in vitro lymphocyte response, plasma cytokine levels, and immunoglobulin autoantibodies. *Environ Health Perspect*, 109(2):173–8. doi:10.1289/ehp.01109173 PMID:11266329

Danzo BJ (1997). Environmental xenobiotics may disrupt normal endocrine function by interfering with the binding of physiological ligands to steroid receptors and binding proteins. *Environ Health Perspect*, 105(3):294–301. doi:10.1289/ehp.97105294 PMID:9171990

Dar SA, Das S, Ramachandran VG, Bhattacharya SN, Mustafa MD, Banerjee BD et al. (2012). Alterations in T-lymphocyte sub-set profiles and cytokine secretion by PBMC of systemic lupus erythematosus patients upon in vitro exposure to organochlorine pesticides. *J Immunotoxicol*, 9(1):85–95. doi:10.3109/1547691X.2011.642103 PMID:22214240

Das SN, Paul BN, Saxena AK, Ray PK (1990). Effect of in utero exposure to hexachlorocyclohexane on the developing immune system of mice. *Immunopharmacol Immunotoxicol*, 12(2):293–310. doi:10.3109/08923979009019674 PMID:1699991

Davies JE, Dedhia HV, Morgade C, Barquet A, Maibach HI (1983). Lindane poisonings. *Arch Dermatol*, 119(2):142–4. doi:10.1001/archderm.1983.01650260050017 PMID:6186202

De Roos AJ, Hartge P, Lubin JH, Colt JS, Davis S, Cerhan JR et al. (2005). Persistent organochlorine chemicals in plasma and risk of non-Hodgkin's lymphoma. *Cancer Res*, 65(23):11214–26. doi:10.1158/0008-5472.CAN-05-1755 PMID:16322272

De Roos AJ, Zahm SH, Cantor KP, Weisenburger DD, Holmes FF, Burmeister LF et al. (2003). Integrative assessment of multiple pesticides as risk factors for non-Hodgkin's lymphoma among men. *Occup Environ Med*, 60(9):E11 doi:10.1136/oem.60.9.e11 PMID:12937207

Detzel A, Patyk A, Fehrenbach H., Franke B., Gingrich J., Lell M., Vogt R, et al. (1998). Investigation of emissions and abatement measures for persistent organic pollutants in the Federal Republic of Germany. Umweltbundesamt. Research report 295 44 365, UBA-FB 98-115, UBA-Texte 74/98. Berlin, Germany: Federal Environmental Agency.

Dewan A, Gupta SK, Jani JP, Kashyap SK (1980). Effect of lindane on antibody response to typhoid vaccine in weanling rats. *J Environ Sci Health B*, 15(4):395–402. doi:10.1080/03601238009372191 PMID:6156966

Dhooge W, Arijs K, D'Haese I, Stuyvaert S, Versonnen B, Janssen C et al. (2006). Experimental parameters affecting sensitivity and specificity of a yeast assay for estrogenic compounds: results of an interlaboratory validation exercise. *Anal Bioanal Chem*, 386(5):1419–28. doi:10.1007/s00216-006-0669-x PMID:16896612

Di Consiglio E, De Angelis G, Traina ME, Urbani E, Testai E (2009). Effect of lindane on CYP-mediated steroid hormone metabolism in male mice following in utero exposure. *J Appl Toxicol*, 29(8):648–55. doi:10.1002/jat.1452 PMID:19557771

Dick IP, Blain PG, Williams FM (1997). The percutaneous absorption and skin distribution of lindane in man. I. In vivo studies. *Hum Exp Toxicol*, 16(11):645–51. doi:10.1177/096032719701601103 PMID:9426365

Drummond L, Gillanders EM, Wilson HK (1988). Plasma gamma-hexachlorocyclohexane concentrations in forestry workers exposed to lindane. *Br J Ind Med*, 45(7):493–7. PMID:2456092

Dubois M, Grosse Y, Thomé JP, Kremers P, Pfohl-Leszkowicz A (1997). Metabolic activation and DNA-adducts detection as biomarkers of chlorinated pesticide exposures. *Biomarkers*, 2(1):17–24. doi:10.1080/135475097231922

Duff RM, Kissel JC (1996). Effect of soil loading on dermal absorption efficiency from contaminated soils. *J Toxicol Environ Health*, 48(1):93–106. doi:10.1080/009841096161492 PMID:8637061

Dunier M, Siwicki AK (1994). Effects of lindane exposure on rainbow trout (Oncorhynchus mykiss) immunity. I. Effect of lindane on antibody-secreting cells (ASC) measured by ELISPOT assay. *Ecotoxicol Environ Saf*, 27(1):1–6. doi:10.1006/eesa.1994.1001 PMID:7525199

Dunier M, Siwicki AK, Scholtens J, Dal Molin S, Vergnet C, Studnicka M (1994). Effects of lindane exposure on rainbow trout (Oncorhynchus mykiss) immunity. III. Effect on nonspecific immunity and B lymphocyte functions. *Ecotoxicol Environ Saf*, 27(3):324–34. doi:10.1006/eesa.1994.1026 PMID:7519551

Eichler D, Heupt W, Paul W (1983). Comparative study on the distribution of alpha- and gamma-hexachlorocyclohexane in the rat with particular reference to the problem of isomerization. *Xenobiotica*, 13(11):639–47. doi:10.3109/00498258309052225 PMID:6201012

Ellero S, Chakhtoura G, Barreau C, Langouët S, Benelli C, Penicaud L et al. (2010). Xenobiotic-metabolizing cytochromes P450 in human white adipose tissue: expression and induction. *Drug Metab Dispos*, 38(4):679–86. doi:10.1124/dmd.109.029249 PMID:20035023

Engst R, Macholz M, Kujawa M (1976a). [Demonstration of metabolites of hexachlorcyclohexane in human urine] *Z Gesamte Hyg*, 22(3):205–8.[German] PMID:61659

Engst R, Macholz RM, Kujawa M (1976b). Occurrence of beta-glucuronide bound metabolites of hexachlorocyclohexane isomers in human urine. *Z Gesamte Hyg*, 22(7):495–8.[German] PMID:60832

Ennaceur S, Gandoura N, Driss MR (2008). Distribution of polychlorinated biphenyls and organochlorine pesticides in human breast milk from various locations in Tunisia: levels of contamination, influencing factors, and infant risk assessment. *Environ Res*, 108(1):86–93. doi:10.1016/j.envres.2008.05.005 PMID:18614165

EPA (2001a). Data evaluation record. Lindane. Study type: oncogenicity feeding – mouse. TX 014564. Washington (DC), USA: United States Environmental Protection Agency.

EPA (2001b). Cancer assessment document. Evaluation of the carcinogenic potential of lindane. TXR No. 0050297. Washington (DC), USA: United States Environmental Protection Agency.

EPA (2015a). Lindane (gamma-hexachlorocyclohexane). Hazard summary. Created in April 1992; revised in January 2000. Health effects notebook for hazardous air pollutants. Washington (DC), USA: United States Environmental Protection Agency. Available from: https://www.epa.gov/sites/production/files/2016-09/documents/lindane.pdf, accessed 29 August 2017.

EPA (2015b) Basic information about lindane in drinking water. Washington (DC), USA: United States Environmental Protection Agency. Available from: http://water.epa.gov/drink/contaminants/basicinformation/lindane.cfm

Etim OE, Farombi EO, Usoh IF, Akpan EJ (2006). The protective effect of aloe vera juice on lindane induced hepatotoxicity and genotoxicity. *Pak J Pharm Sci*, 19(4):337–40.

European Commission (2000a). Directive 2000/60/EC establishing a framework for Community action in the field of water policy; supplemented by the Groundwater Directive (2006/118/EC) and the Environmental Quality Standards Directive (2008/105/EC). Available from: http://ec.europa.eu/health/endocrine_disruptors/docs/wfd_200060ec_directive_en.pdf, accessed June 2016.

European Commission (2000b). Commission decision of 20 December 2000 concerning the non-inclusion of lindane in Annex I to Council Directive 91/414/EEC and the withdrawal of authorisations for plant-protection products containing this active substance. Available from: http://eur-lex.europa.eu/legal-content/EN/TXT/PDF/?uri=CELEX:32000D0801&from=EN, accessed June 2016.

European Commission (2004). Regulation (EC) No 850/2004 of the European Parliament and of the Council of 29 April 2004 on persistent organic pollutants and amending Directive 79/117/EEC. Available from: http://ec.europa.eu/environment/waste/studies/pdf/annex_1_pop_regul_850_2004.pdf, accessed June 2016.

Fahrig R (1974). Comparative mutagenicity studies with pesticides. *IARC Sci Publ*, 10:161–81.

FDA (2015). Food and Drug Administration Public Health Advisory: Safety of topical lindane products for the treatment of scabies and lice. Available from: http://www.fda.gov/drugs/drugsafety/postmarketdrugsafetyinformationforpatientsandproviders/ucm110845.htm, accessed 3 June 2015.

Feldmann RJ, Maibach HI (1974). Percutaneous penetration of some pesticides and herbicides in man. *Toxicol Appl Pharmacol*, 28(1):126–32. doi:10.1016/0041-008X(74)90137-9 PMID:4853576

Filer D, Patisaul HB, Schug T, Reif D, Thayer K (2014). Test driving ToxCast: endocrine profiling for 1858 chemicals included in phase II. *Curr Opin Pharmacol*, 19:145–52. doi:10.1016/j.coph.2014.09.021 PMID:25460227

Fisher BD, Mueller GC (1971). Gamma-hexachlorocyclohexane inhibits the initiation of lymphocyte growth by phytohemagglutinin. *Biochem Pharmacol*, 20(9):2515–8. doi:10.1016/0006-2952(71)90255-3 PMID:4126989

Fitzloff JF, Pan JC (1984). Epoxidation of the lindane metabolite, beta-PCCH, by human- and rat-liver microsomes. *Xenobiotica*, 14(7):599–604. doi:10.3109/00498258409151455 PMID:6209866

Fitzloff JF, Portig J, Stein K (1982). Lindane metabolism by human and rat liver microsomes. *Xenobiotica*, 12(3):197–202. doi:10.3109/00498258209046794 PMID:6180560

Forgue MF, Pinelli E, Beraud M, Souqual MC, Pipy B (1990). Chemiluminescence response and arachidonic acid metabolism of macrophages induced by gamma-hexachlorocyclohexane (lindane). *Food Addit Contam*, 7(supl):Suppl 1: S97–9. doi:10.1080/02652039009373856 PMID:1702068

Franz TJ, Lehman PA, Franz SF, Guin JD (1996). Comparative percutaneous absorption of lindane and permethrin. *Arch Dermatol*, 132(8):901–5. doi:10.1001/archderm.1996.03890320049007 PMID:8712839

Freire C, Koifman RJ, Sarcinelli PN, Rosa AC, Clapauch R, Koifman S (2014). Association between serum levels of organochlorine pesticides and sex hormones in adults living in a heavily contaminated area in Brazil. *Int J Hyg Environ Health*, 217(2–3):370–8. doi:10.1016/j.ijheh.2013.07.012 PMID:23972672

Gencik A (1977). Cytogenetic examination of the bone marrow of rats after administration of lindane. *Bratislavske Lekarske Listy*, 67(5):579–82.[Slovak]

Ginsburg CM, Lowry W, Reisch JS (1977). Absorption of lindane (gamma benzene hexachloride) in infants and children. *J Pediatr*, 91(6):998–1000. doi:10.1016/S0022-3476(77)80915-3 PMID:72815

Giordano A, Richter P, Ahumada I (2011). Determination of pesticides in river water using rotating disk sorptive extraction and gas chromatography-mass spectrometry. *Talanta*, 85(5):2425–9. doi:10.1016/j.talanta.2011.07.087 PMID:21962663

Gopalaswamy UV, Aiyar AS (1986). Biotransformation and toxicity of lindane and its metabolite hexachlorobenzene in mammals. *IARC Sci Publ*, 77:267–76.

Goto M, Hattori M, Miyagawa T, Enomoto M (1972). Beiträge zur ökologischen Chemie: II. Hepatoma-Bildung in Mäusen nach Verabreichung von HCH-Isomeren in hohen dosen. [Contributions to environmental chemistry: II. Hepatoma formation in mice after administration of HCH isomers in high doses]*Chemosphere*, 1(6):279–82. [German] doi:10.1016/0045-6535(72)90033-1

Grabarczyk M, Kopeć-Szlezak J, Szczepańska I, Woźniak J, Podstawka U (1990). The effect of gamma-hexachlorocyclohexane (lindane) on blood cells, kidney and liver tissues in rabbits. *Haematologia (Budap)*, 23(3):171–9. PMID:1706296

Grey WE, Marthrc DE, Rogers SJ (1983). Potential exposure of commercial seed-treating applicators to the pesticides carboxin-thiram and lindane. *Bull Environ Contam Toxicol*, 31(2):244–50. doi:10.1007/BF01607901 PMID:6193827

Gunderson EL (1995). FDA Total Diet Study, July 1986-April 1991, dietary intakes of pesticides, selected elements, and other chemicals. *J AOAC Int*, 78(6):1353–63. PMID:8664570

Guo H, Jin Y, Cheng Y, Leaderer B, Lin S, Holford TR et al. (2014). Prenatal exposure to organochlorine pesticides and infant birth weight in China. *Chemosphere*, 110:1–7. doi:10.1016/j.chemosphere.2014.02.017 PMID:24880592

Halse K, Schlabach M, Eckhardt S, Sweetman A, Jones KC, Breivik K (2011). Spatial variability of POPs in European background air. *Atmos Chem Phys*, 11(4):1549–64. doi:10.5194/acp-11-1549-2011

Hanada M, Yutani C, Miyaji T (1973). Induction of hepatoma in mice by benzene hexachloride. *Gan*, 64(5):511–3. PMID:4129376

Hart LJ, Smith SA, Smith BJ, Robertson J, Holladay SD (1997). Exposure of tilapian fish to the pesticide lindane results in hypocellularity of the primary hematopoietic organ (pronephros) and the spleen without altering activity of phagocytic cells in these organs. *Toxicology*, 118(2-3):211–21. doi:10.1016/S0300-483X(97)03619-6 PMID:9129175

Hassoun E, Bagchi M, Bagchi D, Stohs SJ (1993). Comparative studies on lipid peroxidation and DNA-single strand breaks induced by lindane, DDT, chlordane and endrin in rats. *Comp Biochem Physiolo, C*, 104(3):427–31. doi:10.1016/0742-8413(93)90013-B

Herbst M, Weisse I, Koellmer H (1975). A contribution to the question of the possible hepatocarcinogenic effects of lindane. *Toxicology*, 4(1):91–6. doi:10.1016/0300-483X(75)90025-6 PMID:48294

Hervás JP (1976). Multinucleate plant cells. I. Aneuploidy in a proliferating population. *Exp Cell Res*, 97(1):203–12. doi:10.1016/0014-4827(76)90669-8

Herzel F (1972). Organochlorine insecticides in surface waters in Germany–1970 and 1971. *Pestic Monit J*, 6(3):179–87. PMID:4122187

Hewitt R, Forero A, Luncsford PJ, Martin FL (2007). Enhanced micronucleus formation and modulation of BCL-2:BAX in MCF-7 cells after exposure to binary mixtures. *Environ Health Perspect*, 115(S-1):Suppl 1:129–36. doi:10.1289/ehp.9361 PMID:18174961

Hoar SK, Blair A, Holmes FF, Boysen CD, Robel RJ, Hoover R et al. (1986). Agricultural herbicide use and risk of lymphoma and soft-tissue sarcoma. *JAMA*, 256(9):1141–7. doi:10.1001/jama.1986.03380090081023 PMID:3801091

Hoff RM, Muir DCG, Grift NP (1992a). Annual cycle of polychlorinated biphenyls and organohalogen pesticides in air in southern Ontario. Air concentration data. *Environ Sci Technol*, 26(2):266–75. doi:10.1021/es00026a005

Høyer AP, Jørgensen T, Grandjean P, Hartvig HB (2000). Repeated measurements of organochlorine exposure and breast cancer risk (Denmark). *Cancer Causes Control*, 11(2):177–84. doi:10.1023/A:1008926219539 PMID:10710203

Hrycek A, Kalina Z, Owczarzy I (1984). Cytochemical analysis of selected dehydrogenases in peripheral blood leukocytes of workers exposed to polychlorinated pesticides. *Med Pr*, 35(3):185–9. PMID:6209525 [Polish]

HSDB (2009). Lindane (CAS No. 58-89-9). Hazardous Substances Data Bank [online database]. Bethesda (MD), USA: United States National Library of Medicine. Available from: http://toxnet.nlm.nih.gov/ cgi-bin/sis/htmlgen?HSDB, accessed 18 December 2009.

IARC (1974). Some organochlorine pesticides. *IARC Monogr Eval Carcinog Risk Chem Man*, 5:1–241. Available from: http://monographs.iarc.fr/ENG/Monographs/vol1-42/mono5.pdf

IARC (1976). Some carbamates, thiocarbamates and carbazides. *IARC Monogr Eval Carcinog Risk Chem Man*, 12:1–282. PMID:188751. Available from: http://

monographs.iarc.fr/ENG/Monographs/vol1-42/mono12.pdf

IARC (1979a). Sex hormones (II). *IARC Monogr Eval Carcinog Risk Chem Hum*, 21:1–583. Available from: http://monographs.iarc.fr/ENG/Monographs/vol1-42/mono21.pdf

IARC (1979b). Some halogenated hydrocarbons. *IARC Monogr Eval Carcinog Risk Chem Hum*, 20:1–609. PMID:296120. Available from: http://monographs.iarc.fr/ENG/Monographs/vol1-42/mono20.pdf

IARC (1983). Miscellaneous pesticides. *IARC Monogr Eval Carcinog Risk Chem Hum*, 30:1–424. PMID:6578175. Available from: http://monographs.iarc.fr/ENG/Monographs/vol1-42/mono30.pdf

IARC (1986). Some chemicals used in plastics and elastomers. *IARC Monogr Eval Carcinog Risk Chem Hum*, 39:1–403. Available from: http://monographs.iarc.fr/ENG/Monographs/vol1-42/mono39.pdf

IARC (1987). Overall evaluations of carcinogenicity: an updating of IARC Monographs volumes 1 to 42. *IARC Monogr Eval Carcinog Risks Hum Suppl*, 7:1–440. PMID:3482203. Available from: http://monographs.iarc.fr/ENG/Monographs/suppl7/index.php

IARC (1991). Occupational exposures in insecticide application, and some pesticides. *IARC Monogr Eval Carcinog Risks Hum*, 53:1–612. PMID:1688189 Available from: http://monographs.iarc.fr/ENG/Monographs/vol53/index.php

IARC (2017a). Some organophosphate insecticides and herbicides. *IARC Monogr Eval Carcinog Risks Hum*, 112:1–452. Available from: http://monographs.iarc.fr/ENG/Monographs/vol112/index.php

IARC (2017b). List of ToxCast/Tox21 assay end-points. In: Supplemental Material to *IARC Monographs* Volume 113. Lyon, France: International Agency for Research on Cancer. Available from: http://monographs.iarc.fr/ENG/Monographs/vol113/index.php

Ibarluzea JM, Fernández MF, Santa-Marina L, Olea-Serrano MF, Rivas AM, Aurrekoetxea JJ et al. (2004). Breast cancer risk and the combined effect of environmental estrogens. *Cancer Causes Control*, 15(6):591–600. doi:10.1023/B:CACO.0000036167.51236.86 PMID:15280638

IPCS/ILO (2009). Lindane. ICSC: 0053. International Chemical Safety Cards. International Programme on Chemical Safety & International Labour Organization. Available from: http://www.ilo.org/dyn/icsc/showcard.display?p_card_id=0053, accessed June 2016.

IRPTC (1983). *Lindane, 40.* (Izmerov NF, editor). *Scientific reviews of Soviet literature on toxicity and hazards of chemicals.* Geneva, Switzerland International Registry of Potentially Toxic Chemicals United Nations Environment Programme.

Iverson F, Ryan JJ, Lizotte R, Hierlihy SL (1984). *In vivo* and *in vitro* binding of α- and γ-hexachlorocyclohexane to mouse liver macromolecules. *Toxicol Lett*, 20(3):331–5. doi:10.1016/0378-4274(84)90168-1

Jakszyn P, Goñi F, Etxeandia A, Vives A, Millán E, López R et al. (2009). Serum levels of organochlorine pesticides in healthy adults from five regions of Spain. *Chemosphere*, 76(11):1518–24. doi:10.1016/j.chemosphere.2009.05.048 PMID:19586652

Jaward FM, Farrar NJ, Harner T, Sweetman AJ, Jones KC (2004). Passive air sampling of PCBs, PBDEs, and organochlorine pesticides across Europe. *Environ Sci Technol*, 38(1):34–41. doi:10.1021/es034705n PMID:14740714

Jin L, Tran DQ, Ide CF, McLachlan JA, Arnold SF (1997). Several synthetic chemicals inhibit progesterone receptor-mediated transactivation in yeast. *Biochem Biophys Res Commun*, 233(1):139–46. doi:10.1006/bbrc.1997.6417 PMID:9144411

Joseph P, Viswanatha S, Krishnakumari MK (1992). Role of vitamin A in the haematotoxicity of hexachlorocyclohexane (HCH) in the rat. *J Environ Sci Health B*, 27(3):269–80. doi:10.1080/03601239209372779 PMID:1377731

Joshi PL, Bhattacharya M, Yadava RL, Chand B, Narasimham MV, Nigam DK et al. (1996). A community-based study on the effect of hexachlorocyclo-hexane (HCH) exposure in spraymen and general population. *J Commun Dis*, 28(3):189–98. PMID:8973020

Junqueira VB, Simizu K, Pimentel R, Azzalis LA, Barros SB, Koch O et al. (1991). Effect of phenobarbital and 3-methylcholanthrene on the early oxidative stress component induced by lindane in rat liver. *Xenobiotica*, 21(8):1053–65. doi:10.3109/00498259109039545 PMID:1723229

Kalantzi OI, Hewitt R, Ford KJ, Cooper L, Alcock RE, Thomas GO et al. (2004). Low dose induction of micronuclei by lindane. *Carcinogenesis*, 25(4):613–22. doi:10.1093/carcin/bgh048 PMID:14688026

Kang JJ, Chen IL, Yen-Yang HF (1998). Mediation of γ-hexachlorocyclohexane-induced DNA fragmentation in HL-60 cells through intracellular Ca2+ release pathway. *Food Chem Toxicol*, 36(6):513–20. doi:10.1016/S0278-6915(98)00010-6 PMID:9674959

Kashyap SK (1986). Health surveillance and biological monitoring of pesticide formulators in India. *Toxicol Lett*, 33(1-3):107–14. doi:10.1016/0378-4274(86)90075-5 PMID:2430350

Kassner JT, Maher TJ, Hull KM, Woolf AD (1993). Cholestyramine as an adsorbent in acute lindane poisoning: a murine model. *Ann Emerg Med*, 22(9):1392–7. doi:10.1016/S0196-0644(05)81984-5 PMID:7689801

Kavlock R, Chandler K, Houck K, Hunter S, Judson R, Kleinstreuer N et al. (2012). Update on EPA's ToxCast program: providing high throughput decision support tools for chemical risk management. *Chem*

Res Toxicol, 25(7):1287–302. doi:10.1021/tx3000939 PMID:22519603

Kensler TW, Mueller GC (1978). Effects of hexachlorocyclohexane isomers on the mitogenic response of bovine lymphocytes. Biochem Pharmacol, 27(5):667–71. doi:10.1016/0006-2952(78)90502-6 PMID:77667

Khan FH, Ganesan P, Kumar S (2010). Y Chromosome microdeletion and altered sperm quality in human males with high concentration of seminal hexachlorocyclohexane (HCH). Chemosphere, 80(9):972–7. doi:10.1016/j.chemosphere.2010.05.047

Khanna RN, Das M, Anand M (2002). Influence of phenobarbital and carbon tetrachloride on the modulation of tissue retention profile of hexachlorocyclohexane in rats. Biomed Environ Sci, 15(2):119–29. PMID:12244753

Khanna RN, Kunwar K, Gupta R, Gupta GS (1991). Placental transport of lindane during early and late stages of gestation in rats. Bull Environ Contam Toxicol, 47(4):508–14. doi:10.1007/BF01700938 PMID:1723906

Kogevinas M, Kauppinen T, Winkelmann R, Becher H, Bertazzi PA, Bueno-de-Mesquita HB et al. (1995). Soft tissue sarcoma and non-Hodgkin's lymphoma in workers exposed to phenoxy herbicides, chlorophenols, and dioxins: two nested case-control studies. Epidemiology, 6(4):396–402. doi:10.1097/00001648-199507000-00012 PMID:7548348

Koner BC, Banerjee BD, Ray A (1998). Organochlorine pesticide-induced oxidative stress and immune suppression in rats. Indian J Exp Biol, 36(4):395–8. PMID:9717451

Kopeć-Szlezak J, Szczepańska I, Grabarczyk M, Podstawka U (1990). Late toxic effects of long-term exposure to lindane in peripheral blood cells in rabbits. I. Function impairment and structural disturbances in leucocytes. Mater Med Pol, 22(3):179–83. PMID:1720851

Koutros S, Beane Freeman LE, Lubin JH, Heltshe SL, Andreotti G, Barry KH et al. (2013). Risk of total and aggressive prostate cancer and pesticide use in the Agricultural Health Study. Am J Epidemiol, 177(1):59–74. doi:10.1093/aje/kws225 PMID:23171882

Kroll B, Kunz S, Klein T, Schwarz LR (1999). Effect of lindane and phenobarbital on cyclooxygenase-2 expression and prostanoid synthesis by Kupffer cells. Carcinogenesis, 20(8):1411–6. doi:10.1093/carcin/20.8.1411 PMID:10426785

Kumar D, Khan PK, Sinha SP (1995). Cytogenetic toxicity and no-effect limit dose of pesticides. Food Chem Toxicol, 33(4):309–14. doi:10.1016/0278-6915(94)00147-G

Kutz FW, Wood PH, Bottimore DP (1991). Organochlorine pesticides and polychlorinated biphenyls in human adipose tissue. Rev Environ Contam Toxicol, 120:1–82. doi:10.1007/978-1-4612-3080-9_1 PMID:1899728

La Sala G, Farini D, De Felici M (2009). Proapoptotic effects of lindane on mouse primordial germ cells. Toxicol Sci, 108(2):445–51. doi:10.1093/toxsci/kfp027 PMID:19221147

Lahav N, Kahanovitch Y (1974). Lindane residues in the southern coastal aquifer of Israel. Water Air Soil Pollut, 3(3):253–9. PMID:4142340

Lakkad BC, Nigam SK, Karnik AB, Thakore KN, Aravinda Babu K, Bhatt DK et al. (1982). Dominant-lethal study of technical-grade hexachlorocyclohexane in Swiss mice. Mutat Res, 101(4):315–20. doi:10.1016/0165-1218(82)90124-0

Lange M, Nitzsche K, Zesch A (1981). Percutaneous absorption of lindane in healthy volunteers and scabies patients. Dependency of penetration kinetics in serum upon frequency of application, time and mode of washing. Arch Dermatol Res, 271(4):387–99. doi:10.1007/BF00406683 PMID:6174081

Laws SC, Carey SA, Hart DW, Cooper RL (1994). Lindane does not alter the estrogen receptor or the estrogen-dependent induction of progesterone receptors in sexually immature or ovariectomized adult rats. Toxicology, 92(1–3):127–42. doi:10.1016/0300-483X(94)90172-4 PMID:7524197

Ledirac N, Antherieu S, d'Uby AD, Caron JC, Rahmani R (2005). Effects of organochlorine insecticides on MAP kinase pathways in human HaCaT keratinocytes: key role of reactive oxygen species. Toxicol Sci, 86(2):444–52. doi:10.1093/toxsci/kfi192 PMID:15888667

Lee CH, Edwards AM (2001). Stimulation of DNA synthesis by tumor promoters in primary rat hepatocytes is not mediated by arachidonic acid metabolites. J Cell Physiol, 187(3):336–44. doi:10.1002/jcp.1083 PMID:11319757

Lee HS, Miyauchi K, Nagata Y, Fukuda R, Sasagawa S, Endoh H et al. (2002). Employment of the human estrogen receptor beta ligand-binding domain and co-activator SRC1 nuclear receptor-binding domain for the construction of a yeast two-hybrid detection system for endocrine disrupters. J Biochem, 131(3):399–405. doi:10.1093/oxfordjournals.jbchem.a003115 PMID:11872169

Lee WJ, Cantor KP, Berzofsky JA, Zahm SH, Blair A (2004). Non-Hodgkin's lymphoma among asthmatics exposed to pesticides. Int J Cancer, 111(2):298–302. doi:10.1002/ijc.20273 PMID:15197786

Lemaire G, de Sousa G, Rahmani R (2004). A PXR reporter gene assay in a stable cell culture system: CYP3A4 and CYP2B6 induction by pesticides. Biochem Pharmacol, 68(12):2347–58. doi:10.1016/j.bcp.2004.07.041 PMID:15548381

Lemaire G, Mnif W, Mauvais P, Balaguer P, Rahmani R (2006). Activation of alpha- and beta-estrogen receptors by persistent pesticides in reporter cell lines. Life Sci, 79(12):1160–9. doi:10.1016/j.lfs.2006.03.023 PMID:16626760

Linderholm L, Biague A, Månsson F, Norrgren H, Bergman A, Jakobsson K (2010). Human exposure to persistent organic pollutants in West Africa–a temporal trend study from Guinea-Bissau. *Environ Int*, 36(7):675–82. doi:10.1016/j.envint.2010.04.020 PMID:20537392

Link B, Gabrio T, Zoellner I, Piechotowski I, Paepke O, Herrmann T et al. (2005). Biomonitoring of persistent organochlorine pesticides, PCDD/PCDFs and dioxin-like PCBs in blood of children from South West Germany (Baden-Wuerttemberg) from 1993 to 2003. *Chemosphere*, 58(9):1185–201. doi:10.1016/j.chemosphere.2004.09.061 PMID:15667840

Llabjani V, Trevisan J, Jones KC, Shore RF, Martin FL (2011). Derivation by infrared spectroscopy with multivariate analysis of bimodal contaminant-induced dose-response effects in MCF-7 cells. *Environ Sci Technol*, 45(14):6129–35. doi:10.1021/es200383a PMID:21699185

López FJ, Pitarch E, Egea S, Beltran J, Hernández F (2001). Gas chromatographic determination of organochlorine and organophosphorus pesticides in human fluids using solid phase microextraction. *Anal Chim Acta*, 433(2):217–26. doi:10.1016/S0003-2670(01)00793-0

López-Carrillo L, López-Cervantes M, Torres-Sánchez L, Blair A, Cebrián ME, García RM (2002). Serum levels of beta-hexachlorocyclohexane, hexachlorobenzene and polychlorinated biphenyls and breast cancer in Mexican women. *Eur J Cancer Prev*, 11(2):129–35. doi:10.1097/00008469-200204000-00004 PMID:11984130

Lueken A, Juhl-Strauss U, Krieger G, Witte I (2004). Synergistic DNA damage by oxidative stress (induced by H2O2) and nongenotoxic environmental chemicals in human fibroblasts. *Toxicol Lett*, 147(1):35–43. doi:10.1016/j.toxlet.2003.10.020 PMID:14700526

Macholz RM, Kujawa M (1985). Recent state of lindane metabolism. Part III. *Residue Rev*, 94:119–49. doi:10.1007/978-1-4612-5104-0_4 PMID:2416021

Mahmood A, Malik RN, Li J, Zhang G (2014). Levels, distribution pattern and ecological risk assessment of organochlorines pesticides (OCPs) in water and sediments from two tributaries of the Chenab River, Pakistan. *Ecotoxicology*, 23(9):1713–21. doi:10.1007/s10646-014-1332-5 PMID:25204814

Man YB, Chow KL, Wang HS, Lau KY, Sun XL, Wu SC et al. (2011). Health risk assessment of organochlorine pesticides with emphasis on DDTs and HCHs in abandoned agricultural soils. *J Environ Monit*, 13(8):2250–9. doi:10.1039/c1em10168d PMID:21677982

Mandal A, Chakraborty S, Lahiri P (1986). Hematological changes produced by lindane (gamma-HCH) in six species of birds. *Toxicology*, 40(1):103–11. doi:10.1016/0300-483X(86)90050-8 PMID:2424143

Maranghi F, Rescia M, Macrì C, Di Consiglio E, De Angelis G, Testai E et al. (2007). Lindane may modulate the female reproductive development through the interaction with ER-beta: an in vivo-in vitro approach. *Chem Biol Interact*, 169(1):1–14. doi:10.1016/j.cbi.2007.04.008 PMID:17537412

Martin FL, Cole KJ, Orme MH, Grover PL, Phillips DH, Venitt S (1999). The DNA repair inhibitors hydroxyurea and cytosine arabinoside enhance the sensitivity of the alkaline single-cell gel electrophoresis ('comet') assay in metabolically-competent MCL-5 cells. *Mutat Res*, 445(1):21–43. doi:10.1016/S1383-5718(99)00116-3

Maruyama S, Fujimoto N, Yin H, Ito A (1999). Growth stimulation of a rat pituitary cell line MtT/E-2 by environmental estrogens in vitro and in vivo. *Endocr J*, 46(4):513–20. doi:10.1507/endocrj.46.513 PMID:10580743

Matsuura I, Saitoh T, Tani E, Wako Y, Iwata H, Toyota N et al. (2005). Evaluation of a two-generation reproduction toxicity study adding end-points to detect endocrine disrupting activity using lindane. *J Toxicol Sci*, 30(Spec No):135–61. doi:10.2131/jts.30.S135 PMID:16641539

McDuffie HH, Pahwa P, McLaughlin JR, Spinelli JJ, Fincham S, Dosman JA et al. (2001). Non-Hodgkin's lymphoma and specific pesticide exposures in men: cross-Canada study of pesticides and health. *Cancer Epidemiol Biomarkers Prev*, 10(11):1155–63. PMID:11700263

McGlynn KA, Quraishi SM, Graubard BI, Weber JP, Rubertone MV, Erickson RL (2008). Persistent organochlorine pesticides and risk of testicular germ cell tumors. *J Natl Cancer Inst*, 100(9):663–71. doi:10.1093/jnci/djn101 PMID:18445826

McManus S-L, Coxon CE, Richards KG, Danaher M (2013). Quantitative solid phase microextraction - gas chromatography mass spectrometry analysis of the pesticides lindane, heptachlor and two heptachlor transformation products in groundwater. *J Chromatogr A*, 1284:1–7. doi:10.1016/j.chroma.2013.01.099 PMID:23466207

Meade CJ, Harvey J, Boot JR, Turner GA, Bateman PE, Osborne DJ (1984). gamma-Hexachlorocyclohexane stimulation of macrophage phospholipid hydrolysis and leukotriene production. *Biochem Pharmacol*, 33(2):289–93. doi:10.1016/0006-2952(84)90487-8 PMID:6200116

Meera P, Rao PR, Shanker R, Tripathi O (1992). Immunomodulatory effects of gamma-HCH (Lindane) in mice. *Immunopharmacol Immunotoxicol*, 14(1-2):261–82. doi:10.3109/08923979209009224 PMID:1375957

Meera P, Tripathi O, Kamboj KK, Rao PR (1993). Role of calcium in biphasic immunomodulation by gamma-HCH (lindane) in mice. *Immunopharmacol Immunotoxicol*, 15(1):113–29. doi:10.3109/08923979309066937 PMID:7680676

Michałowicz J, Mokra K, Rosiak K, Sicińska P, Bukowska B (2013). Chlorobenzenes, lindane and dieldrin induce apoptotic alterations in human peripheral blood lymphocytes (in vitro study). *Environ Toxicol Pharmacol*, 36(3):979–88. doi:10.1016/j.etap.2013.08.014 PMID:24077485

Miligi L, Costantini AS, Bolejack V, Veraldi A, Benvenuti A, Nanni O et al. (2003). Non-Hodgkin's lymphoma, leukemia, and exposures in agriculture: results from the Italian multicenter case-control study. *Am J Ind Med*, 44(6):627–36. doi:10.1002/ajim.10289 PMID:14635239

Mills PK, Yang R (2003). Prostate cancer risk in California farm workers. *J Occup Environ Med*, 45(3):249–58. doi:10.1097/01.jom.0000058339.05741.0c PMID:12661182

Mills PK, Yang RC (2007). Agricultural exposures and gastric cancer risk in Hispanic farm workers in California. *Environ Res*, 104(2):282–9. doi:10.1016/j.envres.2006.11.008 PMID:17196584

Miyake Y, Tanaka K, Masuzaki Y, Sato N, Ikeda Y, Chisaki Y et al. (2011). Organochlorine concentrations in breast milk and prevalence of allergic disorders in Japanese women. *Chemosphere*, 85(3):374–8. doi:10.1016/j.chemosphere.2011.07.001 PMID:21802112

Moody RP, Ritter L (1989). Dermal absorption of the insecticide lindane (1 delta, 2 delta, 3 beta, 4 delta, 5 delta, 6 beta-hexachlorocyclohexane) in rats and rhesus monkeys: effect of anatomical site. *J Toxicol Environ Health*, 28(2):161–9. doi:10.1080/15287398909531337 PMID:2477558

Moreno Frías M, Garrido Frenich A, Martínez Vidal JL, Mateu Sánchez M, Olea F, Olea N (2001). Analyses of lindane, vinclozolin, aldrin, p,p'-DDE, o,p'-DDT and p,p'-DDT in human serum using gas chromatography with electron capture detection and tandem mass spectrometry. *J Chromatogr B Biomed Sci Appl*, 760(1):1–15. doi:10.1016/S0378-4347(01)00212-2 PMID:11522051

Munk ZM, Nantel A (1977). Acute lindane poisoning with development of muscle necrosis. *Can Med Assoc J*, 117(9):1050–4. PMID:71942

Murayama H, Takase Y, Mitobe H, Mukai H, Ohzeki T, Shimizu K et al. (2003). Seasonal change of persistent organic pollutant concentrations in air at Niigata area, Japan. *Chemosphere*, 52(4):683–94. doi:10.1016/S0045-6535(03)00105-X PMID:12738282

Naeher LP, Barr DB, Rithmire N, Edwards J, Holmes AK, Needham LL et al. (2009). Pesticide exposure resulting from treatment of lice infestation in school-aged children in Georgia. *Environ Int*, 35(2):358–62. doi:10.1016/j.envint.2008.09.001 PMID:18947873

Nativelle-Serpentini C, Richard S, Séralini GE, Sourdaine P (2003). Aromatase activity modulation by lindane and bisphenol-A in human placental JEG-3 and transfected kidney E293 cells. *Toxicol In Vitro*, 17(4):413–22. doi:10.1016/S0887-2333(03)00046-8 PMID:12849724

NCBI (2015). Lindane (CID 727). PubChem. Open Chemistry Database. Bethesda (MD): USA: National Center for Biotechnology Information. Available from: http://pubchem.ncbi.nlm.nih.gov/summary/summary.cgi?cid=727, accessed 5 March 2015

Nigam SK, Karnik AB, Chattopadhyay P, Lakkad BC, Venkaiah K, Kashyap SK (1993). Clinical and biochemical investigations to evolve early diagnosis in workers involved in the manufacture of hexachlorocyclohexane. *Int Arch Occup Environ Health*, 65:Suppl 1: S193–6. doi:10.1007/BF00381339 PMID:7691758

Nigam SK, Karnik AB, Majumder SK, Visweswariah K, Raju GS, Bai KM et al. (1986). Serum hexachlorocyclohexane residues in workers engaged at a HCH manufacturing plant. *Int Arch Occup Environ Health*, 57(4):315–20. doi:10.1007/BF00406186 PMID:2423460

NTP(1977). Bioassay of lindane for possible carcinogenicity. CAS No. 58-89-9. Technical Report Series NCI-CG-TR-14. *Natl Cancer Inst Carcinog Tech Rep Ser*, 14:1–99. PMID:12844152

Olgun S, Gogal RM Jr, Adeshina F, Choudhury H, Misra HP (2004). Pesticide mixtures potentiate the toxicity in murine thymocytes. *Toxicology*, 196(3):181–95. doi:10.1016/j.tox.2003.09.007 PMID:15036745

OSHA (2017). OSHA law and regulations. Washington (DC), USA: Occupational Safety and Health Administration, United States Department of Labour. Available from: https://www.osha.gov/law-regs.html

Oskarsson A, Ullerås E, Plant KE, Hinson JP, Goldfarb PS (2006). Steroidogenic gene expression in H295R cells and the human adrenal gland: adrenotoxic effects of lindane in vitro. *J Appl Toxicol*, 26(6):484–92. doi:10.1002/jat.1166 PMID:17080404

OSPAR Commission (2002). OSPAR background document on lindane (2004 update). Hazardous Substances Series No. 153. The Convention for the Protection of the Marine Environment of the North-East Atlantic. Available from: https://www.ospar.org/documents?v=6951, accessed 3 June 2015.

Pagès N, Sauviat MP, Bouvet S, Goudey-Perrière F (2002). Reproductive toxicity of lindane. *J Soc Biol*, 196(4):325–38.PMID:12645304 [French]

Parent-Massin D, Thouvenot D, Rio B, Riche C (1994). Lindane haematotoxicity confirmed by in vitro tests on human and rat progenitors. *Hum Exp Toxicol*, 13(2):103–6. doi:10.1177/096032719401300207 PMID:7512355

Parmar D, Yadav S, Dayal M, Johri A, Dhawan A, Seth PK (2003). Effect of lindane on hepatic and brain cytochrome P450s and influence of P450 modulation in lindane induced neurotoxicity. *Food Chem Toxicol*, 41(8):1077–87. doi:10.1016/S0278-6915(03)00045-0 PMID:12842176

Patayová H, Wimmerová S, Lancz K, Palkovičová L, Drobná B, Fabišiková A et al. (2013). Anthropometric, socioeconomic, and maternal health determinants of placental transfer of organochlorine compounds. *Environ Sci Pollut Res Int*, 20(12):8557–66. doi:10.1007/s11356-013-1786-7 PMID:23677752

Pathak R, Mustafa MD, Ahmed T, Ahmed RS, Tripathi AK, Guleria K et al. (2011). Intra uterine growth retardation: association with organochlorine pesticide residue levels and oxidative stress markers. *Reprod*

Toxicol, 31(4):534–9. doi:10.1016/j.reprotox.2011.02.007 PMID:21338667

Paul R, Talukdar A, Bhattacharya R, Santra G (2013). γ-Benzene hexachloride poisoning leading to acute hepatorenal decompensation. *BMJ Case Rep*, 2013:aug07 1: bcr2013009851 doi:10.1136/bcr-2013-009851 PMID:23925679

Pavlíková N, Bláhová L, Klán P, Bathula SR, Sklenář V, Giesy JP et al. (2012). Enantioselective effects of alpha-hexachlorocyclohexane (HCH) isomers on androgen receptor activity in vitro. *Chemosphere*, 86(1):65–9. doi:10.1016/j.chemosphere.2011.08.052 PMID:21962538

Perocco P, Colacci A, Del Ciello C, Grilli S (1995). Cytotoxic and cell transforming effects of the insecticide, lindane (gamma-hexachlorocyclohexane) on BALB/c 3T3 cells. *Res Commun Mol Pathol Pharmacol*, 89(3):329–39. PMID:8680801

Piskac-Collier AL, Smith MA (2009). Lindane-induced generation of reactive oxygen species and depletion of glutathione do not result in necrosis in renal distal tubule cells. *J Toxicol Environ Health A*, 72(19):1160–7. doi:10.1080/15287390903091780 PMID:20077184

Pompa G, Fadini L, Di Lauro F, Caloni F (1994). Transfer of lindane and pentachlorobenzene from mother to newborn rabbits. *Pharmacol Toxicol*, 74(1):28–34. doi:10.1111/j.1600-0773.1994.tb01069.x PMID:7512714

Pool-Zobel BL, Guigas C, Klein R, Neudecker C, Renner HW, Schmezer P (1993). Assessment of genotoxic effects by Lindane. *Food Chem Toxicol*, 31(4):271–83. doi:10.1016/0278-6915(93)90077-C

Pool-Zobel BL, Lotzmann N, Knoll M, Kuchenmeister F, Lambertz R, Leucht U et al. (1994). Detection of genotoxic effects in human gastric and nasal mucosa cells isolated from biopsy samples. *Environ Mol Mutagen*, 24(1):23–45. doi:10.1002/em.2850240105

Prapamontol T, Stevenson D (1991). Rapid method for the determination of organochlorine pesticides in milk. *J Chromatogr A*, 552(1-2):249–57. doi:10.1016/S0021-9673(01)95940-0 PMID:1939438

Purdue MP, Engel LS, Langseth H, Needham LL, Andersen A, Barr DB et al. (2009). Prediagnostic serum concentrations of organochlorine compounds and risk of testicular germ cell tumors. *Environ Health Perspect*, 117(10):1514–9. doi:10.1289/ehp.0800359 PMID:20019899

Purdue MP, Hoppin JA, Blair A, Dosemeci M, Alavanja MC (2007). Occupational exposure to organochlorine insecticides and cancer incidence in the Agricultural Health Study. *Int J Cancer*, 120(3):642–9. doi:10.1002/ijc.22258 PMID:17096337

Rafnsson V (2006a). Risk of non-Hodgkin's lymphoma and exposure to hexachlorocyclohexane, a nested case-control study. *Eur J Cancer*, 42(16):2781–5. doi:10.1016/j.ejca.2006.03.035 PMID:16934973

Rafnsson V (2006b). Cancer incidence among farmers exposed to lindane while sheep dipping. *Scand J Work Environ Health*, 32(3):185–9. doi:10.5271/sjweh.997 PMID:16804620

Raftopoulou EK, Dailianis S, Dimitriadis VK, Kaloyianni M (2006). Introduction of cAMP and establishment of neutral lipids alterations as pollution biomarkers using the mussel *Mytilus galloprovincialis*. Correlation with a battery of biomarkers. *Sci Total Environ*, 368(2–3):597–614. doi:10.1016/j.scitotenv.2006.04.031

Raizada RB, Misra P, Saxena P, Datta KK, Dikshith TS (1980). Weak estrogenic activity of lindane in rats. *J Toxicol Environ Health*, 6(3):483–92. doi:10.1080/15287398009529867 PMID:6158575

Raizada RB, Srivastava MK, Kaushal RA, Singh RP, Gupta KP (2001). Subchronic oral toxicity of a combination of insecticide (HCH) and herbicide (ISP) in male rats. *J Appl Toxicol*, 21(1):75–9. doi:10.1002/jat.733 PMID:11180283

Ramabhatta S, Sunilkumar GR, Somashekhar C (2014). Lindane toxicity following accidental oral ingestion. *Indian J Dermatol Venereol Leprol*, 80(2):181–2. doi:10.4103/0378-6323.129419 PMID:24685871

Rauch AE, Kowalsky SF, Lesar TS, Sauerbier GA, Burkart PT, Scharfman WB (1990). Lindane (Kwell)-induced aplastic anemia. *Arch Intern Med*, 150(11):2393–5. doi:10.1001/archinte.1990.00390220125026 PMID:1700687

Regueiro J, Llompart M, Garcia-Jares C, Cela R (2007). Development of a high-throughput method for the determination of organochlorinated compounds, nitromusks and pyrethroid insecticides in indoor dust. *J Chromatogr A*, 1174(1-2):112–24. doi:10.1016/j.chroma.2007.08.052 PMID:17884065

Regueiro J, Llompart M, Garcia-Jares C, Garcia-Monteagudo JC, Cela R (2008). Ultrasound-assisted emulsification-microextraction of emergent contaminants and pesticides in environmental waters. *J Chromatogr A*, 1190(1-2):27–38. doi:10.1016/j.chroma.2008.02.091 PMID:18359033

Reif DM, Martin MT, Tan SW, Houck KA, Judson RS, Richard AM et al. (2010). Endocrine profiling and prioritization of environmental chemicals using ToxCast data. *Environ Health Perspect*, 118(12):1714–20. doi:10.1289/ehp.1002180 PMID:20826373

Reif DM, Sypa M, Lock EF, Wright FA, Wilson A, Cathey T et al. (2013). ToxPi GUI: an interactive visualization tool for transparent integration of data from diverse sources of evidence. *Bioinformatics*, 29(3):402–3. doi:10.1093/bioinformatics/bts686 PMID:23202747

Reifenrath WG, Chellquist EM, Shipwash EA, Jederberg WW, Krueger GG (1984). Percutaneous penetration in the hairless dog, weanling pig and grafted athymic nude mouse: evaluation of models for predicting skin penetration in man. *Br J Dermatol*, 111(Suppl 27):123–35. doi:10.1111/j.1365-2133.1984.tb15590.x PMID:6204672

Ren A, Qiu X, Jin L, Ma J, Li Z, Zhang L et al. (2011). Association of selected persistent organic pollutants in the placenta with the risk of neural tube defects. *Proc Natl Acad Sci USA*, 108(31):12770–5. doi:10.1073/pnas.1105209108 PMID:21768370

Rippen G (1990/2000). Umweltchemikalien - Stoffdaten. Prüfverfahren. Vorschriften. Ecomed: Landsberg/Lech; 3. Aufl., Loseblattsammlung, Aktualisierung 5/2000; ISBN 3-606-73210-5. [German]

Röllin HB, Sandanger TM, Hansen L, Channa K, Odland JØ (2009). Concentration of selected persistent organic pollutants in blood from delivering women in South Africa. *Sci Total Environ*, 408(1):146–52. doi:10.1016/j.scitotenv.2009.08.049 PMID:19800104

Ronco AM, Valdés K, Marcus D, Llanos M (2001). The mechanism for lindane-induced inhibition of steroidogenesis in cultured rat Leydig cells. *Toxicology*, 159(1–2):99–106. doi:10.1016/S0300-483X(00)00414-5 PMID:11250058

Rotterdam Convention (2004). Rotterdam Convention on the Prior Informed Consent Procedure for Certain Hazardous Chemicals and Pesticides in International Trade. Available from: http://www.pic.int/TheConvention/Overview/TextoftheConvention/RotterdamConventionText/tabid/1160/language/en-US/Default.aspx

Roux F, Treich I, Brun C, Desoize B, Fournier E (1979). Effect of lindane on human lymphocyte responses to phytohemagglutinin. *Biochem Pharmacol*, 28(16):2419–26. doi:10.1016/0006-2952(79)90002-9 PMID:92317

Roy P, Salminen H, Koskimies P, Simola J, Smeds A, Saukko P et al. (2004). Screening of some anti-androgenic endocrine disruptors using a recombinant cell-based in vitro bioassay. *J Steroid Biochem Mol Biol*, 88(2):157–66. doi:10.1016/j.jsbmb.2003.11.005 PMID:15084347

RSC (2015). Lindane. The Merck Index Online [online database]. Royal Society of Chemistry. Available from: https://www.rsc.org/Merck-Index/monograph/m6826/lindane?q=unauthorize, accessed 28 May 2015.

Rugman FP, Cosstick R (1990). Aplastic anaemia associated with organochlorine pesticide: case reports and review of evidence. *J Clin Pathol*, 43(2):98–101. doi:10.1136/jcp.43.2.98 PMID:1690760

Rupa DS, Reddy PP, Reddi OS (1989). Genotoxic effect of benzene hexachloride in cultured human lymphocytes. *Hum Genet*, 83(3):271–3. doi:10.1007/BF00285170

Samanta L, Roy A, Chainy GB (1999a). Changes in rat testicular antioxidant defence profile as a function of age and its impairment by hexachlorocyclohexane during critical stages of maturation. *Andrologia*, 31(2):83–90. doi:10.1046/j.1439-0272.1999.00232.x PMID:10097797

Samanta L, Sahoo A, Chainy GB (1999b). Age-related changes in rat testicular oxidative stress parameters by hexachlorocyclohexane. *Arch Toxicol*, 73(2):96–107. doi:10.1007/s002040050593 PMID:10350190

Sanfeliu C, Camón L, Martínez E, Solà C, Artigas F, Rodríguez-Farré E (1988). Regional distribution of lindane in rat brain. *Toxicology*, 49(1):189–96. doi:10.1016/0300-483X(88)90192-8 PMID:2453938

Saoudi A, Fréry N, Zeghnoun A, Bidondo ML, Deschamps V, Göen T et al. (2014). Serum levels of organochlorine pesticides in the French adult population: the French National Nutrition and Health Study (ENNS), 2006–2007. *Sci Total Environ*, 472:1089–99. doi:10.1016/j.scitotenv.2013.11.044 PMID:24361744

Saradha B, Vaithinathan S, Mathur PP (2008). Single exposure to low dose of lindane causes transient decrease in testicular steroidogenesis in adult male Wistar rats. *Toxicology*, 244(2-3):190–7. doi:10.1016/j.tox.2007.11.011 PMID:18248869

Saradha B, Vaithinathan S, Mathur PP (2009). Lindane induces testicular apoptosis in adult Wistar rats through the involvement of Fas-FasL and mitochondria-dependent pathways. *Toxicology*, 255(3):131–9. doi:10.1016/j.tox.2008.10.016 PMID:19038305

Sauviat MP, Pages N (2002). [Cardiotoxicity of lindane, a gamma isomer of hexachlorocyclohexane] *J Soc Biol*, 196(4):339–48. PMID:12645305. [French]

Sawada N, Iwasaki M, Inoue M, Itoh H, Sasazuki S, Yamaji T et al. ; Japan Public Health Center Based Prospective (JPHC) Study Group(2010). Plasma organochlorines and subsequent risk of prostate cancer in Japanese men: a nested case-control study. *Environ Health Perspect*, 118(5):659–65. doi:10.1289/ehp.0901214 PMID:20435560

Schade G, Heinzow B (1998). Organochlorine pesticides and polychlorinated biphenyls in human milk of mothers living in northern Germany: current extent of contamination, time trend from 1986 to 1997 and factors that influence the levels of contamination. *Sci Total Environ*, 215(1-2):31–9. doi:10.1016/S0048-9697(98)00008-4 PMID:9599454

Schinasi L, Leon ME (2014). Non-Hodgkin lymphoma and occupational exposure to agricultural pesticide chemical groups and active ingredients: a systematic review and meta-analysis. *Int J Environ Res Public Health*, 11(4):4449–527. doi:10.3390/ijerph110404449 PMID:24762670

Schrader TJ, Cooke GM (2000). Examination of selected food additives and organochlorine food contaminants for androgenic activity in vitro. *Toxicol Sci*, 53(2):278–88. doi:10.1093/toxsci/53.2.278 PMID:10696776

Schroeder JC, Olshan AF, Baric R, Dent GA, Weinberg CR, Yount B et al. (2001). Agricultural risk factors for t(14;18) subtypes of non-Hodgkin's lymphoma. *Epidemiology*, 12(6):701–9. doi:10.1097/00001648-200111000-00020 PMID:11679800

Schulte Hermann RS, Koransky W, Leberl C, Noack G (1971). Hyperplasia and hypertrophy of rat liver induced by -hexachlorcyclohexane and butylhydroxytoluene. Retention of the hyperplasia during involution of the enlarged organ. *Virchows Arch B Cell Pathol*, 9(2):125–34. PMID:4108029

Scippo ML, Argiris C, Van De Weerdt C, Muller M, Willemsen P, Martial J et al. (2004). Recombinant human estrogen, androgen and progesterone receptors for detection of potential endocrine disruptors. *Anal Bioanal Chem*, 378(3):664–9. doi:10.1007/s00216-003-2251-0 PMID:14579009

Seiler P, Fischer B, Lindenau A, Beier HM (1994). Effects of persistent chlorinated hydrocarbons on fertility and embryonic development in the rabbit. *Hum Reprod*, 9(10):1920–6. PMID:7531205

Seth V, Banerjee BD, Bhattacharya A, Pasha ST, Chakravorty AK (2001). Pesticide induced alterations in acetylcholine esterase and gamma glutamyl transpeptidase activities and glutathione level in lymphocytes of human poisoning cases. *Clin Biochem*, 34(5):427–9. doi:10.1016/S0009-9120(01)00232-6 PMID:11522282

Sharma BM, Bharat GK, Tayal S, Nizzetto L, Cupr P, Larssen T (2014). Environment and human exposure to persistent organic pollutants (POPs) in India: a systematic review of recent and historical data. *Environ Int*, 66:48–64. doi:10.1016/j.envint.2014.01.022 PMID:24525153

Shirasu Y, Moriya M, Kato K, Furuhashi A, Kada T (1976). Mutagenicity screening of pesticides in the microbial system. *Mutat Res*, 40(1):19–30. doi:10.1016/0165-1218(76)90018-5 PMID:814455

Shunthirasingham C, Oyiliagu CE, Cao X, Gouin T, Wania F, Lee SC et al. (2010). Spatial and temporal pattern of pesticides in the global atmosphere. *J Environ Monit*, 12(9):1650–7. doi:10.1039/c0em00134a PMID:20697628

Siddiqui MK, Anand M, Mehrotra PK, Sarangi R, Mathur N (2005). Biomonitoring of organochlorines in women with benign and malignant breast disease. *Environ Res*, 98(2):250–7. doi:10.1016/j.envres.2004.07.015 PMID:15820732

Siddiqui MK, Nigam U, Kaul PP, Seth TD (1996). Bioaccumulation of HCH isomers in different tissues of young and old rats: a comparison. *Bull Environ Contam Toxicol*, 56(6):896–902. doi:10.1007/s001289900130 PMID:8661878

Siddiqui MK, Saxena MC, Bhargava AK, Seth TD, Murti CR, Kutty D (1981). Agrochemicals in the maternal blood, milk, and cord blood: a source of toxicants for prenates and neonates. *Environ Res*, 24(1):24–32. doi:10.1016/0013-9351(81)90128-6 PMID:6163625

Simić B, Kniewald Z, Davies JE, Kniewald J (1991). Reversibility of the inhibitory effect of atrazine and lindane on cytosol 5 alpha-dihydrotestosterone receptor complex formation in rat prostate. *Bull Environ Contam Toxicol*, 46(1):92–9. doi:10.1007/BF01688260 PMID:1705844

Sipes NS, Martin MT, Kothiya P, Reif DM, Judson RS, Richard AM et al. (2013). Profiling 976 ToxCast chemicals across 331 enzymatic and receptor signaling assays. *Chem Res Toxicol*, 26(6):878–95. doi:10.1021/tx400021f PMID:23611293

Smith MT, Guyton KZ, Gibbons CF, Fritz JM, Portier CJ, Rusyn I et al. (2016). Key characteristics of carcinogens as a basis for organizing data on mechanisms of carcinogenesis. *Environ Health Perspect*, 124(6):713–21. PMID:26600562

Sosan MB, Akingbohungbe AE, Ojo IA, Durosinmi MA (2008). Insecticide residues in the blood serum and domestic water source of cacao farmers in Southwestern Nigeria. *Chemosphere*, 72(5):781–4. doi:10.1016/j.chemosphere.2008.03.015 PMID:18471864

Soto AM, Sonnenschein C, Chung KL, Fernandez MF, Olea N, Serrano FO (1995). The E-SCREEN assay as a tool to identify estrogens: an update on estrogenic environmental pollutants. *Environ Health Perspect*, 103:Suppl 7: 113–22. doi:10.1289/ehp.95103s7113 PMID:8593856

Sreekumaran Nair R, Paulmurugan R, Ranjit Singh AJA (2002). Simple radioactive assay for the estimation of DNA breaks. *J Appl Toxicol*, 22(1):19–23. doi:10.1002/jat.807

SRI (2009). Directory of Chemical Producers [online database]. Menlo Park, CA SRI Consulting database edition, SRI. Available from: http://chemical.ihs.com/nl/Public/2009/0904/0904.html, accessed 3 June 2015.

Srinivasan K, Mahadevappa KL, Radhakrishnamurty R (1991). Effect of maternal dietary hexachlorocyclohexane exposure on pup survival and growth in albino rats. *J Environ Sci Health B*, 26(3):339–49. doi:10.1080/03601239109372740 PMID:1716648

Stellman SD, Djordjevic MV, Muscat JE, Gong L, Bernstein D, Citron ML et al. (1998). Relative abundance of organochlorine pesticides and polychlorinated biphenyls in adipose tissue and serum of women in Long Island, New York. *Cancer Epidemiol Biomarkers Prev*, 7(6):489–96. PMID:9641493

Straif K, Loomis D, Guyton K, Grosse Y, Lauby-Secretan B, El Ghissassi F et al. (2014). Future priorities for the IARC Monographs. *Lancet Oncol*, 15(7):683–4. doi:10.1016/S1470-2045(14)70168-8

Sujatha R, Chitra KC, Latchoumycandane C, Mathur PP (2001). Effect of lindane on testicular antioxidant system and steroidogenic enzymes in adult rats. *Asian J Androl*, 3(2):135–8. PMID:11404799

Sunder Ram Rao CV, Shreenivas R, Singh V, Perez-Atayde A, Woolf A (1988). Disseminated intravascular coagulation in a case of fatal lindane poisoning. *Vet Hum Toxicol*, 30(2):132–4. PMID:2454526

Suwalsky M, Villena F, Marcus D, Ronco AM (2000). Plasma absorption and ultrastructural changes of rat testicular cells induced by lindane. *Hum Exp Toxicol*, 19(9):529–33. doi:10.1191/096032700675826027 PMID:11204556

Tan J, Cheng SM, Loganath A, Chong YS, Obbard JP (2007). Selected organochlorine pesticide and polychlorinated biphenyl residues in house dust in Singapore. *Chemosphere*, 68(9):1675–82. doi:10.1016/j.chemosphere.2007.03.051 PMID:17490710

Thorpe E, Walker AI (1973). The toxicology of dieldrin (HEOD). II. Comparative long-term oral toxicity studies in mice with dieldrin, DDT, phenobarbitone, β-BHC and γ-BHC. *Food Cosmet Toxicol*, 11(3):433–42. doi:10.1016/0015-6264(73)90008-4 PMID:4125578

Tice RR, Austin CP, Kavlock RJ, Bucher JR (2013). Improving the human hazard characterization of chemicals: a Tox21 update. *Environ Health Perspect*, 121(7):756–65. doi:10.1289/ehp.1205784 PMID:23603828

Tomczak S, Baumann K, Lehnert G (1981). Occupational exposure to hexachlorocyclohexane. IV. Sex hormone alterations in HCH-exposed workers. *Int Arch Occup Environ Health*, 48(3):283–7. doi:10.1007/BF00405615 PMID:6166570

Törnkvist A, Glynn A, Aune M, Darnerud PO, Ankarberg EH (2011). PCDD/F, PCB, PBDE, HBCD and chlorinated pesticides in a Swedish market basket from 2005–levels and dietary intake estimations. *Chemosphere*, 83(2):193–9. doi:10.1016/j.chemosphere.2010.12.042 PMID:21269658

Traina ME, Rescia M, Urbani E, Mantovani A, Macrì C, Ricciardi C et al. (2003). Long-lasting effects of lindane on mouse spermatogenesis induced by *in utero* exposure. *Reprod Toxicol*, 17(1):25–35. doi:10.1016/S0890-6238(02)00101-6

Tsatsakis AM, Tzatzarakis MN, Tutudaki M (2008). Pesticide levels in head hair samples of Cretan population as an indicator of present and past exposure. *Forensic Sci Int*, 176(1):67–71. doi:10.1016/j.forsciint.2007.07.017 PMID:17983715

Turner JC, Shanks V (1980). Absorption of some organochlorine compounds by the rat small intestine–in vivo. *Bull Environ Contam Toxicol*, 24(1):652–5. doi:10.1007/BF01608169 PMID:6161652

UCLID (1994). International Uniform Chemical Information Database [online database]. Available from: http://iuclid.eu/, accessed 3 June 2015.

UNECE (2004). Technical review report on lindane. August 20114. Reports on substances scheduled for re-assessments under the UNECE POPs protocol. United Nations Economic Commission for Europe. Available from: http://www.unece.org/fileadmin/DAM/env/lrtap/TaskForce/popsxg/2004/Dossier_Lindane.pdf, accessed June 2016.

UNEP (2002). Stockholm Convention on Persistent Organic Pollutants. Available from: www.pops.int. Geneva, Switzerland: United Nations Environment Programme.

UNEP/WHO (2015). Specific exemptions and acceptable purposes under the Stockholm Convention. Report No. UNEP/POPS/COP.7/4. Geneva, Switzerland: United Nations Environment Programme and the World Health Organization. Available from: http://chm.pops.int/TheConvention/ConferenceoftheParties/Meetings/COP7/tabid/4251/mctl/ViewDetails/EventModID/870/EventID/543/xmid/13075/Default.aspx, accessed 3 June 2015.

USGS (2006). Pesticides in the Nation's Streams and Ground Water, 1992–2001. USGS Circular 1291, Appendix 7A. Statistical summaries of pesticide compounds in stream water, 1992–2001. Indianapolis (IN), USA: United States Geological Survey. Available from: http://water.usgs.gov/nawqa/pnsp/pubs/circ1291/appendix7/7a.html, accessed 26 May 2015.

Vidal M et al. (1997). Analysis of lindane, α- and β-endosulfan and endosulfan sulfate in greenhouse air by gas chromatography. *J Chromatogr A*, 765(1):99–108. doi:10.1016/S0021-9673(96)01088-6 PMID:9035387

Videla LA, Tapia G, Varela P, Cornejo P, Guerrero J, Israel Y et al. (2004). Effects of acute γ-hexachlorocyclohexane intoxication in relation to the redox regulation of nuclear factor-kappaB, cytokine gene expression, and liver injury in the rat. *Antioxid Redox Signal*, 6(2):471–80. doi:10.1089/152308604322899530 PMID:15025948

Videla LA, Troncoso P, Arisi AC, Junqueira VB (1997). Dose-dependent effects of acute lindane treatment on Kupffer cell function assessed in the isolated perfused rat liver. *Xenobiotica*, 27(7):747–57. doi:10.1080/004982597240325 PMID:9253150

Viel JF, Floret N, Deconinck E, Focant JF, De Pauw E, Cahn JY (2011). Increased risk of non-Hodgkin lymphoma and serum organochlorine concentrations among neighbors of a municipal solid waste incinerator. *Environ Int*, 37(2):449–53. doi:10.1016/j.envint.2010.11.009 PMID:21167603

Vijaya Padma V, Sowmya P, Arun Felix T, Baskaran R, Poornima P (2011). Protective effect of gallic acid against lindane induced toxicity in experimental rats. *Food Chem Toxicol*, 49(4):991–8. doi:10.1016/j.fct.2011.01.005 PMID:21219962

Vijgen J, Abhilash PC, Li YF, Lal R, Forter M, Torres J et al. (2011). Hexachlorocyclohexane (HCH) as new Stockholm Convention POPs–a global perspective on the management of Lindane and its waste isomers. *Environ Sci Pollut Res Int*, 18(2):152–62. doi:10.1007/s11356-010-0417-9 PMID:21104204

Volder EC, Li YF (1995). Global usage of selected persistent organochlorines. *Sci Total Environ*, 160/161:201–10. doi:10.1016/0048-9697(95)04357-7

Walorczyk S, Drożdżyński D, Kowalska J, Remlein-Starosta D, Ziółkowski A, Przewoźniak M et al. (2013). Pesticide residues determination in Polish organic crops in 2007–2010 applying gas chromatography-tandem quadrupole mass spectrometry. *Food Chem*, 139(1-4):482–7. doi:10.1016/j.foodchem.2013.01.013 PMID:23561134

Walsh LP, Stocco DM (2000). Effects of lindane on steroidogenesis and steroidogenic acute regulatory protein expression. *Biol Reprod*, 63(4):1024–33. doi:10.1095/biolreprod63.4.1024 PMID:10993823

Ward EM, Schulte P, Grajewski B, Andersen A, Patterson DG Jr, Turner W et al. (2000). Serum organochlorine levels and breast cancer: a nested case-control study of Norwegian women. *Cancer Epidemiol Biomarkers Prev*, 9(12):1357–67. PMID:11142422

Weinhold B (2001). Last call for lindane. *Environ Health Perspect*, 109(6):A254 PMID:11445525

Weisse I, Herbst M (1977). Carcinogenicity study of lindane in the mouse. *Toxicology*, 7(2):233–8. doi:10.1016/0300-483X(77)90069-5 PMID:67663

Welch RM, Levin W, Kuntzman R, Jacobson M, Conney AH (1971). Effect of halogenated hydrocarbon insecticides on the metabolism and uterotropic action of estrogens in rats and mice. *Toxicol Appl Pharmacol*, 19(2):234–46. doi:10.1016/0041-008X(71)90109-8 PMID:4105824

WHO (2004). Lindane in drinking-water. Background document for development of WHO Guidelines for Drinking-water Quality. WHO/SDE/WSH/03.04/102. Geneva, Switzerland: World Health Organization. Available from: http://www.who.int/water_sanitation_health/dwq/chemicals/lindane.pdf

WHO (2015). Human biomonitoring: facts and figures. Copenhagen, Denmark: World Health Organization Regional Office for Europe. Available from: http://www.euro.who.int/en/media-centre/events/events/2015/04/ehp-mid-term-review/publications/human-biomonitoring-facts-and-figures.

Wiles DA, Russell JL, Olson KR, Walson PD, Kelley M (2015). Massive lindane overdose with toxicokinetics analysis. *J Med Toxicol*, 11(1):106–9. doi:10.1007/s13181-014-0403-6 PMID:24805102

Wolff GL, Roberts DW, Morrissey RL, Greenman DL, Allen RR, Campbell WL et al. (1987). Tumorigenic responses to lindane in mice: potentiation by a dominant mutation. *Carcinogenesis*, 8(12):1889–97. doi:10.1093/carcin/8.12.1889 PMID:2445499

Yaduvanshi SK, Srivastava N, Marotta F, Jain S, Yadav H (2012). Evaluation of micronuclei induction capacity and mutagenicity of organochlorine and organophosphate pesticides. *Drug Metabolism Letters*, 6(3):187–97. doi:10.2174/1872312811206030006

Yalçin SS, Orün E, Yalçin S, Aykut O (2014). Organochlorine pesticide residues in breast milk and maternal psychopathologies and infant growth from suburban area of Ankara, Turkey. *Int J Environ Health Res*, 26:1–9. doi:10.1080/09603123.2014.945515 PMID:25155352

Yamamoto T, Egashira T, Yamanaka Y, Yoshida T, Kuroiwa Y (1983). Initial metabolism of gamma-hexachlorocyclohexane (gamma-HCH) by rat liver microsomes. *J Pharmacobiodyn*, 6(10):721–8. doi:10.1248/bpb1978.6.721 PMID:6198504

Yu Y, Li C, Zhang X, Zhang X, Pang Y, Zhang S et al. (2012). Route-specific daily uptake of organochlorine pesticides in food, dust, and air by Shanghai residents, China. *Environ Int*, 50:31–7. doi:10.1016/j.envint.2012.09.007 PMID:23063733

Yu Y, Wang B, Wang X, Liu W, Cao J, Wong M et al. (2013). Temporal trends in daily dietary intakes of DDTs and HCHs in urban populations from Beijing and Shenyang, China. *Chemosphere*, 91(10):1395–400. doi:10.1016/j.chemosphere.2012.12.073 PMID:23427859

Zahm SH, Weisenburger DD, Babbitt PA, Saal RC, Vaught JB, Cantor KP et al. (1990). A case-control study of non-Hodgkin's lymphoma and the herbicide 2,4-dichlorophenoxyacetic acid (2,4-D) in eastern Nebraska. *Epidemiology*, 1(5):349–56. doi:10.1097/00001648-199009000-00004 PMID:2078610

Zhang K, Wei YL, Zeng EY (2013). A review of environmental and human exposure to persistent organic pollutants in the Pearl River Delta, South China. *Sci Total Environ*, 463-464:1093–110. doi:10.1016/j.scitotenv.2012.10.104 PMID:23245873

Zheng T, Holford TR, Mayne ST, Owens PH, Ward B, Carter D et al. (1999). Beta-benzene hexachloride in breast adipose tissue and risk of breast carcinoma. *Cancer*, 85(10):2212–8. doi:10.1002/(SICI)1097-0142(19990515)85:10<2212::AID-CNCR16>3.0.CO;2-F PMID:10326700

Zietz BP, Hoopmann M, Funcke M, Huppmann R, Suchenwirth R, Gierden E (2008). Long-term biomonitoring of polychlorinated biphenyls and organochlorine pesticides in human milk from mothers living in northern Germany. *Int J Hyg Environ Health*, 211(5-6):624–38. doi:10.1016/j.ijheh.2008.04.001 PMID:18550430

Zucchini-Pascal N, de Sousa G, Rahmani R (2009). Lindane and cell death: at the crossroads between apoptosis, necrosis and autophagy. *Toxicology*, 256(1-2):32–41. doi:10.1016/j.tox.2008.11.004 PMID:19041923

2,4-DICHLOROPHENOXYACETIC ACID

1. Exposure Data

1.1 Identification of the agent

1.1.1 Nomenclature

Chem. Abstr. Serv. Reg. No.: 94-75-7

Chem. Abstr. Serv. Name: 2,4-Dichlorophenoxyacetic acid

Preferred IUPAC Name: 2-(2,4-Dichlorophenoxy) acetic acid

Synonyms: 2,4-D; 2,4 dichlorophenoxyacetic acid; 2,4-dichlorophenoxyacetic acid

Trade Names: 2,4-Dichlorophenoxyacetic acid (2,4-D) has been used in many commercial product formulations. Selected trade names include: Hedonal; 2,4-D; Estone; Agrotect; Fernesta; Fernimine; Netagrone; Tributon; Vergemaster; Amoxone; Dicopur; Dormone; Ipaner; Moxone; Phenox; Pielik; Rhodia; Weedone; B-Selektonon.

Additional trade names are available in the PubChem Compound database (NCBI, 2015).

1.1.2 Structural and molecular formulae, and relative molecular mass

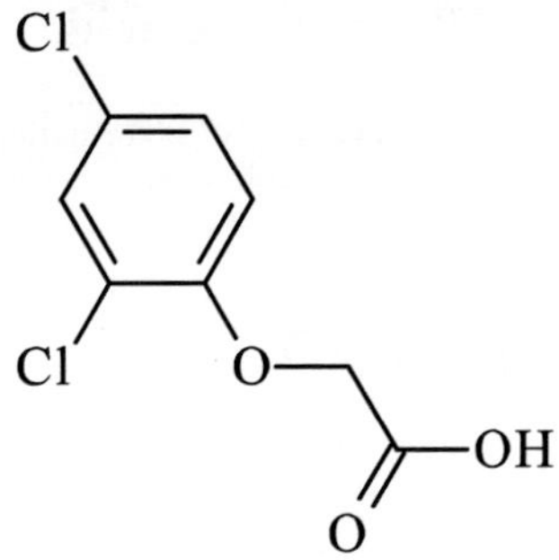

Molecular formula: $C_8H_6Cl_2O_3$

Relative molecular mass: 221.03

1.1.3 Chemical and physical properties of the pure substance

Description: Colourless crystals or white powder

Solubility: Slightly soluble in water (g/100 mL at 25 °C, 0.031). Soluble in organic solvents (ethanol, acetone, dioxane)

Octanol/water partition coefficient: log P_{ow}, 2.81

Conversion factor: 1 ppm = 9.04 mg/m^3, assuming normal temperature (25 °C) and pressure (101 kPa)

See IPCS/ICSC (2015)

Fig. 1.1 Production of 2,4-dichlorophenoxyacetic acid (2,4-D) via 2,4-dichlorophenol

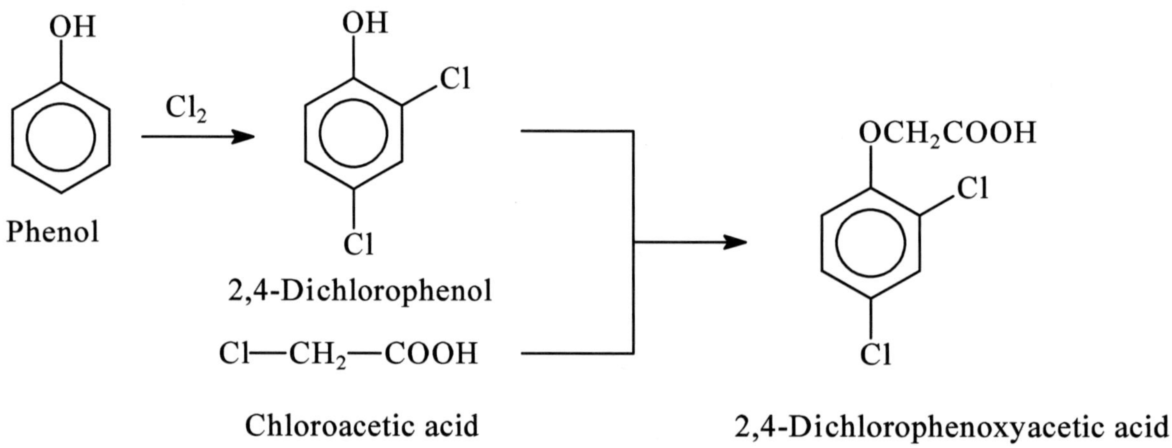

Reprinted from Chemosphere, 92(3), Liu et al. (2013) Formation and contamination of polychlorinated dibenzodioxins/dibenzofurans (PCDD/Fs), polychlorinated biphenyls(PCBs), pentachlorobenzene (PeCBz), hexachlorobenzene (HxCBz), and polychlorophenols in the production of 2,4-D products, pp 304–308, Copyright (2013), with permission from Elsevier

1.1.4 Esters and salts of 2,4-D

Several esters and salts of 2,4-D with various properties have been manufactured and used in herbicide products (NPIC, 2008). In humans, esters and salts of 2,4-D undergo rapid acid or enzymatic hydrolysis in vivo to yield 2,4-D (Garabrant & Philbert, 2002) (see Section 4.1). Esters and salts also undergo hydrolysis to the acid in environmental media at different rates depending on specific conditions of pH, moisture, and other factors (NPIC, 2008). Relevant ester and salt forms of 2,4-D include the following:

- 2,4-D salt (CAS No. 2702-72-9)
- 2,4-D diethanolamine salt (CAS No. 5742-19-8)
- 2,4-D dimethylamine salt (CAS No. 2008-39-1)
- 2,4-D isopropylamine salt (CAS No. 5742-17-6)
- 2,4-D isopropanolamine salt (CAS No. 32341-80-3)
- 2,4-D butoxyethyl ester (CAS No. 1929-73-3)
- 2,4-D butyl ester (CAS No. 94-80-4)
- 2,4-D 2-ethylhexyl ester (CAS No. 1928-43-4)
- 2,4-D isopropyl ester (CAS No. 94-11-1)
- 2,4-D isooctyl ester (CAS No. 25168-26-7)
- 2,4-D choline salt (CAS No. 1048373-72-3)

Physical properties of these 2,4-D salts and esters have been reported elsewhere (NPIC, 2008).

1.2 Production and use

1.2.1 Production

Two processes are currently used for the production of 2,4-D. In the first process, phenol is condensed with chloroacetic acid forming phenoxyacetic acid, which is subsequently chlorinated (Fig. 1.1). In the second process, phenol is chlorinated, generating 2,4-dichlorophenol, which is subsequently condensed with chloroacetic acid (Fig. 1.2).

The butyl ester derivative of 2,4-D is produced by the esterification of the acid with butanol in the presence of a ferric chloride catalyst and chlorine (Liu et al., 2013).

No reliable data on current global production of 2,4-D were available to the Working Group.

Fig. 1.2 Production of 2,4-dichlorophenoxyacetic acid (2,4-D) via phenoxyacetic acid

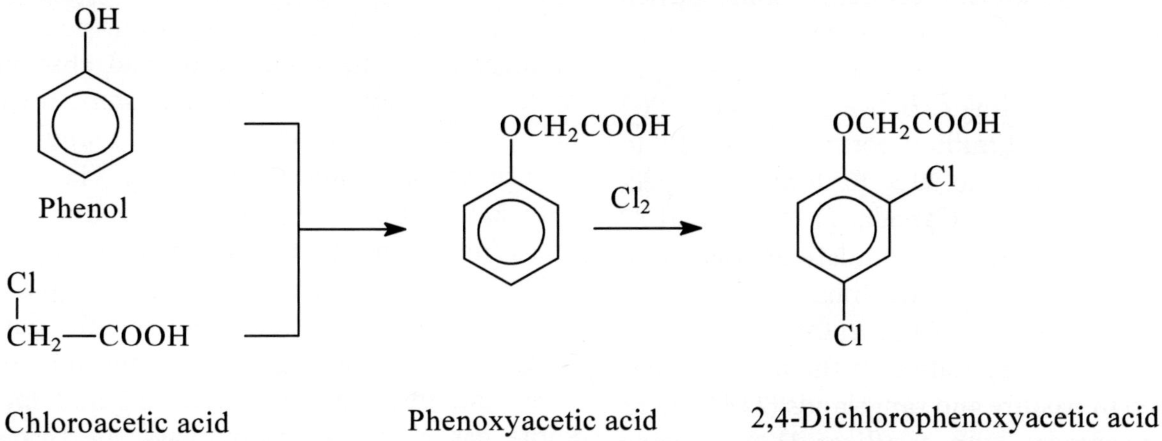

Reprinted from Chemosphere, 92(3), Liu et al. (2013) Formation and contamination of polychlorinated dibenzodioxins/dibenzofurans (PCDD/Fs), polychlorinated biphenyls(PCBs), pentachlorobenzene (PeCBz), hexachlorobenzene (HxCBz), and polychlorophenols in the production of 2,4-D products, pp 304–308, Copyright (2013), with permission from Elsevier

In 2010, the production of 2,4-D reached 40 000 tonnes in China (Liu et al., 2013).

1.2.2 Use

2,4-D is a synthetic auxin, and was the first chemical that could selectively control dicotyledons or broadleaf plants, but spare most monocotyledons, which include grasses and narrow-leaf crops such as wheat, maize (corn), rice, and similar cereal crops (Song, 2014).

2,4-D was first marketed in 1944 and produced by the American Chemical Paint Company. The derivatives of 2,4-D constitute a series of systematic herbicides that are widely used in broad-leaved weeds. 2,4-D is one of the world's most common herbicides because of its general applicability and low cost (Liu et al., 2013)

There are more than 600 products containing 2,4-D currently on the market (Song, 2014). In 2001, the dimethylamine salt and 2-ethylhexyl ester accounted for approximately 90–95% of the total global use of 2,4-D (Charles et al., 2001).

2,4-D is sold in various formulations under a wide variety of brand names and is found, for example, in commercial mixtures of lawn herbicide. 2,4-D can be used alone and is also commonly formulated with other herbicides, for example, dicamba (3,6-dichloro-2-methoxybenzoic acid), mecoprop (methylchlorophenoxypropionic acid, MCPP), mecoprop-P (the (*R*)-(+)-enantiomer of mecoprop), MCPA (2-methyl-4-chlorophenoxyacetic acid), picloram (4-amino-3,5,6 trichloropicolinic acid), and clopyralid (3,6-dichloro pyridine-2-carboxylic acid) (PubChem, 2015). 2,4-D in combination with glyphosate is used as the basis of a herbicide formulation designed for weed control in crops of corn and soybean that have been genetically modified to tolerate 2,4-D and glyphosate via insertion of a bacterial aryloxyalkanoate dioxygenase gene into the plant genome (Wright et al., 2010).

On 18 September 2014, the United States Environmental Protection Agency (EPA) granted registration for a herbicide containing the active ingredients 2,4-D, choline salt, and glyphosate dimethylammonium salt to be used on corn and soybean crops genetically engineered to be resistant to 2,4-D and glyphosate (EPA, 2014).

In the USA, 2,4-D is one of the 10 most commonly used conventional active ingredients of pesticide used in the agricultural sector. Use estimates from 2001 to 2007 ranged from 24 to

35 million pounds [~11×10^3 to 16×10^3 tonnes]. In the non-agricultural sectors, i.e. home/garden and industry/commercial/government, 2,4-D is the most commonly used active herbicide ingredient, with use estimates between 2001 and 2007 of 8–11 and 16–22 million pounds [~3.6×10^3 to 5×10^3 and 7×10^3 to 10×10^3 tonnes], respectively (EPA, 2011). In Canada, 14 tonnes and 87 tonnes of 2,4-D (diverse formulations) were used in British Columbia, and in Ontario respectively, in 2003 (CAREX-CANADA, 2009).

In the USA, application of the herbicide has occurred in pasture and rangelands (24%), lawns by homeowners with fertilizer (12%), spring wheat (8%), winter wheat (7%), lawn/garden without fertilizer (6%), soybean (4%), summer fallow (3%), hay other than alfalfa (3%) and roadways (3%). Other crops on which 2,4-D is used included filberts, sugarcane, barley, seed crops, apples, rye, cherries, oats, millet, rice, soybean, and pears. 2,4-D is also used in forestry, turfgrass management, and in the control of weeds near powerlines, railways, and similar corridors. Rates of application were generally less than 1.7 kg of acid equivalents per hectare, and generally less than 2.2 kg/Ha were applied annually. 2,4-D is predominantly used in the Midwest, Great Plains and Northwestern regions of the USA (EPA, 2005). Low concentrations of 2,4-D are used as plant growth regulators to induce callus formation (Liu et al., 2013). Agricultural use of 2,4-D includes both crop and non-crop applications of primarily liquid formulations, and a variety of application methods ranging from tractor-mounted booms to backpack sprayers. Forestry application ranges from backpack spraying to aerial application. Turf applications may use either liquid spray or granular formulations.

A mixture of roughly equal parts of 2,4-D and 2,4,5-trichlorophenoxyacetic acid (2,4,5-T), known as "agent orange", was used by military forces of the USA as a defoliant in the Viet Nam war (Kahn et al., 1988).

1.3 Measurement and analysis

Exposure to humans may occur as a result of ingestion, inhalation, or dermal absorption of 2,4-D, or any of its salts and esters, through occupational exposure during manufacture or use of herbicide products, or via contact with 2,4-D residues in food, water, air, or soil. Measurement methods have been developed for analysis of 2,4-D and its esters and salts in a wide range of biological, personal air, and dermal samples taken during monitoring for exposure, and in food, and environmental media. Some gas chromatography-mass spectrometry (GC-MS) and liquid chromatography-tandem mass spectrometry (LC-MS/MS) methods have been developed as "multi-residue" methods that can provide simultaneous extraction and analysis of other phenoxy acid herbicides (e.g. MCPA, MCPP, dicamba, 2,4,5-T) or even wider ranges of acidic or otherwise difficult-to-analyse pesticides (Raina-Fulton, 2014).

Analysis of 2,4-D acid in urine is the most widely used approach for biomonitoring of human exposure (Baker et al., 2000; Lindh et al., 2008), because excretion of 2,4-D and its acid-hydrolysable conjugates is almost exclusively in the urine. Esters and salts of 2,4-D are rapidly hydrolysed to the acid in exposed humans (see Section 4.1). This is particularly relevant in occupational settings, where exposure to the ester and salt forms are likely to occur. Methods for analysis of 2,4-D in other biological media, including blood and milk, have been developed and applied primarily in studies of toxicology and metabolism in experimental animals (Dickow et al., 2001; Stürtz et al., 2006). Methods of measurement of exposure for 2,4-D acid and its salt and ester forms have included personal and area air samples, dermal patch and bodysuit samples, and hand-wipe samples that are most often used for assessing occupational exposures (NIOSH, 1994; Gardner et al., 2005). Methods for analysis of 2,4-D in air (Waite et al., 2005), water (EPA,

Table 1.1 Representative methods for the analysis of 2,4-D

Sample matrix	Assay procedure	Limit of detection	Reference
Air, workplace	HPLC-UV	15 µg per filter	NIOSH (1994)
Air, ambient	GC-MSD	0.005 ng/m^3 based on a 2000 m^3 sample volume	Waite et al. (2005)
Ground water	UHPLC-MS/MS	0.0003 µg/L; LOQ, 0.0005 µg/L for 500 mL water samples	McManus et al. (2014)
Drinking-water	GC-ECD	0.055 µg/L	EPA (2000)
Soil	LC-MS/MS	Reporting limit, 0.010 ppm for 20 g of soil sample	Schaner et al. (2007)
Personal exposure (air, hand-wipe, dermal patch)	LC-MS/MS	MDL, 1.1–2.9 µg/L	Gardner et al. (2005)
Urine (human)	LC-MS/MS	0.05 µg/L	Lindh et al. (2008)
Urine (human)	HPLC-MS/MS	0.29 µg/L	Baker et al. (2000)
Plasma (dog)	HPLC-FD	LOQ, 500 µg/L	Dickow et al. (2001)
Serum and milk (rat)	GC-ECD	0.02 ppm [180 µg/L]	Stürtz et al. (2006)
Fruits and vegetables	LC-MS/MS	LOD, not reported; recovery tests performed at 0.01 mg/kg	Shida et al. (2015)
Cereals	LC-MS/MS	LOQ, 0.05 mg/kg	Santilio et al. (2011)
Food (duplicate diet)	GC-MS	MDL, 0.25 ng/g for solid food based on 8 g of homogenized food MDL, 0.20 ng/mL for liquid food based on 30 mL homogenized liquid food	Morgan et al. (2004)
House dust	GC-MS	MDL, 5 ng/g for 0.5 g of dust sample	Colt et al. (2008)

2,4-D, 2,4-dichlorophenoxyacetic acid; ECD, electron capture detector; FD, fluorescence detection; GC, gas chromatography; HPLC, high-performance liquid chromatography; LC, liquid chromatography; LOD, limit of detection; LOQ, limit of quantitation; MDL, method detection limit; MS, mass spectrometry; MS/MS, tandem mass spectrometry; MSD, mass-selective detection; UHPLC, ultra-high performance liquid chromatography

2000; McManus et al., 2014), soil (Schaner et al., 2007), house dust (Colt et al., 2008), and food (Morgan et al., 2004; Santilio et al., 2011; Shida et al., 2015), have primarily (but not exclusively) focused on the acid form of 2,4-D, partly because ester and amine salts of 2,4-D are hydrolysed to the acid at different rates in environmental media, depending on oxygen availability, moisture, and pH levels. In water and aerobic soil and sediment, the half-lives of esters and amines are shorter (in the order of days) than in anaerobic media. 2,4-D undergoes degradation in the outdoor environment, with potentially slower degradation rates in indoor environments (Walters, 1999). Examples of methods of analysis for 2,4-D in a range of media are listed in Table 1.1.

1.4 Occurrence and exposure

2,4-D and its salts and esters do not occur naturally in the environment. Due to widespread production and use of herbicide products containing 2,4-D, there is considerable potential for exposure of humans in occupational and non-occupational settings, as illustrated in Fig 1.3 and Fig. 1.4.

Most of the available data on exposure and environmental occurrence were from North America and Europe. Fewer data were available from other regions of the world. Given the widespread global use of 2,4-D, the lack of data should not be taken as an indicator that human exposures do not occur in other regions.

Fig. 1.3 Urinary concentrations of 2,4-dichlorophenoxyacetic acid (2,4-D)(mean, median, or geometric mean) from studies of occupational or para-occupational exposure, and in the general population

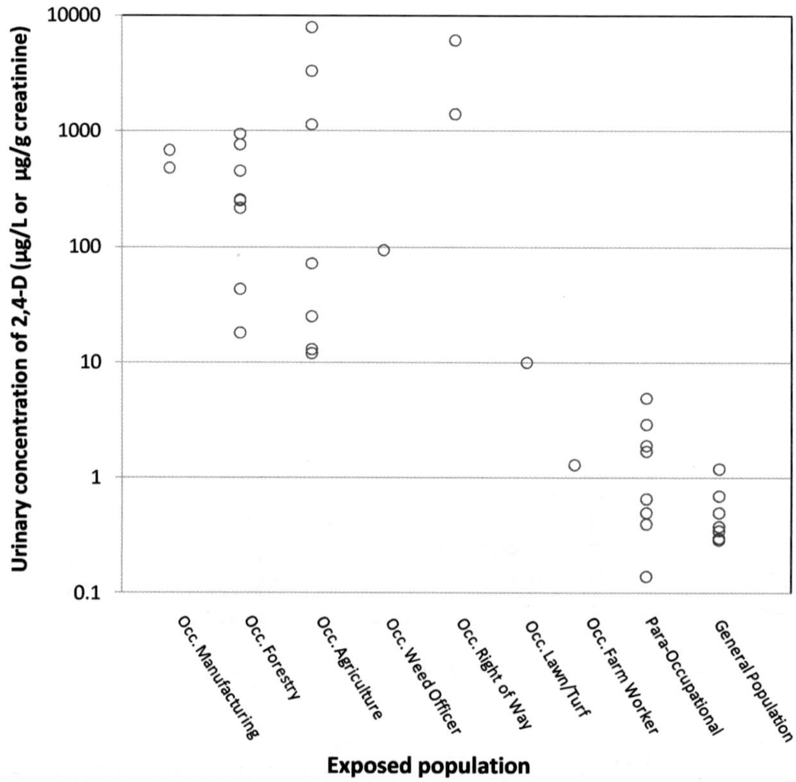

Compiled by the Working Group
Includes multiple subsets of results from several studies: Kolmodin-Hedman & Erne (1980), Draper (1982), Libich et al. (1984), Vural & Burgaz (1984), Knopp (1994), Garry et al. (2001), Hines et al. (2001), Arbuckle et al. (2004, 2005), Curwin et al. (2005a), Alexander et al. (2007), Arcury et al. (2007), Morgan et al. (2008), Bhatti et al. (2010), Thomas et al. (2010a), Zhang et al. (2011), Jurewicz et al. (2012), Rodríguez et al. (2012), Raymer et al. (2014), and CDC (2015)
d, day; occ., occupational

1.4.1 Occupational exposure

Occupational exposure to 2,4-D can result from product manufacturing, agricultural use, forestry, right-of-way, and turf/lawn applications. Indirect or para-occupational exposure may occur in some populations as a result of "take-home" and "drift" pathways. Occupational exposure to 2,4-D typically occurs as a result of dermal absorption and inhalation, although some incidental ingestion may also occur. Some studies cited in a review of dermal absorption of 2,4-D in humans showed that dermal exposure is the primary route of exposure for herbicide-spray applicators (Ross et al., 2005).

(a) Manufacture

In two studies of occupational exposure, workers involved in manufacturing products containing 2,4-D had urinary biomarker concentrations ranging from 35 to 12 693 µg/L, with a mean of 1366 µg/L, in one study as shown in Table 1.2 (Vural & Burgaz, 1984; Knopp, 1994). In one of these studies, values for room air and personal air were 3.2–245 µg/m³ and

Fig. 1.4 Estimated exposure to 2,4-dichlorophenoxyacetic acid (2,4-D) from studies of occupational or para-occupational exposure, and in the general population

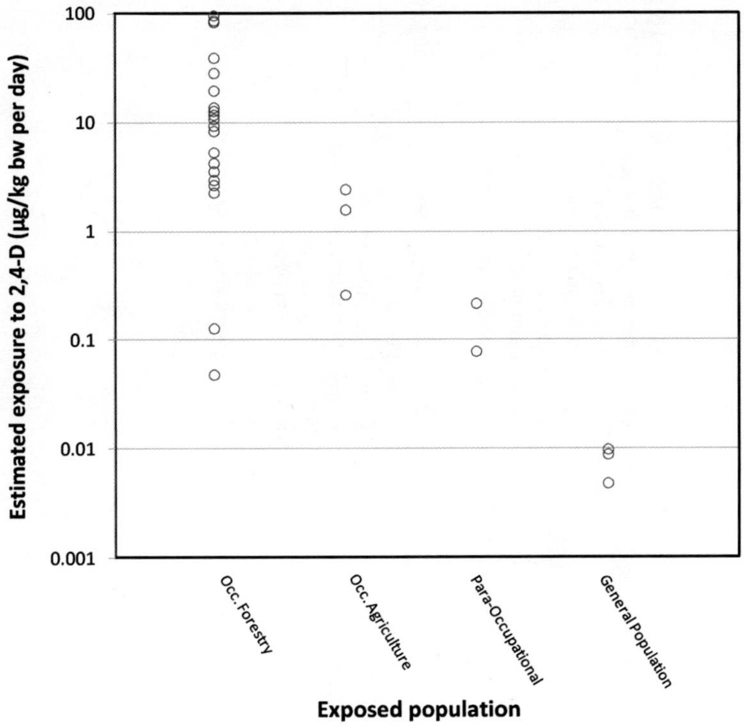

Compiled by the Working Group
Estimates were based on urinary concentrations, except for the general population, for which estimates were derived from residential and dietary measurements. Includes multiple subsets of results from several studies: Lavy et al. (1987), Hines et al. (2001), Alexander et al. (2007), Thomas et al. (2010a), Wilson et al. (2010), Zhang et al. (2011), and Morgan et al. (2014)

23.4–495 µg/m^3, respectively (Vural & Burgaz, 1984).

(b) Application

Many studies have been conducted to measure occupational exposure to 2,4-D from agriculture, forestry, right-of-way, and turf application of herbicidal products (Table 1.2). Both external (dermal, air) and biomonitoring methods have been used for exposure assessment of the applicator. Urinary 2,4-D concentrations for forestry applicators ranged from below the limit of detection (LOD) to 1700 µg/L, with means ranging from 17.6 to 454 µg/L for different job tasks (Garry et al., 2001). Estimated mean values for urinary excretion or the absorbed dose ranged from 2.7 to 98 µg/kg bw per day across several studies of forestry-related job tasks (Lavy et al., 1982; Lavy et al., 1987; Zhang et al., 2011). Professional agricultural applicators had urinary concentrations of 2,4-D ranging from not detected (ND) to 2858 µg/L, with values of 58 (geometric mean, GM) and 94 (median) µg/L (Hines et al., 2003; Bhatti et al., 2010). Many studies reported urinary results for farmer applicators, with 2,4-D concentrations ranging from ND to 14 000 µg/L, with GM values ranging from 5.8 to 715 µg/L, and a mean value of 8000 µg/L reported in one study (Kolmodin-Hedman & Erne, 1980; Draper & Street, 1982; Vural & Burgaz, 1984; Grover et al., 1986; Arbuckle et al., 2005; Curwin et al., 2005a; Alexander et al., 2007; Thomas et al.,

Table 1.2 Occupational exposure to 2,4-D

Country, year	Job/process	Media	No. of exposed individuals	Results		Comments/additional data	Reference
				Mean	Range		
Herbicide production							
Germany, 1985–89	2,4-D herbicide production	Urine	41	–	35–12 963 µg/L		Knopp (1994)
		Serum	41	–	3–3537 µg/L		
		Room air	12	–	3.2–245 µg/m³		
		Personal air	8	–	23.4–495 µg/m³		
Turkey, 1982	2,4-D herbicide production and application	Urine	15	Manufacturing: 1366 µg/L	60–9510 µg/L	15 workers manufacturing 2,4-D esters and amine salt; 6 h work shifts, urine collected on Friday; 13 2,4-D applicator crewmen (pilot, flagman, mixer, supervisor) with urine samples collected at end of 3-month application period	Vural & Burgaz (1984)
			13	Application: 715 µg/L	ND–1920 µg/L		
Forestry workers							
USA, 2002	Forestry backpack applicators	Urine	5	Group A: 768 ± 438 µg/day; 11 ± 5.7 µg/kg bw per day		Mean estimated total absorbed doses estimated for 5 applicators in group A (without protective clothing), 3 applicators in group B (with standard protective clothing), 1 mixer/loader, 1 supervisor; based on daily 24 h urine samples collected for 6 days	Zhang et al. (2011)
			3	Group B: 951 ± 1089 µg/day; 13 ± 14.1 µg/kg bw per day			
			1	Mixer/loader: 217 kg per day ± 103 µg/kg per day; 2.7 ± 1.3 µg/kg bw per day			
			1	Supervisor: 257 ± 117 µg/day; 3.6 ± 1.7 µg/kg per day			

Table 1.2 (continued)

Country, year	Job/process	Media	No. of exposed individuals	Results		Comments/additional data	Reference
				Mean	Range		
USA, year NR	Forestry applicators	Urine	7	Backpack: 454 μg/L	28–1700 μg/L	First void urine collected at end of peak application season	Garry et al. (2001)
			4	Boom spray: 252 μg/L	86–490 μg/L		
			8	Aerial: 42.9 μg/L	ND–97 μg/L		
			5	Skidder: 17.6 μg/L	0.85–58 μg/L		
			15	Controls: 0.5 μg/L	ND–1.8 μg/L		
USA, 1982	Forestry ground workers	Urine, 2,4-D excreted	20	Backpack sprayers: mean, 87.6 (N) and 98 (S) μg/kg per day		24 h urine samples collected; total amount excreted from the application day and 4 following days reported here for normal (N) and special (S) precaution conditions	Lavy et al. (1987)
			20	Injection bar workers: mean, 9.5 (N) and 4.3 (S) μg/kg per day			
			20	Hypohatchet workers: mean, 84.8 (N) and 39.5 (S) μg/kg per day			
			20	Hack/squirt workers: mean, 28.8 (N) and 12.2 (S) μg/kg per day			

Table 1.2 (continued)

Country, year	Job/process	Media	No. of exposed individuals	Results		Comments/additional data	Reference
				Mean	Range		
USA, NR	Aerial crew, forest applications	Urine, 2,4-D excreted	3	Pilots: mean, 19.8 (N) and 8.5 (S) µg/kg per day		24 h urine samples collected; total amount excreted from the application day and the following 5 days, reported here for normal (N) and special (S) precaution conditions	Lavy et al. (1982)
			3	Mechanics: mean, 5.45 (N) and 3.01 (S) µg/kg per day			
			3	Mixer/loaders: mean, 19.6 (N) and 14.0 (S) µg/kg per day			
			3	Supervisors: mean, 2.31 (N) and 0.13 (S) µg/kg per day			
			6	Observers: mean, 0.49 (N) and 0.09 (S) µg/kg per day			
Farmworkers							
USA, 2000–02	Farm applicators	Urine (GM)	68	25 µg/L	1.6–970 µg/L	68 broadcast and hand-spray applicators with 24 h post-application urine; hand-loading, body-loading estimates; air measurements; estimated total absorbed doses for 14 applicators using application day and after 4 days of 24 h urine collection	Thomas et al. (2010a)
		Hand-loading	68	0.39 mg	ND–22 mg		
		Body-loading	68	2.9 mg	0.02–880 mg		
		Personal air	68	0.37 µg/m³	ND–10 µg/m³		
USA, 1996	Custom agricultural applicators	Urine (GM)	15	58 nmol/L [12.8 µg/L]	ND–2600 nmol/L [ND–575 µg/L	5–7 24 h urine samples during 6-wk period; estimated amount excreted in 24 h; air, hand-wipe and body-patch samples for 2,4-D 2-ethylhexyl ester	Hines et al. (2001, 2003)
		Hand-loading	15	–	1.3–4300 µg/sample		
		Body patches	15	–	0.3–6200 µg/sample		
		Personal air	15	–	0.06–2.4 µg/m³		

Table 1.2 (continued)

Country, year	Job/process	Media	No. of exposed individuals	Results		Comments/additional data	Reference
				Mean	Range		
Canada, 1981–82	Farm applicators	Urine, 2,4-D excreted	6	–	215–6258 μg	6 ground-rig spray applicators (one sampled three times); 24 h urine samples collected 4–7 days during/after application; total excreted 2,4-D calculated; hand-wash and dermal-patch samples for estimated dermal exposures	Grover et al. (1986)
		Hand-loading	6	–	10–8840 μg		
		Body-loading	6	–	1.9–1699 mg		
USA, 2000–1	Farm applicators	Urine (GM)	34	71.9 μg/L	1.5–2236	Boom-spray applicators; maximum 24 h urine concentrations during 4-day application and post-application period	Alexander et al. (2007)
USA, 2001	Farm applicators and non-farmers	Urine (GM)	8	Farmers spraying: 2,4-D: 13 μg/L	–	Urine samples collected 1–5 days after application and again 4 wk later	Curwin et al. (2005a)
			14	Farmers not spraying 2,4-D: 0.48 μg/L	–		
			23	Non-farmers: 0.29 μg/L	–		
Canada, 1996	Farm applicators	Urine	43	First 24 h sample: GM, 5.36 μg/L; median, 6.0 μg/L; mean, 27.6 ± 72.5 μg/L	ND–410 μg/L	126 spray applicators using 2,4-D or MCP for first time during growing season; two 24 h urine samples collected from start of application; results reported here for 43 farmers using 2,4-D	Arbuckle et al. (2002, 2005)
			43	Second 24 h sample: GM, 9.9 μg/L; median, 12.0 μg/L; mean, 40.8 ± 91.1 μg/L	ND–514 μg/L		
Sweden, NR	Tractor spray applicators	Urine	4	8000 μg/L	3000–14 000 μg/L	Urine samples during working week and after exposures, personal air samples	Kolmodin-Hedman & Erne (1980)
				24 h excretion: 9 mg	–		
		Air, personal	4	–	100–200 μg/m³		

Table 1.2 (continued)

Country, year	Job/process	Media	No. of exposed individuals	Results		Comments/additional data	Reference
				Mean	Range		
USA, 2003–4	Lawn turf applicators	Urine	135	Mass excreted during 24 h: median, 14.6 µg	0.1–3658 µg	Sprayers sampled across two herbicide and one insecticide spray seasons; two consecutive 24 h urine samples collected during herbicide spraying; not all sprayers used 2,4-D	Harris et al. (2010)
				Creatinine-adjusted concentrations for samples > LOD: median, 10.2 µg/g	0.2–3001 µg/g		
USA, 1994–95	County noxious weed officers	Urine	31	Mean, 259 ± 432 µg/L; median, 94.1 µg/L	0.07–2858 µg/L	Seasonal county agricultural noxious-weed control applicators; overnight (approx. 12 h) urine samples collected every other week during season	Bhatti et al. (2010)
USA, 1980	Pasture spray application	Urine	2	Crew A driver and sprayer: 1000 and 1300 µg/L respectively at 24 h		2 drivers and 2 sprayers using truck-mounted spray system for pasture land; morning void urine collected for 3 days after application; air samples collected in truck cab; hand rinse; crew A had single application, crew B had multiple applications	Draper & Street (1982)
		Urine	2	Crew B driver and sprayer: 4100 and 2800 µg/L respectively at 24 h			
		Hand loading	–	–	1.2–18 mg		
		Truck cab air	–	–	1.2–2.2 µg/m³		

Table 1.2 (continued)

Country, year	Job/process	Media	No. of exposed individuals	Results		Comments/additional data	Reference
				Mean	Range		
Canada, 1979–80	Right-of-way applicators	Urine	12	Roadside gun sprayers: 1.42 ± 1.76 mg/kg	0.04–8.15 mg/kg	Electric right-of-way vehicle or backpack hand-spray applicators; urine collected in morning and afternoon, then combined weekly on Thursdays and daily during air-sampling week	Libich et al. (1984)
		Urine	7	Sprayers in Kapuskasing: 6.16 ± 7.69 mg/kg	0.27–32.74 mg/kg		
		Urine	3	Mist-blower sprayers: 2.55 mg/kg	0.44–5.07 mg/kg		
		Air	12	Roadside gun sprayers: 7.1 ± 4.9 μg/m^3	1.0–19.5 μg/m^3		
		Air	3	Mist-blower sprayers: 55.2 ± 30.7 μg/m^3	16.2–91.3 μg/m^3		
United Kingdom, 1983	Mixing/ loading	Dermal exposure	3	Tractor-mounted: 102, 244, 122 mg		3 tractor-mounted and 2 knapsack sprayers with six replicates each; whole-body dermal dosimetry	Abbott et al. (1987)
			2	Knapsack: 13.2, 11 mg			
	Spraying	Dermal exposure	3	Tractor mounted: 33.7, 38.9, 90.2 mg			
			2	Knapsack: 159, 89 mg			

Table 1.2 (continued)

Country, year	Job/process	Media	No. of exposed individuals	Results		Comments/additional data	Reference
				Mean	Range		
Malaysia, NR	Paddy spray applicators	Personal air	NR	Manual sprayers: 0.027 ± 0.019 µg/L Motorized sprayers: 0.038 ± 0.0028 µg/L		Paddy spray applicators using manual or motorized knapsack sprayers; dermal exposures estimated from DREAM model	Baharuddin et al. (2011)
		Dermal exposure	NR	Manual spray with proper PPE: 37.8 ± 22.9 ppm Manual spray without proper PPE: 86.1 ± 53.4 ppm Motorized spray with proper PPE: 21.8 ± 9.3 ppm Motorized spray without proper PPE: 45.7 ± 20.3 ppm			
USA, 2010	Farmworkers	Urine	361	38.2% with 2,4-D levels > LOD (LOD = 210 µg/L) 16% with levels > LLOQ (LLOQ = 50 µg/L) For 60 people with samples > LLOQ: GM, 1.28 (range, 0.52–18.6) µg/L		Farmworkers exposed to multiple chemicals	Raymer et al. (2014)
Thailand, 2006	Farmers	Urine	136	2,4-D detection for 37.5% [75th percentile, 0.66 µg/L (range, ND–598 µg/L)]		Farmers in two communities; 21 reported use of a 2,4-D product but urine collection was not specifically timed to an application; mixed-crop farmers had higher detection rates for 2,4-D	Panuwet et al. (2008)

2,4-D, 2,4-dichlorophenoxyacetic acid; DREAM, dermal exposure assessment method; GM, geometric mean; LLOQ, lower limit of qualification; LOD, limit of detection; MCP, 4-chloro-2-methylphenoxyacetic acid; NC, not calculated; ND, not detected; NR, data not reported; PPE, protective personal equipment

2010a). Urine samples from farmers in Thailand who were not specifically linked to crop application had a 75th percentile concentration of 0.66 μg/L (median levels were < LOD) (Panuwet et al., 2008). Professional lawn-turf applicators had urinary concentrations ranging from 0.2 to 3001 μg/g creatinine, with a median of 10.2 μg/g creatinine; the range of excreted mass during 24 hours ranged from 0.1 to 3658 μg, with a median value of 14.6 μg (Harris et al., 2010). Professional right-of-way pesticide applicators had mean urinary concentrations ranging from 0.04 to 32.74 mg/kg, with means of 1.42 and 6.16 mg/kg for two groups of sprayers (Libich et al., 1984).

Several studies of different occupations reported measurements of 2,4-D in air, hand-wipe, or body-loading samples (Kolmodin-Hedman & Erne, 1980; Draper & Street, 1982; Kolmodin-Hedman et al., 1983; Libich et al., 1984; Grover et al., 1986; Abbott et al., 1987; Knopp, 1994; Hines et al., 2001; Thomas et al., 2010a; Baharuddin et al., 2011).

Exposures in farm applicators appear to have been higher during the 1980s than during the 2000s, but firm conclusions could not be drawn due to the small number of studies available.

(c) *Para-occupational exposure*

Indirect or para-occupational exposure has been measured in several studies, particularly those involving 2,4-D herbicide spraying on farms, and in farmworker families (Table 1.3). For children of farmers who reported having carried out 2,4-D spray applications, GM urine concentrations ranged from 0.4 to 4.9 μg/L (Arbuckle et al., 2004; Alexander et al., 2007; Rodríguez et al., 2012). For spouses of farmers who reported having carried out 2,4-D spray applications, GM or median urinary concentrations ranged from < LOD to 1.7 μg/L (Arbuckle & Ritter, 2005; Alexander et al., 2007; Jurewicz et al., 2012). In a study of children of farmworkers, the median urinary biomarker concentration was 0.14 μg/L (Arcury et al., 2007). One study measured 2,4-D in house dust at farms where 2,4-D had been applied in the previous 30 days (adjusted GM, 730 ng/g) compared with farms where no spraying had been applied (adjusted GM, 850 ng/g) and with non-farm homes (adjusted GM, 320 ng/g) (Curwin et al., 2005b). [One explanation for the higher concentration of 2,4-D in house dust on farms where no spraying had been carried out compared with farms where spraying was reported to have been undertaken may be that 2,4-D is also widely used on lawns and may be tracked into homes. A second reason is that the ester forms of 2,4-D are often used in agriculture in Iowa, and the study method measured only the acid form.] In the United States Agricultural Health Study (AHS), a longitudinal set of urine samples was collected for 1 year from 30 corn farmers and from 10 non-farmers (controls) (Bakke et al., 2009). For farmers, mean 2,4-D concentrations during pre-planting/off-season, planting, and growing/post-harvest periods were 2.9, 22.9, and 7.8 μg/g creatinine, respectively, while mean 2,4-D concentrations for controls during these periods were 0.5, 1.35, and 0.37 μg/g creatinine, respectively. These data suggested that farmers may be exposed to pesticides at higher levels than controls, even when pesticides are not being actively applied.

1.4.2 *Environmental occurrence and exposure in the general population*

Exposures of the general population may result from the presence of 2,4-D in house dust, food, air, water, and soil. In some areas, residential exposures may be related to use of 2,4-D on lawns, providing a nearby source for direct exposure and tracking into the home. Exposures may occur through inhalation, dermal absorption, and ingestion (Health Canada, 2010).

Table 1.3 Para-occupational exposure to 2,4-D

Country/year	Number of samples/setting	Media	Results	Comments/additional data	Reference
Canada, 1996	92 children (aged 3–18 yrs) of farm 2,4-D or MCPA spray applicators	Urine	First 24 h sample: mean, 0.9 ± 1.4 (max., 12) µg/L Second 24 h sample: mean, 1.9 ± 10.4 (max., 100) µg/L	9.8–14.1% of samples > LOD; data not reported separately for the children of the 43 2,4-D applicators	Arbuckle et al. (2004)
USA, 2004	60 farmworkers' children (aged 1–6 yrs)	Urine	Median, 0.14 µg/L	41.7% of samples > LOD; no information about 2,4-D use	Arcury et al. (2007)
Nicaragua, 2008	Rural schoolchildren; 208 urine samples from 77 children unrelated to 2,4-D application; 3 samples after parental application of 2,4-D	Urine	Unrelated to application: GM, 0.5 (max., 7.4) µg/L Related to application: GM, 0.4 (max., 0.5) µg/L	Study also included data for parental hours and kg a.i. of 2,4-D used for five periods from pre-conception until 8–10 yrs	Rodríguez et al. (2012)
USA, 2000–1	34 spouses and 53 children (aged 4–17 yrs) of farm applicators of 2,4-D spray	Urine	Children: GM, 4.9 µg/L; range, ND–640 µg/L Spouse: GM, 1.7 µg/L; range, 0.5–24.9 µg/L	Maximum 24 h urine concentrations during 4-day application and post-application period	Alexander et al. (2007)
Canada, 1996	125 spouses of farm applicators of 2,4-D or MCPA spray	Urine	First 24 h: GM, 0.6; median, < 1 µg/L; max., 61 µg/L; mean, 1.32 ± 5.6 µg/L Second 24 h: GM, 0.66; median, < LOD (max., 100) µg/L; and mean, 2.0 ± 9.7 µg/L	7.0–14% of samples > LOD; data not reported separately for the spouses of the 43 2,4-D applicators	Arbuckle & Ritter (2005)
Poland, NR	13 spouses of farm applicators of 2,4-D spray	Urine	Day after application: mean, 3.8 (95% CI, 0.6–8.5) µg/L		Jurewicz et al. (2012)
USA, 2002–03	30 farmers, 10 non-farmers; longitudinal collection of urine samples during 1 yr	Urine	Farmer pre-planting/off-season, planting, growing/post-harvest periods: mean, 2.9, 22.9, and 7.8 µg/g creatinine, respectively Non-farmer pre-planting/off-season, planting, growing/post-harvest: mean, 0.5, 1.35, and 0.37 µg/g creatinine, respectively		Bakke et al. (2009)
USA, 2001	House dust collected from 2 farm homes sprayed with 2,4-D in preceding 30 days; 3 farms with no 2,4-D sprayed; 6 non-farm homes	House dust (adjusted GM)	2,4-D detected in 100% of the farm and non-farm home samples: Farms sprayed with 2,4-D: 730 ng/g No 2,4-D sprayed: 850 ng/g Non-farm homes: 320 ng/g	Dust collected from multiple locations in interiors of homes during each of two visits	Curwin et al. (2005b)

a.i., active ingredient; 2,4-D, 2,4-dichlorophenoxyacetic acid; GM, geometric mean; LOD, limit of detection; max., maximum; ND, not detected; NR, data not reported; yr, year

(a) Water

2,4-D may occur in water as a result of direct aquatic uses; from agricultural, forestry, right-of-way, or turfland applications; through application-spray drift; or from atmospheric deposition. Concentrations of 2,4-D in water have been measured for drinking-water supplies, surface water, ground water, and for specific application catchment areas (Table 1.4). In a study of drinking-water supplies in Mexico, 2,4-D concentrations for samples above the detection limit ranged from 0.005 to 0.0038 µg/L (Félix-Cañedo et al., 2013). Detection rates for 2,4-D in surface waters varied widely, with overall concentrations ranging from ND to 14.4 µg/L, and central measures typically < 0.05 µg/L (Phillips & Bode, 2004; Konstantinou et al., 2006; Woudneh et al., 2007; Aulagnier et al., 2008; Loos et al., 2009; Loos et al., 2010b; Glozier et al., 2012; Herrero-Hernández et al., 2013; Tagert et al., 2014). 2,4-D in surface water was found to be associated with agricultural use and, in some cases, was found in urban areas probably as a result of lawn and other turf uses. In a study of ground-water samples from 23 European countries, 2,4-D was detected in only 3.7% of the samples, with a maximum value of 0.012 µg/L (Loos et al., 2010a). Under the United States Geological Survey National Water Quality Assessment Program in 1992–2001, 2,4-D was detected at a frequency of 13% in 1465 samples collected from 62 agricultural surface-water sites, and 13% in 523 samples collected from 19 urban surface-water sites, based on a detection limit of 0.08 µg/L (USGS, 2006). Concentrations at the 90th percentile were 0.11 and 0.16 µg/L, respectively. In ground-water samples in Ireland, the mean 2,4-D concentration was 0.001 µg/L (range, 0.002–0.007 µg/L) (McManus et al., 2014). In a study of surface-water inflow and outflow from a managed turf golf course, the outflow 2,4-D concentration (median, 0.85 µg/L) was significantly higher than the inflow concentration (median, 0.31 µg/L) (King & Balogh, 2010).

(b) Soil

2,4-D is likely to be found in soil in areas where it is applied as an herbicide; however, dissipation/degradation rates have been found to be relatively rapid for most soil types and conditions. While there are many studies published measuring dissipation rates in soils, no publications were found reporting on broad surveys of 2,4-D in soil relevant for assessing the potential for human exposure. One study did report 2,4-D concentrations in soil in 134 home yards in North Carolina and Ohio, USA (Morgan et al., 2008). While most measurements were below the limit of detection, the 95th percentile values were 1.8–4.6 ng/g, with maximum values ranging from 13.3 to 30.5 ng/g.

(c) Residential dust

2,4-D is frequently detected in house dust, with overall concentrations ranging up to 21 700 ng/g, and ranges of GMs, medians, or means from 47.5 to 1035 ng/g (Hartge et al., 2005; Ward et al., 2006; Morgan et al., 2008; Metayer et al., 2013; Deziel et al., 2015; Table 1.5). Surface loading values in homes after lawn applications ranged from 0.05 to 228 µg/m^2 in one study (Nishioka et al., 2001).

(d) Air

2,4-D may occur in outdoor air as a result of application-spray drift, volatilization of applied herbicides, and atmospheric suspension of 2,4-D containing soil and dust. 2,4-D in indoor air may result from tracking-in of 2,4-D in dusts and soils, and from direct intrusion of outdoor air. Concentrations of 2,4-D in outdoor air have been measured in agricultural areas, and in residential indoor and outdoor air (Table 1.6). Several studies in the Canadian prairies measured 2,4-D in outdoor air, with overall ranges of ND–2.73 ng/m^3, and with mean values ranging from 0.059 to 0.44 ng/m^3 (Waite et al., 2005; Yao et al., 2006; Aulagnier et al., 2008). The highest

Table 1.4 Concentration of 2,4-D in water

Country/year of sampling	Number of samples/setting	Results	Comments	Reference
Europe				
Greece, 1988–2000	2,4-D measurement data compiled from literature for 8 rivers	Range of minimum concentrations, ND–0.040 µg/L; range of maximum concentrations, 0.012–1.2 µg/L	2,4-D was detected at least once in 7 out of 8 rivers	Konstantinou et al. (2006)
Ireland, 2012	42 ground-water samples collected from 7 locations	2,4-D: mean, 0.001 (range, 0.002–0.007) µg/L DCP: mean, 0.001 (range, 0.001–0.004) µg/L PAC: mean, 0.456 (range, 0.015–4.15[a]) µg/L	PAC is a transformation product or impurity of 2,4-D and MCPA DCP is a transformation product of 2,4-D	McManus et al. (2014)
Spain, 2011	7 surface-water samples from Ebro river and tributaries; 32 ground-water samples from 3 areas of the La Rioja vineyard region	Rioja Alta: surface water, mean, 0.045 (range, 0.023–0.068) µg/L; ground water, mean, 0.128 (range, 0.046–0.177) µg/L Rioja Baja: surface water, mean, 0.022 (range, 0.020–0.024) µg/L; ground water, mean, 0.031 (range, 0.026–0.034) µg/L Rioja Alavesa: ground water, mean, 0.048 (range, 0.034–0.067) µg/L	2,4-D was detected in 33% of the water samples	Herrero-Hernández et al. (2013)
Europe	122 surface water samples from > 100 European rivers in 27 countries	Detection in 52% of samples; median, 0.003 µg/L; mean, 0.022 µg/L; max., 1.221 µg/L		Loos et al. (2009)
Europe, 2008	164 ground water samples from 23 European Countries	Detection in 3.7% of samples; max., 0.012 µg/L		Loos et al. (2010a)
Europe, 2007	73 Danube River and 23 tributary river surface water samples across 10 countries	Detection in 94% of Danube River samples; median, 0.01 (max., 0.055) µg/L Detection in 72% of tributary rivers; median, 0.003 (max., 0.188) µg/L		Loos et al. (2010b)
Central America				
Mexico, 2008–9	Drinking-water samples from 7 wells, 4 dams, and 15 mixing tanks for surface and ground-water sources supplying 60% of Mexico City water	In mixed water: range, 0.005–0.038 µg/L	2,4-D was found in 20% of the mixed water; 2,4-D was not detected in well and ground-water samples	Félix-Cañedo et al. (2013)

Table 1.4 (continued)

Country/year of sampling	Number of samples/setting	Results	Comments	Reference
North America				
Canada, 2003–5	Surface water collected from 2 reference, 5 agricultural, 2 urban, and 5 mixed agricultural/urban sites	Agricultural sites: range of means, 0–0.044 (overall range, 0–0.345) µg/L Urban sites: range of means, 0.005–0.020 (overall range, 0.002–0.063) µg/L Mixed agricultural/urban sites: range of means, 0.008–0.357 (overall range, 0.002–1.23) µg/L	2,4-D not detected at reference sites	Woudneh et al. (2007)
Canada, 2004	Monthly precipitation samples collected over 5 months at an agricultural site in the Yamaska River Basin, Quebec	2,4-D was detected in one (June) out of 5 monthly samples, at a concentration of 0.007 µg/L		Aulagnier et al. (2008)
Canada, 2007	National survey of 19 sites in 16 urban river watersheds across Canada, including Pacific, prairies, Ontario, Quebec, and Atlantic groupings	2,4-D detected in > 80% of prairie and urban river samples; across all urban samples; mean, 0.172 µg/L; max., > 0.8 µg/L	2,4-D concentrations increased from upstream to downstream across urban sites; highest 2,4-D concentrations were found in summer; 2,4-D concentrations were significantly 2–3 times higher after rain	Glozier et al. (2012)
USA, 2000–1	Surface-water samples from Kisco and Middle Branch of Croton Rivers	Kisco river: 64% of samples > LOD = 0.08 µg/L; 32% > 0.1 µg/L; max., 24 µg/L Middle Branch Croton River: 50% of samples > LOD; 13% > 0.1 µg/L; max., 0.39 µg/L	Highest 2,4-D concentrations measured during stormflow conditions	Phillips & Bode (2004)
USA, 1992–2001	1465 samples from 62 surface-water sites in agricultural areas, 523 samples from 19 surface-water sites in urban areas	Detection frequency of 13% in water from agricultural areas, and 13% in water from urban areas Concentrations at 90th percentile: 0.11 µg/L in water from agricultural areas; and 0.16 µg/L in water from urban areas	Based on LOD of 0.08 µg/L in the USGS National Water Quality Assessment Program	USGS (2006)
USA, 2003–8	Surface water inflow and outflow from a managed turf golf course	Inflow: median, 0.31 µg/L Outflow: median, 0.85 (max., 67.1) µg/L	Outflow concentration was significantly higher than inflow	King & Balogh (2010)
USA, 2002–3	Surface-water samples collected from 7 sites in the upper Pearl River basin	Median, 0.17 (range, 0.10–14.4) µg/L		Tagert et al. (2014)

[a] Extrapolated concentration

2,4-D, 2,4-dichlorophenoxyacetic acid; DCP, 2,4-dichlorophenol; LOD, limit of detection; max., maximum; MCPA, 4-chloro-2-methylphenoxy acetic acid; ND, not detected; PAC, phenoxyacetic acid

Table 1.5 Concentrations of 2,4-D in residential dust

Country/year of sampling	Number of samples/setting	Results	Comments	Reference
USA, 2001–7	House dust collected from 277 homes of children with leukaemia, and 306 control homes	2,4-D was detected in 98% of homes; median, 102 ng/g; 75th percentile, 419 ng/g		Deziel et al., 2015
USA, 2001–7	House dust collected in 333 control homes in a case–control study	2,4-D was detected in > 92% of homes; mean, 831 ± 6041 ng/g		Metayer et al. (2013)
USA, 1998–2000	House dust from 112 home subset of NHL case–control study	2,4-D detected in 95% of homes; GM, 1035 ng/g	Total crop acreage within 750 m of home was significantly associated with increased 2,4-D concentration	Ward et al. (2006)
USA, 2000–6	House dust from 66 homes in NC and 62 homes in OH	OH: median, 156 (range, < LOD–21 700) ng/g NC: median, 47.5 (range, < LOD–7390) ng/g		Morgan et al. (2008)
USA, 1998–2000	House dust from 510 control homes in a NHL case–control study	For control homes: 110 homes < LOD; 161 homes < 500; 59 homes, 500–599; 162 homes, 1000–9999; and 18 homes, > 10 000 ng/g		Hartge et al. (2005)
USA, NR	House indoor-air and surface-wipe and vacuum samples collected at 11 occupied and 2 unoccupied homes during week before application and week after application of 2,4-D	Mean 2,4-D concentrations on particles in air ranged from approx. 1 to 10 ng/m^3, with differences between particle size and collection period; 2,4-D surface loadings ranged from 0.05 to 228 µg/m^2 for carpets, with lower values for bare floors, tables, and window sills	Exposures to young children were estimated to be: median, 1.37 (max., 1.94) µg/day pre-application and 2.42 (max., 8.87) µg/day post-application; track-in factors were important	Nishioka et al. (2001)

2,4-D, 2,4-dichlorophenoxyacetic acid; GM, geometric mean; LOD, limit of detection; OH, Ohio; NC, North Carolina; ND, not detected; NHL, non-Hodgkin lymphoma

Table 1.6 Concentrations of 2,4-D in air

Country/year of sampling	Number of samples/setting	Results	Comments	Reference
Canada, 2004	Weekly and monthly outdoor air samples over 5 months at an agricultural site in the Yamaska River basin	Detection in 38% of the May–June weekly air samples; mean, 0.44 (range, < LOD–1.31) ng/m^3	2,4-D not detected in any of the monthly air samples collected July–September	Aulagnier et al. (2008)
Canada, 2003	Weekly outdoor air samples collected at 8 sites in agricultural and receptor regions over 1 or 3 months	At three prairie sites, means ranged from 0.059 to 0.331 (overall range, ND–1.46) ng/m^3	Highest 2,4-D concentrations were found during typical weeks of application	Yao et al. (2006)
Canada, 2002	6 weekly outdoor air samples at 4 sites on a 500-km north–south transect in Saskatchewan	Across all sites: mean, 0.35 ng/m^3, median, 0.15 ng/m^3; max., 2.73 ng/m^3	Highest 2,4-D concentrations were found during typical weeks of application	Waite et al. (2005)
France, 2001	4 outdoor air samples collected at an urban site, and 5 air samples collected at a rural site	Urban site: range, ND–11 ng/m^3 Rural site: range, ND–37 ng/m^3	Concentrations in gas plus particle phase reported	Baraud et al. (2003)
Netherlands, 2000–1	18 sites nationwide, air and precipitation samples collected once during each 4-week period for 2 yrs; weekly samples collected at three sites	2,4-D was not detected in any air samples Detection in 9% and 31% of precipitation samples in 2000 and 2001 respectively, with means of 0.8 and 1.9 ng/L	Deposition amounts to soil and surface waters were estimated	Duyzer & Vonk (2003)
USA, 2000–1	Home indoor and outdoor air at 66 homes in North Carolina and 67 homes in Ohio	North Carolina, indoor air: 75th percentile, 0.8 ng/m^3; max., 3.7 ng/m^3 Ohio, indoor air: 75th percentile, 0.8 ng/m^3; max., 2.0 ng/m^3 North Carolina, outdoor air: 75th percentile, < LOD; max., 1.7 ng/m^3 Ohio, outdoor air: 75th percentile, 0.3 ng/m^3; max., 3.2 ng/m^3		Morgan et al. (2008)
USA, year NR	Home indoor air collected at 11 occupied and 2 unoccupied homes during pre-application and post-application week	Mean indoor 2,4-D concentrations associated with $PM_{2.5}$ particles ranged from approximately 0.7–3.9 ng/m^3 during application and 0.8–1.5 ng/m^3 3 days after application; mean indoor 2,4-D concentrations associated with PM_{10} particles ranged from approximately 4.2–9.6 ng/m^3 during application, and 1.8–3.4 ng/m^3 3 days after application	Homeowner and pet track-in were significant factors for intrusion; resuspension of floor dust was the major source of 2,4-D in air	Nishioka et al. (2001)

2,4-D, 2,4-dichlorophenoxyacetic acid; LOD, limit of detection; max., maximum; ND, not detected; NR, not reported; $PM_{12.5}$, particulate matter with a diameter of ≤ 2.5 μm; PM_{10}, particulate matter with a diameter of ≤ 10 μm concentrations were observed during weeks when 2,4-D was typically applied. In France, outdoor air concentrations ranged from ND to 11 ng/m³ in an urban location, and ND to 37 ng/m³ in a rural location (Baraud et al., 2003). In a 2-year nationwide monitoring campaign in the Netherlands, 2,4-D was not detected in air, but was detected in precipitation, with mean concentrations of 0.8 and 1.9 ng/L in 2000 and 2001, respectively (Duyzer & Vonk, 2003). In two states in the USA, 75th percentile concentrations of 2,4-D in indoor residential air were each 0.8 ng/m³, with maximum concentrations ranging from 2.0 to 3.7 ng/m³ (Morgan et al., 2008). In the same study, outdoor residential air concentrations of 2,4-D at the 75th percentile ranged from ND to 0.3 ng/m³, with maximum values ranging from 1.7 to 3.2 ng/m³. In other homes with lawn-turf applications, 2,4-D concentrations associated with indoor particulate matter of aerodynamic diameter < 2.5 μm ($PM_{2.5}$) ranged from approximately 0.7 to 3.9 ng/m³ during application, and 0.8 to 1.5 ng/m³ 3 days after application. Mean indoor 2,4-D concentrations associated with particulate matter of aerodynamic diameter < 10 μm (PM_{10}) ranged from approximately 4.2–9.6 ng/m³ during application, and 1.8–3.4 ng/m³ at 3 days after application (Nishioka et al., 2001).

(e) Food

2,4-D residues may be found in some food commodities as a result of use of 2,4-D in agriculture. In 2015, the European Food Safety Authority (EFSA) reported the results of the control activities related to pesticide residues in food carried out in 2013 in the European Union member states, Norway and Iceland (EFSA, 2015). As part of this monitoring programme, 2,4-D was analysed in 2756 food samples and found to be above the LOQ for a single result; the measured concentration of 2,4-D in one lettuce sample was 0.075 mg/kg, and thus higher than the maximum residue level (MRL) of 0.05 mg/kg. EFSA reported results from residue trials for several plant commodities, with all results being less than the proposed MRLs (EFSA, 2011).

Duplicate diet studies provide the most relevant information regarding potential human exposures, accounting for the wide variety of combined foods consumed, and also processing and cooking factors. However, only one duplicate diet study measuring 2,4-D was available (Morgan et al., 2008; Table 1.7). In this study, 2,4-D was not detected in more than half of the duplicate diet samples for children and adult caregivers. The 75th percentile levels for solid foods ranged from 0.4 to 0.9 ng/g, with maximum levels ranging from 3.7 to 20.2 ng/g. The maximum levels for liquid foods/beverages ranged from 0.2 to 0.8 ng/g.

(f) Biological markers

Several studies have examined exposures to 2,4-D in the general population through measurement of 2,4-D in the urine (Table 1.8). The National Health and Nutrition Examination Survey (NHANES) in the USA provides representative population biomonitoring data for 2,4-D for several age groups, with GM values ranging from 0.288 to 0.385 μg/L, and 95th percentile values ranging from 1.12 to 2.08 μg/L (CDC, 2015). A study of children in Thailand found overall concentrations of 2,4-D in the urine ranging from 0.09 to 1.87 μg/g creatinine, and GMs for different groups ranging from 0.17 to 0.21 μg/g creatinine (Panuwet et al., 2009). In a study of children in Puerto Rico, the frequency at which 2,4-D in the urine was found to be greater than the LOD was 11.8%, and the maximum value was 0.9 μg/L (Lewis et al., 2014). For children in two states in the USA, the overall 2,4-D concentration in urine was < LOD – 12.5 μg/L, with median values of 1.2 and 0.5 μg/L in the two states (Morgan et al., 2008). For homeowner lawn/garden applicators and bystanders living in the home, the amount of 2,4-D excreted after

Table 1.7 Concentrations of 2,4-D in food

Country/year of sampling	Number of samples/setting	Results	Comments	Reference
Europe, 2013	2756 food samples analysed for 2,4-D	Only one food (lettuce) had a 2,4-D concentration of > LOQ (the concentration was 0.075 mg/kg)		EFSA (2015)
Europe, NR	Residue trial results for 13 plant commodity types	Median and maximum residue values of less than the proposed MRL of 0.05 mg/kg for food commodities, and greater than proposed MRLs ranging from 0.05 to 50 mg/kg for grass, straw, and maize forage commodities		EFSA (2011)
USA, 2000–1	Solid-food duplicate-diet samples collected from children and adult caregivers at 66 homes in North Carolina, and 69 homes in Ohio	North Carolina, child: 75th percentile, 0.9 ng/g; max., 4.4 ng/g Ohio, child: 75th percentile, 0.4 ng/g; max., 20.2 ng/g North Carolina, adult: 75th percentile, 0.9 ng/g; max., 4.0 ng/g Ohio, adult: 75th percentile, 0.6 ng/g; max., 3.7 ng/g	2,4-D detected in < 10% of liquid-food duplicate-diet samples with maximum values ranging from 0.2 to 0.8 ng/g	Morgan et al. (2008)

2,4-D, 2,4-dichlorophenoxyacetic acid; LOD, limit of detection; LOQ, limit of quantification; max., maximum; MRL, maximum residue limit; ND, not detected

application ranged from ND to 7.1 μg/kg bw (Harris et al., 1992).

1.4.4 Exposure assessment to 2,4-D in epidemiological studies

The key epidemiological studies evaluated for 2,4-D in this *Monograph* can be categorized as occupational studies of applicators in agriculture settings and farmworkers and one study in the general population. For classification of exposure to 2,4-D, cohort and case–control studies of farmers relied on questionnaire-based approaches to collect information regarding pesticide use, work practices, and other important agricultural and lifestyle factors such as smoking. Two studies re-analysed several of the case–control studies, applying techniques to consider the use of multiple pesticides by individuals working in agriculture. A study of farmworkers combined extant geographically based data on pesticide application for specific crops, locations, and dates with union-based records of work history and location information for individuals. The study of the general population examined exposure–outcome relationships using two approaches: (a) questionnaire-based collection of information about residential use of herbicide; and (b) measurement of 2,4-D in house dust as a surrogate indicator of exposure.

There were no epidemiological studies on cancer that relied on measurement of 2,4-D biomarkers for exposure categorization. For 2,4-D, however, the elimination half-life after exposure in humans is short, with ranges of 10–28 hours after an oral dose (Sauerhoff et al., 1977) and 18–68 hours after a dermal dose of 2,4-D, and 18–87 hours for 2,4-D dimethyl amine (Harris & Solomon, 1992). Thus, it is usually impractical to collect adequate numbers of samples to represent long-term exposures in epidemiological cohorts, or to implement collection for large numbers of participants at key time-points such as after herbicide applications. Biomarker measurement therefore has not been the primary means of classifying 2,4-D exposure for research in cancer

Table 1.8 Exposure to 2,4-D in the general population

Country/year of sampling	Subjects/setting	No. of subjects	Age (yrs)	Medium	Results	Comments	Reference
USA, 2009–10	NHANES general population biomonitoring survey	386	6–11	Urine (µg/L)	GM: 0.385; and 95th percentile, 1.59	Representative population sample for USA; data also available for earlier time periods	CDC (2015)
		401	12–19		0.301 and 1.12		
		1309	20–59		0.288 and 1.33		
		651	60–older		0.349 and 2.08		
USA, 2000–1	Children and their adult caregivers in North Carolina and Ohio	66 (North Carolina)	2–5	Urine (µg/L)	Median, 0.5 (range, < LOD–3.3)	Indoor air, outdoor air, house dust, soil, hand wipe, and food data also available from this study	Morgan et al. (2008)
		66 (North Carolina)	Adults		0.7 (< LOD–5.1)		
		69 (Ohio)	2–5		1.2 (< LOD–12.5)		
		69 (Ohio)	Adults		0.7 (< LOD–8.1)		
USA, NR	Homeowner lawn/garden applicators and bystanders living in home	24 (applicators)	NR	Urine (total amount 2,4-D secreted) (µg/kg bw)	Range, < LOD–7.1	Samples collected for 96 h after application	Harris et al. (1992)
		24 (bystanders)	NR		No measurable levels		
Puerto Rico, 2010–12	Pregnant women	152	18–40	Urine (µg/L)	> LOD in 11.8% of samples; 95th percentile, 0.6 Max. value, 0.9	Spot urine samples at approx. 20, 24, and 28 weeks of gestation ORs for 2,4-D detection were significantly elevated for consumption of collard greens and spinach in previous 48 h	Lewis et al. (2014)
Thailand, NR	Children from parents with different occupations:			Urine (µg/g creatinine)		Urine first morning void samples collected No significant differences in urine 2,4-D for agricultural vs non-agricultural families, or boys vs girls	Panuwet et al. (2009)
	Farmer	60	12–13		GM, 0.21 (range, 0.13–1.08)		
	Merchant and trader	39	12–13		0.21 (0.09–0.43)		
	Government and company employee	52	12–13		0.17 (0.10–0.38)		
	Labourer	56	12–13		0.19 (0.12–1.87)		

2,4-D, 2,4-dichlorophenoxyacetic acid; GM, geometric mean; LOD, limit of detection; max., maximum; ND, not detected; NR, data not reported; OR, odds ratio; vs, versus; yr, year

epidemiology. A recent review summarized relevant studies of 2,4-D biomarker measurement, and separately summarized exposure metrics in more recent epidemiological investigations of 2,4-D and cancer (Burns & Swaen, 2012).

This section provides an assessment of the strengths and weaknesses of the exposure assessment and assignment methods used in the key epidemiological studies that were evaluated by the Working Group. The detailed discussions of limitations in exposure assessment in the epidemiological investigations described here should not be construed to suggest that these studies are inferior to others in the literature. In fact, in many ways the studies described here have improved on pesticide exposure assessment in this discipline when compared with many studies that relied on less specific exposure-classification categories, such as "farmer" or any non-specific pesticide use.

(a) Studies of occupational exposure

Several studies were based on reported application of 2,4-D in agricultural settings. Exposure assessment in these studies relied on questionnaire-based approaches for reporting use of 2,4-D, together with reporting of factors potentially affecting exposures. Two cohort studies included farm applicators who reported their lifetime use of 2,4-D (Alavanja et al., 2004; Koutros et al., 2013). Three case–control studies included participants who reported using 2,4-D (Brown et al., 1990; Zahm et al., 1990; McDuffie et al., 2001). Three studies performed additional analyses of data from case–control studies to examine joint exposures to multiple pesticides and pesticide classes (Cantor et al., 1992; De Roos et al., 2003; Hohenadel et al., 2011). One study performed additional analyses of data from case–control studies to examine whether reported use of the insect repellent *N,N*-diethyl-*meta*-toluamide (DEET) might result in increased penetration of 2,4-D through gloves, and thus increase exposure to 2,4-D (McDuffie et al., 2005). Another case–control study examined lympho-haematopoietic cancer in farmworkers, with proxy data on pesticide-use locations and amounts combined with work-location history (Mills et al., 2005).

The two cohort-based studies were from the AHS, which collected detailed information on pesticide use and factors potentially affecting exposure (e.g. spraying techniques, personal protective equipment, etc.). Based on the detailed use information, a semi-quantitative exposure-assessment method was developed in which estimated intensity was combined with years and annual frequency of use (Dosemeci et al., 2002). The intensity score was based on development of an a-priori exposure-intensity algorithm. Several validity evaluations of the exposure assessment process have been carried out. These included: (i) assessment of the reliability of reporting agricultural factors by requiring completion of the enrolment questionnaires twice, approximately 1 year apart; (ii) confirmatory checks correlating the years in which a pesticide was reportedly used with dates of registered use of that particular pesticide; and (iii) comparison of the exposure algorithm with external exposure data. Agreement between reporting of ever/never use of specific pesticides and application practices was high, and generally ranged from 70% to > 90%. Agreement was lower (typically 50–60%) for duration or frequency of use of specific pesticides (Blair et al., 2000). The confirmatory checks on reported usage of specific pesticides established that the majority of respondents provided plausible responses for both decade of first use and total duration of use (Hoppin et al. 2002). The exposure-intensity algorithm was evaluated using measurements of 2,4-D urinary biomarkers in the AHS cohort and in two other studies (Coble et al., 2005; Acquavella et al., 2006; Thomas et al., 2010b). When combined, these studies showed that the AHS algorithm had the capacity to separate the upper tertiles of exposure intensity from the lower.

[The Working Group noted that the AHS had collected very detailed information on pesticide use and practices, and validation studies showed these data to be appropriate for estimating historical exposure to pesticides. In addition, information on exposure was collected before disease diagnosis, eliminating the potential for case-recall bias. It is however important to note that the validity studies were based on information reported at the time the exposure surveys were completed, and would not necessarily reflect the recall of information which had been reported to be good in regard to some but not all aspects, in particular for data regarding frequency and duration of use. The exposure assessment of 2,4-D in the AHS was based on the baseline questionnaire (1993–1997) that relied on recall for recent and previous pesticide-use information. One of the studies also used data from the follow-up questionnaire (1999–2003) that used recent recall to update pesticide-use information. Although use of 2,4-D may be more readily recalled than use of other less widely recognized pesticides, the validity of recalled information is nonetheless unknown. Furthermore, application procedures and other factors in the 1950s–1980s may not necessarily be similar to the parameters addressed in the AHS exposure algorithm. Moreover, it is likely that a certain proportion of exposure of farmers is attributable to non-application circumstances, such as re-entry, and contaminated work and home environments. In this regard, the publication by Bakke et al. (2009) based on corn farmers in Iowa in the AHS is illustrative. In that study, biological samples taken on non-application days revealed elevated levels of 2,4-D among farmers compared with non-farmer controls during the whole year. These levels, although an order of magnitude lower than those recorded when the farmers were applying 2,4-D, still indicated a notable contribution of non-application days to cumulative exposure over the year. This consideration is relevant when accounting for the fact that active application by most farmers amounted to only a few days during the growing season.]

The case–control studies used questionnaire-based approaches to obtain information on participant use of pesticides in agricultural settings, including specific pesticides used, days per year of use, and in one case, year of first use. Analyses were based on ever/never use of specific pesticides, with further analyses using categories of days per year of use for specific pesticides.

[The case–control studies had several strengths, including collection of information on specific pesticides including 2,4-D, using days per year of use to stratify users into categories with higher and lower exposure to 2,4-D, and collection of information from agricultural users likely to be able to accurately identify specific pesticides such as 2,4-D. Exposure assessment in the case–control studies also had several potential limitations. Exposure data were collected after case diagnosis, leading to potential case-recall bias, and general recall bias for both cases and controls is possible. The studies did not collect total years for use of 2,4-D, thus the lifetime days of use metric could not be used to develop a better exposure categorization for distinguishing participants with higher and lower lifetime exposures. Pesticide-use practice information was not collected or used to account for likely differences in exposure intensity. No validation assessments using exposure measurements were performed, nor was questionnaire-response reproducibility examined.]

The farmworker study used available information on the amount of each pesticide used on specific crops at every location in the state of California, USA. The information on pesticide location and amount was used in combination at the county level, with farmworker-union payroll-data records that provided work location information on a monthly level to stratify exposures among the farmworkers over their entire work history (Mills et al., 2005).

[The strengths of this research included the availability of high-quality and very specific records of pesticides and their amount of use for all locations and times across several decades. Use of payroll records that included information of work location removed the possibility of recall bias for workers trying to remember where and when they had worked over the many years of their career. The exposure assessment did, however, have considerable limitations. It was not possible to determine a direct match between participant work location and date with pesticide application and date. It was not possible to ascertain that participants worked in fields where 2,4-D had actually been applied, since the pesticide-use data and work-location data were linked only at a county level. There was no information on re-entry intervals between application and work dates. 2,4-D would not have been applied directly to the crops (it was most likely used as a pre-emergent herbicide) and probably not at any time that crops were present, so the farmworkers were unlikely to have had any exposure during crop picking or exposure to foliar residues. There was no evidence as to whether 2,4-D residues were present and accessible in the soil. There was no information on specific work tasks for specific locations, so the amount of soil contact could not be distinguished within and between individuals over time. There were no measurement studies to demonstrate that the exposure categorization was able to distinguish different exposure intensities between individuals. There was no information to judge whether and to what extent exposure to 2,4-D may have occurred away from the work locations.]

(b) *Studies of the general population*

One case–control study assessed the association between residential use of 2,4-D and non-Hodgkin lymphoma (NHL) in the general population (Hartge et al., 2005). This study used two different approaches for characterizing and classifying exposure to 2,4-D. One approach was based on a questionnaire that collected participant-recalled information on pesticide use in and around each home occupied for more than 2 years since 1970. The categories of use included lawn treatments by the participant, professional applicator, or a third person. The second approach was based on analysis of 2,4-D in house dust in the homes of cases and controls.

[The primary strength of this study was the use of two different approaches for exposure classification, firstly a computer-assisted personal-interview approach, and secondly a measurement-based approach. There were also important limitations in the exposure assessment. There was potential for recall bias among cases and controls, given the difficulty in recalling and accurately reporting lawn-herbicide uses for multiple homes over long periods of time. It was not possible to ascertain that 2,4-D was the active ingredient in the herbicides used for lawn treatments. 2,4-D in house dust may not be a good surrogate for differences in residential occupant exposures at an individual level. In a different study in the same country, there was no significant correlation between 2,4-D concentrations in house dust and in adult urine (Morgan et al., 2008).]

1.5 Regulation

The WHO Guidelines for Drinking-water Quality assign a value of 0.03 mg/L for 2,4-D (WHO, 2011). The guideline value applies to 2,4-D, since salts and esters of 2,4-D are rapidly hydrolysed to the free acid in the water (WHO, 2003, 2011).

This value was determined using an acceptable daily intake (ADI) of 0.01 mg/kg bw for the sum of 2,4-D and its salts and esters, expressed as 2,4-D, on the basis of a no-observed adverse effect level (NOAEL) of 1 mg/kg bw per day in a 1-year study of toxicity in dogs (for a variety of effects, including histopathological lesions in kidneys and liver) and a 2-year study of toxicity and carcinogenicity in rats (for renal lesions) (JMPR, 1996), and assuming a body weight of 60 kg, drinking-water consumption of 2 L/day and a 10% allocation to drinking-water (WHO, 2003, 2011).

Information summarizing regulatory controls and related statutory measures addressing the use of 2,4-D in many regions, particularly countries of Asia, Africa, and South America, was not available to the Working Group.

2,4-D has been registered in the USA since 1948. It was the subject of a registration standard dated 16 February 1988, and a registration standard guidance document dated 1 September 1988. Through the EPA, 2,4-D is registered for use on a variety of food/feed sites, including field, fruit, and vegetable crops, and for use on turf, lawns, rights-of-way, aquatic sites, and forestry applications. 2,4-D is registered for use in terrestrial and aquatic environments (EPA, 2005).

In 2013, the Pest Management Regulatory Agency in Canada updated the re-evaluation notice for 2,4-D, stating that the 2,4-D registrants had provided the required data, which were deemed acceptable (Pest Management Regulatory Agency Canada, 2013).

The European Commission has approved products containing 2,4-D that fulfil safety requirements laid down in Article 5(1)(a) and (b) of Directive 91/414/EEC on the basis of 2,4-D being applied to winter and spring cereals, pasture, green manuring crops, grass-seeds, fallow land, borders of arable land and pastures, and 2,4-D ethylhexyl ether being applied to winter cereals (European Commission, 1991). Extension of use patterns beyond those specified requires evaluation at the member-state level. The European Commission has specified that the theoretical maximum daily intake (TMDI) for an adult of body weight 60 kg is 1.4% of the ADI of 0.05 mg/kg bw per day, based on the European Diet established by the Food and Agriculture Organization of the United Nations and the World Health Organization (FAO/WHO European Diet) (European Commission, 2001).

Concerning water contamination with 2,4-D, the EPA has set an enforceable regulation, called a "maximum contaminant level", at 0.07 mg/L (70 µg/L or ppb) (EPA, 2005).

MRLs for 2,4-D were first set in the European Union in 2002, and re-evaluated by EFSA in 2011. MRLs are specified at 1 mg/kg or less by the European Commission for most of 378 products containing 2,4-D (EFSA, 2011).

In relation to occupational exposure limits, the International Labour Organization (ILO) in collaboration with United States National Institute for Occupational Safety and Health (NIOSH) specified a threshold limit value for 2,4-D of 10 mg/m^3 as a time-weighted average (ILO, 1999). The Occupational Safety and Health Administration in the USA has adopted the following permissible exposure limits for 2,4-D: general industry, 10 mg/m^3; and construction industry, 10 mg/m^3, time-weighted average (OSHA, 2015).

Adequate margins of exposure together with details concerning appropriate personal protective equipment for workers are available through the EPA (EPA, 2005).

In various jurisdictions, statutory authorities exercise decision-making in relation to the marketing and use of various categories of consumer products. A comprehensive listing of such authorities and their respective responsibilities is beyond the scope of this *Monograph*, but examples of such authorities are given. The EPA has registered a herbicide formulation, the active ingredients of which are 2,4-D and glyphosate, to manage "resistant weeds". The EPA has also put in place restrictions to avoid pesticide drift, including a 30-foot in-field "no spray" buffer zone around the application area, no pesticide application when the wind speed is > 15 mph, and only ground applications are permitted (EPA, 2014).

2. Cancer in humans

2.1 Cohort studies

See Table 2.1

This section reviews studies that have specifically assessed exposure to 2,4-D. Studies were excluded from this review if 2,4-D was one of several chemicals or pesticides evaluated, but no specific risk estimate for 2,4-D only was provided (e.g. Kogevinas et al., 1997; Saracci et al., 1991; Boers et al., 2010; Hooiveld et al., 1998; Bueno de Mesquita et al., 1993; Coggon et al., 2015; Lynge, 1985; Lynge, 1998; Becher et al. 1996; 't Mannetje et al., 2005). [The Working Group considered that studies addressing the possible toxic effects of agent orange and related formulations could not be used to indicate the impact of 2,4-D specifically, and such studies are not addressed in the present *Monograph*.] The publications that were informative for assessing the carcinogenicity of 2,4-D are described below.

2.1.1 The International Register of Workers Exposed to Phenoxy Herbicides and Their Contaminants

The International Register of Workers Exposed to Phenoxy Herbicides and Their Contaminants was established in 1980 by IARC in association with the United States National Institute of Environmental Health Sciences to study cancer outcomes associated with these exposures (Saracci et al., 1991; Kogevinas et al., 1997). The register included 21 863 male and female workers in 36 cohorts, in 12 countries, in Europe, Australia, New Zealand, Canada, and the USA. Exposures were classified using individual job records, company-exposure questionnaires developed specifically for this study, and, in some cohorts, measurements of 2,3,7,8-tetrachlorodibenzo-*p*-dioxin (TCDD) and other dioxin and furan congeners in serum and adipose tissue and in the workplace.

Using the IARC International Register, Kogevinas et al. (1995, 1997) conducted nested case–control studes of the incidence of NHL (32 cases) and soft tissue sarcoma (11 cases). Cancer cases were identified by linkage of cohort rosters to death certificates and to cancer registries in each of the countries with cancer-registration systems. Cases were each matched to five controls by age, sex, and country of residence at the time of employment. A panel of three industrial hygienists reviewed company-exposure questionnaires and company records to assess exposure to 21 chemicals or mixtures. Ever exposure to 2,4-D/DP/DB (2,4-D and its precursor compounds, 2,4-dichlorophenoxypropionic acid and 2,4-dichlorophenoxybutyric acid), lagged by 5 years, was associated with an increased risk of soft tissue sarcoma (odds ratio, OR, 5.72; 95% CI, 1.14–28.65) and 1.4 (95% CI, 0.10–15.80) when adjusted for 2,4,5-T and MCPA. In analyses comparing different levels of cumulative exposure to non-exposure, the greatest increase in risk of soft tissue sarcoma in relation to 2,4-D was found for the category of highest exposure (OR, 13.71; 95% CI, 0.90–309.00), with a significant trend across exposure categories ($P = 0.01$). For NHL, the odds ratio was 1.11 (95% CI, 0.46–2.65). After adjustment for 2,4,5-T and MCPA, the odds ratio for NHL was 1.05 (95% CI, 0.26–4.28). No trend was observed for risk of NHL and level of 2,4-D exposure.

2.1.2 United Farm Workers of America

Mills et al. (2005) conducted a case–control study of lympho-haematopoietic cancers (leukaemia, 51 cases; NHL, 60 cases; multiple myeloma, 20 cases) in Hispanic farmworkers, which was nested within a cohort of members of the United Farm Workers of America labour union in California, USA. Incident cases diagnosed between 1988 and 2001 were identified from linkage of union records to the California cancer registry, and for each case, five controls were selected from

Table 2.1 Cohort studies of cancer and exposure to 2,4-D

Reference, location, enrolment/ follow-up period, study design	Population size, description, exposure assessment method	Organ site	Exposure category or level	Exposed cases/ deaths	Risk estimate (95% CI)	Covariates controlled	Comments
Kogevinas et al. (1995) Europe, Australia, New Zealand, Canada, and USA 1939–1992 Nested case–control study	Cases: 11 cases of STS (all men) and 32 cases of NHL (one woman) in a cohort identified from the International Register of Workers Exposed to Phenoxy Herbicides and Their Contaminants) (established jointly with IARC), or national cancer registration records Controls: 55 for STS, 158 for NHL; incidence-density sampling from the cohort; 5 controls per case were matched for age, sex, country of residence at the time of employment Exposure assessment method: company records; exposure information from each plant was retrieved using a questionnaire completed by factory personnel in the presence of an industrial hygienist; mortality records varied between the countries and included records from the company, union, insurance and, physicians, workers and families, and municipal and national vital records	NHL	2,4-D/DP/DB			None	No cases of STS or NHL in the Italian, Canadian, or Austrian cohorts. Risk of STS and of NHL increased with time since first exposure. In workers exposed to phenoxy herbicides with minimal or no contamination from TCDD, mortality was similar to that expected based on national rates Strengths: the study was large and included production workers and herbicide sprayers; the exposure assessment was able to classify workers by ordinal level of exposure to chemicals including 2,4-D; the study examined cancer incidence Limitations: information concerning lifestyle characteristics associated with cancer risk, such as smoking, was not available; there was no quantitative information on exposure
			Ever exposed	12	1.11 (0.46–2.65)		
			Low	4	0.73 (0.22–2.43)		
			Medium	6	2.14 (0.73–6.23)		
			High	2	0.69 (0.11–4.55)		
		NHL	2,4-D/DP/DB			2,4,5-T, MCPA	
			Ever exposed	12	1.05 (0.26–4.28)		
		STS	2.4-D/DP/DB			None	
			Ever exposed	9	5.72 (1.14–28.65)		
			Low	4	4.55 (0.61–53.41)		
			Medium	2	6.13 (0.33–129.70)		
			High	3	13.71 (0.90–309.00)		
			Trend-test *P* value: 0.01				
		STS	2.4-D/DP/DB			2,4,5-T, MCPA	
			Ever exposed	12	1.40 (0.10–15.80)		

Table 2.1 (continued)

Reference, location, enrolment/ follow-up period, study design	Population size, description, exposure assessment method	Organ site	Exposure category or level	Exposed cases/ deaths	Risk estimate (95% CI)	Covariates controlled	Comments
Mills et al. (2005) California USA 1988–1997 Nested case–control study	Cases: 131 cases (response rate, NR) (leukaemia, 51; NHL, 60; multiple myeloma, 20) were identified by linking the cohort to the California cancer registry for 1988–2001 Controls: 655 (response rate, NR); 5 controls for each case were drawn from the cohort, not diagnosed with any cancer, and matched on sex, Hispanic ethnicity, and ± 1 yr of birth Exposure assessment method: crop and pesticide exposures were estimated by linking county/month and crop-specific job-history information from union records with California Department of Pesticide Regulation pesticide-use reports during the 20 yr before cancer diagnosis; "high exposure" can be interpreted as having worked in an area with high use	Leukaemia, total	High (vs low)	NR	1.03 (0.41–2.61)	Age, duration of union affiliation, date of first union affiliation, sex (when appropriate)	United Farm Workers of America Strengths: the study was conducted among farm workers (as opposed to pesticide applicators); included women; used objective exposure-assessment methods not prone to recall bias Limitations: small number of cases; number of cases and controls exposed was not reported; exposure assessment was based on regional pesticide-use data and did not take into account personal use of or exposure to pesticides
			High (men)	NR	0.55 (0.15–2.06)		
			High (women)	NR	3.73 (0.77–18.00)		
		NHL, total	High (vs low)	NR	3.80 (1.85–7.81)		
			High (men)	NR	3.79 (1.58–9.11)		
			High (women)	NR	5.23 (1.30–20.90)		
		Leukaemia, lymphocytic	High (vs low)	NR	1.47 (0.33–6.59)		
		Leukaemia, granulocytic	High (vs low)	NR	1.28 (0.30–5.42)		
		NHL, nodal	High (vs low)	NR	2.29 (0.90–5.82)		
		NHL, extranodal	High (vs low)	NR	9.73 (2.68–35.30)		

Table 2.1 (continued)

Reference, location, enrolment/ follow-up period, study design	Population size, description, exposure assessment method	Organ site	Exposure category or level	Exposed cases/ deaths	Risk estimate (95% CI)	Covariates controlled	Comments
Koutros et al. (2013) Iowa and North Carolina, USA Enrolment, 1993–1997; follow-up to 31 December 2007	54 412 male pesticide applicators; among the 57 310 applicators, 2898 were excluded (1563 women, 1071 prevalent cases of other types of cancer, 264 with no follow-up information) Exposure assessment method: questionnaire; see other details elsewhere; incident prostate cancers were identified from enrolment (1993–1997) until 31 December 2007 (n = 1962)	Prostate: (no family history)	Unexposed to 2,4-D (ref.)	290	1.00	Age, state, smoking, fruit servings, leisure time physical activity in winter, race	Agricultural Health Study Strengths: large cohort study in an agricultural population, thus high exposure prevalence; good exposure assessment
			IWED Q1	262	0.93 (0.78–1.11)		
			IWED Q2	256	0.88 (0.74–1.05)		
			IWED Q3	287	1.03 (0.86–1.22)		
			IWED Q4	260	0.87 (0.73–1.03)		
			Trend-test P value: 0.25				
		Prostate (with family history)	Unexposed to 2,4-D (ref.)	43	1.00	Age, state, smoking, fruit servings, leisure time physical activity in winter, race	
			IWED Q1	60	1.21 (0.8–1.82)		
			IWED Q2	68	1.29 (0.85–1.95)		
			IWED Q3	51	0.86 (0.56–1.31)		
			IWED Q4	73	1.17 (0.78–1.75)		
			Trend-test P value: 0.9				
Flower et al. (2004) Iowa and North Carolina, USA 1993–1997 for enrolment; 1975–1998 for follow-up	17 357 children (aged < 19 yr) of licensed pesticide applicators in Iowa Exposure assessment method: questionnaire; state cancer registries	Childhood cancer	Maternal use of 2,4-D (ever)	7	0.72 (0.32–1.60)	Child's age at enrolment	Agricultural Health Study There was a small number of cases in North Carolina, so these were excluded from all subsequent analyses Strengths: large cohort; assessment of specific chemicals, including 2,4-D Limitations: based on self-reported exposure; potential exposure to multiple pesticides; limited power to study a rare disease such as childhood cancer
			Paternal use of 2,4-D (prenatal)	26	1.29 (0.71–2.35)		

Table 2.1 (continued)

Reference, location, enrolment/follow-up period, study design	Population size, description, exposure assessment method	Organ site	Exposure category or level	Exposed cases/deaths	Risk estimate (95% CI)	Covariates controlled	Comments
Burns et al. (2011) Michigan, USA Enrolment, 1945–1994; follow-up until 2007	1316 men employed by Dow Chemical Company who had ever worked full time for at least 3 days between 1945–1994 in any of four buildings involved in 2,4-D manufacturing, located in Midland, MI, and were alive on 1 January 1985 (corresponding to the initiation of the Michigan state-wide cancer registry) Exposure assessment method: analyses by duration and exposure level based on JEM (from monitoring data) linked to work histories to assign a relative exposure index of 0.005, 0.05, 0.5, or 5 units to each job in in 2,4-D operations; cancer registry (SEER) cancer incidence rates for Michigan white men	All cancers combined	Cohort 3	211	0.96 (0.84–1.10)	Age	Cancer incidence was described only for Michigan. Cohort 3 was limited to persons considered to be Michigan residents for the entire period (the most restrictive but most valid analysis). Cohort 2: continued to accrue person years at risk for persons diagnosed with cancer, beyond the diagnosis date, and whomay have moved out of state. Other respiratory cancers included 4 cases of mesothelioma. Strengths: near complete follow-up; examined cancer incidence; exposures to 2,4-D varied and were assessed for each employee based on the exposure potential of a job during a period of time Limitations: small population; there was exposure to other phenoxyherbicides; lifestyle factors, such as smoking, were not adjusted for
		Other respiratory cancers	Cohort 3	5	4.76 (1.53–11.11)	Age	
		Other cancers of the urinary tract	Cohort 3	3	4.48 (0.90–13.08)	Age	
		NHL	Cohort 3	14	1.71 (0.93–2.87)	Age	
			Cohort 2, ≥ 5 yr duration of employment	4	3.08 (0.84–7.88)		
			Cohort 2, ≥ 5 exposure-years	3	2.16 (0.45–6.31)		

2,4-D/DP/DB, 2,4-dichlorophenoxyacetic acid/2,4-dichlorophenoxypropionic acid/2,4-dichlorophenoxybutyric acid; 2,4,5-T, 2,4,5-trichlorophenoxyacetic acid; approx.., approximately; CI, confidence intervals; IWED, intensity-weighted cumulative exposure days; JEM, job-exposure matrix; MCPA, 2-methyl-4-chlorophenoxyacetic acid; NHL, non-Hodgkin lymphoma; NR, not reported; ref., reference; SEER, Surveillance, Epidemiology and End Results; SMR, standardized mortality ratio; STS, soft tissue sarcoma; TCDD, 2,3,7,8-tetrachlorodibenzo-*p*-dioxin; Q, quintile; vs, versus

cancer-free union members, matched to the case by sex, Hispanic ethnicity, and year of birth. Exposure to 15 pesticides was assessed by linkage of detailed monthly employment union records to the California Department of Pesticide Regulation Pesticide Use Reports, for the two to three decades before cancer diagnosis. Specifically, each subject's employment by month, county, and crop was matched to pesticide application (pounds applied) in the county during that month, and the summed exposures were categorized by median or tertile. Logistic regression was conducted to estimate risk associated with pesticide use, with adjustment for age, sex, date of first contribution to the union, and duration of union membership. The study found that a history of high exposure to 2,4-D was associated with a 3.8-fold (95% CI, 1.85–7.81) increased risk of NHL when compared with a history of low exposure; this association appeared in women (OR, 5.23; 95% CI, 1.30–20.9), and men (OR, 3.79; 95% CI, 1.58–9.11), and was similar after adjustment for the 14 other high-use pesticides examined (OR, 3.58; 95% CI, 1.02–12.56). Leukaemia was not associated with 2,4-D (OR, 1.03; 95% CI, 0.41–2.61). [The semi-ecological exposure assessment in this study limited interpretation concerning individual exposure. On the other hand, the objective manner in which exposures were assigned reduced the possibility of recall bias.]

2.1.3 The Agricultural Health Study

In the USA, the risk of cancer of the prostate (total or aggressive) was examined in the Agricultural Health Study (AHS) for 1993–2007 (Koutros et al., 2013). During this period, 1962 incident cases of cancer of the prostate, including 919 cases of aggressive cancer, were observed among 54 412 pesticide applicators. Rate ratios and 95% confidence limits (CI) were calculated using Poisson regression to evaluate association with lifetime use of 48 pesticides using enrolment-questionnaire data (1993–1997) and follow-up questionnaire data (1999–2003), and incidence of all (total) cancers of the prostate, or aggressive cancers of the prostate (defined by a Gleason score of ≥ 7 or ≥ 8). About three quarters (74.9%) of the cases of cancer of the prostate, and the rest of the cohort (75.7%), had used 2,4-D. No excess incidence of cancer of the prostate was observed in applicators who had used 2,4-D (see Table 2.1). [This was a large study with high-quality exposure assessment.]

Flower et al. (2004) reported the results of analyses of pesticide application by parents and risk of childhood cancer in the AHS. The study included 17 357 children of pesticide applicators in Iowa. The number of cases in North Carolina was insufficient for the analyses, from which they were thus excluded. Parents provided data via questionnaires (1993–1997), and follow-up for cancer (retrospectively and prospectively) was done through the state cancer registries. Fifty incident cases of childhood cancer were identified in 1975–1998 in children aged 0–19 years; roughly half of the fathers had used 2,4-D prenatally (26 exposed cases). For all the children of the pesticide applicators, the incidence of all childhood cancers combined was increased compared with that of the general population, as were the incidence of all lymphomas combined, and of Hodgkin lymphoma. The odds ratio for mothers' use of 2,4-D and risk of childhood cancer was 0.72 (95% CI, 0.32–1.60, 7 exposed cases), and for fathers' use was 1.29 (95% CI, 0.71–2.35; 26 exposed cases). [The Working Group noted that this analysis had limited power to study rare diseases like childhood cancer.]

The risk of cutaneous melanoma was examined in the AHS for 1993–2005 (Dennis et al., 2010). Among 271 incident cases of cutaneous melanoma, no association was seen with exposure to 2,4-D. [The risk estimates for 2,4-D and melanoma were not presented, although the authors stated that "None of the 22 pesticides detailed on the enrollment questionnaire was associated with melanoma" and 2,4-D was one of these pesticides according to the Appendix table.]

2.1.4 The Dow Chemical Company cohort

The Dow Chemical Company cohort in the USA included men involved in the manufacture, formulation, or packaging of 2,4-D and its amines or esters from 1945 until 1982 (Bond et al., 1988). Production of 2,4-D took place in four separate buildings of the plant, each building housing different activities with different levels of exposure. At least one of these buildings, referred to as the "2,4-D plant," also housed formulation and packaging of other herbicides (2,4,5-T, MCPA, and Silvex) at various times during its history. Industrial hygienists developed a job-exposure matrix for the estimation of levels of exposure to 2,4-D among employees based on department, job title, calendar year, available monitoring data and professional judgment, and estimated cumulative exposure based on a time-weighted average of all the jobs held by an individual during the follow-up period. Cumulative exposure to 2,4-D was categorized as: very low (< 0.05 mg/m^3), low (0.05–0.49 mg/m^3), moderate (0.5–4.9 mg/m^3), or high (≥ 5 mg/m^3) (Burns et al., 2011). In an analysis of cancer incidence with follow-up until 2007, there was no increase in the incidence of all cancers combined among employees who were residents in the state for the entire period (n = 1108; "cohort 3") (Burns et al., 2011). For NHL, the standardized incidence ratio (SIR) was 1.71 (95% CI, 0.93–2.87). The highest increases in risk of NHL were observed with duration ≥ 5 years (SIR, 3.08; 95% CI, 0.84–7.88) and in analyses with censoring only at the time employees moved out of the state (n = 1256) with intensity × duration ≥ 5 exposure years (SIR, 2.16; 95% CI, 0.45–6.31). There was no clear pattern in risk of NHL with decade of starting employment. These findings were similar to analyses of mortality with earlier follow-up periods to 1982 (Bond et al., 1988), 1986 (Bloemen et al., 1993), and 1994 (Burns et al., 2001). [The Working Group noted that while analyses were presented for three methods of counting person-time, the method that censored workers at the time of loss to follow-up or diagnosis (state residents for the entire period), provided the most valid estimate ("cohort 3").]

2.2. Case–control studies

2.2.1 Lympho-haematopoietic cancers

See Table 2.2

Many of the case–control studies on lymphoma and leukaemia reported on exposure to phenoxy herbicides as a broad category or report on pesticide mixtures, rather than exposure to 2,4-D, specifically (e.g. Pearce et al., 1986a, b, 1987; Wiklund et al., 1987; Pearce, 1989; Persson et al., 1989; Eriksson & Karlsson, 1992; Fontana et al., 1998; Fritschi et al., 2005; Eriksson et al., 2008; Orsi et al., 2009; Navaranjan et al., 2013). These studies were considered uninformative with regards to carcinogenicity of 2,4-D and are therefore not described further.

In the first of the case–control studies conducted by the National Cancer Institute in the midwest USA in the 1970s and 1980s, Hoar et al. (1986) studied 170 NHL cases diagnosed from 1976 through 1982 and 948 population-based controls in the state of Kansas. Participants were asked in a telephone interview about their years working or living on farmland with wheat, corn, sorghum, or pasture, and specific pesticides handled in these settings. Compared with subjects who had never farmed, ever exposure to phenoxy herbicides was associated with a 2.2-fold increased risk of NHL (OR, 2.2; 95% CI, 1.2–4.1; 24 exposed cases) and exposure to only 2,4-D (eliminating 3 cases who also used 2,4,5-T) was also associated with increased risk (OR, 2.6, 1.4–5.0). There were significant trends in increasing risk of NHL with increasing duration

Table 2.2 Case–control studies of lympho-haematopoietic cancer and exposure to 2,4-D

Reference, location, enrolment/ follow-up period	Population size (response rate), description, exposure assessment method	Organ site	Exposure category or level	Exposed cases/ deaths	Risk estimate (95% CI)	Covariates controlled	Comments
Midwest USA							
Hoar et al. (1986) Kansas, USA 1976–1982	Cases: 170 (response rate, 96%); population-based cancer registry (University of Kansas Cancer Data Service) Controls: 948 (response rate, 94%); random digit dialling (age < 65 years), Medicare records (age ≥ 65 years), death certificates (deceased) Exposure assessment method: questionnaire	NHL	Phenoxyherbicides (synonymous with 2,4-D use)	24	2.2 (1.2–4.1)	Age and vital status (by matched analysis)	Studies in midwest USA This study also enrolled cases of STS and Hodgkin disease, but did not report results associated with 2,4-D exposure for these cancer sites Strengths: excellent response proportion; risk estimated by duration and frequency Limitations: duration and frequency variables were based on reported use of any herbicide (instead of being specific to 2,4-D)
			2,4-D use			Age and vital status (by matched analysis)	
			2,4-D only use (excluding 2,4,5-T)	21	2.6 (1.4–5.0)		
			2,4-D duration (yr):			Age and vital status (by matched analysis)	
			1–5	3	1.3 (0.3–5.1)		
			6–15	7	2.5 (0.9–6.8)		
			16–25	8	3.9 (1.4–10.9)		
			≥ 26	6	2.3 (0.7–6.8)		
			Trend-test *P* value: 0.002				
			2,4-D frequency (days/yr):			Age and vital status (by matched analysis)	
			1–2	6	2.7 (0.9–8.1)		
			3–5	4	1.6 (0.4–5.7)		
			6–10	4	1.9 (0.5–6.7)		
			11–20	4	3.0 (0.7–11.8)		
			≥ 21	5	7.6 (1.8–32.3)		
			Trend-test *P* value: 0.0001				
			2,4-D, first year of use			Age and vital status (by matched analysis)	
			1966 or later	5	1.9 (0.6–6.0)		
			1956–1965	9	2.9 (1.1–7.2)		
			1946–1955	8	2.1 (0.8–5.6)		
			Before 1946	2	6.2 (0.6–65.3)		
			Trend-test *P* value: 0.002				

Table 2.2 (continued)

Reference, location, enrolment/follow-up period	Population size (response rate), description, exposure assessment method	Organ site	Exposure category or level	Exposed cases/deaths	Risk estimate (95% CI)	Covariates controlled	Comments
Zahm et al. (1990) Nebraska, USA 1983–1986	Cases: 201 (response rate, 91%); cases identified by the Nebraska Lymphoma Study Group and area hospitals; white male population Controls: 725 (response rate, 87%); random-digit dialling (age < 65 yr), Medicare records (age ≥ 65 yr), death certificates (deceased) Exposure assessment method: questionnaire; telephone interviews with subjects or next-of-kin	NHL	Mixed or applied 2,4-D	43	1.5 (0.9–2.5)	Age	Studies in midwest USA Strengths: good response proportion; information on duration and frequency, specific to 2,4-D; no evidence of recall bias since ORs for 2,4-D use were similar in subjects who recalled the exposure without prompting and in those who were probed for 2,4-D use Limitations: possibly biased exposure information from proxies
			Frequency mixing or applying 2,4-D (days/yr):			Age	
			1–5	16	1.2 (0.6–2.4)		
			6–20	12	1.6 (0.7–3.6)		
			≥ 21	3	3.3 (0.5–22.1)		
			Trend-test P value: 0.051				
			Duration 2,4-D used on farm (yr):			Age	
			1–5	3	0.9 (0.2–3.6)		
			6–15	11	2.8 (1.1–7.1)		
			16–20	3	0.6 (0.1–2.1)		
			≥ 21	13	1.3 (0.6–2.7)		
			Trend-test P value: 0.274				
			First year of use:			Age	
			Before 1945	8	1.4 (0.5–3.5)		
			1946–1955	13	1.1 (0.5–2.3)		
			1956–1965	5	2.1 (0.6–7.7)		
			1965–1986	4	1.3 (0.3–4.9)		
			Trend-test P value: 0.17				
			By histological subtype for personal use:			Age	
			B-cell/ever 2,4-D	NR	1.5 (0.9–2.6)		
			T-cell/ever 2,4-D	NR	2.0 (0.5–7.3)		

Table 2.2 (continued)

Reference, location, enrolment/ follow-up period	Population size (response rate), description, exposure assessment method	Organ site	Exposure category or level	Exposed cases/ deaths	Risk estimate (95% CI)	Covariates controlled	Comments
Brown et al. (1990) Iowa and Minnesota, USA 1981–1984	Cases: 578 (response rate, 86%); state cancer registry (IA) and surveillance network of hospitals and pathology laboratories (MN) Controls: 1245 (response rate, 77–79%); random-digit dialling (age < 65 yr), Medicare records (age ≥ 65 yr), death certificates (deceased) Exposure assessment method: interviewer-administered questionnaire elicited information about use of specific pesticides	Leukaemia	Ever mixed/ handled/applied 2,4-D	98	1.2 (0.9–1.6)	Vital status, age, state, smoking, family history of lympho-haematopoietic cancer, high-risk occupation, high-risk exposures	Studies in midwest USA Reference group was non-farmers Strengths: agricultural region with high frequency of farming and pesticide use Limitations: no quantitative exposure information; white men only
		Leukaemia (AML)	Ever handled 2,4-D	28	1.5 (0.9–2.5)		
		Leukaemia (CML)	Ever handled 2,4-D	13	1.9 (0.9–3.9)		
		NHL (CLL)	Ever handled 2,4-D	45	1.3 (0.8–2.0)		
Cantor et al. (1992) Iowa and Minnesota, USA 1980–1983	Cases: 622 (response rate, 89%); Iowa State Health Registry and surveillance of Minnesota hospital and pathology laboratory records Controls: 1245 (response rate, 77–79%); random-digit dialling (age < 65 yr), Medicare records (age ≥ 65 yr), death certificates (deceased) Exposure assessment method: questionnaire collected data on use history of specific pesticides	NHL	Handled 2,4-D	115	1.2 (0.9–1.6)	Vital status, age, state, smoking, family history of lympho-haematopoietic cancer, high-risk occupation, high-risk exposures	Studies in midwest USA Reference group was non-farmers; no difference in results by state Strengths: agricultural region with high frequency of farming occupation and pesticide use Limitations: non-quantitative exposure assessment; white men only
			Handled before 1965	86	1.3 (0.9–1.8)		
			Handled without PPE	89	1.2 (0.9–1.7)		
			Iowa – ever handled 2,4-D	51	1.2 (0.8–1.9)		
			Minnesota – ever handled 2,4-D	35	1.4 (0.9–2.3)		

Table 2.2 (continued)

Reference, location, enrolment/ follow-up period	Population size (response rate), description, exposure assessment method	Organ site	Exposure category or level	Exposed cases/ deaths	Risk estimate (95% CI)	Covariates controlled	Comments
De Roos et al. (2003) Iowa, Minnesota, Nebraska, Kansas, USA 1980s (pooled analysis)	Cases: 870 (response rate, NR); white men with NHL Controls: 2569 (response rate, NR); frequency matched on age, race, state of residence using 2 : 1 ratio in Iowa and Minnesota, 4 : 1 in Kansas and Nebraska; random-digit dialling for living cases aged < 65 yr and from the Health Care Financing Administration for those aged ≥ 65 yr; controls for deceased cases from deaths records in each state, matched for age and year of death Exposure assessment method: telephone interviews with subjects or next-of-kin in Kansas and Nebraska, and in-person in Iowa and Minnesota	NHL	Use of 2,4-D (hierarchical regression)	123	0.9 (0.6–1.2)	Age, location, other pesticides	Subjects missing data for any of 47 pesticides were excluded (25% of subjects)

Table 2.2 (continued)

Reference, location, enrolment/ follow-up period	Population size (response rate), description, exposure assessment method	Organ site	Exposure category or level	Exposed cases/ deaths	Risk estimate (95% CI)	Covariates controlled	Comments
Other studies in North America							
Woods et al. (1987) Western Washington, USA 1981–1984	Cases: 576 (response rate, NR); cases identified from population-based tumour registry that covered 13 counties of western Washington State from 1974 Controls: 694 (response rate, NR) chosen by random-digit dialling (aged 20–64 yr) or at random from Health Care Financing Administration records covering social security recipients in the study area (aged 65–79 yr) Exposure assessment method: questionnaire	NHL	Ever occupationally exposed to 2,4-D	NR	0.73 (0.40–1.30)	Age	Strengths: large study population Limitations: number exposed to 2,4-D, NR
			Farmers who "regularly worked with 2,4-D"	NR	0.68 (0.3–1.4)		

Table 2.2 (continued)

Reference, location, enrolment/ follow-up period	Population size (response rate), description, exposure assessment method	Organ site	Exposure category or level	Exposed cases/ deaths	Risk estimate (95% CI)	Covariates controlled	Comments
McDuffie et al. (2001) Six provinces, Canada 1991–1994	Cases: 517 (response rate, 67.1%) men from cancer registries and hospitals; newly diagnosed men (aged ≥ 19 yr) Controls: 1506 (response rate, 48%); random sample from health insurance and voting records; men (aged ≥ 19 yr) frequency-matched on province and age Exposure assessment method: mailed questionnaire followed by telephone interview for subjects reporting ≥ 10 h/yr of pesticide exposure	NHL	Mixed/sprayed 2,4-D	111	1.32 (1.01–1.73)	Age, province, medical history variables, family history of cancer	Men only Strengths: large sample; information on individual pesticides Limitations: low response rate
			Frequency (days/yr):	111	–		
			> 0 and ≤ 2	55	1.17 (0.83–1.64)		
			> 2 and ≤ 5	36	1.39 (0.91–2.13)		
			> 5 and ≤ 7	9	1.38 (0.60–3.15)		
			> 7	11	1.22 (0.60–2.49)		
Hartge et al. (2005) Iowa, Washington (Seattle metropolitan area), Michigan (Detroit metropolitan area), California (Los Angeles county), USA 1998–2000	Cases: 679 (response rate, 59%); population cancer registries (SEER) Controls: 510 (response rate, 44); random-digit dialling (age < 65 yr); Medicare records (≥ age 65 yr), stratified by age, sex, race, centre Exposure assessment method: environmental monitoring; measurement in household dust samples and interview	NHL	2,4-D (ng/g):			Study site, age, sex, race, education	2,4-D half-life in dust is probably shorter than latency period. Frequency and intensity of residential exposures are low compared with those exposed occupationally Strengths: exposure based on measurement (rather than recall) Limitations: low response rate
			Below detection limit	147	1.00		
			< 500	257	1.10 (0.78–1.55)		
			500–999	86	0.91 (0.58–1.45)		
			1000–9999	165	0.66 (0.45–0.98)		
			> 10 000	24	0.82 (0.41–1.66)		

Table 2.2 (continued)

Reference, location, enrolment/ follow-up period	Population size (response rate), description, exposure assessment method	Organ site	Exposure category or level	Exposed cases/ deaths	Risk estimate (95% CI)	Covariates controlled	Comments
Metayer et al. (2013) Northern and central California, USA 1995–2008	Cases: 296 childhood leukaemia, 269 ALL (response rate, 91%); enrolled from paediatric clinical centres Controls: 333 (response rate, 82%); enrolled from birth certificates Exposure assessment method: environmental monitoring; measurement in household dust sample	ALL	2,4-D log-transformed concentration (ng/g)	252	0.96 (0.85–1.08)	Sex, age, Hispanic ethnicity, maternal race, income, season of dust sampling, year of dust sampling, neighbourhood type (urban, rural, etc.)	Strengths: exposure measurement (rather than self-report) Limitations: 2,4-D half-life in dust may be shorter than latency period

Table 2.2 (continued)

Reference, location, enrolment/ follow-up period	Population size (response rate), description, exposure assessment method	Organ site	Exposure category or level	Exposed cases/ deaths	Risk estimate (95% CI)	Covariates controlled	Comments
Mills et al. (2005) California, USA 1987–2001 Nested case–control study	Cases: 131 (response rate, 100%); labour union members, Hispanic population; cases identified from linkage to state-wide population-based cancer registry Controls: 655 (response rate, 100%); 5 controls per case were selected from the union, who had not been diagnosed with any cancer at time of case diagnosis, and matched on sex, Hispanic ethnicity, year of birth Exposure assessment method: detailed month-by-month work histories from union records linked to pesticide-use records from California Department of Pesticide Regulation & Pesticide Use Reports to match pesticide applications given the month/year/ location/crop	NHL	High (vs low)			Age, sex, duration of union affiliation, date of first union affiliation	United Farm Workers of America Strengths: large database with objective exposure assignment (e.g. not based on recall) Limitations: semi-ecological exposure assessment limited interpretation about individual exposure
			High (vs low)	60	3.80 (1.85–7.81)		
			Men	45	3.79 (1.58–9.11)		
			Women	15	5.23 (1.30–20.9)		
		NHL, nodal	High (vs low)	38	2.29 (0.90–5.82)	Same	
		NHL, extranodal	High (vs low)	22	9.73 (2.68–35.30)	Same	
		Leukaemia	High (vs low)	51	1.03 (0.41–2.61)	Same	
			Men	35	0.55 (0.15–2.06)		
			Women	16	3.73 (0.77–18.00)		
		Lymphocytic leukaemia	High (vs low)	23	1.47 (0.33–6.59)	Same	
		Granulocytic leukaemia	High (vs low)	20	1.28 (0.30–5.42)	Same	

Table 2.2 (continued)

Reference, location, enrolment/ follow-up period	Population size (response rate), description, exposure assessment method	Organ site	Exposure category or level	Exposed cases/ deaths	Risk estimate (95% CI)	Covariates controlled	Comments
Sweden							
Hardell et al. (1994) Umea, Sweden 1974–1978	Cases: 105 (response rate, NR); Department of Oncology, Umea Controls: 335 (response rate, NR); national population registry (living) or national registry for causes of death (deceased) Exposure assessment method: questionnaire	NHL	Occupational or leisure-time use of 2,4-D only	3	13.0 (1.2–360.0)	NR	Strengths: separate risk estimate for 2,4-D Limitations: small number exposed to 2,4-D, specifically
Nordström et al. (1998) Sweden 1987–1990	Cases: 121 (response rate, NR); male patients with HCL reported to the Swedish Cancer Registry 1987–1992 Controls: 484 (response rate, NR); age- and county-matched controls from the national population registry Exposure assessment method: mailed, self-administered questionnaire (by subject or next-of-kin) plus supplemental telephone call for clarification of unclear information	NHL (HCL)	2,4-D	2	1.6 (0.3–8.3)	Age	A minimum exposure of 1 working day(8 h) and an induction period of ≥ 1 yr were used in the coding of exposures Strengths: comprehensive population registration Limitations: small number of ever-exposed cases and controls for the analysis; the reference group was not specified

Table 2.2 (continued)

Reference, location, enrolment/ follow-up period	Population size (response rate), description, exposure assessment method	Organ site	Exposure category or level	Exposed cases/ deaths	Risk estimate (95% CI)	Covariates controlled	Comments
Europe							
Miligi et al. (2003) Italy 1990–1993	Cases: 1575 (response rate, 80–83%) Controls: 1232 (response rate, 74%) Exposure assessment method: questionnaire	NHL (ICD 200 & 200) and CLL (ICD 204.1)	Ever occupationally exposed to 2,4-D, men	6	0.7 (0.3–1.9)	Age, area	
			Ever occupationally exposed to 2,4-D, women	7	1.5 (0.4–5.7)		
Miligi et al. (2006a) Italy 1990–1993 Case–control	Cases: 1145 NHL and CLL, 205 MM, 430 leukaemia, 258 HD (80%); all incident cases of malignancies of the haematolymphopoietic system, aged 20–74 yrs, residents of the study area Controls: 1232 (response rate, 74%); age- and sex-matched from the general population Exposure assessment method: questionnaires reviewed by agronomists to assign pesticide-exposure histories	NHL (ICD 200 & 200) and CLL (ICD 204.1)	Ever occupationally exposed to 2,4-D	17	0.9 (0.5–1.8)	Sex, age, area	Association with ever use of 2,4-D did not differ between men and women, as reported in an earlier publication (Miligi et al., 2003) Strengths: expert assessment of exposure
			Probability of use > low and no protective equipment	9	4.4 (1.1–29.1)		

Table 2.2 (continued)

Reference, location, enrolment/ follow-up period	Population size (response rate), description, exposure assessment method	Organ site	Exposure category or level	Exposed cases/ deaths	Risk estimate (95% CI)	Covariates controlled	Comments
Cocco et al. (2013a) Europe (Czech Republic, France, Germany, Italy, Ireland, Spain) 1998–2003	Cases: 2348 (response rate, 88%); referring centres Controls: 2462 (response rate, 52–81%); population controls (Germany, Italy); hospital controls (Czech Republic, France, Ireland, Spain) Exposure assessment method: subjects reporting work in agriculture received job-specific questionnaire eliciting further details on tasks and exposures	B-cell lymphoma	2,4-D	2	0.6 (0.1–3.5)	Age, sex, education, centre	Strengths: detailed exposure questionnaire combined with expert assessment Limitations: low response proportion for population controls
		CLL				Age, sex, education, centre	

ALL, acute lymphoblastic/lymphocytic leukaemia; AML, acute myeloid leukaemia; CLL, chronic lymphocytic leukaemia; CML, chronic myeloid leukaemia; 2,4-D, 2,4-dichlorophenoxyacetic acid; DEET, N,N-Diethyl-meta-toluamide; HCL, hairy cell leukaemia; NHL, non-Hodgkin lymphoma; NR, not reported; OR, odds ratio; PPE, personal protective equipment; SEER, Surveillance, Epidemiology, and End Results; STS, soft tissue sarcoma; vs, versus; yr, year

(years) (trend-test *P* value, 0.002) and frequency (days/year) (trend-test *P* value, 0.0001) of exposure, with the highest risks observed in the third of four exposure categories with 16–25 years exposure (OR, 3.9; 95% CI, 1.4–10.9) or ≥ 21 days/year exposure (OR, 7.6; 95% CI, 1.8–32.3). [The Working Group noted that the information on duration and frequency were based on reported patterns of use of any herbicide, rather than being specific to 2,4-D. This study also enrolled cases of soft tissue sarcoma and Hodgkin, disease, but did not report results associated with 2,4-D exposure for these cancer sites.]

The study conducted by the United States National Cancer Institute (NCI) in Nebraska included 201 cases of NHL and 725 controls, identified between 1983 and 1986 (Zahm et al., 1990). The study was limited to white men. There were elevated, but non-significant increases in risk of NHL observed in association with having having ever mixed or applied 2,4-D (OR, 1.5; 95% CI, 0.9–2.5) and with ≥ 21 days per year use on the farm (OR, 3.3; 95% CI, 0.5–22.1), and a trend in increasing risk by frequency of use (*P* = 0.051). There were no apparent patterns by duration or year of first use. Risk estimates were elevated for T-cell and B-cell NHL, but the trend by days per year was significant only for B-cell NHL (*P* = 0.045). Possible confounding of the results for 2,4-D by use of other pesticides was evaluated. Exclusion of users of 2,4,5-T did not change the risk estimates for handling 2,4-D. The risk associated with having ever mixed or applied 2,4-D was attenuated with exclusion of all organophosphate users (OR, 1.1; CI, not reported), but was increased with adjustment for fungicides (OR, 1.8; 95% CI, 1.1–3.0). For farmers who mixed or applied 2,4-D on > 20 days per year, simultaneous adjustment for organophosphates and fungicides resulted in a risk estimate of 3.1 (CI not reported). Risk estimates for use of 2,4-D were similar in subjects who recalled the exposure without prompting and in subjects whose use of 2,4-D was only reported in response to a specific probe (OR, 1.5 in both groups [CI not reported]), suggesting little recall bias; however, the associations were higher when limited to proxy respondents. [The Working Group noted possible biased exposure information from proxies. This study also enrolled cases of Hodgkin disease, multiple myeloma, and chronic lymphocytic leukaemia, but did not report results associated with 2,4-D exposure for these other cancer sites.]

In Iowa and Minnesota, USA, cases of NHL (*n* = 622) (Cantor et al., 1992) and leukaemia (*n* = 578) (Brown et al., 1990) diagnosed in 1980–1984 were frequency-matched to 1245 population controls. Only white men were included in the analyses. There were no strong associations of 2,4-D use in relation to either NHL or leukaemia in this study, in analyses of ever use, frequency of use, year of first use, use without personal protective equipment, and use by state, or when considering multiple potential confounders including vital status, age, state, smoking, family history of lympho-haematopoietic cancer, high-risk occupation, and high-risk exposures. For subjects who had ever used 2,4-D, the odds ratios were: NHL, 1.2 (95% CI, 0.9–1.6) (Cantor et al., 1992); chronic lymphocytic leukaemia [a subtype of NHL], 1.3 (95% CI, 0.8–2.0) (Brown et al., 1990); and all leukaemia combined, 1.2, (95% CI, 0.9–1.6) (Brown et al., 1990).

A pooled analysis by De Roos et al. (2003) of men from the three United States NCI case–control studies conducted in the midwest USA was limited to men with complete data for 47 pesticides (*n* = 870 cases, and 2569 controls from Hoar et al. (1986); Zahm et al. (1990); Cantor et al. (1992)). About 75% of the original study population was included, but examination of risk factors for NHL, including family history and farming history, indicated no difference in the association of those factors with case status between the original population and the included subset. No association was found between ever use of 2,4-D and risk of NHL (OR, 0.9; 95% CI, 0.6–1.2), with adjustment for other individual pesticides using hierarchical regression. [This estimate was limited to a smaller study population than in the individual studies, due to exclusion of participants with missing data for any of the multiple pesticides examined; nevertheless, the sample size was large. The study suggested that any confounding of the association between NHL and exposure to 2,4-D by exposure to other pesticides is away from the null. The hierarchical regression method "shrinks" (adjusts) unstable estimates towards a prior distribution; however, the shrunken estimate for 2,4-D (OR, 0.9) did not differ substantially from the pooled estimate before shrinkage (OR, 0.8).]

Woods et al. conducted a case–control study of NHL in western Washington State, USA, in the early 1980s, in which a detailed occupational history was obtained, including exposure to phenoxy herbicides (Woods et al., 1987; Woods & Polissar, 1989). In a comparison of 576 cases and 694 controls, there was no observed difference with respect to ever having occupational exposure to 2,4-D (OR, 0.73; 95% CI, 0.4–1.3) or for farmers who reported having "regularly worked with" 2,4-D (OR, 0.68; 95% CI, 0.3–1.4). There were no details provided on the number of cases and controls exposed to 2,4-D. [This study had a strong exposure assessment, including expert review of occupational histories and verification of exposure with a third party in instances where exposure was uncertain. Response proportions were not provided.]

A large study of NHL in men was conducted in six provinces of Canada, between 1991 and 1994 (McDuffie et al., 2001). A total of 517 cases were identified from provincial cancer registries and hospitals, and 1506 controls were identified from provincial health insurance records, computerized telephone listings, and voters' lists. Information about pesticide exposure was collected by means of a mailed questionnaire, followed by a telephone interview for subjects reporting 10 hours per year or more of pesticide exposure. In analyses of individual phenoxyherbicides, the odds ratio for 2,4-D was 1.32, (95% CI, 1.01–1.73). There was no dose–response pattern in analyses of 2,4-D exposure by frequency of use (days/year). Similar effect estimates were presented in a later paper by (Pahwa et al., 2012), in which no interaction was found between exposure to 2,4-D and asthma or allergy in the relationship with NHL. Hohenadel et al. (2011) found no interaction between exposure to 2,4-D and malathion in this study population, and reported no association between exposure to 2,4-D and risk of NHL in the absence of malathion exposure (OR, 0.94; 95% CI, 0.67–1.33) (Hohenadel et al., 2011) [The Working Group noted that the response proportion was 48% for population controls, making selection bias a concern].

Two studies estimated the risk of lympho-haematopoietic cancer associated with exposure to 2,4-D based on measurement of 2,4-D in samples of household dust. Hartge et al. (2005) conducted a study of NHL in Iowa, and the metropolitan region of Seattle, Washington State, Detroit, Michigan, and Los Angeles County, California, USA. Population controls were identified by random-digit telephone dialling and through Medicare records. Dust samples were collected from participating households from the subjects' vacuum-cleaner bag, and 2,4-D concentration (in ng/g) was measured by gas chromatography/mass spectrometry. Comcentrations of 2,4-D and dicamba were higher in the carpets of subjects reporting residential use of herbicides. There was no association between 2,4-D concentration and risk of NHL, and no pattern in the dose–response relationship with increasing concentration. The risk estimate for the highest category of 2,4-D exposure at > 10 μg/g compared with concentrations below the detection limit was 0.82 (95% CI, 0.41–1.66). [The interpretation of this study was limited by the fact that the half-life of 2,4-D in house dust is unknown.]

Metayer et al. (2013) conducted a study of childhood acute lymphocytic leukaemia in northern and central California, USA, in which samples of household dust were collected according to a standardized protocol during a household visit. There was no association between the concentration of 2,4-D in household dust (measured by gas chromatography/mass spectrometry) and risk of childhood acute lymphocytic leukaemia (OR, 0.96; 95% CI, 0.85–1.08). [The half-life of 2,4-D in house dust is unknown.]

Two case–control studies conducted in Sweden used national registration systems for identification of cases and population controls. Hardell et al. (1994) conducted a study on NHL in Umea, Sweden, in which they identified 105 cases and 335 controls from 1974 until 1978. Controls were identified from the national population registry (living) and national registry for causes of death (deceased), and were matched to cases on age, sex, place of residence, vital status and year of death (for the deceased). Exposure information was obtained by questionnaire (completed by proxy for deceased subjects), and exposure was defined as ever use of the pesticide in occupation or during leisure time. There was a significantly increased risk of NHL with use of 2,4-D among those without exposure to any other phenoxy herbicide, based on only 3 exposed cases and 1 exposed control (OR, 13; 95% CI, 1.2–360). There was no estimate presented for exposure to 2,4-D with statistical adjustment for other phenoxy herbicides. [The extreme imprecision of the risk estimate for 2,4-D from this study limited interpretation about the possible magnitude of the association.]

Nordström et al. (1998) identified 121 male patients with hairy cell leukaemia who reported to the Swedish Cancer Registry between 1987–1982 and 484 age- and country-matched population-based controls identified from national registries. Information on pesticide use was collected by mailed questionnaire, with clarification of information by follow-up telephone calls, if needed. Exposure was defined as a minimum of one working day of exposure (8 hours) and a latency period of ≥ 1 year. The association with ever having been exposed to 2,4-D, specifically, was reported for hairy cell leukaemia only, based on two exposed cases and five exposed controls (OR, 1.6; 95% CI, 0.3–8.3).

In a large case–control study on NHL (including chronic lymphocytic leukaemia; $n = 1145$) in Italy in which exposure was assessed through consultation by industrial hygienists and agronomists, there was no association with ever use of 2,4-D (OR, 0.9, 95% CI, 0.5–1.8; 17 exposed cases) (Miligi et al., 2006a). Greater than low probability of 2,4-D use was not associated with risk of NHL in men (OR, 0.7; 0.3–1.9; 6 exposed cases) or women (OR, 1.5; 0.4–5.7; 7 exposed cases) (Miligi et al., 2003); however, an increased risk was observed with greater than low probability of 2,4-D use in combination with lack of protective equipment (OR, 4.4; 95% CI, 1.1–29.1; 9 exposed cases) (Miligi et al., 2006a).

The Epilymph study is a large case–control study of lymphoma (including NHL, Hodgkin lymphoma, chronic lymphocytic leukaemia, and multiple myeloma) conducted in six European countries. Cocco et al. (2013a) evaluated pesticide exposures in the Epilymph study based on expert assessment of detailed work histories coupled with a crop-exposure matrix. Exposure to 2,4-D was not associated with risk of B-cell lymphoma in this study (OR, 0.6; 95% CI, 0.1–3.5), based on two exposed cases and four exposed controls. [Findings for 2,4-D were mentioned in the text of the paper with a reference to Table 4. Therefore, the Working Group interpreted the table entry labelled "2,4-dichlorophenol" as the association between 2,4-dichlorophenoxyacetic acid and B-cell lymphoma.]

2.2.2 Other cancer sites

Most available case–control studies on soft tissue sarcoma evaluating exposure to phenoxy herbicides provided effect estimates for wide exposure categories and did not provide estimates specifically for exposure to 2,4-D (Hardell & Sandström, 1979; Eriksson et al., 1981; Smith et al., 1984; Vineis et al., 1987; Smith & Christophers, 1992). Two studies including soft tissue sarcoma did not provide odds ratios specific for exposure to 2,4-D, but indicated that risks were not associated with exposure to 2,4-D (Hoar et al., 1986; Woods et al., 1987). One case–control study on gastric cancer and one on nasal and nasopharyngeal carcinoma evaluating exposure to phenoxy herbicides provided effect estimates for wide exposure categories and did not provide estimates specifically for exposure to 2,4-D (Hardell et al., 1982; Ekström et al., 1999). These studies were considered to be uninformative about the carcinogenicity of 2,4-D and are not reviewed further here.

A case–control study on glioma was conducted among non-metropolitan residents of Iowa, Michigan, Minnesota and Wisconsin, USA (Yiin et al., 2012). The study included 798 histologically confirmed cases of primary intracranial glioma (45% with proxy respondents) and 1175 population-based controls, aged 18–80 years. Subjects were interviewed face-to-face. Information on exposure from questionnaire responses was evaluated by an experienced industrial hygienist. An inverse association was observed for non-farm occupational use of 2,4-D (OR, 0.56; 95% CI, 0.28–1.10); a similar result was observed when excluding proxy respondents (6 cases) (OR, 0.49; 95% CI, 0.20–1.22). Reported house and garden use of 2,4-D was also inversely associated with risk (OR, 0.64; 95% CI, 0.47–0.88; OR after excluding proxy respondents, 0.76; 95% CI, 0.51–1.11) [The number of exposed subjects was very small. Effect estimates were not presented for farm use of 2,4-D].

Mills et al. (2005) conducted a registry-based case–control study of cancer of the breast in Hispanic members of a farm labour union in California, USA. The study included 128 incident cases of cancer of the breast (1988–2001) diagnosed among past or present members of a large agricultural labour union and identified from the California cancer registry. The controls were 640 cancer-free members of the same trade union matched for ethnicity and the case's attained age at the time of diagnosis. Exposure was determined from detailed employment records that were linked to pesticide information obtained from the Pesticide Databank, a database of historical pesticide-use records collected by the California Department of Food and Agriculture. Exposure to 2,4-D was associated with increased risk of cancer of the breast (OR, 2.14; 95% CI, 1.06–4.32; 21 cases with "high" 2,4-D exposure) among cases diagnosed in the second part of the study period (1995–2001), but not in the first part (1988–1994), after adjusting for age, date of first union affiliation, duration of union affiliation, fertility, and socioeconomic level. [The overall odds ratio for 2,4-D as calculated by the Working Group was 1.40 (95% CI, 0.91–2.17]. Exposure assessment was semi-ecological in this study. There was no adjustment for several potential confounders.]

2.3 Meta-analyses

Schinasi & Leon (2014) conducted a meta-analysis of NHL and exposure to pesticides in agricultural settings. Case–control and cohort studies were included if they were published in English language, used interviews, questionnaires or exposure matrices to assess occupational exposure to agricultural pesticides, and reported quantitative associations of NHL overall or by subtype with specific active ingredients or chemical groups. Five papers on case–control studies contributed to the meta relative-risk estimates for the relationship between occupational exposure to 2,4-D and NHL overall (Zahm et al., 1990; Cantor et al., 1992; Mills et al., 2005; Miligi et al., 2006b; Pahwa et al., 2012). The meta relative-risk for exposure to 2,4-D and NHL was 1.40 (95% CI, 1.03–1.91, I^2 = 61.5%). Sensitivity analyses examining the influence of sex, period of diagnosis, and geographical region did not substantially change the meta relative risk (RR) for 2,4-D.

A meta-analysis of 2,4-D and cancer by Goodman et al. (2015) included a larger number of peer-reviewed observational epidemiological studies of NHL reported before 9 October 2014 that reported quantitative measures of association specifically for 2,4-D, and also estimated cancer of the stomach or prostate in relation to exposure to 2,4-D. In the main analysis, there were nine studies on NHL (Woods & Polissar, 1989; Hardell et al., 1994; Kogevinas et al., 1995; De Roos et al., 2003; Hartge et al., 2005; Mills et al., 2005; Miligi et al., 2006b; Burns et al., 2011; Hohenadel et al., 2011), three studies on cancer of the stomach (Lee et al., 2004; Mills & Yang, 2007; Burns et al., 2011), and two studies on cancer of the prostate (Band et al., 2011; Burns et al., 2011). Risk estimates and confidence intervals extracted from the original studies were log-transformed before analysis. Exposure to 2,4-D was not associated with NHL (RR, 0.97; 95% CI, 0.77–1.22; I^2 = 28.8%, $P_{heterogeneity}$ = 0.19). For cancer of the stomach, the relative risk was 1.14 (95% CI, 0.62–2.10; I^2 = 54.9%, $P_{heterogeneity}$ = 0.11), and for cancer of the prostate it was 1.32 (95% CI, 0.37–4.69; I^2 87.0%, $P_{heterogeneity}$ = 0.01). The tests of heterogeneity of effect by exposure source did not reveal heterogeneity by study design, type of exposure (agricultural or other), geographical location, or sex. Sensitivity analyses indicated that the results were robust to most factors considered. However, after substitution of: (i) a pooled analysis that adjusted for multiple pesticides (De Roos et al., 2003) by the three individual studies (Hoar et al., 1986; Zahm et al., 1990; Cantor et al., 1992) that were not adjusted; and (ii) a study considering 2,4-D in the absence of malathion (Hohenadel et al., 2011) by a study that considered ever exposure to 2,4-D (Pahwa et al., 2012); and (iii) the unadjusted rather than the adjusted estimate from another study (Mills et al., 2005), the meta relative risk for NHL increased to 1.34 (95% CI, 1.04–1.72), and heterogeneity also increased (I^2 = 56.3%, P = 0.011).

The Working Group carried out an additional meta-analysis of 2,4-D and non-Hodgkin lymphoid neoplasms (including NHL, multiple myeloma, hairy cell leukaemia, and chronic lymphocytic leukaemia) (see Table 2.3 for the key characteristics of the studies included in the Working Group meta-analysis and a comparison with the previously published meta-analyses). Thirteen reports were included in the main (primary) analysis and 15 reports were included in a secondary analysis. When the risk estimates for the primary analysis were displayed on a funnel plot, the plot was nearly symmetric [this was interpreted to show no significant evidence of publication bias] (Fig. 2.1). Where available, risk estimates were selected for the primary analysis that adjusted for use of pesticides other than 2,4-D: De Roos et al. (2003) adjusted for more than 40 pesticides, Kogevinas et al. (1995) adjusted for 2,4,5-T and MCPA, Mills et al. (2005) adjusted for 14 pesticides. In addition, a risk estimate for 2,4-D in the absence of exposure to malathion was selected from Hohenadel et al. (2011), the risk estimate for 2,4-D in the absence of exposure to DEET was selected from Pahwa et al. (2006), and a risk estimate in the absence of any other phenoxy herbicide was selected from Hardell et al. (1994). Estimates adjusted for other pesticide use were not available for a further seven studies included in the meta-analysis: Woods & Polissar (1989), Miligi et al. (2006b), Cocco et al. (2013b) (B-cell lymphoma), Nordström et al. (1998) (hairy cell leukaemia), Brown et al. (1993) (multiple myeloma), Brown et al. (1990) (chronic lymphocytic leukaemia), and Burns et al. (2011). Hartge et al. (2005) was not included in the Working Group meta-analysis, in contrast to Goodman et al. (2015), due to the use of exposure measurement via dust, which was qualitatively different from the other studies (Table 2.3). In this analysis, 2,4-D was not associated with risk of NHL (RR, 1.04; 95% CI; 0.88–1.22; I^2 = 6%,

Table 2.3 Selected characteristics of studies included in the Working Group meta-analysis (primary and secondary analysis), and comparison with previously published meta-analyses

Study reference	Relative risk (95% CI)	Included in Working Group analysis[a]		Included in previously published meta-analysis		Exposure–response reported	Study design	Study outcome	Comments
		Primary analysis	Second analysis	Goodman et al. (2015)	Schinasi & Leon (2014)				
Burns et al. (2011)	1.71 (0.93–2.87)					√	Cohort	NHL	Dow cohort of 2,4-D production workers; analysed with three methods of counting person-time. We chose cohort 3 for the primary analysis because it is the most valid since it was censored at lost-to-follow-up or date of diagnosis
	1.36 (0.74–2.29)			√		√	Cohort	NHL	Cohort 2
	0.79 (0.09–2.87)					√	Cohort	MM	Cohort 3
	1.50 (0.85–2.43)	√	√			√	Cohort	NHL + MM	Cohort 3 combined SMR for NHL and MM
Cocco et al. (2013a)	0.60 (0.10–3.50)	√	√				Case–control	NHL	B-cell lymphoma
Nordström et al. (1998)	1.60 (0.30–8.30)	√	√				Case–control	HCL	HCL
Mills et al. (2005)	3.80 (1.85–7.81)		√		√		Cohort	NHL	This estimate was rounded to 3.80 (1.85–7.81) for analysis in Schinasi & Leon (2014)
	3.58 (1.02–12.56)	√		√			Cohort	NHL	Adjusted for use of 14 other pesticides The upper bound of the 95% CI was reported as 12.58 in Goodman et al. (2015)
Kogevinas et al. (1995)	1.11 (0.46–2.65)		√	√		√	Cohort	NHL	
	1.05 (0.26–4.28)	√				√	Cohort	NHL	Adjusted for use of the pesticides 2,4,5-T and MCPA
Pahwa et al. (2012)	1.27 (0.98–1.65)		√		√		Case–control	NHL	This estimate was rounded to 1.30 (1.00–1.70) for analysis in Schinasi & Leon (2014)

Table 2.3 (continued)

Study reference	Relative risk (95% CI)	Included in Working Group analysis[a]		Included in previously published meta-analysis		Exposure–response reported	Study design	Study outcome	Comments
		Primary analysis	Second analysis	Goodman et al. (2015)	Schinasi & Leon (2014)				
Miligi et al. (2006a)	0.90 (0.50–1.80)	√	√	√	√		Case–control	NHL + CLL	
	4.40 (1.10–29.10)						Case–control	NHL + CLL	Probability of use > low (rated by hygienist) and lack of protective equipment. This estimate reflects the highest exposure
Hartge et al. (2005)	0.89 (0.49–1.59)			√		√	Case–control	NHL	Adjusted for use of other pesticides Excluded from the Working Group meta-analysis because 2,4-D was measured in dust, which is different from the exposure assessment for the other studies
Hardell et al. (1994)	13.00 (1.20–360)	√	√	√			Case–control	NHL	Adjusted for use of other pesticides Estimate for "2,4-D only"without exposure to other phenoxyherbicides
Cantor et al. (1992)	1.20 (0.90–1.60)		√		√		Case–control	NHL	Ever handled; 115 exposed cases
	1.20 (0.90–1.70)				√		Case–control	NHL	Highest exposed: handled without protective equipment; 89 exposed cases
Zahm et al. (1990)	1.50 (0.90–2.50)		√		√	√	Case–control	NHL	Adjusted for use of other pesticides Presents a more thorough analysis than Weisenburger (1990), but with complete overlap in patients
	1.50 (0.80–2.60)					√	Case–control	NHL	2,4-D use without 2,4,5-T
Hoar et al. (1986)	2.60 (1.40–5.00)					√	Case–control	NHL	Adjusted for use of other pesticides 2,4-D use without 2,4,5-T
	2.30 (1.30–4.30)		√			√	Case–control	NHL	Overall 2,4-D exposure

Table 2.3 (continued)

Study reference	Relative risk (95% CI)	Included in Working Group analysis[a]		Included in previously published meta-analysis		Exposure–response reported	Study design	Study outcome	Comments
		Primary analysis	Second analysis	Goodman et al. (2015)	Schinasi & Leon (2014)				
Hohenadel et al. (2011)	0.94 (0.67–1.33)	√		√			Case–control	NHL	Adjusted for use of other pesticides. Subset of 2,4-D without malathion exposure (49 exposed cases), but may have been exposed to other pesticides
De Roos et al. (2003)	0.90 (0.60–1.20)	√					Case–control	NHL	Adjusted for use of other pesticides. Hierarchical regression; pooled analysis that replaces Cantor et al. (1992), Zahm et al. (1990), and Hoar et al. (1986)
	0.80 (0.60–1.10)			√			Case–control	NHL	Adjusted for use of other pesticides. Logistic regression
Woods & Polissar (1989)	0.73 (0.40–1.30)	√	√	√			Case–control	NHL	The same estimate was reported in Woods et al. (1987)
Brown et al. (1993)	1.00 (0.60–1.60)	√	√				Case–control	MM	
McDuffie et al. (2001)	1.32 (1.01–1.73)					√	Case–control	NHL	
Brown et al. (1990)	1.30 (0.80–2.00)	√	√				Case–control	CLL	
Pahwa et al. (2006)	1.25 (0.93–1.68)		√				Case–control	MM	
	1.00 (0.62–1.61)	√					Case–control	MM	Adjusted for use of other pesticides.

[a] The "primary analysis" prioritized inclusion of estimates adjusted for other pesticides, when available, whereas the "second analysis" included estimates unadjusted for concomitant use of pesticides from the same studies.

2,4-D, 2,4-dichlorophenoxyacetic acid; 2,4,5-T, 2,4,5-trichlorophenoxyacetic acid; CI, confidence interval; CLL, chronic lymphocytic leukaemia; HCL, hairy cell leukaemia; MCPA, 2-methyl-4-chlorophenoxyacetic acid; MM, multiple myeloma; NHL, non-Hodgkin lymphoma; SMR, standardized mortality ratio

Fig. 2.1 Funnel plot (with pseudo 95% confidence limits) of cohort and case–control studies assessing exposure to 2,4-dichlorophenoxyacetic acid (2,4-D) and non-Hodgkin lymphoid neoplasms: primary analysis

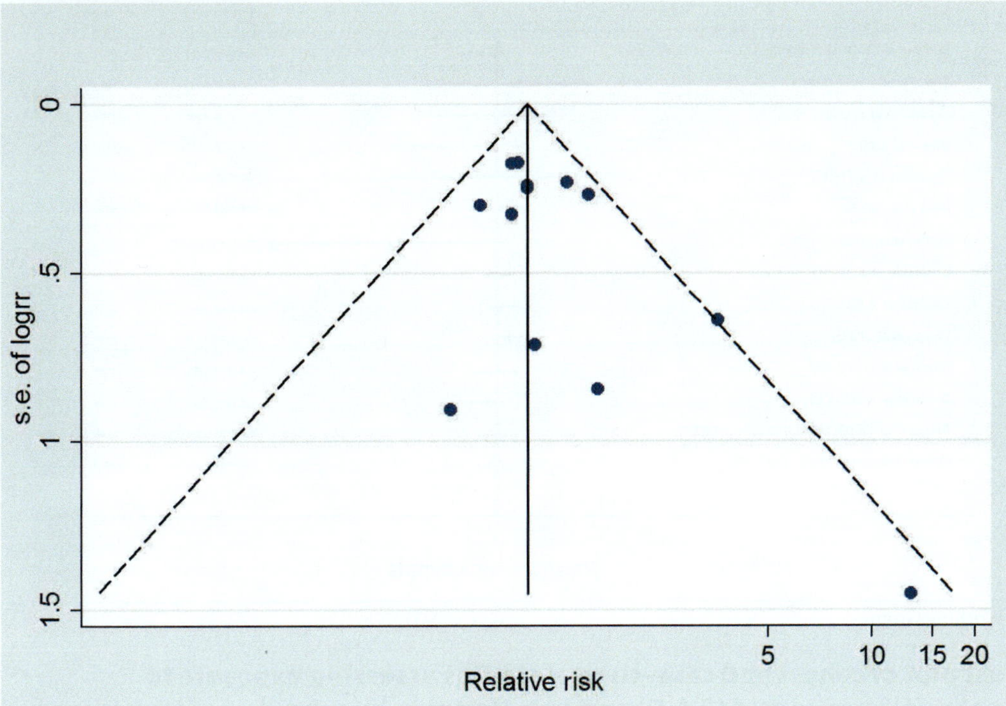

The figure plots the risk estimates for the 13 reports included in the primary analysis
s.e, standard error
Prepared by the Working Group

Table 2.4 Summary of results of the Working Group meta-analysis (primary and secondary analysis) of epidemiological studies on 2,4-dichlorophenoxyacetic acid (2,4-D) and non-Hodgkin lymphoid neoplasms

Analysis	No. of studies included	I^2	Meta-relative risk (95% CI)	Comments
Overall				
Most adjusted (primary analysis)	13	6.00%	1.04 (0.88–1.22)	De Roos et al. (2003) replaces the individual case–control studies in the midwest USA
Least adjusted (secondary analysis)	15	37.60%	1.31 (1.10–1.56)	Least adjusted estimates when possible, compare to primary analysis
By outcome				
Non-Hodgkin lymphoma	9	36.30%	1.06 (0.80–1.40)	Primary analysis restricted to non-Hodgkin lymphoma only, subtypes excluded
Multiple myeloma	3	0.00%	0.99 (0.71–1.39)	Restricted to multiple myeloma
Chronic lymphocytic leukaemia	2	0.00%	1.15 (0.79–1.67)	Restricted to chronic lymphocytic leukaemia

Fig. 2.2 Forest plot of cohort and case–control studies assessing exposure to 2,4-dichlorophenoxyacetic acid (2,4-D) and non-Hodgkin lymphoid neoplasms: primary analysis)

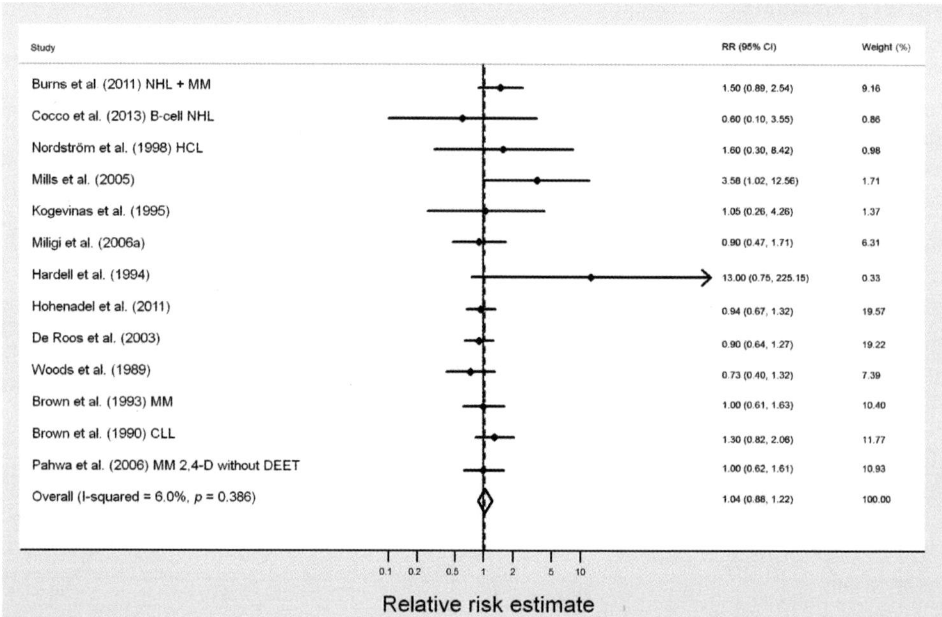

Prepared by the Working Group

Fig. 2.3 Forest plot of cohort and case–control studies assessing exposure to 2,4-dichlorophenoxyacetic acid (2,4-D) and non-Hodgkin lymphoid neoplasms: secondary analysis)

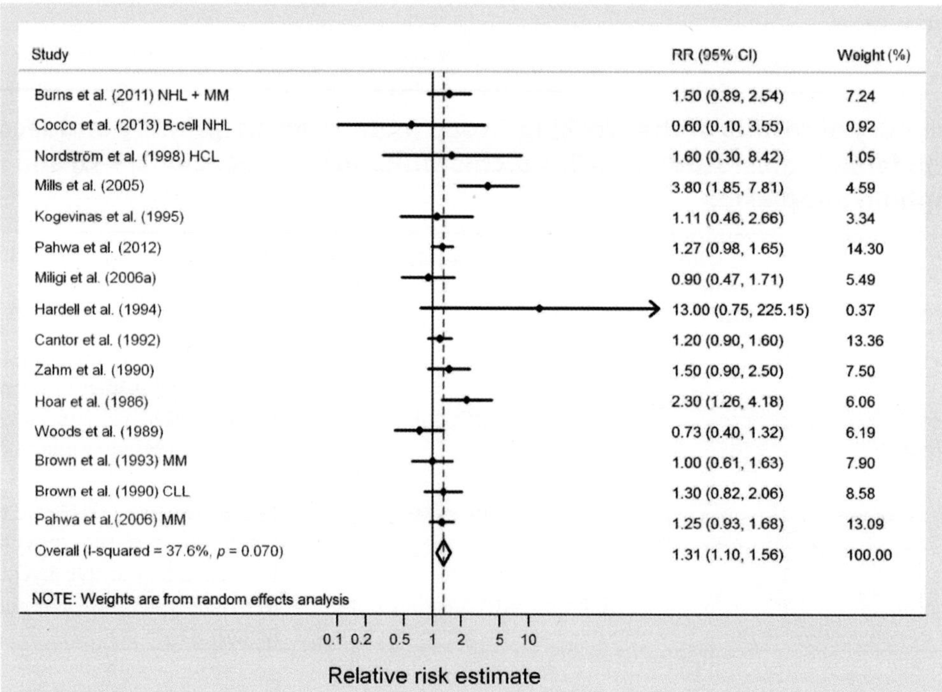

Prepared by the Working Group

$P_{heterogeneity}$ = 0.386) (Fig. 2.2; Table 2.4). A sensitivity analysis showed positive associations and a large heterogeneity when risk estimates that were adjusted for other pesticides (meta relative risk, 1.31; 95% CI, 1.10–1.56; I^2 = 37.6%) (Fig. 2.3; Table 2.4). For both the above, an analysis of the effect of omitting each study in turn did not substantially affect the overall meta relative risk.

3. Cancer in Experimental Animals

2,4-D and its butyl, isopropyl, and isooctyl esters were previously reviewed and evaluated for carcinogenicity by the *IARC Monographs* Working Groups for Volume 15 and Supplement 7 (IARC, 1977, 1987) on the basis of studies of oral administration in rats and mice, and subcutaneous administration in mice. All the studies considered at the time of these evaluations were found to have limitations, inadequate reporting, or inadequate (small) numbers of animals. In addition, only one dose group was used in most of the studies. The previous Working Group (IARC, 1987) concluded that there was *inadequate evidence* in experimental animals for the carcinogenicity of 2,4-D. For the present *Monograph*, the Working Group evaluated all relevant studies of carcinogenicity in experimental animals, including those published since the evaluations made in 1977 and 1987.

3.1 Mouse

See Table 3.1

3.1.1 Oral administration

Groups of 18 male and 18 female (C57BL/6 × C3H/Anf)F_1 mice and groups of 18 male and female (C57BL/6 × AKR)F_1 mice were given commercial 2,4-D (purity, 90%) at a dose of 46.4 mg/kg body weight (bw) per day in 0.5% gelatin (in distilled water) by gavage at age 7–28 days, followed by a diet containing 149 mg/kg bw ad libitum for 18 months. Additional groups of 18 male and 18 female (C57BL/6 × AKR)F_1 mice were given 2,4-D at a dose of 100 mg/kg bw per day by gavage, and subsequently fed a diet containing 2,4-D at 323 mg/kg bw for 18 months. Groups of 18 males and 18 females served as vehicle controls in all experiments. No significant increase in tumour incidences occurred in any of the treatment groups. Similar results were obtained in groups of 18 male and 18 female (C57BL/6 × C3H/Anf)F_1 and (C57BL/6 × AKR) F_1 mice given the 2,4-D isopropyl ester (purity, 99%), butyl ester (purity, 99%), or isooctyl ester (purity, 97%) at a dose of 46.4 mg/kg bw per day in 0.5% gelatin by stomach tube, at age 7–28 days, followed by diets containing the esters at a dose of 111, 149, or 130 mg/kg bw, respectively, for 18 months (NTIS, 1968; Innes et al., 1969). [The Working Group noted the inadequate number of treated and control animals, and the use of a single dose level.]

Groups of 50 male and 50 female B6C3F_1 CRL BR mice were given diets containing 2,4-D (purity, 96.4%) at a concentration of 0, 5, 62.5, or 125 mg/kg bw (males), and 0, 5, 150 or 300 mg/kg bw (females) for 104 weeks. There were no treatment-related deaths or clinical signs of toxicity in either sex. Body weight, body-weight gain, and food consumption were generally similar among the groups of males throughout the study. Body-weight gains of females at 300 mg/kg bw were non-significantly decreased during the first 18 months of the study. There were no treatment-related increases in the incidence of any tumour type in males or females (Charles et al., 1996a).

In a study submitted to the EPA (1997), groups of 50 male and 50 female B6C3F_1 mice were given diets containing 2,4-D (purity, 97.5%) at a dose of 0, 1, 15, or 45 mg/kg bw for 104 weeks. There were no treatment-related effects on survival or body weight. There were no treatment-related increases in tumour incidences in any treated group of males or females.

Table 3.1 Studies[a] of carcinogenicity with 2,4-D and its esters in mice

Strain (sex), age at start Duration Reference	Dosing regimen No. of animals/group at start	Results For each target organ: incidence (%) and/or multiplicity of tumours	Significance	Comments
2,4-D				
(C57BL/6 × C3H/Anf)F_1 (F, M), age 7 days 18 mo NTIS (1968); Innes et al. (1969)	2,4-D (purity, 90%) at a dose of 46.4 mg/kg bw per day in 0.5% gelatin/distilled water by gavage at age 7–28 days, then diet containing 2,4-D at 149 mg/kg bw per day, up to 18 mo; controls were given vehicle only by gavage, followed by untreated diet 18 M and 18 F/group	No increase in the incidence of any tumour type (M or F)	–	Strengths: covered most of the lifespan Limitations: inadequate numbers of animals; only one dose group; study poorly reported Survival, M: 16/18, 15/18 Survival, F: 16/18, 16/18
(C57BL/6 × AKR)F_1 (F, M), age 7 days 18 mo NTIS (1968); Innes et al. (1969)	2,4-D (purity, 90%) at a dose of 46.4 mg/kg bw per day in 0.5% gelatin/distilled water by gavage at age 7–28 days, then diet containing 2,4-D at 149 mg/kg bw per day, up to age 18 mo; controls were given vehicle only by gavage, followed by untreated diet 18 M and 18 F/group	No increase in the incidence of any tumour type (M or F)	–	Strengths: covered most of the lifespan Limitations: inadequate numbers of animals; only one dose group; study poorly reported Survival, M: 18/18, 16/18 Survival, F: 15/18, 16/18
(C57BL/6 × AKR)F_1 (F, M), age 7 days 18 mo NTIS (1968); Innes et al. (1969)	2,4-D (purity, 90%) at a dose of 100 mg/kg bw per day in 0.5% gelatin/distilled water by gavage at age 7–28 days, then diet containing 2,4-D at 323 mg/kg bw per day, up to age 18 mo; controls were given vehicle only by gavage, followed by untreated diet 18 M and 18 F/group	No increase in the incidence of any tumour type (M or F)	–	Strengths: covered most of the lifespan Limitations: inadequate numbers of animals; only one dose group; study poorly reported Survival, M: 18/18, 11/18 Survival, F: 15/18, 13/18
B6C3F_1 CRL BR (F, M), age 6–7 wk 104 wk Charles et al. (1996a)	Diets containing 2,4-D (purity, 96.4%) at a concentration of 0, 5, 62.5, 125 mg/kg bw per day (M) or 0, 5, 150, 300 mg/kg bw per day (F) for 104 wk 50 M and 50 F/group	*Liver* Hepatocellular adenoma: M: 12/50, 9/50, 13/50, 16/50 F: 5/50, 11/50, 8/50, 10/50 Hepatocellular carcinoma: M: 6/50, 3/50, 7/50, 4/50 F: 1/50, 2/50, 0/50, 1/50	 NS NS NS	Strengths: covered most of the lifespan Body-weight gain in females decreased during the first 18 mo of the study, but was not affected in males All mice survived to the end of the study

Table 3.1 (continued)

Strain (sex), age at start Duration Reference	Dosing regimen No. of animals/group at start	Results For each target organ: incidence (%) and/or multiplicity of tumours	Significance	Comments
$B6C3F_1$ (F, M), age NR 104 wk EPA (1997)	Diets containing 2,4-D (purity, 97.5%) at a concentration of 0, 1, 15, or 45 mg/kg bw per day, for 104 wk 50 M and 50 F/group	No increase in the incidence of any tumour type (M or F) Pulmonary adenoma, multiplicity: 14.6 ± 0.8, 15.0 ± 0.8, 18.7 ± 0.9, 14.9 ± 0.8	– NS	Strengths: covered most of the lifespan Survival, NR
(C57BL/6 × C3H/Anf)F1 (F, M), age 28 days 18 mo NTIS (1968)	Single s.c. injection of 2,4-D (purity, 90%) at a dose of 0, or 215 mg/kg bw in DMSO 24 M and 24 F/group (controls) 18 M and 18 F/group (215 mg/kg bw)	No increase in the incidence of any tumour type (M or F)	–	Strengths: covered most of the lifespan Limitations: inadequate numbers of animals; only one dose group; study poorly reported Survival, M: 23/24, 16/18 Survival, F: 23/24, 17/18
(C57BL/6 × AKR)F1 (F, M), age 28 days 18 mo NTIS (1968)	Single s.c. injection of 2,4-D (purity, 90%) at a dose of 0, or 215 mg/kg bw in DMSO 24 M and 24 F/group (controls) 18 M and 18 F/group (215 mg/kg bw)	No increase in the incidence of any tumour type (M or F)	–	Strengths: covered most of the lifespan Limitations: inadequate numbers of animals; only one dose group; study poorly reported Survival, M: 22/24, 18/18 Survival, F: 24/24, 18/18
2,4-D esters				
$(C57BL/6 \times C3H/Anf)F_1$ (F, M), age 7 days 18 mo NTIS (1968); Innes et al. (1969)	2,4-D isopropyl ester (purity, 99%) at a dose of 46.6 mg/kg bw per day in 0.5% gelatin/distilled water by gavage at age 7–28 days, then diet containing 2,4-D isopropyl ester at 111 mg/kg bw per day, up to 18 mo; controls were given vehicle only by gavage, followed by untreated diet 18 M and 18 F/group	No increase in the incidence of any tumour type (M or F)	–	Strengths: covered most of the lifespan Limitations: inadequate numbers of animals; only one dose group; study poorly reported Survival, M: 16/18, 18/18 Survival, F: 16/18, 18/18
$(C57BL/6 \times AKR)F_1$, (F, M), age 7 days 18 mo NTIS (1968); Innes et al. (1969)	2,4-D isopropyl ester (purity, 99%) at a dose of 46.6 mg/kg bw per day in 0.5% gelatin/distilled water by gavage at age 7–28 days, then diet containing 2,4-D isopropyl ester at 111 mg/kg bw per day, up to 18 mo; controls were given vehicle only by gavage, followed by untreated diet 18 M and 18 F/group	No increase in the incidence of any tumour type (M or F)	–	Strengths: covered most of the lifespan Limitations: inadequate numbers of animals; only one dose group; study poorly reported Survival, M: 18/18, 17/18 Survival, F: 15/18, 14/18

Table 3.1 (continued)

Strain (sex), age at start Duration Reference	Dosing regimen No. of animals/group at start	Results For each target organ: incidence (%) and/or multiplicity of tumours	Significance	Comments
$(C57BL/6 \times C3H/Anf)F_1$, (F, M), age 7 days 18 mo NTIS (1968); Innes et al. (1969)	2,4-D butyl ester (purity, 99%) at a dose of 46.4 mg/kg bw per day in 0.5% gelatin/distilled water by gavage at age 7–28 days, then diet containing 2,4-D butyl ester at 149 mg/kg bw per diet, up to 18 mo; controls were given vehicle only by gavage, followed by untreated diet 18 M and 18 F/group	No increase in the incidence of any tumour type (M or F)	–	Strengths: covered most of the lifespan Limitations: inadequate numbers of animals; only one dose group; study poorly reported Survival, M: 16/18, 17/18 Survival, F: 16/18, 17/18
$(C57BL/6 \times AKR)F_1$, (F, M), age 7 days 18 mo NTIS (1968); Innes et al. (1969)	2,4-D butyl ester (purity, 99%) at a dose of 46.4 mg/kg bw per day in 0.5% gelatin/distilled water by gavage at age 7–28 days, then diet containing 2,4-D butyl ester at 149 mg/kg bw per day, up to 18 mo; controls were given vehicle only by gavage, followed by untreated diet 18 M and 18 F/group	No increase in the incidence of any tumour type (M or F)	–	Strengths: covered most of the lifespan Limitations: inadequate numbers of animals; only one dose group; study poorly reported Survival, M: 18/18, 18/18 Survival, F: 15/18, 16/18
$(C57BL/6 \times C3H/Anf)F_1$ (F, M), age 7 days 18 mo NTIS (1968); Innes et al. (1969)	2,4-D isooctyl ester (purity, 97%) at a dose of 46.4 mg/kg bw per day in 0.5% gelatin/distilled water by gavage at age 7–28 days, then diet containing 2,4-D butyl ester at 130 mg/kg bw per day, up to 18 mo; controls were given vehicle only by gavage, followed by untreated diet 18 M and 18 F/group	No increase in the incidence of any tumour type (M or F)	–	Strengths: covered most of the lifespan Limitations: inadequate numbers of animals; only one dose group; study poorly reported Survival, M: 16/18, 16/18 Survival, F: 16/18, 16/18
$(C57BL/6 \times AKR)F_1$ (F, M), age 7 days 18 mo NTIS (1968); Innes et al. (1969)	2,4-D isooctyl ester (purity, 97%) at a dose of 46.4 mg/kg bw per day in 0.5% gelatin/distilled water by gavage at age 7–28 days, then diet containing 2,4-D butyl ester at 130 mg/kg bw per day, up to 18 mo; controls were given vehicle only by gavage, followed by untreated diet 18 M and 18 F/group	No increase in the incidence of any tumour type (M or F)	–	Strengths: covered most of the lifespan Limitations: inadequate numbers of animals; only one dose group; study poorly reported Survival, M: 18/18, 17/18 Survival, F: 15/18, 16/18

Table 3.1 (continued)

Strain (sex), age at start Duration Reference	Dosing regimen No. of animals/group at start	Results For each target organ: incidence (%) and/or multiplicity of tumours	Significance	Comments
(C57BL/6 × C3H/Anf)F_1 (F, M), age 28 days 18 mo NTIS (1968)	Single s.c. injection of 2,4-D isopropyl ester (purity, 99%) at a dose of 0, or 100 mg/kg bw in corn oil 24 M and 24 F/group (controls) 18 M and 18 F/group (100 mg/kg bw)	No increase in the incidence of any tumour type (M or F)	–	Strengths: covered most of the lifespan Limitations: inadequate numbers of animals; only one dose group; study poorly reported Survival, M: 17/24, 16/18 Survival, F: 21/24, 18/18
(C57BL/6 × AKR)F_1 (F, M), age 28 days 18 mo NTIS (1968)	Single s.c. injection of 2,4-D isopropyl ester (purity, 99%) at a dose of 0, or 100 mg/kg bw in corn oil 24 M and 24 F/group (controls) 18 M and 18 F/group (100 mg/kg bw)	No increase in the incidence of any tumour type (M or F)	–	Strengths: covered most of the lifespan Limitations: inadequate numbers of animals; only one dose group; study poorly reported Survival, M: 16/24, 16/18 Survival, F: 19/24, 16/18
(C57BL/6 × C3H/Anf)F_1 (F, M), age 28 days 18 mo NTIS (1968)	Single s.c. injection of 2,4-D butyl ester, (purity, 99%) at a dose of 0, or 21.5 mg/kg bw in corn oil 24 M and 24 F/group (controls) 18 M and 18 F/group (21.5 mg/kg bw)	No increase in the incidence of any tumour type (M or F)	–	Strengths: covered most of the lifespan Limitations: inadequate numbers of animals; only one dose group; study poorly reported Survival, M: 17/24, 16/18 Survival, F: 21/24, 18/18
(C57BL/6 × AKR)F_1 (F, M), age 28 days 18 mo NTIS (1968)	Single s.c. injection of 2,4-D butyl ester, (purity, 99%) at a dose of 0, or 21.5 mg/kg bw in corn oil 24 M and 24 F/group (controls) 18 M and 18 F/group (21.5 mg/kg bw)	No increase in the incidence of any tumour type (M or F)	–	Strengths: covered most of the lifespan Limitations: inadequate numbers of animals; only one dose group; study poorly reported Survival, M: 16/24, 18/18 Survival, F: 19/24, 16/18
(C57BL/6 × C3H/Anf)F_1 (F, M), age 28 days 18 mo NTIS (1968)	Single s.c. injection of 2,4-D isooctyl ester (purity, 97%) at a dose of 0, or 21.5 mg/kg bw in corn oil 24 M and 24 F/group (controls) 18 M and 18 F/group (21.5 mg/kg bw)	No increase in the incidence of any tumour type (M or F)	–	Strengths: covered most of the lifespan Limitations: inadequate numbers of animals; only one dose group; study poorly reported Survival, M: 17/24, 16/18 Survival, F: 21/24, 16/18

Table 3.1 (continued)

Strain (sex), age at start Duration Reference	Dosing regimen No. of animals/group at start	Results For each target organ: incidence (%) and/or multiplicity of tumours	Significance	Comments
(C57BL/6 × AKR)F_1 (F, M), age 28 days 18 mo NTIS (1968)	Single s.c. injection of 2,4-D isooctyl ester (purity, 97%) at a dose of 0, or 21.5 mg/kg bw in corn oil	*Haematopoietic system* Reticulum cell sarcoma: M: 0/16, 0/18 F: 0/19, 5/17	 NS [$P < 0.02$, Fisher exact test]	Strengths: covered most of the lifespan Limitations: inadequate numbers of animals; only one dose group; study poorly reported Survival, M: 16/24, 18/18 Survival, F: 19/24, 17/18
CD-1 (M), age 3 wk (weanling) 15 wk Blakley et al. (1992) Co-carcinogenicity study	Drinking-water containing a commercial amine formulation of 2,4-D (purity, NR) at a dose of 0, 10, 25, or 50 mg/kg bw per day, for 15 wk. After 3 wk of treatment, mice were given a single i.p. injection of urethane at 1.5 g/kg bw in saline, and were killed after the 15-wk exposure period 25/group	*Lung* Pulmonary adenoma, multiplicity: 6.7 ± 0.7, 12.0 ± 1.7*, 11.3 ± 2.1, 9.5 ± 1.6	 $^*P = 0.0207$	Limitations: short duration, inadequate numbers of animals, only included males
CD-1 (F), age 6–7 wk 19 wk Lee et al. (2000) Co-carcinogenicity study	Pregnant mice given drinking-water containing a commercial amine formulation of 2,4-D (purity, NR) at a dose of 0, 15, 65, or 650 mg/kg bw on days 6–16 of gestation. At age 7 wk, female offspring were given a single i.p. injection of urethane at 1.5 g/kg bw in saline, and killed 19 wk after birth 25/group	*Lung* Pulmonary adenoma, multiplicity: 14.6 ± 0.8, 15.0 ± 0.8, 18.7 ± 0.9, 14.9 ± 0.8	 NS	Limitations: short duration, inadequate numbers of animals, only included females In-utero exposure to 2,4-D did not affect the number of urethane-induced adenomas

[a] All studies are full studies of carcinogenicity, unless otherwise stated

–, no significance test performed; 2,4-D, 2,4-dichlorophenoxyacetic acid; bw, body weight; DMSO, dimethyl sulfoxide; F, female; i.p., intraperitoneal; M, male; mo, month; NA, not applicable; NR, not reported; NS, not significant; s.c., subcutaneous; wk, week

3.1.2 Subcutaneous injection

Groups of 18 male and 18 female (C57BL/6 × C3H/Anf)F_1 mice and groups of 18 male and female (C57BL/6 × AKR)F_1 mice were given single subcutaneous injections of 2,4-D (purity, 90%) at a dose of 215 mg/kg bw in dimethyl sulfoxide (DMSO) at age 28 days, and observed for 18 months. At termination of the study, 16–18 mice in the four treated groups were still alive. Groups of 24 male and 24 female mice served as vehicle controls in all experiments. There were no treatment-related increases in tumour incidences in any of the treatment groups.

Tumour incidences were not increased in groups of 18 male and 18 female (C57BL/6 × C3H/Anf)F_1 mice, and 18 male and 18 female (C57BL/6 × AKR)F_1 mice given single subcutaneous injections of 2,4-D butyl esters (purity, 90%) at a dose of 21.5 mg/kg bw, or 2,4-D isopropyl esters (purity, 90%) at a dose of 100 mg/kg bw, in corn oil. However, in (C57BL/6 × AKR)F_1 mice similarly injected with the isooctyl ester of 2,4-D (purity, 97%) at a dose of 21.5 mg/kg bw in corn oil, 5 out of 17 female (C57BL/6 × AKR)F_1 mice developed reticulum cell sarcoma [$P < 0.02$, versus 0 out of 19 controls]; there was no increase in tumour incidence in males treated with the isooctyl ester, or in (C57BL/6 × C3H/Anf)F_1 male and female mice treated with the isooctyl ester (NTIS, 1968). [The Working Group noted the inadequate number of treated and control animals, the use of a single injection and of a single dose. The Working Group also noted that reticulum cell sarcoma may be classified by current terminology as histiocytic sarcoma or as a type of mixed cell malignant lymphoma.]

3.1.3 Coadministration with a known carcinogen

Groups of 25 male CD-1 mice were given drinking-water containing a commercial amine formulation of 2,4-D [containing 2,4-D at 140 g/L; the content of other chemicals was not reported] at a concentration of 0, 10, 25, or 50 mg/kg bw per day for 15 weeks. After 3 weeks, the mice were given a single intraperitoneal injection of urethane at a dose of 1.5 g/kg bw in saline. The effect of 2,4-D on urethane-induced formation of pulmonary adenoma was evaluated 84 days after urethane injection. Treatment with 2,4-D significantly increased the multiplicity of pulmonary adenoma (6.7 ± 0.7, 12.0 ± 1.7*, 11.3 ± 2.1, 9.5 ± 1.6; *$P = 0.0207$) (Blakley et al., 1992).

Pregnant CD-1 mice were given drinking-water containing a commercial amine formulation of 2,4-D [the chemical content was not reported] at a concentration of 0, 15, 65, or 650 mg/kg bw on days 6–16 of gestation. At age 7 weeks, female offspring were given a single intraperitoneal injection of urethane at a dose of 1.5 g/kg bw in saline. The effect of 2,4-D on urethane-induced formation of pulmonary adenoma was evaluated in female offspring 19 weeks after birth. Treatment with 2,4-D did not significantly increase the multiplicity of pulmonary adenoma (14.6 ± 0.8, 15.0 ± 0.8, 18.7 ± 0.9, 14.9 ± 0.8, respectively) (Lee et al., 2000).

3.2 Rat

See Table 3.2

Oral administration

Groups of 25 male and 25 female Osborne-Mendel rats (age, 3 weeks) were given diets containing 2,4-D (purity, 96.7%) at a concentration of 0, 5, 25, 125, 625, or 1250 ppm for 2 years. All rats were killed and necropsied after 2 years,

Table 3.2 Studies[a] of carcinogenicity with 2,4-D in rats

Strain (sex), age at start Duration Reference	Dosing regimen No. animals/group at start	Results For each target organ: incidence (%) and/or multiplicity of tumours	Significance	Comments
Osborne-Mendel (F, M), age 3 wk 2 yr Hansen et al. (1971)	Diets containing 2,4-D (purity, 96.7%) at a concentration of 0, 5, 25, 125, 625, or 1250 ppm for 2 years 25 M and 25 F/group	No increase in the incidence of any tumour type (M or F)	–	Strengths: covered most of the lifespan Limitations: detailed histopathological examination of tissues does not appear to have been consistent across all dose groups: all tissues from 6 males and 6 females from the 1250-ppm and control groups examined, and certain tissues in other groups; inadequate numbers of animals No significant differences in survival, mean body weights, or relative organ weights between controls (M and F) and dosed rats
F344 (F, M), age NR 104 wk EPA (1997)	Diets containing 2,4-D (purity, 97.5%) at a dose of 0, 1, 5, 15, or 45 mg/kg bw per day, for 104 wk 60 F and 60 M/group	*Brain* Astrocytoma: M: 1/60, 0/60, 0/60, 2/58, 6/60 F: 0/60, 1/60, 2/60, 1/60, 1/60	 $P = 0.0026$, trend NS	Strengths: covered most of the lifespan Decrease in body-weight gains in females at the highest dose Survival, NR
F344 (F, M), age 5 wk 104 wk Charles et al. (1996a); EPA (1997)	Diets containing 2,4-D (purity, 96.4%) et a dose of 0, 5, 75, or 150 mg/kg bw per day, for 104 wk 50, 50, 50, 50 28, 25, 33, 36	*Brain* Astrocytoma: M: 0/50, 0/50, 0/50, 1/50 F: 1/50, 0/50, 0/50, 1/50	 NS NS	Strengths: covered most of the lifespan Body-weight gains and average food consumption were decreased throughout the study at the highest dose level (M and F) Survival, M: 28/50, 25/50, 33/50, 36/50 Survival, F: 35/50, 39/50, 40/50, 35/50

[a] All studies are full studies of carcinogenicity, unless otherwise stated

2,4-D, 2,4-dichlorophenoxyacetic acid; bw, body weight; DMSO, dimethyl sulfoxide; F, female; i.p., intraperitoneal; M, male; mo, month; NA, not applicable; NR, not reported; NS, not significant; s.c., subcutaneous; wk, week

except for one rat at the highest dose that died during the experiment. Microscopic examinations were conducted on heart, lung, liver, spleen, kidney, stomach, intestines, pancreas, pituitary, thyroid, adrenal, bone (including marrow), ovary and uterus (or testis and prostate), tumours, and other gross lesions from six males and six females from the group at the highest dose and the control group. Only the liver, kidney, spleen, ovary (or testis), tumours and other gross lesions from rats at other doses were subjected to microscopic examination. Treatment with 2,4-D did not affect survival, body weight, or organ weights at any dose. There were no treatment-related increases in tumour incidences in any treated group of males or females (Hansen et al., 1971). [The Working Group noted that microscopic evaluations were not performed on a standard comprehensive list of tissues and organs across all dose groups].

In a study submitted to the United States EPA (EPA, 1997), groups of 60 male and 60 female F344 rats were given diets containing 2,4-D (purity, 97.5%) at a dose of 0 (control), 1, 5, 15, or 45 mg/kg bw per day for 2 years. Body-weight gains were significantly decreased in females at the highest dose, relative to controls. The incidences of astrocytoma in the brain in the control group, and the groups at 1, 5, 15, and 45 mg/kg bw were: 1/60, 0/60, 0/60, 2/58, and 6/60 in males, respectively; and 0/60, 1/60, 2/60, 1/60, and 1/60 in females, respectively. In males, the incidence of astrocytoma showed a statistically significant positive trend ($P = 0.0026$), but the incidence in the group at the highest dose was not statistically significant.

In a study of combined long-term toxicity and carcinogenicity, groups of 50 male and 50 female F344 rats were given diets containing 2,4-D (purity, 96.4%) at a dose of 0, 5, 75, or 150 mg/kg bw for up to 104 weeks. There were no treatment-related deaths. Body-weight gains and average food consumption were decreased throughout the study in males and females at the highest dose. There was no treatment-related increase in the incidence of any tumour, including astrocytoma of the brain. The incidence of astrocytoma was: 0/50, 0/50, 0/50, 1/50 in males, respectively; and 1/50, 0/50, 0/50, 1/50 in females, respectively (Charles et al. 1996a; EPA, 1997).

[The Working Group noted that the study by Charles et al. (1996a) was designed to address the finding of a significant positive trend in the incidence of astrocytoma of the brain (with no pairwise significance) found in the study submitted to the EPA (1997). In the study by Charles et al. (1996a), the rats were given higher doses, and there was no significant increase in the incidence of astrocytoma of the brain (Charles et al., 1996a; EPA, 1997).]

3.3 Dog

A hospital-based case–control study of pet dogs examined the risk of developing canine malignant lymphoma associated with the use of 2,4-D in and about the dog owner's home. Dogs with histopathologically confirmed malignant lymphoma, newly diagnosed over a 4-year period, were identified using computerized medical-record abstracts from three veterinary medical teaching hospitals in the states of Colorado, Indiana, and Minnesota, USA. Two comparison control groups, matched by age group, but not by sex, were chosen from dogs seen at the same teaching hospital in the same year as the identified case, with one-to-one matching for each control group. The first control group (tumour control) was selected from all other tumour cases diagnosed at the teaching hospital, excluding transitional cell carcinoma of the lower urinary tract because of a potential etiology related to chemical exposures. The second control group (non-tumour control) was selected from dogs seen for any other diagnostic reason, excluding those with conditions possibly related to chemical exposures (e.g. nonspecific dermatitis, neuropathies). Owners of dogs in the case and control groups were sent a standardized questionnaire requesting information about the demographic characteristics of all people living in their home, basic information about the dog's life history, household use of chemicals (in and about the home), including those directly applied to the pet, and personal use of chemicals of whatever kind on the lawn and garden and/or the employment of commercial companies applying such chemicals. In addition, owners were asked about opportunities for exposure of their pets to lawn and garden chemicals, including frequency of access to the yard. The questionnaire did not provide a list of chemicals that the owner could consult in responding to the various questions regarding home, lawn, and gardening chemicals. Information from the self-administered questionnaire and/or a telephone interview of owners of 491 cases, 466 non-tumour controls, and 479 tumour controls indicated that owners of dogs that developed malignant lymphoma had applied 2,4-D herbicides to their lawn and/or employed commercial lawn-care companies to treat their yard significantly more frequently than control owners (OR, 1.3; 95% CI, 1.04–1.67). The risk of canine malignant lymphoma rose to a twofold (OR, 2.0; 95% CI, 0.92–4.15) non-significant excess with four or more lawn applications of 2,4-D per year by the owner. The findings in this study were consistent with those of studies of occupational exposure in humans, which have reported modest associations between agricultural exposure to 2,4-D and increased risk of NHL, the histology and epidemiology of which are similar to those of canine malignant lymphoma (Hayes et al., 1991). [The Working Group noted that details on the assessment procedures were described very briefly, and that information on the type of chemicals applied to the lawns of the respondents by commercial companies was not provided.]

Hayes et al. (1995), responding to a review of their earlier article (Hayes et al., 1991) by Carlo et al. (1992), presented additional data and a subsequent analysis showing that risk estimates did not vary by type of control group (tumour control or non-tumour control), by method of response (self-administered or telephone interview) or by geographical area of recruitment of the subjects.

A re-analysis of the original data was reported by Kaneene & Miller (1999). This re-analysis included a reclassification of exposure to 2,4-D, and a considerable number of the animals characterized as exposed in the original analysis by Hayes et al. (1991) were considered to be non-exposed in the re-analysis. The odds ratio for owner and/or commercial application was slightly lower and non-significant in the re-analysis (OR, 1.2; 95% CI, 0.87–1.65) compared to that in the original article (OR, 1.3; 95% CI, 1.04–1.67) by Hayes et al. (1991). In the re-analysis, there was also no clear dose–response relationship found for level of lawn treatment by the owner, although a non-significant elevated odds ratio was still observed in the highest quartile of exposure (four or more applications per year by the owner) versus non-exposed (OR, 2.4; 95% CI, 0.87–6.72) [no P value for trend was reported]. [The Working Group noted that although Kaneene & Miller (1999) extensively revised the exposure assessment, the re-analysis still lacked information on type of commercial application, as did the original analysis by Hayes et al. (1991). In addition, the classification as non-exposed of many animals that had been previously classified as exposed to applications by commercial companies in the original article may have led to exposure misclassification of the non-exposed group.]

[Overall, results from the original article, the subsequent analysis and the re-analysis were difficult to interpret due to potential exposure misclassification.]

4. Mechanistic and Other Relevant Data

4.1 Toxicokinetics

4.1.1 Humans

(a) Absorption, distribution, and excretion

2,4-D and its salts and esters are readily absorbed after exposure. Dermal absorption has been demonstrated experimentally in humans, with about 5% of the dermally applied dose recovered in the urine in two studies (Feldmann & Maibach, 1974; Harris & Solomon, 1992). However, another study reported that absorption ranged from 7% to 14% depending on the vehicle (water or acetone), and on whether a mosquito repellent (DEET; *N,N*-diethyl-*meta*-toluamide) was also applied (Moody et al., 1992). A review of studies of dermal absorption reported a weighted average (± standard deviation) absorption rate of 5.7% (± 3.4%) (Ross et al., 2005). In skin from human cadavers, the relative percentage of dermal absorption of 2,4-D from contaminated soil increased as soil loadings of 2,4-D decreased (Duff & Kissel, 1996). Absorption after inhalation has not been directly measured experimentally in humans.

Experimental studies in humans demonstrated essentially complete absorption in the gastrointestinal tract after oral exposure (Kohli et al., 1974; Sauerhoff et al., 1977). 2,4-D has been detected in plasma after oral exposure, with terminal plasma half-lives of 7–16 hours, and urinary elimination half-lives of 10–30 hours (Sauerhoff et al., 1977). A somewhat slower elimination half-life of around 33 hours was reported in another study (Kohli et al., 1974). However, both results suggested that 2,4-D would not accumulate with repeated exposure. Additionally, Sauerhoff et al. (1977) reported a distribution volume of around 300 mL/kg, suggesting some distribution to tissues. Like other organic acids of low relative molecular mass, 2,4-D is reversibly bound to plasma proteins, which explains the relatively low distribution volume. For instance, several studies have suggested that 2,4-D binds to human serum albumin (Rosso et al., 1998, Purcell et al., 2001). Rosso et al. (1998) reported that 2,4-D has a poor ability to penetrate membranes.

Available data suggested that excretion of 2,4-D is largely, if not completely, via the urine, regardless of the route of exposure (Feldmann & Maibach, 1974; Kohli et al., 1974; Sauerhoff et al., 1977). The rate of excretion exceeds that which would be expected by passive glomerular filtration, consistent with a role for active transport in the kidney. Based on uptake measured in human kidney slices, organic anion transporter 1 (OAT1) appeared to be largely responsible for the ability of the kidney to excrete 2,4-D (Nozaki et al., 2007). While mainly expressed in the kidney, OAT1 is also expressed at lower levels in the human brain (Burckhardt & Wolff, 2000), and thus may play a role in the blood–brain distribution of 2,4-D (Gao & Meier, 2001).

(b) Metabolism

The salts and esters of 2,4-D are hydrolysed in vivo to 2,4-D, which undergoes a minor amount of metabolism in humans. In one study, 75% of an administered oral dose was excreted unchanged in the urine after 96 hours (Kohli et al., 1974). In another study of oral dosing, 13% of the administered dose was excreted as a 2,4-D hydrolysable conjugates, with about 82% excreted as unchanged 2,4-D (Sauerhoff et al., 1977). The identities of metabolites were not determined in these studies. One study using human CYP3A4 expressed in yeast reported metabolism of 2,4-D to 2,4-dichlorophenol (2,4-DCP) (Mehmood et al., 1996), but no data confirming metabolism of 2,4-D to 2,4-DCP in exposed humans were available.

4.1.2 Experimental systems

The toxicokinetics of 2,4-D has been studied in multiple species, but only studies in mammals are discussed here.

(a) Absorption, distribution, and excretion

2,4-D is readily absorbed by all experimental animal species tested. Absorption from the lung has been measured in rats in one study, although 2,4-D was administered through a tracheal cannula to anaesthetized animals (Burton et al., 1974). The rate of absorption was found to be rapid, with a half-time of 1.4 minutes and no evidence of saturation up to a concentration of 10 mM (Burton et al., 1974). Rapid and nearly complete absorption via the oral route has been reported in multiple species, including mice, rats, hamsters, dogs, pigs, and calves (Erne 1966; Pelletier et al., 1989; Griffin et al., 1997; van Ravenzwaay et al., 2003). Dermal absorption has also been measured in multiple species, including mice, rats, rabbits, and monkeys, with absorption rates between 6% and 36% (Grissom et al., 1985; Pelletier et al., 1989; Moody et al., 1990; Knopp & Schiller 1992). Multiple experimental studies have also reported that 2,4-D absorption is enhanced by ethanol consumption or application of topical sunscreens, moisturizers, or insect repellents (Pelletier et al., 1990; Brand et al., 2003, 2007a, b, c; Pont et al., 2003).

After absorption, 2,4-D readily distributes to tissues via systemic circulation. In all species tested, it is detected in the plasma after oral or dermal exposure. In a study in rats given radiobelled 2,4-D by oral or dermal administration, peak concentrations of radiolabel in the blood and kidney were reached within 30 minutes of administration by either route (Pelletier et al., 1989). In another study in rats given radiolabelled 2,4-D by oral administration, peak concentrations in tissues were reached 6–20 hours after exposure, with the highest levels in the lung, heart, liver, spleen, and kidney, and the lowest levels in adipose tissue and brain (Deregowski et al., 1990). In a study of toxicokinetics in male and female mice and rats given 2,4-D as an oral dose at 5 mg/kg bw, the highest peak concentrations were found in the kidney (Griffin et al., 1997). At a higher oral dose of 200 mg/kg bw, blood concentrations were nearly equal to or higher than kidney concentrations in mice and hamsters (Griffin et al., 1997). In studies in rats, rabbits, and goats, administration of 2,4-D during pregnancy or lactation lead to distribution of 2,4-D to the fetus or to milk (Kim et al., 1996; Sandberg et al., 1996; Stürtz et al., 2000; Barnekow et al., 2001; Stürtz et al., 2006; Saghir et al., 2013).

Like other organic acids of low relative molecular mass, 2,4-D is reversibly bound to plasma proteins (Arnold & Beasley, 1989). In studies in rats, dogs, and goats given 2,4-D orally, the binding fraction was reported to be more than 85% (Örberg, 1980; Griffin et al., 1997; van Ravenzwaay et al., 2003). Plasma binding explains the relatively low apparent volume of distribution (van Ravenzwaay et al., 2003). Binding specifically to bovine serum albumins has been measured in vitro (Mason, 1975; Fang & Lindstrom 1980).

In rats, a disproportionate increase in plasma concentrations of 2,4-D, consistent with saturation of elimination, occurs as doses increase (Saghir et al., 2006). The lowest dietary dose at which this disproportionality has been observed was between 25 and 79 mg/kg bw per day, depending on sex and life-stage (pups, adults, pregnant, lactating) (Saghir et al., 2013).

Several studies in rats and rabbits have examined the distribution of 2,4-D to the brain at higher exposures (Kim et al., 1988; Tyynelä et al., 1990; Kim et al., 1994). OAT1 is expressed in the rat and mouse brain (Burckhardt & Wolff, 2000), and may play a role in the distribution of 2,4-D to the brain (Gao & Meier, 2001). Brain concentrations of 2,4-D at exposures above 100 mg/kg bw appear higher than would be expected based on passive diffusion, given the plasma protein binding of 2,4-D (Tyynelä et al., 1990). Additionally, no increase in permeability of the blood–brain barrier was found (Kim et al., 1988). Mechanistic studies suggest that alterations in OAT1 or other transporters at higher exposures of 2,4-D may be responsible for this accumulation (Pritchard, 1980; Kim et al., 1983).

Excretion of 2,4-D is largely via the urine. After oral administration at 5–50 mg/kg bw, the percentage of 2,4-D recovered in the urine was 81–97% in rats, 71–81% in hamsters, and 54–94% in mice (Griffin et al., 1997; van Ravenzwaay et al., 2003). As in humans, it appears that organic anion transporters, such as OAT1, are responsible for active transport into the kidney, which facilitates excretion (Imaoka et al., 2004). In dogs only 20–38% was recovered in urine after 72 hours, with an additional 10–24% in the faeces (van Ravenzwaay et al., 2003). Moreover, the half-life in dogs appears to be much longer than in other species, even accounting for allometric differences due to differences in body size, leading to much higher body burdens of 2,4-D for a given exposure (van Ravenzwaay et al., 2003; Timchalk, 2004).

(b) Metabolism

The salts and esters of 2,4-D are hydrolysed in vivo to 2,4-D. In most experimental systems, 2,4-D undergoes only limited metabolism to conjugates before excretion. In one study of oral administration, no metabolites were detected in rats, glycine and taurine conjugates of 2,4-D were detected in hamsters and mice, and glucuronide conjugates of 2,4-D were detected in hamsters only (Griffin et al., 1997). In mice, these conjugates accounted for less than 20% of urinary excretion at 5 mg/kg bw, but almost 50% at 200 mg/kg bw. In hamsters, which were only exposed to 2,4-D at 200 mg/kg bw, metabolites accounted for less than 13% of urinary excretion. No metabolites were detected in a study of goats given 2,4-D (Örberg, 1980). In a study comparing rats and dogs, no metabolite in rat urine exceeded 2% of the administered dose. Numerous metabolites were detected in dogs (van Ravenzwaay et al., 2003); the most abundant 2,4-D conjugates were those with glycine (about 34% of the administered dose at 120 hours) and glucuronide (7%), and were more abundant than the parent compound, 2,4-D (1%) (van Ravenzwaay et al., 2003). [The Working Group noted that these data suggested that dogs have a lower capacity to excrete 2,4-D than other species studied.] After administration of 2,4-D, 2,4-DCP has been reported in the rat kidney (Aydin et al., 2005), goat milk and fat, and chicken eggs and liver (Barnekow et al., 2001).

4.1.3 Modulation of metabolic enzymes

No data on modulation of metabolic enzymes in humans were available to the Working Group. At the median lethal dose (LD_{50}, 375 mg/kg), a single gavage dose of 2,4-D induced cytochrome P450 (CYP1A1, CYP1A2, and CYP1B1) mRNAs in the mammary gland, liver, and kidney of female Sprague-Dawley rats (Badawi et al., 2000).

In mouse liver, dietary exposure to 2,4-D at a concentration of 0.125% (w/w) induced total cytochrome oxidase activity and the activities of cytosolic and microsomal epoxide hydrolases (Lundgren et al., 1987). A less pronounced increase in total cytosolic glutathione transferase activity was observed. Total protein levels of CYP450 and cytosolic epoxide hydrolase were induced [probably due to induction of CYP4A mediated by peroxisome proliferator-activated receptor (PPAR).]

4.2 Mechanisms of carcinogenesis

4.2.1 Genotoxicity and related effects

2,4-D has been studied in a variety of assays for genotoxic and related potential. Table 4.1, Table 4.2, Table 4.3, and Table 4.4 summarize the studies carried out in exposed humans, in

Table 4.1 Genetic and related effects of 2,4-D in exposed humans

Tissue	Cell type (if specified)	End-point	Test	Description of exposed and controls	Response, significance	Comments	Reference
Blood	Lymphocytes isolated	Chromosomal damage	Micronucleus formation	12 applicators spraying only 2,4-D and 9 non-exposed controls	–		Figgs et al. (2000)
Blood	Lymphocytes	Chromosomal damage	Micronucleus formation	12 male applicators exposed solely to 2,4-D during a 3-month period	–		Holland et al. (2002)
Blood	Lymphocytes	Chromosomal damage	Chromosomal aberrations, V(D)J region rearrangements	24 forest/roadside pesticide applicators exposed to 2,4-D; 15 non-exposed controls	+ (chromosome translocations, inversions, deletions, $P = 0.003$; breaks, $P = 0.017$; gaps, $P = 0.006$)	Chromosome aberrations associated with the amount of herbicide applied; no association for workers who applied 2,4-D only, or with urinary 2,4-D concentrations V(D)J region rearrangement frequencies positively correlated (r = 0.54) with urinary 2,4-D concentrations ($P = 0.003$)	Garry et al. (2001)
Blood	Lymphocytes	DNA damage	Comet assay	10 subjects exposed to a pesticide mixture (atrazine, alachlor, cyanazine, 2,4-D, malathion) for 22.2 yr (range, 4–30 yr); 10 controls, matched for sex, age, and smoking status	(+)	DNA damage remained elevated, but was significantly decreased 6 mo after exposure cessation [Causative effect of 2,4-D alone could not be demonstrated]	Garaj-Vrhovac & Zeljezic (2000)
Blood	Lymphocytes	DNA and chromosomal damage	Comet assay (DNA damage), chromosomal aberrations, micronucleus formation, sister-chromatid exchanges	20 subjects exposed to a pesticide mixture (atrazine, alachlor, cyanazine, 2,4-D, malathion) for an average of 22.2 yr; 20 controls, matched for sex, age, and smoking status	(+)	Damage remained elevated, but was significantly decreased 8 mo after exposure cessation [Causative effect of 2,4-D alone could be demonstrated]	Zeljezic & Garaj-Vrhovac (2001)

Table 4.1 (continued)

Tissue	Cell type (if specified)	End-point	Test	Description of exposed and controls	Response, significance	Comments	Reference
Blood	Lymphocytes	DNA and chromosomal damage	Comet assay, (DNA damage), chromosomal aberrations, micronucleus formation, sister-chromatid exchanges	10 subjects exposed to a pesticide mixture (atrazine, alachlor, cyanazine, 2,4-D, malathion); 20 controls, matched for sex, age, and smoking status	(+)	[Causative effect of 2,4-D alone could not be demonstrated]	Garaj-Vrhovac & Zeljezic (2002)
Blood	Lymphocytes	Chromosomal damage	Chromosomal aberrations	42 male workers (Idaho, USA) occupationally exposed to 2,4-D, DDT, malathion, ethyl parathion, endosulfan, atrazine, dicamba, among other pesticides; 16 age-matched healthy controls	(+)	[Causative effect of 2,4-D alone could not be demonstrated]	Yoder et al. (1973)
Blood	Lymphocytes	Chromosomal damage	Chromosomal aberrations	19 spraying foliage in forestry workers exposed to 2,4-D and MCPA for 6–28 days	(–)	[Causative effect of 2,4-D alone could not be demonstrated]	Mustonen et al. (1986)
Blood	Lymphocytes	Chromosomal damage	Chromosomal aberrations	19 (2 female, 17 male) herbicide production workers exposed to 2,4-D and 2,4,5-trichlorophenol for 10–30 yr	(+) ($P < 0.05$)	[Causative effect of 2,4-D alone could not be demonstrated]	Kaioumova & Khabutdinova (1998)
Blood	Lymphocytes	Chromosomal damage	Sister-chromatid exchanges	35 males spraying foliage in forestry with 2,4-D and MCPA, and 15 controls not working with herbicides	(–)	[Causative effect of 2,4-D alone could not be demonstrated]	Linnainmaa (1983)

[], comments not provided or present in the original reference
+, positive
–, negative
+/–, equivocal (variable response in several experiments within an adequate study)
(+) or (–), positive or negative result in a study of limited quality
2,4-D, 2,4-dichlorophenoxyacetic acid; DDT, dichlorodiphenyltrichloroethane; MCPA, 2-methyl-4-chlorophenoxyacetic acid; mo, month; yr, year

human cells in vitro, in other mammals in vivo and in vitro, and in non-mammalian systems, respectively.

(a) Humans

(i) Exposed humans

See Table 4.1

No induction of micronucleus formation in circulating blood lymphocytes was reported in applicators exposed only to 2,4-D (Figgs et al., 2000; Holland et al., 2002). In a study of chromosomal aberrations in lymphocytes of forest/roadside herbicide applicators, an association was reported with the amount of herbicide applied, but no association was observed among workers who applied 2,4-D only, or with urinary 2,4-D concentrations (Garry et al., 2001). An association was reported between 2,4-D urine levels and V(D)J rearrangements (Garry et al., 2001). [The Working Group noted that the V(D)J rearrangements may not reflect a genotoxic effect.]

A causative effect of 2,4-D alone could not be demonstrated in several studies of pesticide mixtures. Induction of DNA breaks as measured by the comet assay was seen in lymphocytes from male production workers exposed to a pesticide mixture including 2,4-D (Garaj-Vrhovac & Zeljezic, 2000, 2001, 2002). Chromosomal aberrations were induced in lymphocytes of male agricultural workers (Yoder et al., 1973; Garaj-Vrhovac & Zeljezic, 2001, 2002), but not in male forestry workers (Mustonen et al., 1986). Induction of micronucleus formation in lymphocytes was seen in males exposed to a pesticide mixture including 2,4-D (Garaj-Vrhovac & Zeljezic, 2001, 2002) or in an herbicide plant producing 2,4-D and 2,4,5-trichlorophenol (Kaioumova & Khabutdinova, 1998). For sister-chromatid exchange, positive results were reported in males exposed to a pesticide mixture including 2,4-D (Garaj-Vrhovac & Zeljezic, 2001; Zeljezic & Garaj-Vrhovac, 2002). However, a separate study showed no effect on sister-chromatid exchange in male forestry workers exposed to 2,4-D and MCPA (Linnainmaa, 1983).

(ii) Human cells in vitro

See Table 4.2

No induction of DNA strand breaks by 2,4-D was detected by ^{32}P labelling in exposed leukocytes (Sreekumaran Nair et al., 2002). Comet assay results were positive in lymphocytes isolated from smokers, but not non-smokers, exposed to 2,4-D (Sandal & Yilmaz, 2011). Apurinic/apyrimidinic sites in human fibroblasts were induced by a commercial formulation containing 2,4-D dimethylamine salt, but not by 2,4-D or the 2,4-D trimethylamine salt (Clausen et al., 1990).

Regarding induction of chromosomal aberration by 2,4-D, both positive (Pilinskaia, 1974; Korte & Jalal, 1982) and negative (Mustonen et al., 1986) results have been reported in human lymphocytes. Positive results were reported in human lymphocytes exposed to 2,4-D-based formulations (Mustonen et al., 1986; Zeljezic & Garaj-Vrhovac, 2004).

Regarding micronucleus formation, inconclusive results have also been observed in whole blood or lymphocyte cultures treated with 2,4-D or a 2,4-D-based formulation (Holland et al., 2002). Micronuclei were induced after exposure of lymphocytes to a 2,4-D-based formulation in the presence or absence of metabolic activation (Zeljezic & Garaj-Vrhovac, 2004).

Positive results were seen in an assay for sister-chromatid exchange in human lymphocytes treated with 2,4-D (Korte & Jalal, 1982). Soloneski et al. (2007) reported induction of sister-chromatid exchange in lymphocyte cultures by 2,4-D or by a formulation containing 2,4-D dimethylamine salt, but only when erythrocytes were present in the cultures (Soloneski et al., 2007). [The Working Group noted that the findings of Soloneski et al. (2007) suggest that erythrocytes may play a role in 2,4-D metabolic activation or lipid peroxidation induction.]

Table 4.2 Genetic and related effects of 2,4-D in human cells in vitro

Tissue, cell line	End-point	Test	Results	Concentration (LED or HID)	Comments	Reference
2,4-D						
Leukocytes	DNA damage	^{32}P post labelling	–	80 µg/mL		Sreekumaran Nair et al. (2002)
Lymphocytes, non-smokers	DNA damage	Comet assay	–	10 µM		Sandal & Yilmaz (2011)
Lymphocytes, smokers	DNA damage	Comet assay	+	10 µM		Sandal & Yilmaz (2011)
Fibroblasts	DNA damage	AP sites	–	100 mM		Clausen et al. (1990)
Lymphocytes	Chromosomal damage	Chromosomal aberrations	+	50 µg/mL		Korte & Jalal (1982)
Lymphocytes	Chromosomal damage	Chromosomal aberrations	+	2 µg/mL		Pilinskaia (1974)
Lymphocytes	Chromosomal damage	Chromosomal aberrations	–	0.35 mM		Mustonen et al. (1986)
Whole blood cultures	Chromosomal damage	Micronucleus formation	(+)	0.001–1.0 mM	Only 2 donors; modest dose-dependent ($P = 0.012$) induction in 1 out of 2 donors	Holland et al. (2002)
Lymphocytes	Chromosomal damage	Micronucleus formation	(+)	0.3 mM	Only 2 donors; slight induction at cytotoxic concentration in both donors	Holland et al. (2002)
Lymphocytes	Chromosomal damage	Sister-chromatid exchange	+	10 µg/mL		Korte & Jalal (1982)
Lymphocytes	Chromosomal damage	Sister-chromatid exchange	+	10 µg/mL	Only when erythrocytes were present	Soloneski et al. (2007)
2,4-D-based formulation						
Fibroblasts	DNA damage	AP sites	+	10 mM	2,4-D-based formulation or 2,4-D DMA-HCl salt; 2,4-D TMA-HCl salt was without effect at 100 mM	Clausen et al. (1990)
Lymphocytes	Chromosomal damage	Chromosomal aberrations	+	0.4 µg/mL	With or without metabolic activation	Zeljezic & Garaj-Vrhovac (2004)
Lymphocytes	Chromosomal damage	Chromosomal aberrations	+	0.5 mM		Mustonen et al. (1986)
Lymphocytes	Chromosomal damage	Micronucleus formation	+	0.4 µg/mL	With or without metabolic activation	Zeljezic & Garaj-Vrhovac (2004)

Table 4.2 (continued)

Tissue, cell line	End-point	Test	Results	Concentration (LED or HID)	Comments	Reference
Whole blood cultures or lymphocytes	Chromosomal damage	Micronucleus formation	(–)	0.3 mM	Two 2,4-D-based formulations; only 2 donors; slight induction (from 2/1000 to 6.7/1000) in 1 out of 2 donors that was within the range of baseline variability (3-12/1000)	Holland et al. (2002)
Lymphocytes	Chromosomal damage	Sister-chromatid exchange	+	10 µg/mL	Formulation containing 2,4-D DMA; positive results only when erythrocytes were present	Soloneski et al. (2007)

+, positive
–, negative
+/–, equivocal (variable response in several experiments within an adequate study)
(+) or (–), positive/negative results in a study of limited quality
2,4-D, 2,4-dichlorophenoxyacetic acid; AP sites, apurinic/apyrimidinic sites; DMA, dimethylamine; HID, highest ineffective dose; LED, lowest effective dose; TMA, trimethylamine

Table 4.3 Genetic and related effects of 2,4-D, and its metabolites, salts and esters, in non-human mammals in vivo

Species, strain	Tissue	End-point	Test	Results	Dose (LED or HID)	Route, duration, dosing regimen	Comments	Reference
2,4-D								
Mouse, Swiss	Germ cells	Mutation	Dominant lethal	–	125 mg/kg bw 75 mg/kg bw	i.p. × 1 p.o. × 5	2,4-D	Epstein et al. (1972)
Mouse, B6C3F$_1$	Thymocytes	Mutation	T-cell receptor (*V(D)J*)	–	100 mg/kg bw	p.o. × 1 × 4 days	2,4-D	Knapp et al. (2003)
Mouse, Swiss	Bone-marrow cells, spermatocytes	Chromosomal damage	Chromosomal aberrations	+	3.3 mg/kg bw	p.o. × 3 or 5 days	2,4-D	Amer & Aly (2001)
Mouse, CD-1	Bone-marrow cells	Chromosomal damage	Micronucleus formation	–	400 mg/kg bw	p.o. × 1	2,4-D	Charles et al. (1999b)
Mouse	Bone-marrow cells	Chromosomal damage	Chromosomal aberrations	+	100 mg/kg bw	p.o. × 1	2,4-D	Pilinskaia (1974)
Mouse, C57BL	Bone-marrow cells	Chromosomal damage	Chromosomal aberrations	+	3.5 mg/kg bw	i.p. × 1	2,4-D	Venkov et al. (2000)
Mouse, Swiss	Bone-marrow cells	Chromosomal damage	Chromosomal aberrations	–	3.38 mg/kg bw	i.p. every 3 days for 55 days	2,4-D; offspring of maternal treated mice	Yilmaz & Yuksel (2005)
Mouse, CD-1	Bone-marrow cells	Chromosomal damage	Micronucleus formation	–	1/8 dermal LD_{50}	Topically × 1, 24 h before analysis	2,4-D	Schop et al. (1990)
Mouse, CD-1	Hair follicle	Chromosomal damage	Nuclear aberration assay	+	1/8 dermal LD_{50}	Topically × 1, 24 h before analysis	2,4-D	Schop et al. (1990)
Mouse, CBA	Bone-marrow cells	Chromosomal damage	Micronucleus formation	–	100 mg/kg bw	i.p. × 1	2,4-D Negative results at 24 hours and 7 days after treatment	Jenssen & Renberg (1976)
Mouse, ICR	Bone-marrow cells	Chromosomal damage	Micronucleus formation	–	400 mg/kg bw	p.o. × 1	2,4-D Negative results at 1, 2, and 3 days after treatment	EPA (1990b)

Table 4.3 (continued)

Species, strain	Tissue	End-point	Test	Results	Dose (LED or HID)	Route, duration, dosing regimen	Comments	Reference
Mouse, NIH	Bone-marrow and spermatogonia cells	Chromosomal damage	Sister-chromatid exchange	+	100 mg/kg bw	p.o. × 1	2,4-D	Madrigal-Bujaidar et al. (2001)
Rat, Han Wistar	Hepatocytes	DNA damage	UDS assay	–	1000 mg/kg bw	p.o. × 1	2,4-D	Charles et al. (1999a)
Rat, Wistar	Hepatocytes, kidney cells	DNA damage	Alkaline elution	+	7 mg/kg bw	i.p. × 1	2,4-D	Kornuta et al. (1996)
Rat, Wistar	Spleen, lung, bone-marrow cells	DNA damage	Alkaline elution	+	70 mg/kg bw	i.p. × 1	2,4-D	Kornuta et al. (1996)
Rat, Wistar	Bone-marrow cells	Chromosomal damage	Chromosomal aberrations	+	35 mg/kg bw	i.p. × 1 × 2 days	2,4-D	Adhikari & Grover (1988)
2,4-D metabolites, salts and esters								
Mouse, Swiss	Bone-marrow cells, spermatocytes	Chromosomal damage	Chromosomal aberrations	+	180 mg/kg bw	i.p. × 1 × 3 or 5 days	2,4-DCP	Amer & Aly (2001)
Mouse, CD-1	Bone-marrow cells	Chromosomal damage	Micronucleus formation	–	443 mg/kg bw (	p.o. × 1	2,4-D DEA	Charles et al. (1999b)
					397 mg/kg bw		2,4-D DMA	
					376 mg/kg bw		2,4-D IPA	
					542 mg/kg bw		2,4-D TIPA	
					375 mg/kg bw		2,4-D BEE	
					500 mg/kg bw		2,4-D EHE	
					400 mg/kg bw		2,4,-D IPE	
Mouse, ICR	Bone-marrow cells	Chromosomal damage	Micronucleus formation	–	600 mg/kg bw	p.o. × 1	2,4-D DMA Negative results at 1, 2, and 3 days after treatment	EPA (1990b)

Table 4.3 (continued)

Species, strain	Tissue	End-point	Test	Results	Dose (LED or HID)	Route, duration, dosing regimen	Comments	Reference
Mouse, ICR	Bone-marrow cells	Chromosomal damage	Micronucleus formation	–	500 mg/kg bw	p.o. × 1	2,4-D IOE Negative results at 1, 2, and 3 days after treatment	EPA (1990a)
Mouse, C57BL/6	Bone-marrow cells	Chromosomal damage	Sister-chromatid exchange	–	20–40 mg/kg bw	i.p. × 1	Mixtures of 2,4-D, 2,4,5-T, and TCDD [causative effect of 2,4-D alone could not be demonstrated]	Lamb et al. (1981)
2,4-D-based formulations								
Rat, Wistar	Circulating lymphocytes	Chromosomal damage	Sister-chromatid exchange	–	100 mg/kg bw	p.o. × 1 × 14 day		Linnainmaa (1984)
Hamster, Chinese	Circulating lymphocytes	Chromosomal damage	Sister-chromatid exchange	–	100 mg/kg bw	p.o. × 1 × 9 days		Linnainmaa (1984)
Dog	Malignant lymphoma	Mutation	c-N-*ras* amplification	–	NR	Lawns treated with herbicides containing 2,4-D		Edwards et al. (1993)

+, positive; –, negative; +/–, equivocal (variable response in several experiments within an adequate study); (+) or (–), positive/negative in a study of limited quality
2,4-D, 2,4-dichlorophenoxyacetic acid; 2,4,5-T, 2,4,5-trichlorophenoxyacetic acid; BEE, butoxyethyl ester; DCP, dichlorophenol; DEA, diethanolamine; DMA, dimethylamine; EHE, ethylhexyl ester; HID, highest ineffective dose; IOE, isooctyl ester; i.p, intraperitoneal; IPA, isopropylamine ; IPE, isopropyl ester; LED, lowest effective dose; NR, not reported; p.o., oral administration; TCDD, 2,3,7,8-tetrachlorodibenzo-*p*-dioxin; TIPA, tri-isopropanolamine; UDS, unscheduled DNA synthesis

(b) *Experimental systems*

(i) *Non-human mammals in vivo*

See Table 4.3

In mice, dominant-lethal tests with 2,4-D gave negative results (Epstein et al., 1972). Positive results for chromosomal aberration were reported in bone marrow cells and spermatocytes after oral or intraperitoneal treatment with 2,4-D (Pilinskaia, 1974; Venkov et al., 2000; Amer & Aly, 2001) or after exposure to the metabolite 2,4-DCP (Amer & Aly, 2001). Negative results were seen in the offspring of mice treated with 2,4-D by intraperitoneal administration (Yilmaz & Yuksel, 2005). Micronuclei were not induced in bone marrow cells of mice exposed to 2,4-D topically (Schop et al., 1990), by intraperitoneal administration (Jenssen & Renberg, 1976), or orally (EPA, 1990b; Charles et al., 1999b). 2,4-D salts and esters also gave negative results after oral administration (EPA, 1990a, b; Charles et al., 1999b). Sister-chromatid exchange was induced in bone marrow and spermatogonial cells by 2,4-D administered orally (Madrigal-Bujaidar et al., 2001). Positive results were reported in an assay for nuclear aberration in hair follicles after topical exposure to 2,4-D (Schop et al., 1990).

In rats, inconsistent results were seen for induction of DNA strand breaks. There was no DNA damage induction in hepatocytes evaluated by an assay for unscheduled DNA synthesis after oral exposure to 2,4-D (Charles et al., 1999a). Induction of DNA damage by alkaline elution assay was seen in cells of the liver, kidney, spleen, lung, and bone marrow after intraperitoneal exposure to 2,4-D (Kornuta et al., 1996). Chromosomal aberrations were induced in bone-marrow cells after intraperitoneal exposure to 2,4-D (Adhikari & Grover, 1988). No increase in sister-chromatid exchange in circulating lymphocytes was seen after intragastric exposure to a 2,4-D formulation in rats or Chinese hamsters (Linnainmaa, 1984).

In dogs, no association was reported between exposure to 2,4-D and amplification or mutation of c-N-*ras* in lymphoma specimens (Edwards et al., 1993).

(ii) *Non-human mammalian cells in vitro*

See Table 4.4

In rat cells, no DNA damage was seen in hepatocytes evaluated by assay for unscheduled DNA synthesis after exposure to 2,4-D acid, or to its salts and esters (EPA, 1990a; Charles et al., 1999a).

In Chinese hamster V79 cells, 2,4-D was mutagenic in the hypoxanthine-guanine phosphoribosyl transferase (*HGPRT*) assay (Pavlica et al., 1991). In Chinese hamster ovary (CHO) cells, no mutagenic effect was reported in the *HGPRT* assay after exposure to 2,4-D salts and esters in the presence or absence of metabolic activation (Gollapudi et al., 1999).

Results were variable for DNA strand-break induction by the comet assay. Positive results were observed in Syrian hamster embryo cells exposed to 2,4-D (Maire et al., 2007), and in CHO cells exposed to 2,4-D or 2,4-D DMA (González et al., 2005). Negative results were reported at higher concentrations of 2,4-D in CHO cells (Sorensen et al., 2005).

Sister-chromatid exchange was induced in CHO cells exposed to 2,4-D in the presence (Linnainmaa, 1984), or absence (González et al., 2005) of metabolic activation, or exposed to 2,4-D DMA in the absence of metabolic activation (González et al., 2005). A 2,4-D formulation slightly increased the frequency of sister-chromatid exchange with, but not without, metabolic activation (Linnainmaa, 1984). Negative results were reported in an assay for chromosomal aberration in lymphocytes exposed to 2,4-D salts and esters with and without metabolic activation (Gollapudi et al., 1999).

(iii) *Non-mammalian systems*

See Table 4.5

Table 4.4 Genetic and related effects of 2,4-D in non-human mammalian cells in vitro

Species	Tissue, cell line	End-point	Test	Results		Concentration (LEC or HIC)	Comments	Reference
				Without metabolic activation	With metabolic activation			
2,4-D, esters and salts								
Rat	Hepatocytes	DNA damage	UDS assay	–	NT	2,4-D (96.9 μg/mL) 2,4-D IOE (25 μg/mL) 2,4-D DMA (100 μg/mL)		EPA (1990a)
Rat	Hepatocytes	DNA damage	UDS assay	– – – – – – – –	NT	2,4-D (96.9 μg/mL) 2,4-D DEA (369 μg/mL) 2,4-D DMA (66.2 μg/mL) 2,4-D IPA (250.5 μg/mL) 2,4-D TIPA (354.5 μg/mL) 2,4-D BEE (478 μg/mL) 2,4-D EHE (24.5 μg/mL) 2,4-D IPE (194.2 μg/mL)		Charles et al. (1999a)
Chinese hamster	V79 cells	Mutation	*HGPRT* mutation assay	+	NT	10 μg/mL	2,4-D	Pavlica et al. (1991)
Syrian golden hamster	SHE cells	DNA damage	Comet assay	+	NT	11.5 μM	2,4-D	Maire et al. (2007)
Chinese hamster	CHO cells	DNA damage	Comet assay	+	NT	2 μg/mL	2,4-D or 2,4-D DMA	González et al. (2005)
Chinese hamster	CHO cells	DNA damage	Comet assay	–	NT	1600 μM	2,4-D, alone or after reaction with redox-modified clay	Sorensen et al. (2005)
Chinese hamster	CHO cells	Chromosomal damage	Sister-chromatid exchange	–	+	10^{-4} M	2,4-D, slight increase over control values	Linnainmaa (1984)
Chinese hamster	CHO cells	Chromosomal damage	Sister-chromatid exchanges	+	NT	2 μg/mL	2,4-D or 2,4-D DMA	González et al. (2005)
Rat	Lymphocytes	Chromosomal damage	Chromosomal aberrations	–	–	2,4-D BEE (1400 μg/mL) 2,4-D IPA (3068 μg/mL) 2,4-D TIPA (5000 μg/mL)	For 2,4-D IPA, HIC of 1500 μg/mL with metabolic activation	Gollapudi et al. (1999)

Table 4.4 (continued)

Species	Tissue, cell line	End-point	Test	Results		Concentration (LEC or HIC)	Comments	Reference
				Without metabolic activation	With metabolic activation			
Chinese hamster	CHO cells	Mutation	*HGPRT* assay	–	–	2,4-D BEE (1400µg/mL) 2,4-D IPA (3000 µg/mL) 2,4-D TIPA (5000 µg/mL)	For 2,4-D BEE, HIC of 700 µg/mL without metabolic activation	Gollapudi et al. (1999)
2,4-D-based formulation								
Chinese hamster	CHO cells	Chromosomal damage	Sister-chromatid exchanges	+	–	10^{-5} M	Slight increase over control values	Linnainmaa (1984)

+, positive; –, negative; +/–, equivocal (variable response in several experiments within an adequate study); (+) or (–), positive/negative in a study of limited quality
2,4,5-T, 2,4,5-trichlorophenoxyacetic acid; 2,4-D, 2,4-dichlorophenoxyacetic acid; BEE, butoxyethyl ester; CHO, Chinese hamster ovary; DCP, dichlorophenol; DEA, diethanolamine; DMA, dimethylamine; EHE, ethylhexyl ester; HIC, highest ineffective concentration; IOE, isooctyl ester; IPA, isopropylamine; IPE, isopropyl ester; LEC, lowest effective concentration, NT, not tested; SHE, Syrian hamster embryo; TCDD, 2,3,7,8-tetrachlorodibenzo-*p*-dioxin; TIPA, tri-isopropanolamine; UDS, unscheduled DNA synthesis

In chickens, positive results in sister-chromatid exchanges in embryo cells were reported after exposure to 2,4-D or a 2,4-D formulation (Arias, 2003, 2007).

In fish, DNA strand-breaks, chromosomal aberrations, micronucleus formation, and other abnormalities indicative of genotoxicity were seen after exposure to 2,4-D in *Clarias batrachus* (Ateeq et al., 2002b, 2005), *Channa punctatus* (Farah et al., 2003, 2006), and *Oncorhynchus mykiss* (Martínez-Tabche et al., 2004). DNA strand breaks were induced by 2,4-D or a formulation containing 2,4-D DMA in a carp cell line in vitro (Bokán et al., 2013).

In freshwater snails, exposure to a formulation containing 2,4-D DMA had mutagenic effects in the dominant-lethal assay (Estevam et al., 2006). No induction of micronucleus formation was seen in haemocytes of Mediterranean mussels exposed to 2,4-D (Raftopoulou et al., 2006).

In *Drosophila melanogaster*, mutagenicity was observed with 2,4-D in some assays for sex-linked recessive lethal (Tripathy et al., 1993; Kale et al., 1995) and somatic mutation and recombination (SMART) (Graf & Würgler, 1996), but not others (Vogel & Chandler, 1974; Zimmering et al., 1985). Mutagenicity was observed by the wing-spot test after exposure to 2,4-D (Tripathy et al., 1993; Kaya et al., 1999) or to 2-(2,4-dichlorophenoxy) propionic acid (Surjan, 1989). No mutagenic effect was seen for the white-ivory eye-spot test after 2,4-D exposure (Graf & Würgler, 1996). No chromosomal aberrations were induced in germ-line cells exposed to a 2,4-D-based formulation (Woodruff et al., 1983).

In plants, positive results were reported in assays for point mutation and homologous recombination in *Arabidopsis thaliana* after exposure to 2,4-D (Filkowski et al., 2003). In bean seedlings (*Phaseolus vulgaris*), DNA damage was induced by comet assay and random amplified polymorphic DNA (RAPD) assay after exposure to 2,4-D (Cenkci et al., 2010). 2,4-D induced chromosomal aberrations in *Allium cepa* (Kumari & Vaidyanath, 1989; Ateeq et al., 2002a), and in *A. ascalonicum* (Pavlica et al., 1991), and induced sister-chromatid exchange in *A. sativum* (Doležel et al., 1987). In *Vicia faba*, chromosomal aberration was induced by 2,4-D by when plants were sprayed, but not when seeds were soaked (Amer & Ali, 1974).

A 2,4-D-based formulation (containing butyl ester) was tested in 12 plant species and induced chromosome aberration in three species, giving positive results when applied to roots (*Chrysanthemum leucanthemum*), germinated seeds (*Secale cereale*), or bulbs (*Allium cepa*) (Mohandas & Grant, 1972). Increased frequencies (although slight) of sister-chromatid exchange were seen in *Triticum aestivum* (Murata, 1989).

In *Saccharomyces cerevisiae*, mutagenic effects were reported with 2,4-D in assays for reverse mutation and mitotic gene conversion (Venkov et al., 2000). A separate study reported no mutagenicity in the assay for mitotic gene conversion after exposure to 2,4-D (Fahrig, 1974). 2,4-D-based formulations gave positive results in mitotic gene-conversion assays (Siebert & Lemperle, 1974; Zetterberg et al., 1977), but not in the host-mediated assay (Zetterberg et al., 1977).

In *Salmonella typhimurium*, 2,4-D and its salts and esters did not demonstrate mutagenicity in TA98, TA100, TA1535, TA1537, or TA1538 in the assay for reverse mutation in the presence or absence of metabolic activation (Moriya et al., 1983; Mortelmans et al., 1984; EPA, 1990b; Charles et al., 1999a). A 2,4-D-based formulation gave negative results in TA1535 and TA1538 strains, and in strains TA1530 and TA1531 in the host-mediated assay (Zetterberg et al., 1977). In *Escherichia coli*, no mutagenic effect of 2,4-D was seen in WP2 *hcr* in the reverse mutation assay with or without metabolic activation (Moriya et al., 1983).

In *Bacillus subtilis*, no mutagenic effect was seen with 2,4-D in the Rec mutation assay (Shirasu et al., 1976), and results were inconclusive

Table 4.5 Genetic and related effects of 2,4-D in non-mammalian systems

Phylogenetic class	Species, strain, tissue	End-point	Test	Results		Agent, concentration (LEC or HIC)	Comments	Reference
				Without metabolic activation	With metabolic activation			
Chicken	White Leghorn 288-Shaver strain Chick embryos	Chromosomal damage	Sister-chromatid exchange	+	NT	0–4 mg/embryo	Injection × 1 into the egg air cell Dose-related increase with 2,4-D (borderline significance) or a 2,4-D-based formulation	Arias (2003)
Chicken	White Leghorn 288-Shaver strain Chick embryos	Chromosomal damage	Sister-chromatid exchange	+	NT	4 mg/embryo	Injection × 1 into the egg air cell Induction with 2,4-D (after 10 days) or a 2,4-D formulation (after 4 days)	Arias (2007)
Fish	Catfish (*Clarias batrachus*), circulating erythrocytes	DNA damage	Comet assay	+	NA	25 ppm	2,4-D	Ateeq et al. (2005)
Fish	Catfish (*Clarias batrachus*), circulating erythrocytes	Chromosomal damage	Micronucleus formation	+	NA	25 ppm	2,4-D	Ateeq et al. (2002b)
Fish	Air-breathing *Channa punctatus*, kidney cells	Chromosomal damage	Chromosomal aberrations	+	NA	75 ppm	2,4-D	Farah et al. (2006)
Fish	Air-breathing *Channa punctatus*, circulating erythrocytes	Chromosomal damage	Micronucleus formation	+	NA	25 ppm	2,4-D	Farah et al. (2003)
	Air-breathing *Channa punctatus*, circulating erythrocytes	Chromosomal damage	Micronucleus formation	+	NA	75 ppm	2,4-D	Farah et al. (2006)
Fish	Rainbow trout (*Oncorhynchus mykiss*), gill cells	DNA damage	Comet assay	+	NA	5 mg/l	2,4-D Water not changed during experiment; 1–8 days	Martínez-Tabche et al. (2004)
Snails	*Biomphalaria glabrata*, germ cells	Mutation	Dominant lethal assay	+	NA	75 ppm	2,4-D DMA formulation)	Estevam et al. (2006)

Table 4.5 (continued)

Phylogenetic class	Species, strain, tissue	End-point	Test	Results		Agent, concentration (LEC or HIC)	Comments	Reference
				Without metabolic activation	With metabolic activation			
Mussels	Mediterranean mussel (*Mytilus galloprovincialis*), haemocytes	Chromosomal damage	Micronucleus induction	–	NA	0.03 mg/l	2,4-D	Raftopoulou et al. (2006)
Insects	*Drosophila melanogaster*, germ-line cells	Mutation	Wing-spot test, sex-linked recessive lethal	+	NT	5 mM	2,4-D In feeding media	Tripathy et al. (1993)
	Drosophila melanogaster, germ-line cells	Mutation	Sex-linked recessive lethal	+	NT	10 000 ppm	2,4-D In feeding media	Kale et al. (1995)
	Drosophila melanogaster, germ-line cells	Mutation	Sex-linked recessive lethal	–	NT	9 mM	2,4-D In feeding media	Vogel & Chandler (1974)
	Drosophila melanogaster, germ-line cells	Mutation	Sex-linked recessive lethal	–	NT	10 000 ppm	2,4-D In feeding media	Zimmering et al. (1985)
	Drosophila melanogaster, somatic cells	Mutation	Wing-spot test	+	NT	2.5 mM	2,4-D In feeding media × 2 days	Graf & Würgler (1996)
	Drosophila melanogaster, somatic cells	Mutation	White-ivory eye spot test	–	NT	2.5 mM	2,4-D In feeding media × 3 days	Graf & Würgler (1996)
	Drosophila melanogaster, somatic cells	Mutation	Wing-spot test	+	NT	10 mM	2,4-D In feeding media	Kaya et al. (1999)
	Drosophila melanogaster, somatic cells	Mutation	Wing-spot test	+	NT	0.031 w/w	2-(2,4-D) propionic acid In feeding media	Surjan (1989)
	Drosophila melanogaster with ring-X and doubly marked Y chromosome, germ-line cells	Chromosomal damage	Chromosomal aberrations	–		1000 ppm	2,4-D formulation (not otherwise specified) In feeding media	Woodruff et al. (1983)
Plant systems	*Arabidopsis thaliana* line166 and 166A	Mutation	Point mutation	+	NA	3 µg/l	2,4-D. A→G, but not T→G, reversions	Filkowski et al. (2003)
	Arabidopsis thaliana line 651	Mutation	Homologous recombination assay	+	NA	3 µg/l	2,4-D	Filkowski et al. (2003)
	Common bean, *Phaseolus vulgaris*	DNA damage	Comet assay	+	NA	0.1 ppm	2,4-D	Cenkci et al. (2010)

Table 4.5 (continued)

Phylogenetic class	Species, strain, tissue	End-point	Test	Results		Agent, concentration (LEC or HIC)	Comments	Reference
				Without metabolic activation	With metabolic activation			
Plant systems (cont.)	Onion, *Allium cepa*	Chromosomal damage	Chromosomal aberrations	+	NA	1 ppm	2,4-D	Ateeq et al. (2002a)
	Shallot, *Allium ascalonicum*	Chromosomal damage	Chromosomal aberrations	+	NA	45 µM	2,4-D	Pavlica et al. (1991)
	Garlic, *Allium sativum*	Chromosomal damage	Sister-chromatid exchanges	+	NA	5 µM	2,4-D	Doležel et al. (1987)
Lower eukaryote (yeast, mould, fungi)	*Saccharomyces cerevisiae* D7 tsl	Mutation	Reverse mutation, gene conversion	+	NA	8 mM	2,4-D	Venkov et al. (2000)
	Saccharomyces cerevisiae	Mutation	Mitotic gene conversion	–	NT	Dose, NR	2,4-D	Fahrig (1974)
	Saccharomyces cerevisiae D4	Mutation	Mitotic gene conversion	+	NA	1000 ppm/16 h	2,4-D formulation	Siebert & Lemperle (1974)
	Saccharomyces cerevisiae D4, D5	Mutation	Mitotic gene conversion	+	NA	0.6 mg/mL/3 h (D4)0.3mg/mL/3 h (D5)	2,4-D formulation, in buffer at pH 4.50	Zetterberg et al. (1977)
	Saccharomyces cerevisiae D4	Mutation	Host-mediated assay	–	NA	6 mg/o.p.	Formulated product of 2,4-D as sodium salt	Zetterberg et al. (1977)
Prokaryote (bacteria)	*Salmonella typhimurium* TA98, TA100, TA1535, TA1537, TA1538; *Escherichia coli* WP2 *hcr*	Mutation	Reverse mutation	–	–	5000 µg/plate	2,4-D	Moriya et al. (1983)
	Salmonella typhimurium TA98, TA100, TA1535, TA1537	Mutation	Reverse mutation	–	–	10 mg/plate	2,4-D	Mortelmans et al. (1984)
	Salmonella typhimurium TA98, TA100, TA1535, TA1537, TA1538	Mutation	Reverse mutation	–	–	10 000 µg/plate	Tested were 2,4-D, 2,4-D DMA and 2,4-D IOE	EPA (1990b)

Table 4.5 (continued)

Phylogenetic class	Species, strain, tissue	End-point	Test	Results		Agent, concentration (LEC or HIC)	Comments	Reference
				Without metabolic activation	With metabolic activation			
Prokaryote (bacteria) (cont.)	*Salmonella typhimurium* TA98, TA100, TA1535, TA1537, TA1538	Mutation	Reverse mutation	–	–	2,4-D acid (9610 μg/plate)		Charles et al. (1999a)
				–	–	2,4-D DEA (10 332 μg/plate)		
				–	–	2,4-D DMA (6620 μg/plate)		
				–	–	2,4-D TIPA (7090 μg/plate)		
				–	–	2,4-D BEE (4780 μg/plate)		
				–	–	2,4-D EHE (9800 μg/plate)		
				–	–	2,4-D IPE (4855 μg/plate)		
				–	–	2,4-D IPA (1670 μg/plate)		
	Salmonella typhimurium TA1535, TA1538	Mutation	Reverse mutation	–	–	0.08 mg/mL	2,4-D formulation	Zetterberg et al. (1977)
	Salmonella typhimurium TA1530, TA1531	Mutation	Host-mediated assay	–	NA	6 mg, p.o.	2,4-D formulation	Zetterberg et al. (1977)
	Bacillus subtilis M 45 Rec$^-$, H17 Rec$^+$	Mutation	Rec assay	–	NT	Dose not provided	2,4-D	Shirasu et al. (1976)
	Bacillus subtilis M 45 Rec$^-$, H17 Rec$^+$	Mutation	Rec assay	+/–	NA	2,4-D (10 mg/mL) 2,4-D formulation (7.2 mg/mL)		Grabińska-Sota et al. (2000, 2002)
Acellular systems	Isolated DNA from bacteriophage PM2	DNA damage	DNA single-strand breaks (AP sites)	–	NT	100 mM	2,4-D	Clausen et al. (1990)
	Isolated DNA from bacteriophage PM2	DNA damage	DNA single-strand breaks (AP sites)	+	NT	10 mM	2,4-D DMA formulation	Clausen et al. (1990)

Table 4.5 (continued)

Phylogenetic class	Species, strain, tissue	End-point	Test	Results		Agent, concentration (LEC or HIC)	Comments	Reference
				Without metabolic activation	With metabolic activation			
Acellular systems (cont.)	Isolated DNA from bacteriophage PM2	DNA damage	DNA single-strand breaks (AP sites)	+	NT	25 mM	2,4-D; only positive when DNA was pre-incubated with $CuCl_2$	Jacobi et al. (1992)

+, positive; –, negative; +/–, equivocal (variable response in several experiments within an adequate study); (+) or (–), positive/negative in a study of limited quality
2,4-D, 2,4-dichlorophenoxyacetic acid; 2,4,5-T, 2,4,5-trichlorophenoxyacetic acid; BEE, butoxyethyl ester; DCP, dichlorophenol; DEA, diethanolamine; DMA, dimethylamine; EHE, ethylhexyl ester; HIC, highest ineffective concentration; IOE, isooctyl ester; IPA, isopropylamine; IPE, isopropyl ester; LEC, lowest effective concentration; NA, not applicable; NR, not reported; NT, not tested; TCDD, 2,3,7,8-tetrachlorodibenzo-*p*-dioxin; TIPA, tri-isopropanolamine

at high concentrations of 2,4-D (Grabińska-Sota et al., 2000), and with a 2,4-D-based formulation (Grabińska-Sota et al., 2000, 2002).

2,4-D did not induce apurinic/apyrimidinic sites in bacteriophage PM2 DNA, while positive results were reported with a formulation containing 2,4-D DMA (Clausen et al., 1990). In a separate study, 2,4-D or the 2,4-D DMA formulation only gave positive results after pre-incubation with copper chloride ($CuCl_2$) (Jacobi et al., 1992).

2,4-D interacts with DNA as a groove binder rather than an intercalating agent, based on ultraviolet, fluorescence, and viscosity measurements, and alternating current voltammetry assays (Ahmadi & Bakhshandeh, 2009).

4.2.2 Receptor-mediated effects

(a) Humans

(i) Exposed humans

An increase in hypothyroidism with reported ever-use of 2,4-D and other herbicides has been found in male pesticide applicators in the Agricultural Health Study, USA (Goldner et al., 2013). Garry et al. (2001) reported a correlation between urinary concentrations of 2,4-D and serum concentrations of luteinizing hormone, but not follicle-stimulating hormone, in male pesticide applicators who used a hand-held backpack sprayer. Total testosterone levels in winter were directly correlated with peak levels of 2,4-D in the urine in the application season.

(ii) Human cells in vitro

In human prostate cancer cells (androgen receptor-expressing 22Rv1 and PC3/AR+), 2,4-D alone did not exhibit androgenic activity, but potentiated 5-alpha-dihydroxytestosterone (DHT) androgenic activities. DHT-mediated translocation of the androgen receptor to the nucleus, and androgen-induced transactivation were also increased in the presence of 2,4-D (Kim et al., 2005). [These findings suggested that 2,4-D has the potential to alter DHT-induced transcriptional activity of the androgen receptor.] Similarly, in-vitro reporter gene assays of estrogen receptor, androgen receptor, and thyroid hormone receptor showed no agonist or antagonist activity of 2,4-D against hormone receptors, but 2,4-D enhanced the activity of testosterone through the androgen receptor (Sun et al., 2012).

2,4-D did not interact in vitro with human estrogen, androgen, or steroidogenesis pathways in Tier 1 assays in the Endocrine Disruptor Screening Program run by the United States EPA. 2,4-D gave negative results in assays for estrogen receptor-mediated transcriptional activation (HeLa-9903-ERα transactivation assay), aromatase enzymatic activity inhibition (recombinant human CYP19 aromatase inhibition assay), and interference with steroidogenesis (H295R steroidogenesis assay) (Coady et al., 2014).

In co-transfected Hepa 1 cells, 2,4-D induced transactivation by human PPAR of rat acyl-coenzyme A oxidase (acyl-CoA oxidase) and rabbit CYP4A6 (Pineau et al., 1996).

(b) Experimental systems

(i) Non-human mammals in vivo

In rats, a single dose of 2,4-D has been shown to interfere with thyroid-hormone transport (Malysheva & Zhavoronkov, 1997). 2,4-D has been reported to bind competitively to the thyroxine (T4)-binding site of transthyretin, a carrier of thyroid hormones, and 2,4-D dichlorophenoxybutyric acid reduced plasma total T4 (TT4) in rats (Van den Berg et al., 1991).

In an F1-extended one-generation study of reproductive toxicity with 2,4-D in rats, a slight decrease in follicular size was reported in 3 out of 12 dams at the highest dose, but no other consistent pattern of thyroid effects was evident (Marty et al., 2013).

2,4-D has been reported to cause proliferation of peroxisomes in mouse and rat liver (Kawashima et al., 1984; Lundgren et al., 1987).

(ii) Non-human mammalian cells in vitro

Maloney & Waxman (1999) reported that 2,4-D did not activate mouse PPARα or PPARγ using a transactivation assay in vitro. 2,4-D did not interact in vitro with rodent estrogen or androgen pathways in Tier 1 assays in the Endocrine Disruptor Screening Program run by the EPA. Specifically, 2,4-D gave negative results in assays for estrogen-receptor binding (rat uterine cytosol estrogen-receptor binding assay), and for androgen receptor-binding (rat prostate cytosol androgen-receptor binding assay) (Coady et al., 2014).

(iii) Non-mammalian systems in vivo

Estrogenic activity, as determined by the vitellogenin assay, has been reported after exposure of rainbow trout to 2,4-D (Xie et al., 2005). Plasma vitellogenin levels were 93 times higher in juvenile rainbow trout exposed to 2,4-D (1.64 mg/L) for 7 days than in control untreated fish. No effects of 2,4-D (up to 113 mg acid equivalents/L) were reported in the amphibian metamorphosis assay, and the only effect reported in the fish short-term reproduction assay was decreased fecundity at the highest concentration tested (96.5 mg acid equivalents/L) (Coady et al., 2013).

4.2.3 Oxidative stress

(a) Humans

(i) Exposed humans

No data were available to the Working Group.

(ii) Human cells in vitro

In human erythrocytes in vitro, 2,4-D (10, 50, 100, 250, 500 ppm) induced dose-related decreases in superoxide dismutase activity, and increases in glutathione peroxidase activity. 2,4-D (500 ppm) decreased the level of reduced glutathione in erythrocytes by 18% compared with controls (Bukowska, 2003). In a follow-up study, 2,4-D increased protein carbonyl group content, but had no effect on the denaturation of haemoglobin (Bukowska et al., 2008).

(b) Experimental systems

(i) In vivo

2,4-D increased oxidative stress in ventral prostate, ovary, and mammary gland in the offspring of pregnant rats exposed to 2,4-D by oral gavage at a dose of 70 mg/kg bw per day from day 16 of gestation to 23 days after delivery. The pups were studied on postnatal days 45, 60, or 90. In ventral prostate, 2,4-D increased the concentration of hydroxyl radicals and the rate of lipid and protein oxidation at all ages studied. The activity of certain antioxidant enzymes was increased, but this was insufficient to counteract the oxidative stress. In mammary tissue, 2,4-D promoted oxidative stress, mainly during puberty and adulthood. In the ovary, 2,4-D increased lipid peroxides and altered the activity of several antioxidant enzymes (Pochettino et al., 2013).

2,4-D induced reactive oxygen species and altered antioxidant enzymes in the developing rat brain after exposure in breast milk. Maternal exposure to 2,4-D (100 mg/kg bw per day between postnatal days 9 and 25) had no effect on body weights of pups or lactating mothers. Levels of reactive oxygen species were increased in the neonatal midbrain, striatum, and prefrontal cortex. Glutathione content was significantly decreased in midbrain, and striatum, and there were alterations in levels (either increased or decreased) of antioxidant enzymes (superoxide dismutase, glutathione peroxidase, and catalase) in at least one of the brain regions studied, with the exception of the hypothalamus (Ferri et al., 2007).

2,4-D (600 ppm in drinking-water; from day 14 of pregnancy until 14 days after delivery) induced hepatic oxidative stress and hepatotoxicity in adult and suckling rats (Troudi et al., 2012). In dams and pups, malondialdehyde levels increased, while decreases were seen in the activities of liver antioxidant enzymes (superoxide dismutase, catalase, and glutathione peroxidase). Similarly, 2,4-D induced oxidative stress in the liver of mothers and fetuses after exposure of pregnant rats to 2,4-D (100 mg/kg bw per day) on days 1–19 of gestation. Coadministration of vitamin E (100 mg/kg bw) partially abrogated the oxidative stress (elevated malondialdehyde levels, decreased catalase activity, decreased total antioxidant capacity) induced by 2,4-D. 2,4-D had no effect on the sex ratio, the number of implantations, or the viable and resorbed fetuses. However, lower body weight and a higher rate of morphological and skeletal defects were seen in fetuses of dams treated with 2,4-D (Mazhar et al., 2014).

In rats exposed to a 2,4-D-based formulation (15, 75, and 150 mg/kg via oral gavage for 4 weeks), a significant reduction in the activity of hepatic antioxidant enzymes (glutathione reductase and catalase) was observed (Tayeb et al., 2010). Hepatotoxicity was demonstrated by increased liver weights, histological changes, and elevated levels of serum enzyme markers. Hepatotoxicity and altered lipid metabolism were reported in a follow-up study using the same dosing regime, in which rats showed increased levels of hepatic malondialdehyde and altered levels of liver antioxidant enzyme (Tayeb et al., 2013). Olive oil was found to protect against hepatic oxidative stress induced by the 2,4-D formulation (Nakbi et al., 2010). A further study with the 2,4-D formulation at the same doses (15, 75, and 150 mg/kg bw via oral gavage) reported oxidative stress in the kidney after 28 days (Tayeb et al., 2012). The 2,4-D-based formulation significantly increased levels of malondialdehyde, and altered the activities of renal catalase and superoxide dismutase. Glutathione peroxidase was significantly decreased in rats exposed at 150 mg/kg bw. Dose-dependent increases were seen in the severity of histopathological evidence of tubular and glomerular damage, and the number of pyknotic nuclei. Decreased uric acid and increased plasma levels of urea and creatinine were also reported.

(ii) In vitro

In an in-vitro study using freshly isolated rat hepatocytes, 2,4-D rapidly depleted glutathione and protein thiols, and induced lipid peroxidation (Palmeira et al., 1995).

In *Saccharomyces cerevisiae*, a 2,4-D-based formulation (2,4-D sodium monohydrate) induced concentration-dependent increases in levels of hydroxyl radicals detected by electron paramagnetic resonance spectroscopy (Teixeira et al., 2004). The effect was consistently greater in a mutant lacking the cytosolic Cu-Zn-superoxide dismutase enzyme (*Δsod1*).

4.2.4 Immunosuppression

(a) Humans

(i) Exposed humans

Faustini et al. (1996) reported reductions in immunological parameters in blood samples from 10 farmers 1–12 days after agricultural exposure to chlorophenoxy herbicides. Compared with values before exposure, the following were significantly reduced ($P < 0.05$): circulating helper (CD4) and suppressor T-cells (CD8), CD8 dim, cytotoxic T lymphocytes (CTL), natural killer cells (NK), and CD8 cells expressing the surface antigens HLA-DR (CD8-DR), and lymphoproliferative response to mitogen stimulations. Lymphocyte mitogenic proliferative responses remained significantly decreased 50–70 days after exposure, while other values were similar to pre-exposure levels. Figgs et al. (2000) reported increased proliferation of peripheral blood lymphocytes (replicative index) after spraying in 12 applicators of 2,4-D ($P = 0.016$). Increases were independent of tobacco and alcohol use. 2,4-D concentrations ranged from 1.0 to 1700 μg/g creatinine per L urine), and increased logarithmically with spraying time. No change was seen in complete blood counts or lymphocyte immunophenotypes.

[The Working Group noted that the findings of Faustini et al. (1996) and those of Figgs et al. (2000) appeared to be contradictory, since Faustini *et al.* observed a decrease in lymphocyte proliferation after exposure to chlorophenoxy herbicides, while Figgs et al. observed a small increase.]

(ii) Human cells in vitro

Holland *et al.* evaluated micronuclei (see Section 4.2.1) and reported a dose-dependent inhibition of the replicative index after exposure to 2,4-D in whole blood or isolated lymphocytes from two non-smoking males (aged 31 and 43 years) (Holland et al., 2002). Inhibition was also observed in a separate experiment using isolated lymphocytes from five non-smokers (three females and two males, aged 26–45 years). On the other hand, with low concentrations of a 2,4-D-based formulation (0.005 mM), the replicative index was slightly increased in both experiments (12–15%; $P = 0.052$). No change in mitotic index was seen with either the formulation or pure 2,4-D (Holland et al., 2002).

(b) Experimental systems

(i) Mouse

There were numerous studies of the immunotoxic effects of 2,4-D in mice. Blakley reported immunostimulatory effects with 2,4-D in three studies. In the first study, sheep erythrocyte-stimulated antibody production and lipopolysaccharide-stimulated B-lymphocyte mitogenesis were enhanced in female BDF1 mice (age, 6 weeks) given the n-butyl ester of 2,4-D (2,4-D content, 100 or 200 mg/kg bw). 2,4-D had no effect on T-lymphocyte mitogenesis induced by concanavalin A (Blakley, 1986).

In a second study, production of antibodies to sheep erythrocytes was suppressed at higher single dermal exposures to the n-butyl ester of 2,4-D (2,4-D content, ≤ 500 mg/kg bw) in female CD-1 mice (Blakley & Schiefer, 1986). No effect of acute exposure was seen on T- and B-lymphocyte proliferative responses to concanavalin A or lipopolysaccharide, respectively. However, short-term exposure to 2,4-D n-butyl ester (2,4-D content, up to 300 mg/kg bw, for 3 weeks) enhanced the B- and T-lymphocyte proliferative responses, while having no effect on antibody production.

In the third study (Blakley & Blakley, 1986), the immune response was altered at age 6 weeks in the female offspring of CD-1 mice given 2,4-D n-butyl ester on day 11 of gestation (2,4-D content, up to 200 mg/kg bw). At the highest exposure (200 mg/kg), a slight decrement was reported in the T-lymphocyte proliferative response, and B-lymphocyte stimulation by lipopolysaccharide was significantly reduced. However, when the decrement in background (unstimulated) mitogenic rates was taken into account, no net suppression by 2,4-D was seen. No effect on the humoral immune response (antibody production against sheep erythrocytes) was seen during gestation.

Two studies reported on the immunodepressive effects of 2,4-D in mice (Zhamsaranova et al., 1987; Sapin et al., 2003). [The Working Group noted that the doses and other experimental details were not available for review.]

Lee at al. (2001) demonstrated a suppression of lymphocyte stimulation by concanavalin A in the offspring (age, 7 weeks) of pregnant CD-1 mice exposed on days 6–16 of gestation to a commercial 2,4-D formulation (up to 1.0% in drinking-water; equivalent to 2,4-D amine derivative at 650 mg/kg per day). Body weight and kidney weights were reduced in the offspring of groups at 0.1% and 1.0%. At 1.0% in drinking-water, the formulation increased relative counts of B cells, and reduced counts of T cytotoxic or suppressor cells. No effect was seen on the humoral immune response or peritoneal macrophage phagocytic function.

In contrast, Salazar et al. (2005) reported significant immunosuppressive effects on humoral immunity in C57BL/6 mice treated with 2,4-D. A 2,4-D formulation (dimethylamine salt of 2,4-D, 47.2%; active ingredient, 150 mg/kg bw, given by intraperitoneal administration) decreased by two to three times the number of phosphorylcholine-specific IgM and IgG antibody-secreting B cells in bone marrow, showing an effect on humoral immunity. In serum, titers of phosphorylcholine-specific immunoglobulins IgM, IgG2b, and IgG3 were decreased by three to four times in mice exposed to 2,4-D. In the spleen, however, 2,4-D produced no change in the number of antibody-producing cells in mice treated with 2,4-D [a finding of little relevance because antibodies are mainly produced by bone marrow-derived cells].

In C57Bl/6 female mice, de la Rosa et al. (2003) reported decreased bone marrow pre-B and IgM(+) B-cell populations 7 days after intraperitoneal exposure to a 2,4-D-based formulation (dimethylamine salt, 47.2%; 200 mg/kg bw per day). However, a 1 : 1 mixture of formulations of propanil and 2,4-D decreased pre-B and IgM(+) B cells at a lower dose (each formulation, 50 mg/kg bw) and an earlier time-point (2 and 7 days). De la Rosa et al. went on to demonstrate reduction in thymus-weight to body-weight ratios and thymocyte depletion at 2 days, and inhibition of thymic T-cell repopulation at 7 days, after exposure to the mixture of formulations of propanil and 2,4-D (each formulation, 150 mg/kg bw per day, by intraperitoneal administration) (de la Rosa et al., 2005). Treatment with the 2,4-D-based formulation only (150 mg/kg bw) had no effect on thymus weight. In another study with mixtures, a herbicide formulation containing 2,4-D and picloram (up to 0.42% in drinking-water for 26 days) had an immunosuppressive effect in female CD-1 mice, reducing antibody production in response to sheep erythrocytes (Blakley, 1997).

(ii) Rat

The first of several studies to report an immunosuppressive effect with 2,4-D in experimental animals was that by Kenigsberg (1975), who reported that the amine salt of 2,4-D suppressed the immune response of rats to *Salmonella* bacteria. A separate group of investigators reported that the amino salt of 2,4-D (2.0 and 20 mg/kg bw daily, intragastric administration) decreased the monocytic-precursor count in the bone marrow in 124 non-inbred white rats (Imel'baeva et al., 1999). The capacity for colony formation was increased, monocytopoiesis was activated, and monocyte migration to peripheral blood was increased. In a third study in rats (Mufazalova et al., 2001), a single dose of the 2,4-D amine salt (240 mg/kg bw) induced phasic changes in blood levels of peripheral leukocytes, and alterations in the microbicidal activity of peritoneal macrophages that persisted for 60 days. The activity of polymorphonuclear leukocytes was also affected. In contrast, Blakley et al. reported no alteration in lymphocyte or macrophage function in male Fisher 344 rats exposed to the amine salt of 2,4-D (10.0 mg/kg, by gavage in olive oil vehicle, twice per week, for 28 days) (Blakley et al., 1998). Specifically, there were no changes in lymphocyte cell-surface marker expression or blastogenesis, phagocytic function of peritoneal macrophages, or antibody production (anti-sheep erythrocytes) (Blakley et al., 1998). No changes in body weight, or organ- to body-weight ratios were seen. [The Working Group noted that effects were seen in studies at high doses, but not in a that used lower doses administered less frequently.]

Marty et al. (2013) evaluated developmental immunotoxicity in CD rats fed diets containing 2,4-D (100, 300, or 600 ppm in females, and 800 ppm in males). Adults and F1 offspring were evaluated for immune function using the sheep erythrocyte antibody-forming cell assay, and the NK cell assay. At 600 ppm in females, reductions were seen in antibody plaque-forming cells in the spleen (54% decrease) and the number of antibody plaque-forming cells per 10^6 splenocytes (27% decrease), although neither effect attained statistical significance. [The Working Group noted that the effect on bone-marrow antibody plaque-forming cells was not evaluated; since antibodies are mainly produced by bone marrow-derived cells, this limited the value of this study.]

4.2.5 Inflammation

(a) Humans

No data from studies in exposed humans, or human cells in vitro, were available to the Working Group.

(b) Experimental systems

Fukuyama et al. (2009) evaluated allergic reactions in BALB/c mice topically sensitized (nine times in 3 weeks) and subsequently challenged with 2,4-D (dermal or intratracheal exposure). One day after challenge, immediate-type respiratory reactions were induced by 2,4-D. In bronchoalveolar lavage fluid, there was a rise in total IgE levels and an influx of eosinophils, neutrophils, and chemokines (MCP-1, eotaxin, and MIP-1beta). Serum IgE levels also increased. Additionally, surface antigen expression on B cells increased in lymph nodes, and Th2 cytokine production (IL-4, IL-5, IL-10, and IL-13) was elevated in lymph-node cells. [The Working Group noted that these results indicated that 2,4-D is a respiratory allergen capable of causing inflammatory responses in the respiratory tract of mice.]

In subsequent studies, the same group treated mice (age, 4 weeks) orally with parathion (0, 0.4, or 1.2 mg/kg bw) or methoxychlor (0, 100, or 300 mg/kg), and then 4 weeks later with 2,4-D-butyl (0%, 2.5%, 5%, or 10%) (Fukuyama et al., 2010). Parathion or methoxychlor markedly reduced the concentration of 2,4-D-butyl required to yield a positive response in the local lymph-node assay (i.e. the concentration estimated to yield a stimulation index of 3). Thus, 2,4-D may be a more potent allergen and inducer of inflammation if there is simultaneous exposure to other pesticides.

In a study in BALB/c mice, 2,4-D-specific IgE antibodies were detected after intraperitoneal administration of 2,4-D, but 2,4-D applied epicutaneously did not result in delayed-type hypersensitivity (Cushman & Street, 1982).

4.2.6 Altered cell proliferation or death

(a) Humans

(i) Exposed humans

In a small cohort (n = 12) of applicators spraying 2,4-D, a significant increase in the replicative index (a measure of cell proliferation) in peripheral blood lymphocytes in the absence of micronucleus induction was observed (Figgs et al., 2000; see Section 4.2.1).

(ii) Human cells in vitro

In-vitro exposure of isolated lymphocytes to a low dose of 2,4-D (0.005 mM) increased the replicative index, but not the mitotic index (Holland et al., 2002; see Section 4.2.4). At higher concentrations, cytotoxicity was reported in transformed human haematopoietic cells (Venkov et al., 2000) and isolated human lymphocytes (Soloneski et al., 2007); the latter study also pointed to a delay in cell-cycle progression only when erythrocytes were present. In HepG2 cells, lower concentrations of 2,4-D appeared to induce a G_1-phase arrest, while higher concentrations prolonged S- or G_2-phase; 2,4-D also significantly disrupted mitochondrial membrane potential, and increased the proportion of annexin-positive cells (Tuschl & Schwab, 2003, 2005). In isolated human lymphocytes, the dimethylammonium salt of 2,4-D initiated apoptosis in peripheral blood lymphocytes via a direct effect on mitochondria and disruption of caspase-9 (Kaioumova et al., 2001a). A reduction in HepG2 cell proliferation (determined by incorporation of bromodeoxyuridine) and downregulation of CDC-like kinase 1 (*CLK1*) was noted by Bharadwaj et al. (2005). Several studies have indicated induction of cytotoxicity in human cells after exposure to formulations containing 2,4-D (e.g. Witte et al., 1996; Holland et al., 2002).

(b) Experimental systems

(i) In vivo

No effects on mitotic index or cell proliferation kinetics in murine bone-marrow cells were observed (Madrigal-Bujaidar et al., 2001). Exposure to a formulation containing picloram and 2,4-D caused testicular germ-cell depletion in rats (Oakes et al., 2002). As discussed above (see Section 4.2.3), 2,4-D induced hepatotoxicity and oxidative damage in male Wistar rats (Tayeb et al., 2013). The oxidative damage profile varied among tissues (Pochettino et al., 2013). It has also been reported that 2,4-D is a peroxisome proliferator in rodents (Vainio et al., 1982; Lundgren et al., 1987). Despite observations characteristic of peroxisome proliferators in the liver, 2,4-D appears to primarily induce an architectural change in the outer part of the kidney medulla, characterized by foci of tubules containing basophilic epithelial cells in rodents (primarily in rats, but also in mice) (Ozaki et al., 2001). Renal effects were also noted in fish (Ozcan Oruc et al., 2004). The dimethylammonium salt of 2,4-D caused cell depletion in the white pulp of the spleen and in the cortex of the thymus in rats (Kaioumova et al., 2001b).

(ii) In vitro

At millimolar concentrations in vitro, 2,4-D induced apoptosis associated with mitochondrial cytochrome *c* release and caspase-3 activation in cerebellar granule cells isolated from Wistar rats (age, 8 days) (De Moliner et al., 2002). In CHO cells, an inhibition of protein synthesis associated with polyamine metabolism has been observed, associated with inhibition of cell growth, DNA and protein biosynthesis, and cell accumulation at the G_1-/S-phase boundary (Rivarola et al., 1985, 1992). Ornithine decarboxylase activity was inhibited, and reductions were seen in spermine and spermidine concentrations, but not putrescine (Rivarola & Balegno, 1991).

At concentrations (low micromolar) capable of inducing cell transformation in the Syrian hamster embryo assay, no effect on levels of apoptosis was observed, including unchanged expression of Bcl2 and Bax (Maire et al., 2007). Compared with other herbicides, 2,4-D was the least potent in uncoupling oxidative phosphorylation in rat liver mitochondria (Zychlinski & Zolnierowicz, 1990).

4.2.7 Other mechanisms

Few studies were identified concerning exposure to 2,4-D and immortalization, DNA repair, or epigenetic end-points. Regarding immortalization, the Agricultural Health Study reported that the mean relative telomere length in buccal cells decreased significantly in association with increased lifetime days of use of 2,4-D ($P = 0.004$), among other pesticides (Hou et al., 2013). Epigenetic end-points were addressed in a few studies in plants, in which 2,4-D altered methylation status (Miassod & Cecchini, 1979; Leljak-Levanić et al., 2004).

4.3 Data relevant to comparisons across agents and end-points

4.3.1 General description of the database

The analysis of the in-vitro bioactivity of the agents reviewed in *IARC Monographs* Volume 113 (i.e. 2,4-D, lindane, and DDT) was informed by data from high-throughput screening assays generated by the Toxicity Testing in the

21st Century (Tox21) and Toxicity Forecaster (ToxCast™) research programmes of the government of the USA (Kavlock et al., 2012; Tice et al., 2013). At its meeting in 2014, the Advisory Group To Recommend Priorities for the *IARC Monographs* programme encouraged inclusion of analysis of high-throughput and high-content data (including from curated government databases) (Straif et al., 2014).

Lindane, DDT (*p,p*'-DDT, *o,p*'-DDT, *p,p*'-DDE, *p,p*'-DDD), and 2,4-D were among the approximately 1000 chemicals tested across the full assay battery of the Tox21 and ToxCast research programmes as of 27 April 2015. This assay battery includes 342 assays, for which data on 821 assay end-points (several assays include multiple end-point readouts) are publicly available on the website of the ToxCast research programme (EPA, 2015a). Detailed information about the chemicals tested, assays used, and associated procedures for data analysis is also publicly available (EPA, 2015b). It should be noted that the metabolic capacity of the cell-based assays is variable, and generally limited.

4.3.2 Aligning in-vitro assays to the 10 "key characteristics" of known human carcinogens

In order to explore the bioactivity profiles of the agents being evaluated in *IARC Monographs* Volume 113 with respect to their potential impact on mechanisms of carcinogenesis, the 821 available assay end-points in the ToxCast/Tox21 database were first mapped to the 10 key characteristics of known human carcinogens (Smith et al., 2016). Working Group members and *IARC Monographs* staff made independent assignments for each assay type to one or more "key characteristics." The assignment was based on the biological target being probed by each assay. The consensus assignments comprise 254 assay end-points that mapped to 6 of the 10 "key characteristics" as shown below. Within each key characteristic, the assays were further divided by the Working Group into subsets of similar end-points.

1. *Is electrophilic or can undergo metabolic activation (31 end-points)*: the assay end-points mapped to this characteristic measure CYP450 inhibition (29 end-points) and aromatase inhibition (2 end-points). All 29 assays for CYP inhibition are cell-free. These assay end-points are not direct measures of electrophilicity or metabolic activation.
2. *Is genotoxic (0 end-points):* no assay end-points were mapped to this characteristic.
3. *Alters DNA repair or causes genomic instability (0 end-points):* no assay end-points were mapped to this characteristic.
4. *Induces epigenetic alterations (11 end-points):* assay end-points mapped to this characteristic measure targets associated with DNA binding (e.g. transcription factors) (4 end-points) and transformation catalysts (e.g. histone deacetylase) (7 end-points).
5. *Induces oxidative stress (18 end-points):* the assay end-points mapped to this characteristic measure oxidative stress via cell imaging (7 end-points), markers of oxidative stress (e.g. nuclear factor erythroid 2-related factor, NRF2) (6 end-points), and metalloproteinase (5 end-points).
6. *Induces chronic inflammation (45 end-points):* the assay end-points mapped to this characteristic measure cellular adhesion (14 end-points), cytokines (e.g. IL8) (29 end-points), and nuclear factor kappa-light-chain-enhancer of activated B cells (NF-κB) activity (2 end-points).
7. *Is immunosuppressive (0 end-points):* no assay end-points were mapped to this characteristic.
8. *Modulates receptor-mediated effects (92 end-points):* a large and diverse collection of cell-free and cell-based assay end-points measuring nuclear and other receptor

bioactivity, specifically, AhR (2 end-points), androgen receptor (11 end-points), estrogen receptor (18 end-points), farnesoid X receptor (FXR) (7 end-points), peroxisome proliferator-activated receptor (PPAR) (12 end-points), pregnane X receptor_vitamin D receptor (PXR_VDR) (7 end-points), retinoic acid receptor (RAR) (6 end-points), others (29 end-points), were mapped to this characteristic.

9. *Causes immortalization (0 endpoints):* no assay end-points were mapped to this characteristic.
10. *Alters cell proliferation, cell death, or nutrient supply (68 end-points):* assay end-points mapped to this characteristic measure cytotoxicity (41 end-points), mitochondrial toxicity (7 end-points), cell cycle (16 end-points), and cell proliferation (4 end-points).

By matching assays to key characteristics, additional insights could be obtained on the bioactivity profile for each compound specifically for the purpose of evaluating their potential to interact with or affect mechanisms involved in carcinogenesis. In addition, for each chemical, the results of the in-vitro assays that represent each "key characteristic" can be compared to the results for a larger compendium of substances with similar in-vitro data, so that a particular chemical can be aligned with other chemicals with similar toxicological effects. Nonetheless, the available assays do not cover the full spectrum of targets that may be associated with these mechanisms, and metabolic capacity in many of the assays is limited, which could account for any absence of bioactivity. Conversely, the presence of bioactivity alone does not definitively imply that the agent exhibits that key characteristic, as the assay data are considered along with other information, both in vivo and in vitro.

The Working Group then extracted information from the ToxCast database concerning whether a chemical was "active" or "inactive" for each of the selected assay end-points (Sipes et al., 2013; EPA, 2015b). In the analysis by the Working Group, each "active" was given a value of 1, and each "inactive" was given a value of 0. Thus, by assigning all active compounds a value of 1, the micromolar "potency" estimates from the concentration–response data were not explicitly modelled.

Next, to integrate the data across individual assay end-points into the cumulative score for each "key characteristic," the toxicological prioritization index (ToxPi) approach (Reif et al., 2010) and associated software (Reif et al., 2013; Filer et al., 2014) were used. In the Working Group's analyses, the ToxPi score provides a visual measure of the potential for a chemical to be associated with a "key characteristic" relative to 181 chemicals that have been previously evaluated by the *IARC Monographs* and that have been screened by ToxCast. Assay end-point data were available in ToxCast for these 181 chemicals, and not for other chemicals previously evaluated by IARC. ToxPi is a dimensionless index score that integrates multiple, different, assay results and displays them visually. Within each subset of end-points ("slice"), data are translated into ToxPi slice-wise scores for all compounds as detailed below and in the publications describing the approach and the associated software package (Reif et al., 2013). Within each individual slice for a given chemical, the distance from the origin represents the relative chemical-elicited activity of the component assays (i.e. slices extending farther from the origin were associated with "active" calls on more assays). The overall score for a chemical, visualized as a radial ToxPi profile, is the aggregation of all slice-wise scores.

The list of ToxCast/Tox21 assay end-points included in the analysis by the Working Group, description of the target and/or model system for each end-point (e.g. cell type, species, detection technology, etc.), their mapping to 6 of the 10 "key characteristics" of known human carcinogens, and the decision as to whether each chemical was "active" or "inactive" are available as supplemental material to *Monographs* Volume 113 (IARC, 2016). The output files generated for each "key characteristic" are also provided in the supplemental material, and can be opened using ToxPi software that is freely available for download without a licence (Reif et al., 2013).

4.3.3 Specific effects across 6 of the 10 "key characteristics" based on data from high-throughput screening in vitro

The relative effects of 2,4-D were compared with those of 181 chemicals selected from the more than 800 chemicals previously evaluated by the *IARC Monographs* and also screened by the Tox21/ToxCast programmes, and with those of the other compounds evaluated in the present volume of the *IARC Monographs* (Volume 113) and with their metabolites. Of these 181 chemicals previously evaluated by the *IARC Monographs* and screened in the ToxCast/Tox21 programmes, 8 are classified in Group 1 (*carcinogenic to humans*), 18 are in Group 2A (*probably carcinogenic to humans*), 65 are in Group 2B (*possibly carcinogenic to humans*), 95 are in Group 3 (*not classifiable as to its carcinogenicity to humans*), and 1 is in Group 4 (*probably not carcinogenic to humans*). The results are presented in a dot plot as a rank order of all compounds in the analysis arranged in the order of their relative activity. The relative positions of lindane, DDT (*p,p*'-DDT, *o,p*'-DDT, *p,p*'-DDE, *p,p*'-DDD), and 2,4-D in the ranked list are also shown on the *y*-axis. The colour scheme legend (lower left in each plot) annotates each compound according to its previous *IARC Monographs* group classification. The legend key (lower right graphic in each plot) lists components of the ToxPi chart as subcategories that comprise assay end-points in each characteristic, as well as their respective colour-coding (see Section 4.3.2 and IARC, 2016). The ToxPi profile and numeric score is shown for the highest-ranked chemical in each analysis (directly above the legend key) to represent the maximum ToxPi score and for 2,4-D (upper frame).

Characteristic (1) *Is electrophilic or can undergo metabolic activation*: 2,4-D was tested in all 31 assay end-points mapped to this key characteristic, and found to be active for 1 of the 29 assay end-points related to CYP inhibition. In comparison, the highest-ranked chemical, malathion (IARC Group 2A; IARC, 2017), was active for 20 out of 29 assay end-points for CYP inhibition, and for 1 out of 2 assay end-points related to aromatase inhibition (Fig. 4.1).

Characteristic (4) *Induces epigenetic alterations:* 2,4-D was tested for all 11 assay end-points mapped to this characteristic, and showed activity for 1 of the transformation-catalyst assay end-points. In comparison, the highest-ranked chemical, captan (IARC Group 3; IARC, 1983), was active for 0 out of 4 DNA binding-assay end-points, and 5 out of 7 transformation-catalyst (e.g. histone modification) assay end-points (Fig. 4.2).

Characteristic (5) *Induces oxidative stress*: 2,4-D was tested for all 18 assay end-points mapped to this characteristic, and was not active for any end-point. In comparison, the highest-ranked chemical, carbaryl (IARC Group 3; IARC, 1976), was active for 2 out of 5 metalloproteinase-assay end-points, 3 out of 7 oxidative-stress assay end-points, and 3 out of 6 oxidative-stress marker assay end-points (Fig. 4.3).

Characteristic (6) *Induces chronic inflammation*: 2,4-D was tested for all 45 assay end-points mapped to this characteristic, and was not active for any end-point. In comparison, the highest-ranked chemical, 4,4'-methylenedianiline (IARC Group 2B; IARC, 1986), was active for 2 out of 14 cellular-adhesion assay end-points, and 2 out of 29 cytokine-assay end-points (Fig. 4.4).

Fig. 4.1 ToxPi ranking for 2,4-dichlorophenoxyacetic acid (2,4-D) using ToxCast assay end-points mapped to metabolic activation

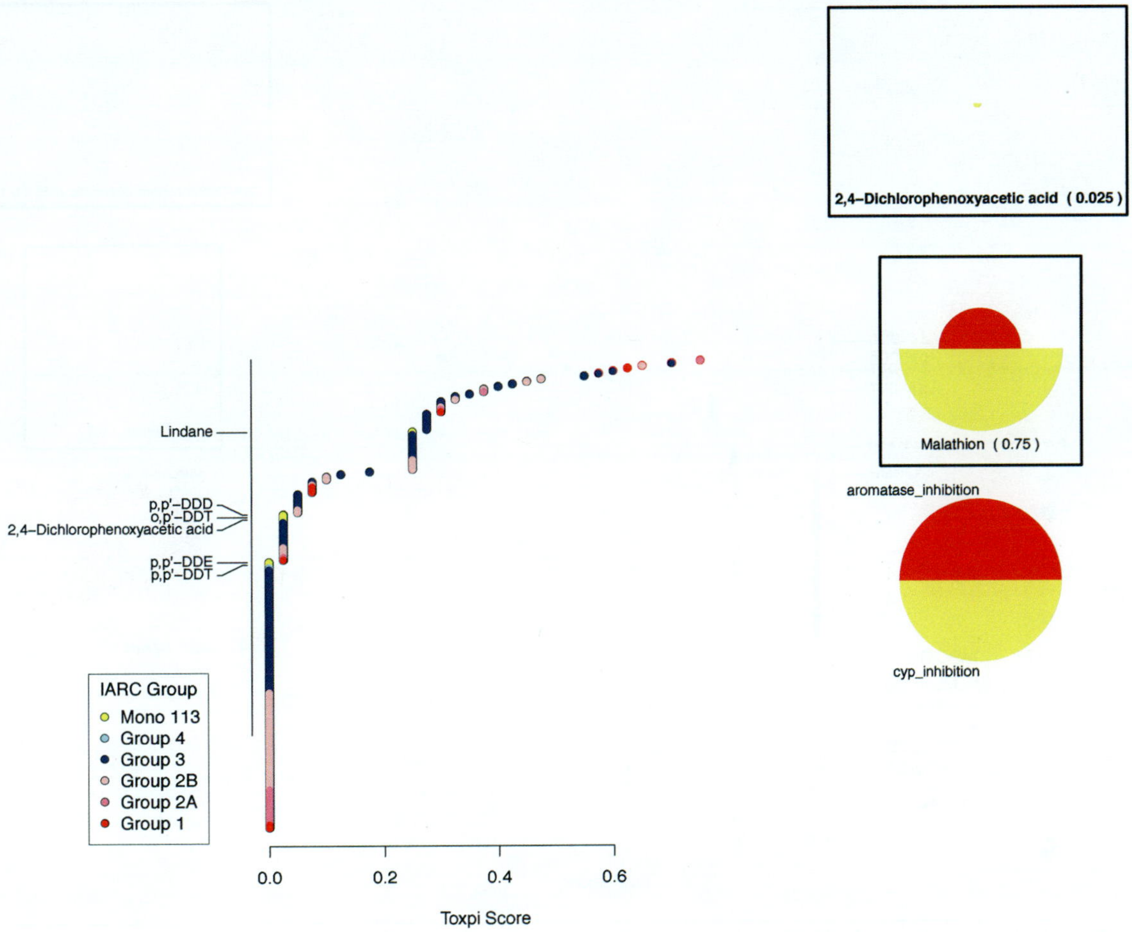

On the left-hand side, the relative rank of 2,4-D is shown (*y*-axis) with respect to its ToxPi score (*x*-axis) compared with the other chemicals evaluated in the present volume (*IARC Monographs* Volume 113) and the 181 chemicals previously evaluated by IARC. The inset in the scatter plot shows subcategories of the ToxPi chart, as well as their respective colour coding. On the right-hand side, the ToxPi charts of the highest-ranked chemical (in this case, malathion) and the target chemical(s) (2,4-D) are shown with their respective ToxPi score in parentheses.

Characteristic (8) *Modulates receptor-mediated effects:* 2,4-D was tested for all 92 assay end-points mapped to this characteristic, and was active for 1 out of 12 PPAR assay end-points. In comparison, the highest-ranked chemical, clomiphene citrate (IARC Group 3; IARC, 1979), was active for 5 out of 11 androgen-receptor assay end-points, 13 out of 18 estrogen-receptor assay end-points, 3 out of 7 FXR assay end-points, 6 out of 29 other nuclear-receptor assay end-points, 2 out of 12 PPAR assay end-points, 5 out of 7 PXR_VDR assay end-points, and 1 out of 6 RAR assay end-points (Fig. 4.5).

Characteristic (10) *Alters cell proliferation, cell death, or nutrient supply:* 2,4-D was tested for 67 out of 68 assay end-points mapped to this characteristic, and was active for 1 of the 41 assay end-points related to cytotoxicity. In comparison, the highest-ranked chemical, ziram (IARC Group 3; IARC, 1991), was active for 2 out of 16 cell-cycle assay end-points,

Fig. 4.2 ToxPi ranking for 2,4-dichlorophenoxyacetic acid (2,4-D) using ToxCast assay end-points mapped to epigenetic alterations

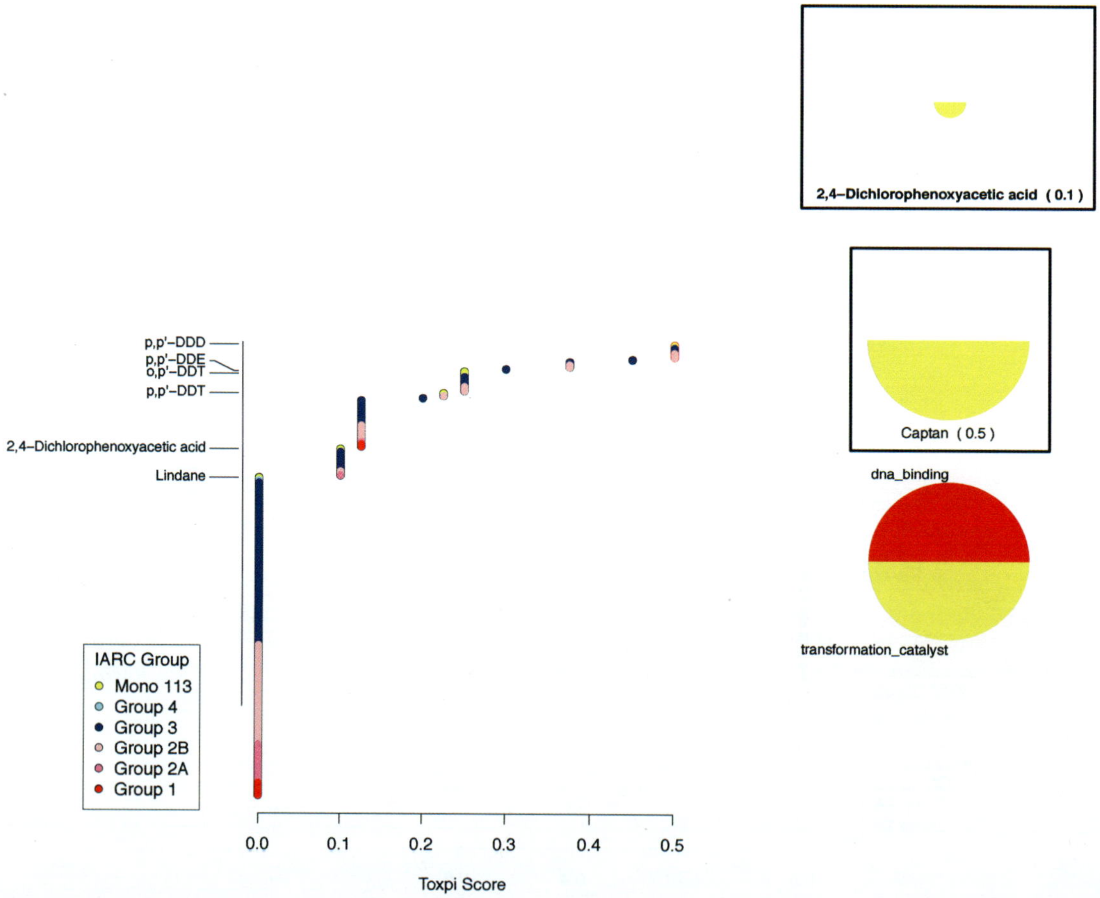

On the left-hand side, the relative rank of 2,4-D is shown (*y*-axis) with respect to their ToxPi score (*x*-axis) compared with the other chemicals evaluated in the present volume (*IARC Monographs* Volume 113) and the 181 chemicals previously evaluated by IARC. The inset in the scatter plot shows subcategories of the ToxPi chart, as well as their respective colour coding. On the right-hand side, the ToxPi charts of the highest-ranked chemical (in this case, captan) and the target chemical(s) (2,4-D) are shown with their respective ToxPi score in parentheses.

33 out of 41 cytotoxicity assay end-points, and 2 out of 7 mitochondrial-toxicity assay end-points (Fig. 4.6).

4.3.4 Summary of all effects across the "key characteristics" based on data from screening in vitro

As a high-level summary of activity, data were recombined into six ToxPi slices, where each slice represents activity across all component assays mapped to a given characteristic. In the figure (Fig. 4.7), slices are labelled "metabolism" (*Is electrophilic or can undergo metabolic activation*), "epigenetic" (*Induces epigenetic alterations*), "stress" (*Induces oxidative stress*), "inflammation" (*Induces chronic inflammation*), "receptor" (*Modulates receptor-mediated effects*), and "cellular" (*Alters cell proliferation, cell death, or nutrient supply*). Overall, 2,4-D was active in four of the assays. In comparison, the highest-ranked chemical, clomiphene citrate (IARC Group 3; IARC, 1979), was active for 104 assay end-points.

Fig. 4.3 ToxPi ranking for 2,4-dichlorophenoxyacetic acid (2,4-D) using ToxCast assay end-points mapped to oxidative stress markers

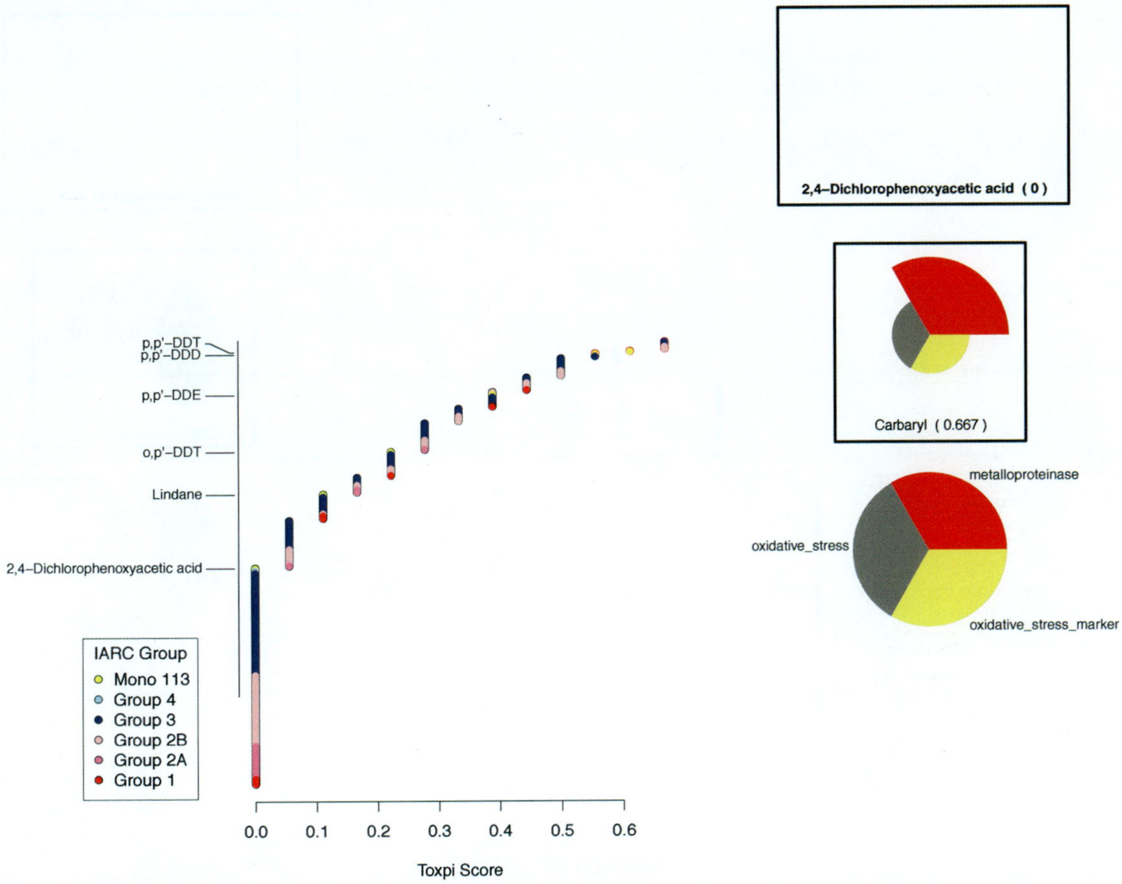

On the left-hand side, the relative rank of 2,4-D is shown (*y*-axis) with respect to its ToxPi score (*x*-axis) compared with the other chemicals evaluated in the present volume (*IARC Monographs* Volume 113) and the 181 chemicals previously evaluated by IARC. The inset in the scatter plot shows subcategories of the ToxPi chart, as well as their respective colour coding. On the right-hand side, the ToxPi charts of the highest-ranked chemical (in this case, carbaryl) and the target chemical(s) (2,4-D) are shown with their respective ToxPi score in parentheses.

4.4 Cancer susceptibility data

Data were not available to the Working Group concerning differential susceptibility due to toxicokinetic or mechanistic factors in humans or experimental systems.

4.5 Other adverse effects

Other adverse effects not addressed in sections 4.1–4.4 that may be relevant to cancer hazard identification for 2,4-D include toxicity in the liver, the lympho-haematopoietic system, or the male reproductive tract.

4.5.1 Humans

One study reported acute hepatitis in a man exposed to 2,4-D through habitual licking of golf balls (Leonard et al., 1997). The patient's liver enzyme levels returned to normal after cessation of exposure, deteriorated again when the behaviour resumed, and then returned back to normal once exposure stopped again.

Fig. 4.4 ToxPi ranking for 2,4-dichlorophenoxyacetic acid (2,4-D) using ToxCast assay end-points mapped to chronic inflammation

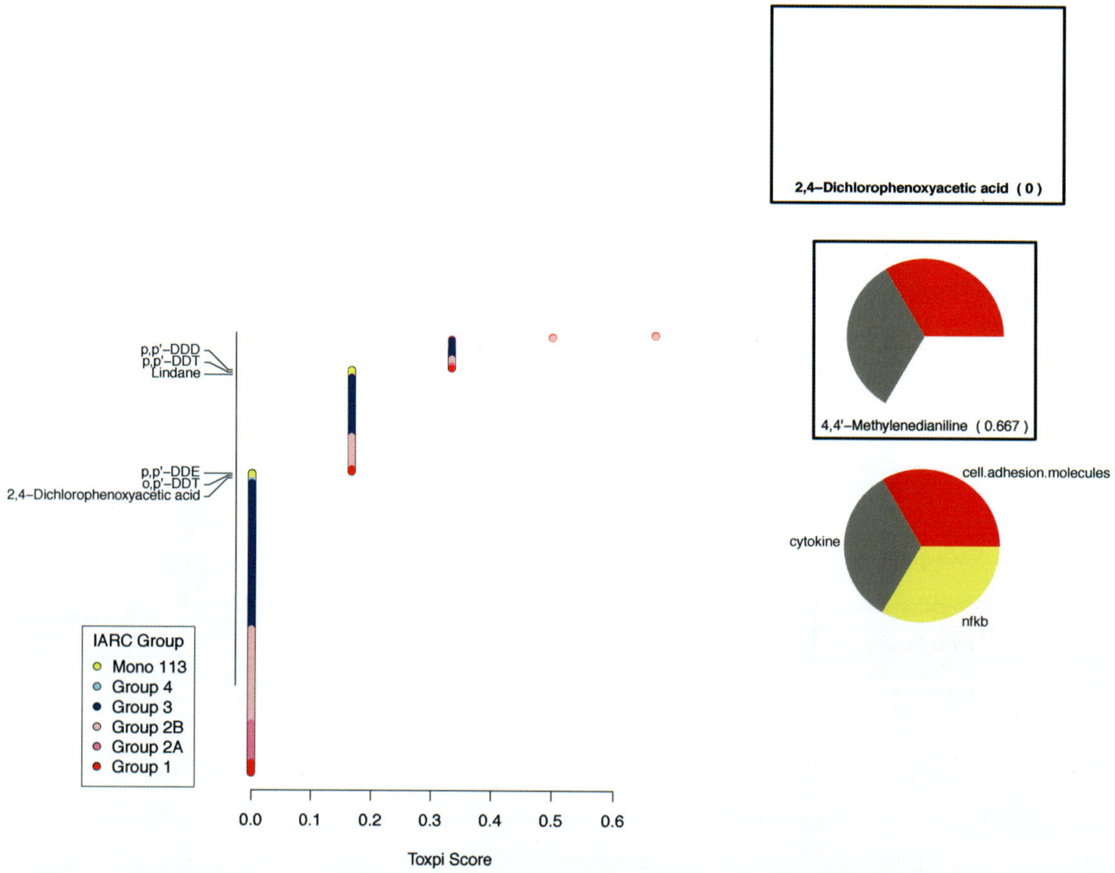

On the left-hand side, the relative rank of 2,4-D is shown (*y*-axis) with respect to its ToxPi score (*x*-axis) compared with the other chemicals evaluated in the present volume (*IARC Monographs* Volume 113) and the 181 chemicals previously evaluated by IARC. The inset in the scatter plot shows subcategories of the ToxPi chart, as well as their respective colour coding. On the right-hand side, the ToxPi charts of the highest-ranked chemical (in this case, 4,4′-methylenedianiline) and the target chemical(s) (2,4-D) are shown with their respective ToxPi score in parentheses.

4.5.2 Experimental systems

Increased liver weights were reported in a short-term study of toxicity with 2,4-D (Charles et al., 1996) and in a short-term study of toxicity with a 2,4-D-based formulation (Tayeb et al., 2010) in rats, but not in single-dose or short-term exposures to mice (Borzelleca et al., 1985). In the short-term study in rats, histological changes, including hepatic cord disruption, focal necrosis, vessel dilation, and pyknotic nuclei, were also reported, with severity increasing with dose (Tayeb et al., 2010). Increases in alkaline phosphatase activity were reported in female mice exposed for 90 days (Borzelleca et al., 1985). No changes in the liver or in any other organs were reported in a long-term study in rats and dogs (Hansen et al., 1971).

With respect to male reproductive toxicity, Charles et al. (1996) reported decreased testes weights in rats after short-term exposure to 2,4-D. In an F1-extended one-generation study of reproductive toxicity with 2,4-D, Marty et al. (2013) reported reduced testicular weights

Fig. 4.5 ToxPi ranking for 2,4-dichlorophenoxyacetic acid (2,4-D) using ToxCast assay end-points mapped to modulation of receptor-mediated effects

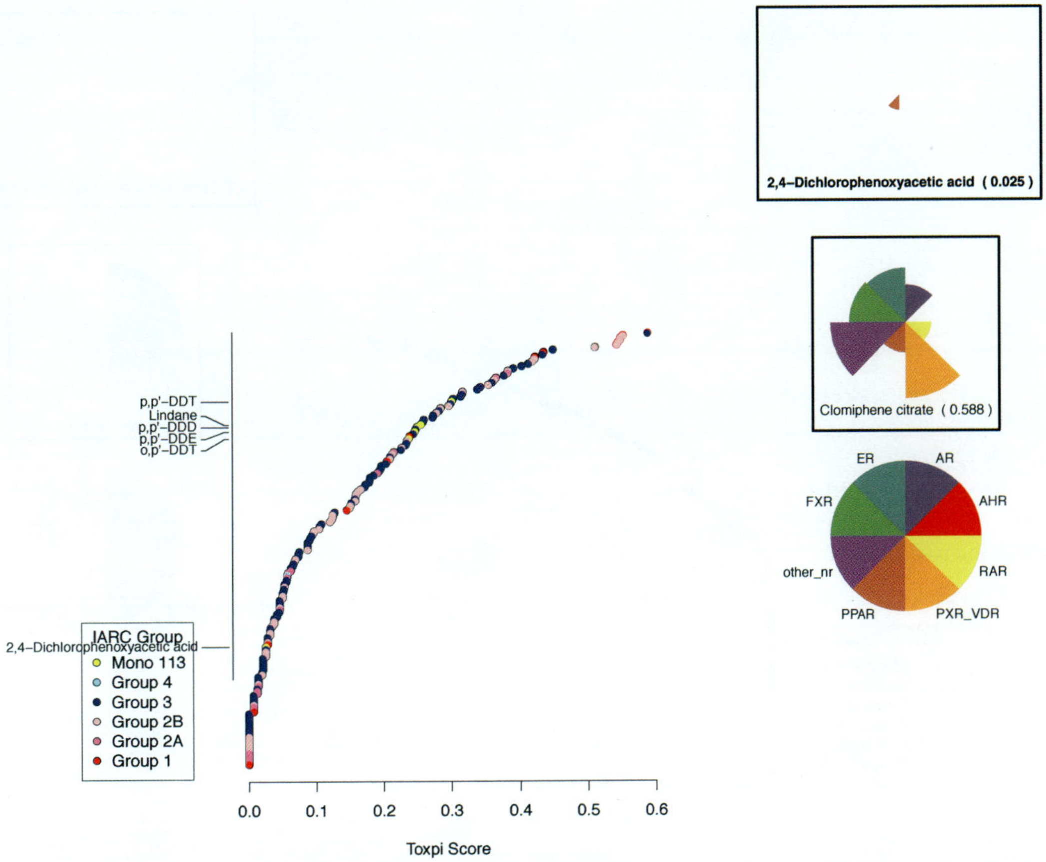

On the left-hand side, the relative rank of 2,4-D is shown (*y*-axis) with respect to its ToxPi score (*x*-axis) compared with the other chemicals evaluated in the present volume (*IARC Monographs* Volume 113) and the 181 chemicals previously evaluated by IARC. The inset in the scatter plot shows subcategories of the ToxPi chart, as well as their respective colour coding. On the right-hand side, the ToxPi charts of the highest-ranked chemical (in this case, clomiphene citrate) and the target chemical(s) (2,4-D) are shown with their respective ToxPi score in parentheses.

accompanied by decreased body weights in weanling rats, and decreased seminal vesicle weights, but no decreases in testicular weights in P1 rats. In rats exposed to a herbicide formulation containing 2,4-D and picloram, Oakes et al. (2002) reported reduced testes weight, shrunken tubules, and germ-cell depletion.

Data were extracted from the Toxicity Reference Database (ToxRefDB), EPA, which contains information on long-term and short-term studies of cancer, developmental and reproductive toxicity in vivo on hundreds of chemicals (Martin et al., 2009). All source files are publicly available from the October 2014 data release (https://www.epa.gov/chemical-research/toxicity-forecaster-toxcasttm-data).

The end-point effects associated with administration of 2,4-D are presented by study type in Table 4.6.

Fig. 4.6 ToxPi ranking for 2,4-dichlorophenoxyacetic acid (2,4-D) using ToxCast assay end-points mapped to metabolic activation

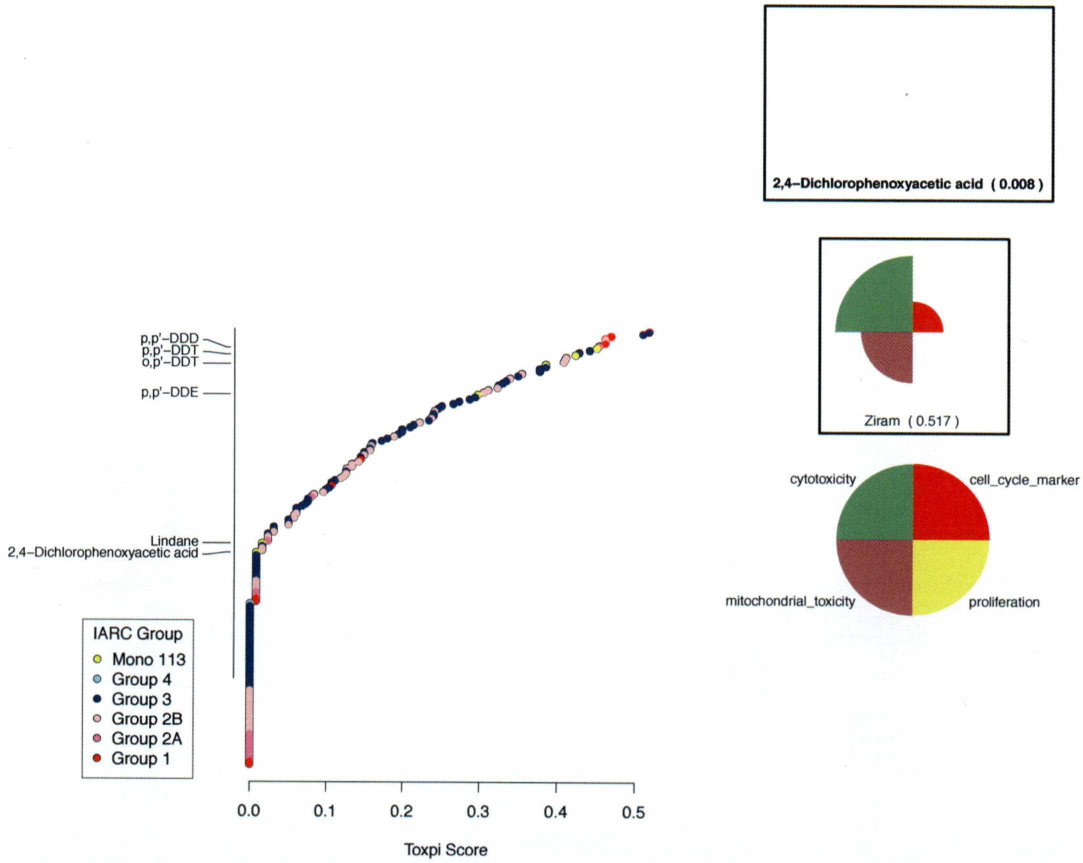

On the left-hand side, the relative rank of 2,4-D is shown (*y*-axis) with respect to its ToxPi score (*x*-axis) compared with the other chemicals evaluated in the present volume (*IARC Monographs* Volume 113) and the 181 chemicals previously evaluated by IARC. The inset in the scatter plot shows subcategories of the ToxPi chart, as well as their respective colour coding. On the right-hand side, the ToxPi charts of the highest-ranked chemical (in this case, ziram) and the target chemical (2,4-D) are shown with its respective ToxPi score in parentheses.

Fig. 4.7 ToxPi ranking for 2,4-dichlorophenoxyacetic acid (2,4-D) using ToxCast assay end-points: summary of key characteristics

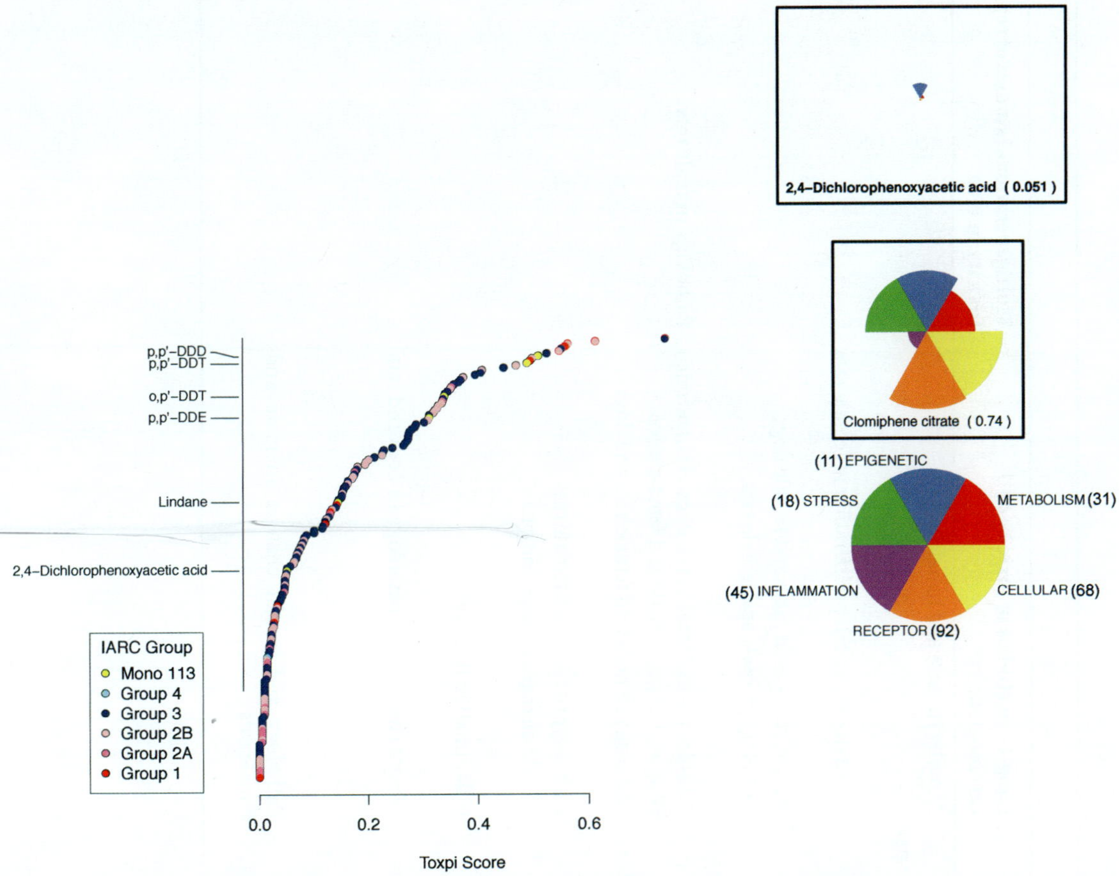

The figure represents a high-level summary of the activity of 2,4-D for end-points covering the seven key characteristics for which it was tested. On the left-hand side, the relative rank of 2,4-D is shown (*y*-axis) with respect to its ToxPi score (*x*-axis) compared with the other chemicals evaluated in the present volume (*IARC Monographs* Volume 113) and the 181 chemicals previously evaluated by IARC. The inset in the scatter plot gives the subcategories and colour coding for the subcategories of the assays. On the right-hand side, the ToxPi charts of the highest-ranked chemical (in this case, clomiphene citrate) and the target chemical (2,4-D) are shown with its respective ToxPi score in parentheses.

Table 4.6 End-point effects associated with administration of 2,4-D reported in the Toxicity Reference Database of the United States Environmental Protection Agency

Target organ	End-point effect		
	13-week studies of toxicity[a]	Long-term studies of toxicity or carcinogenicity[b]	Multigenerational studies of reproductive toxicity in vivo
Adipose tissue		Atrophy in male and female rats	
Adrenal gland	Weight changes in male and female dogs and rats Pathology in male and female rats		
Bone marrow	Haematopoietic hypocellularity in male and female rats	Haematopoietic cell proliferation in female rats	
Brain	Weight changes in male and female rats	Weight changes in male and female dogs	
Eye	Pathology in male and female rats	Pathology in male and female rats	
Heart	Weight changes in male and female dogs and rats		
Kidney	Weight changes and pathology in male and female dogs and rats	Pathology in male and female dogs, rats, and mice Weight changes in male and female mice	Pathology in male rats
Liver	Weight changes in male and female rats and dogs Pathology in male and female rats	Pathology in male and female rats and dogs	
Lung	Pathology in male and female rats	Pathology in male and female rats	
Ovary	Weight changes in rats	Weight changes in dogs and rats	
Pituitary	Weight changes in male and female rats		
Prostate	Inflammation in dogs	Inflammation in dogs	
Spleen	Atrophy in male and female rats		
Testes	Weight changes and pathology in dogs and rats Reduced epididymis size in male rats;	Weight changes and pathology in dogs and rats	
Thymus	Weight changes in male and female rats Atrophy in female rats		
Thyroid	Weight changes in female dogs and male and female rats Pathology in female rats	Pathology and weight changes in male and female rats	

[a] See also Schulze (1991) and EPA (1993)
[b] See also EPA (1995a) and EPA (1995b)
[c] See also Tasker (1985)
2,4-D, 2,4-dichlorophenoxyacetic acid
From EPA (2015c)

5. Summary of data reported

5.1 Exposure data

The common name for 2,4-dichlorophenoxyacetic acid is 2,4-D. In addition to 2,4-D, several 2,4-D ester and salt compounds have been manufactured and used in herbicide products. 2,4-D was commercially introduced in 1944, and has been in continuous production and use worldwide since that time. It is one of the most widely used herbicides around the world for the control of broadleaf weeds and plants in agriculture, forestry, right-of-way (e.g. roadside, rail track, power line), lawn or turf, and aquatic weed control. 2,4-D salts and esters are not persistent under most environmental conditions and are expected to degrade rapidly (within days) to the acid form. Occupational exposures to 2,4-D can result from product manufacturing, and from its use as a herbicide, and occurs primarily via dermal and inhalation routes. Indirect or para-occupational exposure may occur in some populations as a result of take-home and drift exposure pathways, and occurs through dermal absorption, inhalation, and indirect ingestion. The general population may be exposed as a result of the presence of 2,4-D in house dust, food, air, water, and soil. In some areas, residential exposures may be related to use of 2,4-D on lawns. Exposures of the general population may occur through inhalation, dermal absorption, and ingestion. Occupational exposures are often found to be one to three orders of magnitude higher than those in the general population.

5.2 Human carcinogenicity data

Exposure to 2,4-D has been evaluated in relation to cancer risk in population-based case–control studies and in several cohorts of agricultural workers, pesticide applicators, and pesticide manufacturers in the USA, Canada, and Europe. The Working Group also reviewed studies of workers and military personnel who had been exposed to phenoxy herbicides as a class, or to herbicides containing dioxin, but determined that the majority of these studies were uninformative as they did not provide specific risk estimates for 2,4-D. In the studies that were considered to be informative, exposure to 2,4-D was largely assessed by questionnaire, sometimes with expert assessment of work activities and crops grown, and in the industrial cohorts by linkage of work histories to company records of exposure levels. Data were reported in the cohorts for a wide variety of cancers, and in the case–control studies with more detail for non-Hodgkin-lymphoma (NHL), leukaemia, soft tissue sarcoma, and glioma.

More than 10 studies evaluated exposure to 2,4-D in relation to risk of NHL, with NHL classified according to either traditional systems or the more recent WHO classification, which defines lymphoid neoplasms as a broad category that includes lymphoid leukaemia and multiple myeloma. A nested case–control study of NHL within an international cohort of herbicide-manufacturing and -spraying workers, and follow-up of 2,4-D–manufacturing workers in the USA did not observe any strong or consistent increases in risk of NHL in relation to 2,4-D exposure, although the highest risks were observed in the manufacturing cohort in the USA in the highest categories of duration and intensity. A strongly increased risk of NHL, but not leukaemia, was associated with exposure to 2,4-D in a case–control study nested within a cohort of members of a farmworker labour union; nevertheless, the semi-ecological exposure assessment in this study limited the inferences that could be made. Population-based case–control studies of exposure to 2,4-D in relation to risk of lymphoma and leukaemia provided mixed results. Studies in North America found positive associations with exposure to 2,4-D, with evidence for an exposure–response relationship with increasing frequency of use in two studies, but not in a third. Four other studies in the USA and Europe found largely null results. In two studies and a pooled analysis of three studies, associations with 2,4-D were reduced toward the null after adjustment for other pesticides. Studies in which exposure assessment was based on measurement of 2,4-D in house dust did not find an association with risk of NHL or childhood acute lymphocytic lymphoma; however,the validity of dust measurement to reflect exposure during a time frame of etiological interest is unclear.

Two meta-analyses of exposure to 2,4-D and risk of NHL have been published; one included five studies and showed a moderate, statistically significant increase in risk; the other included nine studies and showed no association. The Working Group carried out an additional meta-analysis for exposure to 2,4-D and risk of NHL that included 11 studies, and also showed no association for ever-exposure to 2,4-D. However, sensitivity analyses showed positive associations when risk estimates that were adjusted for other pesticides were replaced by risk estimates from the same studies that were not adjusted for other pesticides, including replacing the pooled analysis that adjusted for multiple pesticides with estimates that were not adjusted for other pesticides from the primary studies.

Risk of soft tissue sarcoma was evaluated in relation to exposure to 2,4-D in three studies. A strong association was found in an international nested case–control study, although the number of cases was small; two case–control studies in the USA reported that they found no association between exposure to 2,4-D and risk of soft tissue sarcoma. There were very few studies examining other cancer sites (e.g. cancers of the prostate, lung, stomach, breast, and melanoma, and glioma), and their findings were inconsistent.

5.3. Animal carcinogenicity data

2,4-D was tested for carcinogenicity by oral administration in two feeding studies in male and female mice, and one study (gavage followed by feeding) in two strains of male and female mice; by single subcutaneous injection in one study in two strains of male and female mice; by oral (drinking-water) administration in two coadministration studies of commercial amine formulations of 2,4-D and the known carcinogen urethane in mice; by oral administration (gavage followed by feeding) in three studies and by single subcutaneous injection of the isopropyl, butyl or isooctyl esters of 2,4-D in three studies in two strains of male and female mice; by oral administration in three feeding studies in male and female rats; and in one epidemiological study in pet dogs.

In mice, subcutaneous administration of the isooctyl ester of 2,4-D resulted in a significantly increased incidence of reticulum cell sarcomas (histiocytic sarcoma/mixed cell malignant lymphoma) in one strain of female mice. This study had limitations in study design, such that a relationship between exposure to 2,4-D and the occurrence of reticulum cell sarcoma could not be established clearly. In one co-carcinogenicity study in male mice, administration of a commercial amine formulation of 2,4-D in drinking-water increased the multiplicity of urethane-induced pulmonary adenoma. The other studies in mice reported negative results.

In one study in rats, oral administration (feeding) of 2,4-D resulted in a significant positive trend in the incidence of astrocytoma of the brain in males. In a second feeding study using higher doses, the incidence and trend in the incidence of astrocytoma of the brain was not increased. The third feeding study reported negative results.

Results of the epidemiological case study in dogs that examined the association between environmental exposure to 2,4-D and risk of canine malignant lymphoma were difficult to evaluate due to potential exposure misclassification.

5.4 Mechanistic and other relevant data

2,4-D, and its salts and esters, are readily absorbed via all routes of exposure. 2,4-D is widely distributed in the body by blood circulation, and is bound reversibly to plasma proteins. The elimination half-life of 2,4-D after cessation of exposure is on the order of 1 day. 2,4-D is largely eliminated unchanged via the urine, with some also eliminated as 2,4-D conjugates. Renal organic anion transporter 1 (OAT1) is involved in the excretion of 2,4-D via the kidneys. Most studies in humans and experimental mammalian systems have reported no evidence of metabolism other than conjugation, but metabolism to 2,4-DCP was reported in human CYP3A4-transfected yeast exposed to 2,4-D.

With respect to the key characteristics of human carcinogens, adequate data to evaluate 2,4-D were available only for oxidative stress, genotoxicity, immunosuppression, receptor-mediated effects, and altered cell proliferation or death.

The evidence that 2,4-D induces oxidative stress that can operate in human is *strong*. In human erythrocytes, 2,4-D induces oxidative stress in vitro. In rats exposed in utero and postnatally through milk, 2,4-D induced oxidative stress in the prostate, ovaries, and mammary tissue up to 90 days after birth. 2,4-D also caused oxidative stress in several regions of the brain in rat pups exposed exclusively via the milk of exposed mothers. 2,4-D increased hepatic oxidative stress in dams and fetuses; this was partially counteracted by co-administration of vitamin E. In yeast, 2,4-D induced hydroxyl-radical formation, and this effect was stronger in superoxide dismutase-deficient mutants. Although there was only one study in human cells in vitro, the results were consistent with available data in rats and yeast.

The evidence that 2,4-D is genotoxic is *weak*. Many of the studies that reported positive results involved mixtures or formulations from which the specific effect of 2,4-D could not be discerned. Several studies in exposed humans found no association between genotoxic effects and exposure to 2,4-D. Evidence for induction of chromosomal aberration, micronucleus formation, and sister-chromatid exchange in human lymphocytes or in vitro was mixed, with experiments using pure 2,4-D giving largely negative results. Data from experimental mammals and non-mammals were mixed. Some studies have reported induction of mutagenic effects in *Drosophila*. 2,4-D does not induce point mutations in bacteria, although positive results have been reported in fish, yeast, and plant systems.

The evidence that 2,4-D causes immunosuppression is *moderate*. There are contradictory results regarding the effects of 2,4-D on lymphocyte proliferation in exposed humans in longitudinal studies; both suppressive and stimulatory effects have been demonstrated, depending on levels of exposure and formulation. In cultures of isolated human lymphocytes exposed to 2,4-D, the lymphocyte proliferation replicative index showed both increases and decreases, depending on the dose and preparation of 2,4-D used. 2,4-D significantly decreased the number of bone-marrow plasma cells in a study in mice, demonstrating suppression of humoral immunity. However, mixed results have been reported in some other studies in rats and mice.

The evidence that 2,4-D modulates receptor activity is *weak*. One study in exposed humans reported a correlation between urinary concentrations of 2,4-D and serum concentrations of luteinizing hormone and testosterone; however, the study subjects were exposed to multiple pesticides and herbicides. Studies in human cells in vitro showed potentiation of androgenic action. In the rat, a single dose of 2,4-D has been shown to interfere with thyroid hormone transport, and to reduce levels of thyroid hormones. Estrogenic activity has been reported in rainbow trout exposed to 2,4-D, according to the vitellogenin assay. 2,4-D causes proliferation of peroxisomes in mouse and rat liver.

The evidence that 2,4-D alters cell proliferation or death is *weak*.

There is inadequate evidence to evaluate whether 2,4-D causes chronic inflammation; however, 2,4-D caused allergy-like hypersensitivity effects in mice.

For the other key characteristics of human carcinogens, the data were insufficient for evaluation.

In high-throughput testing in the Toxicity Testing in the 21st Century (Tox21) and Toxicity Forecaster (ToxCast) research programmes of the government of the USA, 2,4-D gave positive results for 4 assay end-points, including 1 for peroxisome proliferator-activated receptor (PPAR)-related activity, of the 265 assay end-points relevant to the key characteristics of human carcinogens.

There were few data on cancer susceptibility.

2,4-D has been associated with liver effects in a human case report, and in rats and mice, and with reproductive toxicity in males in some studies in rats.

6. Evaluation

6.1 Cancer in humans

There is *inadequate evidence* in humans for the carcinogenicity of 2,4-dichlorophenoxyacetic acid (2,4-D)

6.2 Cancer in experimental animals

There is *limited evidence* in experimental animals for the carcinogenicity of 2,4-dichlorophenoxyacetic acid (2,4-D)

6.3 Overall evaluation

2,4-Dichlorophenoxyacetic acid (2,4-D) is *possibly carcinogenic to humans* (*Group 2B*)

References

Abbott IM, Bonsall JL, Chester G, Hart TB, Turnbull GJ (1987). Worker exposure to a herbicide applied with ground sprayers in the United Kingdom. *Am Ind Hyg Assoc J*, 48(2):167–75. doi:10.1080/15298668791384571 PMID:3565271

Acquavella JF, Alexander BH, Mandel JS, Burns CJ, Gustin C (2006). Exposure misclassification in studies of agricultural pesticides: insights from biomonitoring. *Epidemiology*, 17(1):69–74. doi:10.1097/01.ede.0000190603.52867.22 PMID:16357597

Adhikari N, Grover IS (1988). Genotoxic effects of some systemic pesticides: in vivo chromosomal aberrations in bone marrow cells in rats. *Environ Mol Mutagen*, 12(2):235–42. doi:10.1002/em.2860120209 PMID:3409877

Ahmadi F, Bakhshandeh F (2009). In vitro study of damaging effects of 2,4-dichlorophenoxyacetic acid on DNA structure by spectroscopic and voltammetric techniques. *DNA Cell Biol*, 28(10):527–33. doi:10.1089/dna.2009.0892 PMID:19563251

Alavanja MC, Dosemeci M, Samanic C, Lubin J, Lynch CF, Knott C et al. (2004). Pesticides and lung cancer risk in the Agricultural Health Study cohort. *Am J Epidemiol*, 160(9):876–85. doi:10.1093/aje/kwh290 PMID:15496540

Alexander BH, Mandel JS, Baker BA, Burns CJ, Bartels MJ, Acquavella JF et al. (2007). Biomonitoring of 2,4-dichlorophenoxyacetic acid exposure and dose in farm families. *Environ Health Perspect*, 115(3):370–6. doi:10.1289/ehp.8869 PMID:17431485

Amer SM, Ali EM (1974). Cytological effects of pesticides V. Effects of some herbicides on *Vicia faba*. *Cytologia (Tokyo)*, 39:633–43.

Amer SM, Aly FA (2001). Genotoxic effect of 2,4-dichlorophenoxy acetic acid and its metabolite 2,4-dichlorophenol in mouse. *Mutat Res*, 494(1–2):1–12. doi:10.1016/S1383-5718(01)00146-2 PMID:11423340

Arbuckle TE, Burnett R, Cole D, Teschke K, Dosemeci M, Bancej C et al. (2002). Predictors of herbicide exposure in farm applicators. *Int Arch Occup Environ Health*, 75(6):406–14. doi:10.1007/s00420-002-0323-7 PMID:12070637

Arbuckle TE, Cole DC, Ritter L, Ripley BD (2004). Farm children's exposure to herbicides: comparison of biomonitoring and questionnaire data. *Epidemiology*, 15(2):187–94. doi:10.1097/01.ede.0000112212.01931.81 PMID:15127911

Arbuckle TE, Cole DC, Ritter L, Ripley BD (2005). Biomonitoring of herbicides in Ontario farm applicators. *Scand J Work Environ Health*, 31(Suppl 1):90–7, discussion 63–5. PMID:16190154

Arbuckle TE, Ritter L (2005). Phenoxyacetic acid herbicide exposure for women on Ontario farms. *J Toxicol Environ Health A*, 68(15):1359–70. doi:10.1080/15287390590953635 PMID:16020195

Arcury TA, Grzywacz JG, Barr DB, Tapia J, Chen H, Quandt SA (2007). Pesticide urinary metabolite levels of children in eastern North Carolina farmworker households. *Environ Health Perspect*, 115(8):1254–60. doi:10.1289/ehp.9975 PMID:17687456

Arias E (2003). Sister chromatid exchange induction by the herbicide 2,4-dichlorophenoxyacetic acid in chick embryos. *Ecotoxicol Environ Saf*, 55(3):338–43. doi:10.1016/S0147-6513(02)00131-8 PMID:12798768

Arias E (2007). Cytogenetic effects of short- and long-term exposure of chick embryos to the phenoxyherbicide 2,4-D. *Environ Mol Mutagen*, 48(6):462–6. doi:10.1002/em.20301 PMID:17372986

Arnold EK, Beasley VR (1989). The pharmacokinetics of chlorinated phenoxy acid herbicides: a literature review. *Vet Hum Toxicol*, 31(2):121–5. PMID:2648672

Ateeq B, Abul Farah M, Ahmad W (2005). Detection of DNA damage by alkaline single cell gel electrophoresis in 2,4-dichlorophenoxyacetic-acid- and butachlor-exposed erythrocytes of Clarias batrachus. *Ecotoxicol Environ Saf*, 62(3):348–54. doi:10.1016/j.ecoenv.2004.12.011 PMID:16216628

Ateeq B, Abul Farah M, Niamat Ali M, Ahmad W (2002a). Clastogenicity of pentachlorophenol, 2,4-D and butachlor evaluated by *Allium* root tip test. *Mutat Res*, 514(1–2):105–13. doi:10.1016/S1383-5718(01)00327-8 PMID:11815249

Ateeq B, Abul farah M, Niamat Ali M, Ahmad W (2002b). Induction of micronuclei and erythrocyte alterations in the catfish *Clarias batrachus* by 2,4-dichlorophenoxyacetic acid and butachlor. *Mutat Res*, 518(2):135–44. doi:10.1016/S1383-5718(02)00075-X PMID:12113764

Aulagnier F, Poissant L, Brunet D, Beauvais C, Pilote M, Deblois C et al. (2008). Pesticides measured in air and precipitation in the Yamaska Basin (Québec): occurrence and concentrations in 2004. *Sci Total Environ*, 394(2-3):338–48. doi:10.1016/j.scitotenv.2008.01.042 PMID:18325567

Aydin H, Ozdemir N, Uzunören N (2005). Investigation of the accumulation of 2,4-dichlorophenoxyacetic acid (2,4-D) in rat kidneys. *Forensic Sci Int*, 153(1):53–7. doi:10.1016/j.forsciint.2005.04.018 PMID:15935583

Badawi AF, Cavalieri EL, Rogan EG (2000). Effect of chlorinated hydrocarbons on expression of cytochrome P450 1A1, 1A2 and 1B1 and 2- and 4-hydroxylation of 17beta-estradiol in female Sprague-Dawley rats. *Carcinogenesis*, 21(8):1593–9. doi:10.1093/carcin/21.8.1593 PMID:10910964

Baharuddin MR, Sahid IB, Noor MA, Sulaiman N, Othman F (2011). Pesticide risk assessment: A study on inhalation and dermal exposure to 2,4-D and paraquat among Malaysian paddy farmers. *J Environ Sci Health B*, 46(7):600–7. doi:10.1080/03601234.2011.589309 PMID:21749249

Baker SE, Barr DB, Driskell WJ, Beeson MD, Needham LL (2000). Quantification of selected pesticide metabolites in human urine using isotope dilution high-performance liquid chromatography/tandem mass spectrometry. *J Expo Anal Environ Epidemiol*, 10(6 Pt 2):789–98. doi:10.1038/sj.jea.7500123 PMID:11138671

Bakke B, De Roos AJ, Barr DB, Stewart PA, Blair A, Freeman LB et al. (2009). Exposure to atrazine and selected non-persistent pesticides among corn farmers during a growing season. *J Expo Sci Environ Epidemiol*, 19(6):544–54. doi:10.1038/jes.2008.53 PMID:19052531

Band PR, Abanto Z, Bert J, Lang B, Fang R, Gallagher RP et al. (2011). Prostate cancer risk and exposure to pesticides in British Columbia farmers. *Prostate*, 71(2):168–83. doi:10.1002/pros.21232 PMID:20799287

Baraud L, Tessier D, Aaron JJ, Quisefit JP, Pinart J (2003). A multi-residue method for characterization and determination of atmospheric pesticides measured at two French urban and rural sampling sites. *Anal Bioanal Chem*, 377(7-8):1148–52. doi:10.1007/s00216-003-2196-3 PMID:13680058

Barnekow DE, Hamburg AW, Puvanesarajah V, Guo M (2001). Metabolism of 2,4-dichlorophenoxyacetic acid in laying hens and lactating goats. *J Agric Food Chem*, 49(1):156–63. doi:10.1021/jf000119r PMID:11170571

Becher H, Flesch-Janys D, Kauppinen T, Kogevinas M, Steindorf K, Manz A et al. (1996). Cancer mortality in German male workers exposed to phenoxy herbicides and dioxins. *Cancer Causes Control*, 7(3):312–21. doi:10.1007/BF00052936 PMID:8734824

Bharadwaj L, Dhami K, Schneberger D, Stevens M, Renaud C, Ali A (2005). Altered gene expression in human hepatoma HepG2 cells exposed to low-level 2,4-dichlorophenoxyacetic acid and potassium nitrate. *Toxicol In Vitro*, 19(5):603–19. doi:10.1016/j.tiv.2005.03.011 PMID:15878651

Bhatti P, Blair A, Bell EM, Rothman N, Lan Q, Barr DB et al. (2010). Predictors of 2,4-dichlorophenoxyacetic acid exposure among herbicide applicators. *J Expo Sci*

Environ Epidemiol, 20(2):160–8. doi:10.1038/jes.2009.14 PMID:19319162

Blair A, Tarone R, Sandler D, Lynch C, Rowland A, Wintersteen W et al. (2000). Reliability of reporting on lifestyle and agricultural factors by a sample of participants in the agricultural health study from iowa. *Ann Epidemiol*, 10(7):478 doi:10.1016/S1047-2797(00)00113-7 PMID:11018423

Blakley BR (1986). The effect of oral exposure to the n-butylester of 2,4-dichlorophenoxyacetic acid on the immune response in mice. *Int J Immunopharmacol*, 8(1):93–9. doi:10.1016/0192-0561(86)90077-9 PMID:3485585

Blakley BR (1997). Effect of roundup and tordon 202C herbicides on antibody production in mice. *Vet Hum Toxicol*, 39(4):204–6. PMID:9251167

Blakley BR, Blakley PM (1986). The effect of prenatal exposure to the n-butylester of 2,4-dichlorophenoxyacetic acid (2,4-D) on the immune response in mice. *Teratology*, 33(1):15–20. doi:10.1002/tera.1420330104 PMID:3488604

Blakley BR, Gagnon JM, Rousseaux CG (1992). The effect of a commercial 2,4-D formulation on chemical- and viral-induced tumor production in mice. *J Appl Toxicol*, 12(4):245–9. doi:10.1002/jat.2550120406 PMID:1430774

Blakley BR, Schiefer BH (1986). The effect of topically applied n-butylester of 2,4-dichlorophenoxyacetic acid on the immune response in mice. *J Appl Toxicol*, 6(4):291–5. doi:10.1002/jat.2550060411 PMID:3760456

Blakley BR, Yole MJ, Brousseau P, Boermans H, Fournier M (1998). Effect of 2,4-dicholorophenoxyacetic acid, trifluralin and triallate herbicides on immune function. *Vet Hum Toxicol*, 40(1):5–10. PMID:9467199

Bloemen LJ, Mandel JS, Bond GG, Pollock AF, Vitek RP, Cook RR (1993). An update of mortality among chemical workers potentially exposed to the herbicide 2,4-dichlorophenoxyacetic acid and its derivatives. *J Occup Med*, 35(12):1208–12. PMID:8113924

Boers D, Portengen L, Bueno-de-Mesquita HB, Heederik D, Vermeulen R (2010). Cause-specific mortality of Dutch chlorophenoxy herbicide manufacturing workers. *Occup Environ Med*, 67(1):24–31. doi:10.1136/oem.2008.044222 PMID:19736176

Bokán K, Syberg K, Jensen K, Rank J (2013). Genotoxic potential of two herbicides and their active ingredients assessed with comet assay on a fish cell line, epithelioma papillosum cyprini (EPC). *J Toxicol Environ Health A*, 76(20):1129–37. doi:10.1080/15287394.2013.843068 PMID:24279814

Bond GG, Wetterstroem NH, Roush GJ, McLaren EA, Lipps TE, Cook RR (1988). Cause specific mortality among employees engaged in the manufacture, formulation, or packaging of 2,4-dichlorophenoxyacetic acid and related salts. *Br J Ind Med*, 45(2):98–105. PMID:3342201

Borzelleca JF, Hayes JR, Condie LW, Egle JL Jr (1985). Acute and subchronic toxicity of 2,4-dichlorophenol in CD-1 mice. *Fundam Appl Toxicol*, 5(3):478–86. doi:10.1016/0272-0590(85)90095-8 PMID:4007306

Brand RM, Charron AR, Sandler VL, Jendrzejewski JL (2007a). Moisturizing lotions can increase transdermal absorption of the herbicide 2,4-dichlorophenoxacetic acid across hairless mouse skin. *Cutan Ocul Toxicol*, 26(1):15–23. doi:10.1080/15569520601182791 PMID:17464745

Brand RM, Jendrzejewski JL, Charron AR (2007b). Potential mechanisms by which a single drink of alcohol can increase transdermal absorption of topically applied chemicals. *Toxicology*, 235(3):141–9. doi:10.1016/j.tox.2007.03.008 PMID:17467136

Brand RM, McMahon L, Jendrzejewski JL, Charron AR (2007c). Transdermal absorption of the herbicide 2,4-dichlorophenoxyacetic acid is enhanced by both ethanol consumption and sunscreen application. *Food Chem Toxicol*, 45(1):93–7. doi:10.1016/j.fct.2006.08.005 PMID:17030379

Brand RM, Pike J, Wilson RM, Charron AR (2003). Sunscreens containing physical UV blockers can increase transdermal absorption of pesticides. *Toxicol Ind Health*, 19(1):9–16. doi:10.1191/0748233703th169oa PMID:15462532

Brown LM, Blair A, Gibson R, Everett GD, Cantor KP, Schuman LM et al. (1990). Pesticide exposures and other agricultural risk factors for leukemia among men in Iowa and Minnesota. *Cancer Res*, 50(20):6585–91. PMID:2208120

Brown LM, Burmeister LF, Everett GD, Blair A (1993). Pesticide exposures and multiple myeloma in Iowa men. *Cancer Causes Control*, 4(2):153–6. doi:10.1007/BF00053156 PMID:8481493

Bueno de Mesquita HB, Doornbos G, Van der Kuip DA, Kogevinas M, Winkelmann R (1993). Occupational exposure to phenoxy herbicides and chlorophenols and cancer mortality in The Netherlands. *Am J Ind Med*, 23(2):289–300. doi:10.1002/ajim.4700230206 PMID:8427257

Bukowska B (2003). Effects of 2,4-D and its metabolite 2,4-dichlorophenol on antioxidant enzymes and level of glutathione in human erythrocytes. *Comp Biochem Physiol C Toxicol Pharmacol*, 135(4):435–41. doi:10.1016/S1532-0456(03)00151-0 PMID:12965188

Bukowska B, Rychlik B, Krokosz A, Michałowicz J (2008). Phenoxyherbicides induce production of free radicals in human erythrocytes: oxidation of dichlorodihydrofluorescine and dihydrorhodamine 123 by 2,4-D-Na and MCPA-Na. *Food Chem Toxicol*, 46(1):359–67. doi:10.1016/j.fct.2007.08.011 PMID:17889420

Burckhardt G, Wolff NA (2000). Structure of renal organic anion and cation transporters. *Am J Physiol Renal Physiol*, 278(6):F853–66. PMID:10836973

Burns C, Bodner K, Swaen G, Collins J, Beard K, Lee M (2011). Cancer incidence of 2,4-D production workers. *Int J Environ Res Public Health*, 8(9):3579–90. doi:10.3390/ijerph8093579 PMID:22016704

Burns CJ, Beard KK, Cartmill JB (2001). Mortality in chemical workers potentially exposed to 2,4-dichlorophenoxyacetic acid (2,4-D) 1945–94: an update. *Occup Environ Med*, 58(1):24–30. doi:10.1136/oem.58.1.24 PMID:11119631

Burns CJ, Swaen GM (2012). Review of 2,4-dichlorophenoxyacetic acid (2,4-D) biomonitoring and epidemiology. *Crit Rev Toxicol*, 42(9):768–86. doi:10.3109/10408444.2012.710576 PMID:22876750

Burton JA, Gardiner TH, Schanker LS (1974). Absorption of herbicides from the rat lung. *Arch Environ Health*, 29(1):31–3. doi:10.1080/00039896.1974.10666523 PMID:4841458

Cantor KP, Blair A, Everett G, Gibson R, Burmeister LF, Brown LM et al. (1992). Pesticides and other agricultural risk factors for non-Hodgkin's lymphoma among men in Iowa and Minnesota. *Cancer Res*, 52(9):2447–55. PMID:1568215

CAREX Canada (2009). 2,4-D. Agents cancérogènes. British Columbia: CAREX Canada. Available from: http://www.carexcanada.ca/fr/2,4-d.pdf.

Carlo GL, Cole P, Miller AB, Munro IC, Solomon KR, Squire RA (1992). Review of a study reporting an association between 2,4-dichlorophenoxyacetic acid and canine malignant lymphoma: report of an expert panel. *Regul Toxicol Pharmacol*, 16(3):245–52. doi:10.1016/0273-2300(92)90004-S PMID:1293641

CDC (2015). Fourth national report on human exposure to environmental chemicals, February 2015. Updated tables. Atlanta (GA): Centers for Disease Control and Prevention. Available from: http://www.cdc.gov/biomonitoring/pdf/FourthReport_UpdatedTables_Feb2015.pdf, accessed 9 April 2015.

Cenkci S, Yildiz M, Ciğerci IH, Bozdağ A, Terzi H, Terzi ES (2010). Evaluation of 2,4-D and Dicamba genotoxicity in bean seedlings using comet and RAPD assays. *Ecotoxicol Environ Saf*, 73(7):1558–64. doi:10.1016/j.ecoenv.2010.07.033 PMID:20797789

Charles JM, Bond DM, Jeffries TK, Yano BL, Stott WT, Johnson KA et al. (1996a). Chronic dietary toxicity/oncogenicity studies on 2,4-dichlorophenoxyacetic acid in rodents. *Fundam Appl Toxicol*, 33(2):166–72. doi:10.1006/faat.1996.0154 PMID:8921335

Charles JM, Cunny HC, Wilson RD, Bus JS (1996). Comparative subchronic studies on 2,4-dichlorophenoxyacetic acid, amine, and ester in rats. *Fundam Appl Toxicol*, 33(2):161–5. doi:10.1006/faat.1996.0153 PMID:8921334

Charles JM, Cunny HC, Wilson RD, Bus JS, Lawlor TE, Cifone MA et al. (1999a). Ames assays and unscheduled DNA synthesis assays on 2, 4-dichlorophenoxyacetic acid and its derivatives. *Mutat Res*, 444(1):207–16. doi:10.1016/S1383-5718(99)00074-1 PMID:10477356

Charles JM, Cunny HC, Wilson RD, Ivett JL, Murli H, Bus JS et al. (1999b). In vivo micronucleus assays on 2,4-dichlorophenoxyacetic acid and its derivatives. *Mutat Res*, 444(1):227–34. doi:10.1016/S1383-5718(99)00076-5 PMID:10477358

Charles JM, Hanley TR Jr, Wilson RD, van Ravenzwaay B, Bus JS (2001). Developmental toxicity studies in rats and rabbits on 2,4-dichlorophenoxyacetic acid and its forms. *Toxicol Sci*, 60(1):121–31. doi:10.1093/toxsci/60.1.121 PMID:11222879

Clausen M, Leier G, Witte I (1990). Comparison of the cytotoxicity and DNA-damaging properties of 2,4-D and U 46 D fluid (dimethylammonium salt of 2,4-D). *Arch Toxicol*, 64(6):497–501. doi:10.1007/BF01977633 PMID:2275605

Coady K, Marino T, Thomas J, Sosinski L, Neal B, Hammond L (2013). An evaluation of 2,4-dichlorophenoxyacetic acid in the Amphibian Metamorphosis Assay and the Fish Short-Term Reproduction Assay. *Ecotoxicol Environ Saf*, 90:143–50. doi:10.1016/j.ecoenv.2012.12.025 PMID:23357564

Coady KK, Kan HL, Schisler MR, Gollapudi BB, Neal B, Williams A et al. (2014). Evaluation of potential endocrine activity of 2,4-dichlorophenoxyacetic acid using in vitro assays. *Toxicol In Vitro*, 28(5):1018–25. doi:10.1016/j.tiv.2014.04.006 PMID:24815817

Coble J, Arbuckle T, Lee W, Alavanja M, Dosemeci M (2005). The validation of a pesticide exposure algorithm using biological monitoring results. *J Occup Environ Hyg*, 2(3):194–201. doi:10.1080/15459620590923343 PMID:15764542

Cocco P, Satta G, D'Andrea I, Nonne T, Udas G, Zucca M et al. (2013b). Lymphoma risk in livestock farmers: results of the Epilymph study. *Int J Cancer*, 132(11):2613–8. doi:10.1002/ijc.27908 PMID:23065666

Cocco P, Satta G, Dubois S, Pili C, Pilleri M, Zucca M et al. (2013a). Lymphoma risk and occupational exposure to pesticides: results of the Epilymph study. *Occup Environ Med*, 70(2):91–8. doi:10.1136/oemed-2012-100845 PMID:23117219

Coggon D, Ntani G, Harris EC, Jayakody N, Palmer KT (2015). Soft tissue sarcoma, non-Hodgkin's lymphoma and chronic lymphocytic leukaemia in workers exposed to phenoxy herbicides: extended follow-up of a UK cohort. *Occup Environ Med*, 72(6):435–41. doi:10.1136/oemed-2014-102654 PMID:25694496

Colt JS, Gunier RB, Metayer C, Nishioka MG, Bell EM, Reynolds P et al. (2008). Household vacuum cleaners vs. the high-volume surface sampler for collection of carpet dust samples in epidemiologic studies of children. *Environ Health*, 7(1):6. doi:10.1186/1476-069X-7-6 PMID:18291036

Curwin BD, Hein MJ, Sanderson WT, Barr DB, Heederik D, Reynolds SJ et al. (2005a). Urinary and hand wipe pesticide levels among farmers and nonfarmers in Iowa. *J Expo Anal Environ Epidemiol*, 15(6):500–8. doi:10.1038/sj.jea.7500428 PMID:15841098

Curwin BD, Hein MJ, Sanderson WT, Nishioka MG, Reynolds SJ, Ward EM et al. (2005b). Pesticide contamination inside farm and nonfarm homes. *J Occup Environ Hyg*, 2(7):357–67. doi:10.1080/15459620591001606 PMID:16020099

Cushman JR, Street JC (1982). Allergic hypersensitivity to the herbicide 2,4-D in BALB/c mice. *J Toxicol Environ Health*, 10(4–5):729–41. doi:10.1080/15287398209530291 PMID:7161824

de la Rosa P, Barnett J, Schafer R (2003). Loss of pre-B and IgM(+) B cells in the bone marrow after exposure to a mixture of herbicides. *J Toxicol Environ Health A*, 66(24):2299–313. doi:10.1080/716100638 PMID:14630522

de la Rosa P, Barnett JB, Schafer R (2005). Characterization of thymic atrophy and the mechanism of thymocyte depletion after in vivo exposure to a mixture of herbicides. *J Toxicol Environ Health A*, 68(2):81–98. doi:10.1080/15287390590886072 PMID:15762548

De Moliner KL, Evangelista de Duffard AM, Soto E, Duffard R, Adamo AM (2002). Induction of apoptosis in cerebellar granule cells by 2,4-dichlorophenoxyacetic acid. *Neurochem Res*, 27(11):1439–46. doi:10.1023/A:1021665720446 PMID:12512947

De Roos AJ, Zahm SH, Cantor KP, Weisenburger DD, Holmes FF, Burmeister LF et al. (2003). Integrative assessment of multiple pesticides as risk factors for non-Hodgkin's lymphoma among men. *Occup Environ Med*, 60(9):E11. doi:10.1136/oem.60.9.e11 PMID:12937207

Dennis LK, Lynch CF, Sandler DP, Alavanja MC (2010). Pesticide use and cutaneous melanoma in pesticide applicators in the agricultural heath study. *Environ Health Perspect*, 118(6):812–7. doi:10.1289/ehp.0901518 PMID:20164001

Deregowski K, Sulik M, Kemona A, Stefańska-Sulik E (1990). [Distribution and elimination of C-14–2,4-dichlorophenoxyacetic acid (2,4 D) in rat tissues in acute poisoning]. *Rocz Panstw Zakl Hig*, 41(1-2):71–4. PMID:2244176

Deziel NC, Colt JS, Kent EE, Gunier RB, Reynolds P, Booth B et al. (2015). Associations between self-reported pest treatments and pesticide concentrations in carpet dust. *Environ Health*, 14(1):27. doi:10.1186/s12940-015-0015-x PMID:25889489

Dickow LM, Gerken DF, Sams RA, Ashcraft SM (2001). Simultaneous determination of 2,4-D and MCPA in canine plasma and urine by HPLC with fluorescence detection using 9-anthryldiazomethane (ADAM). *J Anal Toxicol*, 25(1):35–9. doi:10.1093/jat/25.1.35 PMID:11215998

Doležel J, Lucretti S, Novák FJ (1987). The influence of 2,4-dichlorophenoxyacetic acid on cell cycle kinetics and sister-chromatid exchange frequency in garlic (*Allium sativum* L.) meristem cells. *Biol Plant*, 29(4):253–7. doi:10.1007/BF02892785

Dosemeci M, Alavanja MC, Rowland AS, Mage D, Zahm SH, Rothman N et al. (2002). A quantitative approach for estimating exposure to pesticides in the Agricultural Health Study. *Ann Occup Hyg*, 46(2):245–60. doi:10.1093/annhyg/mef011 PMID:12074034

Draper WM (1982). A multiresidue procedure for the determination and confirmation of acidic herbicide residues in human urine. *J Agric Food Chem*, 30(2):227–31. doi:10.1021/jf00110a004 PMID:6853852

Draper WM, Street JC (1982). Applicator exposure to 2,4-D, dicamba, and a dicamba isomer. *J Environ Sci Health B*, 17(4):321–39. doi:10.1080/03601238209372324 PMID:7108143

Duff RM, Kissel JC (1996). Effect of soil loading on dermal absorption efficiency from contaminated soils. *J Toxicol Environ Health*, 48(1):93–106. doi:10.1080/009841096161492 PMID:8637061

Duyzer JH, Vonk AW (2003). Atmospheric deposition of pesticides, PAHs and PCBs in the Netherlands (translation of R2002/606). TNO-report R 2003/255. Available from: http://repository.tudelft.nl/view/tno/uuid%3A9f2f89da-c2e8-4eb1-8c5a-61858d53b9e4/, accessed 23 January 2016.

Edwards MD, Pazzi KA, Gumerlock PH, Madewell BR (1993). c-N-ras is activated infrequently in canine malignant lymphoma. *Toxicol Pathol*, 21(3):288–91. doi:10.1177/019262339302100304 PMID:8248717

EFSA (2011). Reasoned opinion. Review of the existing maximum residue levels (MRLs) for 2,4-D according to Article 12 of Regulation (EC) No 396/2005. *EFSA Journal*, 9(11):2431. Available fromhttp://www.efsa.europa.eu/fr/efsajournal/doc/2431.pdf.

EFSA (2015). The 2013 European Union report on pesticide residues in food. *EFSA Journal*, 13(3):4038.

Ekström AM, Eriksson M, Hansson LE, Lindgren A, Signorello LB, Nyrén O et al. (1999). Occupational exposures and risk of gastric cancer in a population-based case-control study. *Cancer Res*, 59(23):5932–7. PMID:10606238

EPA (1990a).Toxicology data in response to testing requirements of the 2,4-D registration standard - TXR008183. EPA Identification Data. No.0073 *Record No.*, 261:253. Washington (DC): United States Environmental Protection Agency.

EPA (1990b).Toxicology data in response to testing requirements of the 2,4-D registration standard - TXR008183. EPA Identification Data. No.0073 *Record No.*, 261:251. Washington (DC): United States Environmental Protection Agency.

EPA (1993). 13-Week dietary toxicity study of 2,4-D in dogs: Final Report. Lab Project Number: HWA 2184-125. Unpublished study prepared by Hazleton Washington, Inc by Dalgard D. Washington (DC): United States

Environmental Protection Agency. Available from: http://archive.epa.gov/pesticides/chemicalsearch/chemical/foia/web/pdf/030001/030001-1994-04-25a.pdf.

EPA (1995a). Review of Jeffries T, Yano B, Ormand J, Battjes JE. 2,4-Dichlorophenoxyacetic acid: chronic toxicity/oncogenicity study in Fischer 344 rats: Final report. Lab Project Number: K/002372/064. Unpublished study prepared by The Dow Chemical Co., Health and Environment. Washington (DC): United States Environmental Protection Agency.

EPA (1995b). Review of Stott W, Johnson K, Gilbert K, Battjes JE. 2,4-Dichlorophenoxyacetic acid: dietary oncogenicity study in male B6C3F1 mice--Two year final report. Lab Project No: K-002372-063M: 33475: 913. Unpublished study prepared by Dow Chemical Co. Washington (DC): United States Environmental Protection Agency. Available from: http://www3.epa.gov/pesticides/chem_search/cleared_reviews/csr_PC-030001_23-May-96_a.pdf.

EPA (1997). Carcinogenicity peer review (4th) of 2,4-dichlorophenoxyacetic acid (2,4-D). Washington (DC): Office of Prevention, Pesticides and Toxic Substances, United States Environmental Protection Agency.

EPA (2000). Determination of chlorinated acids in drinking water by liquid-liquid microextraction, derivatization, and fast gas chromatography with electron capture detection. Revision 1.0. Method 515.4. Cincinatti (OH): Technical Support Centre, Office of Ground Water and Drinking Water, United States Environmental Protection Agency. Available from: http://nepis.epa.gov/Exe/ZyPDF.cgi/P1008MC8.PDF?Dockey=P1008MC8.PDF, accessed 18 February 2015.

EPA (2005). Reregistration eligibility decision for 2,4-D. EPA 738-R-05-002. Washington (DC): Prevention, pesticides, and Toxic Substances, United States Environmental Protection Agency. Available from: http://archive.epa.gov/pesticides/reregistration/web/pdf/24d_red.pdf.

EPA (2011). Pesticides industry sales and usage, 2006 and 2007 market estimates. Washington (DC): Office of Chemical Safety and Pollution Prevention, United States Environmental Protection Agency.

EPA (2014). Enlist Duo herbicide fact sheet. Washington (DC): United States Environmental Protection Agency. Available from: https://www.epa.gov/sites/production/files/2014-09/documents/enlist-duo-herbicide-fact-sheet.pdf.

EPA (2015a). Interactive Chemical Safety for Sustainability (iCSS) Dashboard. Washington (DC): United States Environmental Protection Agency. Available from: www.actor.epa.gov/dashboard.

EPA (2015b). ToxCastTM Data. Washington (DC): United States Environmental Protection Agency. Available from: http://epa.gov/ncct/toxcast/data.html. Online database

EPA (2015c). Animal toxicity studies: Effects and end-points (Toxicity Reference Database - ToxCast ToxRefDB files). Washington (DC): United States Environmental Protection Agency. Available from: https://www3.epa.gov/research/COMPTOX/animal_toxicity_data.html.

Epstein SS, Arnold E, Andrea J, Bass W, Bishop Y (1972). Detection of chemical mutagens by the dominant lethal assay in the mouse. *Toxicol Appl Pharmacol*, 23(2):288–325. doi:10.1016/0041-008X(72)90192-5 PMID:5074577

Eriksson M, Hardell L, Berg NO, Möller T, Axelson O (1981). Soft-tissue sarcomas and exposure to chemical substances: a case-referent study. *Br J Ind Med*, 38(1):27–33. PMID:7470401

Eriksson M, Hardell L, Carlberg M, Akerman M (2008). Pesticide exposure as risk factor for non-Hodgkin lymphoma including histopathological subgroup analysis. *Int J Cancer*, 123(7):1657–63. doi:10.1002/ijc.23589 PMID:18623080

Eriksson M, Karlsson M (1992). Occupational and other environmental factors and multiple myeloma: a population based case-control study. *Br J Ind Med*, 49(2):95–103. PMID:1536825

Erne K (1966). Distribution and elimination of chlorinated phenoxyacetic acids in animals. *Acta Vet Scand*, 7(3):240–56. PMID:5959182

Estevam EC, Nakano E, Kawano T, de Bragança Pereira CA, Amancio FF, de Albuquerque Melo AMM (2006). Dominant lethal effects of 2,4-D in Biomphalaria glabrata. *Mutat Res*, 611(1–2):83–8. doi:10.1016/j.mrgentox.2006.07.001 PMID:16973407

European Commission (1991). Council Directive of 15 July 1991 concerning the placing of plant protection products on the market (91/414/EEC). Official Journal of the European Communities, No. L230/1–32. Available from: http://eur-lex.europa.eu/legal-content/EN/TXT/PDF/?uri=CELEX:31991L0414&from=EN, accessed 18 February 2016.

European Commission (2001). Review report for the active substance 2,4-D. Report No. 7599/V1/97-final. Commission working document. Brussels: Health & Consumer Protection Directorate-General, Directorate E1 – Food safety, European Commission. Available from: http://24d.org/govtrev/ec20011001.pdf.

Fahrig R (1974). Comparative mutagenicity studies with pesticides. *IARC Sci Publ*, 10:161–81.

Fang SC, Lindstrom FT (1980). In vitro binding of 14C-labeled acidic compounds to serum albumin and their tissue distribution in the rat. *J Pharmacokinet Biopharm*, 8(6):583–97. doi:10.1007/BF01060055 PMID:7229910

Farah MA, Ateeq B, Ahmad W (2006). Antimutagenic effect of neem leaves extract in freshwater fish, Channa punctatus evaluated by cytogenetic tests.

Sci Total Environ, 364(1–3):200–14. doi:10.1016/j.scitotenv.2005.07.008 PMID:16169061

Farah MA, Ateeq B, Ali MN, Ahmad W (2003). Evaluation of genotoxicity of PCP and 2,4-D by micronucleus test in freshwater fish Channa punctatus. *Ecotoxicol Environ Saf*, 54(1):25–9. doi:10.1016/S0147-6513(02)00037-4 PMID:12547631

Faustini A, Settimi L, Pacifici R, Fano V, Zuccaro P, Forastiere F (1996). Immunological changes among farmers exposed to phenoxy herbicides: preliminary observations. *Occup Environ Med*, 53(9):583–5. doi:10.1136/oem.53.9.583 PMID:8882113

Feldmann RJ, Maibach HI (1974). Percutaneous penetration of some pesticides and herbicides in man. *Toxicol Appl Pharmacol*, 28(1):126–32. doi:10.1016/0041-008X(74)90137-9 PMID:4853576

Félix-Cañedo TE, Durán-Álvarez JC, Jiménez-Cisneros B (2013). The occurrence and distribution of a group of organic micropollutants in Mexico City's water sources. *Sci Total Environ*, 454-455:109–18. doi:10.1016/j.scitotenv.2013.02.088 PMID:23542484

Ferri A, Duffard R, de Duffard AM (2007). Selective oxidative stress in brain areas of neonate rats exposed to 2,4-Dichlorophenoxyacetic acid through mother's milk. *Drug Chem Toxicol*, 30(1):17–30. doi:10.1080/01480540601017629 PMID:17364861

Figgs LW, Holland NT, Rothmann N, Zahm SH, Tarone RE, Hill R et al. (2000). Increased lymphocyte replicative index following 2,4-dichlorophenoxyacetic acid herbicide exposure. *Cancer Causes Control*, 11(4):373–80. doi:10.1023/A:1008925824242 PMID:10843448

Filer D, Patisaul HB, Schug T, Reif D, Thayer K (2014). Test driving ToxCast: endocrine profiling for 1858 chemicals included in phase II. *Curr Opin Pharmacol*, 19:145–52. doi:10.1016/j.coph.2014.09.021 PMID:25460227

Filkowski J, Besplug J, Burke P, Kovalchuk I, Kovalchuk O (2003). Genotoxicity of 2,4-D and dicamba revealed by transgenic Arabidopsis thaliana plants harboring recombination and point mutation markers. *Mutat Res*, 542(1–2):23–32. doi:10.1016/j.mrgentox.2003.07.008 PMID:14644350

Flower KB, Hoppin JA, Lynch CF, Blair A, Knott C, Shore DL et al. (2004). Cancer risk and parental pesticide application in children of Agricultural Health Study participants. *Environ Health Perspect*, 112(5):631–5. doi:10.1289/ehp.6586 PMID:15064173

Fontana A, Picoco C, Masala G, Prastaro C, Vineis P (1998). Incidence rates of lymphomas and environmental measurements of phenoxy herbicides: ecological analysis and case-control study. *Arch Environ Health*, 53(6):384–7. doi:10.1080/00039899809605725 PMID:9886156

Fritschi L, Benke G, Hughes AM, Kricker A, Turner J, Vajdic CM et al. (2005). Occupational exposure to pesticides and risk of non-Hodgkin's lymphoma. *Am J Epidemiol*, 162(9):849–57. doi:10.1093/aje/kwi292 PMID:16177143

Fukuyama T, Kosaka T, Tajima Y, Ueda H, Hayashi K, Shutoh Y et al. (2010). Prior exposure to organophosphorus and organochlorine pesticides increases the allergic potential of environmental chemical allergens in a local lymph node assay. *Toxicol Lett*, 199(3):347–56. doi:10.1016/j.toxlet.2010.09.018 PMID:20920556

Fukuyama T, Tajima Y, Ueda H, Hayashi K, Shutoh Y, Harada T et al. (2009). Allergic reaction induced by dermal and/or respiratory exposure to low-dose phenoxyacetic acid, organophosphorus, and carbamate pesticides. *Toxicology*, 261(3):152–61. doi:10.1016/j.tox.2009.05.014 PMID:19467290

Gao B, Meier PJ (2001). Organic anion transport across the choroid plexus. *Microsc Res Tech*, 52(1):60–4. doi:10.1002/1097-0029(20010101)52:1<60::AID-JEMT8>3.0.CO;2-C PMID:11135449

Garabrant DH, Philbert MA (2002). Review of 2,4-dichlorophenoxyacetic acid (2,4-D) epidemiology and toxicology. *Crit Rev Toxicol*, 32(4):233–57. doi:10.1080/20024091064237 PMID:12184504

Garaj-Vrhovac V, Zeljezic D (2000). Evaluation of DNA damage in workers occupationally exposed to pesticides using single-cell gel electrophoresis (SCGE) assay. Pesticide genotoxicity revealed by comet assay. *Mutat Res*, 469(2):279–85. doi:10.1016/S1383-5718(00)00092-9 PMID:10984689

Garaj-Vrhovac V, Zeljezic D (2001). Cytogenetic monitoring of croatian population occupationally exposed to a complex mixture of pesticides. *Toxicology*, 165(2–3):153–62. doi:10.1016/S0300-483X(01)00419-X PMID:11522373

Garaj-Vrhovac V, Zeljezic D (2002). Assessment of genome damage in a population of Croatian workers employed in pesticide production by chromosomal aberration analysis, micronucleus assay and Comet assay. *J Appl Toxicol*, 22(4):249–55. doi:10.1002/jat.855 PMID:12210542

Gardner M, Spruill-McCombs M, Beach J, Michael L, Thomas K, Helburn RS (2005). Quantification of 2,4-D on solid-phase exposure sampling media by LC-MS-MS. *J Anal Toxicol*, 29(3):188–92. doi:10.1093/jat/29.3.188 PMID:15842762

Garry VF, Tarone RE, Kirsch IR, Abdallah JM, Lombardi DP, Long LK et al. (2001). Biomarker correlations of urinary 2,4-D levels in foresters: genomic instability and endocrine disruption. *Environ Health Perspect*, 109(5):495–500. doi:10.1289/ehp.01109495 PMID:11401761

Glozier NE, Struger J, Cessna AJ, Gledhill M, Rondeau M, Ernst WR et al. (2012). Occurrence of glyphosate and acidic herbicides in select urban rivers and streams in Canada, 2007. *Environ Sci Pollut Res Int*, 19(3):821–34. doi:10.1007/s11356-011-0600-7 PMID:21948131

Goldner WS, Sandler DP, Yu F, Shostrom V, Hoppin JA, Kamel F et al. (2013). Hypothyroidism and pesticide use among male private pesticide applicators in the agricultural health study. *J Occup Environ Gar*, 55(10):1171–8. doi:10.1097/JOM.0b013e31829b290b PMID:24064777

Gollapudi BB, Charles JM, Linscombe VA, Day SJ, Bus JS (1999). Evaluation of the genotoxicity of 2,4-dichlorophenoxyacetic acid and its derivatives in mammalian cell cultures. *Mutat Res*, 444(1):217–25. doi:10.1016/S1383-5718(99)00075-3 PMID:10477357

González M, Soloneski S, Reigosa MA, Larramendy ML (2005). Genotoxicity of the herbicide 2,4-dichlorophenoxyacetic and a commercial formulation, 2,4-dichlorophenoxyacetic acid dimethylamine salt. I. Evaluation of DNA damage and cytogenetic endpoints in Chinese Hamster ovary (CHO) cells. *Toxicol In Vitro*, 19(2):289–97. doi:10.1016/j.tiv.2004.10.004 PMID:15649642

Goodman JE, Loftus CT, Zu K (2015). 2,4-Dichlorophenoxyacetic Acid and Non-Hodgkin's Lymphoma, Gastric Cancer, and Prostate Cancer: Meta-analyses of the Published Literature *Ann Epidemiol*, doi:10.1016/j.annepidem.2015.04.002

Grabińska-Sota E, Kalka J, Wiśniowska E (2000). Biodegradation and genotoxicity of some chemical plant protection products. *Cent Eur J Public Health*, 8(Suppl):93–4. PMID:10943490

Grabińska-Sota E, Wiśniowska E, Kalka J, Scieranka B (2002). Genotoxicological effects of some phenoxyherbicides and their metabolites on Bacillus subtilis M45 Rec- and H17 Rec+ strains. *Chemosphere*, 47(1):81–5. doi:10.1016/S0045-6535(01)00211-9 PMID:11996139

Graf U, Würgler FE (1996). The somatic white-ivory eye spot test does not detect the same spectrum of genotoxic events as the wing somatic mutation and recombination test in Drosophila melanogaster. *Environ Mol Mutagen*, 27(3):219–26. doi:10.1002/(SICI)1098-2280(1996)27:3<219::AID-EM7>3.0.CO;2-9 PMID:8625958

Griffin RJ, Godfrey VB, Kim YC, Burka LT (1997). Sex-dependent differences in the disposition of 2,4-dichlorophenoxyacetic acid in Sprague-Dawley rats, B6C3F1 mice, and Syrian hamsters. *Drug Metab Dispos*, 25(9):1065–71. PMID:9311622

Grissom RE Jr, Brownie C, Guthrie FE (1985). Dermal absorption of pesticides in mice. *Pestic Biochem Physiol*, 24(1):119–23. doi:10.1016/0048-3575(85)90121-X

Grover R, Franklin CA, Muir NI, Cessna AJ, Riedel D (1986). Dermal exposure and urinary metabolite excretion in farmers repeatedly exposed to 2,4-D amine. *Toxicol Lett*, 33(1-3):73–83. doi:10.1016/0378-4274(86)90072-X PMID:3775823

Hansen WH, Quaife ML, Habermann RT, Fitzhugh OG (1971). Chronic toxicity of 2,4-dichlorophenoxyacetic acid in rats and dogs. *Toxicol Appl Pharmacol*, 20(1):122–9. doi:10.1016/0041-008X(71)90096-2 PMID:5110820

Hardell L, Eriksson M, Degerman A (1994). Exposure to phenoxyacetic acids, chlorophenols, or organic solvents in relation to histopathology, stage, and anatomical localization of non-Hodgkin's lymphoma. *Cancer Res*, 54(9):2386–9. PMID:8162585

Hardell L, Johansson B, Axelson O (1982). Epidemiological study of nasal and nasopharyngeal cancer and their relation to phenoxy acid or chlorophenol exposure. *Am J Ind Med*, 3(3):247–57. doi:10.1002/ajim.4700030304 PMID:6303119

Hardell L, Sandström A (1979). Case-control study: soft-tissue sarcomas and exposure to phenoxyacetic acids or chlorophenols. *Br J Cancer*, 39(6):711–7. doi:10.1038/bjc.1979.125 PMID:444410

Harris SA, Solomon KR (1992). Percutaneous penetration of 2,4-dichlorophenoxyacetic acid and 2,4-D dimethylamine salt in human volunteers. *J Toxicol Environ Health*, 36(3):233–40. doi:10.1080/15287399209531634 PMID:1629934

Harris SA, Solomon KR, Stephenson GR (1992). Exposure of homeowners and bystanders to 2,4-dichlorophenoxyacetic acid (2,4-D). *J Environ Sci Health B*, 27(1):23–38. doi:10.1080/03601239209372765 PMID:1556388

Harris SA, Villeneuve PJ, Crawley CD, Mays JE, Yeary RA, Hurto KA et al. (2010). National study of exposure to pesticides among professional applicators: an investigation based on urinary biomarkers. *J Agric Food Chem*, 58(18):10253–61. doi:10.1021/jf101209g PMID:20799690

Hartge P, Colt JS, Severson RK, Cerhan JR, Cozen W, Camann D et al. (2005). Residential herbicide use and risk of non-Hodgkin lymphoma. *Cancer Epidemiol Biomarkers Prev*, 14(4):934–7. doi:10.1158/1055-9965.EPI-04-0730 PMID:15824166

Hayes HM, Tarone RE, Cantor KP (1995). On the association between canine malignant lymphoma and opportunity for exposure to 2,4-dichlorophenoxyacetic acid. *Environ Res*, 70(2):119–25. doi:10.1006/enrs.1995.1056 PMID:8674480

Hayes HM, Tarone RE, Cantor KP, Jessen CR, McCurnin DM, Richardson RC (1991). Case-control study of canine malignant lymphoma: positive association with dog owner's use of 2,4-dichlorophenoxyacetic acid herbicides. *J Natl Cancer Inst*, 83(17):1226–31. doi:10.1093/jnci/83.17.1226 PMID:1870148

Health Canada (2010). Report on human biomonitoring of environmental chemicals in Canada. Results of the Canadian Health Measures Survey Cycle 1 (2007–2009). Ottawa (ON): Health Canada. Available from: http://www.hc-sc.gc.ca/ewh-semt/alt_formats/hecs-sesc/pdf/pubs/contaminants/chms-ecms/report-rapport-eng.pdf.

Herrero-Hernández E, Rodríguez-Gonzalo E, Andrades MS, Sánchez-González S, Carabias-Martínez R (2013).

Occurrence of phenols and phenoxyacid herbicides in environmental waters using an imprinted polymer as a selective sorbent. *Sci Total Environ*, 454-455:299–306. doi:10.1016/j.scitotenv.2013.03.029 PMID:23562684

Hines CJ, Deddens JA, Striley CAF, Biagini RE, Shoemaker DA, Brown KK et al. (2003). Biological monitoring for selected herbicide biomarkers in the urine of exposed custom applicators: application of mixed-effect models. *Ann Occup Hyg*, 47(6):503–17. doi:10.1093/annhyg/meg067 PMID:12890659

Hines CJ, Deddens JA, Tucker SP, Hornung RW (2001). Distributions and determinants of pre-emergent herbicide exposures among custom applicators. *Ann Occup Hyg*, 45(3):227–39. doi:10.1093/annhyg/45.3.227 PMID:11295146

Hoar SK, Blair A, Holmes FF, Boysen CD, Robel RJ, Hoover R et al. (1986). Agricultural herbicide use and risk of lymphoma and soft-tissue sarcoma. *JAMA*, 256(9):1141–7. doi:10.1001/jama.1986.03380090081023 PMID:3801091

Hohenadel K, Harris SA, McLaughlin JR, Spinelli JJ, Pahwa P, Dosman JA et al. (2011). Exposure to multiple pesticides and risk of non-Hodgkin lymphoma in men from six Canadian provinces. *Int J Environ Res Public Health*, 8(6):2320–30. doi:10.3390/ijerph8062320 PMID:21776232

Holland NT, Duramad P, Rothman N, Figgs LW, Blair A, Hubbard A et al. (2002). Micronucleus frequency and proliferation in human lymphocytes after exposure to herbicide 2,4-dichlorophenoxyacetic acid in vitro and in vivo. *Mutat Res*, 521(1–2):165–78. doi:10.1016/S1383-5718(02)00237-1 PMID:12438013

Hooiveld M, Heederik DJ, Kogevinas M, Boffetta P, Needham LL, Patterson DG Jr et al. (1998). Second follow-up of a Dutch cohort occupationally exposed to phenoxy herbicides, chlorophenols, and contaminants. *Am J Epidemiol*, 147(9):891–901. doi:10.1093/oxfordjournals.aje.a009543 PMID:9583720

Hoppin JA, Yucel F, Dosemeci M, Sandler DP (2002). Accuracy of self-reported pesticide use duration information from licensed pesticide applicators in the Agricultural Health Study. *J Expo Anal Environ Epidemiol*, 12(5):313–8. doi:10.1038/sj.jea.7500232 PMID:12198579

Hou L, Andreotti G, Baccarelli AA, Savage S, Hoppin JA, Sandler DP et al. (2013). Lifetime pesticide use and telomere shortening among male pesticide applicators in the Agricultural Health Study. *Environ Health Perspect*, 121(8):919–24. doi:10.1289/ehp.1206432 PMID:23774483

IARC (1976). Some carbamates, thiocarbamates and carbazides. *IARC Monogr Eval Carcinog Risk Chem Man*, 12:1–282. Available from: http://monographs.iarc.fr/ENG/Monographs/vol1-42/mono12.pdf PMID:188751

IARC (1977). Some fumigants, the herbicides 2,4-D and 2,4,5-T, chlorinated dibenzodioxins and miscellaneous industrial chemicals. *IARC Monogr Eval Carcinog Risk Chem Man*, 15:1–354. Available from: http://monographs.iarc.fr/ENG/Monographs/vol1-42/mono15.pdf PMID:330387

IARC (1979). Sex hormones (II). *IARC Monogr Eval Carcinog Risk Chem Hum*, 21:1–583. Available from: http://monographs.iarc.fr/ENG/Monographs/vol1-42/mono21.pdf.

IARC (1983). Miscellaneous pesticides. *IARC Monogr Eval Carcinog Risk Chem Hum*, 30:1–424. Available from: http://monographs.iarc.fr/ENG/Monographs/vol1-42/mono30.pdf PMID:6578175

IARC (1986). Some chemicals used in plastics and elastomers. *IARC Monogr Eval Carcinog Risk Chem Hum*, 39:1–403. Available from: http://monographs.iarc.fr/ENG/Monographs/vol1-42/mono39.pdf.

IARC (1987). Overall evaluations of carcinogenicity: an updating of *IARC Monographs* volumes 1 to 42. *IARC Monogr Eval Carcinog Risks Hum*, Suppl 7:1–440. PMID:3482203

IARC (1991). Occupational exposures in insecticide application, and some pesticides. *IARC Monogr Eval Carcinog Risks Hum*, 53:1–612. Available from: http://monographs.iarc.fr/ENG/Monographs/vol53/index.php.

IARC (2017). Malathion. In: Some organophosphate insecticides and herbicides. *IARC Monogr Eval Carcinog Risks Hum*, 112:1–452. Available from http://monographs.iarc.fr/ENG/Monographs/vol112/mono112-07.pdf.

IARC (2016). List of ToxCast/Tox21 assay endpoints. In: Supplemental Material to *IARC Monographs* Volume 113. IARC, Lyon. Available from: http://monographs.iarc.fr/ENG/Monographs/vol113/113-Annex1.pdf.

ILO (1999). 2,4-D. International Chemical Safety Cards. Geneva: International Labour Organization. Available from: http://siri.org/msds/mf/cards/file/0033.html, accessed 23 January 2016

Imaoka T, Kusuhara H, Adachi-Akahane S, Hasegawa M, Morita N, Endou H et al. (2004). The renal-specific transporter mediates facilitative transport of organic anions at the brush border membrane of mouse renal tubules. *J Am Soc Nephrol*, 15(8):2012–22. doi:10.1097/01.ASN.0000135049.20420.E5 PMID:15284287

Imel'baeva EA, Teplova SN, Kamilov FKh, Kaiumova AF (1999). [The effect of the amine salt of 2,4-dichlorophenoxyacetic acid on the cluster- and colony-forming capacity of the bone marrow and on the mononuclear phagocyte system] *Zh Mikrobiol Epidemiol Immunobiol*, (6):67–70. PMID:10876855

Innes JR, Ulland BM, Valerio MG, Petrucelli L, Fishbein L, Hart ER et al. (1969). Bioassay of pesticides and industrial chemicals for tumorigenicity in mice: a preliminary note. *J Natl Cancer Inst*, 42(6):1101–14. PMID:5793189

IPCS/ICSC (2015). 2,4-D. ICSC 0034. International Chemical Safety Card. Geneva: International Programme of Chemical Safety, World Health Organization. Available from: http://www.inchem.org/documents/icsc/icsc/eics0033.htm, accessed 6 November 2015.

Jacobi H, Metzger J, Witte I (1992). Synergistic effects of Cu(II) and dimethylammonium 2,4-dichlorophenoxyacetate (U46 D fluid) on PM2 DNA and mechanism of DNA damage. *Free Radic Res Commun*, 16(2):123–30. doi:10.3109/10715769209049165 PMID:1628858

Jenssen D, Renberg L (1976). Distribution and cytogenetic test of 2,4-D and 2,4,5-T phenoxyacetic acids in mouse blood tissues. *Chem Biol Interact*, 14(3–4):279–89. PMID:954145

JMPR (1996). 2,4-Dichlorophenoxyacetic acid (2,4-D). Geneva: Joint FAO/WHO Meeting on Pesticide Residues World Health Organization. Available from: http://www.inchem.org/documents/jmpr/jmpmono/v96pr04.htm.

Jurewicz J, Hanke W, Sobala W, Ligocka D (2012). Exposure to phenoxyacetic acid herbicides and predictors of exposure among spouses of farmers. *Ann Agric Environ Med*, 19(1):51–6. PMID:22462445

Kahn PC, Gochfeld M, Nygren M, Hansson M, Rappe C, Velez H et al. (1988). Dioxins and dibenzofurans in blood and adipose tissue of Agent Orange-exposed Vietnam veterans and matched controls. *JAMA*, 259(11):1661–7. doi:10.1001/jama.1988.03720110023029 PMID:3343772

Kaioumova D, Kaioumov F, Opelz G, Süsal C (2001b). Toxic effects of the herbicide 2,4-dichlorophenoxyacetic acid on lymphoid organs of the rat. *Chemosphere*, 43(4–7):801–5. doi:10.1016/S0045-6535(00)00436-7 PMID:11372868

Kaioumova D, Süsal C, Opelz G (2001a). Induction of apoptosis in human lymphocytes by the herbicide 2,4-dichlorophenoxyacetic acid. *Hum Immunol*, 62(1):64–74. doi:10.1016/S0198-8859(00)00229-9 PMID:11165716

Kaioumova DF, Khabutdinova LK (1998). Cytogenetic characteristics of herbicide production workers in Ufa. *Chemosphere*, 37(9–12):1755–9. doi:10.1016/S0045-6535(98)00240-9 PMID:9828303

Kale PG, Petty BT Jr, Walker S, Ford JB, Dehkordi N, Tarasia S et al. (1995). Mutagenicity testing of nine herbicides and pesticides currently used in agriculture. *Environ Mol Mutagen*, 25(2):148–53. doi:10.1002/em.2850250208 PMID:7698107

Kaneene JB, Miller R (1999). Re-analysis of 2,4-D use and the occurrence of canine malignant lymphoma. *Vet Hum Toxicol*, 41(3):164–70. PMID:10349709

Kavlock R, Chandler K, Houck K, Hunter S, Judson R, Kleinstreuer N et al. (2012). Update on EPA's ToxCast program: providing high throughput decision support tools for chemical risk management. *Chem Res Toxicol*, 25(7):1287–302. doi:10.1021/tx3000939 PMID:22519603

Kawashima Y, Katoh H, Nakajima S, Kozuka H, Uchiyama M (1984). Effects of 2,4-dichlorophenoxyacetic acid and 2,4,5-trichlorophenoxyacetic acid on peroxisomal enzymes in rat liver. *Biochem Pharmacol*, 33(2):241–5. doi:10.1016/0006-2952(84)90481-7 PMID:6704149

Kaya B, Yanikoglu A, Marcos R (1999). Genotoxicity studies on the phenoxyacetates 2,4-D and 4-CPA in the Drosophila wing spot test. *Teratog Carcinog Mutagen*, 19(4):305–12. doi:10.1002/(SICI)1520-6866(1999)19:4<305::AID-TCM7>3.0.CO;2-X PMID:10406894

Kenigsberg IaE (1975). [Relation of the immune response in rats to the state of the lysosomes and intensity of protein synthesis] *Zh Mikrobiol Epidemiol Immunobiol*, (6):32–5. PMID:1098336

Kim CS, Binienda Z, Sandberg JA (1996). Construction of a physiologically based pharmacokinetic model for 2,4-dichlorophenoxyacetic acid dosimetry in the developing rabbit brain. *Toxicol Appl Pharmacol*, 136(2):250–9. doi:10.1006/taap.1996.0032 PMID:8619233

Kim CS, Gargas ML, Andersen ME (1994). Pharmacokinetic modeling of 2,4-dichlorophenoxyacetic acid (2,4-D) in rat and in rabbit brain following single dose administration. *Toxicol Lett*, 74(3):189–201. doi:10.1016/0378-4274(94)90078-7 PMID:7871543

Kim CS, Keizer RF, Pritchard JB (1988). 2,4-Dichlorophenoxyacetic acid intoxication increases its accumulation within the brain. *Brain Res*, 440(2):216–26. doi:10.1016/0006-8993(88)90989-4 PMID:3359212

Kim CS, O'Tuama LA, Mann JD, Roe CR (1983). Saturable accumulation of the anionic herbicide, 2,4-dichlorophenoxyacetic acid (2,4-D), by rabbit choroid plexus: early developmental origin and interaction with salicylates. *J Pharmacol Exp Ther*, 225(3):699–704. PMID:6864528

Kim H-J, Park YI, Dong M-S (2005). Effects of 2,4-D and DCP on the DHT-induced androgenic action in human prostate cancer cells. *Toxicol Sci*, 88(1):52–9. doi:10.1093/toxsci/kfi287 PMID:16107550

King KW, Balogh JC (2010). Chlorothalonil and 2,4-D losses in surface water discharge from a managed turf watershed. *J Environ Monit*, 12(8):1601–12. doi:10.1039/c0em00030b PMID:20526481

Knapp GW, Setzer RW, Fuscoe JC (2003). Quantitation of aberrant interlocus T-cell receptor rearrangements in mouse thymocytes and the effect of the herbicide 2,4-dichlorophenoxyacetic acid. *Environ Mol Mutagen*, 42(1):37–43. doi:10.1002/em.10168 PMID:12874811

Knopp D (1994). Assessment of exposure to 2,4-dichlorophenoxyacetic acid in the chemical industry: results of a five year biological monitoring study. *Occup*

Environ Med, 51(3):152–9. doi:10.1136/oem.51.3.152 PMID:8130842

Knopp D, Schiller F (1992). Oral and dermal application of 2,4-dichlorophenoxyacetic acid sodium and dimethylamine salts to male rats: investigations on absorption and excretion as well as induction of hepatic mixed-function oxidase activities. *Arch Toxicol*, 66(3):170–4. doi:10.1007/BF01974010 PMID:1497479

Kogevinas M, Becher H, Benn T, Bertazzi PA, Boffetta P, Bueno-de-Mesquita HB et al. (1997). Cancer mortality in workers exposed to phenoxy herbicides, chlorophenols, and dioxins. An expanded and updated international cohort study. *Am J Epidemiol*, 145(12):1061–75. doi:10.1093/oxfordjournals.aje.a009069 PMID:9199536

Kogevinas M, Kauppinen T, Winkelmann R, Becher H, Bertazzi PA, Bueno-de-Mesquita HB et al. (1995). Soft tissue sarcoma and non-Hodgkin's lymphoma in workers exposed to phenoxy herbicides, chlorophenols, and dioxins: two nested case-control studies. *Epidemiology*, 6(4):396–402. doi:10.1097/00001648-199507000-00012 PMID:7548348

Kohli JD, Khanna RN, Gupta BN, Dhar MM, Tandon JS, Sircar KP (1974). Absorption and excretion of 2,4-dichlorophenoxyacetic acid in man. *Xenobiotica*, 4(2):97–100. doi:10.3109/00498257409049349 PMID:4828800

Kolmodin-Hedman B, Erne K (1980). Estimation of occupational exposure to phenoxy acids (2,4-D and 2,4,5-T). *Arch Toxicol Suppl*, 4:318–21. doi:10.1007/978-3-642-67729-8_65 PMID:6933926

Kolmodin-Hedman B, Höglund S, Akerblom M (1983). Studies on phenoxy acid herbicides. I. Field study. Occupational exposure to phenoxy acid herbicides (MCPA, dichlorprop, mecoprop and 2,4-D) in agriculture. *Arch Toxicol*, 54(4):257–65. doi:10.1007/BF01234478 PMID:6667116

Konstantinou IK, Hela DG, Albanis TA (2006). The status of pesticide pollution in surface waters (rivers and lakes) of Greece. Part I. Review on occurrence and levels. *Environ Pollut*, 141(3):555–70. doi:10.1016/j.envpol.2005.07.024 PMID:16226830

Kornuta N, Bagley E, Nedopitanskaya N (1996). Genotoxic effects of pesticides. *J Environ Pathol Toxicol Oncol*, 15(2–4):75–8. PMID:9216788

Korte C, Jalal SM (1982). 2,4-D induced clastogenicity and elevated rates of sister chromatid exchanges in cultured human lymphocytes. *J Hered*, 73(3):224–6. PMID:7096985

Koutros S, Beane Freeman LE, Lubin JH, Heltshe SL, Andreotti G, Barry KH et al. (2013). Risk of total and aggressive prostate cancer and pesticide use in the Agricultural Health Study. *Am J Epidemiol*, 177(1):59–74. doi:10.1093/aje/kws225 PMID:23171882

Kumari TS, Vaidyanath K (1989). Testing of genotoxic effects of 2,4-dichlorophenoxyacetic acid (2,4-D) using multiple genetic assay systems of plants. *Mutat Res*, 226(4):235–8. doi:10.1016/0165-7992(89)90075-4 PMID:2761564

Lamb JC 4th, Marks TA, Gladen BC, Allen JW, Moore JA (1981). Male fertility, sister chromatid exchange, and germ cell toxicity following exposure to mixtures of chlorinated phenoxy acids containing 2,3,7,8-tetrachlorodibenzo-p-dioxin. *J Toxicol Environ Health*, 8(5–6):825–34. doi:10.1080/15287398109530118 PMID:7338944

Lavy TL, Mattice JD, Marx DB, Norris LA (1987). Exposure of forestry ground workers to 2,4-D, picloram and dichlorprop. *Environ Toxicol Chem*, 6(3):209–24. doi:10.1002/etc.5620060306

Lavy TL, Walstad JD, Flynn RR, Mattice JD (1982). (2,4-Dichlorophenoxy)acetic acid exposure received by aerial application crews during forest spray operations. *J Agric Food Chem*, 30(2):375–81. doi:10.1021/jf00110a042

Lee K, Johnson VJ, Blakley BR (2000). The effect of exposure to a commercial 2,4-D herbicide formulation during gestation on urethan-induced lung adenoma formation in CD-1 mice. *Vet Hum Toxicol*, 42(3):129–32. PMID:10839313

Lee K, Johnson VL, Blakley BR (2001). The effect of exposure to a commercial 2,4-D formulation during gestation on the immune response in CD-1 mice. *Toxicology*, 165(1):39–49. doi:10.1016/S0300-483X(01)00403-6 PMID:11551430

Lee WJ, Lijinsky W, Heineman EF, Markin RS, Weisenburger DD, Ward MH (2004). Agricultural pesticide use and adenocarcinomas of the stomach and oesophagus. *Occup Environ Med*, 61(9):743–9. doi:10.1136/oem.2003.011858 PMID:15317914

Leljak-Levanić D, Bauer N, Mihaljević S, Jelaska S (2004). Changes in DNA methylation during somatic embryogenesis in Cucurbita pepo L. *Plant Cell Rep*, 23(3):120–7. doi:10.1007/s00299-004-0819-6 PMID:15221278

Leonard C, Burke CM, O'Keane C, Doyle JS (1997). "Golf ball liver": agent orange hepatitis. *Gut*, 40(5):687–8. doi:10.1136/gut.40.5.687 PMID:9203952

Lewis RC, Cantonwine DE, Anzalota Del Toro LV, Calafat AM, Valentin-Blasini L, Davis MD et al. (2014). Urinary biomarkers of exposure to insecticides, herbicides, and one insect repellent among pregnant women in Puerto Rico. *Environ Health*, 13(1):97 doi:10.1186/1476-069X-13-97 PMID:25409771

Libich S, To JC, Frank R, Sirons GJ (1984). Occupational exposure of herbicide applicators to herbicides used along electric power transmission line right-of-way. *Am Ind Hyg Assoc J*, 45(1):56–62. doi:10.1080/15298668491399370 PMID:6702600

Lindh CH, Littorin M, Amilon A, Jönsson BA (2008). Analysis of phenoxyacetic acid herbicides as biomarkers in human urine using liquid chromatography/triple quadrupole mass spectrometry. *Rapid Commun*

Mass Spectrom, 22(2):143–50. doi:10.1002/rcm.3348 PMID:18059043

Linnainmaa K (1983). Sister chromatid exchanges among workers occupationally exposed to phenoxy acid herbicides 2,4-D and MCPA. *Teratog Carcinog Mutagen*, 3(3):269–79. doi:10.1002/1520-6866(1990)3:3<269::AID-TCM1770030306>3.0.CO;2-F PMID:6137083

Linnainmaa K (1984). Induction of sister chromatid exchanges by the peroxisome proliferators 2,4-D, MCPA, and clofibrate in vivo and in vitro. *Carcinogenesis*, 5(6):703–7. doi:10.1093/carcin/5.6.703 PMID:6722984

Liu W, Li H, Tao F, Li S, Tian Z, Xie H (2013). Formation and contamination of PCDD/Fs, PCBs, PeCBz, HxCBz and polychlorophenols in the production of 2,4-D products. *Chemosphere*, 92(3):304–8. doi:10.1016/j.chemosphere.2013.03.031 PMID:23601123

Loos R, Gawlik BM, Locoro G, Rimaviciute E, Contini S, Bidoglio G (2009). EU-wide survey of polar organic persistent pollutants in European river waters. *Environ Pollut*, 157(2):561–8. doi:10.1016/j.envpol.2008.09.020 PMID:18952330

Loos R, Locoro G, Comero S, Contini S, Schwesig D, Werres F et al. (2010a). Pan-European survey on the occurrence of selected polar organic persistent pollutants in ground water. *Water Res*, 44(14):4115–26. doi:10.1016/j.watres.2010.05.032 PMID:20554303

Loos R, Locoro G, Contini S (2010b). Occurrence of polar organic contaminants in the dissolved water phase of the Danube River and its major tributaries using SPE-LC-MS(2) analysis. *Water Res*, 44(7):2325–35. doi:10.1016/j.watres.2009.12.035 PMID:20074769

Lundgren B, Meijer J, DePierre JW (1987). Induction of cytosolic and microsomal epoxide hydrolases and proliferation of peroxisomes and mitochondria in mouse liver after dietary exposure to p-chlorophenoxyacetic acid, 2,4-dichlorophenoxyacetic acid and 2,4,5-trichlorophenoxyacetic acid. *Biochem Pharmacol*, 36(6):815–21. doi:10.1016/0006-2952(87)90169-9 PMID:3032197

Lynge E (1985). A follow-up study of cancer incidence among workers in manufacture of phenoxy herbicides in Denmark. *Br J Cancer*, 52(2):259–70. doi:10.1038/bjc.1985.186 PMID:4027168

Lynge E (1998). Cancer incidence in Danish phenoxy herbicide workers, 1947–1993. *Environ Health Perspect*, 106(Suppl 2):683–8. doi:10.1289/ehp.98106683 PMID:9599717

Madrigal-Bujaidar E, Hernández-Ceruelos A, Chamorro G (2001). Induction of sister chromatid exchanges by 2,4-dichlorophenoxyacetic acid in somatic and germ cells of mice exposed in vivo. *Food Chem Toxicol*, 39(9):941–6. doi:10.1016/S0278-6915(01)00037-0 PMID:11498271

Maire MA, Rast C, Landkocz Y, Vasseur P (2007). 2,4-Dichlorophenoxyacetic acid: effects on Syrian hamster embryo (SHE) cell transformation, c-Myc expression, DNA damage and apoptosis. *Mutat Res*, 631(2):124–36. doi:10.1016/j.mrgentox.2007.03.008 PMID:17540612

Maloney EK, Waxman DJ (1999). trans-Activation of PPARalpha and PPARgamma by structurally diverse environmental chemicals. *Toxicol Appl Pharmacol*, 161(2):209–18. doi:10.1006/taap.1999.8809 PMID:10581215

Malysheva LN, Zhavoronkov AA (1997). Morphological and histochemical changes in the thyroid gland after a single exposure to 2,4-DA herbicide. *Bull Exp Biol Med*, 124(6):1223–4. doi:10.1007/BF02445126

Martin MT, Judson RS, Reif DM, Kavlock RJ, Dix DJ (2009). Profiling chemicals based on chronic toxicity results from the U.S. EPA ToxRef Database. *Environ Health Perspect*, 117(3):392–9. doi:10.1289/ehp.0800074 PMID:19337514

Martínez-Tabche L, Madrigal-Bujaidar E, Negrete T (2004). Genotoxicity and lipoperoxidation produced by paraquat and 2,4-dichlorophenoxyacetic acid in the gills of rainbow trout (Oncorhynchus mikiss). *Bull Environ Contam Toxicol*, 73(1):146–52. doi:10.1007/s00128-004-0406-0 PMID:15386085

Marty MS, Neal BH, Zablotny CL, Yano BL, Andrus AK, Woolhiser MR et al. (2013). An F1-extended one-generation reproductive toxicity study in Crl:CD(SD) rats with 2,4-dichlorophenoxyacetic acid. *Toxicol Sci*, 136(2):527–47. doi:10.1093/toxsci/kft213 PMID:24072463

Mason RW (1975). Binding of some phenoxyalkanoic acids to bovine serum albumin in vitro. *Pharmacology*, 13(2):177–86. doi:10.1159/000136898 PMID:1170578

Mazhar FM, Moawad KM, El-Dakdoky MH, Am AS (2014). Fetotoxicity of 2,4-dichlorophenoxyacetic acid in rats and the protective role of vitamin E. *Toxicol Ind Health*, 30(5):480–8. doi:10.1177/0748233712459915 PMID:22949405

McDuffie HH, Pahwa P, McLaughlin JR, Spinelli JJ, Fincham S, Dosman JA et al. (2001). Non-Hodgkin's lymphoma and specific pesticide exposures in men: cross-Canada study of pesticides and health. *Cancer Epidemiol Biomarkers Prev*, 10(11):1155–63. PMID:11700263

McDuffie HH, Pahwa P, Robson D, Dosman JA, Fincham S, Spinelli JJ et al. (2005). Insect repellents, phenoxyherbicide exposure, and non-Hodgkin's lymphoma. *J Occup Environ Med*, 47(8):806–16. PMID:16093930

McManus SL, Moloney M, Richards KG, Coxon CE, Danaher M (2014). Determination and occurrence of phenoxyacetic acid herbicides and their transformation products in groundwater using ultra high performance liquid chromatography coupled to tandem mass spectrometry. *Molecules*, 19(12):20627–49. doi:10.3390/molecules191220627 PMID:25514054

Mehmood Z, Williamson MP, Kelly DE, Kelly SL (1996). Human cytochrome P450 3A4 is involved in the biotransformation of the herbicide 2,4-dichlorophenoxyacetic acid. *Environ Toxicol Pharmacol*, 2(4):397–401. doi:10.1016/S1382-6689(96)00077-4 PMID:21781748

Metayer C, Colt JS, Buffler PA, Reed HD, Selvin S, Crouse V et al. (2013). Exposure to herbicides in house dust and risk of childhood acute lymphoblastic leukemia. *J Expo Sci Environ Epidemiol*, 23(4):363–70. doi:10.1038/jes.2012.115 PMID:23321862

Miassod R, Cecchini JP (1979). Partial base-methylation and other structural differences in the 17 S ribosomal RNA of sycamore cells during growth in cell culture. *Biochim Biophys Acta*, 562(2):292–301. doi:10.1016/0005-2787(79)90174-6 PMID:444529

Miligi L, Costantini AS, Bolejack V, Veraldi A, Benvenuti A, Nanni O et al. (2003). Non-Hodgkin's lymphoma, leukemia, and exposures in agriculture: results from the Italian multicenter case-control study. *Am J Ind Med*, 44(6):627–36. doi:10.1002/ajim.10289 PMID:14635239

Miligi L, Costantini AS, Veraldi A, Benvenuti A, Vineis P (2006b). Living in a Chemical World: Framing the Future in Light of the Past. Volume 1076. Oxford, UK: Blackwell Publishing. pp. 366–377.

Miligi L, Costantini AS, Veraldi A, Benvenuti A, Vineis P; WILL (2006a). Cancer and pesticides: an overview and some results of the Italian multicenter case-control study on hematolymphopoietic malignancies. *Ann N Y Acad Sci*, 1076(1):366–77. doi:10.1196/annals.1371.036 PMID:17119216

Mills PK, Yang R, Riordan D (2005). Lymphohematopoietic cancers in the United Farm Workers of America (UFW), 1988–2001. *Cancer Causes Control*, 16(7):823–30. doi:10.1007/s10552-005-2703-2 PMID:16132792

Mills PK, Yang RC (2007). Agricultural exposures and gastric cancer risk in Hispanic farm workers in California. *Environ Res*, 104(2):282–9. doi:10.1016/j.envres.2006.11.008 PMID:17196584

Mohandas T, Grant WF (1972). Cytogenetic effects of 2,4-D and amitrole in relation to nuclear volume and DNA content in some higher plants. *Can J Genet Cytol*, 14(4):773–83. doi:10.1139/g72-095

Moody RP, Franklin CA, Ritter L, Maibach HI (1990). Dermal absorption of the phenoxy herbicides 2,4-D, 2,4-D amine, 2,4-D isooctyl, and 2,4,5-T in rabbits, rats, rhesus monkeys, and humans: a cross-species comparison. *J Toxicol Environ Health*, 29(3):237–45. doi:10.1080/15287399009531387 PMID:2313737

Moody RP, Wester RC, Melendres JL, Maibach HI (1992). Dermal absorption of the phenoxy herbicide 2,4-D dimethylamine in humans: effect of DEET and anatomic site. *J Toxicol Environ Health*, 36(3):241–50. doi:10.1080/15287399209531635 PMID:1629935

Morgan MK, Sheldon LS, Croghan CW, Chuang JC, Lordo RA, Wilson NK, et al. (2004). A pilot study of children's total exposure to persistent pesticides and other persistent organic pollutants (CTEPP). United States Environmental Protection Agency. EPA/600/R-041/193.

Morgan MK, Sheldon LS, Thomas KW, Egeghy PP, Croghan CW, Jones PA et al. (2008). Adult and children's exposure to 2,4-D from multiple sources and pathways. *J Expo Sci Environ Epidemiol*, 18(5):486–94. doi:10.1038/sj.jes.7500641 PMID:18167507

Morgan MK, Wilson NK, Chuang JC (2014). Exposures of 129 preschool children to organochlorines, organophosphates, pyrethroids, and acid herbicides at their homes and daycares in North Carolina. *Int J Environ Res Public Health*, 11(4):3743–64. doi:10.3390/ijerph110403743 PMID:24705361

Moriya M, Ohta T, Watanabe K, Miyazawa T, Kato K, Shirasu Y (1983). Further mutagenicity studies on pesticides in bacterial reversion assay systems. *Mutat Res*, 116(3–4):185–216. doi:10.1016/0165-1218(83)90059-9 PMID:6339892

Mortelmans K, Haworth S, Speck W, Zeiger E (1984). Mutagenicity testing of agent orange components and related chemicals. *Toxicol Appl Pharmacol*, 75(1):137–46. doi:10.1016/0041-008X(84)90084-X PMID:6379990

Mufazalova NA, Medvedev IuA, Basyrova NK (2001). [Corrective effects of tocopherol on changes in indices of the protective activity of phagocytes under the action of the herbicide 2,4-DA] *Gig Sanit*, (6):61–3. PMID:11810914

Murata M (1989). Effects of auxin and cytokinin on induction of sister chromatid exchanges in cultured cells of wheat (Triticum aestivum L.). *Theor Appl Genet*, 78(4):521–4. doi:10.1007/BF00290836 PMID:24225679

Mustonen R, Kangas J, Vuojolahti P, Linnainmaa K (1986). Effects of phenoxyacetic acids on the induction of chromosome aberrations in vitro and in vivo. *Mutagenesis*, 1(4):241–5. doi:10.1093/mutage/1.4.241 PMID:3331666

Nakbi A, Tayeb W, Grissa A, Issaoui M, Dabbou S, Chargui I et al. (2010). Effects of olive oil and its fractions on oxidative stress and the liver's fatty acid composition in 2,4-Dichlorophenoxyacetic acid-treated rats. *Nutr Metab (Lond)*, 7(1):80 doi:10.1186/1743-7075-7-80 PMID:21034436

Navaranjan G, Hohenadel K, Blair A, Demers PA, Spinelli JJ, Pahwa P et al. (2013). Exposures to multiple pesticides and the risk of Hodgkin lymphoma in Canadian men. *Cancer Causes Control*, 24(9):1661–73. doi:10.1007/s10552-013-0240-y PMID:23756639

NCBI (2015). Compound summary for CID 1486. 2,4-dichlorophenoxyacetic acid. PubChem Open Chemistry Database. Bethesda (MD): National Center for Biotechnology Information. Available from: https://pubchem.ncbi.nlm.nih.gov/compound/1486#section=3D-Conformer, accessed 5 March 2015.

NIOSH (1994). 2,4-D. Method 5001, Issue 2, dated 15 August 1994. In: NIOSH Manual of Analytical Methods, Fourth Edition. Atlanta (GA): National Institute for Occupational Safety and Health, Centers for Disease Control and Prevention. Available from: http://www.cdc.gov/niosh/docs/2003-154/pdfs/500124-d.pdf. Accessed May 27, 2015.

Nishioka MG, Lewis RG, Brinkman MC, Burkholder HM, Hines CE, Menkedick JR (2001). Distribution of 2,4-D in air and on surfaces inside residences after lawn applications: comparing exposure estimates from various media for young children. *Environ Health Perspect*, 109(11):1185–91. doi:10.1289/ehp.011091185 PMID:11713005

Nordström M, Hardell L, Magnuson A, Hagberg H, Rask-Andersen A (1998). Occupational exposures, animal exposure and smoking as risk factors for hairy cell leukaemia evaluated in a case-control study. *Br J Cancer*, 77(11):2048–52. doi:10.1038/bjc.1998.341 PMID:9667691

Nozaki Y, Kusuhara H, Kondo T, Hasegawa M, Shiroyanagi Y, Nakazawa H et al. (2007). Characterization of the uptake of organic anion transporter (OAT) 1 and OAT3 substrates by human kidney slices. *J Pharmacol Exp Ther*, 321(1):362–9. doi:10.1124/jpet.106.113076 PMID:17255469

NPIC (2008). 2,4-D Technical Fact Sheet. National Pesticide Information Center, Oregon State University. Available from: http://npic.orst.edu/factsheets/archive/2,4-DTech.html, accessed 18 February 2016.

NTIS (1968). Evaluation of carcinogenic, teratogenic and mutagenic activities of selected pesticides and industrial chemicals. Volume I. Carcinogenic study. Prepared by Bionetics Research Labs, Incorporated. Report No. PB 223 159, Washington (DC): National Technical Information Service, United States Department of Commerce. Available from: http://www.nal.usda.gov/exhibits/speccoll/files/original/81a28fea12a8e39a2c7fa354bf29737d.pdf.

Oakes DJ, Webster WS, Brown-Woodman PD, Ritchie HE (2002). Testicular changes induced by chronic exposure to the herbicide formulation, Tordon 75D (2,4-dichlorophenoxyacetic acid and picloram) in rats. *Reprod Toxicol*, 16(3):281–9. doi:10.1016/S0890-6238(02)00015-1 PMID:12128102

Örberg J (1980). Observations on the 2,4-dichlorophenoxyacetic acid (2,4-D) excretion in the goat. *Acta Pharmacol Toxicol (Copenh)*, 46(1):78–80. doi:10.1111/j.1600-0773.1980.tb02424.x PMID:7361563

Orsi L, Delabre L, Monnereau A, Delval P, Berthou C, Fenaux P et al. (2009). Occupational exposure to pesticides and lymphoid neoplasms among men: results of a French case-control study. *Occup Environ Med*, 66(5):291–8. doi:10.1136/oem.2008.040972 PMID:19017688

OSHA (2015). 2,4-D. Chemical sampling information. Washington (DC): Occupational Safety and Health Administration, United States Department of Labor. Available from: https://www.osha.gov/dts/chemicalsampling/data/CH_231150.html.

Ozaki K, Mahler JF, Haseman JK, Moomaw CR, Nicolette ML, Nyska A (2001). Unique renal tubule changes induced in rats and mice by the peroxisome proliferators 2,4-dichlorophenoxyacetic acid (2,4-D) and WY-14643. *Toxicol Pathol*, 29(4):440–50. doi:10.1080/01926230152499791 PMID:11560249

Ozcan Oruc E, Sevgiler Y, Uner N (2004). Tissue-specific oxidative stress responses in fish exposed to 2,4-D and azinphosmethyl. *Comp Biochem Physiol C Toxicol Pharmacol*, 137(1):43–51. doi:10.1016/j.cca.2003.11.006 PMID:14984703

Pahwa M, Harris SA, Hohenadel K, McLaughlin JR, Spinelli JJ, Pahwa P et al. (2012). Pesticide use, immunologic conditions, and risk of non-Hodgkin lymphoma in Canadian men in six provinces. *Int J Cancer*, 131(11):2650–9. doi:10.1002/ijc.27522 PMID:22396152

Pahwa P, McDuffie HH, Dosman JA, McLaughlin JR, Spinelli JJ, Robson D et al. (2006). Hodgkin lymphoma, multiple myeloma, soft tissue sarcomas, insect repellents, and phenoxyherbicides. *J Occup Environ Med*, 48(3):264–74. doi:10.1097/01.jom.0000183539.20100.06 PMID:16531830

Palmeira CM, Moreno AJ, Madeira VMC (1995). Thiols metabolism is altered by the herbicides paraquat, dinoseb and 2,4-D: a study in isolated hepatocytes. *Toxicol Lett*, 81(2–3):115–23. doi:10.1016/0378-4274(95)03414-5 PMID:8553365

Panuwet P, Prapamontol T, Chantara S, Barr DB (2009). Urinary pesticide metabolites in school students from northern Thailand. *Int J Hyg Environ Health*, 212(3):288–97. doi:10.1016/j.ijheh.2008.07.002 PMID:18760967

Panuwet P, Prapamontol T, Chantara S, Thavornyuthikarn P, Montesano MA, Whitehead RD Jr et al. (2008). Concentrations of urinary pesticide metabolites in small-scale farmers in Chiang Mai Province, Thailand. *Sci Total Environ*, 407(1):655–68. doi:10.1016/j.scitotenv.2008.08.044 PMID:18954893

Pavlica M, Papes D, Nagy B (1991). 2,4-Dichlorophenoxyacetic acid causes chromatin and chromosome abnormalities in plant cells and mutation in cultured mammalian cells. *Mutat Res*, 263(2):77–81. doi:10.1016/0165-7992(91)90063-A PMID:2046706

Pearce N (1989). Phenoxy herbicides and non-Hodgkin's lymphoma in New Zealand: frequency and duration of herbicide use. *Br J Ind Med*, 46(2):143–4. PMID:2923826

Pearce NE, Sheppard RA, Smith AH, Teague CA (1987). Non-Hodgkin's lymphoma and farming: an expanded case-control study. *Int J Cancer*, 39(2):155–61. doi:10.1002/ijc.2910390206 PMID:3804490

Pearce NE, Smith AH, Howard JK, Sheppard RA, Giles HJ, Teague CA (1986a). Case-control study of multiple myeloma and farming. *Br J Cancer*, 54(3):493–500. doi:10.1038/bjc.1986.202 PMID:3756085

Pearce NE, Smith AH, Howard JK, Sheppard RA, Giles HJ, Teague CA (1986b). Non-Hodgkin's lymphoma and exposure to phenoxyherbicides, chlorophenols, fencing work, and meat works employment: a case-control study. *Br J Ind Med*, 43(2):75–83. PMID:3753879

Pelletier O, Ritter L, Caron J (1990). Effects of skin preapplication treatments and postapplication cleansing agents on dermal absorption of 2,4-dichlorophenoxyacetic acid dimethylamine by Fischer 344 rats. *J Toxicol Environ Health*, 31(4):247–60. doi:10.1080/15287399009531454 PMID:2254951

Pelletier O, Ritter L, Caron J, Somers D (1989). Disposition of 2,4-dichlorophenoxyacetic acid dimethylamine by Fischer 344 rats dosed orally and dermally. *J Toxicol Environ Health*, 28(2):221–34. doi:10.1080/15287398909531342 PMID:2795703

Persson B, Dahlander AM, Fredriksson M, Brage HN, Ohlson CG, Axelson O (1989). Malignant lymphomas and occupational exposures. *Br J Ind Med*, 46(8):516–20. PMID:2775671

Pest Management Regulatory Agency (Canada) (2013). Re-evaluation update 2,4-D REV2013-02. Ottawa. Available from: http://www.hc-sc.gc.ca/cps-spc/alt_formats/pdf/pubs/pest/decisions/rev2013-02/rev2013-02-eng.pdf.

Phillips PJ, Bode RW (2004). Pesticides in surface water runoff in south-eastern New York State, USA: seasonal and stormflow effects on concentrations. *Pest Manag Sci*, 60(6):531–43. doi:10.1002/ps.879 PMID:15198325

Pilinskaia MA (1974). [The cytogenetic effect of herbicide 2,4-D on human and animal chromosomes]. *Tsitol Genet*, 8(3):202–6. PMID:4464590

Pineau T, Hudgins WR, Liu L, Chen LC, Sher T, Gonzalez FJ et al. (1996). Activation of a human peroxisome proliferator-activated receptor by the antitumor agent phenylacetate and its analogs. *Biochem Pharmacol*, 52(4):659–67. doi:10.1016/0006-2952(96)00340-1 PMID:8759039

Pochettino AA, Bongiovanni B, Duffard RO, Evangelista de Duffard AM (2013). Oxidative stress in ventral prostate, ovary, and breast by 2,4-dichlorophenoxyacetic acid in pre- and postnatal exposed rats. *Environ Toxicol*, 28(1):1–10. doi:10.1002/tox.20690 PMID:21374790

Pont AR, Charron AR, Wilson RM, Brand RM (2003). Effects of active sunscreen ingredient combinations on the topical penetration of the herbicide 2,4-dichlorophenoxyacetic acid. *Toxicol Ind Health*, 19(1):1–8. doi:10.1191/0748233703th172oa PMID:15462531

Pritchard JB (1980). Accumulation of anionic pesticides by rabbit choroid plexus in vitro. *J Pharmacol Exp Ther*, 212(2):354–9. PMID:7351648

PubChem (2015). PubChem Open Chemistry Database. Bethesda (MD): National Center for Biotechnology Information. Available from: http://pubchem.ncbi.nlm.nih.gov, accessed November 2015.

Purcell M, Neault JF, Malonga H, Arakawa H, Carpentier R, Tajmir-Riahi HA (2001). Interactions of atrazine and 2,4-D with human serum albumin studied by gel and capillary electrophoresis, and FTIR spectroscopy. *Biochim Biophys Acta*, 1548(1):129–38. doi:10.1016/S0167-4838(01)00229-1 PMID:11451446

Raftopoulou EK, Dailianis S, Dimitriadis VK, Kaloyianni M (2006). Introduction of cAMP and establishment of neutral lipids alterations as pollution biomarkers using the mussel Mytilus galloprovincialis. Correlation with a battery of biomarkers. *Sci Total Environ*, 368(2–3):597–614. doi:10.1016/j.scitotenv.2006.04.031 PMID:16780930

Raina-Fulton R (2014). A review of methods for the analysis of orphan and difficult pesticides: glyphosate, glufosinate, quaternary ammonium and phenoxy acid herbicides, and dithiocarbamate and phthalimide fungicides. *J AOAC Int*, 97(4):965–77. doi:10.5740/jaoacint.SGERaina-Fulton PMID:25145125

Raymer JH, Studabaker WB, Gardner M, Talton J, Quandt SA, Chen H et al. (2014). Pesticide exposures to migrant farmworkers in Eastern NC: detection of metabolites in farmworker urine associated with housing violations and camp characteristics. *Am J Ind Med*, 57(3):323–37. doi:10.1002/ajim.22284 PMID:24273087

Reif DM, Martin MT, Tan SW, Houck KA, Judson RS, Richard AM et al. (2010). Endocrine profiling and prioritization of environmental chemicals using ToxCast data. *Environ Health Perspect*, 118(12):1714–20. doi:10.1289/ehp.1002180 PMID:20826373

Reif DM, Sypa M, Lock EF, Wright FA, Wilson A, Cathey T et al. (2013). ToxPi GUI: an interactive visualization tool for transparent integration of data from diverse sources of evidence. *Bioinformatics*, 29(3):402–3. doi:10.1093/bioinformatics/bts686 PMID:23202747

Rivarola V, Mori G, Balegno H (1992). 2,4-Dichlorophenoxyacetic acid action on in vitro protein synthesis and its relation to polyamines. *Drug Chem Toxicol*, 15(3):245–57. doi:10.3109/01480549209014154 PMID:1425363

Rivarola VA, Balegno HF (1991). Effects of 2,4-dichlorophenoxyacetic acid on polyamine synthesis in Chinese hamster ovary cells. *Toxicol Lett*, 56(1–2):151–7. doi:10.1016/0378-4274(91)90101-B PMID:2017772

Rivarola VA, Bergesse JR, Balegno HF (1985). DNA and protein synthesis inhibition in Chinese hamster ovary cells by dichlorophenoxyacetic acid. *Toxicol Lett*, 29(2–3):137–44. doi:10.1016/0378-4274(85)90034-7 PMID:4089882

Rodríguez T, van Wendel de Joode B, Lindh CH, Rojas M, Lundberg I, Wesseling C (2012). Assessment of long-term and recent pesticide exposure among rural school children in Nicaragua. *Occup Environ Med*, 69(2):119–25. doi:10.1136/oem.2010.062539 PMID:21725072

Ross JH, Driver JH, Harris SA, Maibach HI (2005). Dermal absorption of 2,4-D: a review of species differences. *Regul Toxicol Pharmacol*, 41(1):82–91. doi:10.1016/j.yrtph.2004.10.001 PMID:15649830

Rosso SB, Gonzalez M, Bagatolli LA, Duffard RO, Fidelio GD (1998). Evidence of a strong interaction of 2,4-dichlorophenoxyacetic acid herbicide with human serum albumin. *Life Sci*, 63(26):2343–51. doi:10.1016/S0024-3205(98)00523-2 PMID:9877224

Saghir SA, Marty MS, Zablotny CL, Passage JK, Perala AW, Neal BH et al. (2013). Life-stage-, sex-, and dose-dependent dietary toxicokinetics and relationship to toxicity of 2,4-dichlorophenoxyacetic acid (2,4-D) in rats: implications for toxicity test dose selection, design, and interpretation. *Toxicol Sci*, 136(2):294–307. doi:10.1093/toxsci/kft212 PMID:24105888

Saghir SA, Mendrala AL, Bartels MJ, Day SJ, Hansen SC, Sushynski JM et al. (2006). Strategies to assess systemic exposure of chemicals in subchronic/chronic diet and drinking water studies. *Toxicol Appl Pharmacol*, 211(3):245–60. doi:10.1016/j.taap.2005.06.010 PMID:16040073

Salazar KD, de la Rosa P, Barnett JB, Schafer R (2005). The polysaccharide antibody response after Streptococcus pneumoniae vaccination is differentially enhanced or suppressed by 3,4-dichloropropionanilide and 2,4-dichlorophenoxyacetic acid. *Toxicol Sci*, 87(1):123–33. doi:10.1093/toxsci/kfi244 PMID:15976183

Sandal S, Yilmaz B (2011). Genotoxic effects of chlorpyrifos, cypermethrin, endosulfan and 2,4-D on human peripheral lymphocytes cultured from smokers and nonsmokers. *Environ Toxicol*, 26(5):433–42. doi:10.1002/tox.20569 PMID:20196147

Sandberg JA, Duhart HM, Lipe G, Binienda Z, Slikker W Jr, Kim CS (1996). Distribution of 2,4-dichlorophenoxyacetic acid (2,4-D) in maternal and fetal rabbits. *J Toxicol Environ Health*, 49(5):497–509. doi:10.1080/009841096160718 PMID:8968410

Santilio A, Stefanelli P, Girolimetti S, Dommarco R (2011). Determination of acidic herbicides in cereals by QuEChERS extraction and LC/MS/MS. *J Environ Sci Health B*, 46(6):535–43. PMID:21726153

Sapin MR, Lebedeva SN, Zhamsaranova SD, Erofeeva LM (2003). [Comparative analysis of disorders in duodenal lymphoid tissue of mice treated with azathioprine and herbicide 2,4-dichlorophenoxyacetic acid and their correction by plant and animal origin remedies] *Morfologiia*, 124(4):70–3. PMID:14628561

Saracci R, Kogevinas M, Bertazzi PA, Bueno de Mesquita BH, Coggon D, Green LM et al. (1991). Cancer mortality in workers exposed to chlorophenoxy herbicides and chlorophenols. *Lancet*, 338(8774):1027–32. doi:10.1016/0140-6736(91)91898-5 PMID:1681353

Sauerhoff MW, Braun WH, Blau GE, Gehring PJ (1977). The fate of 2,4-dichlorophenoxyacetic acid (2,4-D) following oral administration to man. *Toxicology*, 8(1):3–11. doi:10.1016/0300-483X(77)90018-X PMID:929615

Schaner A, Konecny J, Luckey L, Hickes H (2007). Determination of chlorinated acid herbicides in vegetation and soil by liquid chromatography/electrospray-tandem mass spectrometry. *J AOAC Int*, 90(5):1402–10. PMID:17955986

Schinasi L, Leon ME (2014). Non-Hodgkin lymphoma and occupational exposure to agricultural pesticide chemical groups and active ingredients: a systematic review and meta-analysis. *Int J Environ Res Public Health*, 11(4):4449–527. doi:10.3390/ijerph110404449 PMID:24762670

Schop RN, Hardy MH, Goldberg MT (1990). Comparison of the activity of topically applied pesticides and the herbicide 2,4-D in two short-term in vivo assays of genotoxicity in the mouse. *Fundam Appl Toxicol*, 15(4):666–75. doi:10.1016/0272-0590(90)90183-K PMID:2086312

Schulze G (1991). Subchronic Toxicity Study in Rats with 2,4-Di- chlorophenoxyacetic Acid. Lab Project Number: 2184-116. Unpublished study prepared by Hazleton Laboratories America, 529 p.

Shida SS, Nemoto S, Matsuda R (2015). Simultaneous determination of acidic pesticides in vegetables and fruits by liquid chromatography–tandem mass spectrometry. *J Environ Sci Health B*, 50(3):151–62. doi:10.1080/03601234.2015.982381 PMID:25602148

Shirasu Y, Moriya M, Kato K, Furuhashi A, Kada T (1976). Mutagenicity screening of pesticides in the microbial system. *Mutat Res*, 40(1):19–30. doi:10.1016/0165-1218(76)90018-5 PMID:814455

Siebert D, Lemperle E (1974). Genetic effects of herbicides: induction of mitotic gene conversion in Saccharomyces cerevisiae. *Mutat Res*, 22(2):111–20. doi:10.1016/0027-5107(74)90090-6 PMID:4601758

Sipes NS, Martin MT, Kothiya P, Reif DM, Judson RS, Richard AM et al. (2013). Profiling 976 ToxCast chemicals across 331 enzymatic and receptor signalling assays. *Chem Res Toxicol*, 26(6):878–95. doi:10.1021/tx400021f PMID:23611293

Smith MT, Guyton KZ, Gibbons CF, Fritz JM, Portier CJ, Rusyn I et al. (2016). Key Characteristics of Carcinogens as a Basis for Organizing Data on Mechanisms of Carcinogenesis. *Environ Health Perspect*, 124(6):713–21. PMID:26600562

Smith AH, Pearce NE, Fisher DO, Giles HJ, Teague CA, Howard JK (1984). Soft tissue sarcoma and exposure to phenoxyherbicides and chlorophenols in New Zealand. *J Natl Cancer Inst*, 73(5):1111–7. PMID:6593487

Smith JG, Christophers AJ (1992). Phenoxy herbicides and chlorophenols: a case control study on soft tissue sarcoma and malignant lymphoma. *Br J Cancer*, 65(3):442–8. doi:10.1038/bjc.1992.90 PMID:1558802

Soloneski S, González NV, Reigosa MA, Larramendy ML (2007). Herbicide 2,4-dichlorophenoxyacetic acid (2,4-D)-induced cytogenetic damage in human lymphocytes in vitro in presence of erythrocytes. *Cell Biol Int*, 31(11):1316–22. doi:10.1016/j.cellbi.2007.05.003 PMID:17606385

Song Y (2014). Insight into the mode of action of 2,4-dichlorophenoxyacetic acid (2,4-D) as an herbicide. *J Integr Plant Biol*, 56(2):106–13. doi:10.1111/jipb.12131 PMID:24237670

Sorensen KC, Stucki JW, Warner RE, Wagner ED, Plewa MJ (2005). Modulation of the genotoxicity of pesticides reacted with redox-modified smectite clay. *Environ Mol Mutagen*, 46(3):174–81. doi:10.1002/em.20144 PMID:15920753

Sreekumaran Nair R, Paulmurugan R, Ranjit Singh AJA (2002). Simple radioactive assay for the estimation of DNA breaks. *J Appl Toxicol*, 22(1):19–23. doi:10.1002/jat.807 PMID:11807925

Straif K, Loomis D, Guyton K, Grosse Y, Lauby-Secretan B, El Ghissassi F et al. (2014). Future priorities for *IARC Monographs*. *Lancet Oncol*, 15(7):683–4.

Stürtz N, Bongiovanni B, Rassetto M, Ferri A, de Duffard AM, Duffard R (2006). Detection of 2,4-dichlorophenoxyacetic acid in rat milk of dams exposed during lactation and milk analysis of their major components. *Food Chem Toxicol*, 44(1):8–16. doi:10.1016/j.fct.2005.03.012 PMID:16216402

Stürtz N, Evangelista de Duffard AM, Duffard R (2000). Detection of 2,4-dichlorophenoxyacetic acid (2,4-D) residues in neonates breast-fed by 2,4-D exposed dams. *Neurotoxicology*, 21(1-2):147–54. PMID:10794394

Sun H, Si C, Bian Q, Chen X, Chen L, Wang X (2012). Developing in vitro reporter gene assays to assess the hormone receptor activities of chemicals frequently detected in drinking water. *J Appl Toxicol*, 32(8):635–41. doi:10.1002/jat.1790 PMID:22912978

Surjan A (1989). Analysis of genotoxic activity of 16 compounds and mixtures by the Drosophila mosaic test. *Ann Ist Super Sanita*, 25(4):569–72. PMID:2517189

't Mannetje A, McLean D, Cheng S, Boffetta P, Colin D, Pearce N (2005). Mortality in New Zealand workers exposed to phenoxy herbicides and dioxins. *Occup Environ Med*, 62(1):34–40. doi:10.1136/oem.2004.015776 PMID:15613606

Tagert MLM, Massey JH, Shaw DR (2014). Water quality survey of Mississippi's Upper Pearl River. *Sci Total Environ*, 481:564–73. doi:10.1016/j.scitotenv.2014.02.084 PMID:24631619

Tasker E (1985). A Dietary Two-Generation Reproduction Study in Fischer 344 Rats with 2,4-Dichlorophenoxyacetic Acid: Final Report. Project No. WIL-81137. Unpublished study prepared by Wil Research Laboratories, Inc. 1402 p.

Tayeb W, Nakbi A, Cheraief I, Miled A, Hammami M (2013). Alteration of lipid status and lipid metabolism, induction of oxidative stress and lipid peroxidation by 2,4-dichlorophenoxyacetic herbicide in rat liver. *Toxicol Mech Methods*, 23(6):449–58. doi:10.3109/15376516.2013.780275 PMID:23464821

Tayeb W, Nakbi A, Trabelsi M, Attia N, Miled A, Hammami M (2010). Hepatotoxicity induced by sub-acute exposure of rats to 2,4-Dichlorophenoxyacetic acid based herbicide "Désormone lourd". *J Hazard Mater*, 180(1–3):225–33. doi:10.1016/j.jhazmat.2010.04.018 PMID:20447766

Tayeb W, Nakbi A, Trabelsi M, Miled A, Hammami M (2012). Biochemical and histological evaluation of kidney damage after sub-acute exposure to 2,4-dichlorophenoxyacetic herbicide in rats: involvement of oxidative stress. *Toxicol Mech Methods*, 22(9):696–704. doi:10.3109/15376516.2012.717650 PMID:22894658

Teixeira MC, Telo JP, Duarte NF, Sá-Correia I (2004). The herbicide 2,4-dichlorophenoxyacetic acid induces the generation of free-radicals and associated oxidative stress responses in yeast. *Biochem Biophys Res Commun*, 324(3):1101–7. doi:10.1016/j.bbrc.2004.09.158 PMID:15485668

Thomas KW, Dosemeci M, Coble JB, Hoppin JA, Sheldon LS, Chapa G et al. (2010b). Assessment of a pesticide exposure intensity algorithm in the agricultural health study. *J Expo Sci Environ Epidemiol*, 20(6):559–69. doi:10.1038/jes.2009.54 PMID:19888312

Thomas KW, Dosemeci M, Hoppin JA, Sheldon LS, Croghan CW, Gordon SM et al. (2010a). Urinary biomarker, dermal, and air measurement results for 2,4-D and chlorpyrifos farm applicators in the Agricultural Health Study. *J Expo Sci Environ Epidemiol*, 20(2):119–34. doi:10.1038/jes.2009.6 PMID:19240759

Tice RR, Austin CP, Kavlock RJ, Bucher JR (2013). Improving the human hazard characterization of chemicals: a Tox21 update. *Environ Health Perspect*, 121(7):756–65. doi:10.1289/ehp.1205784 PMID:23603828

Timchalk C (2004). Comparative inter-species pharmacokinetics of phenoxyacetic acid herbicides and related organic acids. evidence that the dog is not a relevant species for evaluation of human health risk. *Toxicology*, 200(1):1–19. doi:10.1016/j.tox.2004.03.005 PMID:15158559

Tripathy NK, Routray PK, Sahu GP, Kumar AA (1993). Genotoxicity of 2,4-dichlorophenoxyacetic acid tested in somatic and germ-line cells of Drosophila. *Mutat Res*, 319(3):237–42. doi:10.1016/0165-1218(93)90083-P PMID:7694145

Troudi A, Ben Amara I, Samet AM, Zeghal N (2012). Oxidative stress induced by 2,4-phenoxyacetic acid in liver of female rats and their progeny: biochemical and histopathological studies. *Environ Toxicol*, 27(3):137–45. doi:10.1002/tox.20624 PMID:20607813

Tuschl H, Schwab C (2003). Cytotoxic effects of the herbicide 2,4-dichlorophenoxyacetic acid in HepG2 cells. *Food Chem Toxicol*, 41(3):385–93. doi:10.1016/S0278-6915(02)00238-7 PMID:12504171

Tuschl H, Schwab CE (2005). The use of flow cytometric methods in acute and long-term in vitro testing. *Toxicol In Vitro*, 19(7):845–52. doi:10.1016/j.tiv.2005.06.026 PMID:16081244

Tyynelä K, Elo HA, Ylitalo P (1990). Distribution of three common chlorophenoxyacetic acid herbicides into the rat brain. *Arch Toxicol*, 64(1):61–5. doi:10.1007/BF01973378 PMID:2306196

USGS (2006). Appendix 7A. Statistical summaries of pesticide compounds in stream water, 1992–2001. In: Pesticides in the nation's streams and ground water, 1992–2001. USGS Circular 1291. United States Geological Survey. Available from: http://water.usgs.gov/nawqa/pnsp/pubs/circ1291/appendix7/7a.html, accessed 26 May 2015.

Vainio H, Nickels J, Linnainmaa K (1982). Phenoxy acid herbicides cause peroxisome proliferation in Chinese hamsters. *Scand J Work Environ Health*, 8(1):70–3. doi:10.5271/sjweh.2494 PMID:7134925

Van den Berg KJ, van Raaij JAGM, Bragt PC, Notten WRF (1991). Interactions of halogenated industrial chemicals with transthyretin and effects on thyroid hormone levels in vivo. *Arch Toxicol*, 65(1):15–9. doi:10.1007/BF01973497 PMID:2043046

van Ravenzwaay B, Hardwick TD, Needham D, Pethen S, Lappin GJ (2003). Comparative metabolism of 2,4-dichlorophenoxyacetic acid (2,4-D) in rat and dog. *Xenobiotica*, 33(8):805–21. doi:10.1080/0049825031000135405 PMID:12936702

Venkov P, Topashka-Ancheva M, Georgieva M, Alexieva V, Karanov E (2000). Genotoxic effect of substituted phenoxyacetic acids. *Arch Toxicol*, 74(9):560–6. doi:10.1007/s002040000147 PMID:11131037

Vineis P, Terracini B, Ciccone G, Cignetti A, Colombo E, Donna A et al. (1987). Phenoxy herbicides and soft-tissue sarcomas in female rice weeders. A population-based case-referent study. *Scand J Work Environ Health*, 13(1):9–17. doi:10.5271/sjweh.2077 PMID:3576149

Vogel E, Chandler JL (1974). Mutagenicity testing of cyclamate and some pesticides in Drosophila melanogaster. *Experientia*, 30(6):621–3. doi:10.1007/BF01921506 PMID:4209504

Vural N, Burgaz S (1984). A gas chromatographic method for determination of 2,4-D residues in urine after occupational exposure. *Bull Environ Contam Toxicol*, 33(5):518–24. doi:10.1007/BF01625578 PMID:6498355

Waite DT, Bailey P, Sproull JF, Quiring DV, Chau DF, Bailey J et al. (2005). Atmospheric concentrations and dry and wet deposits of some herbicides currently used on the Canadian Prairies. *Chemosphere*, 58(6):693–703. doi:10.1016/j.chemosphere.2004.09.105 PMID:15621183

Walters J (1999). Environmental fate of 2,4-dichlorophenoxyacetic acid. Sacramento (CA): California Department of Pesticide Regulation and CalEPA. Available from: http://www.cdpr.ca.gov/docs/emon/pubs/fatememo/24-d.pdf, accessed 23 January 2015

Ward MH, Lubin J, Giglierano J, Colt JS, Wolter C, Bekiroglu N et al. (2006). Proximity to crops and residential exposure to agricultural herbicides in iowa. *Environ Health Perspect*, 114(6):893–7. doi:10.1289/ehp.8770 PMID:16759991

Weisenburger DD (1990). Environmental epidemiology of non-Hodgkin's lymphoma in eastern Nebraska. *Am J Ind Med*, 18(3):303–5. doi:10.1002/ajim.4700180310 PMID:2220835

WHO (2003). 2,4-D in drinking-water. Background document for development of WHO Guidelines for Drinking-water Quality. Report No. WHO/SDE/WSH/03.04/70. Geneva: World Health Organization. Available from: http://www.who.int/water_sanitation_health/dwq/chemicals/24D.pdf.

WHO (2011). *Guidelines for drinking water quality*. Geneva: Global Malaria Programme, World Health Organization.

Wiklund K, Dich J, Holm LE (1987). Risk of malignant lymphoma in Swedish pesticide appliers. *Br J Cancer*, 56(4):505–8. doi:10.1038/bjc.1987.234 PMID:3689667

Wilson NK, Strauss WJ, Iroz-Elardo N, Chuang JC (2010). Exposures of preschool children to chlorpyrifos, diazinon, pentachlorophenol, and 2,4-dichlorophenoxyacetic acid over 3 years from 2003 to 2005: A longitudinal model. *J Expo Sci Environ Epidemiol*, 20(6):546–58. doi:10.1038/jes.2009.45 PMID:19724304

Witte I, Jacobi H, Juhl-Strauss U (1996). Suitability of different cytotoxicity assays for screening combination effects of environmental chemicals in human fibroblasts. *Toxicol Lett*, 87(1):39–45. doi:10.1016/0378-4274(96)03697-1 PMID:8701443

Woodruff RC, Phillips JP, Irwin D (1983). Pesticide-induced complete and partial chromosome loss in screens with repair-defective females of Drosophila melanogaster. *Environ Mutagen*, 5(6):835–46. doi:10.1002/em.2860050608 PMID:6418539

Woods JS, Polissar L (1989). Non-Hodgkin's lymphoma among phenoxy herbicide-exposed farm workers in western Washington State. *Chemosphere*, 18(1-6):401–6. doi:10.1016/0045-6535(89)90148-3

Woods JS, Polissar L, Severson RK, Heuser LS, Kulander BG (1987). Soft tissue sarcoma and non-Hodgkin's lymphoma in relation to phenoxyherbicide and chlorinated phenol exposure in western Washington. *J Natl Cancer Inst*, 78(5):899–910. PMID:3471999

Woudneh MB, Sekela M, Tuominen T, Gledhill M (2007). Acidic herbicides in surface waters of Lower Fraser Valley, British Columbia, Canada. *J Chromatogr A*, 1139(1):121–9. doi:10.1016/j.chroma.2006.10.081 PMID:17118381

Wright TR, Shan G, Walsh TA, Lira JM, Cui C, Song P et al. (2010). Robust crop resistance to broadleaf and grass herbicides provided by aryloxyalkanoate dioxygenase transgenes. *Proc Natl Acad Sci USA*, 107(47):20240–5. doi:10.1073/pnas.1013154107 PMID:21059954

Xie L, Thrippleton K, Irwin MA, Siemering GS, Mekebri A, Crane D et al. (2005). Evaluation of estrogenic activities of aquatic herbicides and surfactants using an rainbow trout vitellogenin assay. *Toxicol Sci*, 87(2):391–8. doi:10.1093/toxsci/kfi249 PMID:16049272

Yao Y, Tuduri L, Harner T, Blanchard P, Waite D, Poissant L et al. (2006). Spatial and temporal distribution of pesticide air concentrations in Canadian agricultural regions. *Atmos Environ*, 40(23):4339–51. doi:10.1016/j.atmosenv.2006.03.039

Yiin JH, Ruder AM, Stewart PA, Waters MA, Carreón T, Butler MA et al.; Brain Cancer Collaborative Study Group (2012). The Upper Midwest Health Study: a case-control study of pesticide applicators and risk of glioma. *Environ Health*, 11(1):39. doi:10.1186/1476-069X-11-39 PMID:22691464

Yilmaz HR, Yuksel E (2005). Effect of 2,4-dichlorophenoxyacetic acid on the activities of some metabolic enzymes for generating pyridine nucleotide pool of cells from mouse liver. *Toxicol Ind Health*, 21(9):231–7. doi:10.1191/0748233705th231oa PMID:16342474

Yoder J, Watson M, Benson WW (1973). Lymphocyte chromosome analysis of agricultural workers during extensive occupational exposure to pesticides. *Mutat Res*, 21(6):335–40. doi:10.1016/0165-1161(73)90057-5 PMID:4779319

Zahm SH, Weisenburger DD, Babbitt PA, Saal RC, Vaught JB, Cantor KP et al. (1990). A case-control study of non-Hodgkin's lymphoma and the herbicide 2,4-dichlorophenoxyacetic acid (2,4-D) in eastern Nebraska. *Epidemiology*, 1(5):349–56. doi:10.1097/00001648-199009000-00004 PMID:2078610

Zeljezic D, Garaj-Vrhovac V (2001). Chromosomal aberration and single cell gel electrophoresis (Comet) assay in the longitudinal risk assessment of occupational exposure to pesticides. *Mutagenesis*, 16(4):359–63. doi:10.1093/mutage/16.4.359 PMID:11420406

Zeljezic D, Garaj-Vrhovac V (2002). Sister chromatid exchange and proliferative rate index in the longitudinal risk assessment of occupational exposure to pesticides. *Chemosphere*, 46(2):295–303. doi:10.1016/S0045-6535(01)00073-X PMID:11827288

Zeljezic D, Garaj-Vrhovac V (2004). Chromosomal aberrations, micronuclei and nuclear buds induced in human lymphocytes by 2,4-dichlorophenoxyacetic acid pesticide formulation. *Toxicology*, 200(1):39–47. doi:10.1016/j.tox.2004.03.002 PMID:15158562

Zetterberg G, Busk L, Elovson R, Starec-Nordenhammar I, Ryttman H (1977). The influence of pH on the effects of 2,4-D (2,4-dichlorophenoxyacetic acid, Na salt) on Saccharomyces cerevisiae and Salmonella typhimurium. *Mutat Res*, 42(1):3–17. doi:10.1016/S0027-5107(77)80003-1 PMID:15215

Zhamsaranova SD, Lebedeva SN, Liashenko VA (1987). [Immunodepressive effects of the herbicide 2,4-D on mice]. *Gig Sanit*, (5):80–1. PMID:3609786

Zhang X, Acevedo S, Chao Y, Chen Z, Dinoff T, Driver J et al. (2011). Concurrent 2,4-D and triclopyr biomonitoring of backpack applicators, mixer/loader and field supervisor in forestry. *J Environ Sci Health B*, 46(4):281–93. doi:10.1080/03601234.2011.559424 PMID:21500074

Zimmering S, Mason JM, Valencia R, Woodruff RC (1985). Chemical mutagenesis testing in Drosophila. II. Results of 20 coded compounds tested for the National Toxicology Program. *Environ Mutagen*, 7(1):87–100. doi:10.1002/em.2860070105 PMID:3917911

Zychlinski L, Zolnierowicz S (1990). Comparison of uncoupling activities of chlorophenoxy herbicides in rat liver mitochondria. *Toxicol Lett*, 52(1):25–34. doi:10.1016/0378-4274(90)90162-F PMID:2356568

LIST OF ABBREVIATIONS

2,4-D	2,4-dichlorophenoxyacetic acid
2,4-DB	4-(2,4-dichlorophenoxy)butyric acid
2,4-DCP	2,4-dichlorophenol
2,4,5-T	2,4,5-trichlorophenoxyacetic acid
8-OHdG	8-hydroxy-2′-deoxyguanosine
AAF	2-acetylaminofluorene
ACTH	adrenocorticotropic hormone
ADI	acceptable daily intake
AHS	Agricultural Health Study
BMI	body mass index
bw	body weight
CHO	Chinese hamster ovary
CLL	chronic lymphocytic leukaemia
CYP	cytochrome P450
DDT	dichlorodiphenyltrichloroethane
DEET	*N,N*-diethyl-*meta*-toluamide
DHT	5α-dihydroxytestosterone
dicamba	3,6-dichloro-2-methoxybenzoic acid
DoDSR	Department of Defense Serum Repository
E_2	17β-estradiol
ECD	electron capture detector
EFSA	European Food Safety Authority
EGF	epidermal growth factor
EPA	United States Environmental Protection Agency
ER	estrogen receptor
EROD	ethoxyresorufin-*O*-deethylase
FAO	Food and Agriculture Organization of the United Nations
FXR	farnesoid X receptor
GC	gas chromatography
GC-ECD	gas chromatography-electron capture detector
GC-MS	gas chromatography-mass spectrometry
GC-MS-EI	gas chromatography-mass spectrometry-electron ionization
GCR	glucocorticoid receptor

$GdCl_3$	gadolinium chloride
GM	geometric mean
HBsAg	hepatitis B surface antigen
HBV	hepatitis B virus
HCC	hepatocellular carcinoma
HCCH	hexachlorocyclohex-1-ene
HCH	hexachlorocyclohexane
HGPRT	hypoxanthine-guanine phosphoribosyl transferase
HRMS	high-resolution mass spectrometry
ILO	International Labour Organization
IP3	inositol 1,4,5-triphosphate
IQR	interquartile range
JECFA	Joint FAO/WHO Expert Committee on Food Additives
LD_{50}	median lethal dose
LH	luteinizing hormone
LOD	limit of detection
MCPA	2-methyl-4-chlorophenoxyacetic acid
MCPP	mecoprop or methylchlorophenoxypropionic acid
meta-RR	meta relative risk
MHCIα	α-cardiac-like myosin heavy chain
MHCIIα	myosin heavy chain gene II α
MRL	maximum residue level
MS	mass spectrometry
MTT	3-(4,5-dimethylthiazol-2-yl)-2,5-diphenyltetrazolium bromide
Mx	MX dynamin-like GTPase
NAC	*N*-acetyl-*L*-cysteine
ND	not detected
NDMA-*d*	*N*-nitrosodimethylamine demethylase
NF-κβ	nuclear factor kappa β
NIOSH	United States National Institute for Occupational Safety and Health
NK	natural killer
NOAEL	no-observed-adverse-effect level
NR	not reported
NT	not tested
OAT1	organic anion transporter 1
OCP	organochlorine pesticide
OR	odds ratio
PCB	polychlorinated biphenyl
PCCH	pentachlorocyclohex-1-ene
PGD2	prostaglandin D2
PGE2	prostaglandin E2
PGF2α	prostaglandin F2α
PI	phosphatidylinositol
PIP	phosphatidylinositol 4-phosphate
PIP2	phosphatidylinositol 4,5-bisphosphate
$PM_{2.5}$	particulate matter with aerodynamic diameter $\leq 2.5\ \mu m$
PM_{10}	particulate matter with aerodynamic diameter $\leq 10\ \mu m$
PPAR	peroxisome proliferator-activated receptor
ppb	parts per billion
ppm	parts per million

PROD	7-pentoxyresorufin-*O*-dealkylase
PSA	prostate-specific antigen
PXR	pregnane X receptor
ROS	reactive oxygen species
SD	standard deviation
SEER	Surveillance, Epidemiology and End Results Program
SHBG	sex hormone-binding globulin
SIR	standardized incidence ratio
SMR	standardized mortality ratio
T3	triiodothyronine
T4	thyroxine
TCDD	2,3,7,8-tetrachlorodibenzo-*p*-dioxin
TCP	trichlorophenol
TMDI	theoretical maximum daily intake
TNFα	tumour necrosis factor α
ToxCast™	Toxicity Forecaster
ToxPi	toxicological prioritization index
TPA	12-O-tetradecanoyl phorbol-13-acetate
TSH	thyroid-stimulating hormone
USA	United States of America
WHO	World Health Organization
w/w	weight per weight
ww	wet weight
vs	versus

ANNEX 1. SUPPLEMENTAL MATERIAL FOR TOXCAST/TOX21

This supplemental material (which is available online at: http://publications.iarc.fr/550) contains a spreadsheet (.xlsx) and a zip folder containing several ToxPi data files (.csv) analysed by the Working Group for Volume 113 of the *IARC Monographs*. The spreadsheet lists the ToxCast/Tox21 assay end-points, the associated target and/or model system (e.g. cell type, species, detection technology, etc.), their mapping to 6 of the 10 "key characteristics" of known human carcinogens, and whether each chemical was "active" or "inactive" (EPA, 2015). The ToxPi files integrate the results by "key characteristic" and can be accessed using ToxPi software that is freely available for download without a licence (Reif et al., 2013).

References

EPA (2015). ToxCastTM Data. Washington (DC): United States Environmental Protection Agency. Available from: https://www.epa.gov/chemical-research/toxicity-forecaster-toxcasttm-data. Data released December 2014.

Reif DM, Sypa M, Lock EF, Wright FA, Wilson A, Cathey T et al. (2013). ToxPi GUI: an interactive visualization tool for transparent integration of data from diverse sources of evidence. *Bioinformatics*, 29(3):402–3. doi:10.1093/bioinformatics/bts686 PMID:23202747